MATRICES

ALGEBRA, ANALYSIS AND APPLICATIONS

MATRICES

ALGEBRA, ANALYSIS AND APPLICATIONS

SHMUEL FRIEDLAND

University of Illinois at Chicago, USA

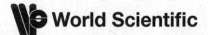

World Scientific

NEW JERSEY · LONDON · SINGAPORE · BEIJING · SHANGHAI · HONG KONG · TAIPEI · CHENNAI · TOKYO

Published by

World Scientific Publishing Co. Pte. Ltd.

5 Toh Tuck Link, Singapore 596224

USA office: 27 Warren Street, Suite 401-402, Hackensack, NJ 07601

UK office: 57 Shelton Street, Covent Garden, London WC2H 9HE

Library of Congress Cataloging-in-Publication Data
Friedland, Shmuel.
   Matrices : algebra, analysis, and applications / by Shmuel Friedland (University of Illinois at Chicago, USA).
      pages cm
   Includes bibliographical references and index.
   ISBN 978-9814667968 (alk. paper) -- ISBN 978-9813141032 (pbk)
   1. Matrices. 2. Eigenvalues. 3. Non-negative matrices. I. Title.
   QA188.F75 2015
   512.9'434--dc23

                        2015020408

British Library Cataloguing-in-Publication Data
A catalogue record for this book is available from the British Library.

In-house Editors: Kwong Lai Fun/Dipasri Sardar

Typeset by Stallion Press
Email: enquiries@stallionpress.com

Printed in Singapore

To the memory of my parents: Aron and Golda Friedland

Preface

Linear algebra and matrix theory are closely related subjects that are used extensively in pure and applied mathematics, bioinformatics, computer science, economy, engineering, physics and social sciences. Some results in these subjects are quite simple and some are very advanced and technical. This book reflects a very personal selection of the topics in matrix theory that the author was actively working on in the past 40 years. Some of the topics are very classical and available in a number of books. Other topics are not available in the books that are currently on the market. The author lectured several times certain parts of this book in graduate courses in Hebrew University of Jerusalem, University of Illinois at Chicago, Technion, and TU-Berlin.

The book consists of seven chapters which are somewhat interdependent. Chapter 1 discusses the fundamental notions of Linear Algebra over general and special integral domains. Chapter 2 deals with well-known canonical form: Jordan canonical form, Kronecker canonical form, and their applications. Chapter 3 discusses functions of matrices and analytic similarity with respect to one complex variable. Chapter 4 is devoted to linear operators over finite dimensional inner product spaces. Chapter 5 is a short chapter on elements of multilinear algebra. Chapter 6 deals with non-negative matrices. Chapter 7 discusses various topics as norms, complexity problem of the convex hull of a tensor product of certain two convex sets, variation of tensor power and spectra, inverse eigenvalue problems for non-negative matrices, and cones.

This book started as an MRC report "Spectral theory of matrices", 1980, University of Madison, Wisconsin. I continued to work on this book in Hebrew University of Jerusalem, University of Illinois at Chicago and in Technion and Technical University of Berlin during my Sabbaticals in 2000 and 2007–2008 respectively.

I thank Eleanor Smith for reading parts of this book and for her useful remarks.

Chicago, January 1, 2015

Contents

Chapter 1

Domains, Modules and Matrices

1.1 Rings, Domains and Fields

Definition 1.1.1 *A nonempty set* R *is called a ring if* R *has two binary operations, called addition and multiplication and denoted by* $a + b$ *and* ab *respectively, such that for all* $a, b, c \in$ R *the following holds:*

$$a + b \in \text{R};\tag{1.1.1}$$

$$a + b = b + a \quad \text{(the commutative law)};\tag{1.1.2}$$

$$(a + b) + c = a + (b + c) \quad \text{(the associative law)};\tag{1.1.3}$$

$$\exists\, 0 \in \text{R such that } a + 0 = 0 + a = a, \ \forall\, a \in \text{R};\tag{1.1.4}$$

$$\forall\, a \in \text{R}, \ \exists -a \in \text{R such that } a + (-a) = 0;\tag{1.1.5}$$

$$ab \in \text{R};\tag{1.1.6}$$

$$a(bc) = (ab)c \quad \text{(the associative law)};\tag{1.1.7}$$

$$a(b + c) = ab + ac, \quad (b + c)a = ba + ca, \quad \text{(the distributive laws)}.\tag{1.1.8}$$

R has an *identity* element 1 if $a1 = 1a$ for all $a \in$ R. R is called *commutative* if

$$ab = ba, \quad \text{for all } a, b \in \text{R}.\tag{1.1.9}$$

1

Note that the properties $(1.1.2)-(1.1.8)$ imply that $a0 = 0a = 0$. If a and b are two nonzero elements such that

$$ab = 0, \qquad\qquad (1.1.10)$$

then a and b are called *zero divisors*.

Definition 1.1.2 $\mathbb{D}$ *is called an integral domain if* $\mathbb{D}$ *is a commutative ring without zero divisors which contains an identity element 1.*

The classical example of an integral domain is the ring of integers $\mathbb{Z}$. In this book we shall use the following example of an integral domain.

Example 1.1.3 *Let* $\Omega \subset \mathbb{C}^n$ *be a nonempty set. Then* $\mathrm{H}(\Omega)$ *denotes the ring of functions* $f(z_1, \ldots, z_n)$ *such that for each* $\zeta \in \Omega$ *there exists an open neighborhood* $O(f, \zeta)$ *of* ζ *on which* f *is analytic. If* Ω *is open we assume that* f *is defined only on* Ω. *If* Ω *consists of one point* ζ *then* H_ζ *stands for* $\mathrm{H}(\{\zeta\})$.

Recall that $\Omega \subset \mathbb{C}^n$ is called *connected*, if in the relative topology on Ω, induced by the standard topology on $\mathbb{C}^n$, the only subsets of Ω which are both open in Ω and closed in Ω are $\emptyset$ and Ω. Note that the zero element is the zero function of $\mathrm{H}(\Omega)$ and the identity element is the constant function which is equal to 1. The properties of analytic functions imply that $\mathrm{H}(\Omega)$ is an integral domain if and only if Ω is a *connected* set. In this book, we shall assume that Ω is connected unless otherwise stated. See [Rud74] and [GuR65] for properties of analytic functions in one and several complex variables.

Definition 1.1.4 *A nonempty* $\Omega \subset \mathbb{C}^n$ *is called a domain if* Ω *is an open connected set.*

For $a, b \in \mathbb{D}$, a *divides* b, (or a is a *divisor* of b), denoted by $a|b$, if $b = ab_1$ for some $b_1 \in \mathbb{D}$. An element a is called *invertible*, (unit, unimodular), if $a|1$. $a, b \in \mathbb{D}$ are *associates*, denoted by $a \equiv b$, if $a|b$ and $b|a$. Let $\{\{b\}\} = \{a \in \mathbb{D} : a \equiv b\}$. The associates of a and units are called *improper* divisors of a. For an invertible a denote by a^{-1} the unique element such that

$$aa^{-1} = a^{-1}a = 1. \tag{1.1.11}$$

$f \in H(\Omega)$ is invertible if and only if f does not vanish at any point of Ω.

Definition 1.1.5 *A field* $\mathbb{F}$ *is an integral domain* $\mathbb{D}$ *such that any nonzero element is invertible. A field* $\mathbb{F}$ *has characteristic 0 if for any nonzero integer n and a nonzero element* $f \in \mathbb{F}$ $nf \neq 0$.

The familiar examples of fields are the set of rational numbers $\mathbb{Q}$, the set of real numbers $\mathbb{R}$, and the set of complex numbers $\mathbb{C}$. Note that the characteristic of $\mathbb{Q}, \mathbb{R}, \mathbb{C}$ is 0. Given an integral domain $\mathbb{D}$ there is a standard way to construct the *field* $\mathbb{F}$ of its *quotients*. $\mathbb{F}$ is formed by the set of equivalence classes of all quotients $\frac{a}{b}, b \neq 0$ such that

$$\frac{a}{b} + \frac{c}{d} = \frac{ad+bc}{bd}, \quad \frac{a}{b}\frac{c}{d} = \frac{ac}{bd}, \quad b, d \neq 0. \tag{1.1.12}$$

Definition 1.1.6 *For* $\Omega \subset \mathbb{C}^n, \zeta \in \mathbb{C}^n$ *let* $\mathcal{M}(\Omega), \mathcal{M}_\zeta$ *denote the quotient fields of* $H(\Omega), H_\zeta$ *respectively.*

Definition 1.1.7 *Let* $\mathbb{D}[x_1, \ldots, x_n]$ *be the ring of all polynomials in n variables with coefficients in* $\mathbb{D}$:

$$p(x_1, \ldots, x_n) = \sum_{|\alpha| \leq m} a_\alpha x^\alpha, \text{ for some } m \in \mathbb{N}, \tag{1.1.13}$$

where $\alpha = (\alpha_1, \ldots, \alpha_n) \in \mathbb{Z}_+^n, \quad |\alpha| := \sum_{i=1}^n \alpha_i, \quad x^\alpha := x_1^{\alpha_1} \cdots x_n^{\alpha_n}.$

Sometimes we denote $\mathbb{D}[x_1, \ldots, x_n]$ *and* $p(x_1, \ldots, x_n)$ *by* $\mathbb{D}[\mathbf{x}]$ *and* $p(\mathbf{x})$ *respectively.*

The *degree* of $p(x_1, \ldots, x_n) \neq 0$ (denoted $\deg p$) is the largest natural number d such that there exists $a_\alpha \neq 0$ with $|\alpha| = d$. ($\deg 0 = -\infty$.) A polynomial p is called *homogeneous* if $a_\alpha = 0$ for all $|\alpha| < \deg p$. It is a standard fact that $\mathbb{D}[x_1, \ldots, x_n]$ is an integral domain. (See Problems 2–3 below.) As usual $\mathbb{F}(x_1, \ldots, x_n)$ denotes the quotient field of $\mathbb{F}[x_1, \ldots, x_n]$.

Problems

1. Let $C[a, b]$ be the set of real valued continuous functions on the interval $[a, b], a < b$. Show that $C[a, b]$ is a commutative ring with identity and zero divisors.

2. Let $\mathbb{D}$ be an integral domain. Prove that $\mathbb{D}[x]$ is an integral domain.

3. Prove that $\mathbb{D}[x_1, \ldots, x_n]$ is an integral domain. (Use the previous problem and the identity $\mathbb{D}[x_1, \ldots, x_n] = \mathbb{D}[x_1, \ldots, x_{n-1}][x_n]$.)

4. Let $p(x_1, \ldots, x_n) \in \mathbb{D}[x_1, \ldots, x_n]$. Show that $p = \sum_{i \leq \deg p} p_i$, where each p_i is either a zero polynomial or a homogeneous polynomial of degree i for $i \geq 0$. If p is not a constant polynomial then $m = \deg p \geq 1$ and $p_m \neq 0$. The polynomial p_m is called the *principal part* of p and is denoted by p_π. (If p is a constant polynomial then $p_\pi = p$.)

5. Let $p, q \in \mathbb{D}[x_1, \ldots, x_n]$. Show $(pq)_\pi = p_\pi q_\pi$.

6. Let $\mathbb{F}$ be a field with two elements at least which does not have characteristic 0. Show that there exists a unique prime integer $p \geq 2$ such that $pf = 0$ for each $f \in \mathbb{F}$. p is called the characteristic of $\mathbb{F}$.

1.2   Bezout Domains

Let $a_1, \ldots, a_n \in \mathbb{D}$. Assume first that not all of $a_1, \ldots, a_n$ are equal to zero. An element $d \in \mathbb{D}$ is a *greatest common divisor* (g.c.d.) of $a_1, \ldots, a_n$ if $d|a_i$ for $i = 1, \ldots, n$, and for any d' such that $d'|a_i, i = 1, \ldots, n$, $d'|d$. Denote by $(a_1, \ldots, a_n)$ any g.c.d. of $a_1, \ldots, a_n$. Then $\{\{(a_1, \ldots, a_n)\}\}$ is the equivalence class of all g.c.d. of $a_1, \ldots, a_n$.

For $a_1 = \cdots = a_n = 0$, we define 0 to be the GCD of $a_1, \ldots, a_n$, i.e. $(a_1, \ldots, a_n) = 0$. The elements $a_1, \ldots, a_n$ are called *coprime* if $\{\{(a_1, \ldots, a_n)\}\} = \{\{1\}\}$.

Definition 1.2.1 $\mathbb{D}$ *is called a greatest common divisor domain, or simply GCD domain and denoted by* $\mathbb{D}_G$, *if any two elements in* $\mathbb{D}$ *have a GCD.*

A simple example of $\mathbb{D}_G$ is $\mathbb{Z}$. See Problem 5 for a nonGCD domain.

Definition 1.2.2 *Let* $\mathbb{D}$ *be a commutative ring. A subset $I \subset \mathbb{D}$ is called an ideal if for any $a, b \in I$ and $p, q \in \mathbb{D}$ the element $pa + qb$ belongs to I.*

In $\mathbb{Z}$ any nontrivial ideal is the set of all numbers divisible by an integer $k \neq 0$. In $H(\Omega)$, the set of functions which vanishes on a prescribed set $U \subset \Omega$, i.e.

$$I(U) := \{f \in H(\Omega): \quad f(\zeta) = 0, \quad \text{for all } \zeta \in U\}, \qquad (1.2.1)$$

is an ideal. An ideal I is called *prime* if $ab \in I$ implies that either a or b is in I. $I \subset \mathbb{Z}$ is a prime ideal if and only if I is the set of integers divisible by some prime number p. An ideal I is called *maximal* if the only ideals which contain I are I and $\mathbb{D}$. I is called *finitely generated* if there exists k elements (*generators*) $p_1, \ldots, p_k \in I$ such that any $i \in I$ is of the form

$$i = a_1 p_1 + \cdots + a_k p_k \qquad (1.2.2)$$

for some $a_1, \ldots, a_k \in \mathbb{D}$. For example, in $\mathbb{D}[x, y]$ the set of all polynomials $p(x, y)$ such that

$$p(0, 0) = 0, \qquad (1.2.3)$$

is an ideal generated by x and y. An ideal is called a *principal* ideal if it is generated by one element p.

Definition 1.2.3 $\mathbb{D}$ *is called a Bezout domain, or simply BD and denoted by* $\mathbb{D}_B$, *if any two elements* $a, b \in \mathbb{D}$ *have g.c.d.* (a, b) *such that*

$$(a, b) = pa + qb, \qquad (1.2.4)$$

for some $p, q \in \mathbb{D}$.

It is easy to show by induction that for $a_1, \ldots, a_n \in \mathbb{D}_B$

$$(a_1, \ldots, a_n) = \sum_{i=1}^{n} p_i a_i, \quad \text{for some } p_1, \ldots, p_n \in \mathbb{D}_B. \qquad (1.2.5)$$

Lemma 1.2.4 *An integral domain is a Bezout domain if and only if any finitely generated ideal is principal.*

Proof. Assume that an ideal of $\mathbb{D}_B$ is generated by $a_1, \ldots, a_n$. Then (1.2.5) implies that $(a_1, \ldots, a_n) \in I$. Clearly $(a_1, \ldots, a_n)$ is a generator of I. Assume now that any finitely generated ideal of $\mathbb{D}$ is principal. For given $a, b \in \mathbb{D}$ let I be the ideal generated by a and b. Let d be a generator of I. So

$$d = pa + qb. \qquad (1.2.6)$$

Since d generates I, d divides a and b. (1.2.6) implies that if d' divides a and b then $d'|d$. Hence, $d = (a, b)$ and $\mathbb{D}$ is $\mathbb{D}_B$. $\qquad \square$

Let $I \subset \mathbb{D}[x, y]$ be the ideal given by (1.2.3). Clearly $(x, y) = 1$. As $1 \notin I$, I is not principal. As x, y generate I we obtain that $\mathbb{D}[x, y]$ is not $\mathbb{D}_B$. In particular $\mathbb{F}[x_1, \ldots, x_n]$ is not $\mathbb{D}_B$ for $n \geq 2$. The same argument shows that $H(\Omega)$ is not $\mathbb{D}_B$ for $\Omega \subset \mathbb{C}^n$ and $n \geq 2$. It is a standard fact that $\mathbb{F}[x]$ is a Bezout domain [Lan67]. (See §1.3.) For a connected set $\Omega \subset \mathbb{C}$ $H(\Omega)$ is $\mathbb{D}_B$. This result is implied by the following interpolation theorem [Rud74, Theorems 15.11 and 15.15]:

Theorem 1.2.5 *Let* $\Omega \subset \mathbb{C}$ *be an open set,* $A \subset \Omega$ *be a countable set with no accumulation point in* Ω. *Assume that for each*

$\zeta \in A$, $m(\zeta)$ and $w_{0,\zeta}, \ldots, w_{m(\zeta),\zeta}$ *are a non-negative integer and* $m(\zeta) + 1$ *complex numbers, respectively. Then there exists* $f \in \mathrm{H}(\Omega)$ *such that*

$$f^{(n)}(\zeta) = n! w_{n,\zeta}, \quad n = 0, \ldots, m(\zeta), \quad \text{for all } \zeta \in A.$$

Furthermore, if all $w_{n,\zeta} = 0$ *then there exists* $g \in \mathrm{H}(\Omega)$ *such that all zeros of g are in A and g has a zero of order* $m(\zeta) + 1$ *at each* $\zeta \in A$.

Theorem 1.2.6 *Let* $\Omega \subset \mathbb{C}$ *be a domain. Then for* $a, b \in \mathrm{H}(\Omega)$ *there exists* $p \in H(\Omega)$ *such that* $(a, b) = pa + b$.

Proof. If $a = 0$ or $b = 0$ then $(a, b) = 1a + 1b$. Assume that $ab \neq 0$. Let A be the set of common zeros of $a(z)$ and $b(z)$. For each $\zeta \in A$ let $m(\zeta) + 1$ be the minimum multiplicity of the zero $z = \zeta$ of $a(z)$ and $b(z)$. Theorem 1.2.5 implies the existence of $f \in \mathrm{H}(\Omega)$ which has its zeros at A, such that at each $\zeta \in A$ $f(z)$ has a zero of order $m(\zeta) + 1$. Hence,

$$a = \hat{a}f, \; b = \hat{b}f, \; \hat{a}, \hat{b} \in \mathrm{H}(\Omega).$$

Thus $\hat{a}$ and $\hat{b}$ do not have common zeros. If A is empty then $\hat{a} = a$, $\hat{b} = b$. Let $\hat{A}$ be the set of zeros of $\hat{a}$. Assume that for each $\zeta \in \hat{A}$, $\hat{a}$ has a zero of multiplicity $n(\zeta) + 1$. Since $b(\zeta) \neq 0$ for any $\zeta \in \hat{A}$, Theorem 1.2.5 implies the existence of a function $g \in \mathrm{H}(\Omega)$ which satisfies the interpolation conditions:

$$\frac{d^k}{dz^k}(e^{g(z)})|_{z=\zeta} = \frac{d^k}{dz^k}\hat{b}(z)|_{z=\zeta}, \quad k = 0, \ldots, n(\zeta), \quad \zeta \in \hat{A}.$$

Then

$$p = \frac{e^g - \hat{b}}{\hat{a}}, \quad (a, b) = fe^g = pa + b$$

and the theorem follows. $\qquad\qquad\square$

Corollary 1.2.7 *Let* $\Omega \subset \mathbb{C}$ *be a connected set. Then* $\mathrm{H}(\Omega)$ *is a Bezout domain.*

Problems

1. Let $a, b, c \in \mathbb{D}_B$. Assume that $(a, b) = 1$, $(a, c) = 1$. Show that $(a, bc) = 1$.

2. Let I be a prime ideal in $\mathbb{D}$. Show that $\mathbb{D}/I$ (the set of all cosets of the form $I + a$) is an integral domain.

3. Let I be an ideal in $\mathbb{D}$. For $p \in \mathbb{D}$ denote by $I(p)$ the set:

$$I(p) := \{a \in \mathbb{D} : \quad a = bp + q, \text{ for all } b \in \mathbb{D}, \ q \in I\}.$$

   Show that $I(p)$ is an ideal. Prove that I is a maximal ideal if and only if for any $p \notin I$, $I(p) = \mathbb{D}$.

4. Show that an ideal I is maximal if and only if $\mathbb{D}/I$ is a field.

5. Let $\mathbb{Z}[\sqrt{-3}] = \{a \in \mathbb{C}, \ a = p + q\sqrt{-3}, \ p, q \in \mathbb{Z}\}$. Show

   (a) $\mathbb{Z}[\sqrt{-3}]$, viewed as a subset of $\mathbb{C}$, is a domain with respect to the addition and multiplication in $\mathbb{C}$.

   (b) Let $z = a + b\sqrt{-3} \in \mathbb{Z}[\sqrt{-3}]$. Then

   $$|z| = 1 \iff z = \pm 1, \quad |z| = 2 \iff z = \pm 2 \text{ or}$$
   $$z = \pm 1 \pm \sqrt{-3}.$$

   $|z| \geq \sqrt{7}$ for all other values of $z \neq 0$. In particular if $|z| = 2$ then z is a prime.

   (c) Let

   $$a = 4 = 2 \cdot 2 = (1 + \sqrt{-3})(1 - \sqrt{-3}),$$
   $$b = (1 + \sqrt{-3}) \cdot 2 = -(1 - \sqrt{-3})^2.$$

   Then any d that divides a and b divides one of the following primes $d_1 := 1 + \sqrt{-3}$, $\bar{d}_1 = 1 - \sqrt{-3}$, $d_2 := 2$.

   (d) $\mathbb{Z}[\sqrt{-3}]$ is not GCD domain.

1.3 $\mathbb{D}_U, \mathbb{D}_P$ and $\mathbb{D}_E$ Domains

$p \in \mathbb{D}$ is *irreducible* (*prime*) if it is not a unit and every divisor of p is improper. A positive integer $p \in \mathbb{Z}$ is irreducible if and only if p is prime. A linear polynomial in $\mathbb{D}[x_1, \ldots, x_n]$ is irreducible.

Lemma 1.3.1 *Let $\Omega \subset \mathbb{C}$ be a connected set. Then all irreducible elements of $H(\Omega)$ (up to multiplication by invertible elements) are of the form $z - \zeta$ for each $\zeta \in \Omega$.*

Proof. Let $f \in H(\Omega)$ be noninvertible. Then there exists $\zeta \in \Omega$ such that $f(\zeta) = 0$. Hence, $z - \zeta | f(z)$. Therefore, the only irreducible elements are $z - \zeta$. Clearly $\frac{z-\eta}{z-\zeta}$ is analytic in Ω if and only if $\eta = \zeta$. □

For $\zeta \in \mathbb{C}$, H_ζ has one irreducible element, namely $z - \zeta$.

Definition 1.3.2 $\mathbb{D}$ *is a unique factorization domain, or simply UFD and denoted by $\mathbb{D}_U$, if any nonzero, noninvertible element a can be factored as a product of irreducible elements*

$$a = p_1, \ldots, p_r, \tag{1.3.1}$$

and these are uniquely determined up to order and invertible factors.

$\mathbb{Z}$ and H_ζ, $\zeta \in \mathbb{C}$ are $\mathbb{D}_U$. $\mathbb{F}[x_1, \ldots, x_n]$ is $\mathbb{D}_U$ [Lan67].

Lemma 1.3.3 *Let $\Omega \subset \mathbb{C}$ be a domain. Then $H(\Omega)$ is not a unique factorization domain.*

Proof. Theorem 1.2.6 yields the existence of a nonzero function $a(z) \in H(\Omega)$ which has a countably infinite number of zeros Ω (which do not accumulate in Ω). Use Lemma 1.3.1 to deduce that a cannot be a product of a finite number of irreducible elements. □

A straightforward consequence of this lemma is that for any domain $\Omega \subset \mathbb{C}^n$, $H(\Omega)$ is not $\mathbb{D}_U$. See Problem 2.

Definition 1.3.4 $\mathbb{D}$ *is principal ideal domain, or simply PID and denoted by $\mathbb{D}_P$, if every ideal of $\mathbb{D}$ is principal.*

$\mathbb{Z}$ and $\mathbb{F}[z]$ are $\mathbb{D}_P$. It is known that any $\mathbb{D}_P$ is $\mathbb{D}_U$ ([Lan67] or [vdW59]). Thus, $H(\Omega)$ is not $\mathbb{D}_P$ for any open connected set $\Omega \subset \mathbb{C}^n$.

Definition 1.3.5 $\mathbb{D}$ *is a Euclidean domain, or simply ED and denoted by* $\mathbb{D}_E$, *if there exists a function* $d : \mathbb{D}\backslash\{0\} \to \mathbb{Z}_+$ *such that:*

$$\text{for all } a, b \in \mathbb{D}, \ ab \neq 0 \quad d(a) \leq d(ab); \tag{1.3.2}$$

$$\text{for any } a, b \in \mathbb{D}, \ ab \neq 0, \text{ there exists } t, r \in \mathbb{D} \text{ such that}$$
$$a = tb + r, \text{ where either } r = 0 \text{ or } d(r) < d(b). \tag{1.3.3}$$

We define $d(0) = -\infty$.

Standard examples of Euclidean domains are $\mathbb{Z}$ and $\mathbb{F}[x]$, see Problem 1.

Lemma 1.3.6 *Any ideal* $\{0\} \neq I \subset \mathbb{D}_E$ *is principal.*

Proof. Let $\min_{x \in I \backslash \{0\}} d(x) = d(a)$. Then I is generated by a. $\square$

Lemma 1.3.7 *Let* $\Omega \subset \mathbb{C}$ *be a compact connected set. Then* $H(\Omega)$ *is* $\mathbb{D}_E$. *Here,* $d(a)$ *is the number of zeros of a nonzero function* $a \in H(\Omega)$ *counted with their multiplicities.*

Proof. Let a be a nonzero analytic functions on a domain $O \supset \Omega$. Since each zero of a is an isolated zero of finite multiplicity, the assumption that Ω is compact yields that a has a finite number of zeros in Ω. Hence, $d(a) < \infty$. Let p_a be a nonzero polynomial of degree $d(a)$ such that $a_0 := \frac{a}{p_a}$ does not vanish on Ω. By definition, $d(a) = d(p_a) = \deg p$. Let $a, b \in H(\Omega)$, $ab \neq 0$. Since $\mathbb{C}[z]$ is $\mathbb{D}_E$ we deduce that

$$p_a(z) = t(z)p_b(z) + r(z), \quad r = 0 \text{ or } d(r) < d(p_b).$$

Hence

$$a = \frac{a_0 t}{b_0}b + a_0 r, \quad a_0 r = 0 \text{ or } d(a_0 r) = d(r) < d(p_b) = d(b).$$

$\square$

The Weierstrass preparation theorem [GuR65] can be used to prove the following extension of the above lemma to several complex variables:

Lemma 1.3.8 *Let $\Omega \subset \mathbb{C}^n$ be a compact connected set. Then $H(\Omega)$ is $\mathbb{D}_U$.*

Let $a_1, a_2 \in \mathbb{D}_E \setminus \{0\}$. Assume that $d(a_1) \geq d(a_2)$. The *Euclidean algorithm* consists of a sequence $a_1, \ldots, a_{k+1}$ which is defined recursively as follows:

$$a_i = t_i a_{i+1} + a_{i+2}, \quad a_{i+2} = 0 \quad \text{or} \quad d(a_{i+2}) < d(a_{i+1}). \quad (1.3.4)$$

Since $d(a) \geq 0$ the Euclidean algorithm terminates $a_1 \neq 0, \ldots, a_k \neq 0$ and $a_{k+1} = 0$. Hence

$$(a_1, a_2) = a_k. \quad (1.3.5)$$

See Problem 3 below.

Problems

1. Show that the following domains are Euclidean.

   (a) $\mathbb{Z}$, where $d(a) = |a|$ for any $a \in \mathbb{Z}$.

   (b) $\mathbb{F}[x]$, where $d(p(x)) = \deg p(x)$ for each nonzero polynomial $p(x) \in \mathbb{F}[x]$.

2. Let $\Omega \subset \mathbb{C}^n$ be a domain. Construct a nonzero function f depending on one variable in Ω, which has an infinite number of zeros in Ω. Prove that f cannot be decomposed to a finite product of irreducible elements. Hence, $H(\Omega)$ is not $\mathbb{D}_U$.

3. Consider the equation in (1.3.3) for $r \neq 0$. Show that $(a, b) = (a, r)$. Using this result prove (1.3.5).

1.4   Factorizations in $\mathbb{D}[x]$

Let $\mathbb{F}$ be the field of quotients of $\mathbb{D}$. Assume that $p(x) \in \mathbb{D}[x]$. Suppose that

$$p(x) = p_1(x)p_2(x), \quad for \ some \ p_1(x), p_2(x) \in \mathbb{F}[x].$$

We discuss the problem of determining when $p_1(x), p_2(x) \in \mathbb{D}[x]$. One has to take into account that for any $q(x) \in \mathbb{F}[x]$,

$$q(x) = \frac{p(x)}{a}, \quad for \ some \ p(x) \in \mathbb{D}[x], \quad and \ some \ a \in \mathbb{D}. \quad (1.4.1)$$

Definition 1.4.1 *Let*

$$p(x) = a_0 x^m + \cdots + a_m \in \mathbb{D}[x]. \qquad (1.4.2)$$

$p(x)$ *is called normalized if $a_0 = 1$. Let $\mathbb{D}$ be a GCD domain and denote $c(p) = (a_0, \ldots, a_m)$. $p(x)$ is called primitive if $c(p) = 1$.*

The following result follows from Problem 2.

Lemma 1.4.2 *Let $\mathbb{F}$ be the quotient field of $\mathbb{D}_G$. Then for any $q(x) \in \mathbb{F}[x]$ there exists a decomposition (1.4.1) where $(c(p), a) = 1$. The polynomial $p(x)$ is uniquely determined up to an invertible factor in $\mathbb{D}_G$. Furthermore,*

$$q(x) = \frac{b}{a} r(x), \quad r(x) \in \mathbb{D}_G[x], \ a, b \in \mathbb{D}_G, \qquad (1.4.3)$$

where $(a, b) = 1$ and $r(x)$ is primitive.

Lemma 1.4.3 (Gauss's lemma) *Let $p(x), q(x) \in \mathbb{D}_U[x]$ be primitive. Then $p(x)q(x)$ is primitive.*

The proof of Gauss's lemma follows from the following proposition.

Proposition 1.4.4 *Let $p, q \in \mathbb{D}_G[x]$. Assume that $\pi \in \mathbb{D}$ is a prime element which divides $c(pq)$. Then π divides either $c(p)$ or $c(q)$.*

Proof. Clearly, it is enough to assume that $p, q \neq 0$. We prove the proposition by induction on $k = \deg p + \deg q$. For $k = 0$ $p(x) = a_0$, $q(x) = b_0$. Hence, $c(pq) = a_0 b_0$. Since $\pi | a_0 b_0$, we deduce that π divides either $a_0 = c(p)$ or $b_0 = c(q)$.

Assume that the proposition holds for $k \leq l$ and assume that $k = l + 1$. Let $p = \sum_{i=0}^{m} a_i x^i, q = \sum_{j=0}^{n} b_j x^j$, where $a_m b_n \neq 0$ and $l + 1 = m + n$. So $\pi | a_m b_n$. Without loss of generality we may assume the nontrivial case $\pi | a_m$ and $m > 0$. Let $r(x) = \sum_{i=0}^{m-1} a_i x^i$. Since $\pi | c(pq)$ it is straightforward to show that $\pi | c(rq)$. As $\deg r + \deg q \leq l$ we deduce that $\pi | c(r) c(q)$. If $\pi | c(q)$ the proposition follows. If $\pi | c(r)$ then $\pi | c(p)$ and the proposition follows in this case too. $\square$

Corollary 1.4.5 *Let $p(x) \in \mathbb{D}_U[x]$ be primitive. Assume that $p(x)$ is irreducible in $\mathbb{F}[x]$, where $\mathbb{F}$ is the quotient field of $\mathbb{D}_U$. Then $p(x)$ is irreducible in $\mathbb{D}_U[x]$.*

Theorem 1.4.6 *Let $\mathbb{F}$ be the quotient field of $\mathbb{D}_U$. Then any $p(x) \in \mathbb{D}_U[x]$ has unique decomposition (up to invertible elements in $\mathbb{D}_U$) :*

$$p(x) = a q_1(x) \cdots q_s(x), \quad q_1, \ldots, q_s \in \mathbb{D}_U[x], \; a \in \mathbb{D}_U, \qquad (1.4.4)$$

where $q_1(x), \ldots, q_s(x)$ are primitive and irreducible in $\mathbb{F}[x]$, and a has decomposition (1.3.1). Hence, $\mathbb{D}_U[x]$ is a UFD.

See [Lan67] and Problems 3–5.

Normalization 1.4.7 *Let $\mathbb{F}$ be a field and assume that $p(x) \in \mathbb{F}[x]$ is a nonconstant normalized polynomial in $\mathbb{F}[x]$. Let (1.4.4) be a decomposition to irreducible factors. Normalize the decomposition (1.4.4) by letting $q_1(x), \ldots, q_s(x)$ be normalized irreducible polynomials in $\mathbb{F}[x]$. (Then $a = 1$.)*

Lemmas 1.4.3 and 1.4.5 yield (see Problem 5):

Theorem 1.4.8 *Let $p(x)$ be a normalized nonconstant polynomials in $\mathbb{D}_U[x]$. Let (1.4.4) be a normalized decomposition in $\mathbb{F}[x]$, where $\mathbb{F}$ is the quotient field of $\mathbb{D}_U$. Then $q_1(x), \ldots, q_s(x)$ are irreducible polynomials in $\mathbb{D}_U[x]$.*

Theorem 1.4.9 *Let $\Omega \subset \mathbb{C}^n$ be a connected set. Assume that $p(x)$ is a normalized nonconstant polynomial in $H(\Omega)[x]$. Let $(1.4.4)$ be a normalized decomposition in $\mathcal{M}[x]$, where $\mathcal{M}$ is the field of meromorphic functions in Ω. Then each $q_j(x)$ is an irreducible polynomial in $H(\Omega)[x]$.*

Proof. By the definition of $H(\Omega)$ we may assume that $p(x) \in H(\Omega_0)[x]$, $q_j(x) \in \mathcal{M}(\Omega_0)[x]$, $j = 1, \ldots, s$ for some domain $\Omega_0 \supset \Omega$. Let

$$q(x, z) = x^t + \sum_{r=1}^{t} \frac{\alpha_r(z)}{\beta_r(z)} x^{t-r}, \tag{1.4.5}$$

$$x \in \mathbb{C}, \ z \in \Omega_0, \ \alpha_r(z), \beta_r(z) \in H(\Omega_0), \ r = 1, \ldots, t.$$

Then $q(x, z)$ is analytic on $\Omega_0 \backslash \Gamma$, where Γ is an analytic variety given by

$$\Gamma = \{z \in \Omega_0 : \prod_{r=1}^{t} \beta_r(z) = 0\}.$$

Let $x_1(z), \ldots, x_t(z)$ be the roots of $q(x, z) = 0$, which is well defined as an unordered set of functions $\{x_1(z), \ldots, x_t(z)\}$ on $\Omega \backslash \Gamma$. Suppose that each $x_k(z)$ is bounded on some neighborhood O of a point $\zeta \in \Gamma$. Then each $\frac{\alpha_j(z)}{\beta_j(z)}$, which is the j symmetric function of $\{x_1(z), \ldots, x_t(z)\}$, is bounded on O. The Riemann extension theorem [GrH78] implies that $\frac{\alpha_j(z)}{\beta_j(z)}$ is analytic in O. If each $x_k(z)$ is bounded in the neighborhood of each $\zeta \in \Gamma$ it follows that $\frac{\alpha_k(z)}{\beta_k(z)} \in H(\Omega_0)$, $k = 1, \ldots, t$.

The assumption that $p(x, z)$ is a normalized polynomial in $H(\Omega_0)$ yields that all the roots of $p(x, z) = 0$ are bounded on any compact set $S \subset \Omega$. The above arguments show that each $q_j(x, z)$ in the decomposition $(1.4.4)$ of $p(x, z)$ is an irreducible polynomial in $H(\Omega)[x]$. $\square$

Problems

1. $a_1, \ldots, a_k \in \mathbb{D}\setminus\{0\}$ are said to have a least common multiple, denoted by $\text{lcm}(a_1, \ldots, a_k)$ and abbreviated as lcm, if the following conditions hold. Assume that $b \in \mathbb{D}$ is divisible by each $a_i, i = 1, \ldots, k$. Then $\text{lcm}(a_1, \ldots, a_k) | b$. (Note that the lcm is defined up to an invertible element.) Let $\mathbb{D}$ be a GCD domain. Show

   (a) $\text{lcm}(a_1, a_2) = \frac{a_1 a_2}{(a_1, a_2)}$.

   (b) For $k > 2$ $\text{lcm}(a_1, \ldots, a_k) = \frac{\text{lcm}(a_1, \ldots, a_{k-1}) a_k}{(\text{lcm}(a_1, \ldots, a_{k-1}), a_k)}$.

2. Let $\mathbb{F}$ be the quotient field of $\mathbb{D}_G$. Assume that $0 \neq q(x) \in \mathbb{F}[x]$. Write $q(x) = \sum_{i \in I} \frac{b_i}{a_i} x^i$ where $a_i, b_i \in \mathbb{D}_G\setminus\{0\}$ for each $i \in I$, and $I = \{0 \leq i_1 < \cdots < i_k\}$ is a finite subset of $\mathbb{Z}_+$. Let $a_i' = \frac{a_i}{(a_i, b_i)}, b_i' = \frac{b_i}{(a_i, b_i)}$ for $i \in I$. Then (1.4.1) holds, where $a = \text{lcm}(a_{i_1}', \ldots, a_{i_k}')$ and $p(x) = \sum_{i \in I} \frac{b_i' a}{a_i'} x^i$. Show that $(c(p), a) = 1$. Furthermore, if $q(x) = \frac{r(x)}{c}$ for some $r(x) \in \mathbb{D}_G[x], c \in \mathbb{D}_G$ then $c = ea, r(x) = ep(x)$ for some $e \in \mathbb{D}_G\setminus\{0\}$.

3. Let $p(x)$ be given by (1.4.2) and put

   $$q(x) = b_0 x^n + \cdots + b_n, \quad r(x) = p(x)q(x) = c_0 x^{m+n} + \cdots + c_{m+n}.$$

   Assume that $p(x), q(x) \in \mathbb{D}_U[x]$. Let π be an irreducible element in $\mathbb{D}_U$ such that

   $$\pi | a_i, \ i = 0, \ldots, \alpha, \quad \pi | b_j, \ j = 0, \ldots, \beta, \quad \pi | c_{\alpha+\beta+2}.$$

   Then either $\pi | a_{\alpha+1}$ or $\pi | b_{\beta+1}$.

4. Prove that if $p(x), q(x) \in \mathbb{D}_U[x]$ then $c(pq) = c(p)c(q)$.

   Deduce from the above equality Lemma 1.4.3. Also, show that if $p(x)$ and $q(x)$ are normalized polynomials then $p(x)q(x)$ is primitive.

5. Prove Theorem 1.4.8.

6. Using the equality $\mathbb{D}[x_1, \ldots, x_{n-1}][x_n] = \mathbb{D}[x_1, \ldots, x_n]$, prove that $\mathbb{D}_U[x_1, \ldots, x_n]$ is a UFD. Deduce that $\mathbb{F}[x_1, \ldots, x_n]$ is a UFD.

1.5   Elementary Divisor Domains

Definition 1.5.1 $\mathbb{D}_G$ *is an elementary divisor domain, or simply EDD and denoted by* $\mathbb{D}_{ED}$, *if for any three elements* $a, b, c \in \mathbb{D}$ *there exists* $p, q, x, y \in \mathbb{D}$ *such that*

$$(a, b, c) = (px)a + (py)b + (qy)c. \qquad (1.5.1)$$

By letting $c = 0$ we obtain that (a, b) is a linear combination of a and b. Hence, an elementary divisor domain is a Bezout domain.

Theorem 1.5.2 *Let* $\mathbb{D}$ *be a principal ideal domain. Then* $\mathbb{D}$ *is an elementary divisor domain.*

Proof. Without loss of generality we may assume that $abc \neq 0$, $(a, b, c) = 1$. Let $(a, c) = d$. Since $\mathbb{D}$ is $\mathbb{D}_U$ ([Lan67]), we decompose $a = a'a''$, where in the prime decomposition (1.3.1) of a, a' contains all the irreducible factors of a, which appear in the decomposition of d into irreducible factors. Thus

$$a = a'a'', \ (a', a'') = 1, \ (a', c) = (a, c), \ (a'', c) = 1, \qquad (1.5.2)$$

and if a', f are not coprime then c, f are not coprime.

Hence, there exist q and α such that

$$b - 1 = -qc + \alpha a''. \qquad (1.5.3)$$

Let $d' = (a, b+qc)$. The above equality implies that $(d', a'') = 1$. Suppose that d' is not coprime with a'. Then there exists a noninvertible f such that f divides d' and a'. According to (1.5.2) $(f, c) = f'$ and f' is not invertible. Thus $f'|b$ which implies that f' divides a, c and b, which is contradictory to our assumption that $(a, b, c) = 1$. So $(d', a') = 1$ which implies $(d', a) = 1$. Therefore, there exists $x, y \in \mathbb{D}$ such that $xa + y(b + qc) = 1$. This shows (1.5.1) with $p = 1$. $\square$

Theorem 1.5.3 *Let $\Omega \subset \mathbb{C}$ be a connected set. Then $H(\Omega)$ is an elementary divisor domain.*

Proof. Given $a, b, c \in H(\Omega)$ we may assume that $a, b, c \in H(\Omega_0)$ for some domain $\Omega_0 \supset \Omega$. Theorem 1.2.6 yields

$$(a, b, c) = (a, (b, c)) = a + y(b, c) = a + y(b + qc). \qquad (1.5.4)$$

$\square$

Problems

1. $\mathbb{D}$ is called *adequate* if for any $0 \neq a, c \in \mathbb{D}$ (1.5.2) holds. Use the proof of Theorem 1.5.2 to show that any adequate $\mathbb{D}_B$ is $\mathbb{D}_{ED}$.

2. Prove that for any connected set $\Omega \subset \mathbb{C}$, $H(\Omega)$ is an adequate domain ([Hel43]).

1.6 Modules

Definition 1.6.1 M *is an abelian group if it has a binary operation, denoted by $+$, which satisfies the conditions (1.1.1)–(1.1.5).*

Definition 1.6.2 *Let R be a ring with identity. An abelian group* **M** *is called a (left) R-module if for each $r \in R$,* **v** $\in$ **M** *the product $r\mathbf{v}$ is an element of* **M** *such that the following properties hold:*

$$r(\mathbf{v}_1 + \mathbf{v}_2) = r\mathbf{v}_1 + r\mathbf{v}_2, \quad (r_1 + r_2)\mathbf{v} = r_1\mathbf{v} + r_2\mathbf{v},$$

$$(1.6.1)$$

$$(rs)\mathbf{v} = r(s\mathbf{v}), \quad 1\mathbf{v} = \mathbf{v}.$$

N $\subset$ **M** *is called a submodule if* **N** *is an R-module.*

Assume that R does not have zero divisors. (That is, if $r, s \in R$ and $rs = 0$ then either $r = 0$ or $s = 0$.) Then **M** *does not have zero divisors if*

$$r\mathbf{v} = \mathbf{0} \text{ if and only if } \mathbf{v} = \mathbf{0} \text{ for any } r \neq 0. \qquad (1.6.2)$$

Assume that $\mathbb{D}$ is an integral domain. Then **M** *is a called a $\mathbb{D}$-module if in addition to being a module,* **M** *does not have zero divisors.*

Let $\mathbb{F}$ be a field. Then an $\mathbb{F}$-module is called a vector space $\mathbf{V}$ over $\mathbb{F}$. A submodule of $\mathbf{V}$ is called a subspace.

A standard example of an R-module is

$$\mathrm{R}^m := \{\mathbf{v} = (v_1, \ldots, v_m)^\top : \quad v_i \in \mathrm{R}, \; i = 1, \ldots, m\}, \qquad (1.6.3)$$

where

$$\mathbf{u} + \mathbf{v} = (u_1 + v_1, \ldots, u_m + v_m)^\top,$$

$$(1.6.4)$$

$$r\mathbf{u} = (ru_1, \ldots, ru_m)^\top, \quad r \in \mathrm{R}.$$

Note that if R does not have zero divisors then R^n is an R-module with no zero divisors.

One of the standard examples of submodules in $\mathbb{D}^n$ is as follows: Consider the linear homogeneous system

$$\sum_{j=1}^{n} a_{ij}x_j = 0, \quad a_{ij}, x_j \in \mathbb{D}, \; i = 1, \ldots, m, \; j = 1, \ldots, n. \qquad (1.6.5)$$

Then the set of solutions $\mathbf{x} = (x_1, \ldots, x_n)^\top$ is a submodule of $\mathbb{D}^n$.

Definition 1.6.3 *A $\mathbb{D}$-module $\mathbf{M}$ is finitely generated if there exist n-elements (generators) $\mathbf{v}_1, \ldots, \mathbf{v}_n \in \mathbf{M}$ such that any $\mathbf{v} \in \mathbf{M}$ is of the form*

$$\mathbf{v} = \sum_{i=1}^{n} a_i \mathbf{v}_i, \quad a_i \in \mathbb{D}, \; i = 1, \ldots, n. \qquad (1.6.6)$$

If each $\mathbf{v}$ can be expressed uniquely in the above form then $\mathbf{v}_1, \ldots, \mathbf{v}_n$ is called a basis in $\mathbf{M}$, and $\mathbf{M}$ is said to have a finite basis. We denote a basis in $\mathbf{M}$ by $[\mathbf{v}_1, \ldots, \mathbf{v}_n]$.

Note that $\mathbb{D}^m$ has a standard basis $\mathbf{v}_i = (\delta_{i1}, \ldots, \delta_{in})^\top, i = 1, \ldots, n.$

We now bring a short proof to the well-known fact that a finitely generated vector space has a finite basis. We start with the following well-known result.

Lemma 1.6.4 *Let $\mathbb{D}$ be an integral domain. Assume that $n > m > 0$ are integers. Consider the submodule $\mathbf{N} \subset \mathbb{D}^n$ given by (1.6.5). Then $\mathbf{N}$ contains a nonzero element.*

Proof. We first prove this result where $\mathbb{D}$ is a field $\mathbb{F}$. We prove this result by induction on n. For $n = 2$ we have $m = 1$ and the lemma follows immediately. Suppose that lemma holds for $n = N$. Assume that $n = N + 1$. If $a_{m1} = \cdots = a_{mn} = 0$, let $x_n = 0$ and use the induction hypothesis. If $a_{mj} \neq 0$ then $x_j = \frac{-1}{a_{mj}} \sum_{i \neq j} a_{mi} x_i$. Substitute x_j by this expression in the first $m - 1$ equations in (1.6.5) and use induction to deduce the lemma.

Consider now the case $\mathbb{D}$ is an integral domain. Let $\mathbb{F}$ be its field of quotients. By previous arguments there exist a solution $\mathbf{x} = (x_1, \ldots, x_n)^\top \in \mathbb{F}^n \backslash \{\mathbf{0}\}$. Hence, there exists $a \in \mathbb{D} \backslash \{0\}$ such that $a\mathbf{x} \in \mathbb{D}^n$. So $a\mathbf{x} \in \mathbf{N} \backslash \{\mathbf{0}\}$. $\qquad\square$

Definition 1.6.5 *Let $\mathbb{D}$ be an integral domain and $\mathbf{M}$ be a $\mathbb{D}$-module. Assume that $\mathbf{v}_1, \ldots, \mathbf{v}_n \in \mathbf{M}$.*

1. $a_1 \mathbf{v}_1 + \cdots + a_n \mathbf{v}_n$ *is called a linear combination of $\mathbf{v}_1, \ldots, \mathbf{v}_n$.*

2. $\mathbf{v}_1, \ldots, \mathbf{v}_n$ *are called linearly independent if the equality $\sum_{i=1}^n a_i \mathbf{v}_i = \mathbf{0}$ implies that $a_1 = \ldots = a_n = 0$.*

3. $\mathbf{v}_1, \ldots, \mathbf{v}_n$ *are called linearly dependent if there exist $a_1, \ldots, a_n \in \mathbb{D}$ not all 0 such that $\sum_{i=1}^n a_i \mathbf{v}_i = \mathbf{0}$.*

Theorem 1.6.6 *Let $\mathbf{V} \neq \{\mathbf{0}\}$ be a vector space over $\mathbb{F}$. Then the following are equivalent.*

1. $\mathbf{V}$ *has a basis $[\mathbf{v}_1, \ldots, \mathbf{v}_n]$.*

2. *Any $n+1$ elements in $\mathbf{V}$ are linearly dependent and any n linearly independent elements in $\mathbf{V}$ form a basis in $\mathbf{V}$. (n is called the dimension of $\mathbf{V}$ and denoted as $\dim V$.)*

3. $\mathbf{V}$ *is finitely generated.*

Proof. $1 \Rightarrow 2$. Let $\mathbf{u}_1, \ldots, \mathbf{u}_{n+1} \in \mathbf{V}$. So $\mathbf{u}_j = \sum_{i=1}^{n} b_{ij} \mathbf{v}_i$ for $j = 1, \ldots, n+1$. Clearly

$$\sum_{j=1}^{n+1} x_j \mathbf{u}_j = \sum_{i=1}^{n} \left(\sum_{j=1}^{n+1} b_{ij} x_j \right) \mathbf{v}_i.$$

Lemma 1.6.4 yields that the system $\sum_{j=1}^{n+1} b_{ij} x_j = 0$ for $i = 1, \ldots, n$ has a nontrivial solution. Hence, $\mathbf{u}_1, \ldots, \mathbf{u}_{n+1}$ are linearly dependent. Assume now that $\mathbf{u}_1, \ldots, \mathbf{u}_n \in \mathbf{V}$ linearly independent. Let $\mathbf{u} \in \mathbf{V}$. So $\mathbf{u}_1, \ldots, \mathbf{u}_n, \mathbf{u}$ are linearly dependent. Hence, $x_1 \mathbf{u}_1 + \cdots + x_n \mathbf{u}_n + a\mathbf{u} = \mathbf{0}$ for some $(x_1, \ldots, x_n, a)^\top \neq \mathbf{0}$. As $\mathbf{u}_1, \ldots, \mathbf{u}_n$ are linearly independent $a \neq 0$. So, $\mathbf{u} = \sum_{i=1}^{n} y_i \mathbf{u}_i$. Suppose $\mathbf{u} = \sum_{i=1}^{n} z_i \mathbf{u}_i$. So, $\mathbf{u} - \mathbf{u} = \mathbf{0} = \sum_{i=1}^{n} (y_i - z_i) \mathbf{u}_i$. As $\mathbf{u}_1, \ldots, \mathbf{u}_n$ are linearly independent it follows that $y_i = z_i$ for $i = 1, \ldots, n$. Hence, $\mathbf{u}_1, \ldots, \mathbf{u}_n$ is a basis of $\mathbf{V}$.

Clearly, $2 \Rightarrow 1$ and $1 \Rightarrow 3$ We now show $3 \Rightarrow 1$. Assume that $\mathbf{u}_1, \ldots, \mathbf{u}_m$ generate $\mathbf{V}$. Suppose first that $\mathbf{u}_1, \ldots, \mathbf{u}_m$ are linearly independent. Then the above arguments show that $\mathbf{u}_1, \ldots, \mathbf{u}_m$ is a basis in $\mathbf{V}$. Assume next that $\mathbf{u}_1, \ldots, \mathbf{u}_m$ are linearly dependent. Hence, $\mathbf{u}_i$ is a linear combination of $\mathbf{u}_1, \ldots, \mathbf{u}_{i-1}, \mathbf{u}_{i+1}, \ldots, \mathbf{u}_m$ for some i. In particular, these $m - 1$ vectors generate $\mathbf{V}$. Continuing in this manner we find a subset of n vectors out of $\mathbf{u}_1, \ldots, \mathbf{u}_m$ which generate $\mathbf{V}$ and are linearly independent. These n vectors form a basis in $\mathbf{V}$. $\qquad\square$

A finitely generated vector space is called *finite dimensional*.

Lemma 1.6.7 *Let $\mathbb{D}$ be an integral domain and assume that $\mathbf{M}$ is a $\mathbb{D}$-module. Let $\mathbb{F}$ be the quotient field of $\mathbb{D}$. Let $\tilde{\mathbf{V}}$ be the set of all elements of the form $(a, \mathbf{v})$ where $a \in \mathbb{F} \setminus \{0\}$ and $\mathbf{v} \in \mathbf{M}$. Define a relation on $\tilde{\mathbf{V}}$: $(a, \mathbf{v}) \sim (b, \mathbf{w})$ if there exists $c \in \mathbb{D} \setminus \{0\}$ such that $ca, cb \in \mathbb{D}$ and $(ca)\mathbf{v} = (cb)\mathbf{w}$. Then*

1. $\sim$ *is an equivalence relation.*

2. *Let $\mathbf{V} := \tilde{\mathbf{V}} / \sim$. Define $\phi : \mathbf{M} \to \mathbf{V}$ such that $\phi(\mathbf{v}) = (1, \mathbf{v})$. Then*

(a) *One can define uniquely the addition on* **V** *and multiplication from the left by* $r \in \mathrm{R}$ *such that*

$$\phi(r(\mathbf{u}+\mathbf{v})) = r(\phi(\mathbf{u})+\phi(\mathbf{v})), \quad \phi((r+s)\mathbf{u}) = (r+s)\phi(\mathbf{u}).$$

(b) **V** *is a vector space over* **F**.

(c) *If* **M** *is finitely generated then* **V** *is finitely generated.*

We leave the proof of this lemma as an exercise.

Corollary 1.6.8 *Any two finite bases of a* $\mathbb{D}$ *module contain the same number of elements* dim **V**, *where* **V** *is defined in Lemma* 1.6.7.

In §1.13, we show that the above module has a basis if $\mathbb{D}$ is a Bezout domain.

Notation 1.6.9 *For a set* S *denote by* $\mathrm{S}^{m \times n}$ *the set of all* $m \times n$ *matrices* $A = [a_{ij}]_{i=j=1}^{i=m,j=n}$, *where each* $a_{ij} \in \mathrm{S}$. *In what follows assume that* $A \in \mathrm{S}^{m \times n}$ *as above.*

1. *The vectors* $\mathbf{r}_i := (a_{i1}, \ldots, a_{in})$ *and* $\mathbf{c}_j := (a_{1j}, \ldots, a_{mj})^\top$ *are called the* ith *row and the* jth *columns of* A *respectively.*

2. *Denote by* $A^\top \in \mathrm{S}^{n \times m}$ *the matrix* $B = [b_{pq}]_{p=q=1}^{n,m}$ *where* $b_{pq} = a_{qp}$ *for* $p = 1, \ldots, n, q = 1, \ldots, m$.

3. A *is called upper triangular or lower triangular if* $a_{ij} = 0$ *for* $i > j$ *or* $j > i$ *respectively.*

4. A *is called diagonal if* $a_{ij} = 0$ *for* $i \neq j$.

Definition 1.6.10 *Let* R *be a ring with identity. Then:*

1. *Assume that* $m, n, p \in \mathbb{N}$. *Then for* $A = [a_{ij}]_{i=j=1}^{m,n} \in \mathrm{R}^{m \times n}$ *and* $B = [b_{jk}]_{j=k=1}^{n,p} \in \mathrm{R}^{n \times p}$ *define* $AB = [c_{ik}]_{i=k=1}^{m,p}$ *by the formula:*

$$c_{ik} = \sum_{j=1}^{n} a_{ij} b_{jk}, \quad i = 1, \ldots, m, \ k = 1, \ldots, p.$$

2. *Define* $I_n := [\delta_{ij}]_{i=j=1}^{n} \in \mathrm{R}^{n \times n}$.

3. $A \in \mathbb{R}^{n \times n}$ *is called invertible, or unimodular, if there exists $B \in$*
   $\mathbb{R}^{n \times n}$ *such that $AB = BA = I_n$.*

Lemma 1.6.11 *Let $\mathbb{R}$ be a ring with identity. Then*

1. $\mathbb{R}^{n \times n}$ *is a ring with identity I_n.*

2. $\mathbb{R}^{m \times n}$ *is an $\mathbb{R}$-left and $\mathbb{R}$-right module. (That is rA and Ar are defined for each $A \in \mathbb{R}^{m \times n}$ and $r \in \mathbb{R}$, and the corresponding distribution properties apply.)*

3. $\mathbb{R}^{m \times n}$ *is a left $\mathbb{R}^{m \times m}$-module and a right $\mathbb{R}^{n \times n}$-module.*

4. *For $n \geq 2$ $\mathbb{R}^{n \times n}$ has zero divisors even if $\mathbb{R}$ is a field.*

5. *Assume in addition that $\mathbb{R}$ is commutative. Then $\mathbb{R}^{n \times n}$ is an algebra: $r(AB) = (rA)B = A(rB)$.*

The proof of the lemma is left to the reader.

Definition 1.6.12 *Let $\mathbf{M}$ be a module over $\mathbb{D}$. Assume that $\mathbf{M}_i$ is a submodule of $\mathbf{M}$ for $i = 1, \ldots, k$. Then $\mathbf{M}$ is called a direct sum of $\mathbf{M}_1, \ldots, \mathbf{M}_k$, and denoted as $\mathbf{M} = \oplus_{i=1}^k \mathbf{M}_i$, if every element $\mathbf{m} \in \mathbf{M}$ can be expressed in unique way as a sum $\mathbf{m} = \sum_{i=1}^k \mathbf{m}_i$, where $\mathbf{m}_i \in \mathbf{M}_i$ for $i = 1, \ldots, k$.*

Definition 1.6.13 *The ring of quaternions $\mathbb{H}$ is a four-dimensional vector space over $\mathbb{R}$ with the basis $1, \mathbf{i}, \mathbf{j}, \mathbf{k}$, i.e. vectors of the form*

$$\mathbf{q} = a + b\mathbf{i} + c\mathbf{j} + d\mathbf{k}, \quad a, b, c, d \in \mathbb{R}, \tag{1.6.7}$$

where

$$\mathbf{i}^2 = \mathbf{j}^2 = \mathbf{k}^2 = -1, \ \mathbf{ij} = -\mathbf{ji} = \mathbf{k}, \ \mathbf{jk} = -\mathbf{kj} = \mathbf{i}, \ \mathbf{ki} = -\mathbf{ik} = \mathbf{j}.$$
$$\tag{1.6.8}$$

It is known that $\mathbb{H}$ is a noncommutative division algebra over $\mathbb{R}$. See Problem 7.

Problems

1. Prove Lemma 1.6.7.

2. Let $\mathbf{M}$ be a finitely generated module over $\mathbb{D}$. Let $\mathbb{F}$ be the quotient field of $\mathbb{D}$. Show

   (a) Assume that $\mathbf{M}$ is generated by $\mathbf{a}_1, \ldots, \mathbf{a}_m$. Let $\mathbf{N} = \{\mathbf{x} = (x_1, \ldots, x_m)^\top \in \mathbb{D}^m, \sum_{i=1}^m x_i \mathbf{a}_i = \mathbf{0}\}$. Then $\mathbf{N}$ is a $\mathbb{D}$-module.

   (b) Let $\mathbf{U} \subseteq \mathbb{F}^m$ be the subspace generated by all vectors in $\mathbf{N}$. (Any vector in $\mathbf{U}$ is a finite linear combination of vectors in $\mathbf{N}$.)

   (i) Show that any vector $\mathbf{u} \in \mathbf{U}$ is of the form $\frac{1}{c}\mathbf{b}$, where $\mathbf{b} \in \mathbf{N}$.

   (ii) Assume that $\dim \mathbf{U} = l$. Pick a basis $[\mathbf{u}_1, \ldots, \mathbf{u}_l]$ in $\mathbf{U}$ and complete this basis to a basis in $\mathbb{F}^m$. So $[\mathbf{u}_1, \ldots, \mathbf{u}_l, \mathbf{w}_1, \ldots, \mathbf{w}_{m-l}]$ is a basis in $\mathbb{F}^m$. Let $\mathbf{W} = \mathrm{span}\,(\mathbf{w}_1, \ldots, \mathbf{w}_{m-l})$. Let $\mathbf{V}$ be the quotient space $\mathbb{F}^m/\mathbf{U}$. Show that any vector in $\mathbf{V}$ is of the form of a coset $\mathbf{w} + \mathbf{U}$ for a unique vector $\mathbf{w} \in \mathbf{W}$.

   (c) Define $\phi : \mathbf{M} \to \mathbf{V}$ as follows. Let $\mathbf{a} \in \mathbf{M}$ and write $\mathbf{a} = \sum_{i=1}^m a_i \mathbf{a}_i$. Set $\phi(\mathbf{a}) = (a_1, \ldots, a_m)^\top + \mathbf{U}$. Then

   (i) ϕ is well defined, i.e. does not depend on a particular representation of $\mathbf{a}$ as a linear combination of $\mathbf{a}_1, \ldots, \mathbf{a}_m$.

   (ii) $\phi(\mathbf{a}) = \phi(\mathbf{b}) \iff \mathbf{a} = \mathbf{b}$.

   (iii) $\phi(a\mathbf{a} + b\mathbf{b}) = a\phi(\mathbf{a}) + b\phi(\mathbf{b})$ for any $a, b \in \mathbb{D}$ and $\mathbf{a}, \mathbf{b} \in \mathbf{M}$.

   (iv) For any $\mathbf{v} \in \mathbf{V}$ there exists $a \in \mathbb{D}$ and $\mathbf{a} \in \mathbf{M}$ such that $\phi(\mathbf{a}) = a\mathbf{v}$.

   (d) Let $\mathbf{Y}$ be a finite dimensional vector space over $\mathbb{F}$ with the following properties.

   (i) There is an injection $\phi : \mathbf{M} \to \mathbf{Y}$, i.e. ϕ is one to one, such that $\phi(a\mathbf{m} + b\mathbf{n}) = a\phi(\mathbf{m}) + b\phi(\mathbf{n})$ for any $a, b \in \mathbb{D}$ and $\mathbf{m}, \mathbf{n} \in \mathbf{M}$.

(ii) For any $\mathbf{y} \in \mathbf{Y}$ there exists $a \in \mathbb{D}$ and $\mathbf{m} \in \mathbf{M}$ such that $\phi(\mathbf{m}) = a\mathbf{y}$.

Show that dim $\mathbf{Y} = \dim \mathbf{V}$, where $\mathbf{V}$ is defined in part b.

3. Let $\mathbf{M}$ be a $\mathbb{D}$-module with a finite basis. Let $\mathbf{N}$ be a submodule of $\mathbf{M}$. Show that if $\mathbb{D}$ is $\mathbb{D}_P$ then $\mathbf{N}$ has a finite basis.

4. Let $\mathbf{M}$ be a $\mathbb{D}$-module with a finite basis. Assume that $\mathbf{N}$ is a finitely generated submodule of $\mathbf{M}$. Show that if $\mathbb{D}$ is $\mathbb{D}_B$ then $\mathbf{N}$ has a finite basis.

5. Prove Lemma 1.6.11.

6. Let $\mathbf{M}$ be a module over $\mathbb{D}$. Assume that $\mathbf{M}_i$ is a submodule of $\mathbf{M}$ for $i = 1, \ldots, k$. Then $\mathbf{N} := M_1 + \cdots + M_k$ is the set of all elements $\mathbf{m} \in \mathbf{M}$ of the form $\mathbf{m}_1 + \cdots + \mathbf{m}_k$, where $\mathbf{m}_i \in \mathbf{M}_i$ for $i = 1, \ldots, k$. Show

(a) $\mathbf{N}$ is a submodule of $\mathbf{M}$.

(b) $\cap_{i=1}^k \mathbf{M}_i$ is a submodule of $\mathbf{M}$.

(c) Assume that $\mathbf{M}_1, \ldots, \mathbf{M}_k$ are finitely generated. Then $\mathbf{N}$ is finitely generated and dim $\mathbf{N} \leq \sum_{i=1}^k \dim \mathbf{M}_i$.

(d) Assume that $\mathbf{M}_1, \ldots, \mathbf{M}_k$ have bases and $\mathbf{N} = \oplus_{i=1}^k \mathbf{M}_i$. Then $\mathbf{N}$ has a basis and dim $\mathbf{N} = \sum_{i=1}^k \dim \mathbf{M}_i$.

7. Show

(a) The ring of quaternions $\mathbb{H}$ can be viewed as $\mathbb{C}^2$, where each $\mathbf{q} \in \mathbb{H}$ of the form (1.6.7) can be written as $\mathbf{q} = z + w\mathbf{j}$, where $z = a + bi, w = c + di \in \mathbb{C}$. Furthermore, for any $z \in \mathbb{C}$, $\mathbf{j}z = \bar{z}\mathbf{j}$.

(b) $\mathbb{H}$ is a ring with the identity $1 = 1 + 0\mathbf{i} + 0\mathbf{j} + 0\mathbf{k}$.

(c) $(r\mathbf{q})\mathbf{s} = \mathbf{q}(r\mathbf{s})$ for any $\mathbf{q}, \mathbf{s} \in \mathbb{H}$ and $r \in \mathbb{R}$. Hence, $\mathbb{H}$ is an algebra over $\mathbb{R}$.

(d) Denote $|\mathbf{q}| = \sqrt{a^2 + b^2 + c^2 + d^2}$, $\bar{\mathbf{q}} = a - b\mathbf{i} - c\mathbf{j} - d\mathbf{k}$ for any $\mathbf{q}$ of the form (1.6.7). Then $\bar{\mathbf{q}}\mathbf{q} = \mathbf{q}\bar{\mathbf{q}} = |\mathbf{q}|^2$. Hence, $|\mathbf{q}|^{-2}\bar{\mathbf{q}}$ is the right and the left inverse of $\mathbf{q} \neq 0$.

1.7 Determinants

Definition 1.7.1

1. *For a positive integer n denote $[n] := \{1, \ldots, n\}$.*

2. *For $k \in [n]$ denote by $[n]_k$ the set of all subsets of $[n]$ of cardinality k. Each $\alpha \in [n]_k$ is represented by $\boldsymbol{\alpha} := (\alpha_1, \ldots, \alpha_k)$, where $\alpha_1, \ldots, \alpha_k$ are integers satisfying $1 \leq \alpha_1 < \cdots < \alpha_k \leq n$. Denote $\|\boldsymbol{\alpha}\|_1 := \sum_{j=1}^{k} \alpha_j$.*

3. *Let $A = [a_{ij}]_{i=j=1}^{m,n} S^{m \times n}$. Assume that $\boldsymbol{\alpha} = (\alpha_1, \ldots, \alpha_k) \in [m]_k, \boldsymbol{\beta} = (\beta_1, \ldots, \beta_l) \in [n]_l$. Denote by $A[\boldsymbol{\alpha}, \boldsymbol{\beta}] = [a_{\alpha_p \beta_q}]_{p=q=1}^{k,l} \in S^{k \times l}$ and by $A(\boldsymbol{\alpha}, \boldsymbol{\beta}) \in S^{(m-k) \times (n-l)}$ the matrix obtained from A by deleting $\alpha_1, \ldots, \alpha_k$ rows and $\beta_1, \ldots, \beta_l$ columns respectively.*

4. *Let Σ_n the group of bijections (permutations) $\sigma : [n] \to [n]$.*

5. *id $\in \Sigma_n$ is the identity element, i.e. $\mathrm{id}(i) = i$ for all $i \in [n]$.*

6. *An element $\tau \in \Sigma_n$ is called a transposition if there exists $i, j \in [n], i \neq j$ such that $\tau(i) = j, \tau(j) = i$ and $\tau(k) = k$ for $k \in [n] \setminus \{i, j\}$.*

7. *Let $\mathbb{D}$ be an integral domain and let $\mathbb{F}$ be its field of quotients. For $n \geq 2$ define $\mathrm{sign} : \Sigma_n \to \{-1, 1\} \subset \mathbb{D}$ as follows: Let $x_1, \ldots, x_n \in \mathbb{F}(x_1, \ldots, x_n)$. Then*

$$\mathrm{sign}\ \sigma = \frac{\prod_{1 \leq i < j \leq n}(x_{\sigma(j)} - x_{\sigma(i)})}{\prod_{1 \leq i < j \leq n}(x_j - x_i)}.$$

For $n = 1$ let $\mathrm{sign}\ \mathrm{id} = 1$.

8. *σ is called even or odd if either $\mathrm{sign}\ \sigma = 1$ or $\mathrm{sign}\ \sigma = -1$ respectively.*

The following lemma is well known and its proof is left as Problem 1:

Lemma 1.7.2

1. *Each $\sigma \in \Sigma_n$ is a product of transpositions*

$$\sigma = \tau_1 \circ \tau_2 \circ \cdots \circ \tau_m. \tag{1.7.1}$$

2. sign id $= 1$.

3. sign $\tau = -1$ *for each transposition.*

4. *The map* sign *is a homomorphism:* sign $\sigma \circ \omega = (\text{sign } \sigma)(\text{sign } \omega)$ *for all $\sigma, \omega \in \Sigma_n$.*

5. *Assume that* (1.7.1) *holds. Then* sign $\sigma = (-1)^m$.

Definition 1.7.3 *Let $\mathbb{D}$ be an integral domain. Let $A = [a_{ij}]_{i=j=1}^n \in \mathbb{D}^{n \times n}$. Then the determinant of A, denoted as* det A, *is defined as follows:*

$$\det A = \sum_{\sigma \in \Sigma_n} \text{sign } \sigma \, a_{1\sigma(1)} a_{2\sigma(2)} \cdots a_{n\sigma(n)}. \tag{1.7.2}$$

Theorem 1.7.4 *Let $\mathbb{D}$ be an integral domain. Assume that $A = [a_{ij}], B = [b_{ij}] \in \mathbb{D}^{n \times n}$. Then the folowing conditions hold:*

1. det A *is a multilinear function on the rows or columns of A.*

2. det $I_n = 1$.

3. det $A^\top = $ det A.

4. *The determinant of lower triangular or upper triangular matrix A is a product of its diagonal entries.*

5. *Let B be obtained from A by permuting two rows or columns of A. Then* det $B = -$det A.

6. *If A has two equal rows or columns then* det $A = 0$.

7. (*Laplace row and column expansion for determinants*) *For* $i, j \in$ $[n]$ *denote by* $A(i, j) \in \mathbb{D}^{(n-1) \times (n-1)}$ *the matrix obtained from A by deleting its ith row and jth column. Then*

$$\det A = \sum_{j=1}^{n} a_{ij}(-1)^{i+j} \det A(i, j) \qquad (1.7.3)$$

$$= \sum_{j=1}^{n} a_{ji}(-1)^{i+j} \det A(j, i),$$

for $i \in [n]$.

8. (*Laplace mulitple row or column expansion*) *Fix* $k \in [n-1]$ *and* $\boldsymbol{\alpha} = (\alpha_1, \ldots, \alpha_k) \in [n]_k$. *Then*

$$\det A = \sum_{\boldsymbol{\beta} \in [n]_k} (-1)^{\|\boldsymbol{\alpha}\|_1 + \|\boldsymbol{\beta}\|_1} \det A[\boldsymbol{\alpha}, \boldsymbol{\beta}] \det A(\boldsymbol{\alpha}, \boldsymbol{\beta}) =$$

$$(1.7.4)$$

$$\det A = \sum_{\boldsymbol{\beta} \in [n]_k} (-1)^{\|\boldsymbol{\alpha}\|_1 + \|\boldsymbol{\beta}\|_1} \det A[\boldsymbol{\beta}, \boldsymbol{\alpha}] \det A(\boldsymbol{\beta}, \boldsymbol{\alpha})$$

9. $\det AB = (\det A)(\det B)$.

10. (*Cauchy–Binet identity*) *Let* $n \geq 2$ *and* $m \in [n]$. *Assume that* $F \in \mathbb{D}^{m \times n}, G \in \mathbb{D}^{n \times m}$. *Then*

$$\det FG = \sum_{\boldsymbol{\alpha} \in [n]_m} \det F[[m], \boldsymbol{\alpha}] \det G[\boldsymbol{\alpha}, [m]]. \qquad (1.7.5)$$

Proof.

1. Immediate.

2. Follows from sign id $= 1$.

3. Note that if $\sigma(i) = j$ then $i = \sigma^{-1}(j)$. Hence

$$\det A = \sum_{\sigma \in \Sigma_n} \text{sign } \sigma \prod_{j \in [n]} a_{\sigma^{-1}(j)j}.$$

Let $\omega := \sigma^{-1}$ and recall that sign $\omega =$ sign σ. This proves the equality $\det A^\top = \det A$.

4. Assume that A is upper triangular. Note that $\prod_{i\in[n]} a_{i\sigma(i)} = 0$ if $\sigma(i) < i$ for some $i \in [n]$. Hence, $\prod_{i\in[n]} a_{i\sigma(i)} = 0$ if $\sigma \neq$ id. Therefore, $\det A = a_{11}, \ldots, a_{nn}$. Similar result holds for lower triangular matrices.

5. Let B obtained from A by permuting two columns. Hence, there exists a transposition $\tau \in \Sigma_n$ such that $b_{ij} = a_{i\tau(j)}$. Thus

$$\det B = \sum_{\sigma\in\Sigma_n} \text{sign } \sigma \prod_{i\in[n]} b_{i\sigma(i)} = \sum_{\sigma\in\Sigma_n} \text{sign } \sigma \prod_{i\in[n]} a_{i(\sigma\circ\tau)(i)}.$$

Let $\omega = \sigma \circ \tau$. Recall that sign $\omega = $ sign σ sign $\tau = -$sign σ. Hence, $\det B = -\det A$. Use 3 to deduce this equality if B is obtained from A by permuting two rows.

6. Assume that A has two identical rows. Suppose first that $-1 \neq 1$. Permute the two identical rows to deduce that $\det A = -\det A$. So $2\det A = 0$, hence $\det A = 0$. Suppose now that $-1 = 1$. So sign $\sigma = 1$ for each $\sigma \in \Sigma_n$. To show that $\det A = 0$ we may assume without loss of generality that the first and the second rows of A are the same: $a_{1j} = a_{2j} = b_j$ for $j \in [n]$. In the formula for $\det A$ consisting of $n!$ terms we have exactly two identical terms of the form $(b_i b_j)x, (b_j b_i)x$. Hence, $(b_i b_j)x + (b_j b_i)x = 2b_i b_j x = 0$. Therefore, $\det A = 0$. The same results follow if A has two identical columns.

7. We first show the validity of the first identity of (1.7.4). It is clear that the right-hand side of (1.7.4) contains exactly $n!$ terms appearing in $\det A$. It is left to show that the product $a_{1\sigma(1)}, \ldots, a_{n\sigma(n)}$ appears in the right-hand side of (1.7.4) with the sign sign σ. Assume first that $\beta = \alpha$. Then in the right-hand side of (1.7.4) we have the expression $\det A[\alpha, \alpha] \det A(\alpha, \alpha)$. Consider in Σ_n the subgroup $\Sigma_n(\alpha)$ of all permutations on $[n]$ which leave invariant α. This subgroup is isomorphic to the direct product of $\Sigma_k \times \Sigma_{n-k}$. So each element of this product is $\sigma = (\mu, \nu)$. It is straightforward to show that sign $\sigma = $ sign μ sign ν. Hence, the expression $\det A[\alpha, \alpha] \det A(\alpha, \alpha)$ contains $k!(n-k)!$ terms with the right signs. For a general β

we interchange the column β_i with the column α_i for $i = 1, \ldots, k$. Now use the previous case to deduce that the correct sign of the product $\det A[\boldsymbol{\alpha}, \boldsymbol{\beta}] \det A(\boldsymbol{\alpha}, \boldsymbol{\beta})$ is $(-1)^{\|\boldsymbol{\alpha}\|_1 + \|\boldsymbol{\beta}\|_1}$.

8. Follows from 7.

9. Let $\mathbf{a}_1, \ldots, \mathbf{a}_n$ and $\mathbf{b}_1, \ldots, \mathbf{b}_n$ be the columns of the matrix A and B respectively. So $A = [\mathbf{a}_1, \ldots, \mathbf{a}_n]$ and $B = [\mathbf{b}_1, \ldots, \mathbf{b}_n]$. It is straightforward to show that $BA = [B\mathbf{a}_1, \ldots, B\mathbf{a}_n]$. Let $\mathbf{x} = (x_1, \ldots, x_n)^\top \in \mathbb{D}^n$. Then $B\mathbf{x} = x_1\mathbf{b}_1 + \cdots + x_n\mathbf{b}_n$. Use the multilinearity of $\det BA$ in columns to deduce that $\det BA$ is a sum of n^n terms of the form: $(a_{i_1 1}, \ldots, a_{i_n n})\det [\mathbf{b}_{i_1} \ \mathbf{b}_{i_2}, \ldots, \mathbf{b}_{i_n}]$. Recall that $\det [\mathbf{b}_{i_1} \ \mathbf{b}_{i_2}, \ldots, \mathbf{b}_{i_n}] = 0$ if $i_p = i_q$ for some $p \neq q$. Hence, $\det BA$ is a sum of $n!$ terms of the form $(a_{i_1 1}, \ldots, a_{i_n n})\det [\mathbf{b}_{i_1} \ \mathbf{b}_{i_2}, \ldots, \mathbf{b}_{i_n}]$, where $\{i_1, \ldots, i_n\} = [n]$. There exists a unique $\sigma \in \Sigma_n$ such that $\sigma(j) = i_j$ for $j = 1, \ldots, n$. Note that $\det [\mathbf{b}_{i_1} \ \mathbf{b}_{i_2}, \ldots, \mathbf{b}_{i_n}] = \text{sign } \sigma \det B$. Hence

$$\det BA = \sum_{\sigma \in \Sigma_n} \det B \Big(\sum_{\sigma \in \Sigma_n} \text{sign } \sigma \prod a_{\sigma(i)i} \Big)$$

$$= \det B \det A^\top = \det B \det A.$$

10. Is proved similarly to 9. $\qquad\qquad\square$

Definition 1.7.5 *Let* $A = [a_{ij}] \in \mathbb{D}^{n \times n}$. *Then the adjoint of* A, *sometimes called the adjugate of* A, *is defined as* $\text{adj } A := [b_{ij}] \in \mathbb{D}^{n \times n}$, *where* $b_{ij} = (-1)^{i+j} A(j, i)$ *for* $i, j \in [n]$.

Lemma 1.7.6 *Let* $\mathbb{D}$ *be an integral domain. Assume that* $A \in \mathbb{D}^{n \times n}$. *Then*

1. $$A \text{ adj } A = (\text{adj } A)A = (\det A)I_n. \qquad (1.7.6)$$

2. *A is invertible if and only if* $\det A$ *is an invertible element in* $\mathbb{D}$. *Furthermore, if* A *is invertible then* $A^{-1} = \frac{1}{\det A} \text{adj } A$.

Proof. 1. Let adj $A = [b_{ij}]_{i,j \in [n]}$. The equality $\sum_{j=1}^n a_{ij} b_{ji} = \sum_{j=1}^n b_{ij} a_{ji} = \det A$ follows from the Laplace expansion. The equality $\sum_{j=1}^n a_{ij} b_{jk} = \sum_{j=1}^n b_{ij} a_{jk} = 0$ for $i \neq k$ follows from the observation that these sums are row or column expansion of the determinants of corresponding matrices having two identical rows or columns.

2. Suppose that $I_n = AB$. Then $1 = \det AB = (\det A)(\det B)$. Hence, $\det A$ is invertible. Vice versa assume that $\det A$ is invertible. Then (1.7.6) implies that $A^{-1} = \frac{1}{\det A} \text{adj } A$. Suppose that $AB = I_n$. Multiply from the left by $\frac{1}{\det A} \text{adj } A$ to deduce that $B = \frac{1}{\det A} \text{adj } A$. $\qquad\square$

Problems

1. Prove Lemma 1.7.2.

2. Let $m > n$ be positive integers and assume that $\mathbb{D}$ is an integral domain. Assume that $F \in \mathbb{D}^{m \times n}, G \in \mathbb{D}^{n \times m}$. Show that $\det AB = 0$.

1.8   Algebraically Closed Fields

Definition 1.8.1 *A field* $\mathbb{F}$ *is algebraically closed if any polynomial* $p(x) \in \mathbb{F}[x]$ *of the form* (1.4.2) *splits into linear factors in* $\mathbb{F}[x]$:

$$p(x) = a_0 \prod_{i=1}^m (x - \xi_i), \quad \xi_i \in \mathbb{F}, \; i = 1, \ldots, m, \; a_0 \neq 0. \qquad (1.8.1)$$

The classical example of an algebraically closed field is the field of complex numbers $\mathbb{C}$. The field of real numbers $\mathbb{R}$ is not algebraically closed.

Definition 1.8.2 *Let* $\mathbb{K} \supset \mathbb{F}$ *be fields. Then* $\mathbb{K}$ *is called an extension field of* $\mathbb{F}$. $\mathbb{K}$ *is called a finite extension of* $\mathbb{F}$ *if* $\mathbb{K}$ *is a finite dimensional vector space over* $\mathbb{F}$. *The dimension of the vector space* $\mathbb{K}$ *over* $\mathbb{F}$ *is called the degree of* $\mathbb{K}$ *over* $\mathbb{F}$ *and is denoted by* $[\mathbb{K} : \mathbb{F}]$.

Thus $\mathbb{C}$ is a finite extension of $\mathbb{R}$ of degree 2. It is known ([Lan67], see Problems 1–2) that:

Theorem 1.8.3 *Let $p(x) \in \mathbb{F}[x]$. Then there exists a finite extension $\mathbb{K}$ of $\mathbb{F}$ such that $p(x)$ splits into linear factors in $\mathbb{K}[x]$.*

The classical Weierstrass preparation theorem in two complex variables is an explicit example of the above theorem. We state the Weierstrass preparation theorem in a form needed later [GuR65].

Theorem 1.8.4 *Let H_0 be the ring of analytic functions in one variable in the neighborhood of the origin $0 \in \mathbb{C}$. Let $p(\lambda) \in H_0[\lambda]$ be a normalized polynomial of degree n*

$$p(\lambda, z) = \lambda^n + \sum_{j=1}^{n} a_j(z)\lambda^{n-j}, \quad a_j(z) \in H_0, \ j = 1, \ldots, n.$$

$$(1.8.2)$$

Then there exists a positive integer $s|n!$ such that

$$p(\lambda, w^s) = \prod_{j=1}^{n} (\lambda - \lambda_j(w)), \quad \lambda_j(w) \in H_0, \ j = 1, \ldots, n. \quad (1.8.3)$$

In this particular case the extension field $\mathbb{K}$ of $\mathbb{F} = \mathcal{M}_0$ is the set of multivalued functions in z, which are analytic in $z^{\frac{1}{s}}$ in the neighborhood of the origin. Thus $\mathbb{K} = \mathcal{M}_0(w)$, where

$$w^s = z. \quad (1.8.4)$$

The degree of $\mathbb{K}$ over $\mathbb{F}$ is s.

Problems

1. Let $\mathbb{F}$ be a field and assume that $p(x) = x^d + a_d x^{d-1} + \cdots + a_1 \in \mathbb{F}[x]$, where $d > 1$. On the vector space $\mathbb{F}^d$ define a product as follows. Let $b(x) = \sum_{i=1}^{d} b_i x^{i-1}, c(x) = \sum_{i=1}^{d} c_i x^{i-1}$. Then $(b_1, \ldots, b_d)(c_1, \ldots, c_d) = (r_1, \ldots, r_d)$, where $r(x) = \sum_{i=1}^{d} r_i x^{i-1}$

is the remainder of $b(x)c(x)$ after dividing by $p(x)$. That is, $b(x)c(x) = g(x)p(x) + r(x)$ where deg $r(x) < d$. Let $\mathcal{P}_d$ be $\mathbb{F}^d$ with the above product.

Show

(a) $\mathcal{P}_d$ is a commutative ring with identity $\mathbf{e}_1 = (1, 0, \ldots, 0)$.

(b) $\mathbb{F}$ is isomorphic to span $(\mathbf{e}_1)$, where $f \mapsto f\mathbf{e}_1$.

(c) Let $\mathbf{e}_i = (\delta_{1i}, \ldots, \delta_{di}), i = 2, \ldots, d$. Then

$$\mathbf{e}_2^i = \mathbf{e}_{1+i}, i = 0, \ldots, d-1, \quad p(\mathbf{e}_2) = 0.$$

(d) $\mathcal{P}_d$ is a domain if and only if $p(x)$ is an irreducible polynomial over $\mathbb{F}[x]$.

(e) $\mathcal{P}_d$ is a field if and only if $p(x)$ is an irreducible polynomial over $\mathbb{F}[x]$.

(f) Assume that $p(x) \in \mathbb{F}[x]$ is irreducible. Then $\mathbb{K} := \mathcal{P}_d$ is an extension field of $\mathbb{F}$ with $[\mathbb{K} : \mathbb{F}] = d$. Furthermore $p(x)$ viewed as $p(x) \in \mathbb{K}[x]$ decomposes to $p(x) = (x - \mathbf{e}_2)q(x)$, where $q(x) = x^{d-1} + \sum_{i=1}^{d} g_i x^{i-1} \in \mathbb{K}[x]$.

2. Let $\mathbb{F}$ be a field and $p(x) \in \mathbb{F}[x]$. Show that there exists a finite extension field $\mathbb{K}$ such that $p(x)$ splits in $\mathbb{K}$. Furthermore $[\mathbb{K} : \mathbb{F}] \le (\deg p)!$

1.9   The Resultant and the Discriminant

Let $\mathbb{D}$ be an integral domain. Suppose that

$$p(x) = a_0 x^m + \cdots + a_m, \quad q(x) = b_0 x^n + \cdots + b_n \in \mathbb{D}[x]. \qquad (1.9.1)$$

Assume furthermore that $m, n \geq 1$ and $a_0 b_0 \neq 0$. Let $\mathbb{F}$ be the quotient field of $\mathbb{D}$ and assume that $\mathbb{K}$ is a finite extension of $\mathbb{F}$ such that $p(x)$ and $q(x)$ split to linear factors in $\mathbb{K}$. That is

$$p(x) = a_0 \prod_{i=1}^{m} (x - \xi_i), \qquad \xi_i \in \mathbb{K}, \ i = 1, \ldots, m, \ a_0 \neq 0.$$

(1.9.2)

$$q(x) = b_0 \prod_{j=1}^{n} (x - \eta_j), \qquad \eta_j \in \mathbb{K}, \ j = 1, \ldots, n, \ b_0 \neq 0.$$

Then the *resultant* $R(p, q)$ of p, q and the *discriminant* $D(p)$ of p are defined as follows:

$$R(p, q) = a_0^n b_0^m \prod_{i,j=1}^{m,n} (\xi_i - \eta_j),$$

(1.9.3)

$$D(p) = a_0^{m(m-1)} \prod_{1 \leq i < j \leq m} (\xi_i - \xi_j)^2.$$

It is a classical result that $R(p, q) \in \mathbb{D}[a_0, \ldots, a_m, b_0, \ldots, b_n]$ and $D(p) \in \mathbb{D}[a_0, \ldots, a_m]$, (see [vdW59]). More precisely, we have the following.

Theorem 1.9.1 *Let*

$$\mathbf{a} = (a_0, \ldots, a_m) \in \mathbb{D}^{m+1}, \mathbf{b} = (b_0, \ldots, b_n) \in \mathbb{D}^{n+1},$$

$$p(x) = \sum_{i=0}^{m} a_i x^{m-i}, q(x) = \sum_{j=0}^{n} b_j x^{n-j}.$$

Then $R(p, q) = \det C(\mathbf{a}, \mathbf{b})$, where

$$
C(\mathbf{a}, \mathbf{b}) = \begin{bmatrix}
a_0 & a_1 & a_2 & \cdots & a_m & 0 & 0 & \cdots & 0 \\
0 & a_0 & a_1 & \cdots & a_{m-1} & a_m & 0 & \cdots & 0 \\
\vdots & \vdots & \vdots & \ddots & \vdots & \vdots & \vdots & \ddots & \vdots \\
0 & 0 & 0 & \cdots & a_0 & a_1 & a_2 & \cdots & a_m \\
b_0 & b_1 & \cdots & b_{n-2} & b_{n-1} & b_n & 0 & \cdots & 0 \\
0 & b_0 & b_1 & \cdots & b_{n-2} & b_{n-1} & b_n & 0 & \cdots \\
\vdots & \vdots & \vdots & \ddots & \vdots & \vdots & \vdots & \ddots & \vdots \\
0 & 0 & 0 & \cdots & b_0 & b_1 & b_2 & \cdots & b_n
\end{bmatrix}
$$

is an $(m + n) \times (m + n)$ matrix.

Proof. Let $\mathbb{F}$ be the quotient field of $\mathbb{D}$, and assume that $p, q \in \mathbb{D}[x]$ split in a finite extension field $\mathbb{K}$ of $\mathbb{F}$. Let $c(x) = \sum_{i=0}^{n-1} c_i x^{n-1-i}, d(x) = \sum_{j=0}^{m-1} d_j x^{m-1-j} \in \mathbb{F}[x]$. Then $c(x)p(x) + d(x)q(x) = \sum_{l=0}^{m+n-1} g_l x^{m+n-1-l}$.

Denote by $\mathbf{f} = (c_0, \ldots, c_{n-1}, d_0, \ldots, d_{m-1}), \mathbf{g} = (g_0, \ldots, g_{m+n-1}) \in \mathbb{D}^{m+n}$. A straightforward calculation shows that $\mathbf{f}C(\mathbf{a}, \mathbf{b}) = \mathbf{g}$.

Assume that $\det C(\mathbf{a}, \mathbf{b}) \neq 0$. Let $\mathbf{f} = (0, \ldots, 0, 1)C(\mathbf{a}, \mathbf{b})^{-1}$. Hence there exist $c(x), d(x) \in \mathbb{F}[x]$ of the above form such that $c(x)p(x) + d(x)q(x) = 1$. Thus $p(x), q(x)$ do not have common zeros in $\mathbb{K}$.

We now show that if $a_0 b_0 \neq 0$ then $R(p, q) = \det C(\mathbf{a}, \mathbf{b})$. Divide the first n rows of $C(\mathbf{a}, \mathbf{b})$ by a_0 and the last m rows of $C(\mathbf{a}, \mathbf{b})$ by b_0, to deduce that it is enough to show the equality $R(p, q) = \det C(\mathbf{a}, \mathbf{b})$ in the case $a_0 = b_0 = 1$. Then $p(x) = \prod_{i=1}^{m}(x - u_i), q(x) = \prod_{j=1}^{n}(x - v_j) \in \mathbb{K}[x]$. Recall that $(-1)^i a_i$ and $(-1)^j b_j$ are the ith and jth elementary symmetric polynomials in $u_1, \ldots, u_m$ and $v_1, \ldots, v_n$, respectively:

$$a_i = (-1)^i \sum_{1 \le l_1 < \cdots < l_i \le m} u_{l_1} \cdots u_{l_i}, \quad i = 1, \ldots, m,$$

$$b_j = (-1)^j \sum_{1 \le l_1 < \cdots < l_j \le n} v_{l_1} \cdots v_{l_j}, \quad j = 1, \ldots, n. \qquad (1.9.4)$$

Then $C(\mathbf{a}, \mathbf{b})$ is a matrix with polynomial entries in $\mathbf{u} = (u_1, \ldots, u_m), \mathbf{v} = (v_1, \ldots, v_n)$. Hence, $s(\mathbf{u}, \mathbf{v}) := \det C(\mathbf{a}, \mathbf{b})$ is a polynomial in $m + n$ variables, the coordinates of $\mathbf{u}$ and $\mathbf{v}$ respectively.

Assume that $u_i = v_j$ for some $i \in [1, m], j \in [1, n]$. Then $p(x)$ and $q(x)$ have a common factor $x - u_i = x - v_j$. The above arguments shows that $s(\mathbf{u}, \mathbf{v}) = 0$. Hence, $s(\mathbf{u}, \mathbf{v})$ is divisible by $t(\mathbf{u}, \mathbf{v}) = \prod_{i=1, j=1}^{m,n}(u_i - v_j)$. So $s(\mathbf{u}, \mathbf{v}) = h(\mathbf{u}, \mathbf{v})t(\mathbf{u}, \mathbf{v})$, for some polynomial $h(\mathbf{u}, \mathbf{v})$.

Consider $s(\mathbf{u}, \mathbf{v}), t(\mathbf{u}, \mathbf{v}), h(\mathbf{u}, \mathbf{v})$ as polynomials in $\mathbf{v}$ with coefficients in $\mathbb{D}[\mathbf{u}]$. Then $\deg_{\mathbf{v}} t(\mathbf{u}, \mathbf{v}) = nm$ and the term of the highest degree is $(-1)^{mn} v_1^m, \ldots, v_n^m$. Observe next that the contribution of the variables $\mathbf{v}$ in $\det C(\mathbf{a}, \mathbf{b})$ comes from it last m rows. The term of the maximal degree in each such row is n which comes only from $b_n = (-1)^n v_1, \ldots, v_n$. Hence, the coefficient of the product b_n^m comes from the minor of $C(\mathbf{a}, \mathbf{b})$ based on the first n rows and columns. Clearly, this minor is equal to $a_0^n = 1$. So $h(\mathbf{u}, \mathbf{v})$ is only polynomial in $\mathbf{u}$, i.e. its does not depend on the coordinates of $\mathbf{v}$. Furthermore $h(\mathbf{u}) = 1$, i.e. $h(\mathbf{u}, \mathbf{v})$ is a constant polynomial whose value is 1. $\qquad \square$

If $\mathbb{F}$ is a field of characteristic 0 then

$$D(p) = \pm a_0^{-1} R(p, p'). \qquad (1.9.5)$$

Note that if a_i, b_i are given the weight i for $i = 0, \ldots, \max(m, n)$ then $R(p, q)$ and $D(p)$ are polynomials with total degrees mn and $m(m - 1)$ respectively. See Problem 4 below.

Problems

1. Let $\mathbb{D}$ be an integral domain and assume that $p(x) = x^m, q(x) = (x + 1)^n$. Show

   (a) $R(p, q) = 1$.

   (b) Let $\mathbf{a} = (1, 0, \ldots, 0) \in \mathbb{D}^{m+1}, \mathbf{b} = \left(\binom{n}{0}, \binom{n}{1}, \ldots, \binom{n}{n}\right) \in \mathbb{D}^{n+1}$. Let $C(\mathbf{a}, \mathbf{b})$ be defined as in Theorem 1.9.1. Then $\det C(\mathbf{a}, \mathbf{b}) = 1$.

2. Let $\mathbf{u} = (u_1, \ldots, u_m), \mathbf{v} = (v_1, \ldots, v_n)$. Assume that each $a_i \in \mathbb{D}[\mathbf{u}], b_j \in \mathbb{D}[\mathbf{v}]$, is a multilinear polynomial for $i = 0, \ldots, m, j = 0, \ldots, n$, where the degree of a_i, b_j with respect to any variable is at most 1. Let $C(\mathbf{a}, \mathbf{b})$ be defined as in Theorem 1.9.1. Show that $\det C(\mathbf{a}, \mathbf{b})$ is a polynomial of degree at most n and m with respect to u_i and v_j respectively, for any $i = 1, \ldots, m$ and $j = 1, \ldots, n$.

3. Let the assumptions of Theorem 1.9.1 hold. Show

   (a) If $a_0 = b_0$ then $\det C(\mathbf{a}, \mathbf{b}) = 0$.

   (b) Assume that $p(x)$ is not the zero polynomial and $a_0 = 0$, $b_0 \neq 0$. Then $\det C(\mathbf{a}, \mathbf{b}) = 0$ if and only if $p(x)$ and $q(x)$ have a common root in an extension field $\mathbb{K}$ of $\mathbb{F}$, where $p(x)$ and $q(x)$ split.

4. Let $C(\mathbf{a}, \mathbf{b})$ be defined as in Theorem 1.9.1. View $\det C(\mathbf{a}, \mathbf{b})$ as a polynomial $F(\mathbf{a}, \mathbf{b})$. Assume that the weight $\omega(a_i) = i, \omega(b_j) = j$. Then the weight of a monomial in the variables $\mathbf{a}, \mathbf{b}$ is the sum of the weights of each variable times the number of times in appears in this monomial. Show

   (a) Each nontrivial monomial in $F(\mathbf{a}, \mathbf{b})$ is of weight mn.

   (b) Assume as in the proof of Theorem 1.9.1 that $a_0 = b_0 = 1$ and a_i and b_j are the ith and jth elementary symmetric polynomials in $\mathbf{u}$ and $\mathbf{v}$ respectively. Then each nontrivial monomial in $\mathbf{u}, \mathbf{v}$ appearing in $F(\mathbf{a}(\mathbf{u}), \mathbf{b}(\mathbf{v}))$ is of total degree mn.

1.10 The Ring $\mathbb{F}[x_1, \ldots, x_n]$

Definition 1.10.1 *Let U, V be two subsets of W. Then $U \backslash V$ denotes the set of all points of U which are not in V. ($U \backslash V$ may be empty.)*

In §1.2, we pointed out that $\mathbb{F}[x_1, \ldots, x_n]$ is not $\mathbb{D}_B$ for $n \geq 2$. It is known (see [Lan67]) that $\mathbb{F}[x_1, \ldots, x_n]$ is Noetherian:

Definition 1.10.2 $\mathbb{D}$ *is Noetherian, denoted by $\mathbb{D}_N$, if any ideal of $\mathbb{D}$ is finitely generated.*

In what follows we assume that $\mathbb{F}$ is algebraically closed. Let $p_1, \ldots, p_k \in \mathbb{F}[x_1, \ldots, x_n]$. Denote by $U(p_1, \ldots, p_k)$ the common set of zeros of $p_1, \ldots, p_k$:

$$U(p_1, \ldots, p_k) = \{\mathbf{x} = (x_1, \ldots, x_n)^T : \quad p_j(\mathbf{x}) = 0, \ j = 1, \ldots, k\}.$$
$$(1.10.1)$$

$U(p_1, \ldots, p_k)$ may be an empty set. A set of the form $U(p_1, \ldots, p_k)$ is called an algebraic variety (in $\mathbb{F}^n$). It is known (see [Lan67]) that any nonempty variety in $\mathbb{F}^n$ splits as

$$U = \cup_{i=1}^{k} V_i, \qquad (1.10.2)$$

where each V_i is an irreducible algebraic variety, which is not contained in any other V_j. Over $\mathbb{C}$ each irreducible variety $V \subset \mathbb{C}^n$ is a closed connected set. Furthermore, there exists a strict subvariety $W \subset V$ (of singular points of V) such that $V \backslash W$ is a connected analytic manifold of complex dimension d in $\mathbb{C}^n$. $\dim V := d$ is called the dimension of V. If $d = 0$ then V consists of one point. For any set $U \subset \mathbb{F}^n$ let $I(U)$ be the ideal of polynomials vanishing on U:

$$I(U) = \{p \in \mathbb{F}[x_1, \ldots, x_n] : \quad p(\mathbf{x}) = 0, \ \forall \mathbf{x} \in U\}. \qquad (1.10.3)$$

Theorem 1.10.3 (Hilbert Nullstellensatz) *Let $\mathbb{F}$ be an algebraically closed field. Let $I \subseteq \mathbb{F}[x_1, \ldots, x_n]$ be the ideal generated by $p_1, \ldots, p_k$. Assume that $g \in \mathbb{F}[x_1, \ldots, x_n]$. Then $g^j \in I$ for some positive integer j if and only if $g \in I(U(p_1, \ldots, p_k))$.*

Corollary 1.10.4 *Let* $p_1, \ldots, p_k \in \mathbb{F}[x_1, \ldots, x_n]$, *where* $\mathbb{F}$ *is an algebraically closed field. Then* $p_1, \ldots, p_k$ *generate* $\mathbb{F}[x_1, \ldots, x_n]$ *if and only if* $U(p_1, \ldots, p_k) = \emptyset$.

1.11   Matrices and Homomorphisms

Definition 1.11.1 *Let* **M**, **N** *be* $\mathbb{D}$-*modules. Let* $T : \mathbf{N} \to \mathbf{M}$. *T is a homomorphism if*

$$T(a\mathbf{u} + b\mathbf{v}) = aT\mathbf{u} + bT\mathbf{v}, \quad \text{for all } \mathbf{u}, \mathbf{v} \in \mathbf{N}, \ a, b \in \mathbb{D}. \quad (1.11.1)$$

Let

$$\text{Range } T = \{\mathbf{u} \in \mathbf{M} : \quad \mathbf{u} = T\mathbf{v}, \ \mathbf{v} \in \mathbf{N}\},$$
$$\text{Ker } T = \{\mathbf{v} \in \mathbf{N} : \quad T\mathbf{v} = \mathbf{0}\},$$

be the range and the kernel of T. *Denote by* $\text{Hom}(\mathbf{N}, \mathbf{M})$ *the set of all homomorphisms of* **N** *to* **M**.

$\mathbf{N}' := \text{Hom}(\mathbf{N}, \mathbb{D})$ *is called the dual module of* **N**. *For a vector space* **V** *over* $\mathbb{F}$ *the vector space* $\mathbf{V}'$ *is called the dual vector space of* **V**.

$T \in \text{Hom}(\mathbf{N}, \mathbf{M})$ *is an isomorphism if there exists* $Q \in \text{Hom}(\mathbf{M}, \mathbf{N})$ *such that* QT *and* TQ *are the identity maps on* **M** *and* **N** *respectively.* **M** *and* **N** *are isomorphic if there exists an isomorphism* $T \in \text{Hom}(\mathbf{N}, \mathbf{M})$.

$\text{Hom}(\mathbf{N}, \mathbf{M})$ is a $\mathbb{D}$-module with

$$(aS + bT)\mathbf{v} = aS\mathbf{v} + bT\mathbf{v}, \quad a, b \in \mathbb{D}, \ S, T \in \text{Hom}(\mathbf{N}, \mathbf{M}), \ \mathbf{v} \in \mathbf{N}.$$

Assume that **M** and **N** have finite bases. Let $[\mathbf{u}_1, \ldots, \mathbf{u}_m]$ and $[\mathbf{v}_1, \ldots, \mathbf{v}_n]$ be bases in **M** and **N** respectively. Then there exists a natural isomorphism between $\text{Hom}(\mathbf{N}, \mathbf{M})$ and $\mathbb{D}^{m \times n}$. For each

$T \in \mathrm{Hom}(\mathbf{N}, \mathbf{M})$ let $A = [a_{ij}] \in \mathbb{D}^{m \times n}$ be defined as follows:

$$Tv_j = \sum_{i=1}^{m} a_{ij} u_i, \quad j = 1, \ldots, n. \tag{1.11.2}$$

Conversely, for each $A = [a_{ij}] \in \mathbb{D}^{m \times n}$ there exists a unique $T \in \mathrm{Hom}(\mathbf{N}, \mathbf{M})$ which satisfies (1.11.2). The matrix A is called the representation matrix of T in the bases $[\mathbf{u}_1, \ldots, \mathbf{u}_m]$ and $[\mathbf{v}_1, \ldots, \mathbf{v}_n]$.

Definition 1.11.2 *Let $\mathbb{D}$ be an integral domain and $A = [a_{ij}]_{i \in [m], j \in [n]} \in \mathbb{D}^{m \times n}$.*

(1) *Assume that $\alpha = (\alpha_1, \ldots, \alpha_k) \in [m]_k, \beta = (\beta_1, \ldots, \beta_k) \in [n]_k$. Then $\det A[\alpha, \beta]$ is called an (α, β) minor, k-minor, or simply a minor of A.*

(2) *The rank of A, denoted by $\mathrm{rank}\, A$, is the maximal size of a nonvanishing minor of A. (The rank of the zero matrix is 0.)*

(3) *The nullity of A, denoted by $\mathrm{nul}\, A$, is $n - \mathrm{rank}\, A$.*

Any $A \in \mathbb{D}^{m \times n}$ can be viewed as $T \in \mathrm{Hom}(\mathbb{D}^n, \mathbb{D}^m)$, where $T\mathbf{x} := A\mathbf{x}$, $\mathbf{x} = (x_1, \ldots, x_n)^\top$. We will sometime denote T by A. If $\mathbb{D}$ is $\mathbb{D}_B$ then $\mathrm{Range}\, A$ has a finite basis of dimension $\mathrm{rank}\, A$ (see Problem 1).

We now study the relations between the representation matrices of a fixed $T \in \mathrm{Hom}(\mathbf{N}, \mathbf{M})$ with respect to different bases in $\mathbf{M}$ and $\mathbf{N}$.

Notation 1.11.3 *Denote by $\mathbf{GL}(n, \mathbb{D})$ the group of invertible matrices in $\mathbb{D}^{n \times n}$.*

Lemma 1.11.4 *Let $\mathbf{M}$ be a $\mathbb{D}$-module with a finite basis $[\tilde{\mathbf{u}}_1, \ldots, \tilde{\mathbf{u}}_m]$. Then $[\mathbf{u}_1, \ldots, \mathbf{u}_m]$ is a basis in $\mathbf{M}$ if and only if the matrix $Q = [q_{ki}] \in \mathbb{D}^{m \times m}$ given by the equalities*

$$\mathbf{u}_i = \sum_{k=1}^{m} q_{ki} \tilde{\mathbf{u}}_k, \quad i = 1, \ldots, m, \tag{1.11.3}$$

is an invertible matrix.

Proof. Suppose first that $[\mathbf{u}_1, \ldots, \mathbf{u}_m]$ is a basis in $\mathbf{M}$. Then

$$\tilde{\mathbf{u}}_k = \sum_{l=1}^{m} r_{lk} \mathbf{u}_l, \quad k = 1, \ldots, m. \tag{1.11.4}$$

Let $R = [r_{lk}]_1^m$. Insert (1.11.4) to (1.11.3) and use the assumption that $[\mathbf{u}_1, \ldots, \mathbf{u}_m]$ is a basis to obtain that $RQ = I$. Hence, $\det R \det Q = 1$. Lemma 1.7.6 yields that $Q \in \mathbf{GL}(m, \mathbb{D})$. Assume now that Q is invertible. Let $R = Q^{-1}$. Hence (1.11.4) holds. It is straightforward to deduce that $[\tilde{\mathbf{u}}_1, \ldots, \tilde{\mathbf{u}}_m]$ is a basis in $\mathbf{M}$. $\qquad \square$

Definition 1.11.5 *Let* $A, B \in \mathbb{D}^{m \times n}$. *Then* A *and* B *are right equivalent, left equivalent and equivalent if the following conditions hold respectively:*

$$B = AP \quad \text{for some } P \in \mathbf{GL}(n, \mathbb{D}) \quad (A \sim_r B), \tag{1.11.5}$$

$$B = QA \quad \text{for some } Q \in \mathbf{GL}(m, \mathbb{D}) \quad (A \sim_l B), \tag{1.11.6}$$

$$B = QAP \quad \text{for some } P \in \mathbf{GL}(n, \mathbb{D}), \ Q \in \mathbf{GL}(m, \mathbb{D}) \quad (A \sim B). \tag{1.11.7}$$

Clearly, all the above relations are equivalence relations.

Theorem 1.11.6 *Let* $\mathbf{M}$ *and* $\mathbf{N}$ *be* $\mathbb{D}$-*modules with finite bases having* m *and* n *elements respectively. Then* $A, B \in \mathbb{D}^{m \times n}$ *represent some* $T \in \mathrm{Hom}(\mathbf{N}, \mathbf{M})$ *in certain bases as follows:*
(l) $A \sim_l B$ *if and only if* A *and* B *represent* T *in the corresponding bases of* $\mathbf{U}$ *and* $\mathbf{V}$ *respectively*

$$[\mathbf{u}_1, \ldots, \mathbf{u}_m], \ [\mathbf{v}_1, \ldots, \mathbf{v}_n] \quad \text{and} \quad [\tilde{\mathbf{u}}_1, \ldots, \tilde{\mathbf{u}}_m], \ [\mathbf{v}_1, \ldots, \mathbf{v}_n].$$
$$\tag{1.11.8}$$

(r) $A \sim_r B$ *if and only if* A *and* B *represent* T *in the corresponding bases of* $\mathbf{U}$ *and* $\mathbf{V}$ *respectively*

$$[\mathbf{u}_1, \ldots, \mathbf{u}_m], \ [\mathbf{v}_1, \ldots, \mathbf{v}_n] \quad \text{and} \quad [\mathbf{u}_1, \ldots, \mathbf{u}_m], \ [\tilde{\mathbf{v}}_1, \ldots, \tilde{\mathbf{v}}_n].$$
$$\tag{1.11.9}$$

(e) $A \sim B$ *if and only if A and B represent T in the corresponding bases of $\mathbf{U}$ and $\mathbf{V}$ respectively*

$$[\mathbf{u}_1, \ldots, \mathbf{u}_m], \ [\mathbf{v}_1, \ldots, \mathbf{v}_n] \quad \text{and} \quad [\tilde{\mathbf{u}}_1, \ldots, \tilde{\mathbf{u}}_m], \ [\tilde{\mathbf{v}}_1, \ldots, \tilde{\mathbf{v}}_n].$$
$$(1.11.10)$$

Sketch of a proof. Let A be the representation matrix of T in the bases $[\mathbf{u}_1, \ldots, \mathbf{u}_m]$ and $[\mathbf{v}_1, \ldots, \mathbf{v}_n]$ given in (1.11.2). Assume that the relation between the bases $[\mathbf{u}_1, \ldots, \mathbf{u}_m]$ and $[\tilde{\mathbf{u}}_1, \ldots, \tilde{\mathbf{u}}_m]$ is given by (1.11.3). Then

$$T\mathbf{v}_j = \sum_{i=1}^{m} a_{ij}\mathbf{u}_i = \sum_{i=k=1}^{m} q_{ki}a_{ij}\tilde{\mathbf{u}}_k, \quad j = 1, \ldots, n.$$

Hence, the representation matrix B in bases $[\tilde{\mathbf{u}}_1, \ldots, \tilde{\mathbf{u}}_m]$ and $[\mathbf{v}_1, \ldots, \mathbf{v}_n]$ is given by (1.11.6).

Change the basis $[\mathbf{v}_1, \ldots, \mathbf{v}_n]$ to $[\tilde{\mathbf{v}}_1, \ldots, \tilde{\mathbf{v}}_n]$ according to

$$\tilde{\mathbf{v}}_j = \sum_{l=1}^{n} p_{lj}\mathbf{v}_l, \ j = 1, \ldots, n, \quad P = [p_{lj}] \in \mathbf{GL}(n, \mathbb{D}).$$

Then a similar computation shows that T is represented in the bases $[\mathbf{u}_1, \ldots, \mathbf{u}_m]$ and $[\tilde{\mathbf{v}}_1, \ldots, \tilde{\mathbf{v}}_n]$ by AP. Combine the above results to deduce that the representation matrix B of T in bases $[\tilde{\mathbf{u}}_1, \ldots, \tilde{\mathbf{u}}_m]$ and $[\tilde{\mathbf{v}}_1, \ldots, \tilde{\mathbf{v}}_n]$ is given by (1.11.7). $\qquad\square$

Problems

1. Let $A \in \mathbb{D}_B^{m \times n}$. View A a as linear transformation from $A : \mathbb{D}_B^n \to \mathbb{D}_B^m$ to show that Range A is a module with basis of dimension rank A. (*Hint*: Use Problem 1.6.4.)

2. For $A, B \in \mathbb{D}^{m \times n}$ show:

   (a) If $A \sim_l B$ then Ker $A =$ Ker B and Range A and Range B are isomorphic.

   (b) $A \sim_r B$ then Range $A =$ Range B and Ker A and Ker B are isomorphic.

1.12   Hermite Normal Form

We start this section with two motivating problems.

Problem 1.12.1 *Given $A, B \in \mathbb{D}^{m \times n}$, when are they*
(l) *left equivalent;*
(r) *right equivalent;*
(e) *equivalent?*

Problem 1.12.2 *For a given $A \in \mathbb{D}^{m \times n}$ characterize the equivalence classes corresponding to the relations of left equivalence, right equivalence and equivalence as defined in Problem 1.12.1.*

For $\mathbb{D}_G$ the equivalence relation has the following natural invariants:

Lemma 1.12.3 *For $A \in \mathbb{D}_G^{m \times n}$ let*

$$\mu(\alpha, A) := \text{g.c.d.} \, (\{\det A[\alpha, \theta], \ \theta \in [n]_k\}), \quad \alpha \in [m]_k,$$
$$\nu(\beta, A) := \text{g.c.d.} \, (\{\det A[\phi, \beta], \ \phi \in [m]_k\}), \quad \beta \in [n]_k,$$

$$(1.12.1)$$

$$\delta_k(A) := \text{g.c.d.} \, (\{\det A[\phi, \theta], \ \phi \in [m]_k, \ \theta \in [n]_k\}),$$

($\delta_k(A)$ is called the kth determinant invariant of A). Then

$$\mu(\alpha, A) \equiv \mu(\alpha, B) \quad \text{for all } \alpha \in [m]_k \quad \text{if } A \sim_r B,$$
$$\nu(\beta, A) \equiv \nu(\beta, B) \quad \text{for all } \beta \in [n]_k, \quad \text{if } A \sim_l B, \quad (1.12.2)$$
$$\delta_k(A) \equiv \delta_k(B) \quad \text{if } A \sim B,$$

for $k = 1, \ldots, \min(m, n)$. (Recall that for $a, b \in \mathbb{D}$, $a \equiv b$ if $a = bc$ for some invertible $c \in \mathbb{D}$.)

Proof. Suppose that $B = AP$ for some $P \in \mathbf{GL}(n, \mathbb{D})$. Then the Cauchy–Binet formula (1.7.5) yields:

$$\det B[\alpha, \gamma] = \sum_{\theta \in [n]_k} \det A[\alpha, \theta] \det P[\theta, \gamma]. \qquad (1.12.3)$$

Hence, $\mu(\alpha, A)$ divides $\mu(\alpha, B)$. As $A = BP^{-1}$ we get $\mu(\alpha, B) | \mu(\alpha, A)$. Thus $\mu(\alpha, A) \equiv \mu(\alpha, B)$. The other equalities in (1.12.2) are established in a similar way. $\square$

Clearly

$$A \sim_l B \iff A^T \sim_r B^T, \quad A, B \in \mathbb{D}^{m \times n}. \tag{1.12.4}$$

Hence, it is enough to consider the left equivalence relation. We characterize the left equivalence classes for Bezout domains $\mathbb{D}_B$. To do that we need some notation.

Recall that $P \in \mathbb{D}^{n \times n}$ is called a permutation matrix if P is a matrix having in each row and each column exactly one nonzero element which is equal to the identity element 1. A permutation matrix is invertible since $P^{-1} = P^T$.

Definition 1.12.4 *Let* $\mathcal{P}_n \subset \mathbf{GL}(n, \mathbb{D})$ *be the group of* $n \times n$ *permutation matrices.*

Definition 1.12.5 *An invertible matrix* $U \in \mathbf{GL}(n, \mathbb{D})$ *is called simple if there exists* $P, Q \in \mathcal{P}_n$ *such that*

$$U = P \begin{bmatrix} V & 0 \\ 0 & I_{n-2} \end{bmatrix} Q, \tag{1.12.5}$$

where

$$V = \begin{bmatrix} \alpha & \beta \\ \gamma & \delta \end{bmatrix} \in \mathbf{GL}(2, \mathbb{D}), \quad (\alpha\delta - \beta\gamma \text{ is invertible}). \tag{1.12.6}$$

U *is called elementary if* U *is of the form* (1.12.5) *and*

$$V = \begin{bmatrix} \alpha & \beta \\ 0 & \delta \end{bmatrix} \in \mathbf{GL}(2, \mathbb{D}), \quad \text{and } \alpha, \delta \text{ are invertible}. \tag{1.12.7}$$

Definition 1.12.6 *Let* $A \in \mathbb{D}^{m \times n}$. *The following row (column) operations are called elementary:*

(a) *interchanging any two rows (columns) of* A;
(b) *multiplying row (column)* i *by an invertible element* a;
(c) *adding* b *times row (column)* i *to row (column)* j $(i \neq j)$.
    The following row (column) operation is called simple;
(d) *replacing row (column)* i *by* a *times row (column)* i *plus* b *times row (column)* j, *and row (column)* j *by* c *times row (column)* i *plus* d *times row (column)* j, *where* $i \neq j$ *and* $ad - bc$ *is invertible in* $\mathbb{D}$.

It is straightforward to see that the elementary row (column) operations can be carried out by multiplication of A by a suitable elementary matrix from left (right), and the simple row (column) operations are carried out by multiplication of A by a simple matrix U from left (right).

Theorem 1.12.7 *Let $\mathbb{D}_B$ be a Bezout domain. Let $A \in \mathbb{D}_B^{m \times n}$. Assume that rank $A = r$. Then there exists $B = [b_{ij}] \in \mathbb{D}_B^{m \times n}$ which is equivalent to A and satisfies the following conditions:*

the ith row of B is nonzero if and only if $i \le r$.   (1.12.8)

Let b_{in_i} be the first nonzero entry in ith row for $i = 1, \ldots, r$. Then

$$1 \le n_1 < n_2 < \cdots < n_r \le n. \qquad (1.12.9)$$

The numbers $n_1, \ldots, n_r$ are uniquely determined and the elements b_{in_i}, $i = 1, \ldots, r$, which are called pivots, are also uniquely determined, up to invertible factors, by the conditions

$$\nu((n_1, \ldots, n_i), A) = b_{1n_1}, \ldots, b_{in_i}, \quad i = 1, \ldots, r,$$
$$(1.12.10)$$
$$\nu(\alpha, A) = 0, \ \alpha \in [n_i - 1]_i, \quad i = 1, \ldots, r.$$

For $1 \le j < i \le r$, adding to the row j a multiple of the row i does not change the above form of B. Assume that $B = [b_{ij}], C = [c_{ij}] \in \mathbb{D}^{m \times n}$ are left equivalent to A and satisfy the above conditions. If $b_{jn_i} = c_{jn_i}$, $j = 1, \ldots, i, i = 1, \ldots, r$ then $B = C$. The invertible matrix Q which satisfies $B = QA$ can be given by a finite product of simple matrices.

Proof. Clearly, it is enough to consider the case $A \ne 0$, i.e. $r \ge 1$. Our proof is by induction on n and m. For $n = m = 1$ the theorem is obvious. Let $n = 1$ and assume that for a given $m \ge 1$ there exists a matrix Q, which is a finite product of simple matrices, such that the entries $(i, 1)$ of Q are zero for $i = 2, \ldots, m$ if $m \ge 2$. Let $A_1 = [a_{i1}] \in \mathbb{D}_B^{(m+1) \times 1}$ and denote by A

the submatrix $[a_{i1}]_{i=1}^m$. Set

$$Q_1 := \begin{bmatrix} Q & 0 \\ 0 & 1 \end{bmatrix}.$$

Then the $(i,1)$ entries of $A_2 = [a_{i1}^{(2)}] = Q_1 A_1$ are equal to zero for $i = 2, \ldots, m$. Interchange the second and the last row of A_2 to obtain A_3. Clearly $A_3 = [a_{i1}^{(3)}] = Q_2 A_2$ for some permutation matrix Q_2. Let $A_4 = (a_{11}^{(3)}, a_{21}^{(3)})^\top$. As $\mathbb{D}_B$ is a Bezout domain there exist $\alpha, \beta \in \mathbb{D}_B$ such that

$$\alpha a_{11}^{(3)} + \beta a_{21}^{(3)} = (a_{11}^{(3)}, a_{21}^{(3)}) = d. \qquad (1.12.11)$$

As $(\alpha, \beta) = 1$ there exists $\gamma, \delta \in \mathbb{D}_B$ such that

$$\alpha\delta - \beta\gamma = 1. \qquad (1.12.12)$$

Let V be a 2×2 invertible matrix given by (1.12.6). Then

$$A_5 = V A_4 = \begin{bmatrix} d \\ d' \end{bmatrix}.$$

Lemma 1.12.3 implies $\nu((1), A_5) = \nu((1), A_4) = d$. Hence, $d' = pd$ for some $p \in \mathbb{D}_B$. Thus

$$\begin{bmatrix} d \\ 0 \end{bmatrix} = W A_5, \quad W = \begin{bmatrix} 1 & 0 \\ -p & 1 \end{bmatrix} \in \mathbf{GL}(2, \mathbb{D}_B).$$

Let

$$Q_3 = \begin{bmatrix} W & 0 \\ 0 & I_{m-1} \end{bmatrix} \begin{bmatrix} V & 0 \\ 0 & I_{m-1} \end{bmatrix}.$$

Then the last m rows of $A_6 = [a_{i1}^{(6)}] = Q_3 A_3$ are zero rows. So $a_{11}^{(6)} = \nu((1), A_6) = \nu((1), A_1)$ and the theorem is proved in this case.

Assume now that we proved the theorem for all $A_1 \in \mathbb{D}_B^{m \times n}$ where $n \le p$. Let $n = p + 1$ and $A \in \mathbb{D}_B^{m \times (p+1)}$. Let $A_1 = [a_{ij}]_{i=j=1}^{m,p}$. The induction hypothesis implies the existence of $Q_1 \in \mathbf{GL}(m, \mathbb{D}_B)$, which is a finite product of simple matrices, such that $B_1' =$

$[b_{ij}^{(1)}]_{i=j=1}^{m,p} = Q_1 A_1$ satisfies the assumptions of our theorem. Let $n_1', \ldots, n_s'$ be the integers defined by A_1. Let $B_1 = [b_{ij}^{(1)}]_{i=j=1}^{m,n} = Q_1 A$. If $b_{in}^{(1)} = 0$ for $i > s$ then $n_i = n_i'$, $i = 1, \ldots, s$ and B_1 is in the right form. Suppose now that $b_{in}^{(1)} \neq 0$ for some $s < i \leq m$. Let $B_2 = [b_{in}^{(1)}]_{i=s+1}^{m} \in \mathbb{D}_B^{(m-s) \times 1}$. We proved above that there exists $Q_2 \in \mathbf{GL}(m - s, \mathbb{D}_B)$ such that $Q_2 B_2 = (c, 0, \ldots, 0)^T$, $c \neq 0$. Then

$$B_3 = \begin{bmatrix} I_s & 0 \\ 0 & Q_2 \end{bmatrix} B_1$$

is in the right form with

$$s = r - 1, \ n_1 = n_1', \ldots, n_{r-1} = n_{r-1}', \ n_r = n.$$

We now show (1.12.10). First if $\alpha \in [n_i - 1]_i$ then any matrix $B[\beta, \alpha]$, $\beta \in [m]_i$ has at least one zero row. Hence, $\det B[\beta, \alpha] = 0$. Therefore, $\nu(\alpha, B) = 0$. Lemma 1.12.3 yields that $\nu(\alpha, A) = 0$. Let $\alpha = (n_1, \ldots, n_i)$. Then $B[\beta, \alpha]$ has at least one zero row unless β is equal to $\gamma = (1, 2, \ldots, i)$. Therefore

$$\nu(\alpha, A) = \nu(\alpha, B) = \det B[\gamma, \alpha] = b_{1n_1} \cdots b_{in_i} \neq 0.$$

This establishes (1.12.10).

It is obvious that $b_{1n_1}, \ldots, b_{rn_r}$ are determined up to invertible elements. For $1 \leq j < i \leq r$ we can perform the following elementary row operation on B: add to row j a multiple of row i. The new matrix C will satisfy the assumption of the theorem. It is left to show that if $B = [b_{ij}], C = [c_{ij}] \in \mathbb{D}_B^{m \times n}$ are left equivalent to A, have the same form given by the theorem and satisfying

$$b_{jn_i} = c_{jn_i}, \quad j = 1, \ldots, i, \ i = 1, \ldots, r, \tag{1.12.13}$$

then $B = C$. See Problem 1. $\square$

A matrix $B \in \mathbb{D}_B^{m \times n}$ is said to be in a *Hermite normal form*, abbreviated as *HNF*, if it satisfies conditions (1.12.8) and (1.12.9).

Normalization 1.12.8 *Let $B = [b_{ij}] \in \mathbb{D}_B^{m \times n}$ be in a Hermite normal form. If b_{in_i} is invertible we set $b_{in_i} = 1$ and $b_{jn_i} = 0$ for $i < j$.*

Theorem 1.12.9 *Let U be an invertible matrix over a Bezout domain. Then U is a finite product of simple matrices.*

Proof. Since $\det U$ is invertible, Theorem 1.12.7 yields that b_{ii} is invertible. Normalization 1.12.8 implies that the Hermite normal form of U is I. Hence, the inverse of U is a finite product of simple matrices. Therefore, U itself is a finite product of simple matrices. $\qquad\qquad\qquad\qquad\qquad\qquad\qquad\qquad\Box$

Normalization 1.12.10 *For Euclidean domains assume*

$$\text{either } b_{jn_i} = 0 \quad \text{or } d(b_{jn_i}) < d(b_{in_i}) \text{ for } j < i. \qquad (1.12.14)$$

For $\mathbb{Z}$ we assume that $b_{in_i} \geq 1$ and $0 \leq b_{jn_i} < b_{in_i}$ for $j < i$. For $\mathbb{F}[x]$ we assume that b_{in_i} is a normalized polynomial.

Corollary 1.12.11 *Let $\mathbb{D}_E = \mathbb{Z}, \mathbb{F}[x]$. Under Normalization 1.12.10 any $A \in \mathbb{D}_E^{m \times n}$ has a unique Hermite normal form.*

It is a well-known fact that over Euclidean domains Hermite normal form can be achieved by performing elementary row operations.

Theorem 1.12.12 *Let $A \in \mathbb{D}_E^{m \times n}$. Then there exists $Q \in \mathbf{GL}(m, \mathbb{D}_E)$ such that $B = QA$, where B is in a Hermite normal form satisfying Normalization 1.12.10 and Q is a product of finite elementary matrices.*

Proof. From the proof of Theorem 1.12.7 it follows that it is enough to show that any $A \in \mathbf{GL}(2, \mathbb{D}_E)$ is a finite product of elementary invertible matrices in $\mathbf{GL}(2, \mathbb{D}_E)$. As I_2 is the Hermite normal form of any 2×2 invertible matrix, it suffices to show that any $A \in \mathbb{D}_E^{2 \times 2}$ can be brought to its Hermite form by a finite number of elementary row operations. Let

$$A_i = \begin{bmatrix} a_i & b_i \\ a_{i+1} & b_{i+1} \end{bmatrix}, \quad A_1 = PA,$$

where P is a permutation matrix such that $d(a_1) \geq d(a_2)$. Suppose first that $a_2 \neq 0$. Compute a_{i+2} by (1.3.4). Then

$$A_{i+1} = \begin{bmatrix} 0 & 1 \\ 1 & 0 \end{bmatrix} \begin{bmatrix} 1 & -t_i \\ 0 & 1 \end{bmatrix} A_i, \quad i = 1, \dots$$

As the Euclidean algorithm terminates after a finite number of steps we obtain that $a_{k+1} = 0$. Then A_k is the Hermite normal form of A. If $b_{k+1} = 0$ we are done. If $b_{k+1} \neq 0$ subtract from the first row of A_k a corresponding multiple of the second row of A_k to obtain the matrix

$$B' = \begin{bmatrix} a_k & b'_k \\ 0 & b_{k+1} \end{bmatrix}, \quad d(b_{k+1}) > d(b'_k).$$

Multiply each row of B' by an invertible element if necessary to obtain the Hermite normal form of B according to Normalization 1.12.10. We obtained B by a finite number of elementary row operations. If $a_1 = a_2 = 0$ perform the Euclid algorithm on the second column of A. $\qquad \square$

Corollary 1.12.13 *Let $U \in \mathbf{GL}(n, \mathbb{D}_E)$. Then U is a finite product of elementary invertible matrices.*

Corollary 1.12.14 *Let $\mathbb{F}$ be a field. Then $A \in \mathbb{F}^{m \times n}$ can be brought to its unique reduced row echelon form given by Theorem 1.12.7 with*

$$b_{in_i} = 1, \; b_{jn_i} = 0, \quad j = 1, \dots, i-1, \; i = 1, \dots, r,$$

by a finite number of elementary row operations.

Problems

1. Show

    (a) Let $A, B \in \mathbb{D}^{m \times m}$ be two upper triangular matrices with the same nonzero diagonal entries. Assume that $QA = B$ for some $Q \in \mathbb{D}^{m \times m}$. Then Q is un upper triangular matrix with 1 on the main diagonal. (*Hint*: First prove this claim for the quotient field $\mathbb{F}$ of $\mathbb{D}$.)

    (b) Let $Q \in \mathbb{D}^{m \times m}$ be an upper triangular matrix with 1 on the main diagonal. Show that $Q = R_2 \ldots R_m = T_m \ldots T_2$ where $R_i - I_m, Q_i - I_m$ may have nonzero entries only in the places (j, i) for $j = 1, \ldots, i - 1$ and $i = 2, \ldots, m$.

    (c) Let $A, B \in \mathbb{D}^{m \times n}$. Assume that $A \sim_l B$, and A and B are in *HNF* and have the same pivots. Then B can be obtained from A, by adding multiples of the row b_{in_i} to the rows $j = 1, \ldots, i - 1$ for $i = 2, \ldots, r$.

    (d) Let $B = [b_{ij}], C = [c_{ij}] \in \mathbb{D}^{m \times n}$. Assume that B, C are in Hermite normal form with the same r numbers $1 \le n_1 < \cdots < n_r \le n$. Suppose furthermore (1.12.13) holds and $B = QC$ for some $Q \in \mathbb{D}^{m \times m}$. Then

$$Q = \begin{bmatrix} I_r & * \\ 0 & * \end{bmatrix} \quad \Rightarrow \quad B = C.$$

    (Here $*$ denotes a matrix of the corresponding size.)

    (e) Let $\mathbf{M}$ be a $\mathbb{D}_B$-module, $\mathbf{N} = \mathbb{D}_B^n$ and $T \in \mathrm{Hom}\,(\mathbf{N}, \mathbf{M})$. Let Range (T) be the range of T in $\mathbf{M}$. Then the module Range (T) has a basis $T\mathbf{u}_1, \ldots, T\mathbf{u}_k$ such that

$$\mathbf{u}_i = \sum_{j=1}^{i} c_{ij}\mathbf{v}_j, \quad c_{ii} \ne 0, \quad i = 1, \ldots, k, \qquad (1.12.15)$$

where $\mathbf{v}_1, \ldots, \mathbf{v}_n$ is a permutation of the standard basis

$$\mathbf{e}_i = (\delta_{i1}, \ldots, \delta_{in})^T, \quad i = 1, \ldots, n. \qquad (1.12.16)$$

2. Let $A \in \mathbb{D}_B^{m \times n}$ and assume that B is its Hermite normal form. Assume that $n_i < j < n_{i+1}$. Prove that

$$\nu(\alpha, A) = b_{1n_1} \cdots b_{(i-1)n_{i-1}} b_{ij}, \quad \text{for } \alpha = (n_1, \ldots, n_{i-1}, j).$$

3. **Definition 1.12.15** *Let* $\mathbb{F}$ *be a field and* $\mathbf{V}$ *a vector space over* $\mathbb{F}$ *of dimension* n. *A flag* $\mathbf{F}_*$ *on* $\mathbf{V}$ *is a strictly increasing sequence of subspaces*

$$0 = \mathbf{F}_0 \subset \mathbf{F}_1 \subset \cdots \subset \mathbf{F}_n = \mathbf{V},$$

$$(1.12.17)$$

$$\dim \mathbf{F}_i = i, \quad i = 1, \ldots, n = \dim V.$$

Show

(a) Let $\mathbf{L}$ be a subspace of $\mathbf{V}$ of dimension ℓ. Then

$$\dim \mathbf{L} \cap \mathbf{F}_{i-1} \leq \dim \mathbf{L} \cap \mathbf{F}_i \leq \dim \mathbf{L} \cap \mathbf{F}_{i-1} + 1, \ i = 1, \ldots, n.$$

$$(1.12.18)$$

(b) Let $\mathrm{Gr}(\ell, \mathbf{V})$ be the space of all ℓ-dimensional subspaces of $\mathbf{V}$. Let $J = \{1 \leq j_1 < \cdots < j_\ell \leq n\}$ be a subset of $[n]$. Then

$$\Omega^o(J, \mathbf{F}_*) := \{\mathbf{L} \in \mathrm{Gr}(\ell, \mathbf{V}) : \quad \dim \mathbf{L} \cap \mathbf{F}_{j_i} = i, \ i = 1, \ldots, \ell\},$$

$$(1.12.19)$$

$$\Omega(J, \mathbf{F}_*) := \{\mathbf{L} \in \mathrm{Gr}(\ell, \mathbf{V}) : \quad \dim \mathbf{L} \cap \mathbf{F}_{j_i} \geq i, \ i = 1, \ldots, \ell\},$$

are called the open and the closed Schubert cells in $\mathrm{Gr}(\ell, \mathbf{V})$ respectively. Show that a given $\mathbf{L} \in \mathrm{Gr}(\ell, \mathbf{V})$ belongs to the smallest open Schubert cell Ω_J^o, where $J = J(\mathbf{L}, \mathbf{F}_*)$ given by the condition

$$\dim \mathbf{L} \cap \mathbf{F}_{j_i} = i, \quad \dim \mathbf{L} \cap \mathbf{F}_{j_i - 1} = i - 1, \quad i = 1, \ldots, \ell.$$

$$(1.12.20)$$

(c) Let $\mathbf{V} = \mathbb{F}^n$ and assume that $\mathbf{e}_1, \ldots, \mathbf{e}_n$ is the standard basis of $\mathbb{F}^n$. Let

$$\mathbf{F}_i = \mathrm{span}\,(\mathbf{e}_n, \mathbf{e}_{n-1}, \ldots, \mathbf{e}_{n-i+1}), \quad i = 1, \ldots, n \quad (1.12.21)$$

be the *reversed* standard flag in $\mathbb{F}^n$. Let $A \in \mathbb{F}^{m \times n}$. Assume that $\ell = \operatorname{rank} A \geq 1$. Let $\mathbf{L} \in \operatorname{Gr}(\ell, \mathbb{F}^n)$ be the vector space spanned by the columns of A^T. Let $N = \{1 \leq n_1 < \cdots < n_\ell \leq n\}$ be the integers given by the row echelon form of A. Then $J(\mathbf{L}, \mathbf{F}_*) = N$.

1.13 Systems of Linear Equations over Bezout Domains

Consider a system of m linear equations in n unknowns:

$$\sum_{j=1}^{n} a_{ij} x_j = b_i, \quad i = 1, \ldots, m,$$

$$(1.13.1)$$

$$a_{ij}, b_i \in \mathbb{D}, \ i = 1, \ldots, m, \ j = 1, \ldots, n.$$

In matrix notation (1.13.1) is equivalent to

$$A\mathbf{x} = \mathbf{b}, \quad A \in \mathbb{D}^{m \times n}, \ \mathbf{x} \in \mathbb{D}^n, \ \mathbf{b} \in \mathbb{D}^m. \qquad (1.13.2)$$

Let

$$\hat{A} = [A, \mathbf{b}] \in \mathbb{D}^{m \times (n+1)}. \qquad (1.13.3)$$

The matrix A is called the *coefficient* matrix and the matrix $\hat{A}$ is called the *augmented coefficient* matrix. If $\mathbb{D}$ is a field, the classical Kronecker–Capelli theorem states [Gan59] that (1.13.1) is solvable if and only if

$$\operatorname{rank} A = \operatorname{rank} \hat{A}. \qquad (1.13.4)$$

Let $\mathbb{F}$ be the quotient field of $\mathbb{D}$. If (1.13.1) is solvable over $\mathbb{D}$ it is also solvable over $\mathbb{F}$. Therefore, (1.13.4) is a necessary condition for the solvability of (1.13.1) over $\mathbb{D}$. Clearly, even in the case $m = n = 1$ this condition is not sufficient. In this section, we give necessary and sufficient conditions on $\hat{A}$ for the solvability of (1.13.1) over a Bezout domain. First we need the following lemma:

Lemma 1.13.1 *Let* $0 \neq A \in \mathbb{D}_B^{m \times n}$. *Then there exist* $P \in \Pi_m$, $U \in \mathbf{GL}(n, \mathbb{D}_B)$ *such that*

$$
\begin{aligned}
C &= [c_{ij}] = PAU, \\
c_{ii} &\neq 0, \quad i = 1, \ldots, \text{rank A}, \\
c_{ij} &= 0 \quad \text{if either } j > i \text{ or } j > \text{rank A}.
\end{aligned}
\tag{1.13.5}
$$

Proof. Consider the matrix $A^\top$. By interchanging the columns of A^T, i.e. multiplying $A^\top$ from the right by some permutation matrix P^T, we can assume that the Hermite normal form of $A^\top P^\top$ satisfies $n_i = i$, $i = 1, \ldots, \text{rank A}$. $\qquad\square$

Theorem 1.13.2 *Let* $\mathbb{D}$ *be a Bezout domain. Then the system* (1.13.1) *is solvable if and only if*

$$
r = \text{rank A} = \text{rank } \hat{A}, \quad \delta_r(A) \equiv \delta_r(\hat{A}).
\tag{1.13.6}
$$

Proof. Assume first the existence of $\mathbf{x} \in \mathbb{D}^n$ which satisfies (1.13.2). Hence (1.13.4) holds, i.e. the first part of (1.13.6) holds. As any minor $r \times r$ of A is a minor of $\hat{A}$ we deduce that $\delta_r(\hat{A}) | \delta_r(A)$. (1.13.2) implies that $\mathbf{b}$ is a linear combination of the columns of A. Consider any $r \times r$ minor of $\hat{A}$ which contains the $(n+1)$st column $\mathbf{b}$. Since $\mathbf{b}$ is a linear combination of columns of A it follows that $\delta_r(A)$ divides this minor. Hence $\delta_r(A) | \delta_r(\hat{A})$, which establishes the second part of (1.13.6). (Actually we showed that if (1.13.1) is solvable over $\mathbb{D}_G$ then (1.13.6) holds.)

Assume now that (1.13.6) holds. Let

$$
V\hat{A} = \hat{B} = [B, \hat{\mathbf{b}}] \in \mathbb{D}^{m \times (n+1)}, \ V \in \mathbf{GL}(m, \mathbb{D})
$$

be Hermite's normal form of $\hat{A}$. Hence, B is Hermite's normal form of A. Furthermore

$$
\begin{aligned}
VA &= B, \quad \text{rank B} = \text{rank A} = \text{rank } \hat{A} = \text{rank } \hat{B} = r, \\
&\delta_r(B) \equiv \delta_r(A) \equiv \delta_r(\hat{A}) \equiv \delta_r(\hat{B}).
\end{aligned}
$$

Hence, n_r in Hermite's normal form of $\hat{A}$ is at most n. Note that the last $m - r$ equations of $B\mathbf{x} = \hat{\mathbf{b}}$ are the trivial equations $0 = 0$.

That is, it is enough to show the solvability of the system (1.13.2) under the assumptions (1.13.6) with $r = m$. By changing the order of equations in (1.13.1) and introducing a new set of variables

$$\mathbf{y} = U^{-1}\mathbf{x}, \quad U \in \mathbf{GL}(n, \mathbb{D}), \tag{1.13.7}$$

we may assume that the system (1.13.2) is

$$C\mathbf{y} = \mathbf{d}, \quad C = PAU, \, \mathbf{d} = (d_1, \ldots, d_m)^T = P\mathbf{b}, \tag{1.13.8}$$

where C is given as in Lemma 1.13.5 with $r = m$. Let $\hat{C} = [C, \mathbf{d}]$. It is straightforward to see that $A \sim C$, $\hat{A} \sim \hat{C}$. Hence

$$\text{rank } C = \text{rank } A = \text{rank } \hat{A} = \text{rank } \hat{C} = m,$$
$$\delta_m(C) \equiv \delta_m(A) \equiv \delta_m(\hat{A}) \equiv \delta_m(\hat{C}).$$

Thus it is enough to show that the system (1.13.8) is solvable. In view of the form of C the solvability of the system (1.13.8) over $\mathbb{D}$ is equivalent the solvability of the system

$$\tilde{C}\tilde{\mathbf{y}} = \mathbf{d}, \quad \tilde{C} = [c_{ij}]_{i=j=1}^m \in \mathbb{D}^{m \times m}, \, \tilde{\mathbf{y}} = (y_1, \ldots, y_m)^T. \tag{1.13.9}$$

Note that $\delta_m(C) = \delta_m(\tilde{C}) = \det \tilde{C}$. Cramer's rule for the above system in the quotient field $\mathbb{F}$ of $\mathbb{D}$ yields

$$y_i = \frac{\det \tilde{C}_i}{\det \tilde{C}}, \quad i = 1, \ldots, m.$$

Here, $\tilde{C}_i$ is obtained by replacing column i of $\tilde{C}$ by $\mathbf{d}$. Clearly $\det \tilde{C}_i$ is an $m \times m$ minor of $\hat{C}$ up to the factor ± 1. Hence, it is divisible by $\delta_m(\hat{C}) \equiv \delta_m(C) = \det(\tilde{C})$. Therefore, $y_i \in \mathbb{D}$, $i = 1, \ldots, m$. $\square$

Theorem 1.13.3 *Let $A \in \mathbb{D}_B^{m \times n}$. Then* Range A *and* Ker A *are modules in $\mathbb{D}_B^m$ and $\mathbb{D}_B^n$ having finite bases with* rank A *and* nul A *elements respectively. Moreover, the basis of* Ker A *can be completed to a basis of $\mathbb{D}_B^n$.*

Proof. As in the proof of Theorem 1.13.2 we may assume that rank $A = m$ and $A = C$, where C is given by (1.13.5) with $r = m$. Let $\mathbf{e}_1, \ldots, \mathbf{e}_n$ be the standard basis of $\mathbb{D}_B^n$. Then $C\mathbf{e}_1, \ldots, C\mathbf{e}_m$ is a basis in Range C and $\mathbf{e}_{m+1}, \ldots, \mathbf{e}_n$ is a basis for Ker A. $\square$

Let $A \in \mathbb{D}_G^{m \times n}$. Expand any $q \times q$ minor of A by any $q - p$ rows, where $1 \leq p < q$. We then deduce

$$\delta_p(A) | \delta_q(A) \quad \text{for any } 1 \leq p \leq q \leq \min(m, n). \tag{1.13.10}$$

Definition 1.13.4 *For* $A \in \mathbb{D}_G^{m \times n}$ *let*

$$i_j(A) := \frac{\delta_j(A)}{\delta_{j-1}(A)}, \quad j = 1, \ldots, \text{rank } A, \quad (\delta_0(A) = 1),$$

$$i_j(A) = 0 \quad \text{for rank } A < j \leq \min(m, n),$$

be the invariant factors of A. $i_j(A)$ *is called a trivial factor if* $i_j(A)$ *is invertible in* $\mathbb{D}_G$.

Suppose that (1.13.1) is solvable over $\mathbb{D}_B$. Using the fact that **b** is a linear combination of the columns of A and Theorem 1.13.2 we get an equivalent version of Theorem 1.13.2. (See Problem 2.)

Corollary 1.13.5 *Let* $A \in \mathbb{D}_B^{m \times n}$, $\mathbf{b} \in \mathbb{D}_B^m$. *Then the system* (1.13.1) *is solvable over* $\mathbb{D}_B$ *if and only if*

$$r = \text{rank } A = \text{rank } \hat{A}, \quad i_k(A) \equiv i_k(\hat{A}), \quad k = 1, \ldots, r. \tag{1.13.11}$$

Problems

1. Let $A \in \mathbb{D}_G^{n \times n}$. Assume that $r = \text{rank } A$. Show

   (a)
   $$\delta_j(A) = \bar{\omega}_j i_1(A) \ldots i_j(A), \quad \text{where } \omega_j \text{ is invertible in } \mathbb{D}_G$$
   $$\text{for } j = 1, \ldots, n. \tag{1.13.12}$$

   (b) $i_1(A) | i_j(A)$ for $j = 2, \ldots, r$. (*Hint*: Expand any minor of order j by any row.)

   (c) Let $2 \leq k, 2k - 1 < j \leq ' r$. Then $i_1(A) \ldots i_k(A) | i_{j-k+1}(A) \ldots i_j(A)$.

2. Give a complete proof of Corollary 1.13.5.

3. Let $A \in \mathbb{D}_B^{m \times n}$. Assume that all the pivots in HNF of $A^\top$ are invertible elements. Show

    (a) Any basis of Range A can be completed to a basis in $\mathbb{D}_B^m$.

    (b) $i_1(A) = \cdots = i_{\text{rank } A}(A) = 1$.

4. Assume that $\mathbb{D} = \mathbb{D}_B$, $\mathbf{M}$ is a $\mathbb{D}$-module with a basis, $\mathbf{M}_1, \mathbf{M}_2$ are finitely generated modules of $\mathbf{M}$. Show

    (a) $\mathbf{M}_1 \cap \mathbf{M}_2$ has a basis which can be completed to bases in $\mathbf{M}_1$ and $\mathbf{M}_2$.

    (b) $\mathbf{M}_i = \mathbf{M}_1 \cap \mathbf{M}_2 \oplus N_i$ for $i = 1, 2$, where each N_i has a basis.

$$\dim \mathbf{M}_i = \dim (\mathbf{M}_1 \cap \mathbf{M}_2) + \dim \mathbf{N}_i, i = 1, 2,$$
$$\mathbf{M}_1 + \mathbf{M}_2 = (\mathbf{M}_1 \cap \mathbf{M}_2) \oplus \mathbf{N}_1 \oplus \mathbf{N}_2.$$

    In particular, $\dim (\mathbf{M}_1 + \mathbf{M}_2) = \dim \mathbf{M}_1 + \dim \mathbf{M}_2 - \dim (\mathbf{M}_1 \cap \mathbf{M}_2)$.

1.14   Smith Normal Form

A matrix $D = [d_{ij}] \in \mathbb{D}^{m \times n}$ is called a diagonal matrix if $d_{ij} = 0$ for all $i \neq j$. The entries $d_{11}, \ldots, d_{\ell\ell}$, $\ell = \min(m, n)$ are called the diagonal entries of D. D is denoted by $D = \text{diag}(d_{11}, \ldots, d_{\ell\ell}) = \text{diag}(d_1, \ldots, d_l)$.

Theorem 1.14.1 *Let $0 \neq A \in \mathbb{D}^{m \times n}$. Assume that $\mathbb{D}$ is an elementary divisor domain. Then A is equivalent to a diagonal matrix*

$$B = \text{diag}(i_1(A), \ldots, i_r(A), 0, \ldots, 0), \quad r = \text{rank A}. \qquad (1.14.1)$$

Furthermore

$$i_{j-1}(A) | i_j(A), \quad \text{for } j = 2, \ldots, \text{rank A}. \qquad (1.14.2)$$

Proof. Recall that an elementary divisor domain is a Bezout domain. For $n = 1$ the Hermite normal form of A is a diagonal matrix with $i_1(A) = \delta_1(A)$. Next we consider the case $m = n = 2$. Let

$$A_1 = WA = \begin{bmatrix} a & b \\ 0 & c \end{bmatrix}, \quad W \in \mathbf{GL}(2, \mathbb{D}),$$

be the Hermite normal form of A. As $\mathbb{D} = \mathbb{D}_{ED}$ there exists $p, q, x, y \in \mathbb{D}$ such that

$$(px)a + (py)b + (qy)c = (a, b, c) = \delta_1(A).$$

Clearly $(p, q) = (x, y) = 1$. Hence, there exist $\bar{p}, \bar{q}, \bar{x}, \bar{y}$ such that

$$p\bar{p} - q\bar{q} = x\bar{x} - y\bar{y} = 1.$$

Let

$$V = \begin{bmatrix} p & q \\ \bar{q} & \bar{p} \end{bmatrix}, \quad U = \begin{bmatrix} x & \bar{y} \\ y & \bar{x} \end{bmatrix}.$$

Thus

$$G = VAU = \begin{bmatrix} \delta_1(A) & g_{12} \\ g_{21} & g_{22} \end{bmatrix}.$$

Since $\delta_1(G) \equiv \delta_1(A)$ we deduce that $\delta_1(A)$ divides g_{12} and g_{21}. Apply appropriate elementary row and column operations to deduce that A is equivalent to a diagonal matrix $C = \text{diag}(i_1(A), d_2)$. As $\delta_2(C) = i_1(A)d_2 \equiv \delta_2(A)$ we see that C is equivalent to the matrix of the form (1.14.1), where we can assume that $d_2 = i_2(A)$. Since $i_1(A)|d_2$ we have that $i_1(A)|i_2(A)$. We now prove the theorem in the case $m \geq 3$, $n = 2$ by induction starting from $m = 2$. Let $A = [a_{ij}] \in \mathbb{D}^{m \times 2}$ and denote by $\bar{A} = [a_{ij}]_{i=j=1}^{m-1,2}$. Use the induction hypothesis to assume that $\bar{A}$ is in the form (1.14.1). Interchange the second row of A with the last one to obtain $A_1 \in \mathbb{D}^{m \times 2}$. Apply simple row and column operations on the first two rows and columns of A_1 to obtain $A_2 = [a_{ij}^{(2)}] \in \mathbb{D}^{m \times 2}$, where $a_{11}^{(2)} = i_1(A)$. Use the elementary row and column operations to obtain A_3 of the form

$$A_3 = \begin{bmatrix} i_1(A) & 0 \\ 0 & A_4 \end{bmatrix}, \quad A_4 \in \mathbb{D}^{(m-1)\times 1}. \tag{1.14.3}$$

Recall that $i_1(A)$ divides all the entries of A_4. Hence, $A_4 = i_1(A)B_4$ and $i_1(A_4) = i_1(A)i_1(B_4)$. Use simple row operations on the rows $2, \ldots, m$ of A_3 to bring B_4 to a diagonal form. Thus A is equivalent to the diagonal matrix

$C = \operatorname{diag}(i_1(A), i_1(A)i_1(B_4)) = \operatorname{diag}(i_1(A), i_1(A_4)) \in \mathbb{D}^{m\times 2}$. Recall that

$$i_1(A) = \delta_1(A) \equiv \delta_1(C), \quad \delta_2(A) \equiv \delta_2(C) = i_1(A)i_1(A_4)$$
$$= i_1(A)i_1(A)i_1(B_4).$$

Thus $i_1(A)|i_1(A_4)$ so $i_1(A) = i_1(C)$ and $i_2(A) \equiv i_1(A_4)$. Hence, C is equivalent to B of the form (1.14.1) and $i_1(A)|i_2(A)$.

By considering $A^\top$ we deduce that we proved the theorem in the case $\min(m, n) \le 2$. We now prove the remaining cases by a double induction on $m \ge 3$ and $n \ge 3$. Assume that the theorem holds for all matrices in $\mathbb{D}^{(m-1)\times n}$ for $n = 2, 3, \ldots$ Assume that $m \ge 3$ and is fixed, and theorem holds for any $E \in \mathbb{D}^{m\times(n-1)}$ for $n \ge 3$. Let $A = [a_{ij}] \in \mathbb{D}^{m\times n}$, $\bar{A} = [a_{ij}]_{i=j=1}^{m,n-1}$. Use the induction hypothesis to assume that $\bar{A} = \operatorname{diag}(d_1, \ldots, d_l)$, $l = \min(m, n-1)$. Here, $d_1|d_i$, $i = 2, \ldots, l$. Interchange the second and the last column of A to obtain $A_1 = [a_{ij}^{(1)}] \in \mathbb{D}^{m\times n}$. Perform simple row operations on the rows of A_1 and simple column operations on the the first $n-1$ columns of A_1 to obtain the matrix $A_2 = [a_{ij}^{(2)}] \in \mathbb{D}^{m\times n}$ such that $\bar{A}_2 = [a_{ij}^{(2)}]_{i=j=1}^{m,n-1} = \operatorname{diag}(a_{11}^{(2)}, \ldots, a_{ll}^{(2)})$ is the Smith normal form of $\bar{A}_1 = [a_{ij}^{(1)}]_{i=j=1}^{m,n-1}$. The definition of A_2 yields that $i_1(A) \equiv a_{11}^{(2)}$. Use elementary row operations to obtain an equivalent matrix to A_2:

$$A_3 = \begin{bmatrix} i_1(A) & 0 \\ 0 & A_4 \end{bmatrix}, \quad A_4 \in \mathbb{D}^{(m-1)\times(n-1)}.$$

As $i_1(A) = \delta_1(A) \equiv \delta_1(A_3)$ it follows that $i_1(A)$ divides all the entries of A_4. So $A_4 = i_1(A)B_4$. Hence $i_j(A_4) = i_1(A)i_j(B_4)$ use simple row and column operations on the last $m - 1$ rows and the last $n - 1$ columns of A_3 to bring B_4 to Smith normal form using the induction hypothesis:

$$A \sim A_5 = \begin{bmatrix} i_1(A) & 0 \\ 0 & i_1(A)\,\mathrm{diag}(i_1(B_4), \ldots, i_l(B_4)) \end{bmatrix}.$$

By induction hypothesis

$$i_j(B_4)|i_{j+1}(B_4), \ j = 1, \ldots, \mathrm{rank}\,A - 1, \quad i_j(B_4) = 0, \ j > \mathrm{rank}\,A - 1.$$

A similar claim holds for A_4. Hence

$$\delta_k(A) \equiv \delta_k(A_5) = i_1(A)i_1(A_4)\cdots i_{k-1}(A_4), \quad k = 2, \ldots, \mathrm{rank}\,A.$$

Thus

$$i_j(A_4) \equiv i_{j+1}(A), \quad j = 1, \ldots, \mathrm{rank}\,A - 1$$

and A_5 is equivalent to B given by (1.14.1). Furthermore, we showed (1.14.2). $\qquad\qquad\square$

The matrix (1.14.1) is called the *Smith normal* form of A.

Corollary 1.14.2 *Let* $A, B \in \mathbb{D}_{ED}^{m\times n}$. *Then* A *and* B *are equivalent if and only if* A *and* B *have the same rank and the same invariant factors.*

Over an elementary divisor domain, the system (1.13.2) is equivalent to a simple system

$$i_k(A)y_k = c_k, \quad k = 1, \ldots, \mathrm{rank}\,A,$$

$$\hspace{6cm}(1.14.4)$$

$$0 = c_k, \quad k = \mathrm{rank}\,A + 1, \ldots, m,$$

$$\mathbf{y} = P^{-1}\mathbf{x}, \quad \mathbf{c} = Q\mathbf{b}. \hspace{2cm}(1.14.5)$$

Here, P and Q are the invertible matrices appearing in (1.11.7) and B is of the form (1.14.1). For the system (1.14.4) Theorems 1.13.2 and 1.13.3 are straightforward. Clearly

Theorem 1.14.3 *Let $A \in \mathbb{D}_{ED}^{m \times n}$. Assume that all of the invariant factors of A are invertible elements in $\mathbb{D}_{ED}$. Then the basis of* Range A *can be completed to a basis of* $\mathbb{D}_{ED}^n$.

In what follows we adopt the following:

Normalization 1.14.4 *Let $A \in \mathbb{F}[x]^{m \times n}$. Then the invariant polynomials (the invariant factors) of A are assumed to be normalized polynomials.*

Notation 1.14.5 *Let $A_i \in \mathbb{D}^{m_i \times n_i}$ for $i = 1, \ldots, k$. Then $\oplus_{i=1}^k A_i = \operatorname{diag}(A_1, \ldots, A_k)$ denotes the block diagonal matrix $B = [B_{ij}]_{i,j=1}^k \in \mathbb{D}^{m \times n}$, where $B_{ij} \in \mathbb{D}^{m_i \times n_j}$ for $i, j = 1, \ldots, k$, $m = \sum_{i=1}^k m_i$, $n = \sum_{j=1}^k n_j$, such that $B_{ii} = A_i$ and $B_{ij} = 0$ for $i \neq j$.*

Problems

1. Let $A = \begin{bmatrix} p & 0 \\ 0 & q \end{bmatrix} \in \mathbb{D}_B^{2 \times 2}$. Show that A is equivalent to $\operatorname{diag}\left((p, q), \frac{pq}{(p,q)}\right)$.

2. Let $A \in \mathbb{D}_G^{m \times n}$, $B \in \mathbb{D}_G^{p \times q}$. Suppose that either $i_s(A)|i_t(B)$ or $i_t(B)|i_s(A)$ for $s = 1, \ldots, \operatorname{rank} A = \alpha$, $t = 1, \ldots, \operatorname{rank} B = \beta$. Show that the set of the invariant factors $A \oplus B$ is $\{i_1(A), \ldots, i_\alpha(A), i_1(B), \ldots, i_\beta(B)\}$.

3. Let $\mathbf{M} \subset \mathbf{N}$ be $\mathbb{D}_{ED}$ modules with finite bases. Prove that there exists a basis $\mathbf{u}_1, \ldots, \mathbf{u}_n$ in $\mathbf{N}$ such that $i_1 \mathbf{u}_1, \ldots, i_r \mathbf{u}_r$ is a basis in $\mathbf{M}$, where $i_1, \ldots, i_r \in \mathbb{D}_{ED}$ and $i_j | i_{j+1}$ for $j = 1, \ldots, r - 1$.

4. Let $\mathbf{M}$ be a $\mathbb{D}$-module and $\mathbf{N}_1, \mathbf{N}_2 \subset \mathbf{M}$ be submodules. $\mathbf{N}_1$ and $\mathbf{N}_2$ are called equivalent if there exists an isomorphism $T \in \operatorname{Hom}(\mathbf{M}, \mathbf{M})$ $(T^{-1} \in \operatorname{Hom}(\mathbf{M}, \mathbf{M}))$ such that $T\mathbf{N}_1 = \mathbf{N}_2$. Suppose that $\mathbf{M}$, $\mathbf{N}_1$, $\mathbf{N}_2$ have bases $[\mathbf{u}_1, \ldots, \mathbf{u}_m]$, $[\mathbf{v}_1, \ldots, \mathbf{v}_n]$

and $[\mathbf{w}_1, \ldots, \mathbf{w}_n]$ respectively. Let

$$\mathbf{v}_j = \sum_{i=1}^{m} a_{ij} \mathbf{u}_i, \quad \mathbf{w}_j = \sum_{i=1}^{m} b_{ij} \mathbf{u}_i, \quad j = 1, \ldots, n,$$

(1.14.6)

$$A = [a_{ij}]_{i=j=1}^{m,n}, \quad B = [b_{ij}]_{i=j=1}^{m,n}.$$

Show that $\mathbf{N}_1$ and $\mathbf{N}_2$ are equivalent if and only if $A \sim B$.

5. Let $\mathbf{N} \subset \mathbf{M}$ be $\mathbb{D}$ modules with bases. Assume that $\mathbf{N}$ has the division property: if $a\mathbf{x} \in \mathbf{N}$ for $0 \neq a \in \mathbb{D}$ and $\mathbf{x} \in \mathbf{M}$ then $\mathbf{x} \in \mathbf{N}$. Show that if $\mathbb{D}$ is an elementary divisor domain and $\mathbf{N}$ has the division property then any basis in $\mathbf{N}$ can be completed to a basis in $\mathbf{M}$.

6. Let $\mathbb{D}$ be an elementary divisor domain. Assume that $\mathbf{N} \subset \mathbb{D}^m$ is a submodule with basis of dimension $k \in [1, m]$. Let $\mathbf{N}' \subset \mathbb{D}^m$ be the following set: $\mathbf{n} \in \mathbf{N}'$ if there exists $0 \neq a \in \mathbb{D}$ such that $a\mathbf{n} \in \mathbf{N}$. Show that $\mathbf{N}'$ is a submodule of $\mathbb{D}^m$ that has the division property. Furthermore, $\mathbf{N}'$ has a basis of dimension k which can be obtained from a basis of $\mathbf{N}$ as follows. Let $\mathbf{w}_1, \ldots, \mathbf{w}_k$ be a basis of $\mathbf{N}$. Let $W \in \mathbb{D}^{m \times k}$ be the matrix whose columns are $\mathbf{w}_1, \ldots, \mathbf{w}_k$. Assume that $D = \mathrm{diag}(n_1, \ldots, n_k)$ is the Smith normal form of W. So $W = UDV, U \in \mathbf{GL}(m, \mathbb{D}), V \in \mathbf{GL}(k, \mathbb{D})$. Let $\mathbf{u}_1, \ldots, \mathbf{u}_k$ be the first k columns of U. Then $\mathbf{u}_1, \ldots, \mathbf{u}_k$ is a basis of $\mathbf{N}'$.

1.15   Local Analytic Functions in One Variable

In this section, we consider applications of the Smith normal form to a system of linear equations over H_0, the ring of local analytic functions in one variable at the origin. In §1.3, we showed that the only noninvertible irreducible element in H_0 is z. Let $A \in H_0^{m \times n}$. Then $A = A(z) = [a_{ij}(z)]_{i=j=1}^{m,n}$ and $A(z)$ has the

MacLaurin expansion

$$A(z) = \sum_{k=0}^{\infty} A_k z^k, \quad A_k \in \mathbb{C}^{m \times n}, \; k = 0, \ldots, \tag{1.15.1}$$

which converges in some disk $|z| < R(A)$. Here, $R(A)$ is a positive number which depends on A. That is, each entry $a_{ij}(z)$ has convergent MacLaurin series for $|z| < R(A)$.

Notations and Definitions 1.15.1 *Let $A \in \mathrm{H}_0^{m \times n}$. The local invariant polynomials (the invariant factors) of A are normalized to be*

$$i_k(A) = z^{\iota_k(A)}, \quad 0 \le \iota_1(A) \le \iota_2(A) \le \cdots \le \iota_r(A), \quad r = \mathrm{rank}\, A. \tag{1.15.2}$$

The number $\iota_r(A)$ is called the index of A and is denoted by $\eta = \eta(A)$. For a non-negative integer p denote the number of local invariant polynomials of A whose degree is equal to p by $\kappa_p = \kappa_p(A)$.

We start with the following perturbation result.

Lemma 1.15.2 *Let $A, B \in \mathrm{H}_0^{m \times n}$. Let*

$$C(z) = A(z) + z^{k+1} B(z), \tag{1.15.3}$$

where k is a non-negative integer. Then A and C have the same local invariant polynomials up to degree k. Moreover, if k is equal to the index of A, and A and C have the same ranks then A is equivalent to C.

Proof. Since H_0 is a Euclidean domain we may already assume that A is in Smith normal form

$$A = \mathrm{diag}(z^{\iota_1}, \ldots, z^{\iota_r}, 0, \ldots, 0). \tag{1.15.4}$$

Let $s = \sum_{j=0}^{k} \kappa_j(A)$. Assume first that $s \ge t \in \mathbb{N}$. Consider any any $t \times t$ submatrix $D(z)$ of $C(z) = [c_{ij}(z)]$. View $\det D(z)$ as a sum of $t!$ products. As $k + 1 > \iota_t$ it follows each such product is divisible by $z^{\iota_1 + \cdots + \iota_t}$. Let $D(z) = [c_{ij}(z)]_{i=j=1}^{t}$. Then the product of the diagonal entries is of the form $z^{\iota_1 + \cdots + \iota_t}(1 + zO(z))$. All other $t! - 1$ products appearing in det $D(z)$ are divisible by $z^{\iota_1 + \cdots \iota_t - 2 + 2(k+1)}$. Hence

$$\delta_t(C) = z^{\iota_1 + \cdots + \iota_t} = \delta_t(A), \quad t = 1, \ldots, s, \tag{1.15.5}$$

which implies that

$$\iota_t(C) = \iota_t(A), \quad t = 1, \ldots, s. \tag{1.15.6}$$

As $s = \sum_{j=0}^{k} \kappa_j(A)$ it follows that

$$\kappa_j(C) = \kappa_j(A), \quad j = 0, \ldots, k - 1, \quad \kappa_k(A) \le \kappa_k(C).$$

Write $A = C - z^{k+1}B$ and deduce from the above arguments that

$$\kappa_j(C) = \kappa_j(A), \quad j = 0, \ldots, k. \tag{1.15.7}$$

Hence, A and C have the same local invariant polynomials up to degree k. Suppose that rank A = rank C. Then (1.15.6) implies that A and C have the same local invariant polynomials. Hence, $A \sim B$. $\square$

Consider a system of linear equations over H_0

$$A(z)\mathbf{u} = \mathbf{b}(z), \quad A(z) \in H_0^{m \times n}, \ \mathbf{b}(z) \in H_0^m, \tag{1.15.8}$$

where we look for a solution $\mathbf{u}(z) \in H_0^n$. Theorem 1.13.2 claims that the above system is solvable if and only if rank A = rank $\hat{A} = r$ and the g.c.d. of all $r \times r$ minors of A and $\hat{A}$ are equal. In the area of analytic functions it is common to try to solve (1.15.8) by the method of power series. Assume that $A(z)$ has the expansion (1.15.1) and $\mathbf{b}(z)$ has the expansion

$$\mathbf{b}(z) = \sum_{k=0}^{\infty} \mathbf{b}_k z^k, \quad \mathbf{b}_k \in \mathbb{C}^m, \ k = 0, \ldots. \tag{1.15.9}$$

Then one looks for a formal solution

$$\mathbf{u}(z) = \sum_{k=0}^{\infty} \mathbf{u}_k z^k, \quad \mathbf{u}_k \in \mathbb{C}^n, \ k = 0, \ldots, \tag{1.15.10}$$

which satisfies

$$\sum_{j=0}^{k} A_{k-j} \mathbf{u}_j = \mathbf{b}_k, \qquad (1.15.11)$$

for $k = 0, \ldots$. A vector $\mathbf{u}(z)$ is called a formal solution of (1.15.8) if (1.15.11) holds for any $k \in \mathbb{Z}_+$. A vector $\mathbf{u}(z)$ is called an analytic solution if $\mathbf{u}(z)$ is a formal solution and the series (1.15.10) converges in some neighborhood of the origin, i.e. $\mathbf{u}(z) \in H_0^n$. We now give the exact conditions for which (1.15.11) is solvable for $k = 0, \ldots, q$.

Theorem 1.15.3 *Consider the system* (1.15.11) *for* $k = 0, \ldots, q$ $\in \mathbb{Z}_+$. *Then this system is solvable if and only if* $A(z)$ *and* $\hat{A}(z)$ *have the same local invariant polynomials up to degree* q:

$$\kappa_j(A) = \kappa_j(\hat{A}), \quad j = 0, \ldots, q. \qquad (1.15.12)$$

Assume that the system (1.15.8) *is solvable over* H_0. *Let* $q = \eta(A)$ *and suppose that* $\mathbf{u}_0, \ldots, \mathbf{u}_q$ *satisfies* (1.15.11) *for* $k = 0, \ldots, q$. *Then there exists* $\mathbf{u}(z) \in H_0^n$ *satisfying* (1.15.8) *and* $\mathbf{u}(0) = \mathbf{u}_0$.

Let $q \in \mathbb{Z}_+$ *and* $\mathbf{W}_q \subset \mathbb{C}^n$ *be the subspace of all vectors* $\mathbf{w}_0$ *such that* $\mathbf{w}_0, \ldots, \mathbf{w}_q$ *is a solution to the homogeneous system*

$$\sum_{j=0}^{k} A_{k-j} \mathbf{w}_j = 0, \quad k = 0, \ldots, q. \qquad (1.15.13)$$

Then

$$\dim \mathbf{W}_q = n - \sum_{j=0}^{q} \kappa_j(A). \qquad (1.15.14)$$

In particular, for $\eta = \eta(A)$ *and any* $\mathbf{w}_0 \in \mathbf{W}_\eta$ *there exists* $\mathbf{w}(z) \in H_0^n$ *such that*

$$A(z)\mathbf{w}(z) = \mathbf{0}, \quad \mathbf{w}(0) = \mathbf{w}_0. \qquad (1.15.15)$$

Proof. Let

$$\mathbf{u}_k = (u_{k,1}, \ldots, u_{k,n})^\top, \quad k = 0, \ldots, q.$$

We first establish the theorem when $A(z)$ is in Smith normal form (1.15.4). In that case the system (1.15.11) reduces to

$$u_{k-\iota_s,s} = b_{k,s} \quad \text{if } \iota_s \leq k,$$

$$(1.15.16)$$

$$0 = b_{k,s} \quad \text{if either } \iota_s > k \text{ or } s > \text{rank A}.$$

The above equations are solvable for $k = 0, \ldots, q$ if and only if z^{ι_s} divides $b_s(z)$ for all $\iota_s \leq q$, and for $\iota_s > q$, z^{q+1} divides $b_s(z)$. If $\iota_s \leq q$ then subtract from the last column of $\hat{A}$ the s-column times $\frac{b_s(z)}{z^{\iota_s}}$. So $\hat{A}$ is equivalent to the matrix

$$A_1(z) = \text{diag}(z^{\iota_1}, \ldots, z^{\iota_l}) \oplus z^{q+1} A_2(z),$$

$$l = \sum_{j=0}^{q} \kappa_j(A), \quad A_2 \in \text{H}_0^{(m-l) \times (n+1-l)}.$$

According to Problem 2, the local invariant polynomials of $A_1(z)$ whose degrees do not exceed q are $z^{\iota_1}, \ldots, z^{\iota_l}$. So $A(z)$ and $A_1(z)$ have the same local invariant polynomials up to degree q. Assume now that A and $\hat{A}$ have the same local invariant polynomial up to degree q. Hence

$$z^{\iota_1 + \cdots + \iota_k} = \delta_k(A) = \delta_k(\hat{A}), \quad k = 1, \ldots, l,$$

$$z^{\iota_1 + \cdots + \iota_l + q + 1} | \delta_{l+1}(\hat{A}).$$

The first set of the equalities implies that $z^{\iota_s} | b_s(z)$, $s = 1, \ldots, l$. The last equality yields that for $s > l$, $z^{q+1} | b_s(z)$. Hence, (1.15.11) is solvable for $k = 0, \ldots, q$ if and only if A and $\hat{A}$ have the same local invariant polynomials up to the degree q.

Assume next that (1.15.8) is solvable. Since $A(z)$ is of the form (1.15.4) the general solution of (1.15.8) in that case is

$$u_j(z) = \frac{b_i(z)}{z^{\iota_j}}, \quad j = 1, \ldots, \text{rank A},$$

where $u_j(z)$ is an arbitrary function in H_0, $j = \text{rank A} + 1, \ldots, n$.

Hence

$$u_j(0) = b_{i_j}, \quad j = 1, \ldots, \text{rank A},$$

(1.15.17)

where $u_j(0)$ is an arbitrary complex number,

$$j = \text{rank A} + 1, \ldots, n.$$

Clearly (1.15.16) implies that $u_{0,s} = u_s(0)$ for $k = i_s$. The solvability of (1.15.8) implies that $b_s(z) = 0$ for $s > \text{rank A}$. So $u_{0,s}$ is not determined by (1.15.16) for $s > \text{rank A}$. This proves the existence of $\mathbf{u}(z)$ satisfying (1.15.8) such that $\mathbf{u}(0) = \mathbf{u}_0$. Consider the homogeneous system (1.15.13) for $k = 0, \ldots, q$. Then $w_{0,s} = 0$ for $i_s \leq$ and otherwise $u_{0,s}$ is a free variable. Hence, (1.15.14) holds. Let $q = \eta = \eta(A)$. Then the system (1.15.13) implies that the coordinates of $\mathbf{w}_0$ satisfy the conditions (1.15.17). Hence, the system (1.15.15) is solvable.

Assume now that $A(z) \in \mathrm{H}_0^{m \times n}$ is an arbitrary matrix. Theorem 1.14.1 implies the existence of $P(z) \in \mathbf{GL}(n, \mathrm{H}_0)$, $Q(z) \in \mathbf{GL}(m, \mathrm{H}_0)$ such that

$$Q(z)A(z)P(z) = B(z) = \text{diag}(z^{\iota_1}, \ldots, z^{\iota_r}, 0, \ldots, 0), 0 \leq \iota_1 \leq \cdots \leq \iota_r,$$
$$r = \text{rank A}.$$

It is straightforward to show that $P(z) \in \mathbf{GL}(n, \mathrm{H}_0)$ if and only if $P(z) \in \mathrm{H}_0^{n \times n}$ and $P(0)$ is invertible. To this end let

$$P(z) = \sum_{k=0}^{\infty} P_k z^k, \quad P_k \in \mathbb{C}^{n \times n}, \ k = 0, \ldots, \quad \det P_0 \neq 0,$$

$$Q(z) = \sum_{k=0}^{\infty} Q_k z^k, \quad Q_k \in \mathbb{C}^{n \times n}, \ k = 0, \ldots, \quad \det Q_0 \neq 0.$$

Introduce a new set of variables $\mathbf{v}(z)$ and $\mathbf{v}_0, \mathbf{v}_1, \ldots, \mathbf{v}_k$ such that

$$\mathbf{u}(z) = P(z)\mathbf{v}(z),$$
$$\mathbf{u}_k = \sum_{j=0}^{k} P_{k-j}\mathbf{v}_j, \quad k = 0, \ldots.$$

Since $\det P_0 \neq 0$ $\mathbf{v}(z) = P(z)^{-1}\mathbf{u}(z)$ and we can express each $\mathbf{v}_k$ in terms of $\mathbf{u}_k, \ldots, \mathbf{u}_0$ for $k = 0, 1, \ldots$, we have that (1.15.8) and (1.15.11) are respectively equivalent to

$$B(z)\mathbf{v}(z) = \mathbf{c}(z), \quad \mathbf{c}(z) = Q(z)\mathbf{b}(z),$$

$$\sum_{j=0}^{k} B_{k-j}\mathbf{v}_j = \mathbf{c}_k, \quad k = 0, \ldots, q.$$

As $B \sim A$ and $\hat{B} = Q\hat{A}(P \oplus I_1) \sim \hat{A}$, we deduce the theorem. $\square$

Problems

1. The system (1.15.8) is called solvable in the punctured disc if the system

$$A(z_0)\mathbf{u}(z_0) = \mathbf{b}(z_0), \tag{1.15.18}$$

is solvable for any point $0 < |z_0| < R$ as a linear system over $\mathbb{C}$ for some $R > 0$, i.e.

$$\operatorname{rank} A(z_0) = \operatorname{rank} \hat{A}(z_0), \quad \text{for all } 0 < |z_0| < \text{R.} \tag{1.15.19}$$

Show that (1.15.8) is solvable in the punctured disk if and only if (1.15.8) is solvable over $\mathcal{M}_0$, the quotient field of H_0.

2. The system (1.15.8) is called pointwise solvable if (1.15.18) is solvable for all z_0 in some open disk $|z_0| < R$. Show that (1.15.8) is pointwise solvable if and only if (1.15.8) is solvable over $\mathcal{M}_0$ and

$$\operatorname{rank} A(0) = \operatorname{rank} \hat{A}(0). \tag{1.15.20}$$

3. Let $A(z) \in H_0^{m \times n}$. $A(z)$ is called *generic* if, whenever the system (1.15.8) is pointwise solvable, it is also analytically solvable, i.e. solvable over H_0. Prove that $A(z)$ is generic if and only if $\eta(A) \leq 1$.

4. Let $\Omega \subset \mathbb{C}$ be a domain and consider the system

$$A(z)\mathbf{u} = \mathbf{b}(z), \quad A(z) \in \mathrm{H}(\Omega)^{m \times n}, \quad \mathbf{b}(z) \in \mathrm{H}(\Omega)^{m}. \quad (1.15.21)$$

Show that the above system is solvable over $\mathrm{H}(\Omega)$ if and only if for each $\zeta \in \Omega$ this system is solvable in H_{ζ}. (*Hint*: As $\mathrm{H}(\Omega)$ is $\mathbb{D}_{ED}$ it suffices to analyze the case where $A(z)$ is in Smith normal form.)

5. Let $A(z)$ and $\mathbf{b}(z)$ satisfy the assumptions of Problem 4. $A(z)$ is called *generic* if whenever (1.15.21) is pointwise solvable it is solvable over $\mathrm{H}(\Omega)$. Show that $A(z)$ is generic if and only the invariant functions (factors) of $A(z)$ have only simple zeros. (ζ is a simple zero of $f \in \mathrm{H}(\Omega)$ if $f(\zeta) = 0$ and $f'(\zeta) \neq 0$.)

6. Let $A(z) \in \mathrm{H}(\Omega)^{m \times n}$, where Ω is a domain in $\mathbb{C}$. Prove that the invariant factors of $A(z)$ are invertible in $\mathrm{H}(\Omega)$ if and only if

$$\mathrm{rank}\, \mathrm{A}(\zeta) = \mathrm{rank}\, \mathrm{A}, \quad \text{for all } \zeta \in \Omega. \quad (1.15.22)$$

7. Let $A(z) \in \mathrm{H}(\Omega)^{m \times n}$, where Ω is a domain in $\mathbb{C}$. Assume that (1.15.22) holds. View $A(z) \in \mathrm{Hom}\,(\mathrm{H}(\Omega)^{n}, \mathrm{H}(\Omega)^{m})$. Show that Range $A(z)$ has a basis which can be completed to a basis in $\mathrm{H}(\Omega)^{m}$. (*Hint*: Use Theorem 1.15.5.)

1.16 The Local–Global Domains in $\mathbb{C}^{p}$

Let p be a positive integer and assume that $\Omega \subseteq \mathbb{C}^{p}$ is a domain. Consider the system of m nonhomogeneous equations in n unknowns:

$$A(\mathbf{z})\mathbf{u} = \mathbf{b}(\mathbf{z}), \quad A(\mathbf{z}) \in \mathrm{H}(\Omega)^{m \times n}, \quad \mathbf{b}(\mathbf{z}) \in \mathrm{H}(\Omega)^{m}. \quad (1.16.1)$$

In this section, we are concerned with the problem of the existence of a solution $\mathbf{u}(\mathbf{z}) \in \mathrm{H}(\Omega)^{m}$ to the above system. Clearly a necessary condition for the solvability is the local condition:

Condition 1.16.1 *Let* $\Omega \subseteq \mathbb{C}^{p}$ *be a domain. Then for each* $\zeta \in \Omega$ *the system* $A(\mathbf{z})\mathbf{u} = \mathbf{b}(\mathbf{z})$ *has a solution* $\mathbf{u}_{\zeta}(\mathbf{z}) \in \mathrm{H}_{\zeta}^{m}$.

Definition 1.16.2 *A domain $\Omega \subseteq \mathbb{C}^p$ is called a local–global domain, if any system of the form (1.16.1), satisfying the condition 1.16.1, has a solution $\mathbf{u}(\mathbf{z}) \in \mathrm{H}(\Omega)^m$.*

Problem 1.15.4 implies that any domain $\Omega \subset \mathbb{C}$ is a local–global domain. In this section, we assume that $p > 1$. Problem 1 shows that not every domain in $\mathbb{C}^p$ is a local–global domain. We give a sufficient condition on domain Ω to be a local–global domain.

Definition 1.16.3 *A domain $\Omega \subset \mathbb{C}^p$ is called a domain of holomorphy, if there exists $f \in \mathrm{H}(\Omega)$ such that for any larger domain $\Omega_1 \subset C^p$, strictly containing Ω, there is no $f_1 \in \mathrm{H}(\Omega_1)$ which coincides with f on Ω.*

The following theorem is a very special case of Hartog's theorem [GuR65].

Theorem 1.16.4 *Let $\Omega \subseteq \mathbb{C}^p, p > 1$ be a domain. Assume that $\zeta \in \Omega$ and $f \in \mathrm{H}(\Omega \backslash \{\zeta\})$. Then $f \in \mathrm{H}(\Omega)$.*

Thus $\Omega \backslash \{\zeta\}$ is not a domain of holomorphy. A simple example of a domain of holomorphy is:

Example 1.16.5 *Let $\Omega \subseteq \mathbb{C}^p$ be an open convex set. Then Ω is a domain of holomorphy [GuR65].*

(Recall that $\Omega \subseteq \mathbb{C}^p$ is convex if for any two points $\xi, \zeta \in \Omega$ the point $(1-t)\xi + t\zeta$ is in Ω for each $t \in (0,1)$.) The main result of this section is:

Theorem 1.16.6 *Let $\Omega \subseteq \mathbb{C}^p, p > 1$ be a domain of holomorphy. Then Ω is a local–global domain.*

The proof needs basic knowledge of sheaves and is given below for the reader who has been exposed to the basic concepts in this field. See for example [GuR65]. We discuss only very special types of sheaves which are needed for the proof of Theorem 1.16.6.

Definition 1.16.7 *Let $\Omega \subseteq \mathbb{C}^p$ be an open set. Then*

1. *$\mathcal{F}(\Omega)$, called the sheaf of rings of holomorphic functions on Ω, is the union all $H(U)$, where U ranges over all open subsets of Ω. Then for each $\zeta \in \Omega$ the local ring H_ζ is viewed as a subset of $\mathcal{F}(\Omega)$ and is called the stalk of $\mathcal{F}(\Omega)$ over ζ. A function $f \in H(U)$ is called a section of $\mathcal{F}(\Omega)$ on U.*

2. *For an integer $n \geq 1$, $\mathcal{F}_n(\Omega)$, called an $\mathcal{F}(\Omega)$–sheaf of modules, is the union all $H(U)^n$, where U ranges over all open subsets of Ω. Then for each $\zeta \in \Omega$ the local module H_ζ^n is viewed as a subset of $\mathcal{F}_n(\Omega)$ and is called the stalk of $\mathcal{F}_n(\Omega)$ over ζ. (Note: H_ζ^n is an H_ζ module.) A vector $\mathbf{u} \in H(U)^n$ is called a section of $\mathcal{F}_n(\Omega)$ on U. (If $U = \emptyset$ then $H(U)^n$ consists of the zero element $\mathbf{0}$.)*

3. *$\mathcal{F} \subseteq \mathcal{F}_n(\Omega)$ is called a subsheaf if the following conditions holds.*

   (a) *$\mathcal{F} \cap H^n(U)$ contains the trivial section $\mathbf{0}$ for each open set $U \subseteq \Omega$.*

   (b) *Assume that $\mathbf{u} \in H(U)^n \cap \mathcal{F}, \mathbf{v} \in H(V)^n \cap \mathcal{F}$ and $W \subseteq U \cap V$ is an open nonempty set.*

      (i) *For any $f, g \in H(W)$ the vector $f\mathbf{u} + g\mathbf{v} \in \mathcal{F} \cap H(W)^n$. (This property is called the **Restriction property**.)*

      (ii) *If $\mathbf{u} = \mathbf{v}$ on W then the section $\mathbf{w} \in H^n(U \cup V)$, which coincides with $\mathbf{u}, \mathbf{v}$ on U, V respectively, belongs to $\mathcal{F} \cap H^n(U \cup V)$. (This property is called the **Extension property**.)*

4. *Assume that $\mathcal{F} \subseteq \mathcal{F}_n(\Omega)$ is a subsheaf. Then*

   (a) *$\mathcal{F}_\zeta := \mathcal{F} \cap H_\zeta^n$ is called the stalk of $\mathcal{F}$ over $\zeta \in \Omega$.*

   (b) *Let U be an open subset of Ω. Then $\mathcal{F}(U) := \mathcal{F} \cap \mathcal{F}_n(U)$ is called the restriction of the subsheaf $\mathcal{F}$ to U. The sections $\mathbf{u}_1, \ldots, \mathbf{u}_k \in \mathcal{F} \cap H(U)^n$ are said to generate $\mathcal{F}$ on U, if for any $\zeta \in U$ $\mathcal{F}_\zeta$ is generated by $\mathbf{u}_1, \ldots, \mathbf{u}_k$ over H_ζ. $\mathcal{F}$ is called finitely generated over U if such $\mathbf{u}_1, \ldots, \mathbf{u}_k \in \mathcal{F}(U)$ exists.*

(c) $\mathcal{F}$ is called *finitely generated* if it is finitely generated over Ω. $\mathcal{F}$ is called *of finite type* if for each $\zeta \in \Omega$ there exists an open set $U_\zeta \subset \Omega$, containing ζ, such that $\mathcal{F}$ is finitely generated over U_ζ. (*That is, each $\mathcal{F}_\zeta$ is finitely generated.*)

(d) $\mathcal{F}$ is called a *coherent sheaf* if the following two conditions hold. First $\mathcal{F}$ is of finite type. Second, for each open set $U \subseteq \Omega$ and for any $q \geq 1$ sections $\mathbf{u}_1, \ldots, \mathbf{u}_q \in \mathcal{F} \cap \mathrm{H}(\Omega)^n$ let $\mathcal{G} \subseteq \mathcal{F}_q(U)$ be a subsheaf generated by the condition $\sum_{i=1}^q f_i \mathbf{u}_i = 0$. That is, $\mathcal{G}$ is a union of all $(f_1, \ldots, f_q)^\top \in \mathrm{H}(V)^q$ satisfying the condition $\sum_{i=1}^q f_i \mathbf{u}_i = 0$ for all open $V \subseteq U$. Then $\mathcal{G}$ is of finite type.

The following result is a straight consequence of Oka's coherence theorem [GuR65].

Theorem 1.16.8 *Let $\Omega \subseteq \mathbb{C}^P$ be an open set. Then*

- *The sheaf $\mathcal{F}_n(\Omega)$ is coherent.*

- *Let $A \in \mathrm{H}(\Omega)^{m \times n}$ be given. Let $\mathcal{F} \subseteq \mathcal{F}(\Omega)$ be the subsheaf consisting of all $\mathbf{u} \in \mathrm{H}(U)^n$ satisfying $A\mathbf{u} = 0$ for all open sets $U \subseteq \Omega$. Then $\mathcal{F}$ is coherent.*

Note that $\mathcal{F}_n(\Omega)$ is generated by n constant sections $\mathbf{u}_i := (\delta_{i1}, \ldots, \delta_{in})^\top \in \mathrm{H}(\Omega)^n, i = 1, \ldots, n$. The following theorem is a special case of Cartan's Theorem A.

Theorem 1.16.9 *Let $\Omega \subseteq \mathbb{C}^p$ be a domain of holomorphy. Let $\mathcal{F} \subset \mathcal{F}_n(\Omega)$ be a subsheaf defined in Definition 1.16.7. If $\mathcal{F}$ is coherent then $\mathcal{F}$ is finitely generated.*

Corollary 1.16.10 *Let $\Omega \subseteq \mathbb{C}^p$ be a domain of holomorphy and $A \in \mathrm{H}(\Omega)^{m \times n}$. Then there exist $\mathbf{u}_1, \ldots, \mathbf{u}_l \in \mathrm{H}(\Omega)^n$, such that for any $\zeta \in \Omega$, every solution of the system $A\mathbf{u} = \mathbf{0}$ over H_ζ^n is of the form $\sum_{i=1}^l f_i \mathbf{u}_i$ for some $f_1, \ldots, f_l \in \mathrm{H}_\zeta$.*

We now introduce the notion of *sheaf cohomology* of $\mathcal{F} \subseteq \mathcal{F}_n(\Omega)$.

Definition 1.16.11 *Let $\Omega \subseteq \mathbb{C}^p$ be an open set. Let $\mathcal{U} := \{U_i \subseteq \Omega, \; i \in \mathcal{I}\}$ be an open cover of Ω. (That is, each U_i is open, and $\cup_{i \in \mathcal{I}} U_i = \Omega$.) For each integer $p \geq 0$ and $p + 1$ tuples of indices $(i_0, \ldots, i_p) \in \mathcal{I}^{p+1}$ denote $U_{i_0, \ldots, i_p} := \cap_{j=0}^{p} U_{i_j}$.*

Assume that $\mathcal{F} \subseteq \mathcal{F}_n(\Omega)$ is a subsheaf. A p-cochain c is a map carrying each $p + 1$-tuples of indices $(i_0, \ldots, i_p)$ to a section $\mathcal{F} \cap \mathrm{H}^n(U_{i_0, \ldots, i_p})$ satisfying the following properties.

1. *$c(i_0, \ldots, i_p) = 0$ if $U_{i_0 \ldots i_p} = \emptyset$.*

2. *$c(\pi(i_0), \ldots, \pi(i_p)) = \mathrm{sgn}(\pi) c(i_0, \ldots, i_p)$ for any permutation $\pi : \{0, \ldots, p\} \to \{0, \ldots, p\}$. (Note that $c(i_0, \ldots, i_p)$ is the trivial section if $i_j = i_k$ for $j \neq k$.)*

The zero cochain is the cochain which assigns a zero section to any $(i_0, \ldots, i_p)$. Two cochains c, d are added and subtracted by the identity $(c \pm d)(i_0, \ldots, i_p) = c(i_0, \ldots, i_p) \pm d(i_0, \ldots, i_p)$. Denote by $\mathrm{C}^p(\Omega, \mathcal{F}, \mathcal{U})$ the group of $p + 1$ cochains.

The pth coboundary operator $\delta_p : \mathrm{C}^p(\Omega, \mathcal{F}, \mathcal{U}) \to \mathrm{C}^{p+1}(\Omega, \mathcal{F}, \mathcal{U})$ is defined as follows:

$$(\delta_p c)(i_0, \ldots, i_{p+1}) = \sum_{j=0}^{p+1} (-1)^j c(i_0, \ldots, \hat{i}_j, \ldots, i_{p+1}),$$

where $\hat{i}_j$ is a deleted index. Then pth cohomology group is given by

1. *$\mathrm{H}^0(\Omega, \mathcal{F}, \mathcal{U}) := \mathrm{Ker}\, \delta_0$.*

2. *For $p \geq 1$ $\mathrm{H}^p(\Omega, \mathcal{F}, \mathcal{U}) := \mathrm{Ker}\, \delta_p / \mathrm{Range}\, \delta_{p-1}$. (See Problem 2.)*

Lemma 1.16.12 *Let the assumptions of Definition 1.16.11 hold. Let $c \in \mathrm{C}^0(\Omega, \mathcal{F}, \mathcal{U})$. Then $c \in \mathrm{H}^0(\Omega, \mathcal{F}, \mathcal{U})$ if and only if c represents a global section $\mathbf{u} \in \mathcal{F} \cap \mathrm{H}(\Omega)^n$.*

Proof. Let $c \in \mathrm{C}^0(\Omega, \mathcal{F}, \mathcal{U})$. Assume that $c \in \mathrm{H}^0(\Omega, \mathcal{F}, \mathcal{U})$. Let U_0, U_1 be two open sets in $\mathcal{U}$. Then $c(i_0) - c(i_1)$ is the zero section on $U_0 \cap U_1$. Thus for each $\zeta \in U_0 \cap U_1$ $c(i_0)(\zeta) = c(i_1)(\zeta)$. Let $\mathbf{u}(\mathbf{z}) := c(i_0)(\mathbf{z}) \in \mathbb{C}^n$. It follows that $\mathbf{u} \in \mathrm{H}(\Omega)^n$. The extension property of subsheaf $\mathcal{F}$ yields that $\mathbf{u} \in \mathcal{F} \cap \mathrm{H}^n(\Omega)$. Vice versa, assume that $\mathbf{u} \in \mathcal{F} \cap \mathrm{H}^n(\Omega)$. Define $c(i_0) = \mathbf{u}|U_0$. Then $c \in \mathrm{H}^0(\Omega, \mathcal{F}, \mathcal{U})$. $\square$

We identify $\mathrm{H}^0(\Omega, \mathcal{F}, \mathcal{U})$ with the set of global sections $\mathcal{F} \cap \mathrm{H}(\Omega)^n$. The cohomology groups $\mathrm{H}^p(\Omega, \mathcal{F}, \mathcal{U}), p \geq 1$ depend on the open cover $\mathcal{U}$ of Ω. By refining the covers of Ω one can define the cohomology groups $\mathrm{H}^p(\Omega, \mathcal{F}), p \geq 0$. See Problem 3. Cartan's Theorem B claims [GuR65]:

Theorem 1.16.13 *Let $\Omega \subseteq \mathbb{C}^p$ be domain of holomorphy. Assume that the sheaf $\mathcal{F}$ given in Definition 1.16.7 is coherent. Then $\mathrm{H}^p(\Omega, \mathcal{F})$ is trivial for any $p \geq 1$.*

Proof of Theorem 1.16.6. Consider the system (1.16.1). Let $\mathcal{F}$ be the coherent sheaf defined in Theorem 1.16.8. Assume that the system (1.16.1) is locally solvable over Ω. Let $\zeta \in \Omega$. Then there exists an open set $U_\zeta \subseteq \Omega$ such that there exists $\mathbf{u}_\zeta \in \mathrm{H}^n(U_\zeta)$ satisfying (1.16.1) over $\mathrm{H}(U_\zeta)$. Let $\mathcal{U} := \{U_\zeta, \zeta \in \Omega\}$ be an open cover of Ω. Define $c \in \mathrm{C}^1(\Omega, \mathcal{F}, \mathcal{U})$ by $c(\zeta, \eta) = \mathbf{u}_\zeta - \mathbf{u}_\eta$. Note that

$$(\delta_1 c)(\zeta, \eta, \theta) = c(\eta, \theta) - c(\zeta, \theta) + c(\zeta, \eta) = 0.$$

Hence, $c \in \mathrm{Ker}\, \delta_1$. Since $\mathcal{F}$ is coherent Cartan's Theorem B yields that $\mathrm{H}^1(\Omega, \mathcal{F})$ is trivial. Hence, $\mathrm{H}^1(\Omega, \mathcal{F}, \mathcal{U})$ is trivial. (See Problem 3c below.) Thus, there exists an element $d \in \mathrm{C}^0(\Omega, \mathcal{F}, \mathcal{U})$ such that $\delta_0 d = c$. Thus for each $\zeta, \eta \in \Omega$ such that $U_\zeta \cap U_\eta$ there exist sections $\mathbf{d}(\zeta) \in \mathcal{F} \cap \mathrm{H}^n(U_\zeta), \mathbf{d}(\eta) \in \mathcal{F} \cap \mathrm{H}^n(U_\eta)$ such that $\mathbf{d}(\eta) - \mathbf{d}(\zeta) = \mathbf{u}_\zeta - \mathbf{u}_\eta$ on $U_\zeta \cap U_\eta$. Hence, $\mathbf{d}(\eta) + \mathbf{u}_\eta = \mathbf{d}(\zeta) + \mathbf{u}_\zeta$ on $U_\zeta \cap U_\eta$. Since $A\mathbf{d}_\zeta = \mathbf{0} \in \mathrm{H}(U_\zeta)^m$ it follows that $A(\mathbf{d}_\zeta + \mathbf{u}_\zeta) = \mathbf{b} \in \mathrm{H}(U_\zeta)^m$. As $\mathbf{d}(\eta) + \mathbf{u}_\eta = \mathbf{d}(\zeta) + \mathbf{u}_\zeta$ on $U_\zeta \cap U_\eta$ it follows that all these section can be patched to the vector $\mathbf{v} \in \mathrm{H}(\Omega)^n$ which is a global solution of (1.16.1). $\square$

Problems

1. Consider a system of one equation over $\mathbb{C}^p, p > 1$

$$\sum_{i=1}^{p} z_i u_i = 1, \quad \mathbf{u} = (u_1, \ldots, u_p)^\top, \ \mathbf{z} = (z_1, \ldots, z_p).$$

Let $\Omega := \mathbb{C}^p \backslash \{\mathbf{0}\}$.

(a) Show that Condition 1.16.1 holds for Ω.

(b) Show that the system is not solvable at $\mathbf{z} = \mathbf{0}$. (Hence, it does not have a solution $\mathbf{u}(\mathbf{z}) \in \mathrm{H}_{\mathbf{0}}^p$.)

(c) Show that the system does not have a solution $\mathbf{u}(\mathbf{z}) \in \mathrm{H}(\Omega)^p$. (*Hint*: Prove by contradiction using Hartog's theorem.)

2. Let the assumptions of Definition 1.16.11 hold. Show for any $p \geq 0$.

(a) $\delta_{p+1}\delta_p = 0$.

(b) Range $\delta_p \subseteq \mathrm{Ker}\ \delta_{p+1}$.

3. Let $\mathcal{U} = \{U_i, i \in \mathcal{I}\}, \mathcal{V} = \{V_j, j \in \mathcal{J}\}$ be two open covers of an open set $\Omega \subset \mathbb{C}^p$. $\mathcal{V}$ is called a refinement of $\mathcal{U}$, denoted $\mathcal{V} \prec \mathcal{U}$, if each V_j is contained in some U_i. For each V_j we fix an arbitrary U_i with $V_j \subseteq U_i$, and write it as $U_{i(j)} : V_j \subseteq U_{i(j)}$. Let $\mathcal{F}$ be a subsheaf as in Definition 1.16.11. Show

(a) Define $\phi : \mathrm{C}^p(\Omega, \mathcal{F}, \mathcal{U}) \to \mathrm{C}^p(\Omega, \mathcal{F}, \mathcal{V})$ as follows. For $c \in \mathrm{C}^p(\Omega, \mathcal{F}, \mathcal{U})$ let $(\phi(c))(j_0, \ldots, j_p) \in \mathrm{C}^p(\Omega, \mathcal{F}, \mathcal{V})$ be the restriction of the section $c(i(j_0), \ldots, i(j_p))$ to $V_{j_0, \ldots, j_p}$. Then ϕ is a homomorphism.

(b) ϕ induces a homomorphism $\tilde{\phi} : \mathrm{H}^p(\Omega, \mathcal{F}, \mathcal{U}) \to \mathrm{H}^p(\Omega, \mathcal{F}, \mathcal{V})$. Furthermore, $\tilde{\phi}$ depends only on the covers $\mathcal{U}, \mathcal{V}$. (That is, the choice of $i(j)$ is irrelevant.)

(c) By refining the covers one obtains the pth cohomology group $\mathrm{H}^p(\Omega, \mathcal{F})$ with the following property. The homomorphism $\tilde{\phi}$ described in 3b induces an injective homomorphism $\tilde{\phi} : \mathrm{H}^p(\Omega, \mathcal{F}, \mathcal{U}) \to \mathrm{H}^p(\Omega, \mathcal{F})$ for $p \geq 1$. (Recall that $\mathrm{H}^0(\Omega, \mathcal{F}, U) \equiv \mathcal{F} \cap \mathrm{H}^n(\Omega)$.) In particular, $\mathrm{H}^p(\Omega, \mathcal{F})$ is trivial, i.e. $\mathrm{H}^p(\Omega, \mathcal{F}) = \{0\}$, if and only if each $\mathrm{H}^p(\Omega, \mathcal{F}, \mathcal{U})$ is trivial.

1.17   Historical Remarks

Most of the material in §1.1–1.11 is standard. See [Lan58], [Lan67] and [vdW59] for the algebraic concepts. Consult [GuR65] and [Rud74] for the concepts and results concerning the analytic functions. See [Kap49] for the properties of elementary divisor domains. It is not known if there exists a Bezout domain which is not an elementary divisor domain. Theorem 1.5.3 for $\Omega = \mathbb{C}$ is due to [Hel40]. For §1.11 see [CuR62] or [McD33]. Most of §1.12 is well known, e.g. [McD33]. §1.13 seems to be new since the underlying ring is assumed to be only a Bezout domain. Theorems 1.13.2 and 1.13.3 are well known for an elementary divisor domain, since A is equivalent to a diagonal matrix. It would be interesting to generalize Theorem 1.13.2 for $\mathbb{D} = \mathbb{F}[x_1, \ldots, x_p]$ for $p \geq 2$. The fact that the Smith normal form can be achieved for $\mathbb{D}_{ED}$ is due to Helmer [Hel43]. More can be found in [Kap49].

Most of the results of §1.15 are from [Fri80b]. I assume that Theorem 1.16.6 is known to the experts.

Chapter 2

Canonical Forms for Similarity

2.1 Strict Equivalence of Pencils

Definition 2.1.1 *A matrix* $A(x) \in \mathbb{D}[x]^{m \times n}$ *is a pencil if*

$$A(x) = A_0 + x A_1, \quad A_0, A_1 \in \mathbb{D}^{m \times n}. \tag{2.1.1}$$

A pencil $A(x)$ *is regular if* $m = n$ *and* $\det A(x) \neq 0$. *Otherwise* $A(x)$ *is a singular pencil. Two pencils* $A(x), B(x) \in \mathbb{D}[x]^{m \times n}$ *are strictly equivalent if*

$$A(x) \overset{s}{\sim} B(x) \iff B(x) = Q A(x) P, \quad P \in \mathbf{GL}(n, \mathbb{D}), \quad Q \in \mathbf{GL}(m, \mathbb{D}). \tag{2.1.2}$$

The classical works of Weierstrass [Wei67] and Kronecker [Kro90] classify the equivalence classes of pencils under the strict equivalence relation in the case $\mathbb{D}$ is a field $\mathbb{F}$. We give a short account of their main results.

First note that the strict equivalence of $A(x), B(x)$ implies the equivalence of $A(x), B(x)$ over the domain $\mathbb{D}[x]$. Furthermore let

$$B(x) = B_0 + x B_1. \tag{2.1.3}$$

75

Then the condition (2.1.2) is equivalent to

$$B_0 = QA_0P, \quad B_1 = QA_1P, \quad P \in \mathbf{GL}(n, \mathbb{D}), \quad Q \in \mathbf{GL}(m, \mathbb{D}).$$
$$(2.1.4)$$

Thus we can interchange A_0 with A_1 and B_0 with B_1 without affecting the strict equivalence relation. Hence, it is natural to consider a *homogeneous* pencil

$$A(x_0, x_1) = x_0 A_0 + x_1 A_1. \tag{2.1.5}$$

Assume that $\mathbb{D}$ is $\mathbb{D}_U$. Then $\mathbb{D}_U[x_0, x_1]$ is also $\mathbb{D}_U$ (Problem 1.4.6) In particular $\mathbb{D}_U[x_0, x_1]$ is $\mathbb{D}_G$. Let $\delta_k(x_0, x_1)$, $i_k(x_0, x_1)$ be the invariant determinants and factors of $A(x_0, x_1)$ respectively for $k = 1, \ldots, \operatorname{rank} A(x_0, x_1)$.

Lemma 2.1.2 *Let $A(x_0, x_1)$ be a homogeneous pencil over $\mathbb{D}_U[x_0, x_1]$. Then its invariant determinants and the invariant polynomials $\delta_k(x_0, x_1)$, $i_k(x_0, x_1)$, $k = 1, \ldots, \operatorname{rank} A(x_0, x_1)$ are homogeneous polynomials. Moreover, if $\delta_k(x)$ and $i_k(x)$ are the invariant determinants and factors of the pencil $A(x)$ for $k = 1, \ldots, \operatorname{rank} A(x)$, then*

$$\delta_k(x) = \delta_k(1, x), \quad i_k(x) = i_k(1, x), \quad k = 1, \ldots, \operatorname{rank} A(x). \quad (2.1.6)$$

Proof. Clearly any $k \times k$ minor of $A(x_0, x_1)$ is either zero or a homogeneous polynomial of degree k. In view of Problem 1 we deduce that the g.c.d. of all nonvanishing $k \times k$ minors is a homogeneous polynomial $\delta_k(x_0, x_1)$. As $i_k(x_0, x_1) = \frac{\delta_k(x_0, x_1)}{\delta_{k-1}(x_0, x_1)}$ Problem 1 implies that $i_k(x_0, x_1)$ is a homogeneous polynomial. Consider the pencil $A(x) = A(1, x)$. So $\delta_k(x)$ — the g.c.d. of $k \times k$ minors of $A(x)$ is obviously divisible by $\delta_k(1, x)$. On the other hand, we have the following relation between the minors of $A(x_0, x_1)$ and $A(x)$

$$\det A(x_0, x_1)[\alpha, \beta] = x_0^k \det A\left(\frac{x_1}{x_0}\right)[\alpha, \beta], \quad \alpha, \beta \in [n]_k. \quad (2.1.7)$$

This shows that $x_0^{\rho_k} \delta_k(\frac{x_1}{x_0})$ ($\rho_k = \deg \delta_k(x)$) divides any $k \times k$ minor of $A(x_0, x_1)$. So $x_0^{\rho_k} \delta_k(\frac{x_1}{x_0}) | \delta_k(x_0, x_1)$. This proves the first part of

(2.1.6). So

$$\delta_k(x_0, x_1) = x_0^{\phi_k}\left(x_0^{\rho_k}\delta_k\left(\frac{x_1}{x_0}\right)\right), \quad \rho_k = \deg \delta_k(x), \quad \phi_k \geq 0.$$

(2.1.8)

The equality

$$i_k(x_0, x_1) = \frac{\delta_k(x_0, x_1)}{\delta_{k-1}(x_0, x_1)}$$

implies

$$i_k(x_0, x_1) = x_0^{\psi_k}\left(x_0^{\sigma_k}i_k\left(\frac{x_1}{x_0}\right)\right), \quad \sigma_k = \deg i_k(x), \quad \psi_k \geq 0. \quad (2.1.9)$$

$\square$

$\delta_k(x_0, x_1)$ and $i_k(x_0, x_1)$ are called the invariant *homogeneous* determinants and the invariant *homogeneous* polynomials (factors) respectively. The classical result due to Weierstrass [Wei67] states:

Theorem 2.1.3 *Let $A(x) \in \mathbb{F}[x]^{n \times n}$ be a regular pencil. Then a pencil $B(x) \in \mathbb{F}[x]^{n \times n}$ is strictly equivalent to $A(x)$ if and only if $A(x)$ and $B(x)$ have the same invariant homogeneous polynomials.*

Proof. The necessary part of the theorem holds for any $A(x)$, $B(x)$ which are strictly equivalent. Suppose now that $A(x)$ and $B(x)$ have the same invariant homogeneous polynomials. According to (1.4.4) the pencils $A(x)$ and $B(x)$ have the same invariant polynomials. So $A(x) \sim B(x)$ over $\mathbb{F}[x]$. Therefore

$$W(x)B(x) = A(x)U(x), \quad U(x), \; W(x) \in \mathbf{GL}(n, \mathbb{F}[x]). \quad (2.1.10)$$

Assume first that A_1 and B_1 are nonsingular. Then (see Problem 2) it is possible to divide $W(x)$ by $A(x)$ from the right and to divide $U(x)$ by $B(x)$ from the left

$$W(x) = A(x)W_1(x) + R, \quad U(x) = U_1(x)B(x) + P, \quad (2.1.11)$$

where P and R are constant matrices. So

$$A(x)(W_1(x) - U_1(x))B(x) = A(x)P - RB(x).$$

Since $A_1, B_1 \in \mathbf{GL}(n, \mathbb{F})$ we must have that $W_1(x) = U_1(x)$, otherwise the left-hand side of the above equality would be of degree 2 at least (see Definition 2.1.5), while the right-hand side of this equality is at most 1. So

$$W_1(x) = U_1(x), \quad RB(x) = A(x)P. \tag{2.1.12}$$

It is left to show that P and Q are nonsingular. Let $V(x) = W(x)^{-1} \in \mathbf{GL}(n, \mathbb{F}[x])$. Then $I = W(x)V(x)$. Let $V(x) = B(x)V_1(x) + S$. Use the second identity of (2.1.12) to obtain

$$
\begin{aligned}
I &= (A(x)W_1(x) + R)V(x) = A(x)W_1(x)V(x) + RV(x) \\
&= A(x)W_1(x)V(x) + RB(x)V_1(x) + RS \\
&= A(x)W_1(x)V(x) + A(x)PV_1(x) + RS \\
&= A(x)(W_1(x)V(x) + PV_1(x)) + RS.
\end{aligned}
$$

Since $A_1 \in \mathbf{GL}(n, \mathbb{F})$ the above equality implies

$$W_1(x)V(x) + PV_1(x) = 0, \quad RS = I.$$

Hence, R is invertible. Similar arguments show that P is invertible. Thus $A(x) \overset{s}{\sim} B(x)$ if $\det A_1, \det B_1 \neq 0$.

Consider now the general case. Introduce new variables y_0, y_1:

$$y_0 = ax_0 + bx_1, \quad y_1 = cx_0 + dx_1, \quad ad - cb \neq 0.$$

Then

$$A(y_0, y_1) = y_0 A_0' + y_1 A_1', \quad B(y_0, y_1) = y_0 B_0' + y_1 B_1'.$$

Clearly $A(y_0, y_1)$ and $B(y_0, y_1)$ have the same invariant homogeneous polynomials. Also $A(y_0, y_1) \overset{s}{\sim} B(y_0, y_1) \iff A(x_0, x_1) \overset{s}{\sim} B(x_0, x_1)$. Since $A(x_0, x_1)$ and $B(x_0, x_1)$ are regular pencils it is possible to choose a, b, c, d such that A_1' and B_1' are nonsingular. This shows that $A(y_0, y_1) \overset{s}{\sim} B(y_0, y_1)$. Hence, $A(x) \overset{s}{\sim} B(x)$. $\qquad\square$

Using the proof of Theorem 2.1.3 and Problem 2 we obtain:

Theorem 2.1.4 *Let* $A(x), B(x) \in \mathbb{D}[x]^{n \times n}$. *Assume that* $A_1, B_1 \in \mathbf{GL}(n, \mathbb{D})$. *Then* $A(x) \overset{s}{\sim} B(x) \iff A(x) \sim B(x)$.

For singular pencils the invariant homogeneous polynomials alone do not determine the class of strictly equivalent pencils. We now introduce the notion of column and row indices for $A(x) \in \mathbb{F}[x]^{m \times n}$. Consider the system (1.15.15). The set of all solutions $\mathbf{w}(x)$ is an $\mathbb{F}[x]$-module $\mathbf{M}$ with a finite basis $\mathbf{w}_1(x), \ldots, \mathbf{w}_s(x)$ (Theorem 1.13.3). To specify a choice of a basis we need the following definition.

Definition 2.1.5 *Let $A \in \mathbb{D}[x_1, \ldots, x_k]^{m \times n}$. So*

$$A(x_1, \ldots, x_k) = \sum_{|\alpha| \le d} A_\alpha x^\alpha, \quad A_\alpha \in \mathbb{D}^{m \times n},$$

$$\alpha = (\alpha_1, \ldots, \alpha_k) \in \mathbb{Z}_+^k, \quad |\alpha| = \sum_{i=1}^{k} \alpha_i, \quad x^\alpha = x_1^{\alpha_1}, \ldots, x_k^{\alpha_k}.$$

$$(2.1.13)$$

Then the degree of $A(x_1, \ldots, x_k) \ne 0$ (deg A) is d if there exists $A_\alpha \ne 0$ with $|\alpha| = d$. Let $\deg 0 = 0$.

Definition 2.1.6 *Let $A \in \mathbb{F}[x]^{m \times n}$ and consider the module $\mathbf{M} \subset \mathbb{F}[x]^n$ of all solutions of (1.15.15). Choose a basis $\mathbf{w}_1(x), \ldots, \mathbf{w}_s(x)$, $s = n - \operatorname{rank} A$ in $\mathbf{M}$ such that $\mathbf{w}_k(x) \in \mathbf{M}$ has the lowest degree among all $\mathbf{w}(x) \in \mathbf{M}$ which are linearly independent over $\mathbb{F}(x)$ of $\mathbf{w}_1, \ldots, \mathbf{w}_{k-1}(x)$ for $k = 1, \ldots, s$. Then the column indices $0 \le \alpha_1 \le \alpha_2 \le \cdots \le \alpha_s$ of $A(x)$ are given as*

$$\alpha_k = \deg \mathbf{w}_k(x), \quad k = 1, \ldots, s. \qquad (2.1.14)$$

The row indices $0 \le \beta_1 \le \beta_2 \le \cdots \le \beta_t$, $t = m - \operatorname{rank} A$, of $A(x)$ are the column indices of $A(x)^\top$.

It can be shown [Gan59] that the column (row) indices are independent of a particular allowed choice of a basis $\mathbf{w}_1(x), \ldots, \mathbf{w}_s(x)$. We state the Kronecker result [Kro90]. (See [Gan59] for a proof.)

Theorem 2.1.7 *The pencils $A(x), B(x) \in \mathbb{F}[x]^{m \times n}$ are strictly equivalent if and only if they have the same invariant homogeneous polynomials and the same row and column indices.*

For a canonical form of a singular pencil under the strict equivalence see Problems 8–12.

Problems

1. Using the fact that $\mathbb{D}_U[x_1, \ldots, x_n]$ is $\mathbb{D}_U$ and the equality (1.14.5) show that if $a \in \mathbb{D}_U[x_1, \ldots, x_n]$ is a homogeneous polynomial then in the decomposition (1.3.1) each p_i is a homogeneous polynomial.

2. Let

$$W(x) = \sum_{k=0}^{q} W_k x^k, \quad U(x) = \sum_{k=0}^{p} U_k x^k \in \mathbb{D}[x]^{n \times n}. \quad (2.1.15)$$

Assume that $A(x) = A_0 + xA_1$ such that $A_0 \in \mathbb{D}^{n \times n}$ and $A_1 \in \mathbf{GL}(n, \mathbb{D})$. Show that if $p, q \geq 1$ then

$$W(x) = A(x)A_1^{-1}(W_q x^{q-1}) + \tilde{W}(x),$$
$$U(x) = (U_p x^{p-1})A_1^{-1}A(x) + \tilde{U}(x),$$

where

$$\deg \tilde{W}(x) < q, \quad \deg \tilde{U}(x) < p.$$

Prove the equalities (2.1.11) where R and P are constant matrices. Suppose that $A_1 = I$. Show that R and P in (2.1.11) can be given as

$$R = \sum_{k=0}^{q}(-A_0)^k W_k, \quad P = \sum_{k=0}^{p} U_k(-A_0)^k. \quad (2.1.16)$$

3. Let $A(x) \in \mathbb{D}_U[x]^{n \times n}$ be a regular pencil such that $\det A_1 \neq 0$. Prove that in (2.1.8) and (2.1.9) $\phi_k = \psi_k = 0$ for $k = 1, \ldots, n$. (Use equality (1.13.12) for $A(x)$ and $A(x_0, x_1)$.)

4. Consider the following two pencils

$$A(x) = \begin{bmatrix} 2+x & 1+x & 3+3x \\ 3+x & 2+x & 5+2x \\ 3+x & 2+x & 5+2x \end{bmatrix},$$

$$B(x) = \begin{bmatrix} 2+x & 1+x & 1+x \\ 1+x & 2+x & 2+x \\ 1+x & 1+x & 1+x \end{bmatrix}$$

over $\mathbb{R}[x]$. Show that $A(x)$ and $B(x)$ are equivalent but not strictly equivalent.

5. Let

$$A(x) = \sum_{k=0}^{p} A_k x^k \in \mathbb{F}[x]^{m \times n}.$$

Put

$$A(x_0, x_1) = \sum_{k=0}^{q} A_k x_0^{q-k} x_1^k, \quad q = \deg A(x).$$

Prove that $i_k(x_0, x_1)$ is a homogeneous polynomial for $k = 1, \ldots,$ rank $A(x)$. Show that $i_1(1, x), \ldots, i_k(1, x)$ are the invariant factors of $A(x)$.

6. Let $A(x), B(x) \in \mathbb{F}[x]^{m \times n}$. $A(x)$ and $B(x)$ are called strictly equivalent $(A(x) \overset{s}{\sim} B(x))$ if

$$B(x) = PA(x)Q, \quad P \in \mathbf{GL}(m, \mathbb{F}), \quad Q \in \mathbf{GL}(n, \mathbb{F}).$$

Show that if $A(x) \overset{s}{\sim} B(x)$ then $A(x_0, x_1)$ and $B(x_0, x_1)$ have the same invariant factors.

7. Let $A(x), B(x) \in \mathbb{F}[x]^{m \times n}$. Show that $A(x) \overset{s}{\sim} B(x) \iff A(x)^\top \overset{s}{\sim} B(x)^\top$.

8. (a) Let $L_m(x) \in \mathbb{F}[x]^{m \times (m+1)}$ be matrix with 1 on the main diagonal and x on the diagonal above it, and all other entries 0:

$$L_m(x) = \begin{bmatrix} 1 & x & 0 & \cdots & 0 \\ 0 & 1 & x & \cdots & 0 \\ \vdots & \vdots & \ddots & \ddots & \vdots \\ 0 & 0 & \cdots & 1 & x \end{bmatrix}.$$

Show that rank $L_m = m$ and $\alpha_1 = m$.

(b) Let $1 \leq \alpha_1 \leq \cdots \leq \alpha_s, 1 \leq \beta_1 \leq \cdots \leq \beta_t$ be integers. Assume that $B(x) = B_0 + xB_1 \in \mathbb{F}[x]^{l \times l}$ is a regular pencil. Show that $A(x) = B(x) \oplus_{i=1}^s L_{\alpha_i} \oplus_{j=1}^t L_{\beta_j}^\top$ has the column and the row indices $1 \leq \alpha_1 \leq \cdots \leq \alpha_s, 1 \leq \beta_1 \leq \cdots \leq \beta_t$ respectively.

9. Show if a pencil $A(x)$ is a direct sum of pencils of the below form, where one of the summands of the form (9a)–(9b) appears, it is a singular pencil.

   (a) $L_m(x)$.

   (b) $L_m(x)^\top$.

   (c) $B(x) = B_0 + xB_1 \in \mathbb{F}[x]^{l \times l}$ is a regular pencil.

   Furthermore, one of the summands of the form 1–2 appears if and only if $A(x)$ a singular pencil.

10. Show that a singular pencil $A(x)$ is strictly equivalent to the singular pencil given in Problem 9, if and only if there are no column and row indices equal to 0.

11. Assume that $A(x) \in \mathbb{F}[x]^{m \times n}$ is a singular pencil.

    (a) Show that $A(x)$ has exactly k column indices equal to 0, if and only if it is strict equivalent to $[0_{m \times k} \ A_1(x)]$, $A_1(x) \in \mathbb{F}[x]^{m \times (n-k)}$, where either $A_1(x)$ is regular or singular. If $A_1(x)$ is singular then the row indices of $A_1(x)$ are the row indices of $A(x)$, and the column indices of $A_1(x)$ are the nonzero column indices of of $A(x)$.

    (b) By considering $A(x)^\top$ state and prove similar result for the row indices of $A(x)$.

12. Use Problems 8–11 to find a canonical from for a singular pencil $A(x)$ under the strict equivalence.

2.2   Similarity of Matrices

Definition 2.2.1 *Let $A, B \in \mathbb{D}^{m \times m}$. Then A and B are similar $(A \approx B)$ if*

$$B = QAQ^{-1}, \tag{2.2.1}$$

for some $Q \in \mathbf{GL}(m, \mathbb{D})$.

Clearly, the similarity relation is an equivalence relation. So $\mathbb{D}^{m \times m}$ is divided into equivalences classes which are called the similarity classes. For a $\mathbb{D}$ module $\mathbf{M}$ we let Hom $(\mathbf{M}) :=$ Hom $(\mathbf{M}, \mathbf{M})$. It is a standard fact that each similarity class corresponds to all possible representations of some $T \in$ Hom $(\mathbf{M})$, where $\mathbf{M}$ is a $\mathbb{D}$-module having a basis of m elements. Indeed, let $[\mathbf{u}_1, \dots, \mathbf{u}_m]$ be a basis in $\mathbf{M}$. Then T is represented by $A = [a_{ij}] \in \mathbb{D}^{m \times m}$, where

$$T\mathbf{u}_j = \sum_{i=1}^{m} a_{ij}\mathbf{u}_i, \quad j = 1, \dots, m. \tag{2.2.2}$$

Let $[\tilde{\mathbf{u}}_1, \dots, \tilde{\mathbf{u}}_m]$ be another basis in $\mathbf{M}$. Assume that $Q \in \mathbf{GL}(m, \mathbb{D})$ is given by (1.11.3). According to (2.2.2) and the arguments of §1.11, the representation of T in the basis $[\tilde{\mathbf{u}}_1, \dots, \tilde{\mathbf{u}}_m]$ is given by the matrix B of the form (2.2.1).

The similarity notion of matrices is closely related to strict equivalence of certain regular pencils.

Lemma 2.2.2 *Let $A, B \in \mathbb{D}^{m \times m}$. Associate with these matrices the following regular pencils*

$$A(x) = -A + xI, \quad B(x) = -B + xI. \tag{2.2.3}$$

Then A and B are similar if and only if the pencils $A(x)$ and $B(x)$ are strictly equivalent.

Proof. Assume first that $A \approx B$. Then (2.2.1) implies (2.1.2) where $P = Q^{-1}$. Suppose now that $A(x) \overset{s}{\sim} B(x)$. So $B = QAP$, $I = QP$. That is $P = Q^{-1}$ and $A \approx B$. $\qquad \square$

Clearly $A(x) \overset{s}{\sim} B(x) \Rightarrow A(x) \sim B(x)$.

Corollary 2.2.3 *Let $A, B \in \mathbb{D}_U^{m \times m}$. Assume that A and B are similar. Then the corresponding pencils $A(x)$, $B(x)$ given by (2.2.3) have the same invariant polynomials.*

In the case $\mathbb{D}_U = \mathbb{F}$ the above condition is also a sufficient condition in view of Lemma 2.2.2 and Corollary 2.1.4.

Theorem 2.2.4 *Let $A, B \in \mathbb{F}^{m \times m}$. Then A and B are similar if and only if the pencils $A(x)$ and $B(x)$ given by (2.2.3) have the same invariant polynomials.*

It can be shown (see Problem 1) that even over Euclidean domains the condition that $A(x)$ and $B(x)$ have the same invariant polynomials does not imply in general that $A \approx B$.

Problems

1. Let

$$A = \begin{bmatrix} 1 & 0 \\ 0 & 5 \end{bmatrix}, \quad B = \begin{bmatrix} 1 & 1 \\ 0 & 5 \end{bmatrix} \in \mathbb{Z}^{2 \times 2}.$$

   Show that $A(x)$ and $B(x)$ given by (2.2.3) have the same invariant polynomials over $\mathbb{Z}[x]$. Show that A and B are not similar over $\mathbb{Z}$.

2. Let $A(x) \in \mathbb{D}_U[x]^{n \times n}$ be given by (2.2.3). Let $i_1(x), \ldots, i_n(x)$ be the invariant polynomial of $A(x)$. Using the equality (1.13.12) show that each $i_k(x)$ can be assumed to be normalized polynomial and

$$\sum_{k=1}^{n} \deg i_k(x) = n. \tag{2.2.4}$$

3. Let $A \in \mathbb{F}^{n \times n}$. Show that $A \approx A^\top$.

2.3 The Companion Matrix

Theorem 2.2.4 shows that if $A \in \mathbb{F}^{n \times n}$ then the invariant polynomials determine the similarity class of A. We now show that any set of normalized polynomials $i_1(x), \ldots, i_n(x) \in \mathbb{D}_U[x]$, such that $i_j(x)|i_{j+1}(x)$, $j = 1, \ldots, n-1$ and which satisfy (2.2.4), are invariant polynomials of $xI - A$ for some $A \in \mathbb{D}_U^{n \times n}$. To do so we introduce the notion of a companion matrix.

Definition 2.3.1 *Let $p(x) \in \mathbb{D}[x]$ be a normalized polynomial*

$$p(x) = x^m + a_1 x^{m-1} + \cdots + a_m.$$

Then $C(p) = [c_{ij}]_1^m \in \mathbb{D}^{m \times m}$ *is the companion matrix of* $p(x)$ *if*

$$c_{ij} = \delta_{(i+1)j}, \quad i = 1, \ldots, m-1, \ j = 1, \ldots, m,$$

$$c_{mj} = -a_{m-j+1}, \quad j = 1, \ldots, m,$$

$$C(p) = \begin{bmatrix} 0 & 1 & 0 & \cdots & 0 & 0 \\ 0 & 0 & 1 & \cdots & 0 & 0 \\ \vdots & \vdots & \vdots & \cdots & \vdots & \vdots \\ 0 & 0 & 0 & \cdots & 0 & 1 \\ -a_m & -a_{m-1} & -a_{m-2} & \cdots & -a_2 & -a_1 \end{bmatrix}. \quad (2.3.1)$$

Lemma 2.3.2 *Let* $p(x) \in \mathbb{D}_U[x]$ *be a normalized polynomial of degree* m. *Consider the pencil* $C(x) = xI - C(p)$. *Then the invariant polynomials of* $C(x)$ *are*

$$i_1(C) = \cdots = i_{m-1}(C) = 1, \quad i_m(C) = p(x). \quad (2.3.2)$$

Proof. For $k < m$ consider a minor of $C(x)$ composed of the rows $1, \ldots, k$ and the columns $2, \ldots, k+1$. Since this minor is the determinant of a lower triangular matrix with -1 on the main diagonal we deduce that its value is $(-1)^k$. So $\delta_k(C) = 1$, $k = 1, \ldots, m-1$. This establishes the first equality in (2.3.2). Clearly $\delta_m(C) = \det(xI - C)$. Expand the determinant of $C(x)$ by the first row and use induction to prove that $\det(xI - C) = p(x)$. This shows that $i_m(C) = \frac{\delta_m(C)}{\delta_{m-1}(C)} = p(x)$. $\qquad \square$

Using the results of Problem 2.1.14 and Lemma 2.3.2 we get:

Theorem 2.3.3 *Let* $p_j(x) \in \mathbb{D}_U[x]$ *be normalized polynomials of positive degrees such that* $p_j(x) | p_{j+1}(x)$, $j = 1, \ldots, k-1$. *Consider the matrix*

$$C(p_1, \ldots, p_k) = \oplus_{j=1}^k C(p_j). \quad (2.3.3)$$

Then the nontrivial invariant polynomials of $xI - C(p_1, \ldots, p_k)$ *(i.e. those polynomials which are not the identity element) are* $p_1(x), \ldots, p_k(x)$.

Combining Theorems 2.2.4 and 2.3.3 we obtain a canonical representation for the similarity class in $\mathbb{F}^{n \times n}$.

Theorem 2.3.4 *Let $A \in \mathbb{F}^{n \times n}$ and assume that $p_j(x) \in \mathbb{F}[x]$, $j = 1, \ldots, k$ are the nontrivial normalized invariant polynomials of $xI - A$. Then A is similar to $C(p_1, \ldots, p_k)$.*

Definition 2.3.5 *For $A \in \mathbb{F}^{n \times n}$ the matrix $C(p_1, \ldots, p_k)$ is called the rational canonical form of A.*

Let $\mathbb{F}$ be the quotient field of $\mathbb{D}$. Assume that $A \in \mathbb{D}^{n \times n}$. Let $C(p_1, \ldots, p_k)$ be the rational canonical form of A in $\mathbb{F}^{n \times n}$. We now discuss the case when $C(p_1, \ldots, p_k) \in \mathbb{D}^{n \times n}$. Assume that $\mathbb{D}$ is $\mathbb{D}_U$. Let δ_k be the g.c.d. of $k \times k$ minors of $xI - A$. So δ_k divides the minor $p(x) = \det (xI - A)[\alpha, \alpha]$, $\alpha = \{1, \ldots, k\}$. Clearly $p(x)$ is normalized polynomial in $\mathbb{D}_U[x]$. Recall that $\mathbb{D}_U[x]$ is also $\mathbb{D}_U$ (§1.4).

According to Theorem 1.4.8 the decomposition of $p(x)$ into irreducible factors in $\mathbb{D}_U[x]$ is of the form (1.4.4), where $a = 1$ and each $q_i(x)$ is a nontrivial normalized polynomial in $\mathbb{D}_U[x]$. Hence, $i_k = \frac{\delta_k}{\delta_{k-1}}$ is either identity or a nontrivial polynomial in $\mathbb{D}_U[x]$. Thus

Theorem 2.3.6 *Let $A \in \mathbb{D}_U^{n \times n}$. Then the rational canonical form $C(p_1, \ldots, p_k)$ of A over the quotient field $\mathbb{F}$ of $\mathbb{D}_U$ belongs to $\mathbb{D}_U^{n \times n}$.*

Corollary 2.3.7 *Let $A \in \mathbb{C}[x_1, \ldots, x_m]^{n \times n}$. Then the rational canonical form of A over $\mathbb{C}(x_1, \ldots, x_n)$ belongs to $\mathbb{C}[x_1, \ldots, x_m]^{n \times n}$.*

Using the results of Theorem 1.4.9 we deduce that Theorem 2.3.6 applies to the ring of analytic functions in several variables although this ring is not $\mathbb{D}_U$ (§1.3).

Theorem 2.3.8 *Let $A \in \mathrm{H}(\Omega)^{n \times n}$ $(\Omega \subset \mathbb{C}^m)$. Then the rational canonical form of A over the field of meromorphic functions in Ω belongs to $\mathrm{H}(\Omega)^{n \times n}$.*

Problems

1. Let $p(x) \in \mathbb{D}_U[x]$ be a normalized nontrivial polynomial. Assume that $p(x) = p_1(x)p_2(x)$, where $p_i(x)$ is a normalized nontrivial polynomial in $\mathbb{D}_U[x]$ for $i = 1, 2$. Using Problems 1.14.1

and 1.14.2 show that $xI - C(p_1, p_2)$ given by (2.3.3) have the same invariant polynomials as $xI - C(p)$ if and only if $(p_1, p_2) = 1$.

2. Let $A \in \mathbb{D}_U^{n \times n}$ and assume that $p_1(x), \ldots, p_k(x)$ are the nontrivial normalized invariant polynomials of $xI - A$. Let

$$p_j(x) = (\phi_1(x))^{m_{1j}} \cdots (\phi_l(x))^{m_{lj}}, \quad j = 1, \ldots, k, \qquad (2.3.4)$$

where $\phi_1(x), \ldots, \phi_l(x)$ are nontrivial normalized irreducible polynomials in $\mathbb{D}_U[x]$ such that $(\phi_i, \phi_j) = 1$ for $i \neq j$. Prove that

$$m_{ik} \geq 1, \quad m_{ik} \geq m_{i(k-1)} \geq \cdots \geq m_{i1} \geq 0, \quad \sum_{i,j=1}^{l,k} m_{ij} = n.$$

$$(2.3.5)$$

The polynomials $\phi_i^{m_{ij}}$ for $m_{ij} > 0$ are called the *elementary divisors* of $xI - A$. Let

$$E = \oplus_{m_{ij} > 0} C\left(\phi_i^{m_{ij}}\right). \qquad (2.3.6)$$

Show that $xI - A$ and $xI - E$ have the same invariant polynomials. Hence $A \approx E$ over the quotient field $\mathbb{F}$ of $\mathbb{D}_U$. (In some references E is called the rational canonical form of A.)

2.4   Splitting to Invariant Subspaces

Let $\mathbf{V}$ be an m dimensional vector space over $\mathbb{F}$ and let $T \in \text{Hom}(\mathbf{V})$. In §2.2 we showed that the set of all matrices $\mathcal{A} \subset \mathbb{F}^{m \times m}$, which represents T in different bases, is an equivalence class of matrices with respect to the similarity relation. Theorem 2.2.4 shows that the class $\mathcal{A}$ is characterized by the invariant polynomials of $xI - A$ for some $A \in \mathcal{A}$. Since $xI - A$ and $xI - B$ have the same invariant polynomials if and only if $A \approx B$ we define:

Definition 2.4.1 *Let $T \in \text{Hom}(\mathbf{V})$ and let $A \in \mathbb{F}^{m \times m}$ be a representation matrix of T given by the equality (2.2.2) in some basis $\mathbf{u}_1, \ldots, \mathbf{u}_m$ of $\mathbf{V}$. Then the invariant polynomials $i_1(x), \ldots, i_m(x)$ of T are defined as the invariant polynomials of $xI - A$. The characteristic polynomial of T is the polynomial $\det(xI - A)$.*

The fact that the characteristic polynomial of T is independent of a representation matrix A follows from the identity (1.13.12)

$$\det (xI - A) = p_1(x) \cdots p_k(x), \qquad (2.4.1)$$

where $p_1(x), \ldots, p_k(x)$ are the nontrivial invariant polynomials of $xI - A$. In §2.3, we showed that the matrix $C(p_1, \ldots, p_k)$ is a representation matrix of T. In this section, we consider another representation of T which is closely related to the matrix E given in (2.3.6). This form is achieved by splitting $\mathbf{V}$ to a direct sum

$$\mathbf{V} = \oplus_{i=1}^{l} \mathbf{U}_i, \qquad (2.4.2)$$

where each $\mathbf{U}_i$ is an invariant subspace of T defined as follows:

Definition 2.4.2 *Let* $\mathbf{V}$ *be a finite dimensional vector space over* $\mathbb{F}$ *and* $T \in \mathrm{Hom}\,(\mathbf{V})$. *A subspace* $\mathbf{U} \subseteq \mathbf{V}$ *is an invariant subspace of* T *(T-invariant) if*

$$T\mathbf{U} \subseteq \mathbf{U}. \qquad (2.4.3)$$

$\mathbf{U}$ *is called trivial if* $\mathbf{U} = \{\mathbf{0}\}$ *or* $\mathbf{U} = \mathbf{V}$. $\mathbf{U}$ *is called nontrivial, (proper), if* $\{\mathbf{0}\} \neq \mathbf{U} \neq \mathbf{V}$. $\mathbf{U}$ *is called irreducible if* $\mathbf{U}$ *cannot be expressed a direct sum of two nontrivial invariant subspaces of* T. *The restriction of* T *to a* T-*invariant subspace* $\mathbf{U}$ *is denoted by* $T|\mathbf{U}$.

Thus if $\mathbf{V}$ splits into a direct sum of nontrivial invariant subspaces of T, then a direct sum of matrix representations of T on each $\mathbf{U}_j$ gives a representation of T. So, a simple representation of T can be achieved by splitting $\mathbf{V}$ into a direct sum of irreducible invariant subspaces. To do so we need to introduce the notion of the minimal polynomial of T. Consider the linear operators $I = T^0$, $T, T^2, \ldots, T^{m^2}$, where I is the identity operator $(I\mathbf{v} = \mathbf{v})$. As $\dim \mathrm{Hom}\,(\mathbf{V}) = m^2$, these $m^2 + 1$ operators are linearly dependent. So there exists an integer $q \in [0, m^2]$ such that $I, T, \ldots, T^{q-1}$ are linearly independent and $I, T, \ldots, T^q$ are linearly dependent. Let $0 \in \mathrm{Hom}\,(\mathbf{V})$ be the zero operator: $0\mathbf{v} = \mathbf{0}$. For $\phi \in \mathbb{F}[x]$ let $\phi(T)$ be the operator

$$\phi(T) = \sum_{i=0}^{l} c_i T^i, \quad \phi(x) = \sum_{i=1}^{l} c_i x^i.$$

ϕ is annihilated by T if $\phi(T) = 0$.

Definition 2.4.3 *A polynomial* $\psi(x) \in \mathbb{F}[x]$ *is a minimal polynomial of* $T \in \mathrm{Hom}\,(\mathbf{V})$ *if* $\psi(x)$ *is a normalized polynomial of the smallest degree annihilated by* T.

Lemma 2.4.4 *Let* $\psi(x) \in \mathbb{F}[x]$ *be the minimal polynomial* $T \in \mathrm{Hom}\,(\mathbf{V})$. *Assume that* T *annihilates* ϕ. *Then* $\psi | \phi$.

Proof. Divide ϕ by ψ:

$$\phi(x) = \chi(x)\psi(x) + \rho(x), \quad \deg \rho < \deg \psi.$$

As $\phi(T) = \psi(T) = 0$ it follows that $\rho(T) = 0$. From the definition of $\psi(x)$ it follows that $\rho(x) = 0$. □

Since $\mathbb{F}[x]$ is a unique factorization domain, let

$$\psi(x) = \phi_1^{s_1} \cdots \phi_l^{s_l},$$

$$(\phi_i, \phi_j) = 1 \quad \text{for } 1 \le i < j \le l, \ \deg \phi_i \ge 1, \ i = 1, \ldots, l,$$

$$(2.4.4)$$

where each ϕ_i is a normalized irreducible polynomial if $\mathbb{F}[x]$.

Theorem 2.4.5 *Let* ψ *be the minimal polynomial of* $T \in \mathrm{Hom}\,(\mathbf{V})$. *Assume that* ψ *splits to a product of coprime factors given in* (2.4.4). *Then the vector space* $\mathbf{V}$ *splits to a direct sum* (2.4.2), *where each* $\mathbf{U}_j$ *is a nontrivial invariant subspace of* $T|\mathbf{U}_j$. *Moreover,* $\phi_j^{s_j}$ *is the minimal polynomial of* $T|\mathbf{U}_j$.

The proof of the theorem follows immediately from the Lemma 2.4.6.

Lemma 2.4.6 *Let ψ be the minimal polynomial of $T \in \mathrm{Hom}\,(\mathbf{V})$. Assume that ψ splits to a product of two nontrivial coprime factors*

$$\psi(x) = \psi_1(x)\psi_2(x), \quad \deg \psi_i \geq 1, \quad i = 1, 2, \quad (\psi_1, \psi_2) = 1, \quad (2.4.5)$$

where each ψ_i is normalized. Then

$$\mathbf{V} = \mathbf{U}_1 \oplus \mathbf{U}_2, \tag{2.4.6}$$

where each $\mathbf{U}_j$ is a nontrivial T-invariant subspace and ψ_j is the minimal polynomial of $T_j := T|\mathbf{U}_j$.

Proof. The assumptions of the lemma imply the existence of polynomials $\theta_1(x)$ and $\theta_2(x)$ such that

$$\theta_1(x)\psi_1(x) + \theta_2(x)\psi_2(x) = 1. \tag{2.4.7}$$

Define

$$\mathbf{U}_j = \{\mathbf{u} \in \mathbf{V} : \; \psi_j(T)\mathbf{u} = \mathbf{0}\}, \quad j = 1, 2. \tag{2.4.8}$$

Since any two polynomials in T commute (i.e. $\mu(T)\nu(T) = \nu(T)\mu(T)$) it follows that each $\mathbf{U}_j$ is T-invariant. The equality (2.4.7) implies

$$I = \psi_1(T)\theta_1(T) + \psi_2(T)\theta_2(T).$$

Hence, for any $\mathbf{u} \in \mathbf{V}$ we have

$$\mathbf{u} = \mathbf{u}_1 + \mathbf{u}_2, \quad \mathbf{u}_1 = \psi_2(T)\theta_2(T)\mathbf{u} \in \mathbf{U}_1, \quad \mathbf{u}_2 = \psi_1(T)\theta_1(T)\mathbf{u} \in \mathbf{U}_2.$$

So $\mathbf{V} = \mathbf{U}_1 + \mathbf{U}_2$. Suppose that $\mathbf{u} \in \mathbf{U}_1 \cap \mathbf{U}_2$. Then $\psi_1(T)\mathbf{u} = \psi_2(T)\mathbf{u} = \mathbf{0}$. Hence, $\theta_1(T)\psi_1(T)\mathbf{u} = \theta_2(T)\psi_2(T)\mathbf{u} = \mathbf{0}$. Thus

$$\mathbf{u} = \psi_1(T)\mathbf{u} + \psi_2(T)\mathbf{u} = 0.$$

So $\mathbf{U}_1 \cap \mathbf{U}_2 = \{\mathbf{0}\}$ and (2.4.6) holds. Clearly T_j annihilates ψ_j. Let $\bar{\psi}_j$ be the minimal polynomial of T_j. So $\bar{\psi}_j | \psi_j$, $j = 1, 2$. Now

$$\bar{\psi}_1(T)\bar{\psi}_2(T)\mathbf{u} = \bar{\psi}_1(T)\bar{\psi}_2(T)(\mathbf{u}_1 + \mathbf{u}_2)$$
$$= \bar{\psi}_2(T)\bar{\psi}_1(T)\mathbf{u}_1 + \bar{\psi}_1(T)\bar{\psi}_2(T)\mathbf{u}_2 = 0.$$

Hence, T annihilates $\bar{\psi}_1\bar{\psi}_2$. Since ψ is the minimal polynomial of T we have $\psi_1\psi_2|\bar{\psi}_1\bar{\psi}_2$. Therefore $\bar{\psi}_j = \psi_j$, $j = 1, 2$. As deg $\psi_j \geq 1$ it follows that dim $\mathbf{U}_j \geq 1$. $\qquad\square$

Problems

1. Assume that (2.4.6) holds, where $T\mathbf{U}_j \subseteq \mathbf{U}_j$, $j = 1, 2$. Let ψ_j be the minimal polynomial of $T_j = T|\mathbf{U}_j$ for $j = 1, 2$. Show that the minimal polynomial ψ of T is equal to $\frac{\psi_1\psi_2}{(\psi_1,\psi_2)}$.

2. Let the assumptions of Problem 1 hold. Assume furthermore that $\psi = \phi^s$, where ϕ is irreducible over $\mathbb{F}[x]$. Then either $\psi_1 = \psi$ or $\psi_2 = \psi$.

3. Let $C(p) \in \mathbb{D}^{m \times m}$ be the companion matrix given by (2.3.1). Let $\mathbf{e}_i = (\delta_{i1}, \ldots, \delta_{im})^\top$, $i = 1, \ldots, m$ be the standard basis in $\mathbb{D}^m$. Show

$$C(p)\mathbf{e}_i = \mathbf{e}_{i-1} - a_{m-i+1}\mathbf{e}_m, \quad i = 1, \ldots, m, \quad (\mathbf{e}_0 = \mathbf{0}). \quad (2.4.9)$$

Prove that $p(C) = 0$ and that any polynomial $0 \neq q \in \mathbb{D}[x]$, deg $q < m$ is not annihilated by C. (Consider $q(C)\mathbf{e}_i$ and use (2.4.9).) That is: p is the minimal polynomial of $C(p)$.

Hint: Use the induction on m as follows. Set $\mathbf{f}_i = \mathbf{e}_{m-i+1}$ for $i = 1, \ldots, m$. Let $q = x^{m-1} + a_1 x^{m-1} + \cdots + a_{m-1}$. Set $Q = \begin{bmatrix} 0 & \mathbf{0}_{m-1}^\top \\ \mathbf{0}_{m-1} & C(q) \end{bmatrix}$. Use the induction hypothesis on $C(q)$, i.e. $q(C(q)) = 0$, and the facts that $C(p)\mathbf{f}_i = Q\mathbf{f}_i$ for $i = 1, \ldots, m-1$, $C(p)\mathbf{f}_m = \mathbf{f}_{m+1} + Q\mathbf{f}_m$ to obtain that $p(C(p))\mathbf{f}_1 = \mathbf{0}$. Now use (2.4.9) to show that $p(C(p))\mathbf{f}_i = \mathbf{0}$ for $i = 2, \ldots, m+1$.

4. Let $A \in \mathbb{F}^{m \times m}$. Using Theorem 2.3.4 and Problems 1 and 3 show that the minimal polynomial ψ of A is the last invariant polynomial of $xI - A$. That is:

$$\psi(x) = \frac{\det (xI - A)}{\delta_{m-1}(x)}, \quad (2.4.10)$$

where $\delta_{m-1}(x)$ is the g.c.d. of all $(m - 1) \times (m - 1)$ minors of $xI - A$.

5. Show that the results of Problem 4 apply to $A \in \mathbb{D}_U^{m \times m}$. In particular, if $A \approx B$ then A and B have the same minimal polynomials.

6. Deduce from Problem 4 the *Cayley–Hamilton* theorem which states that $T \in \mathrm{Hom}\,(\mathbf{V})$ annihilates its characteristic polynomial.

7. Let $A \in \mathbb{D}^{m \times m}$. Prove that A annihilates its characteristic polynomial. (Consider the quotient field $\mathbb{F}$ of $\mathbb{D}$.)

8. Use Problem 6 and Lemma 2.4.4 to show

$$\deg \psi \le \dim \mathbf{V}. \tag{2.4.11}$$

9. Let $\psi = \phi^s$, where ϕ is irreducible in $\mathbb{F}[x]$. Assume that $\deg \psi = \dim \mathbf{V}$. Use Problems 2 and 8 to show that $\mathbf{V}$ is an irreducible invariant subspace of T.

10. Let $p(x) \in \mathbb{F}[x]$ be a nontrivial normalized polynomial such that $p = \phi^s$, where ϕ is a normalized irreducible in $\mathbb{F}[x]$. Let $T \in \mathrm{Hom}\,(\mathbf{V})$ be represented by $C(p)$. Use Problem 9 to show that $\mathbf{V}$ is an irreducible invariant subspace of T.

11. Let $T \in \mathrm{Hom}\,(\mathbf{V})$ and let E be the matrix given by (2.3.6), which is determined by the elementary divisors of T. Using Problem 10 show that the representation E of T corresponds to a splitting of V to a direct sum of irreducible invariant subspaces of T.

12. Deduce from Problems 9 and 11 that $\mathbf{V}$ is an irreducible invariant subspace of T if and only if the minimal polynomial ψ of T satisfies the assumptions of Problem 9.

2.5   An Upper Triangular Form

Definition 2.5.1 *Let* $\mathbf{M}$ *be a* $\mathbb{D}$*-module and assume that* $T \in \mathrm{Hom}\,(\mathbf{M})$. $\lambda \in \mathbb{D}$ *is an eigenvalue of* T *if there exists* $\mathbf{0} \ne \mathbf{u} \in \mathbf{M}$ *such that*

$$T\mathbf{u} = \lambda\mathbf{u}. \tag{2.5.1}$$

The element, (vector), **u** *is an eigenelement, (eigenvector), corresponding to* λ. *An element* $\mathbf{0} \neq \mathbf{u}$ *is a generalized eigenelement, (eigenvector), if*

$$(\lambda I - T)^k \mathbf{u} = \mathbf{0} \tag{2.5.2}$$

for some positive integer k, *where* λ *is an eigenvalue of* T. *For* $T \in \mathbb{D}^{m \times m}$ λ *is an eigenvalue if* (2.5.1) *holds for some* $\mathbf{0} \neq \mathbf{u} \in \mathbb{D}^m$. *The element* **u** *is eigenelement, (eigenvector), or generalized eigenelement, (eigenvector), if either* (2.5.1) *or* (2.5.2) *holds respectively.*

Lemma 2.5.2 *Let* $T \in \mathbb{D}^{m \times m}$. *Then* λ *is an eigenvalue of* T *if and only if* λ *is a root of the characteristic polynomial* $\det (xI - T)$.

Proof. Let $\mathbb{F}$ be the quotient field of $\mathbb{D}$. Assume first that λ is an eigenvalue of T. As (2.5.1) is equivalent to $(\lambda I - T)\mathbf{u} = \mathbf{0}$ and $\mathbf{u} \neq \mathbf{0}$, then above system has a nontrivial solution. Therefore $\det (\lambda I - T) = 0$. Vice versa, if $\det (\lambda I - T) = 0$ then the system $(\lambda I - T)\mathbf{v} = \mathbf{0}$ has a nontrivial solution $\mathbf{v} \in \mathbb{F}^m$. Then there exists $0 \neq a \in \mathbb{D}$ such that $\mathbf{u} := a\mathbf{v} \in \mathbb{D}^m$ and $T\mathbf{u} = \lambda \mathbf{u}$. $\square$

Definition 2.5.3 *A matrix* $A = [a_{ij}] \in \mathbb{D}^{m \times m}$ *is an upper, (lower), triangular if* $a_{ij} = 0$ *for* $j < i$, $(j > i)$. *Let* $\mathbf{UT}(m, \mathbb{D})$, $\mathbf{LT}(m, \mathbb{D}), \mathbf{D}(m, \mathbb{D}) \subset \mathbb{D}^{m \times m}$ *be the ring of upper triangular, lower triangular, diagonal* $m \times m$ *matrices. Let* $\mathbf{UTG}(m, \mathbb{D}) = \mathbf{UT}(m, \mathbb{D}) \cap \mathbf{GL}(m, \mathbb{D})$, $\mathbf{LTG}(m, \mathbb{D}) = \mathbf{LT}(m, \mathbb{D}) \cap \mathbf{GL}_m(\mathbb{D})$.

Theorem 2.5.4 *Let* $T \in \mathbb{D}^{m \times m}$. *Assume that the characteristic polynomial of* T *splits to linear factors over* $\mathbb{D}$

$$\det (xI - T) = \prod_{i=1}^{m} (x - \lambda_i), \quad \lambda_i \in \mathbb{D}, \ i = 1, \ldots, m. \tag{2.5.3}$$

Assume furthermore that $\mathbb{D}$ *is a Bezout domain. Then*

$$T = QAQ^{-1}, \quad Q \in \mathbf{GL}(m, \mathbb{D}), \quad A = [a_{ij}]_1^m \in \mathbf{UT}(m, \mathbb{D}), \tag{2.5.4}$$

such that $a_{11}, \ldots, a_{mm}$ *are the eigenvalues* $\lambda_1, \ldots, \lambda_m$ *appearing in any specified order.*

Proof. Let λ be an eigenvalue of T and consider the set of all $\mathbf{u} \in \mathbb{D}^m$ which satisfies (2.5.1). Clearly this set is a $\mathbb{D}$-module $\mathbf{M}$. Lemma 2.5.2 yields that $\mathbf{M}$ contains nonzero vectors. Assume that $\mathbb{D}$ is $\mathbb{D}_B$. According to Theorem 1.13.3 $\mathbf{M}$ has a basis $\mathbf{u}_1, \ldots, \mathbf{u}_k$ which can be completed to a basis $\mathbf{u}_1, \ldots, \mathbf{u}_m$ in $\mathbb{D}^m$. Let

$$T\mathbf{u}_i = \sum_{j=1}^{m} b_{ji}\mathbf{u}_j, \quad i = 1, \ldots, m, \quad B = [b_{ij}] \in \mathbb{D}^{m \times m}. \quad (2.5.5)$$

A straightforward computation shows that $T \approx B$. As $T\mathbf{u}_i = \lambda\mathbf{u}_i$, $i = 1, \ldots, k$ we have that $b_{j1} = 0$ for $j > 1$. So

$$\det (xI - T) = \det (xI - B) = (x - \lambda)\det (xI - \tilde{B}),$$

where $\tilde{B} = [b_{ij}]_{i,j=2}^n \in \mathbb{D}^{(m-1)\times(m-1)}$. Here, the last equality is achieved by expanding $\det (xI - B)$ by the first column. Use the induction hypothesis to obtain that $\tilde{B} \approx A_1$, where $A_1 \in \mathbf{UT}(m - 1, \mathbb{D})$, with the eigenvalues of $\tilde{B}$ on the main diagonal of A_1 appearing in any prescribed order. Hence, $T \approx C = [c_{ij}]_1^n$, where $C \in \mathbf{UT}(m, \mathbb{D})$ with $c_{11} = \lambda$, $[c_{ij}]_{i,j=2}^n = A_1$.   □

The upper triangular form of A is not unique unless A is a *scalar matrix*: $A = aI$. See Problem 1.

Definition 2.5.5 *Let $T \in \mathbb{D}^{m \times m}$ and assume that (2.5.3) holds. Then the eigenvalue multiset of T is the set $\mathrm{S}(T) = \{\lambda_1, \ldots, \lambda_m\}$. The multiplicity of $\lambda \in \mathrm{S}(T)$, denoted by $m(\lambda)$, is the number of elements in $\mathrm{S}(T)$ which are equal to λ. λ is called a simple eigenvalue if $m(\lambda) = 1$. The spectrum of T, denoted by $\mathrm{spec}\,(T)$, is the set of all distinct eigenvalues of T:*

$$\sum_{\lambda \in \mathrm{spec}\,(T)} m(\lambda) = m. \quad (2.5.6)$$

For $T \in \mathbb{C}^{m \times m}$ arrange the eigenvalues of T in the decreasing order of their absolute values (unless otherwise stated):

$$|\lambda_1| \geq \cdots \geq |\lambda_m| \geq 0, \quad (2.5.7)$$

The spectral radius of T, denoted by $\rho(T)$, is equal to $|\lambda_1|$.

Problems

1. Let Q correspond to the elementary row operation described in Definition 1.12.6(iii). Assume that $A \in \mathbf{UT}(m, \mathbb{D})$. Show that if $j < i$ then $QAQ^{-1} \in \mathbf{UT}(m, \mathbb{D})$ with the same diagonal as A. More general, for any $Q \in \mathbf{UTG}_m(\mathbb{D})$ $QAQ^{-1} \in \mathbf{UT}(m, \mathbb{D})$ with the same diagonal as A.

2. Show that if $T \in \mathbb{D}^{m \times m}$ is similar to $A \in \mathbf{UT}(m, \mathbb{D})$ then the characteristic polynomial of T splits to linear factors over $\mathbb{D}[x]$.

3. Let $T \in \mathbb{D}^{m \times m}$ and put

$$\det (xI - T) = x^m + \sum_{j=1} a_{m-j} x^j. \qquad (2.5.8)$$

Assume that the assumptions of Theorem 2.5.4 hold. Show that

$$(-1)^k a_k = \sum_{\alpha \in [m]_k} \det T[\alpha, \alpha] = s_k(\lambda_1, \ldots, \lambda_m), \quad k = 1, \ldots, m.$$

$$(2.5.9)$$

Here $s_k(x_1, \ldots, x_m)$ is the kth elementary symmetric polynomial of $x_1, \ldots, x_m$. The coefficient $-a_1$ is called the *trace* of A:

$$\operatorname{tr} A = \sum_{i=1}^m a_{ii} = \sum_{i=1}^m \lambda_i. \qquad (2.5.10)$$

4. Let $T \in \mathbb{D}^{m \times m}$ and assume the assumptions of Theorem 2.5.4. Suppose furthermore that $\mathbb{D}$ is $\mathbb{D}_U$. Using the results of Theorem 2.5.4 and Problem 2.4.5 show that the minimal polynomial $\psi(x)$ of T is of the form

$$\psi(x) = \prod_{i=1}^l (x - \alpha_i)^{s_i},$$

$$\alpha_i \neq \alpha_j \text{ for } i \neq j, \ 1 \leq s_i \leq m_i := m(\alpha_i), \ i = 1, \ldots, l,$$

$$(2.5.11)$$

where spec $(T) = \{\alpha_1, \ldots, \alpha_l\}$. (*Hint*: Consider the diagonal elements of $\psi(A)$.)

5. Let $T \in \mathbb{D}_U^{m \times m}$ and assume that the minimal polynomial of T is given by (2.5.11). Using Problem 2.4.4 and the equality (2.4.1) show

$$\det (xI - T) = \prod_{i=1}^{l} (x - \alpha_i)^{m_i}. \qquad (2.5.12)$$

2.6   Jordan Canonical Form

Theorem 2.5.4 and Problem 2.5.2 show that $T \in \mathbb{D}^{m \times m}$ is similar to an upper triangular matrix if and only if the characteristic polynomial of T splits to linear factors. Unfortunately, the upper triangular form of T is not unique. If $\mathbb{D}$ is a field then there is a special upper triangular form in the similarity class of T which is essentially unique. For convenience we state the theorem for an operator $T \in \text{Hom} \, (\mathbf{V})$.

Theorem 2.6.1 *Let* $\mathbf{V}$ *be a vector space over the field* $\mathbb{F}$. *Let* $T \in \text{Hom} \, (\mathbf{V})$ *and assume that the minimal polynomial* $\psi(x)$ *of* T *splits to a product of linear factors as given by (2.5.11). Then* $\mathbf{V}$ *splits to a direct sum of nontrivial irreducible invariant subspaces of* T

$$\mathbf{V} = \mathbf{W}_1 \oplus \cdots \oplus \mathbf{W}_q. \qquad (2.6.1)$$

In each invariant subspace $\mathbf{W}(= \mathbf{W}_j)$ *it is possible to choose a basis consisting of generalized eigenvectors* $\mathbf{x}_1, \ldots, \mathbf{x}_r$ *such that*

$$\begin{aligned} T\mathbf{x}_1 &= \lambda_0 \mathbf{x}_1, \\ T\mathbf{x}_{k+1} &= \lambda_0 \mathbf{x}_{k+1} + \mathbf{x}_k, \quad k = 1, \ldots, r - 1, \end{aligned} \qquad (2.6.2)$$

where λ_0 *is equal to some* α_i *and* $r \le s_i$. *(For* $r = 1$ *the second part of (2.6.2) is void.) Moreover for each* α_i *there exists an invariant subspace* $\mathbf{W}$ *whose basis satisfies (2.6.2) with* $\lambda_0 = \alpha_i$ *and* $r = s_i$.

Proof. Assume first that the minimal polynomial of T is

$$\psi(x) = x^s. \qquad (2.6.3)$$

Recall that $\psi(x)$ is the last invariant polynomial of T. Hence, each nontrivial invariant polynomial of T is of the form x^r for $1 \le r \le s$.

Theorem 2.3.4 implies that $\mathbf{V}$ has a basis in which T is presented by its rational canonical form

$$C(x^{r_1}) \oplus \cdots \oplus C(x^{r_k}), \quad 1 \leq r_1 \leq r_2 \leq \cdots \leq r_k = s.$$

Hence, $\mathbf{V}$ splits to a direct sum of T-invariant subspaces (2.6.1). Let $\mathbf{W}$ be an invariant subspace in the decomposition (2.6.1). Then $\tilde{T} := T|\mathbf{W}$ has the minimal polynomial x^r, $1 \leq r \leq s$. Furthermore, $\mathbf{W}$ has a basis $\mathbf{x}_1, \ldots, \mathbf{x}_r$ so that $\tilde{T}$ is represented in this basis by the companion matrix $C(x^r)$. It is straightforward to show that $x_1, \ldots, x_r$ satisfies (2.6.2) with $\lambda_0 = 0$. As $\mathbf{W}$ is spanned by $\mathbf{x}_r, \tilde{T}\mathbf{x}_r, \ldots, \tilde{T}^{r-1}\mathbf{x}_r$ it follows that $\mathbf{W}$ is an irreducible invariant subspace of T. Assume now that the minimal polynomial of T is $(x - \lambda_0)^s$. Let $T_0 = T - \lambda_0 I$. Clearly x^s is the minimal polynomial of T_0. Let (2.6.1) be the decomposition of $\mathbf{V}$ to invariant subspaces of T_0 as above. In each invariant subspace $\mathbf{W}$ choose a basis for T_0 as above. Then our theorem holds in this case too.

Assume now that the minimal polynomial of T is given by (2.5.11). Use Theorem 2.4.5 and the above arguments to deduce the theorem. $\qquad\square$

Let

$$H_n := C(x^n) = \begin{bmatrix} 0 & 1 & 0 & \ldots & 0 & 0 \\ 0 & 0 & 1 & \ldots & 0 & 0 \\ \vdots & \vdots & \vdots & \ldots & \vdots & \vdots \\ 0 & 0 & 0 & \ldots & 0 & 1 \\ 0 & 0 & 0 & \ldots & 0 & 0 \end{bmatrix}. \qquad (2.6.4)$$

Sometimes we denote H_n by H when the dimension of H is well defined.

Let $\mathbf{W} = \operatorname{span}(\mathbf{x}_1, \mathbf{x}_2, \ldots, \mathbf{x}_r)$. Let $T \in \operatorname{Hom}(\mathbf{W})$ be given by (2.6.2). Then T is presented in the basis $\mathbf{x}_1, \ldots, \mathbf{x}_r$ by the *Jordan block* $\lambda_0 I_r + H_r$. Theorem 2.6.1 yields:

Theorem 2.6.2 *Let* $A \in \mathbb{F}^{n \times n}$. *Assume that the minimal polynomial* $\psi(x)$ *of* A *splits to linear factors as in (2.5.11). Then there* *exists $P \in \mathbf{GL}(n, \mathbb{F})$ such that*

$$P^{-1}AP = J,$$

$$J = \oplus_{i=1}^{l} \oplus_{j=1}^{q_i} (\alpha_i I_{m_{ij}} + H_{m_{ij}}), \tag{2.6.5}$$

$$1 \leq m_{iq_i} \leq m_{iq_{i-1}} \leq \cdots \leq m_{i1} = s_i, \quad i = 1, \ldots, l. \tag{2.6.6}$$

Definition 2.6.3 *Let $A \in \mathbb{F}^{n \times n}$ satisfy the assumptions of Theorem 2.6.2. The matrix J in (2.6.5) is called the Jordan canonical form of A. Let $T \in \mathrm{Hom}\,(\mathbf{V})$ and assume that its minimal polynomial splits over $\mathbb{F}$. Then a representation matrix J (2.6.5) is called the Jordan canonical form of T.*

Remark 2.6.4 *Let $A \in \mathbb{F}^{n \times n}$ and suppose that the minimal polynomial ψ of A does not split over $\mathbb{F}$. Then there exists a finite extension $\mathbb{K}$ of $\mathbb{F}$ such that ψ splits over $\mathbb{K}$. Then (2.6.5) holds for some $P \in \mathbf{GL}(n, \mathbb{K})$. J is referred as the Jordan canonical form of A.*

Corollary 2.6.5 *Let $A \in \mathbb{F}^{n \times n}$. Assume that the minimal polynomial of A is given by (2.5.11). Let J be the Jordan canonical form of A given by (2.6.5). Set*

$$m_{iq_i+1} = \cdots = m_{in} = 0, \quad i = 1, \ldots, l. \tag{2.6.7}$$

Then the elementary polynomials of $xI - A$, which are the elementary divisors of $xI - A$ defined in Problem 2, are

$$\phi_{ij} = (x - \alpha_i)^{m_{ij}}, \quad j = 1, \ldots, n, \ i = 1, \ldots, l. \tag{2.6.8}$$

Hence, the invariant polynomials $i_1(x), \ldots, i_n(x)$ of $xI - A$ are

$$i_r(x) = \prod_{i=1}^{l} (x - \alpha_i)^{m_{i(n-r+1)}}, \quad r = 1, \ldots, n. \tag{2.6.9}$$

The above corollary shows that the Jordan canonical form is unique up to a permutation of Jordan blocks.

Problems

1. Show directly that to each eigenvalue λ_0 of a companion matrix $C(p) \in \mathbb{F}^{n \times n}$ corresponds one-dimensional eigenvalues subspace spanned by the vector $(1, \lambda_0, \lambda_0^2, \ldots, \lambda_0^{n-1})^{\top}$.

2. Let $A \in \mathbb{F}^{n \times n}$ and assume that the minimal polynomial of A splits in $\mathbb{F}$. Let $\mathbf{U}_1, \mathbf{U}_2 \subset \mathbb{F}^n$ be the subspaces of all generalized eigenvectors of A, $A^\top$ respectively corresponding to $\lambda \in \operatorname{spec}(A)$. Show that there exists bases $\mathbf{x}_1, \ldots, \mathbf{x}_m$ and $\mathbf{y}_1, \ldots, \mathbf{y}_m$ in $\mathbf{U}_1$ and $\mathbf{U}_2$ respectively so that

$$\mathbf{y}_i^\top \mathbf{x}_j = \delta_{ij}, \quad i, j = 1, \ldots, m.$$

   (*Hint*: Assume first that A is in its Jordan canonical form.)

3. Let $A \in \mathbb{F}^{n \times n}$. Let $\lambda, \mu \in \mathbb{F}$ be two distinct eigenvalues of A. Let $\mathbf{x}, \mathbf{y} \in \mathbb{F}^n$ be two generalized eigenvectors of $A, A^\top$ corresponding to λ, μ respectively. Show that $\mathbf{y}^\top \mathbf{x} = 0$.

4. Verify directly that J (given in (2.6.5)) annihilates its characteristic polynomial. Using the fact that any $A \in \mathbb{F}^{n \times n}$ is similar to its Jordan canonical form over the finite extension field $\mathbb{K}$ of $\mathbb{F}$ deduce the Cayley–Hamilton theorem.

5. Let $A, B \in \mathbb{F}^{n \times n}$. Show that $A \approx B$ if and only if A and B have the same Jordan canonical form.

2.7　Some Applications of Jordan Canonical Form

Definition 2.7.1 *Let $A \in \mathbb{F}^{n \times n}$ and assume that* $\det(xI - A)$ *splits in $\mathbb{F}$. Let λ_0 be an eigenvalue of A. Then the number of factors of the form $x - \lambda_0$ appearing in the minimal polynomial $\psi(x)$ of A is called the index of λ_0 and is denoted by* index λ_0. *The dimension of the eigenvalue subspace of A corresponding to λ_0 is called the geometric multiplicity of λ_0.*

Using the results of the previous section we obtain.

Lemma 2.7.2 *Let the assumptions of Definition 2.7.1 hold. Then* index λ_0 *is the size of the largest Jordan block corresponding to λ_0, and the geometric multiplicity of λ_0 is the number of the Jordan blocks corresponding to λ_0.*

Let $T \in \mathrm{Hom}\ (\mathbf{V})$, $\lambda_0 \in \mathrm{spec}\ (T)$ and consider the invariant subspaces

$$\mathbf{X}_r = \{\mathbf{x} \in \mathbf{V} :\ (\lambda_0 I - T)^r \mathbf{x} = \mathbf{0}\}, \quad r = 0, 1, \ldots, \quad (2.7.1)$$

$$\mathbf{Y}_r = (\lambda_0 I - T)^r \mathbf{V}, \quad r = 0, 1, \ldots.$$

Theorem 2.7.3 *Let $T \in \mathrm{Hom}\ (\mathbf{V})$ and assume that λ_0 is the eigenvalue of T. Let index $\lambda_0 = m_1 \geq m_2 \geq \cdots \geq m_p \geq 1$ be the dimensions of all Jordan blocks corresponding to λ_0 which appear in the Jordan canonical form of T. Then*

$$\dim \mathbf{X}_r = \sum_{i=1}^{p} \min(r, m_i), \quad r = 0, 1, \ldots, \quad (2.7.2)$$

$$\dim \mathbf{Y}_r = \dim \mathbf{V} - \dim \mathbf{X}_r, \quad r = 0, 1, \ldots.$$

In particular

$$[\mathbf{0}] = \mathbf{X}_0 \subsetneqq \mathbf{X}_1 \subsetneqq \mathbf{X}_2 \subsetneqq \cdots \subsetneqq \mathbf{X}_m,$$

$$\mathbf{X}(\lambda_0) := \mathbf{X}_m = \mathbf{X}_{m+1} = \cdots, \quad m = \mathrm{index}\ \lambda_0.$$

$$\mathbf{V} = \mathbf{Y}_0 \supsetneqq \mathbf{Y}_1 \supsetneqq \mathbf{Y}_2 \supsetneqq \cdots \supsetneqq \mathbf{Y}_m, \quad (2.7.3)$$

$$\mathbf{Y}(\lambda_0) := \mathbf{Y}_m = \mathbf{Y}_{m+1} = \cdots,$$

$$\mathbf{V} = \mathbf{X}(\lambda_0) \oplus \mathbf{Y}(\lambda_0).$$

Let

$$\nu_i = \dim \mathbf{X}_i - \dim \mathbf{X}_{i-1}, \quad i = 1, \ldots, m + 1, \quad m := \mathrm{index}\ \lambda_0.$$
$$(2.7.4)$$

Then ν_i is the number of Jordan block of at least size i corresponding to λ_0. In particular

$$\nu_1 \geq \nu_2 \geq \cdots \geq \nu_m > \nu_{m+1} = 0. \quad (2.7.5)$$

Furthermore

$\nu_i - \nu_{i+1}$ *is the number of Jordan blocks of order*
i in the Jordan canonical form of T corresponding to λ_0. (2.7.6)

Proof. Assume first that det $(xI - T) = \psi(x) = (x - \lambda_0)^m$. That is T has one Jordan block of order m corresponding to λ_0. Then the theorem follows straightforward. Observe next that for

$$\text{Ker } (\lambda I - T) = 0, \quad \text{Range } (\lambda I - T) = \mathbf{V}, \quad \lambda \neq \lambda_0.$$

Assume now that det $(xI - T)$ splits in $\mathbb{F}$ and $\mathbf{V}$ has the decomposition (2.6.1). Apply the above arguments to each $T|\mathbf{W}_i$ for $i = 1, \ldots, q$ to deduce the theorem in this case. In the general case, where det $(xI - T)$ does not split to linear factors, use the rational canonical form of T to deduce the theorem. $\square$

Thus (2.7.3) gives yet another characterization of the index λ_0. Note that in view of Definition 2.5.1 each $0 \neq \mathbf{x} \in \mathbf{X}_k$ is a generalized eigenvector of T. The sequence (2.7.4) is called the Weyr sequence corresponding to λ_0.

Definition 2.7.4 *A transformation $T \in \text{Hom }(\mathbf{V})$ is diagonable if there exists a basis in $\mathbf{V}$ which consists entirely of eigenvectors of T. That is any representation matrix A of T is diagonable, i.e. A is similar to a diagonal matrix.*

For such T we have that $\mathbf{X}_1 = \mathbf{X}_{m_1}$ for each $\lambda_0 \in \text{spec }(T)$. Theorem 2.6.1 yields.

Theorem 2.7.5 *Let $T \in \text{Hom }(\mathbf{V})$. Then T is diagonable if and only if the minimal polynomial ψ of T splits to linear, pairwise different factors. That is the index of any eigenvalue of T equals to 1.*

Definition 2.7.6 *Let $\mathbf{M}$ be a $\mathbb{D}$-module and let $T \in \text{Hom }(\mathbf{M})$. T is nilpotent if $T^s = 0$ for some positive integer s.*

Let $T \in \text{Hom }(\mathbf{V})$ and assume that det $(xI - T)$ splits in $\mathbb{F}$. For $\lambda_0 \in \text{spec }(T)$ let $\mathbf{X}(\lambda_0) \subset \mathbf{V}$ be the T-invariant subspace defined in (2.7.3). Then the decomposition (2.6.1) yields the spectral decomposition of $\mathbf{V}$:

$$\mathbf{V} = \oplus_{\lambda \in \text{spec }(T)} \mathbf{X}(\lambda). \tag{2.7.7}$$

The above decomposition is courser then the fine decomposition (2.6.1). The advantage of the spectral decomposition is that it is uniquely defined. Note that each $\mathbf{X}(\lambda)$, $\lambda \in \text{spec}\,(T)$ is direct sum of irreducible T-invariant subspaces corresponding to the eigenvalue λ in the decomposition (2.6.1). Clearly $T - \lambda I | \mathbf{X}(\lambda)$ is a nilpotent operator. In the following theorem we address the problem of the choices of irreducible invariant subspaces in the decomposition (2.6.1) for a nilpotent transformation T.

Theorem 2.7.7 *Let* $T \in \text{Hom}\,(\mathbf{V})$ *be nilpotent. Let* index $0 = m = m_1 \geq m_2 \geq \cdots \geq m_p \geq 1$ *be the dimensions of all Jordan blocks appearing in the Jordan canonical form of* T. *Let* (2.6.1) *be a decomposition of* $\mathbf{V}$ *to a direct sum of irreducible* T-*invariant subspaces such that*

$$\dim \mathbf{W}_1 = m_1 \geq \dim \mathbf{W}_2 = m_2 \geq \cdots \geq \dim \mathbf{W}_q = m_q \geq 1,$$

$$m_1 = \cdots = m_{i_1} > m_{i_1+1} = \cdots = m_{i_2} > \cdots > m_{i_{p-1}+1}$$

$$= \cdots = m_{i_p} = m_q. \tag{2.7.8}$$

Assume that each $\mathbf{W}_i$ *has a basis* $\mathbf{y}_{i,1}, \ldots, \mathbf{y}_{i,m_i}$ *satisfying* (2.6.2), *with* $\lambda_0 = 0$. *Let* $\mathbf{X}_i, \mathbf{Y}_i, i = 0, \ldots$ *be defined as in* (2.7.1) *for* $\lambda_0 = 0$. *Then the above bases in* $\mathbf{W}_1, \ldots, \mathbf{W}_q$ *can be chosen recursively as follows:*

(a) $\mathbf{y}_{1,1}, \ldots, \mathbf{y}_{i_1,1}$ *is an arbitrary basis in* $\mathbf{Y}_{m-1}$.
(b) *Let* $1 \leq k < m$.

Assume that $\mathbf{y}_{l,j}$ *are given for all* l *such that* $m_l \geq m - k + 1$ *and all* j *such that* $1 \leq j \leq m_l - m + k$. *Then each* $\mathbf{y}_{l,(k+1)}$ *is any element in* $T^{-1}\mathbf{y}_{l,k} \cap \mathbf{Y}_{m-k-1}$, *which is a coset of the subspace* $\mathbf{Y}_{m-k-1} \cap \mathbf{X}_1 = \text{Ker}\,T | \mathbf{Y}_{m-k-1}$. *If* $m - k = m_t$ *for some* $1 < t \leq i_p$ *then* $\mathbf{y}_{i_{t-1}+1,1}, \ldots, \mathbf{y}_{i_t,1}$ *is any set of linearly independent vectors in* $\mathbf{Y}_{m-k-1} \cap \mathbf{X}_1$, *which complements the above chosen vectors* $\mathbf{y}_{l,j}$, $m_l \geq m - k + 1$, $m_l - m + k + 1 \geq j$ *to a basis in* $\mathbf{Y}_{m-k-1}$.

See Problem 1 for the proof of the Theorem.

Corollary 2.7.8 *Let the assumptions of Theorem 2.7.7 hold. Suppose furthermore that* $\mathbf{Z} \subset \mathbf{V}$ *is an eigenspace of* T. *Then there exists a decomposition* (2.6.1) *of* $\mathbf{V}$ *to a direct sum of irreducible*

T-invariant subspaces such that $\mathbf{Z}$ *has a basis consisting of* $l = \dim \mathbf{Z}$ *eigenvectors of the restrictions of* $T|\mathbf{W}_{j_1}, \ldots, T|\mathbf{W}_{j_l}$ *for* $1 \leq j_1 < \cdots < j_l \leq q.$

Proof. Let $\mathbf{Z}_1 := \mathbf{Z} \cap \mathbf{Y}_{m-1} \subset \cdots \subset \mathbf{Z}_m := \mathbf{Z} \cap \mathbf{Y}_0$ and denote $l_i = \dim \mathbf{Z}_i$ for $i = 1, \ldots, m$. We then construct bases in $\mathbf{W}_1, \ldots, \mathbf{W}_q$ as in Theorem 2.7.7 in the following way. If $l_1 = \dim \mathbf{Z}_1 > 0$ we pick $\mathbf{y}_{1,1}, \ldots, \mathbf{y}_{l_1,1}$ to be from $\mathbf{Z}_1$. In general, for each $k = 1, \ldots, m-1$, $1 < t \leq i_p$ and $m_t = m - k$ such that $l_{k+1} > l_k$ we let $\mathbf{y}_{i_{t-1}+1,1}, \ldots, \mathbf{y}_{i_{t-1}+l_{k+1}-l_k,1}$ be any set of linearly independent vectors in $\mathbf{Z}_{k+1}$, which form a basis in $\mathbf{Z}_{k+1}/\mathbf{Z}_k$. $\qquad \square$

Problem

1. Prove Theorem 2.7.7.

2.8   The Matrix Equation $AX - XB = 0$

Let $A, B \in \mathbb{D}^{n \times n}$. A possible way to determine if A and B are similar over $\mathbf{GL}(n, \mathbb{D})$ is to consider the matrix equation

$$AX - XB = 0. \tag{2.8.1}$$

Then $A \approx B$ if and only if there exists a solution $X \in \mathbb{D}^{n \times n}$ such that $\det X$ is an invertible element in $\mathbb{D}$. For $X = [x_{ij}] \in \mathbb{D}^{m \times n}$ let $\hat{X} \in \mathbb{D}^{mn}$ be the column vector composed of the n columns of X:

$$\hat{X} = (x_{11}, \ldots, x_{m1}, x_{12}, \ldots, x_{m2}, \ldots, x_{m(n-1)}, x_{1n}, \ldots, x_{mn})^\top. \tag{2.8.2}$$

Then Eq. (2.8.1), where $A \in \mathbb{D}^{m \times m}, B \in \mathbb{D}^{n \times n}$, has a simple form in tensor notation [MaM64]. (See also Problems 1 and 2.8.13.)

$$(I \otimes A - B^\top \otimes I)\hat{X} = 0. \tag{2.8.3}$$

Assume that $\mathbb{D}$ is a Bezout domain. Then the set of all $X \in \mathbb{D}^{n \times n}$ satisfying (2.8.1) forms a $\mathbb{D}$-module with a basis $X_1, \ldots, X_\nu,$

(Theorem 1.13.3). So any matrix X which satisfies (2.8.1) is of the form

$$X = \sum_{i=1}^{\nu} x_i X_i, \quad x_i \in \mathbb{D}, \quad i = 1, \ldots, \nu.$$

It is "left" to find whether a function

$$\delta(x_1, \ldots, x_\nu) := \det \left(\sum_{i=1}^{\nu} x_i X_i \right)$$

has an invertible value. In such a generality this is a difficult problem. A more modest task is to find the value of ν and to determine if $\delta(x_1, \ldots, x_\nu)$ vanish identically. For that purpose it is enough to assume that $\mathbb{D}$ is actually a field $\mathbb{F}$ (for example, the quotient field of $\mathbb{D}$). Also we may replace $\mathbb{F}$ by a finite extension field $\mathbb{K}$ in which the characteristic polynomial of A and B split. Finally, we are going to study Eq. (2.8.1) where

$$A \in \mathbb{K}^{m \times m}, \quad B \in \mathbb{K}^{n \times n}, \quad X \in \mathbb{K}^{m \times n}.$$

Let $\psi(x)$, $\phi(x)$ and J, K be the minimal polynomials and the Jordan canonical forms of A, B respectively.

$$\psi(x) = \prod_{i=1}^{l} (x - \lambda_i)^{s_i}, \quad \text{spec}(A) = \{\lambda_1, \ldots, \lambda_l\},$$

$$\phi(x) = \prod_{j=1}^{k} (x - \mu_j)^{t_j}, \quad \text{spec}(B) = \{\mu_1, \ldots, \mu_k\},$$

$P^{-1}AP = J = \oplus_{i=1}^{l} J_i,$

$J_i = \oplus_{r=1}^{q_i} (\lambda_i I_{m_{ir}} + H_{m_{ir}}), 1 \le m_{iq_i} \le \cdots \le m_{i1} = s_i, \quad i = 1, \ldots, l,$

$Q^{-1}BQ = K = \oplus_{j=1}^{k} K_j,$

$K_j = \oplus_{r=1}^{p_j} (\mu_j I_{n_{jr}} + H_{n_{jr}}), 1 \le n_{jp_j} \le \cdots \le n_{j1} = t_j, \quad j = 1, \ldots, k.$

$$(2.8.4)$$

Let $Y = P^{-1}XQ$. Then the system (2.8.1) is equivalent to $JY - YK = 0$. Partition Y according to the partitions of J and K as given in (2.8.4). So

$$Y = [Y_{ij}], \quad Y_{ij} \in \mathbb{K}^{m_i \times n_j},$$

$$m_i = \sum_{r=1}^{q_i} m_{ir}, \quad n_j = \sum_{r=1}^{p_j} n_{jr}, \quad i = 1, \ldots, l, \quad j = 1, \ldots, k.$$

Then the matrix equation for Y reduces to lk matrix equations

$$J_i Y_{ij} - Y_{ij} K_j = 0, \quad i = 1, \ldots, l, \quad j = 1, \ldots, k. \qquad (2.8.5)$$

The following two lemmas analyze the above matrix equations.

Lemma 2.8.1 *Let $i \in [l]$, $j \in [k]$. If $\lambda_i \neq \mu_j$ then the corresponding matrix equation in (2.8.5) has the unique trivial solution $Y_{ij} = 0$.*

Proof. Let

$$J_i = \lambda_i I_{m_i} + \bar{J}_i, \quad \bar{J}_i = \oplus_{r=1}^{q_i} H_{m_{ir}},$$

$$K_j = \mu_j I_{n_j} + \bar{K}_j, \quad \bar{K}_j = \oplus_{r=1}^{p_j} H_{n_{jr}}.$$

Note that $\bar{J}^u = \bar{K}^v = 0$ for $u \geq m_i$ and $v \geq n_j$. Then (2.8.5) becomes

$$(\lambda_i - \mu_j)Y_{ij} = -\bar{J}Y_{ij} + Y_{ij}\bar{K}.$$

Thus

$$\begin{aligned}
(\lambda_i - \mu_j)^2 Y_{ij} &= -\bar{J}_i(\lambda_i - \mu_j)Y_{ij} + (\lambda_i - \mu_j)Y_{ij}\bar{K}_j \\
&= -\bar{J}_i(-\bar{J}_i Y_{ij} + Y_{ij}\bar{K}_j) + (-\bar{J}_i Y_{ij} + Y_{ij}\bar{K}_j)\bar{K}_j \\
&= (-\bar{J}_i)^2 Y_{ij} + 2(-\bar{J}_i)Y_{ij}\bar{K}_j + Y_{ij}\bar{K}_j^2.
\end{aligned}$$

Continuing this procedure we get

$$(\lambda_i - \mu_j)^r Y_{ij} = \sum_{u=0}^{r} \binom{r}{u}(-\bar{J}_i)^u Y_{ij}\bar{K}_j^{r-u}.$$

Hence, for $r = m_i + n_j$ either $\bar{J}^u$ or $\bar{K}_j^{r-u}$ is a zero matrix. Since $\lambda_i \neq \mu_j$ we deduce that $Y_{ij} = 0$. $\qquad \square$

Lemma 2.8.2 *Let* $Z = [z_{\alpha\beta}] \in \mathbb{F}^{m \times n}$ *satisfy the equation*

$$H_m Z = Z H_n. \tag{2.8.6}$$

Then the entries of Z *are of the form*

$$\begin{aligned} z_{\alpha\beta} &= 0 \quad \text{for } \beta < \alpha + n - \min(m, n), \\ z_{\alpha\beta} &= z_{(\alpha+1)(\beta+1)} \quad \text{for } \beta \geq \alpha + n - \min(m, n). \end{aligned} \tag{2.8.7}$$

In particular, the subspace of all $m \times n$ *matrices* Z *satisfying* (2.8.6) *has dimension* $\min(m, n)$.

Proof. Note that the first column and the last row of H_l are equal to zero. Hence, the first column and the last row of $Z H_n = H_m Z$ are equal to zero. That is

$$z_{\alpha 1} = z_{m\beta} = 0, \quad \alpha = 2, \ldots, m, \quad \beta = 1, \ldots, n - 1.$$

In all other cases, equating the (α, β) entries of $H_m Z$ and $Z H_n$ we obtain

$$z_{(\alpha+1)\beta} = z_{\alpha(\beta-1)}, \quad \alpha = 1, \ldots, m - 1, \quad \beta = 2, \ldots, n.$$

The above two sets of equalities yield (2.8.7). $\qquad \square$

Combine the above two lemmas to obtain.

Theorem 2.8.3 *Consider the system of* (2.8.5). *If* $\lambda_i \neq \mu_j$ *then* $Y_{ij} = 0$. *Assume that* $\lambda_i = \mu_j$. *Partition* Y_{ij} *according to the partitions of* J_i *and* K_j *as given in* (2.8.4):

$$Y_{ij} = \left[Y_{ij}^{(uv)} \right], \quad Y_{ij}^{(uv)} \in \mathbb{K}^{m_{iu} \times n_{jv}}, \quad u = 1, \ldots, q_i, \quad v = 1, \ldots, p_j.$$

Then each $Y_{ij}^{(uv)}$ *is of the form given in Lemma 2.8.2 with* $m = m_{iu}$ *and* $n = n_{jv}$. *Assume that*

$$\begin{aligned} \lambda_i &= \mu_i, \quad i = 1, \ldots, t, \\ \lambda_i &\neq \mu_j, \quad i = t+1, \ldots, l, \quad j = t+1, \ldots, k. \end{aligned} \tag{2.8.8}$$

Then the dimension of the subspace $\mathcal{Y} \subset \mathbb{K}^{m \times n}$ *of block matrices* $Y = [Y_{ij}]_{i,j=1}^{lk}$ *satisfying (2.8.5) is given by the formula*

$$\dim \mathcal{Y} = \sum_{i=1}^{t} \sum_{u,v=1}^{q_i,p_i} \min(m_{iu}, n_{iv}). \tag{2.8.9}$$

Consider a special case of (2.8.1)

$$C(A) = \{X \in \mathbb{D}^{n \times n} : AX - XA = 0\}. \tag{2.8.10}$$

Then $C(A)$ is an *algebra* over $\mathbb{D}$ with the identity I. In case $\mathbb{D}$ is a field $\mathbb{F}$, or more generally $\mathbb{D}$ is a Bezout domain, $C(A)$ has a finite basis. Theorem 2.8.3 yields

$$\dim C(A) = \sum_{i=1}^{l} \sum_{u,v=1}^{q_i} \min(m_{iu}, m_{iv}).$$

(Note that the dimension of $C(A)$ does not change if we pass from $\mathbb{F}$ to a finite extension field $\mathbb{K}$ in which the characteristic polynomial of A splits.) As $\{m_{iu}\}_{u=1}^{p_i}$ is a decreasing sequence we have

$$\sum_{v=1}^{q_i} \min(m_{iu}, m_{iv}) = u m_{iu} + \sum_{v=u+1}^{q_i} m_{iv}.$$

So

$$\dim C(A) = \sum_{i=1}^{l} \sum_{u=1}^{q_i} (2u - 1) m_{iu}. \tag{2.8.11}$$

Let $i_1(x), \ldots, i_n(x)$ be the invariant polynomials of $xI - A$. Use (2.6.7)–(2.6.9) to deduce

$$\dim C(A) = \sum_{u=1}^{n} (2u - 1) \deg i_{n-u+1}(x). \tag{2.8.12}$$

The above formula enables us to determine when any commuting matrix with A is a polynomial in A. Clearly, the dimension of the subspace spanned by the powers of A is equal to the degree of the minimal polynomial of A.

Corollary 2.8.4 *Let $A \in \mathbb{F}^{n \times n}$. Then each commuting matrix with A can be expressed as a polynomial in A if and only if the minimal and the characteristic polynomial of A are equal. That is, A is similar to a companion matrix $C(p)$, where $p(x) = \det(xI - A)$.*

A matrix for which the minimal and characteristic polynomial coincide is called *nonderogatory*. If the minimal polynomial of A is a strict factor of the characteristic polynomial of A, i.e. the degree of the minimal polynomial is strictly less than the degree of the characteristic polynomial, then A is called *derogatory*.

Problems

1. For $A = [a_{ij}] \in \mathbb{D}^{m \times p}$, $B = [b_{kl}] \in \mathbb{D}^{n \times q}$ let

$$A \otimes B := [a_{ij}B] \in \mathbb{D}^{mn \times pq}, \qquad (2.8.13)$$

be the tensor (Kronecker) product of A and B. Show

$$(A_1 \otimes A_2)(B_1 \otimes B_2) = (A_1 B_1) \otimes (A_2 B_2),$$
$$A_i \in \mathbb{D}^{m_i \times n_i}, \quad B_i \in \mathbb{D}^{n_i \times p_i}, \quad i = 1, 2.$$

2. Let $\mu : \mathbb{D}^{m \times n} \to \mathbb{D}^{mn}$ be given by $\mu(X) = \hat{X}$, where $\hat{X}$ is defined be (2.8.2). Show that

$$\mu(AX) = (I_n \otimes A)\mu(X), \quad \mu(XB) = (B^\top \otimes I_m)\mu(X),$$
$$A \in \mathbb{D}^{m \times m}, \quad B \in \mathbb{D}^{n \times n}.$$

3. Let $P \in \mathbb{F}^{m \times m}$, $Q \in \mathbb{F}^{n \times n}$, $R \in \mathbb{F}^{m \times n}$. Let

$$A = \begin{bmatrix} P & R \\ 0 & Q \end{bmatrix}, \quad B = \begin{bmatrix} P & 0 \\ 0 & Q \end{bmatrix} \in \mathbb{F}^{(m+n) \times (m+n)}.$$

Assume that the characteristic polynomials of P and Q are coprime. Show that there exists $X = \begin{bmatrix} I_m & Y \\ 0 & I_m \end{bmatrix}$ which satisfies (2.8.1). Hence, $A \approx B$.

4. Let $A = \oplus_{i=1}^{\ell} A_i \in \mathbb{F}^{n \times n}$. Show that

$$\dim C(A) \geq \sum_{i=1}^{\ell} \dim C(A_i), \qquad (2.8.14)$$

and the equality holds if and only if

$$(\det (xI - A_i), \det (xI - A_j)) = 1$$
$$\text{for } i = 1, \ldots, \ell, \ j = 1, \ldots, \ell - 1.$$

5. Let $A \in \mathbb{D}^{n \times n}$. Show that the ring $C(A)$ is a commutative ring if and only if A satisfies the conditions of Corollary 2.8.4, where $\mathbb{F}$ is the quotient field of $\mathbb{D}$.

6. Let $A \in \mathbb{D}^{n \times n}$, $B \in C(A)$. Then B is an invertible element in the ring $C(A)$ if and only if B is a unimodular matrix.

7. Let $A \in \mathbb{D}^{m \times m}$, $B \in \mathbb{D}^{n \times n}$. Define

$$C(A, B) := \{X \in \mathbb{D}^{m \times n} : \ AX - XB = 0\}. \qquad (2.8.15)$$

Show that $C(A, B)$ is a left (right) module of $C(A)$ $(C(B))$ under the matrix multiplication.

8. Let $A, B \in \mathbb{D}^{n \times n}$. Show that $A \approx B$ if and only if the following two conditions hold:

(a) $C(A, B)$ is a $C(A)$-module with a basis consisting of one element U;

(b) any basis element U is a unimodular matrix.

2.9   A Criterion for Similarity of Two Matrices

Definition 2.9.1 *Let $A \in \mathbb{D}^{m \times m}$, $B \in \mathbb{D}^{n \times n}$. Denote by $r(A, B)$ and $\nu(A, B)$ the rank and the nullity of the matrix $I_n \otimes A - B^{\top} \otimes I_m$ viewed as a matrix acting on the vector space $\mathbb{F}^{m \times n}$, where $\mathbb{F}$ is the quotient field of $\mathbb{D}$.*

According to Theorem 2.8.3 we have

$$\nu(A, B) = \sum_{i=1}^{t} \sum_{u,v=1}^{q_i, p_i} \min(m_{iu}, n_{iv}),$$

$$r(A, B) = mn - \sum_{i=1}^{t} \sum_{u,v=1}^{q_i, p_i} \min(m_{iu}, n_{iv}).$$

(2.9.1)

Theorem 2.9.2 *Let* $A \in \mathbb{D}^{m \times m}$, $B \in \mathbb{D}^{n \times n}$. *Then*

$$\nu(A, B) \leq \frac{1}{2}(\nu(A, A) + \nu(B, B)).$$

Equality holds if and only if $m = n$ *and* A *and* B *are similar over the quotient field* $\mathbb{F}$.

Proof. Without loss of generality we may assume that $\mathbb{D} = \mathbb{F}$ and the characteristic polynomials of A and B split over $\mathbb{F}[x]$. For $x, y \in \mathbb{R}$ let $\min(x, y)$ $(\max(x, y))$ be the minimum (maximum) of the values of x and y. Clearly $\min(x, y)$ is a homogeneous concave function on $\mathbb{R}^2$. Hence,

$$\min(a + b, c + d) \geq \frac{\min(a, c) + \min(b, d) + \min(a, d) + \min(b, c)}{2}.$$

(2.9.2)

A straightforward calculation shows that if $a = c$ and $b = d$ then equality holds if and only if $a = b$. Let

$$N = \max(m, n), \quad m_{iu} = n_{jv} = 0,$$

$$\text{for } q_i < u \leq N, \ p_i < v \leq N, \ i = 1, \ldots, \ell, \ j = 1, \ldots, k.$$

Then

$$\nu(A, A) + \nu(B, B) = \sum_{i,u=1}^{\ell, N} (2u - 1)m_{iu} + \sum_{j,u=1}^{k, N} (2u - 1)n_{ju}$$

$$\geq \sum_{i,u=1}^{t, N} (2u - 1)(m_{iu} + n_{iu}),$$

and the equality holds if and $\ell = k = t$. Next consider the inequality

$$\sum_{i,u=1}^{t,N} (2u-1)(m_{iu} + n_{iu})$$

$$= \sum_{i=1}^{t} \sum_{u,v=1}^{N} \min(m_{iu} + n_{iu}, m_{iv} + n_{iv})$$

$$\geq \frac{1}{2} \sum_{i=1}^{t} \sum_{u,v=1}^{N} (\min(m_{iu}, m_{iv}) + \min(m_{iu}, n_{iv})$$

$$+ \min(n_{iu}, m_{iv}) + \min(n_{iu}, n_{iv}))$$

$$= \frac{1}{2} \sum_{i,u=1}^{t,N} (2u-1)(m_{iu} + n_{iu}) + \sum_{i}^{t} \sum_{u,v=1}^{q_i,p_i} \min(m_{iu}, n_{iv}).$$

Combine the above results to obtain the inequality (2.9.2). Equality sign holds in (2.9.2) if and only if A and B have the same Jordan canonical forms. That is $m = n$ and A is similar to B over $\mathbb{F}$.   □

Suppose that $A \approx B$. Hence, (2.2.1) holds. The rules for the tensor product (Problem 2.8.1) imply

$$I \otimes A - B^\top \otimes I = ((Q^\top)^{-1} \otimes I)(I \otimes A - A^\top \otimes I)(Q^\top \otimes I),$$
$$I \otimes A - B^\top \otimes I = ((Q^\top)^{-1} \otimes Q)(I \otimes A - A^\top \otimes I)(Q^\top \otimes Q^{-1}).$$
$$(2.9.3)$$

Hence, the three matrices

$$I \otimes A - A^\top \otimes I, \quad I \otimes A - B^\top \otimes I, \quad I \otimes B - B^\top \otimes I \quad (2.9.4)$$

are similar. In particular, these matrices are equivalent. Over a field $\mathbb{F}$ the above matrices are equivalent if and only if the have the same nullity. Hence, Theorem 2.9.2 yields.

Theorem 2.9.3 *Let $A, B \in \mathbb{F}^{n \times n}$. Then A and B are similar if and only if the three matrices in (2.9.4) are equivalent.*

The obvious part of Theorem 2.9.3 extends trivially to any integral domain $\mathbb{D}$.

Proposition 2.9.4 *Let* $A, B \in \mathbb{D}^{n \times n}$. *If* A *and* B *are similar over* $\mathbb{D}$ *then the three matrices in* (2.9.4) *are equivalent over* $\mathbb{D}$.

However, this condition is not sufficient for the similarity of A and B even in the case $\mathbb{D} = \mathbb{Z}$. (See Problem 1.) The disadvantage of the similarity criterion stated in Theorem 2.9.3 is due to the appearance of the matrix $I \otimes A - B^\top \otimes I$, which depends on A and B. It is interesting to note that the equivalence of just two matrices in (2.9.4) does not imply the similarity of A and B. Indeed

$$I \otimes A - A^\top \otimes I = I \otimes (A + \lambda I) - (A + \lambda I)^\top \otimes I$$

for any $\lambda \in \mathbb{F}$. If $\mathbb{F}$ has zero characteristic then $A \not\approx A + \lambda I$ for any $\lambda \neq 0$. (Problem 2) Also if $A = H_n$ and $B = 0$ then $\nu(A, A) = \nu(A, B) = n$. (Problem 3) However, under certain assumptions the equality $\nu(A, A) = \nu(A, B)$ implies $A \approx B$.

Theorem 2.9.5 *Let* $A \in \mathbb{C}^{n \times n}$. *Then there exists a neighborhood of* $A = [a_{ij}]$

$$D(A, \rho) := \left\{ B = [b_{ij}] \in \mathbb{C}^{n \times n} : \sum_{i,j=1}^{n} |b_{ij} - a_{ij}|^2 < \rho^2 \right\}, \qquad (2.9.5)$$

for some positive ρ *depending on* A, *such that if*

$$\nu(A, A) = \nu(A, B), \quad B \in D(A, \rho), \qquad (2.9.6)$$

then B *is similar to* A.

Proof. Let r be the rank of $I \otimes A - A^\top \otimes I$. So there exist indices

$$\alpha = \{(\alpha_{11}, \alpha_{21}), \dots, (\alpha_{1r}, \alpha_{2r})\},$$

$$\beta = \{(\beta_{11}, \beta_{21}), \dots, (\beta_{1r}, \beta_{2r})\} \subset [n] \times [n],$$

viewed as elements of $[n^2]_r$, such that $\det (I \otimes A - A^\top \otimes I)[\alpha, \beta] \neq 0$. Also $\det (I \otimes A - A^\top \otimes I)[\gamma, \delta] = 0$ for any $\gamma, \delta \in [n^2]_{r+1}$. First choose a positive ρ' such that

$$\det (I \otimes A - B^\top \otimes I)[\alpha, \beta] \neq 0, \quad \text{for all } B \in D(A, \rho'). \quad (2.9.7)$$

Consider the system (2.8.1) as a system in n^2 variables, which are the entries of $X = [x_{ij}]_1^n$. In the system (2.8.1) consider the subsystem of r equations corresponding to the set α:

$$\sum_{k=1}^n a_{ik} x_{kj} - x_{ik} b_{kj} = 0, \quad i = \alpha_{1p}, \ j = \alpha_{2p}, \ p = 1, \ldots, r. \quad (2.9.8)$$

Let

$$x_{kj} = \delta_{kj} \quad \text{for } (k, j) \neq (\beta_{1p}, \beta_{2p}), \ p = 1, \ldots, r. \quad (2.9.9)$$

The condition (2.9.7) yields that the system (2.9.8) and (2.9.9) has a unique solution $X(B)$ for any $B \in D(A, \rho')$. Also $X(A) = I$. Use the continuity argument to deduce the the existence of $\rho \in (0, \rho']$ so that $\det X(B) \neq 0$ for all $B \in D(A, \rho)$. Let $\mathcal{V}$ be the algebraic variety of all matrices $B \in \mathbb{C}^{n \times n}$ satisfying

$$\det (I \otimes A - B^\top \otimes I)[\gamma, \delta] = 0 \quad \text{for any } \gamma, \delta \in Q_{(r+1), n^2}. \quad (2.9.10)$$

We claim that $\mathcal{V} \cap D(A, \rho)$ is the set of matrices of the form (2.9.6). Indeed, let $B \in \mathcal{V} \cap D(A, \rho)$. Then (2.9.7) and (2.9.10) yield that $\nu(A, B) = \nu(A, A) = n^2 - r$. Assume that B satisfies (2.9.6). Hence, (2.9.10) holds and $B \in \mathcal{V} \cap D(A, \rho)$. Assume that $B \in \mathcal{V} \cap D(A, \rho)$. Then

$$AX(B) - X(B)B = 0, \quad \det X(B) \neq 0 \quad \Rightarrow \quad A \approx B. \qquad \square$$

Problems

1. Show that for A and B given in Problem 2.2.1 the three matrices in (2.9.4) are equivalent over $\mathbb{Z}$, but A and B are not similar over $\mathbb{Z}$. (See Problem 2.2.1.)

2. Show that if $\mathbb{F}$ has an infinite characteristic then for any $A \in \mathbb{F}^{n \times n}$ $A \approx A + \lambda I$ if and only if $\lambda = 0$. (Compare the traces of A and $A + \lambda I$.)

3. Show that if $A = H_n$ and $B = 0$ then $\nu(A, A) = \nu(A, B) = n$.

4. Let $A, B \in \mathbb{D}^{n \times n}$. Assume that the three matrices in (2.9.4) are equivalent. Let $\mathcal{I}$ be a maximal ideal in $\mathbb{D}$. Let $\mathbb{F} = \mathbb{D}/\mathcal{I}$ and view A, B as matrices over $\mathbb{F}$. Prove that A and B are similar over $\mathbb{F}$. (Show that the matrices in (2.9.4) are equivalent over $\mathbb{F}$.)

2.10   The Matrix Equation $AX - XB = C$

A related equation to (2.8.1) is the nonhomogeneous equation

$$AX - XB = C, \quad A \in \mathbb{F}^{m \times m}, \quad B \in \mathbb{F}^{n \times n}, \quad C, X \in \mathbb{F}^{m \times n}. \quad (2.10.1)$$

This equation can be written in the tensor notation as

$$(I_n \otimes A - B^\top \otimes I_m)\hat{X} = \hat{C}. \quad (2.10.2)$$

The necessary and sufficient condition for the solvability of (2.10.2) can be stated in the dual form as follows. Consider the homogeneous system whose coefficient matrix is the transposed coefficient matrix of (2.10.2), (see Problem 1),

$$(I_n \otimes A^\top - B \otimes I_m)\hat{Y} = 0.$$

Then (2.10.2) is solvable if and only if any solution $\hat{Y}$ of the above system is orthogonal to $\hat{C}$ (e.g. Problem 2). In matrix form the above equation is equivalent to

$$A^\top Y - YB^\top = 0, \quad Y \in \mathbb{F}^{m \times n}.$$

The orthogonality of $\hat{Y}$ and $\hat{C}$ are written as $\operatorname{tr} Y^\top C = 0$. (See Problem 3.) Thus we showed:

Theorem 2.10.1 *Let* $A \in \mathbb{F}^{m \times m}$, $B \in \mathbb{F}^{n \times n}$. *Then* (2.10.1) *is solvable if and only if*

$$\operatorname{tr} ZC = 0 \quad (2.10.3)$$

for all $Z \in \mathbb{F}^{m \times n}$ satisfying

$$ZA - BZ = 0. \tag{2.10.4}$$

Using Theorem 2.10.1 we can obtain a stronger version of Problem 4.

Theorem 2.10.2 *Let*

$$G = [G_{ij}]_1^\ell, \quad G_{ij} \in \mathbb{F}^{n_i \times n_j}, \quad G_{ij} = 0 \ for \ j < i, \ i, j = 1, \ldots, \ell.$$

Then

$$\dim C(G) \geq \sum_{i=1}^{\ell} \dim C(G_{ii}). \tag{2.10.5}$$

Proof. Consider first the case $\ell = 2$. Let $G = \begin{bmatrix} A & E \\ 0 & B \end{bmatrix}$. Assume that $T = \begin{bmatrix} U & X \\ 0 & V \end{bmatrix}$ commutes with G. So

$$AU - UA = 0, \quad BV - VB = 0, \quad AX - XB = UE - EV. \tag{2.10.6}$$

Theorem 2.10.1 implies that $U \in C(A)$, $V \in C(B)$ satisfy the last equation of (2.10.6) if and only if $\operatorname{tr} Z(UE - EV) = 0$ for all Z satisfying (2.10.4). Thus the dimension of pairs (U, V) satisfying (2.10.6) is at least

$$\dim C(A) + \dim C(B) - \dim C(B, A).$$

On the other hand, for a fixed (U, V) satisfying (2.10.6), the set of all X satisfying the last equation of (2.10.6) is of the form $X_0 + C(A, B)$. The equality (2.8.9) yields $\dim C(A, B) = \dim C(B, A)$. Hence, (2.10.5) holds for $\ell = 2$. The general case follows straightforward by induction on ℓ. □

We remark that contrary to the results given in Problem 2.8.4 the equality in (2.10.5) may occur even if $G_{ii} = G_{jj}$ for some $i \neq j$. (See Problem 4.)

Theorem 2.10.3 *Let* $A \in \mathbb{F}^{m \times m}$, $B \in \mathbb{F}^{n \times n}$, $C \in \mathbb{F}^{m \times n}$. *Let*

$$F = \begin{bmatrix} A & 0 \\ 0 & B \end{bmatrix}, \quad G = \begin{bmatrix} A & C \\ 0 & B \end{bmatrix} \in \mathbb{F}^{(m+n) \times (m+n)}.$$

Then $F \approx G$ *if and only if the matrix equation* (2.10.1) *is solvable.*

Proof. Assume that (2.10.1) solvable. Then $U = \begin{bmatrix} I_m & X \\ 0 & I_n \end{bmatrix} \in$ $\mathbf{GL}(m+n, \mathbb{F})$ and $G = U^{-1}FU$.

Assume now that $F \approx G$. We prove the solvability of (2.10.1) by induction on $m + n$, where $m, n \geq 1$. Let $\mathbb{K}$ be a finite extension of $\mathbb{F}$ such that the characteristic polynomial of A and B split to linear factors. Clearly it is enough to prove the solvability of (2.10.1) for the field $\mathbb{K}$. Suppose first that A and B do not have common eigenvalues. Then Problem 2.8.3 yields that $F \approx G$. Assume now that A and B have a common eigenvalue λ_1. For $m = n = 1$ it means that $A = B = \lambda_1 \in \mathbb{F}$. Then the assumption that $F \approx G$ implies that $C = 0$ and (2.10.1) is solvable with $X = 0$.

Assume now that the theorem holds for all $2 \leq m + n < L$. Let $m + n = L$. The above arguments yield that it is enough to consider the case where the characteristic polynomials of A and B split to linear factors and λ_1 is a common eigenvalue of A and B. By considering the matrices $F - \lambda_1 I_{m+n}$, $G - \lambda_1 I_{m+n}$ we may assume that 0 is an eigenvalue of A and B. By considering the similar matrices $U^{-1}FU$, $U^{-1}GU$ where $U = U_1 \oplus U_2$, $U_1 \in \mathbf{GL}(m, \mathbb{F})$, $U_2 \in \mathbf{GL}(n, \mathbb{F})$ we may assume that A and B are in a Jordan canonical form of the form

$$A = A_1 \oplus A_2, \quad B = B_1 \oplus B_2, \quad A_1^m = 0, \quad B_1^n = 0,$$

$$0 \notin (\operatorname{spec}(A_2) \cup \operatorname{spec}(B_2)).$$

(It is possible that either $A = A_1$ or $B = B_1$.) Let

$$U = \begin{bmatrix} I_m & X \\ 0 & I_n \end{bmatrix}, \quad X = \begin{bmatrix} 0 & X_{12} \\ X_{21} & 0 \end{bmatrix}, \quad C = \begin{bmatrix} C_{11} & C_{12} \\ C_{21} & C_{22} \end{bmatrix}.$$

Use Problem 2.8.3 to deduce that one can choose X_{12}, X_{21} such that $G' = U^{-1}GU = G = \begin{bmatrix} A & C' \\ 0 & B \end{bmatrix}$ and $C'_{12} = 0$, $C'_{21} = 0$. For simplicity of notation we will assume that $C_{12} = 0$, $C_{21} = 0$. Permute second and third blocks in F, G to obtain that F, G are permutationally similar to

$$\hat{F} = \begin{bmatrix} A_1 & 0 & 0 & 0 \\ 0 & B_1 & 0 & 0 \\ 0 & 0 & A_2 & 0 \\ 0 & 0 & 0 & B_2 \end{bmatrix}, \quad \hat{G} = \begin{bmatrix} A_1 & C_{11} & 0 & 0 \\ 0 & B_1 & 0 & 0 \\ 0 & 0 & A_2 & C_{22} \\ 0 & 0 & 0 & B_2 \end{bmatrix},$$

respectively. So the Jordan canonical form of $\hat{F}, \hat{G}$ corresponding to 0 are determined by the Jordan canonical forms of $\begin{bmatrix} A_1 & 0 \\ 0 & B_1 \end{bmatrix}, \begin{bmatrix} A_1 & C_{11} \\ 0 & B_1 \end{bmatrix}$ respectively. The Jordan canonical form of $\hat{F}, \hat{G}$ corresponding to other eigenvalues are determined by the Jordan canonical forms of $\begin{bmatrix} A_2 & 0 \\ 0 & B_2 \end{bmatrix}, \begin{bmatrix} A_2 & C_{22} \\ 0 & B_2 \end{bmatrix}$ respectively. Hence

$$\begin{bmatrix} A_1 & 0 \\ 0 & B_1 \end{bmatrix} \approx \begin{bmatrix} A_1 & C_{11} \\ 0 & B_1 \end{bmatrix}, \quad \begin{bmatrix} A_2 & 0 \\ 0 & B_2 \end{bmatrix} \approx \begin{bmatrix} A_2 & C_{22} \\ 0 & B_2 \end{bmatrix}.$$

Thus if either A or B are not nilpotent the theorem follows by induction.

It is left to consider the case where A and B are nilpotent matrices, which are in their Jordan canonical form. If $A = 0$, $B = 0$ then $C = 0$ and the theorem follows. So we assume that either at least one of the matrices in $\{A, B\}$ is not a zero matrix. Since dim Ker F = dim Ker G Problem 6 yields that (after the upper triangular similarity applied to G) we may assume that Ker F = Ker G. Let

$$A = \oplus_{i=1}^p A_i, \quad B = \oplus_{j=1}^q B_j,$$

where each A_i, B_j is an upper triangular Jordan block of dimension m_i, n_j respectively. Let

$$C = [C_{ij}], \quad C_{ij} \in \mathbb{C}^{m_i \times n_j}, \quad i = 1, \ldots, p, \ j = 1, \ldots, q,$$

be the block partition of C induced by the block partition of A and B respectively. The assumption that Ker F = Ker G is equivalent to the assumption that the first column of each C_{ij} is zero. Consider $\mathbf{V} = \mathbb{F}^{m+n}/\mathrm{Ker}\, F$. Then F, G induce the operators $\hat{F}, \hat{G}$ on $\mathbf{V}$ which are obtained from F, G by deleting the rows and columns corresponding to the vectors in the kernels of A and B. (These vectors are formed by some of the vectors in the canonical basis of $\mathbb{F}^{m+n}$.) Note that the Jordan canonical forms of $\hat{F}, \hat{G}$ are direct sums of reduced Jordan blocks (obtained by deleting the first row and column in each Jordan block) corresponding to $\hat{F}, \hat{G}$ respectively. As F and G have the same Jordan blocks it follows that $\hat{F}, \hat{G}$ have the same Jordan blocks, i.e. $\hat{F} \approx \hat{G}$. It is easy to see that

$$\hat{F} = \begin{bmatrix} \hat{A} & 0 \\ 0 & \hat{B} \end{bmatrix}, \quad \hat{G} = \begin{bmatrix} \hat{A} & \hat{C} \\ 0 & \hat{B} \end{bmatrix},$$

$$\hat{A} = \oplus_{i=1}^{p} \hat{A}_i, \quad \hat{B} = \oplus_{j=1}^{q} \hat{B}_j, \quad \hat{C} = (\hat{C}_{ij}),$$

$$\hat{A}_i \in \mathbb{F}^{(m_i-1)\times(m_i-1)}, \quad \hat{B}_j \in \mathbb{F}^{(n_j-1)\times(n_j-1)}, \quad \hat{C}_{ij} \in \mathbb{F}^{(m_i-1)\times(n_j-1)}.$$

Here $\hat{A}_i, \hat{B}_j, \hat{C}_{ij}$ obtained from A_i, B_j, C_{ij} be deleting the first row and column respectively. Since $\hat{F} \approx \hat{G}$ we can use the induction hypothesis. That is there exists $X = (X_{ij}) \in \mathbb{F}^{m\times n}$ partitioned as C with the following properties: The first row and the column of each X_{ij} is zero. $A_i X_{ij} - X_{ij} B_j - C_{ij}$ have zero entries in the last $m_i - 1$ rows and in the first column. By considering $U^{-1} G U$ with $U = \begin{bmatrix} I_m & X \\ 0 & I_n \end{bmatrix}$ we already may assume that the last $m_i - 1$ rows and the first column of each C_{ij} are zero. Finally, we observe that if A_i and B_j are Jordan blocks that Eq. (2.10.1) is solvable by letting X_{ij} be a corresponding matrix with the last $m_i - 1$ rows equal to zero. $\qquad\Box$

Problems

1. Let $A \otimes B$ be defined as in (2.8.13). Prove that $(A \otimes B)^{\top} = A^{\top} \otimes B^{\top}$.

2. Consider the system

$$Ax = b, \quad A \in \mathbb{F}^{m \times n}, \ b \in \mathbb{F}^n.$$

Show the above system is solvable if and only any solution of $A^\top y = 0$ satisfies $y^\top b = 0$. (Change variables to bring A to its diagonal form as in §1.12.)

3. Let $X, Y \in \mathbb{D}^{m \times n}$. Let $\mu(X), \mu(Y) \in \mathbb{D}^{mn}$ be defined as in Problem 2.8.2. Show that

$$\mu(X)^\top \mu(Y) = \operatorname{tr} Y^\top X.$$

4. Assume in Theorem 2.10.2 $\ell = 2$, $G_{11} = G_{22} = 0$, $G_{12} = I$. Show that in this case the equality sign holds in (2.10.5).

5. Let $A_i \in \mathbb{F}^{n_i \times n_i}$, $i = 1, 2$ and suppose that A_1 and A_2 do not have a common eigenvalue. Assume that $A = A_1 \oplus A_2$. Let

$$C = [C_{ij}]_1^2, \quad X = [X_{ij}]_1^2, \quad C_{ij}, X_{ij} \in \mathbb{F}^{n_i \times n_j}, \quad i, j = 1, 2.$$

Using Problem 2.8.4 prove that the equation $AX - XA = C$ is solvable if and only if the equations $A_i X_{ii} - X_{ii} A_i = C_{ii}$, $i = 1, 2$ are solvable.

6. Let $A \in \mathbb{F}^{m \times m}$, $B \in \mathbb{F}^{n \times n}$ be two nilpotent matrix. Let $C \in \mathbb{F}^{m \times n}$ and define the matrices $F, G \in \mathbb{F}^{(m+n) \times (m+n)}$ as in Theorem 2.10.3. Show that $\dim \operatorname{Ker} F \geq \dim \operatorname{Ker} G$. Equality holds if and only if $C \operatorname{Ker} B \subset \operatorname{Range} A$. Equivalently, equality holds if and only if there exists $X \in \mathbb{F}^{m \times n}$ such that

$$\operatorname{Ker} F = \operatorname{Ker} U^{-1} G U, \quad U = \begin{bmatrix} I_m & X \\ 0 & I_n \end{bmatrix}.$$

2.11 A Case of Two Nilpotent Matrices

Theorem 2.11.1 *Let $T \in \operatorname{Hom}(\mathbf{V})$ be nilpotent. Let* index $0 = m = m_1 \geq m_2 \geq \cdots \geq m_p \geq 1$ *be the dimensions of all Jordan blocks appearing in the Jordan canonical form of T. Let $\mathbf{Z} \subset \mathbf{V}$ be an eigenspace of T corresponding to the eigenvalue 0. Denote by*

$\mathbf{W} = \mathbf{V}/\mathbf{Z}$. *Then T induces a nilpotent operator $T' \in \mathrm{Hom}\,(\mathbf{W})$. The dimension of Jordan blocks of T' correspond to the positive integers in the sequence $m'_1, m'_2, \ldots, m'_p$, where m'_i is either m_i or $m_i - 1$. Furthermore, exactly $\dim \mathbf{Z}$ of indices of m'_i are equal to $m_i - 1$.*

Proof. Suppose first that $p = 1$, i.e. $\mathbf{W}$ is an irreducible invariant subspace of T. Then $\mathbf{Z}$ is the eigenspace of T and the theorem is straightforward. Use Corollary 2.7.8 in the general case to deduce the theorem. $\qquad\qquad\square$

Theorem 2.11.2 *Let $A \in \mathbb{F}^{n \times n}$ be a nilpotent matrix. Put*

$$\mathbf{X}_k = \{\mathbf{x} \in \mathbb{F}^n :\ A^k \mathbf{x} = \mathbf{0}\}, \quad k = 0, \ldots$$

Then

$$\mathbf{X}_0 = \{\mathbf{0}\} \quad and \quad \mathbf{X}_i \subset \mathbf{X}_{i+1}, \quad i = 0, \ldots \qquad (2.11.1)$$

Assume that

$$\mathbf{X}_i \neq \mathbf{X}_{i+1} \quad for\ i = 0, \ldots, p-1, \quad and \quad \mathbf{X}_p = \mathbb{F}^n, \quad p = \mathrm{index}\ 0.$$

Suppose that $B \in \mathbb{F}^{n \times n}$ satisfies

$$B\mathbf{X}_{i+1} \subset \mathbf{X}_i, \quad for\ i = 1, \ldots, p-1. \qquad (2.11.2)$$

Then B is nilpotent and

$$\nu(A, A) \leq \nu(B, B). \qquad (2.11.3)$$

Equality holds if and only if B is similar to A.

Proof. Clearly (2.11.2) holds for any $A \in \mathbb{F}^{n \times n}$. As $B^p \mathbb{F}^n = B^p \mathbf{X}_p \subset \mathbf{X}_0 = \{\mathbf{0}\}$, it follows that B is nilpotent. We prove the claim by induction on p. For $p = 1$ $A = B = 0$ and equality holds in (2.11.3). Suppose that the theorem holds for $p = q - 1$. Let $p = q$.

Assume that the Jordan blocks of A and B are the sizes $q = m_1 \geq \cdots \geq m_j \geq 1$ and $l_1 \geq \cdots \geq l_k \geq 1$ respectively. Recall that $\mathbf{X}_1$ is the eigenspace of A corresponding to $\lambda = 0$. Hence, $j = \dim \mathbf{X}_1$. Since $B\mathbf{X}_1 = \mathbf{X}_0 = \{\mathbf{0}\}$ it follows that the dimension of the eigenspace of B is at least j. Hence, $k \geq j$.

Let $\mathbf{W} := \mathbf{V}/\mathbf{X}_1$. Since $A\mathbf{X}_1 = \{0\}$ A induces a nilpotent operator $A' \in \mathrm{Hom}\,(\mathbf{W})$. Let $\mathbf{X}'_i = \ker(A')^i, i = 1, \ldots$. Then $\mathbf{X}'_i = \mathbf{X}_{i+1}/\mathbf{X}_1, i = 0, 1, \ldots$. Hence, the index of $A' = q-1$. Furthermore the Jordan blocks of A' correspond to the positive numbers in the sequence $m'_1 = m_1 - 1 \geq \cdots \geq m'_j = m_j - 1$. Since $B\mathbf{X}_1 = \{0\}$ it follows that B induces the operator $B' \in \mathrm{Hom}\,(\mathbf{W})$. The equality $\mathbf{X}'_i = \mathbf{X}_{i+1}/\mathbf{X}_1$ implies that $B'\mathbf{X}'_i \subset \mathbf{X}'_{i-1}$ for $i = 1, \ldots$.

Theorem 2.11.1 implies that the Jordan blocks of B' correspond to nonzero $l'_1, \ldots, l'_k$, where l'_i is either l_i or l_i-1. Furthermore exactly j of l'_i are equal to $l_i - 1$. Recall (2.8.11) that

$$\nu(A, A) = \sum_{i=1}^{j}(2i - 1)m_i$$

$$= \sum_{i=1}^{j}(2i - 1)m'_i + \sum_{i=1}^{j}(2i - 1) = \nu(A', A') + j^2.$$

Assume that

$$l_1 = \cdots = l_{k_1} > l_{k_1+1} = \cdots = l_{k_2} > l_{k_2+1}$$

$$= \cdots > l_{k_{r-1}+1} = \cdots = l_{k_r},$$

where $k_r = k$. let $k_0 = 0$. Suppose that in the set of $\{k_{s-1}+1, \ldots, k_s\}$ we have exactly $i \leq k_s - k_{s-1}$ indices such that $l'_r = l_r - 1$ for $r \in \{k_{s-1} + 1, \ldots, k_s\}$. We then assume that $l'_r = l_r - 1$ for $r = k_s, k_s - 1, \ldots, k_s - i + 1$. Hence, $l'_1 \geq \cdots \geq l'_k \geq 0$. Thus $\nu(B', B') = \sum_{i=1}^{k}(2i - 1)l'_i$. So

$$\nu(B, B) = \sum_{i=1}^{k}(2i - 1)l_i$$

$$= \nu(B', B') + \sum_{i=1}^{k}(2i - 1)(l_i - l'_i) \geq \nu(B', B') + j^2$$

equality holds if and only if $l'_i = l_i - 1$ for $i = 1, \ldots, j$. The induction hypothesis implies that $\nu(A', A') \leq \nu(B', B')$ and equality holds if and only if $A' \sim B'$, i.e. A' and B' have the same Jordan blocks. Hence, $\nu(A, A) \leq \nu(B, B)$ and equality holds if and only if $A \sim B$.   □

2.12   Historical Remarks

The exposition of §2.1 is close to [Gan59]. The content of §2.2 is standard. Theorem 2.3.4 is well known [Gan59]. Other results of §2.3 are not common and some of them may be new. §2.4 is standard and its exposition is close to [Gan59]. Theorem 2.5.4 is probably known for $\mathbb{D}_{ED}$ (see [Lea48] for the case $\mathbb{D} = H(\Omega)$, $\Omega \subset \mathbb{C}$.) Perhaps it is new for Bezout domains. The results of §2.6 are standard. Most of §2.7 is standard. The exposition of §2.8 is close to [Gan59]. For additional properties of tensor product see [MaM64]. Problem 2.8.8 is close to the results of [Fad66]. See also [Gur80, Gur81] for an arbitrary integral domain $\mathbb{D}$. Theorems 2.9.2 and 2.9.3 are taken from [Fri80b]. See [GaB77] for a weaker version of Theorem 2.9.3. Some of the results of §2.10 may be new. Theorem 2.10.1 was taken from [Fri80a]. Theorem 2.10.3 is called Roth's theorem [Rot52]. Theorem 2.11.2 is taken from [Fri80b].

Chapter 3

Functions of Matrices and Analytic Similarity

3.1 Components of a Matrix and Functions of Matrices

In this chapter, we assume that all the matrices are complex valued ($\mathbb{F} = \mathbb{C}$) unless otherwise stated. Let $\phi(x)$ be a polynomial ($\phi \in \mathbb{C}[x]$). The following relations are easily established

$$\phi(B) = P\phi(A)P^{-1}, \quad B = PAP^{-1}, \ A, B \in \mathbb{C}^{n \times n}, \ P \in \mathbf{GL}_n(\mathbb{C}),$$

$$(3.1.1)$$

$$\phi(A_1 \oplus A_2) = \phi(A_1) \oplus \phi(A_2).$$

It often pays to know the explicit formula for $\phi(A)$ in terms of the Jordan canonical form of A. In view of (3.1.1) it is enough to consider the case where J is composed of one Jordan block.

Lemma 3.1.1 *Let* $J = \lambda_0 I + H \in \mathbb{C}^{n \times n}$, *where* $H = H_n$. *Then for any* $\phi \in \mathbb{C}[x]$

$$\phi(J) = \sum_{k=0}^{n-1} \frac{\phi^{(k)}(\lambda_0)}{k!} H^k.$$

Proof. For any ϕ we have the Taylor expansion

$$\phi(x) = \sum_{k=0}^{N} \frac{\phi^{(k)}(\lambda_0)}{k!}(x - \lambda_0)^k, \quad N = \max(\deg \phi, n).$$

As $H^\ell = 0$ for $\ell \geq n$ from the above equality we deduce the lemma.

$\square$

Using the Jordan canonical form of A we obtain.

Theorem 3.1.2 *Let $A \in \mathbb{C}^{n\times n}$. Assume that the Jordan canonical form of A is given by (2.6.5). Then for $\phi \in \mathbb{C}[x]$ we have*

$$\phi(A) = P\left(\oplus_{i=1}^{\ell} \oplus_{j=1}^{q_i} \sum_{k=0}^{m_{ij}-1} \frac{\phi^{(k)}(\lambda_i)}{k!}H_{m_{ij}}^k\right)P^{-1}. \tag{3.1.2}$$

Definition 3.1.3 *Let the assumptions of Theorem 3.1.2 hold. Then $Z_{ik} = Z_{ik}(A)$ is called the (i,k) component of A and is given by*

$$Z_{ik} = P(0 \oplus \cdots \oplus 0 \oplus_{j=1}^{q_i} H_{m_{ij}}^k \oplus 0 \cdots \oplus 0)P^{-1},$$
$$k = 0, \ldots, s_i - 1, \quad s_i = m_{i1}, \, i = 1, \ldots, \ell. \tag{3.1.3}$$

Compare (3.1.2) with (3.1.3) to deduce

$$\phi(A) = \sum_{i=1}^{\ell} \sum_{j=0}^{s_i-1} \frac{\phi^{(j)}(\lambda_i)}{j!}Z_{ij}. \tag{3.1.4}$$

Definition 3.1.4 *Let $A \in \mathbb{C}^{n\times n}$ and assume that $\Omega \subset \mathbb{C}$ contains $\mathrm{spec}\,(A)$. Then for $\phi \in \mathrm{H}(\Omega)$ define $\phi(A)$ by (3.1.4).*

Using (3.1.3) it is easy verify that the components of A satisfy

$Z_{ij}, \, i = 1, \ldots, \ell, \, j = 1, \ldots, s_i - 1, \quad$ are linearly·independent,

$Z_{ij}Z_{pq} = 0, \quad$ if either $i \neq p$, or $i = p$ and $j + q \geq s_i$,

$Z_{ij}Z_{iq} = Z_{i(j+q)}, \quad$ for $j + q \leq s_i - 1$,

$$A = P\left(\sum_{i=1}^{\ell} \lambda_i Z_{i0} + Z_{i1}\right)P^{-1}. \tag{3.1.5}$$

Consider the component $Z_{i(s_i-1)}$. The above relations imply

$$AZ_{i(s_i-1)} = Z_{i(s_i-1)}A = \lambda_i Z_{i(s_i-1)}. \qquad (3.1.6)$$

Thus the nonzero columns of $Z_{i(s_i-1)}$, $Z_{i(s_i-1)}^\top$ are the eigenvectors of A, $A^\top$ respectively corresponding to λ_i. (Note that $Z_{i(s_i-1)} \neq 0$.)

Lemma 3.1.5 *Let $A \in \mathbb{C}^{n \times n}$. Assume that λ_i is an eigenvalue of A. Let $\mathbf{X}_i$ be the generalized eigenspace of A corresponding to λ_i:*

$$\mathbf{X}_i = \{\mathbf{x} \in \mathbb{C}^n : \quad (\lambda_i I - A)^{s_i}\mathbf{x} = \mathbf{0}\}. \qquad (3.1.7)$$

Then

$$\text{rank } Z_{i(s_i-1)} = \dim \ (\lambda_i I - A)^{s_i-1}\mathbf{X}_i. \qquad (3.1.8)$$

Proof. It is enough to assume that A is in its Jordan form. Then $\mathbf{X}_i$ is the subspace of all $\mathbf{x} = (x_1, \ldots, x_n)^\top$, where the first $\sum_{p=1}^{i-1} \sum_{j=1}^{q_p} m_{pj}$ coordinates and the last $\sum_{p=i+1}^{\ell} \sum_{j=1}^{q_p} m_{pj}$ coordinates vanish. So $(\lambda_i I - A)^{s_i-1}\mathbf{X}_i$ contains only those eigenvectors which correspond to Jordan blocks of the length s_i. Clearly, the rank of $Z_{i(s_i-1)}$ is exactly the number of such blocks. $\square$

Definition 3.1.6 *Let $A \in \mathbb{C}^{n \times n}$. Then the spectral radius $\rho(A)$, the peripheral spectrum* spec $_{\text{peri}}(A)$ *and the index A of A are given by*

$$\rho(A) = \max_{\lambda \in \text{spec } (A)} |\lambda|,$$

$$\text{spec }_{\text{peri}} = \{\lambda \in \text{spec } (A) : \quad |\lambda| = \rho(A)\}, \qquad (3.1.9)$$

$$\text{index } A = \max_{\lambda \in \text{spec }_{\text{peri}}(A)} \text{index } \lambda.$$

Problems

1. Let $A \in \mathbb{C}^{n \times n}$ and let $\psi \in \mathbb{C}[x]$ be the minimal polynomial of A. Assume that $\Omega \subset \mathbb{C}$ is an open set in $\mathbb{C}$ such that spec $(A) \subset \Omega$. Let $\phi \in H(\Omega)$. Then the values

$$\phi^{(k)}(\lambda), \quad k = 0, \ldots, \text{index } \lambda - 1, \ \lambda \in \text{spec } (A) \qquad (3.1.10)$$

are called the *values* of ϕ on the *spectrum* of A. Two functions $\phi, \theta \in \mathrm{H}(\Omega)$ are said to *coincide* on spec (A) if they have the same values on spec (A). Assume that $\phi \in \mathbb{C}[x]$ and let

$$\phi = \omega \psi + \theta, \quad \deg \theta < \deg \psi.$$

Show that θ coincide with ϕ on spec (A). Let

$$\frac{\phi(x)}{\psi(x)} = \omega(x) + \frac{\theta(x)}{\psi(x)}$$

$$= \omega(x) + \sum_{i=1}^{\ell} \sum_{j=1}^{s_i} \frac{\alpha_{ij}}{(x - \lambda_i)^j}, \quad s_i = \text{index } \lambda_i, \ i = 1, \ldots, \ell,$$

where ψ is given by (2.5.11). Show that α_{ij}, $j = s_i, \ldots, s_i - p$ are determined recursively by $\phi^{(j)}$, $j = 0, \ldots, p$. (Multiply the above equality by $\psi(x)$ and evaluate this identity at λ_i.) For any $\phi \in \mathrm{H}(\Omega)$ define θ by the equality

$$\theta(x) = \psi(x) \sum_{i=1}^{\ell} \sum_{j=1}^{s_i} \frac{\alpha_{ij}}{(x - \lambda_i)^j}. \tag{3.1.11}$$

The polynomial θ is called the *Lagrange–Sylvester* (L–S) *interpolation* polynomial of ϕ (corresponding to ψ). Prove that

$$\phi(A) = \theta(A). \tag{3.1.12}$$

Let θ_j be the L–S polynomials of $\phi_j \in \mathrm{H}(\Omega)$ for $j = 1, 2$. Show that $\theta_1 \theta_2$ coincides with L–S polynomial of $\phi_1 \phi_2$ on spec (A). Use this fact to prove the identity

$$\phi_1(A)\phi_2(A) = \phi(A), \quad \phi = \phi_1 \phi_2. \tag{3.1.13}$$

2. Prove (3.1.13) by using the definition (3.1.4) and the relation (3.1.5).

3. Let the assumptions of Problem 1 hold. Assume that a sequence $\{\phi_m\}_1^{\infty} \subset \mathrm{H}(\Omega)$ converges to $\phi \in \mathrm{H}(\Omega)$. That is $\{\phi_m\}_1^{\infty}$ converges

uniformly on any compact set of Ω. Hence

$$\lim_{m\to\infty} \phi_m^{(j)}(\lambda) = \phi^{(j)}(\lambda), \quad \text{for any } j \in \mathbb{Z}_+ \text{ and } \lambda \in \Omega.$$

Use the definition (3.1.4) to show

$$\lim_{m\to\infty} \phi_m(A) = \phi(A). \tag{3.1.14}$$

Apply this result to prove

$$e^A = \sum_{m=0}^{\infty} \frac{A^m}{m!} \quad \left(= \lim_{N\to\infty} \sum_{m=0}^{N} \frac{A^m}{m!} \right). \tag{3.1.15}$$

$$(\lambda I - A)^{-1} = \sum_{m=0}^{\infty} \frac{A^m}{\lambda^{m+1}} \quad \text{for } |\lambda| > \rho(A). \tag{3.1.16}$$

3.2   Cesaro Convergence of Matrices

Let

$$A_k = \left[a_{ij}^{(k)} \right] \in \mathbb{C}^{m\times n}, \quad k = 0, 1, \ldots \tag{3.2.1}$$

be a sequence of matrices. The pth *Cesaro sequence* is defined as follows. First $A_{k,0} = A_k$ for each $k \in \mathbb{Z}_+$. Then for $p \in \mathbb{N}$ $A_{k,p}$ defined recursively by

$$A_{k,p} = \left[a_{ij}^{(k,p)} \right] := \frac{1}{k+1} \sum_{j=0}^{k} A_{j,p-1}, \quad k \in \mathbb{Z}_+, \ p \in \mathbb{N}. \tag{3.2.2}$$

Definition 3.2.1 *A sequence* $\{A_k\}_0^\infty$ *converges to* $A = [a_{ij}] \in \mathbb{C}^{m\times n}$ *if*

$$\lim_{k\to\infty} a_{ij}^{(k)} = a_{ij}, \quad i = 1, \ldots, m, \ j = 1, \ldots, n \quad \Longleftrightarrow \quad \lim_{k\to\infty} A_k = A.$$

A sequence $\{A_k\}_0^\infty$ *converges* p*-Cesaro to* $A = [a_{ij}]$ *if* $\lim_{k\to\infty} A_{k,p} = A$ *for* $p \in \mathbb{Z}_+$. *A sequence* $\{A_k\}_0^\infty$ *converges* p*-Cesaro exactly to* $A = [a_{ij}]$ *if* $\lim_{k\to\infty} A_{k,p} = A$ *and* $\{A_{k,p-1}\}_{k=0}^\infty$ *does not converge.*

It is known (e.g. [Har49]) that if $\{A_k\}$ is p-Cesaro convergent then $\{A_k\}$ is also $p + 1$-Cesaro convergent. A simple example of exact 1-Cesaro convergent sequence is the sequence $\{\lambda^k\}$, where $|\lambda| = 1$, $\lambda \neq 1$. More generally, see [Har49] or Problem 1:

Lemma 3.2.2 *Let* $|\lambda| = 1$, $\lambda \neq 1$. *Then for* $p \in \mathbb{N}$ *the sequence* $\{\binom{k}{p-1}\lambda^k\}_{k=0}^{\infty}$ *is exactly p-Cesaro convergent.*

We now show how to recover the component $Z_{\alpha(s_\alpha-1)}(A)$ for $0 \neq \lambda_\alpha \in \text{spec}_{\text{peri}}(A)$ using the notion of Cesaro convergence.

Theorem 3.2.3 *Let* $A \in \mathbb{C}^{n \times n}$. *Assume that* $\rho(A) > 0$ *and* $\lambda_\alpha \in \text{spec}_{\text{peri}}(A)$. *Let*

$$A_k = \frac{(s_\alpha - 1)!}{k^{s_\alpha-1}} \left(\frac{\bar{\lambda}_\alpha A}{|\lambda_\alpha|^2} \right)^k, \quad s_\alpha = \text{index } \lambda_\alpha. \tag{3.2.3}$$

Then

$$\lim_{k \to \infty} A_{k,p} = Z_{\alpha(s_\alpha-1)}, \quad p = \text{index } A - \text{index } \lambda_\alpha + 1. \tag{3.2.4}$$

The sequence A_k *is exactly p-Cesaro convergent unless either* $\text{spec}_{\text{peri}}(A) = \{\lambda_\alpha\}$ *or* $\text{index } \lambda < \text{index } \lambda_\alpha$ *for any* $\lambda \neq \lambda_\alpha$ *in* $\text{spec}_{\text{peri}}(A)$. *In these exceptional cases* $\lim_{k \to \infty} A_k = Z_{\alpha(s_\alpha-1)}$.

Proof. It is enough to consider the case where $\lambda_\alpha = \rho(A) = 1$. By letting $\phi(x) = x^k$ in (3.1.4) we get

$$A^k = \sum_{i=1}^{\ell} \sum_{j=0}^{s_i-1} \binom{k}{j} \lambda_i^{k-j} Z_{ij}. \tag{3.2.5}$$

So

$$A_k = \sum_{i=1}^{\ell} \sum_{j=0}^{s_i-1} \frac{(s_\alpha - 1)!}{k^{s_\alpha-1}} \frac{k(k-1)\dots(k-j+1)}{j!} \lambda_i^{k-j} Z_{ij}.$$

Since the components Z_{ij}, $i = 1, \dots, \ell$, $j = 0, \dots, s_i - 1$ are linearly independent it is enough to analyze the sequence $\frac{(s_\alpha-1)!}{k^{s_\alpha-1}} \binom{k}{j} \lambda_i^{k-j}$, $k = j, j+1, \dots$ Clearly for $|\lambda| < 1$ and any j or for $|\lambda| = 1$ and $j < s_\alpha - 1$

this sequence converges to zero. For $\lambda_i = 1$ and $j = s_\alpha - 1$ the above sequence converges to 1. For $|\lambda_i| = 1$, $\lambda_i \neq 1$ and $j \geq s_\alpha - 1$ the given sequence is exactly $j - s_\alpha + 2$ convergent to 0 in view of Lemma 3.2.2. From these arguments the theorem easily follows. $\square$

The proof of Theorem 3.2.3 yields:

Corollary 3.2.4 *Let the assumptions of Theorem 3.2.3 hold. Then*

$$\lim_{N \to \infty} \frac{1}{N+1} \sum_{k=0}^{N} \frac{(s-1)!}{k^{s-1}} \left(\frac{A}{\rho(A)} \right)^k = Z, \quad s = \text{index } A. \quad (3.2.6)$$

If $\rho(A) \in \text{spec}(A)$ and index $\rho(A) = s$ then $Z = Z_{\rho(A)(s-1)}$. Otherwise $Z = 0$.

Problems

1. Let $|\lambda| = 1$, $\lambda \neq 1$ be fixed. Differentiate the formula

$$\sum_{j=0}^{k-1} \lambda^j = \frac{\lambda^k - 1}{\lambda - 1}$$

r times with respect to λ and divide by $r!$ to obtain

$$\sum_{j=0}^{k-1} \binom{j}{r} \lambda^j = k \sum_{\ell=0}^{r-1} f(\lambda, r, \ell) \binom{k-1}{\ell} \lambda^{k-1} + (-1)^r \frac{\lambda^k - 1}{(\lambda - 1)^{r+1}},$$

where $f(\lambda, r, \ell)$ are some fixed nonzero functions. Use the induction on r to prove Lemma 3.2.2.

2. Let $\phi(x)$ be a normalized polynomial of degree $p - 1$. Prove that the sequence $\{\phi(k)\lambda^k\}_{k=0}^{\infty}$ for $|\lambda| = 1$, $\lambda \neq 1$ is exactly p-Cesaro convergent.

3. Let $A \in \mathbb{C}^{n \times n}$. For $\lambda_i \in \text{spec}(A)$ let

$$Z_{ij}(A) = \left[z_{\mu\nu}^{(ij)} \right]_{\mu,\nu=1}^{n}, \quad j = 0, \ldots, \text{index } \lambda_i - 1. \quad (3.2.7)$$

Let

$$\text{index}_{\mu\nu}\lambda_i := 1 + \max\{j : z_{\mu\nu}^{(ij)} \neq 0, \ j = 0,\dots,\text{index}; \lambda_i - 1\}, \tag{3.2.8}$$

where $\text{index}_{\mu\nu}\lambda_i = 0$ if $z_{\mu\nu}^{(ij)} = 0$ for $j = 0,\dots,\text{index};\lambda_i - 1$.

$$\rho_{\mu\nu}(A) = \max\{|\lambda_i| : \text{index}_{\mu\nu}\lambda_i > 0\}, \tag{3.2.9}$$

where $\rho_{\mu\nu}(A) = -\infty$ if $\text{index}_{\mu\nu}\lambda_i = 0$ for all $\lambda_i \in \text{spec}\,(A)$. The quantities $\text{index}_{\mu\nu}\lambda_i$, $\rho_{\mu\nu}(A)$ are called the (μ,ν) index of λ_i and the (μ,ν) spectral radius respectively. Alternatively these quantities are called the local index and the local spectral radius respectively. Show that Theorem 3.2.3 and Corollary 3.2.4 could be stated in a local form. That is for $1 \leq \mu, \nu \leq n$ assume that

$$\lambda_\alpha = \rho_{\mu\nu}(A), \ s_\alpha = \text{index}_{\mu\nu}\lambda_\alpha, \ A^k = \left(a_{\mu\nu}^{(k)}\right),$$
$$A_k = [a_{\mu\nu,k}], \ A_{k,p} = [a_{\mu\nu,kp}],$$

where A_k and $A_{k,p}$ are given by (3.2.3) and (3.2.2) respectively. Prove

$$\lim_{k\to\infty} a_{\mu\nu,kp} = z_{\mu\nu}^{(\alpha(s_\alpha-1))}, \quad p = \text{index}_{\mu\nu}A - \text{index}_{\mu\nu}\lambda_\alpha + 1,$$

$$\lim_{N\to\infty} \frac{1}{N+1} \sum_{k=0}^{N} \frac{(s-1)!}{k^{s-1}} \left(\frac{a_{\mu\nu,k}}{\rho_{\mu\nu}(A)}\right)^k = z_{\mu\nu}, \quad s = \text{index}_{\mu\nu}A,$$

$$\rho_{\mu\nu}(A) > 0,$$

where $z_{\mu\nu} = 0$ unless $\lambda_1 = \rho_{\mu\nu}(A) \in \text{spec}\,(A)$ and $\text{index}_{\mu\nu}\lambda_1 = \text{index}_{\mu\nu} = s$. In this exceptional case $z_{\mu\nu} = z_{\mu\nu}^{(1(s-1))}$.

Finally A is called irreducible if $\rho_{\mu\nu}(A) = \rho(A)$ for each $\mu, \nu = 1,\dots,n$. Thus for an irreducible A the local and the global versions of Theorem 3.2.3 and Corollary 3.2.4 coincide.

3.3 An Iteration Scheme

Consider an iteration given by

$$\mathbf{x}^{i+1} = A\mathbf{x}^i + \mathbf{b}, \quad i = 0, 1, \ldots, \tag{3.3.1}$$

where $A \in \mathbb{C}^{n \times n}$ and $\mathbf{x}^i, \mathbf{b} \in \mathbb{C}^n$. Such an iteration can be used to solve a system

$$\mathbf{x} = A\mathbf{x} + \mathbf{b}. \tag{3.3.2}$$

Assume that $\mathbf{x}$ is the unique solution of (3.3.2) and let $\mathbf{y}^i := \mathbf{x}^i - \mathbf{x}$. Then

$$\mathbf{y}^{i+1} = A\mathbf{y}^i, \quad i = 0, 1, \ldots \tag{3.3.3}$$

Definition 3.3.1 *The system* (3.3.3) *is called stable if the sequence* $\mathbf{y}^i$, $i = 0, 1, \ldots$ *converges to zero for any choice of* $\mathbf{y}^0$. *The system* (3.3.3) *is called bounded if the sequence* $\mathbf{y}^i$, $i = 0, 1, \ldots$ *is bounded for any choice of* $\mathbf{y}^0$.

Clearly, the solution to (3.3.3) is $\mathbf{y}^i = A^i \mathbf{y}^0$, $i = 0, 1, \ldots$ So (3.3.3) is stable if and only if

$$\lim_{i \to \infty} A^i = 0. \tag{3.3.4}$$

Furthermore (3.3.3) is bounded if and only if

$$\|A^i\| \leq M, \quad i = 0, 1, \ldots, \tag{3.3.5}$$

for some (or any) vector norm $\| \cdot \| : \mathbb{C}^{n \times n} \to \mathbb{R}_+$ and some $M > 0$. For example one can choose the l_∞ norm on $\mathbb{C}^{m \times n}$ to obtain the induced matrix norm:

$$\|B\| = \max_{1 \leq i \leq m, 1 \leq j \leq n} |b_{ij}|, \quad B = [b_{ij}] \in \mathbb{C}^{m \times n}. \tag{3.3.6}$$

See §7.1 and 7.4 for definitions and properties of vector and operator norms.

Theorem 3.3.2 *Let $A \in \mathbb{C}^{n \times n}$. Then*

1. *The condition (3.3.4) holds if and only if $\rho(A) < 1$.*

2. *The condition (3.3.5) holds if either $\rho(A) < 1$ or $\rho(A) = 1$ and index $A = 1$.*

3. *The condition*

$$\lim_{i \to \infty} A^i = B \tag{3.3.7}$$

holds if and only if either $\rho(A) < 1$ or $\rho(A) = 1$ and the only eigenvalue on the unit circle is $\lambda = 1$ whose index is 1. Furthermore, if (3.3.7) holds then $B^2 = B$, i.e. B is a projection.

Proof. Consider the identity (3.2.5). Recall that all the components of A are linearly independent.

1. Clearly, (3.3.4) is equivalent to

$$\lim_{k \to \infty} \binom{k}{j} \lambda_i^{k-j} = 0, \quad \lambda_i \in \text{spec}(A), \ j = 0, 1, \dots, \text{index } \lambda_i - 1.$$

Hence, the above conditions are equivalent to $\rho(A) < 1$.

2. Since all vector norms on $\mathbb{C}^{m \times n}$ are equivalent, the condition (3.3.5) is equivalent to the statement that the sequence $\binom{k}{j} \lambda_i^{k-j}$, $k = 0, \dots$, is bounded for each $\lambda_i \in \text{spec}(A)$ and each $j \in [0, \text{index } \lambda_i - 1]$. Hence, $\rho(A) \le 1$. Furthermore if $|\lambda_i| = 1$ then index $\lambda_i = 1$.

3. Clearly, (3.3.7) holds if and only if the sequence $\binom{k}{j} \lambda_i^{k-j}$, $k = 0, \dots$, is convergent for each $\lambda_i \in \text{spec}(A)$ and each $j \in [0, \text{index } \lambda_i - 1]$. Hence, $\rho(A) \le 1$. Furthermore if $|\lambda_i| = 1$ then $\lambda_i = 1$ index $\lambda_i = 1$. Assume that (3.3.7) holds. So $A^i A^i = A^{2i}$. Let $i \to \infty$ to deduce that $B^2 = B$. $\qquad \square$

Problems

1. Let $A \in \mathbb{C}^{n \times n}$ and ψ be the minimal polynomial of A given by (2.5.11). Verify

$$e^{At} = \sum_{i=1}^{\ell} \sum_{j=0}^{s_i - 1} \frac{t^j e^{\lambda_i t}}{j!} Z_{ij}. \tag{3.3.8}$$

Use (3.1.5) or (3.1.15) to show

$$\frac{d}{dt}e^{At} = Ae^{At} = e^{At}A. \qquad (3.3.9)$$

(In general t may be complex valued, but in this problem we assume that t is real.) Verify that the system

$$\frac{d\mathbf{x}}{dt} = A\mathbf{x}, \quad x(t) \in \mathbb{C}^n \qquad (3.3.10)$$

has the unique solution

$$\mathbf{x}(t) = e^{A(t-t_0)}\mathbf{x}(t_0). \qquad (3.3.11)$$

The system (3.3.10) is called stable if $\lim_{t\to\infty} \mathbf{x}(t) = 0$ for any solution (3.3.11). The system (3.3.10) is called bounded if any solution $\mathbf{x}(t)$ (3.3.11) is bounded on $[t_0, \infty)$. Prove that (3.3.10) is stable if and only if

$$\Re\lambda < 0 \quad \text{for each } \lambda \in \text{spec}\,(A). \qquad (3.3.12)$$

Furthermore (3.3.10) is bounded if and only if each $\lambda \in \text{spec}\,(A)$ satisfies

$$\Re\lambda \leq 0 \quad \text{and} \quad \text{index } \lambda = 1 \text{ if } \Re\lambda = 0. \qquad (3.3.13)$$

3.4 Cauchy Integral Formula for Functions of Matrices

Let $A \in \mathbb{C}^{n\times n}$, $\phi \in \mathrm{H}(\Omega)$, where Ω is an open set in $\mathbb{C}$. If spec $(A) \subset \Omega$ it is possible to define $\phi(A)$ by (3.1.4). The aim of this section is to give an integral formula for $\phi(A)$ using the Cauchy integration formula for $\phi(\lambda)$. The resulting expression is simply looking and very useful in theoretical studies of $\phi(A)$. Moreover, this formula remains valid for bounded operators in Banach spaces (e.g. [Kat80], [Kat82]).

Consider the function $\phi(x, \lambda) = (\lambda - x)^{-1}$. The domain of analyticity of $\phi(x, \lambda)$ (with respect to x) is the punctured complex plane

$\mathbb{C}$ at λ. Thus if $\lambda \notin \text{spec}(A)$ (3.1.4) yields

$$(\lambda I - A)^{-1} = \sum_{i=1}^{\ell} \sum_{j=0}^{s_i-1} (\lambda - \lambda_i)^{-(j+1)} Z_{ij}. \tag{3.4.1}$$

Definition 3.4.1 *The function* $(\lambda I - A)^{-1}$ *is called the resolvent of A and is denoted by*

$$R(\lambda, A) = (\lambda I - A)^{-1}. \tag{3.4.2}$$

We call a curve in $\mathbb{C}$ a simple closed curve if it is a rectifiable nonintersecting closed curve. It is well known that the interior of a simple closed curve in $\mathbb{C}$ bounds a simply connected domain. Let $\Gamma = \{\Gamma_1, \ldots, \Gamma_k\}$ be a set of disjoint simple closed curves such that Γ forms the boundary ∂D of an open set D and

$$D \cup \Gamma \subset \Omega, \quad \Gamma = \partial D. \tag{3.4.3}$$

For $\phi \in H(\Omega)$ the classical Cauchy integration formula states (e.g. [Rud74])

$$\phi(\zeta) = \frac{1}{2\pi\sqrt{-1}} \int_{\Gamma} (\lambda - \zeta)^{-1} \phi(\lambda) d\lambda, \quad \zeta \in D. \tag{3.4.4}$$

Differentiate the above equality j times to obtain

$$\frac{\phi^{(j)}(\zeta)}{j!} = \frac{1}{2\pi\sqrt{-1}} \int_{\Gamma} (\lambda - \zeta)^{-(j+1)} \phi(\lambda) d\lambda, \quad \zeta \in D, \, j = 0, 1, 2, \ldots \tag{3.4.5}$$

Theorem 3.4.2 *Let* Ω *be an open set in* $\mathbb{C}$. *Assume that* $\Gamma = \{\Gamma_1, \ldots, \Gamma_k\}$ *is a set of disjoint simple closed curves such that* Γ *is a boundary of an open set* D, *and* $\Gamma \cup D \subset \Omega$. *Assume that* $A \in \mathbb{C}^{n \times n}$ *and* $\text{spec}(A) \subset D$. *Then for any* $\phi \in H(\Omega)$

$$\phi(A) = \frac{1}{2\pi\sqrt{-1}} \int_{\Gamma} R(\lambda, A) \phi(\lambda) d\lambda. \tag{3.4.6}$$

Proof. Insert the expression (3.4.1) into the above integral to obtain

$$\frac{1}{2\pi\sqrt{-1}}\int_\Gamma R(\lambda, A)\phi(\lambda)d\lambda$$

$$= \sum_{i=1}^{\ell}\sum_{j=0}^{s_i-1}\left(\frac{1}{2\pi\sqrt{-1}}\int_\Gamma (\lambda - \lambda_i)^{-(j+1)}\phi(\lambda)d\lambda\right)Z_{ij}.$$

Use the identity (3.4.5) to deduce

$$\frac{1}{2\pi\sqrt{-1}}\int_\Gamma R(\lambda, A)\phi(\lambda)d\lambda = \sum_{i=1}^{\ell}\sum_{j=0}^{s_i-1}\frac{\phi^{(j)}(\lambda_i)}{j!}Z_{ij}.$$

The definition (3.1.4) yields the equality (3.4.6). □

We generalize the above theorem as follows.

Theorem 3.4.3 *Let Ω be an open set in $\mathbb{C}$. Assume that $\Gamma = \{\Gamma_1,\ldots,\Gamma_k\}$ is a set of disjoint simple closed curves such that Γ is a boundary of an open set D, and $\Gamma \cup D \subset \Omega$. Assume that $A \in \mathbb{C}^{n\times n}$ and $\mathrm{spec}\,(A)\cap\Gamma = \emptyset$. Let $\mathrm{spec}\,_D(A) := \mathrm{spec}\,(A)\cap D$. Then for any $\phi \in H(\Omega)$*

$$\sum_{\lambda_i\in\mathrm{spec}\,_D(A)}\sum_{j=0}^{s_i-1}\frac{\phi^{(j)}(\lambda_i)}{j!}Z_{ij} = \frac{1}{2\pi\sqrt{-1}}\int_\Gamma R(\lambda, A)\phi(\lambda)d\lambda. \quad (3.4.7)$$

If $\mathrm{spec}\,_D(A) = \emptyset$ then the left-hand side of the above identity is zero.

See Problem 1.

We illustrate the usefulness of Cauchy integral formula by two examples.

Theorem 3.4.4 *Let $A \in \mathbb{C}^{n\times n}$ and assume that $\lambda_p \in \mathrm{spec}\,(A)$. Suppose that D and Γ satisfy the assumptions of Theorem 3.4.3*

$(\Omega = \mathbb{C})$. *Assume furthermore that* spec $(A) \cap D = \{\lambda_p\}$. *Then the* (p, q) *component of* A *is given by*

$$Z_{pq}(A) = \frac{1}{2\pi\sqrt{-1}} \int_\Gamma R(\lambda, A)(\lambda - \lambda_p)^q d\lambda. \qquad (3.4.8)$$

$(Z_{pq} = 0 \ for \ q > s_p - 1.)$

See Problem 2.

Our next example generalizes the first part of Theorem 3.3.2 to a compact set of matrices.

Definition 3.4.5 *A set* $\mathcal{A} \subset \mathbb{C}^{n \times n}$ *is called power stable if*

$$\lim_{k \to \infty} \left(\sup_{A \in \mathcal{A}} \|A^k\| \right) = 0, \qquad (3.4.9)$$

for some vector norm on $\mathbb{C}^{n \times n}$. *A set* $\mathcal{A} \subset \mathbb{C}^{n \times n}$ *is called power bounded if*

$$\|A^k\| \le K, \quad \text{for any } A \in \mathcal{A} \text{ and } k = 0, 1, \ldots, \qquad (3.4.10)$$

for some positive K *and some vector norm on* $\mathbb{C}^{n \times n}$.

Theorem 3.4.6 *Let* $\mathcal{A} \subset \mathbb{C}^{n \times n}$ *be a compact set. Then* $\mathcal{A}$ *is power stable if and only if* $\rho(A) < 1$ *for any* $A \in \mathcal{A}$.

To prove the theorem we need a well-known result on the roots of normalized polynomials in $\mathbb{C}[x]$ (e.g. [Ost66]).

Lemma 3.4.7 *Let* $p(x) = x^m + \sum_{i=1}^{m} a_i x^{m-i} \in \mathbb{C}[x]$. *Then the zeros* $\xi_1, \ldots, \xi_m$ *of* $p(x)$ *are continuous functions of its coefficients. That is for a given* $a_1, \ldots, a_n$ *and* $\epsilon > 0$ *there exists* $\delta(\epsilon)$, *depending on* $a_1, \ldots, a_n$, *such that if* $|b_i - a_i| < \delta(\epsilon)$, $i = 1, \ldots, m$ *it is possible to enumerate the zeros of* $q(x) = x^m + \sum_{i=1}^{m} b_i x^{m-i}$ *by* $\eta_1, \ldots, \eta_m$, *such that* $|\eta_i - \xi_i| < \epsilon$. $i = 1, \ldots, m$. *In particular the function*

$$\rho(p) = \max_{1 \le i \le m} |\xi_i| \qquad (3.4.11)$$

is a continuous function of $a_1, \ldots, a_m$.

Corollary 3.4.8 *The function $\rho : \mathbb{C}^{n \times n} \to \mathbb{R}_+$, which assigns to $A \in \mathbb{C}^{n \times n}$ its spectral radius $\rho(A)$ is a continuous function.*

Proof of Theorem 3.4.6. Suppose that (3.4.9) holds. Then by Theorem 3.3.2 $\rho(A) < 1$ for each $A \in \mathcal{A}$. Assume that $\mathcal{A}$ is compact and $\rho(A) < 1$. Corollary 3.4.8 yields

$$\rho := \max_{A \in \mathcal{A}} \rho(A) = \rho(\tilde{A}) < 1, \quad \tilde{A} \in \mathcal{A}.$$

Recall that $(\lambda I - A)^{-1} = [\frac{p_{ij}(\lambda)}{\det (\lambda I - A)}]_1^n$, where $p_{ij}(\lambda)$ is the (j, i) cofactor of $\lambda I - A$. Let $\lambda_1, \ldots, \lambda_n$ be the eigenvalues of A counted with multiplicities. Then for $|\lambda| > \rho$

$$|\det (\lambda I - A)| = \left| \prod_{i=1}^n (\lambda - \lambda_i) \right| \geq \prod_{i=1}^n (|\lambda| - \rho)^n.$$

Let $\rho < r < 1$. Since $\mathcal{A}$ is a bounded set, the above arguments yield that there exists a positive constant K such that $||(\lambda I - A)^{-1}|| \leq K$ for each $A \in \mathcal{A}$, $|\lambda| = r$. Apply (3.4.6) to obtain

$$A^p = \frac{1}{2\pi \sqrt{-1}} \int_{|\lambda| = r} (\lambda I - A)^{-1} \lambda^p d\lambda, \qquad (3.4.12)$$

for each $A \in \mathcal{A}$. Combine this equality with the estimate $||\lambda I - A)^{-1}|| \leq K$ for $|\lambda| = r$ to obtain $||A^p|| \leq K r^{p+1}$ for any $A \in \mathcal{A}$. As $r < 1$ the theorem follows. $\qquad \square$

Theorem 3.4.9 *Let $\mathcal{A} \subset \mathbb{C}^{n \times n}$. Then $\mathcal{A}$ is power bounded if and only*

$$||(\lambda I - A)^{-1}|| \leq \frac{K}{|\lambda| - 1}, \quad \text{for all } A \in \mathcal{A} \text{ and } |\lambda| > 1, \qquad (3.4.13)$$

for some vector norm $|| \cdot ||$ on $\mathbb{C}^{n \times n}$ and $K \geq ||I_n||$.

Proof. For $|\lambda| > \rho(A)$ we have the Neumann series

$$(\lambda I - A)^{-1} = \sum_{i=0}^{\infty} \frac{A^i}{\lambda^{i+1}}. \qquad (3.4.14)$$

Hence, for any vector norm on $\mathbb{C}^{n \times n}$

$$\|(\lambda I - A)^{-1}\| \leq \sum_{i=0}^{\infty} \frac{\|A^i\|}{|\lambda|^{i+1}}, \quad |\lambda| > \rho(A). \tag{3.4.15}$$

(See Problem 3.) Assume first that (3.4.10) hold. As $A^0 = I_n$ it follows that $K \geq \|I_n\|$. Furthermore as each $A \in \mathcal{A}$ is power bounded Theorem 3.3.2 yields that $\rho(A) \leq 1$ for each $A \in \mathcal{A}$. Combine (3.4.10) and (3.4.15) to obtain (3.4.13).

Assume now that (3.4.13) holds. Since all vector norms on $\mathbb{C}^{n \times n}$ are equivalent we assume that the norm in (3.4.13) is the l_∞ norm given in (3.3.6). Let $A \in \mathcal{A}$. Note that $(\lambda I - A)$ in invertible for each $|\lambda| > 1$. Hence, $\rho(A) \leq 1$. Let $(\lambda I - A)^{-1} = [\frac{p_{ij}(\lambda)}{p(\lambda)}]$. Here, $p(\lambda) = \det(\lambda I - A)$ is a polynomial of degree n and $p_{ij}(\lambda)$ (the (j,i) cofactor of $\lambda I - A$) is a polynomial of degree $n - 1$ at most. Let $A^p = [a_{ij}^{(p)}]$, $p = 0, 1, \ldots$ Then for any $r > 1$ the equality (3.4.12) yields that

$$a_{ij}^{(p)} = \frac{1}{2\pi\sqrt{-1}} \int_{|\lambda|=r} \frac{p_{ij}(\lambda)}{p(\lambda)} \lambda^p d\lambda$$

$$= \frac{1}{2\pi} \int_0^{2\pi} \frac{p_{ij}(re^{\sqrt{-1}\theta})}{p(re^{\sqrt{-1}\theta})} r^{p+1} e^{(p+1)\sqrt{-1}\theta} d\theta.$$

Problem 6 implies that

$$\left|a_{ij}^{(p)}\right| \leq \frac{4(2n-1)r^{p+1}}{\pi(p+1)} \max_{|\lambda|=r} \left|\frac{p_{ij}(\lambda)}{p(\lambda)}\right| \leq \frac{4(2n-1)r^{p+1}K}{\pi(p+1)(r-1)}.$$

Choose $r = 1 + \frac{1}{p+1}$ to obtain

$$\left|a_{ij}^{(p)}\right| \leq \frac{4(2n-1)eK}{\pi}, \quad i, j = 1, \ldots, n, \ p = 0, 1, \ldots, \ A \in \mathcal{A}. \tag{3.4.16}$$

Hence, $\|A^p\| \leq \frac{4(2n-1)eK}{\pi}$. $\square$

Problems

1. Use the proof of Theorem 3.4.2 to prove Theorem 3.4.3.

2. Prove Theorem 3.4.4.

3. Let $A \in \mathbb{C}^{n \times n}$. Show the Neumann series converge to the resolvent (3.4.14) for any $|\lambda| > \rho(A)$. (You may use (3.4.1).) Prove (3.4.15) for any vector norm on $\mathbb{C}^{n \times n}$.

4. Let $f(x)$ be a real continuous periodic function on $\mathbb{R}$ with period 2π. Assume furthermore that f' is a continuous function on $\mathbb{R}$. (f' is periodic of period 2π.) Then the Fourier series of f converge to f (e.g. [Pin09, Corollary 1.2.28]).

$$f(\theta) = \sum_{k \in \mathbb{Z}} a_k e^{\sqrt{-1}k\theta},$$

$$a_k = \bar{a}_k = \frac{1}{2\pi} \int_{\theta_0}^{\theta_0 + 2\pi} f(\theta) e^{-\sqrt{-1}k\theta} d\theta, \ k \in \mathbb{Z}, \ \theta_0 \in \mathbb{R}.$$
$$\tag{3.4.17}$$

Use integration by parts to conclude that

$$a_k = \frac{1}{2\pi\sqrt{-1}k} \int_{\theta_0}^{\theta_0 + 2\pi} f'(\theta) e^{-\sqrt{-1}k\theta} d\theta, \ k \in \mathbb{Z}\backslash\{0\}. \tag{3.4.18}$$

Assume that $f'(\theta)$ vanishes exactly at $m(\geq 2)$ points on the interval $[0, 2\pi)$. Show that

$$|a_0| \leq \max_{\theta \in [0,2\pi)} |f(\theta)|,$$

$$|a_k| \leq \frac{m}{\pi|k|} \max_{\theta \in [0,2\pi)} |f(\theta)|, \ \text{for all } k \in \mathbb{Z}\backslash\{0\}. \tag{3.4.19}$$

(*Hint*: The first inequality of (3.4.19) follows immediately from (3.4.17). Assume that f' vanishes at $0 \leq \theta_0 < \cdots < \theta_{m-1} <$

$2\pi \leq \theta_m = \theta_0 + 2\pi$. Then

$$\left| \int_{\theta_{i-1}}^{\theta_i} f'(\theta) e^{-2\pi\sqrt{-1}k\theta} d\theta \right| \leq \int_{\theta_{i-1}}^{\theta_i} |f'(\theta)| d\theta$$

$$= |f(\theta_i) - f(\theta_{i-1})|$$

$$\leq 2 \max_{\theta \in [0, 2\pi)} |f(\theta)|, \quad i = 1, \ldots, m.$$

Use (3.4.18) to deduce the second part of (3.4.19).)

5. A real periodic function f is called a trigonometric polynomial of degree n if f has the expansion (3.4.17), where $a_k = 0$ for $|k| > n$ and $a_n \neq 0$. Show

  (a) A nonzero trigonometric polynomial $f(\theta)$ of degree n vanishes at most $2n$ points on the interval $[0, 2\pi)$. (*Hint:* Let $z = e^{\sqrt{-1}\theta}$. Then $f = z^{-n} p(z)|_{|z|=1}$ for a corresponding polynomial p of degree $2n$.)

  (b) Let $f(\theta) = \frac{g(\theta)}{h(\theta)}$ be a nonconstant function, where g is a nonzero trigonometric polynomial of degree m at most and h is a nowhere vanishing trigonometric polynomial of degree n. Show that f' has at most $2(m + n)$ zeros on $[0, 2\pi)$.

6. Let $p(z), q(z)$ be nonconstant polynomials of degree m, n respectively. Suppose that $q(z)$ does not vanish on the circle $|z| = r > 0$. Let $M := \max_{|z|=r} \frac{|p(z)|}{|q(z)|}$. Show that for all $k \in \mathbb{Z}$

$$\left| \frac{1}{2\pi} \int_0^{2\pi} \frac{p(re^{\sqrt{-1}\theta})}{q(re^{\sqrt{-1}\theta})} e^{\sqrt{-1}k\theta} d\theta \right| \leq \frac{4M \max(m+n, 2n-1)}{\pi \max(|k|, 1)}.$$

$$(3.4.20)$$

Hint: Let $F(z) = \frac{p(z)}{q(z)} = \frac{p(z)\bar{q}(z)}{q(z)\bar{q}(z)}$ be a nonconstant rational function. Then $F(re^{\sqrt{-1}\theta}) = f_1(\theta) + \sqrt{-1}f_2(\theta)$, where f_1, f_2 as in

Problem 5. Clearly $|f_1(\theta)|, |f_2(\theta)| \leq M$. Observe next that

$$f_1'(\theta) + \sqrt{-1}f_2'(\theta) = \frac{\sqrt{-1}re^{\sqrt{-1}\theta}(p'(re^{\sqrt{-1}\theta})q(re^{\sqrt{-1}\theta})}{|q(re^{\sqrt{-1}\theta})|^2} \cdot \frac{-p(re^{\sqrt{-1}\theta})q'(re^{\sqrt{-1}\theta})(q(re^{\sqrt{-1}\theta}))^2}{|q(re^{\sqrt{-1}\theta})|^2}.$$

Hence, f_1', f_2' vanish at most $2\max(m+n, 2n-1)$ points on $[0, 2\pi)$. Use (3.4.19) for f_1, f_2 to deduce (3.4.20).

7. Let $\alpha \geq 0$ be fixed and assume that $\mathcal{A} \subset \mathbb{C}^{n \times n}$. Show that the following statements are equivalent:

$$\|A^k\| \leq k^\alpha K, \quad \text{for any } A \in \mathcal{A} \text{ and } k = 0, 1, \dots, \qquad (3.4.21)$$

$$\|(\lambda I - A)^{-1}\| \leq \frac{K'|\lambda|^\alpha}{(|\lambda| - 1)^{1+\alpha}}, \quad \text{for all } A \in \mathcal{A} \text{ and } |\lambda| > 1.$$
$$(3.4.22)$$

Hint: Use the fact that $(-1)^k \binom{-(1+\alpha)}{k} k^{-\alpha} \in [a, b], \ k = 1, \dots$ for some $0 < a < b$, (e.g. [Olv74], p. 119).

8. Let $A \in \mathbb{C}^{n \times n}$. Using (3.4.1) deduce

$$Z_{i(s_i-1)} = \lim_{x \to \lambda_i} (x - \lambda_i)^{s_i} (xI - A)^{-1}, \quad i = 1, \dots, \ell. \quad (3.4.23)$$

Let $R(x, A) = [r_{\mu\nu}]$. Using the definitions of Problem 3 show

$$z_{\mu\nu}^{i(s_i-1)} = \lim_{x \to \lambda_i} (x - \lambda_i)^s \, r_{\mu\nu}(x), \quad \text{if } s = \text{index}_{\mu\nu}\lambda_i > 0.$$
$$(3.4.24)$$

9. A set $\mathcal{A} \subset \mathbb{C}^{n \times n}$ is called *exponentially* stable if

$$\lim_{T \to \infty} \sup_{t \geq T} \|e^{At}\| = 0. \qquad (3.4.25)$$

Show that a compact set $\mathcal{A}$ is exponentially stable if and only if $\Re\lambda < 0$ for each $\lambda \in \text{spec}(A)$ and each $A \in \mathcal{A}$.

10. A matrix $B \in \mathbb{C}^{n \times n}$ is called projection (idempotent) if $B^2 = B$. Let Γ be a set of simply connected rectifiable curves such that Γ

from a boundary of an open bounded set $D \subset \mathbb{C}$. Let $A \in \mathbb{C}^{n \times n}$ and assume that $\Gamma \cap \operatorname{spec}(A) = \emptyset$. Define

$$P_D(A) := \frac{1}{2\pi\sqrt{-1}} \int_\Gamma R(x, A)dx, \qquad (3.4.26)$$

$$A(D) := \frac{1}{2\pi\sqrt{-1}} \int_\Gamma R(x, A)xdx.$$

Show that $P_D(A)$ is a projection. $P_D(A)$ is called the projection of A on D, and $A(D)$ is called the restriction of A to D. Prove

$$P_D(A) = \sum_{\lambda_i \in \operatorname{spec}_D(A)} Z_{i0}, \quad A(D) = \sum_{\lambda_i \in \operatorname{spec}_D(A)} (\lambda_i Z_{i0} + Z_{i1}).$$

$$(3.4.27)$$

Show that the rank of $P_D(A)$ is equal to the number of eigenvalues of A in D counted with their multiplicities. Prove that there exists a neighborhood of A such that $P_D(B)$ and $B(D)$ are analytic functions in B in this neighborhood. In particular, if D satisfies the assumptions of Theorem 3.4.4 then $P_D(A)$ is called the projection of A on λ_p: $P_D(A) = Z_{p0}$.

11. Let $B = QAQ^{-1} \in \mathbb{C}^{n \times n}$. Assume that D satisfies the assumptions of Problem 10. Show that $P_D(B) = QP_D(A)Q^{-1}$.

12. Let $A \in \mathbb{C}^{n \times n}$ and assume that the minimal polynomial $\psi(x)$ of A is given by (2.5.11). Let $\mathbb{C}^n = \mathbf{U}_1 \oplus \cdots \oplus \mathbf{U}_\ell$, where each $\mathbf{U}_p$ is an invariant subspace of A ($A\mathbf{U}_p \subset \mathbf{U}_p$), such that the minimal polynomial of $A|\mathbf{U}_p$ is $(x - \lambda_p)^{s_p}$. Show that

$$\mathbf{U}_p = Z_{p0}\mathbb{C}^n. \qquad (3.4.28)$$

Hint: It is enough to consider the case where A is in the Jordan canonical form.

13. Let D_i satisfy the assumptions of Problem 10 for $i = 1, \ldots, k$. Assume that $D_i \cap D_j = \emptyset$ for $i \neq j$. Show that $P_{D_i}(A)\mathbb{C}^n \cap P_{D_j}(A)\mathbb{C}^n = \{0\}$ for $i \neq j$. Assume furthermore that

$D_i \cap \operatorname{spec}(A) \neq \emptyset$, $i = 1, \ldots, k$, and $\operatorname{spec}(A) \subset \cup_{i=1}^{k} D_i$. Let

$$P_{D_i}(A)\mathbb{C}^n = \operatorname{span}\left(\mathbf{y}_1^{(i)}, \ldots, \mathbf{y}_{n_i}^{(i)}\right), \quad i = 1, \ldots, k,$$

$$X = \left[\mathbf{y}_1^{(1)}, \ldots, \mathbf{y}_{n_1}^{(1)}, \ldots, \mathbf{y}_{n_k}^{(k)}\right] \in \mathbb{C}^{n \times n}.$$

Show that

$$X^{-1}AX = \sum_{i=1}^{k} \oplus B_i, \quad \operatorname{spec}(B_i) = D_i \cap \operatorname{spec}(A), \quad i = 1, \ldots, k.$$

$$(3.4.29)$$

14. Let $A \in \mathbb{C}^{n \times n}$ and $\lambda_p \in \operatorname{spec}(A)$. Show that if index $\lambda_p = 1$ then

$$Z_{p0} = \prod_{\lambda_j \in \operatorname{spec}(A), \lambda_j \neq \lambda_p} \frac{(A - \lambda_j I)^{s_j}}{(\lambda_p - \lambda_j)^{s_j}}, \quad s_j = \text{index } \lambda_j.$$

Hint: Use the Jordan canonical form of A.

3.5 A Canonical Form over $\mathbf{H}_A$

Consider the space $\mathbb{C}^{n \times n}$. Clearly $\mathbb{C}^{n \times n}$ can be identified with $\mathbb{C}^{n^2}$. As in Example 1.1.3 denote by $\mathbf{H}_A$ the set of analytic functions $f(B)$, where B ranges over a neighborhood $D(A, \rho)$ of the form (2.9.5) $(\rho = \rho(f) > 0)$. Thus $B = [b_{ij}]$ is an element in $\mathbf{H}_A^{n \times n}$. Let $C \in \mathbf{H}_A^{n \times n}$ and assume that $C = C(B)$ is similar to B over $\mathbf{H}_A$. Then

$$C(B) = X^{-1}(B)BX(B), \qquad (3.5.1)$$

where $X(B) \in \mathbf{H}_A^{n \times n}$ and $\det X(A) \neq 0$. We want to find a "simple" form for $C(B)$ (simpler than B!). Let $\mathcal{M}_A$ be the quotient field of $\mathbf{H}_A$ (the set of meromorphic functions in the neighborhood of A). If we let $X \in \mathcal{M}_A^{n \times n}$ then we may take $C(B)$ to be $R(B)$ — the rational canonical form of B (2.3.3). According to Theorem 2.3.8 $R(B) \in \mathbf{H}_A^{n \times n}$. However, B and $R(B)$ are not similar over $\mathbf{H}_A$ in general. (We shall give below the necessary and sufficient conditions for $B \approx R(B)$ over $\mathbf{H}_A$.) For $C(B) = [c_{ij}(B)]$ we may ask how many independent variables are among $c_{ij}(B)$, $i, j = 1, \ldots, n$. For

$X(B) = I$ the number of independent variables in $C(B) = B$ is n^2. Thus we call $C(B)$ to be simpler than B if $C(B)$ contains less independent variable than B. For a given $C(B)$ we can view $C(B)$ as a map

$$C(\cdot) : D(A, \rho) \to \mathbb{C}^{n \times n}, \tag{3.5.2}$$

where $D(A, \rho)$ is given by (2.9.5), for some $\rho > 0$. It is well known, e.g. [GuR65], that the number of independent variables is equal to the rank of the Jacobian matrix $DC(\cdot)$ over $\mathcal{M}_A$

$$DC(B) := \left(\frac{\partial \mu(C)}{\partial b_{ij}}(B) \right) \in H_A^{n^2 \times n^2}, \tag{3.5.3}$$

where μ is the map given in Problem 2.8.2.

Definition 3.5.1 *Let* rank DC, rank $DC(A)$ *be the ranks of* $DC(\cdot)$, $DC(A)$ *over the fields* $\mathcal{M}_A$, $\mathbb{C}$ *respectively.*

Lemma 3.5.2 *Let* $C(B)$ *be similar to* B *over* H_A. *Then*

$$\text{rank } DC(A) \geq \nu(A, A). \tag{3.5.4}$$

Proof. Differentiating the relation $X^{-1}(B)X(B) = I$ with respect to b_{ij} we get

$$\frac{\partial X^{-1}}{\partial b_{ij}} = -X^{-1} \frac{\partial X}{\partial b_{ij}} X^{-1}.$$

So

$$\frac{\partial C}{\partial b_{ij}} = X^{-1} \left(-\left(\frac{\partial X}{\partial b_{ij}} X^{-1} \right) B + B \left(\frac{\partial X}{\partial b_{ij}} X^{-1} \right) + E_{ij} \right) X, \tag{3.5.5}$$

where

$$E_{ij} = [\delta_{i\alpha} \delta_{j\beta}]_{\alpha, \beta=1}^{m,n} \in \mathbb{C}^{m \times n}, \quad i = 1, \ldots, m, \; j = 1, \ldots, n, \tag{3.5.6}$$

and $m = n$. So

$$X(A)\frac{\partial C}{\partial b_{ij}}(A)X^{-1}(A) = AP_{ij} - P_{ij}A + E_{ij}, \quad P_{ij} = \frac{\partial X}{\partial b_{ij}}(A)X^{-1}(A).$$

Clearly, $AP_{ij} - P_{ij}A$ is in Range $\tilde{A}$, where

$$\tilde{A} = (I \otimes A - A^\top \otimes I) : \mathbb{C}^{n \times n} \to \mathbb{C}^{n \times n}. \tag{3.5.7}$$

According to Definition 2.9.1 dim Range $\tilde{A} = r(A, A)$. Let

$$\mathbb{C}^{n \times n} = \text{Range } \tilde{A} \oplus \text{span } (\Gamma_1, \dots, \Gamma_{\nu(A,A)}). \tag{3.5.8}$$

As E_{ij}, $i, j = 1, \dots, n$ is a basis in $\mathbb{C}^{n \times n}$

$$\Gamma_p = \sum_{i,j=1}^n \alpha_{ij}^{(p)} E_{ij}, \quad p = 1, \dots, \nu(A, A).$$

Let

$$T_p := \sum_{i,j=1}^n \alpha_{ij}^{(p)} \frac{\partial C}{\partial b_{ij}}(A) = X^{-1}(A)(Q_p + \Gamma_p)X(A),$$

$$Q_p \in \text{Range } (\tilde{A}), \quad p = 1, \dots, \nu(A, A).$$

According to (3.5.8) $T_1, \dots, T_{\nu(A,A)}$ are linearly independent. Hence, (3.5.4) holds. $\qquad\square$

Clearly rank $DC \geq$ rank $DC(A) \geq \nu(A, A)$.

Theorem 3.5.3 *Let $A \in \mathbb{C}^{n \times n}$ and assume that $\Gamma_1, \dots, \Gamma_{\nu(A,A)}$ be any $\nu(A, A)$ matrices satisfying (3.5.8). Then for any nonsingular matrix $P \in \mathbb{C}^{n \times n}$ it is possible to find $X(B) \in \mathrm{H}_A^{n \times n}$, $X(A) = P$, such that*

$$X^{-1}(B)BX(B) = P^{-1}AP + \sum_{i=1}^{\nu(A,A)} f_i(B)P^{-1}\Gamma_i P,$$

$$\tag{3.5.9}$$

$$f_i \in \mathrm{H}_A, \ f_i(A) = 0, \quad i = 1, \dots, \nu(A, A).$$

Proof. Let $R_1, \ldots, R_{r(A,A)}$ be a basis in Range $\tilde{A}$. So there exist T_i such that $AT_i - T_iA = R_i$ for $i = 1, \ldots, r(A, A)$. Assume that $X(B)$ is of the form

$$X(B)P^{-1} = I + \sum_{j=1}^{r(A,A)} g_j(B)T_j, \quad g_j \in \mathrm{H}_A, \ g_j(A) = 0,$$

$$j = 1, \ldots, r(A, A). \quad (3.5.10)$$

The theorem will follow if we can show that the system

$$B\left(I + \sum_{j=1}^{r(A,A)} g_jT_j\right) = \left(I + \sum_{j=1}^{r(A,A)} g_jT_j\right)\left(A + \sum_{i=1}^{\nu(A,A)} f_i\Gamma_i\right)$$

$$(3.5.11)$$

is solvable for some $g_1, \ldots, g_{r(A,A)}, f_1, \ldots, f_{\nu(A,A)} \in \mathrm{H}_A$ which vanish at A. Clearly, the above system is trivially satisfied at $B = A$. The implicit function theorem implies that the above system is solved uniquely if the Jacobian of this system is nonsingular. Let $B = A + F$, $F = [f_{ij}] \in \mathbb{C}^{n \times n}$. Let $\alpha_i(F), \beta_j(F)$ be the linear terms of the Taylor expansions of $f_i(A + F), g_j(A + F)$. The linear part of (3.5.11) reduces to

$$F + \sum_{j=1}^{r(A,A)} \beta_j A T_j = \sum_{j=1}^{r(A,A)} \beta_j T_j A + \sum_{i=1}^{\nu(A,A)} \alpha_i \Gamma_i.$$

That is

$$F = \sum_{j=1}^{r(A,A)} \beta_j R_j + \sum_{i=1}^{\nu(A,A)} \alpha_i \Gamma_i.$$

In view of (3.5.8) $\alpha_1, \ldots, \alpha_{\nu(A,a)}, \beta_1, \ldots, \beta_{r(A,A)}$ are uniquely determined by F. $\qquad\square$

Note that if $A = aI$ then the form (3.5.10) is not simpler than B. Also by mapping $T \to P^{-1}TP$ we get

$$\mathbb{C}^{n \times n} = \mathrm{Range}\ \widetilde{P^{-1}AP} \oplus \mathrm{span}\ (P^{-1}\Gamma_1 P, \ldots, P^{-1}\Gamma_{\nu(A,A)}P).$$

$$(3.5.12)$$

Lemma 3.5.4 *Let $B \in \mathrm{H}_A^{n \times n}$. Then the rational canonical form of B over $\mathcal{M}_A$ is a companion matrix $C(p)$, where $p(x) = \det (xI - B)$.*

Proof. The rational canonical form of B is $C(p_1, \ldots, p_k)$ is given by (2.3.3). We claim that $k = 1$. Otherwise $p(x)$ and $p'(x)$ have a common factor over $\mathcal{M}_A$. In view of Theorem 2.1.9 implies that $p(x)$ and $p'(x)$ have a common factor over H_A. That is any $B \in D(A, \rho)$ has at least one multiple eigenvalue. Evidently this is false. Consider $C = P^{-1}BP$ where $P \in \mathbb{C}^{n \times n}$ and $J = P^{-1}AP$ is the Jordan canonical form of A. So, $C \in D(J, \rho')$. Choose C to be an upper diagonal. (This is possible since J is an upper diagonal matrix.) So, the eigenvalues of C are the diagonal elements of C, and we can choose them to be pairwise distinct. Thus $p(x)$ and $p'(x)$ are coprime over $\mathcal{M}_A$, hence $k = 1$. Furthermore $p_1(x) = \det (xI - C(p)) = \det (xI - B)$. $\qquad \square$

Theorem 3.5.5 *Let $A \in \mathbb{C}^{n \times n}$. Then $B \in \mathrm{H}_A^{n \times n}$ is similar to the companion matrix $C(p)$, $p(x) = \det (xI - B)$ over H_A if and only if $\nu(A, A) = n$. That is the minimal and the characteristic polynomial of A coincide, i.e. A is nonderogatory.*

Proof. Assume first that $C(B)$ in (3.5.1) can be chosen to be $C(p)$. Then for $B = A$ we obtain that A is similar to the companion matrix. Corollary 2.8.4 yields $\nu(A, A) = n$. Assume now that $\nu(A, A) = n$. According to (2.8.12) we have that $i_1(x) = i_2(x) = \cdots = i_{n-1}(x) = 1$. That is, the minimal and the characteristic polynomials of A coincide, i.e. A is similar to a companion matrix. Use (3.5.10) to see that we may assume that A is a companion matrix. Choose $\Gamma_i = E_{ni}$, $i = 1, \ldots, n$, where E_{ni} are defined in (3.5.6).

It is left to show that Range $\tilde{A} \cap$ span $(E_{n1}, \ldots, E_{nn}) = \{0\}$. Suppose that $\Gamma = \sum_{i=1}^{n} \alpha_i E_{ni} \in$ Range $(\tilde{A})$. Theorem 2.10.1 and Corollary 2.8.4 yield that $\mathrm{tr}\,\Gamma A^k = 0$, $k = 0, 1, \ldots, n - 1$. Let $\alpha = (\alpha_1, \ldots, \alpha_n)$. Since the first $n - 1$ rows of Γ are zero rows we have

$$0 = \operatorname{tr} \Gamma A^k = \alpha A^k \mathbf{e}_n, \quad \mathbf{e}_j = (\delta_{j1}, \ldots, \delta_{jn})^\top, \ j = 1, \ldots, n.$$

For $k = 0$ the above equality implies that $\alpha_n = 0$. Suppose that we already proved that these equalities for $k = 0, \ldots, \ell$ imply that $\alpha_n = \cdots = \alpha_{n-l} = 0$. Consider the equality $\operatorname{tr} \Gamma A^{\ell+1} = 0$. Use Problem 2.4.9 to deduce

$$A^{\ell+1} \mathbf{e}_n = \mathbf{e}_{n-\ell-1} + \sum_{j=0}^{\ell} f_{(\ell+1)j} \mathbf{e}_{n-j}.$$

So $\operatorname{tr} \Gamma A^{\ell+1} = \alpha_{n-\ell-1}$ as $\alpha_n = \cdots = \alpha_{n-\ell} = 0$. Thus $\alpha_{n-\ell-1} = 0$, which implies that $\Gamma = 0$.

Theorem 3.5.3 yields that

$$C(B) = X^{-1}(B) B X(B) = A + \sum_{i=1}^{n} f_i(B) E_{ni}.$$

So $C(B)$ is a companion matrix. As $\det (xI - C(B)) = \det (xI - B)$ it follows that $C(B) = C(p)$. $\qquad\square$

Problem 5 yields.

Lemma 3.5.6 *Let $A_i \in \mathbb{C}^{n_i \times n_i}$ $i = 1, 2$ and assume that*

$$\mathbb{C}^{n_i \times n_i} = \operatorname{Range} \tilde{A}_i \oplus \operatorname{span} \left(\Gamma_1^{(i)}, \ldots, \Gamma_{\nu(A_i, A_i)}^{(i)} \right), \quad i = 1, 2.$$

Suppose that A_1 and A_2 do not have a common eigenvalue. Then

$$\mathbb{C}^{(n_1+n_2) \times (n_1+n_2)} = \operatorname{Range} \widetilde{A_1 \oplus A_2} \oplus$$
$$\operatorname{span} \left(\Gamma_1^{(1)} \oplus 0, \ldots, \Gamma_{\nu(A_1, A_1)}^{(1)} \oplus 0, 0 \oplus \Gamma_1^{(2)}, \ldots, 0 \oplus \Gamma_{\nu(A_2, A_2)}^{(2)} \right).$$

Theorem 3.5.7 *Let $A \in \mathbb{C}^{n \times n}$. Assume that* spec (A) *consists of ℓ distinct eigenvalues $\lambda_1, \ldots, \lambda_\ell$, where the multiplicity of λ_i is n_i*

for $i = 1, \ldots, \ell$. *Then* B *is similar over* H_A *to the matrix*

$$C(B) = \sum_{i=1}^{\ell} \oplus C_i(B), \ C_i(B) \in H_A^{n_i \times n_i},$$

$$(\lambda_i I_{n_i} - C_i(A))^{n_i} = 0, \ i = 1, \ldots, \ell. \tag{3.5.13}$$

Moreover, $C(A)$ *is the Jordan canonical form of* A.

Proof. Choose P in the equality (3.5.10) such that $P^{-1}AP$ is the Jordan canonical of A and each $P^{-1}\Gamma_i P$ is of the form $\sum_{j=1}^{\ell} \Gamma_i^{(j)}$ as follows from Lemma 3.5.6. Then (3.5.10) yields the theorem. $\square$

Problems

1. Let $A = \sum_{i=1}^{k} \oplus H_{n_i}$, $n = \sum_{i=1}^{k} n_i$. Partition any $B \in \mathbb{C}^{n \times n}$ as a block matrix as A: $B = [B_{ij}]$, $B_{ij} \in \mathbb{C}^{n_i \times n_j}$, $i, j = 1, \ldots, k$. Using the results of Theorem 2.8.3 and Theorem 2.10.1 show that the matrices

$$\Gamma_{\alpha,\beta,\gamma} = [\Gamma_{ij}^{(\alpha,\beta,\gamma)}]_1^k \in \mathbb{C}^{n \times n},$$

$$\Gamma_{ij}^{(\alpha,\beta,\gamma)} = 0 \in \mathbb{C}^{n_i \times n_j}, \quad \text{if } (\alpha,\beta) \neq (i,j),$$

$$\Gamma_{\alpha\beta}^{(\alpha,\beta,\gamma)} = E_{n_\alpha \gamma} \in \mathbb{C}^{n_\alpha \times n_\beta},$$

$$\gamma = 1, \ldots, \min(n_\alpha, n_\beta), \ \alpha, \beta = 1, \ldots, k,$$

   satisfy (3.5.8).

2. Let A be a matrix given by (2.8.4). Use Theorem 3.5.7 and Problem 1 to find a set of matrices $\Gamma_1, \ldots, \Gamma_{\nu(A,A)}$ which satisfy (3.5.8).

3. Let $A \in \mathbb{C}^{n \times n}$ and assume that λ_i is a simple eigenvalue of A, i.e. λ_i is a simple root of the characteristic polynomial of A. Use Theorem 2.8.3 to show the existence of $\lambda(B) \in H_A$ such that $\lambda(B)$ is an eigenvalue of B and $\lambda(A) = \lambda_i$.

4. Let A satisfy the assumptions of Theorem 3.5.7. Denote by D_p an open set satisfying the assumptions of Theorem 3.4.4 for

$p = 1, \ldots, \ell$. Let $P_k(B)$ be the projection of $B \in \mathrm{H}_A^{n \times n}$ on D_k, $k = 1, \ldots, \ell$. Problem 10 implies that $P_k(B) \in \mathrm{H}_A^{n \times n}$, $k = 1, \ldots, \ell$. Let $P_k(A)\mathbb{C}^n = \mathrm{span}\,(\mathbf{x}^{k1}, \ldots, \mathbf{x}^{kn_k})$, $k = 1, \ldots, \ell$, $B \in D(A, \rho)$, where ρ is some positive number. Let $X(B) \in \mathrm{H}_A^{n \times n}$ be formed ·by the columns $P_k(B)\mathbf{x}^{k1}, \ldots, P_k(B)\mathbf{x}^{kn_k}$, $k = 1, \ldots, \ell$. Show that $C(B)$ given by (3.5.1) satisfies (3.5.13). (This yields another proof of Theorem 3.5.7.)

3.6 Analytic, Pointwise and Rational Similarity

Definition 3.6.1 *Let $\Omega \subset \mathbb{C}^m$ and $A, B \in \mathrm{H}(\Omega)^{n \times n}$. Then*

(a) *A and B are called analytically similar, denoted by $A \overset{a}{\approx} B$, if A and B are similar over $\mathrm{H}(\Omega)$.*

(b) *A and B are called pointwise similar, denoted by $A \overset{p}{\approx} B$, if $A(x)$ and $B(x)$ are similar over $\mathbb{C}$ for all $x \in \Omega_0$, for some open set $\Omega_0 \supset \Omega$.*

(c) *A and B are called rationally similar, denoted by $A \overset{r}{\approx} B$, if A and B are similar over the field of meromorphic functions $\mathcal{M}(\Omega)$.*

Theorem 3.6.2 *Let $\Omega \subset \mathbb{C}^m$ and assume that $A, B \in \mathrm{H}(\Omega)^{n \times n}$. Then*

$$A \overset{a}{\approx} B \Rightarrow A \overset{p}{\approx} B \Rightarrow A \overset{r}{\approx} B.$$

Proof. Suppose that

$$B(x) = P^{-1}(x)A(x)P(x), \tag{3.6.1}$$

where $P, P^{-1} \in \mathrm{H}(\Omega)^{n \times n}$. Let $x_0 \in \Omega$. Then (3.6.1) holds in some neighborhood of x_0. So $A \overset{p}{\approx} B$. Assume now that $A \overset{p}{\approx} B$. Let $C(p_1, \ldots, p_k)$ and $C(q_1, \ldots, q_\ell)$ be the rational canonical forms of A

and B respectively over $\mathcal{M}(\Omega)$. Then

$$C(p_1, \ldots, p_k) = S(x)^{-1} A(x) S(x),$$
$$C(q_1, \ldots, q_\ell) = T(x)^{-1} B(x) T(x),$$
$$S(x), T(x) \in \mathrm{H}(\Omega)^{n \times n}, \quad \det A(x) \not\equiv 0, \ \det B(x) \not\equiv 0.$$

Theorem 2.3.8 yields that $C(p_1, \ldots, p_k), C(q_1, \ldots, q_\ell) \in \mathrm{H}(\Omega)^{n \times n}$. Let $\Omega_0 \supset \Omega$ be an open set such that $A, B, S, T \in \mathrm{H}(\Omega_0)^{n \times n}$ and $A(x)$ and $B(x)$ are similar over $\mathbb{C}$ for any $x \in \Omega_0$. Let $x_0 \in \Omega_0$ be a point such that $\det S(x_0) T(x_0) \neq 0$. Then for all $x \in D(x_0, \rho)$ $C(p_1, \ldots, p_k) = C(q_1, \ldots, q_\ell)$. The analyticity of $C(p_1, \ldots, p_k)$ and $C(q_1, \ldots, q_\ell)$ imply that these matrices are identical in $\mathrm{H}(\Omega)$, i.e. $A \overset{r}{\approx} B$. $\qquad\square$

Assume that $A \overset{a}{\approx} B$. Then according to Lemma 2.9.4 the three matrices

$$I \otimes A(x) - A(x)^\top \otimes I, \quad I \otimes A(x) - B(x)^\top \otimes I,$$
$$I \otimes B(x) - B(x)^\top \otimes I \qquad (3.6.2)$$

are equivalent over $\mathrm{H}(\Omega)$. Theorem 2.9.3 yields.

Theorem 3.6.3 *Let* $A, B \in \mathrm{H}(\Omega)^{n \times n}$. *Assume that the three matrices in* (3.6.2) *are equivalent over* $\mathrm{H}(\Omega)$. *Then* $A \overset{p}{\approx} B$.

Assume that $\Omega \subset \mathbb{C}$ is a domain. Then $\mathrm{H}(\Omega)$ is EDD. Hence, we can determine when these matrices are equivalent.

The problem of finding a canonical form of $A \in \Omega^{n \times n}$ under analytic similarity is a very hard problem. This problem for the ring of local analytic functions in one variable will be discussed in the next sections. We now determine when A is analytically similar to its rational canonical form over H_ζ, the ring of local analytic functions in the neighborhood of $\zeta \in \mathbb{C}^m$.

For $A, B \in \mathrm{H}(\Omega)^{n \times n}$ denote by $r(A, B)$ and $\nu(A, B)$ the rank and the nullity of the matrix $C = I \otimes A - B^\top \otimes I$ over the field $\mathcal{M}(\Omega)$. Denote by $r(A(x), B(x))$ and $\nu(A(x), B(x))$ the rank of $C(x)$ over $\mathbb{C}$.

As the rank of $C(x)$ is the largest size of a nonvanishing minor, we deduce

$$r(A(\zeta), B(\zeta)) \leq r(A(x), B(x)) \leq r(A, B)$$
$$\nu(A, B) \leq \nu(A(x), B(x)) \leq \nu(A(\zeta), B(\zeta)), \quad x \in D(\zeta, \rho) \tag{3.6.3}$$

for some positive ρ. Moreover, for any $\rho > 0$ there exists at least one $x_0 \in D(\zeta, \rho)$ such that

$$r(A(x_0), B(x_0)) = r(A, B), \quad \nu(A(x_0, B(x_0)) = \nu(A, B). \tag{3.6.4}$$

Theorem 3.6.4 *Let $\zeta \in \mathbb{C}^m$ and $A \in H_\zeta^{n \times n}$. Assume that $C(p_1, \ldots, p_k)$ is the rational canonical form of A over $\mathcal{M}_\zeta$ and $C(\sigma_1, \ldots, \sigma_\ell)$ is the rational canonical form of $A(\zeta)$ over $\mathbb{C}$. That is $p_i = p_i(\lambda, x)$ and $\sigma_j(\lambda)$ are normalized polynomials in λ belonging to $H_\zeta[\lambda]$ and $\mathbb{C}[\lambda]$ respectively for $i = 1, \ldots, k$ and $j = 1, \ldots, \ell$. Then*

(a) $\ell \geq k$;

(b) $\prod_{i=0}^{q} \sigma_{\ell-i}(\lambda) | \prod_{i=0}^{q} p_{k-i}(\lambda, \zeta)$ *for $q = 0, 1, \ldots, k-1$.*

Moreover, $\ell = k$ and $p_i(\lambda, \zeta) = \sigma_i(\lambda)$ for $i = 1, \ldots, k$ if and only if

$$r(A(\zeta), B(\zeta)) = r(A, B), \quad \nu(A(\zeta), B(\zeta)) = \nu(A, B), \tag{3.6.5}$$

which is equivalent to the condition

$$r(A(\zeta), B(\zeta)) = r(A(x), B(x)), \tag{3.6.6}$$
$$\nu(A(\zeta), B(\zeta)) = \nu(A(x), B(x)), \quad x \in D(\zeta, \rho),$$

for some positive ρ.

Proof. Let

$$u_{n-k+i}(\lambda, x) = \prod_{\alpha=1}^{i} p_\alpha(\lambda, x), \quad v_{n-\ell+j}(\lambda) = \prod_{\beta=1}^{j} \sigma_\beta(\lambda),$$

$$i = 1, \ldots, k, \ j = 1, \ldots, \ell,$$

$$u_\alpha(\lambda, x) = v_\beta(\lambda) = 1, \quad \text{for} \quad \alpha \leq n - k, \ \beta \leq n - \ell.$$

So $u_i(\lambda, x)$ and $v_i(\lambda)$ are the g.c.d. of all minors of order i of matrices $\lambda I - A$ and $\lambda I - A(\zeta)$ over the rings $\mathcal{M}_\zeta[\lambda]$ and $\mathbb{C}[\lambda]$ respectively. As $u_i(\lambda, x) \in H_\zeta[\lambda]$ it is clear that $u_i(\lambda, \zeta)$ divides all the minors of

$\lambda I - A(\zeta)$ of order i. So $u_i(\lambda, \zeta)|v_i(\lambda)$ for $i = 1, \ldots, n$. Since $v_{n-\ell} = 1$ it follows that $u_{n-\ell}(\lambda, x) = 1$. Hence, $k \leq \ell$. Furthermore

$$u_n(\lambda, x) = \det(\lambda I - A(x)), \quad v_n(\lambda) = \det(\lambda I - A(\zeta)).$$

Therefore, $u_n(\lambda, \zeta) = v_n(\lambda)$ and $\frac{v_n(\lambda)}{v_i(\lambda)} | \frac{u_n(\lambda, \zeta)}{u_i(\lambda, \zeta)}$. This establishes claims (a) and (b) of the theorem. Clearly if $C(q_1, \ldots, q_\ell) = C(p_1, \ldots, p_k)(\zeta)$ then $k = \ell$ and $p_i(\lambda, \zeta) = q_i(\lambda)$ for $i = 1, \ldots, \ell$. Assume now that (3.6.5) holds. According to (2.8.12)

$$\nu(A, A) = \sum_{i=1}^{k} (2i - 1) \deg p_{k-i+1}(\lambda, x),$$

$$\nu(A(\zeta), A(\zeta)) = \sum_{j=1}^{\ell} (2j - 1) \deg q_{\ell-j+1}(\lambda).$$

Note that the degrees of the invariant polynomials of $\lambda I - A$ and $\lambda I - A(\zeta)$ satisfy the assumptions of Problem 2. From the results of Problem 2 it follows that the second equality in (3.6.5) holds if and only if $k = \ell$ and $\deg p_i(\lambda, x) = \deg q_i(\lambda)$ for $i = 1, \ldots, k$. Finally, (3.6.3) and (3.6.4) imply the equivalence of the conditions of (3.6.5) and (3.6.6). $\square$

Corollary 3.6.5 *Let* $A \in H_\zeta^{n \times n}$. *Assume that* (3.6.6) *holds. Then* $A \overset{a}{\approx} B$ *if and only if* $A \overset{p}{\approx} B$.

Proof. According to Theorem 3.6.2 it is enough to show that $A \overset{p}{\approx} B$ implies that $A \overset{a}{\approx} B$. Since A satisfies (3.6.6) the assumption that $A \overset{p}{\approx} B$ implies that B satisfies (3.6.6) too. According to Theorem 3.6.4, A and B are analytically similar to their canonical rational form. From Theorem 3.6.2 it follows that A and B have the same rational canonical form. $\square$

Problems

1. Let

$$A(x) = \begin{bmatrix} 0 & x \\ 0 & 0 \end{bmatrix}, \quad B(x) = \begin{bmatrix} 0 & x^2 \\ 0 & 0 \end{bmatrix}.$$

Show that $A(x)$ and $B(x)$ are rationally similar over $\mathbb{C}(x)$ to $H_2 = A(1)$. Prove that

$$A \overset{p}{\not\gtrsim} H_2, \quad B \overset{p}{\not\gtrsim} H_2, \quad A \overset{p}{\approx} B, \quad A \overset{a}{\not\gtrsim} B$$

over $\mathbb{C}[x]$.

2. Let n be a positive integer and assume that $\{\ell_i\}_1^n$, $\{m_i\}_1^n$ are two nonincreasing sequences of non-negative integers satisfying

$$\sum_{i=1}^k \ell_i \le \sum_{i=1}^k m_i, \quad k = 1, \ldots, n-1,$$

$$\sum_{i=1}^n \ell_i = \sum_{i=1}^n m_i.$$

Show (by induction) that

$$\sum_{i=1}^n (2i-1)m_i \le \sum_{i=1}^n (2i-1)\ell_i$$

and equality holds if and only if $\ell_i = m_i$, $i = 1, \ldots, n$.

3. Let $\zeta_n \in \mathbb{C}$, $n = 1, \ldots$, and $\lim_{n \to \infty} \zeta_n = \zeta$. Suppose that $\Omega \subset \mathbb{C}$ is a connected set and $\zeta_n \in \Omega$, $n = 1, \ldots, \zeta \in \Omega$. Recall that if $f \in H(\Omega)$ and $f(\zeta_n) = 0$, $n = 1, \ldots$, then $f = 0$. Show that for $A, B \in H(\Omega)^{n \times n}$ the assumption that $A(\zeta_n) \approx B(\zeta_n)$, $n = 1, \ldots$, implies that $A \overset{r}{\approx} B$.

3.7   A Global Splitting

From this section to the end of the chapter we assume that Ω is a domain in $\mathbb{C}$. We now give a global version of Theorem 3.5.7.

Theorem 3.7.1 *Let $A \in H(\Omega)^{n \times n}$. Suppose that*

$$\det (\lambda I - A(x)) = \phi_1(\lambda, x)\phi_2(\lambda, x), \qquad (3.7.1)$$

where ϕ_1, ϕ_2 are two nontrivial normalized polynomials in $H(\Omega)[\lambda]$ of positive degrees n_1 and n_2 respectively. Assume that $(\phi_1(\lambda, x_0),$

$\phi_2(\lambda, x_0)) = 1$ *for each* $x_0 \in \Omega$. *Then there exists* $X \in \mathbf{GL}(n, \mathrm{H}(\Omega))$ *such that*

$$X^{-1}(x)C(x)X(x) = C_1(x) \oplus C_2(x),$$

$$C_i(x) \in \mathrm{H}(\Omega)^{n_i \times n_i}, \quad \det(\lambda I - C_i(x)) = \phi_i(\lambda, x), \quad i = 1, 2.$$
$$(3.7.2)$$

Proof. Let $P_i(x)$ be the projection of $A(x)$ on the eigenvalues of $A(x)$ satisfying $\phi_i(\lambda, x) = 0$. Since $(\phi_1(\lambda, x_0), \phi_2(\lambda, x_0)) = 1$ it follows that $P_i(x) \in \mathrm{H}(\Omega)^{n \times n}$ for $i = 1, 2$. (See Problem 3.4.10.) Also for any x_0 the rank of $P_i(x_0)$ is n_i. Since $\mathrm{H}(\Omega)$ is *EDD* each $P_i(x)$ can be brought to the Smith normal form

$$P_i(x) = U_i(x) \operatorname{diag}\left(\epsilon_1^{(i)}(x), \ldots, \epsilon_{n_i}^{(i)}(x), 0, \ldots, 0\right) V_i(x)),$$

$$U_i, V_i \in \mathbf{GL}(n_i, \mathrm{H}(\Omega)), \ i = 1, 2.$$

As rank $P_i(x_0) = n_i$ for any $x_0 \in \Omega$ we deduce that $\epsilon_j^{(i)} = 1$, $j = 1, \ldots, n$, $i = 1, 2$. Let $\mathbf{u}_1^{(i)}(x), \ldots, \mathbf{u}_n^{(i)}(x)$ be the columns of $U_i(x)$ for $i = 1, 2$. As $V \in \mathbf{GL}(n, \mathrm{H}(\Omega))$ we obtain

$$P_i(x)\mathbb{C}^n = \operatorname{span}(\mathbf{u}_1^{(i)}(x), \ldots, \mathbf{u}_{n_i}^{(i)}(x)), \qquad (3.7.3)$$

for any $x \in \Omega$. Let

$$X(x) = \left[\mathbf{u}_1^{(1)}(x), \ldots, \mathbf{u}_{n_1}^{(1)}(x), \mathbf{u}_1^{(2)}(x), \ldots, \mathbf{u}_{n_2}^{(2)}(x)\right] \in \mathrm{H}(\Omega)^{n \times n}.$$

According to Problem 3.4.13 $\det X(x_0) \neq 0$ for any $x_0 \in \mathrm{H}(\Omega)$. So $X(x) \in \mathbf{GL}(n, \mathrm{H}(\Omega))$. Then (3.7.2) follows from (3.4.29). $\square$

3.8   First Variation of a Geometrically Simple Eigenvalue

Theorem 3.8.1 *Let* $A(x)$ *be a continuous family of* $n \times n$ *complex valued matrices for* $|x - x_0| < \delta$, *where the parameter* x *is either* *real or complex. Suppose that*

$$A(x) = A_0 + (x - x_0)A_1 + |x - x_0|o(1). \qquad (3.8.1)$$

Assume furthermore that λ_0 is a geometrically simple eigenvalue of A_0 of multiplicity m. Let $\mathbf{x}_1, \ldots, \mathbf{x}_m$ and $\mathbf{y}_1, \ldots, \mathbf{y}_m$ be eigenvectors of A_0 and $A_0^\top$ respectively corresponding to λ_0, which form a biorthonormal system $\mathbf{y}_i^\top \mathbf{x}_j = \delta_{ij}$, $i, j = 1, \ldots, m$. Then it is possible to enumerate the eigenvalues of $A(x)$ by $\lambda_i(x)$, $i = 1, \ldots, n$, such that

$$\lambda_i(x) = \lambda_0 + (x - x_0)\mu_i + |x - x_0|o(1), \quad i = 1, \ldots, m, \qquad (3.8.2)$$

where $\mu_1, \ldots, \mu_m$ are the eigenvalues of the matrix

$$S = [s_{ij}] \in \mathbb{C}^{m \times m}, \quad s_{ij} = \mathbf{y}_i^\top A_1 \mathbf{x}_j, \ i, j = 1, \ldots, m. \qquad (3.8.3)$$

Proof. By considering the matrix $P^{-1}A(x)P$, for an appropriate $P \in \mathbf{GL}(n, \mathbb{C})$, we can assume that A_0 is in the Jordan canonical form such that the first m diagonal entries of A_0 are λ_0. The proofs of Theorems 3.5.3 and 3.5.7 imply the existence of

$$X(B) = I + Z(B), \quad Z \in \mathrm{H}_0^{n \times n}, \quad Z(0) = 0,$$

such that

$$X^{-1}(B)(A_0 + B)X(B) = \oplus_{i=1}^{\ell} C_i(B), \quad C_1(0) = \lambda_0 I_m. \qquad (3.8.4)$$

Substituting

$$B(x) = A(x) - A_0 = (x - x_0)A_1 + |x - x_0|o(1),$$
$$X(x) = X(B(x)) = I + (x - x_0)X_1 + |x - x_0|o(1),$$

we get

$$C(X) = X^{-1}A(x)X(x)$$
$$= A_0 + (x - x_0)(A_1 + A_0X_1 - X_1A_0) + |x - x_0|o(1).$$

According to (3.8.4) $\lambda_1(x), \ldots, \lambda_m(x)$ are the eigenvalues of $C_1(B(x))$. As $C_1(B(x_0)) = \lambda_0 I_m$, by considering $(C_1(B(x)) - \lambda_0 I_m)/(x - x_0)$ we deduce that $(\lambda_i(x) - \lambda_0)/(x - x_0)$ are continuous functions at x_0. Also

$$(C_1(B(x)) - \lambda_0 I_m)/(x - x_0)$$
$$= \left[\mathbf{v}_i^\top (A_1 + A_0 X_1 - X_1 A_0) \mathbf{u}_j \right]_{i,j=1}^m + o(1),$$

where $\mathbf{u}_i = \mathbf{v}_i = (\delta_{i1}, \ldots, \delta_{in})^\top$ for $i = 1, \ldots, m$. Since $\mathbf{u}_i$ and $\mathbf{v}_i$ are the eigenvectors of A_0 and $A_0^\top$ respectively corresponding to λ_0 for $i = 1, \ldots, m$, it follows that $\mathbf{v}_i^\top (A_0 X_1 - X_1 A_0) \mathbf{u}_j = 0$ for $i, j = 1, \ldots, m$. This establishes the result for a particular choice of eigenvectors $\mathbf{u}_1, \ldots, \mathbf{u}_m$ and $\mathbf{v}_1, \ldots, \mathbf{v}_m$. It is left to note that any other choice of the eigenvectors $\mathbf{x}_1, \ldots, \mathbf{x}_m$ and $\mathbf{y}_1, \ldots, \mathbf{y}_m$, which form a biorthonormal system amounts to a new matrix S_1 which is similar to S. In particular S and S_1 have the same eigenvalues. $\square$

Problem

1. Let $A(x) = \begin{bmatrix} 0 & 1 \\ x & 0 \end{bmatrix}$. Find the eigenvalues and the eigenvectors of $A(x)$ in terms of $\sqrt{x}$. Show that (3.8.2) does not apply for $x_0 = 0$ in this case. Let $B(x) = A(x^2)$. Show that (3.8.2) holds for x_0 even though $\lambda_0 = 0$ is not geometrically simple for $B(0)$.

3.9 Analytic Similarity over H_0

Let $A, B \in H_0^{n \times n}$. That is

$$A(x) = \sum_{k=0}^\infty A_k x^k, \quad |x| < r(A),$$

$$B(x) = \sum_{k=0}^\infty B_k x^k, \quad |x| < r(B).$$

$$(3.9.1)$$

Definition 3.9.1 *For $A, B \in H_0^{n \times n}$ let $\eta(A, B)$ and $\kappa_p(A, B)$ be the index and the number of local invariant polynomials of degree p of the matrix $I_n \otimes A(x) - B(x)^\top \otimes I_n$ respectively.*

Theorem 3.9.2 *Let $A, B \in H_0^{n \times n}$. Then A and B are analytically similar over H_0 if and only if A and B are rationally similar over H_0 and there exists $\eta(A, A) + 1$ matrices $T_0, \dots, T_\eta \in \mathbb{C}^{n \times n}$ $(\eta = \eta(A, A))$, such that $\det T_0 \neq 0$ and*

$$\sum_{i=0}^{k} A_i T_{k-i} - T_{k-i} B_i = 0, \quad k = 0, \dots, \eta(A, A). \tag{3.9.2}$$

Proof. The necessary part of the theorem is obvious. Assume now that $A(x) \overset{r}{\approx} B(x)$ and the matrices $T_0, \dots, T_\eta$ satisfy (3.9.2), where $T_0 \in \mathbf{GL}(n, \mathbb{C})$. Put

$$C(x) = T(x)B(x)T^{-1}(x), \quad T(x) = \sum_{k=0}^{\eta} T_k x^k.$$

As $\det T_0 \neq 0$ we deduce that $B(x) \overset{a}{\approx} C(x)$. Hence, $A(x) \overset{r}{\approx} C(x)$. In particular $r(A, A) = r(A, C)$. Also (3.9.2) is equivalent to $A(x) - C(x) = x^{\eta+1}O(1)$. Thus

$$(I_n \otimes A(x) - A(x)^\top \otimes I_n) - (I_n \otimes A(x) - C(x)^\top \otimes I_n) = x^{\eta+1}O(1).$$

In view of Lemma 1.15.2 the matrices $(I_n \otimes A(x) - A(x)^\top \otimes I_n), (I_n \otimes A(x) - C(x)^\top \otimes I_n)$ are equivalent over H_0. In particular $\eta(A, A) = \eta(A, C)$. Also $I, 0, \dots, 0$ satisfy the system (3.9.2) where $B_i = C_i, i = 0, 1, \dots, \eta$. Theorem 1.15.3 yields the existence $P(x) \in H_0^{n \times n}$ such that

$$A(x)P(x) - P(x)C(x) = 0, \quad P(0) = I.$$

Hence, $A(x) \overset{a}{\approx} C(x)$. By the definition $C(x) \overset{a}{\approx} B(x)$. Therefore, $A(x) \overset{a}{\approx} B(x)$. $\square$

Note that if $\eta(A, A) = 0$ the assumptions of Theorem 3.9.2 are equivalent to $A(x) \overset{r}{\approx} B(x)$. Then the implication that $A(x) \overset{a}{\approx} B(x)$ follows from Corollary 3.6.5.

Suppose that the characteristic polynomial of $A(x)$ splits over H_0. That is

$$\det (\lambda I - A(x)) = \prod_{i=1}^{n} (\lambda - \lambda_i(x)), \quad \lambda_i(x) \in H_0, \ i = 1, \ldots, n.$$

$$(3.9.3)$$

As H_0 is *ED* Theorem 2.5.4 yields that $A(x)$ is similar to an upper triangular matrix. Using Theorems 2.5.4 and 3.5.7 we obtain that $A(x)$ is analytically similar to

$$C(x) = \oplus_{i=1}^{\ell} C_i(x), \quad C_i(x) \in H_0^{n_i \times n_i},$$

$$(\alpha_i I_{n_i} - C_i(0))^{n_i} = 0, \ \alpha_i = \lambda_{n_i}(0), \ \alpha_i \neq \alpha_j \text{ for } i \neq j, \ i, j = 1, \ldots, \ell.$$

$$(3.9.4)$$

Furthermore each $C_i(x)$ is an upper triangular matrix. In what follows we are more specific on the form of the upper triangular matrix.

Theorem 3.9.3 *Let $A(x) \in H_0^{n \times n}$. Assume that the characteristic polynomial of $A(x)$ splits in H_0. Then $A(x)$ is analytically similar to a block diagonal matrix $C(x)$ of the form (3.9.4) such that each $C_i(x)$ is an upper triangular matrix whose off-diagonal entries are polynomial in x. Moreover, the degree of each polynomial entry above the diagonal in the matrix $C_i(x)$ does not exceed $\eta(C_i, C_i)$ for $i = 1, \ldots, \ell$.*

Proof. In view of Theorem 3.5.7 we may assume that $\ell = 1$. That is, $A(0)$ has one eigenvalue α_0. Furthermore, by considering $A(x) - \alpha_0 I$ we may assume that $A(0)$ is nilpotent. Also in view of Theorem 3 we may assume that $A(x)$ is already in the upper triangular form. Suppose in addition to all the above assumptions $A(x)$ is nilpotent. Define

$$\mathbf{X}_k = \{\mathbf{y} : \ A^k \mathbf{y} = \mathbf{0}, \ \mathbf{y} \in H_0^n\}, \quad k = 0, 1, \ldots$$

Then

$$\{\mathbf{0}\} = \mathbf{X}_0 \subsetneq \mathbf{X}_1 \subsetneq \mathbf{X}_2 \subsetneq \ldots \subsetneq \mathbf{X}_p = H_0^n.$$

Using Theorem 1.13.3 one can show the existence of a basis $\mathbf{y}_1(x), \ldots, \mathbf{y}_n(x)$ in H_0^n, such that $\mathbf{y}_1(x), \ldots, \mathbf{y}_{\psi_k}(x)$ is a basis in $\mathbf{X}_k$ for $k = 1, \ldots, p$. As $A(x)\mathbf{X}_{k+1} \subset \mathbf{X}_k$ we have

$$A(x)\mathbf{y}_j = \sum_{i=1}^{\psi_k} g_{ij}\mathbf{y}_i(x), \quad \psi_k < j \le \psi_{k+1}.$$

Define $g_{ij} = 0$ for $i > \psi_k$ and $\psi_k < j \le \psi_{k+1}$. Put

$$G(x) = [g_{ij}]_1^n, \ T(x) = [\mathbf{y}_1(x), \ldots, \mathbf{y}_n(x)] \in \mathrm{H}_0^{n \times n}.$$

Since $\mathbf{y}_1(x), \ldots, \mathbf{y}_n(x)$ is a basis in H_0^n we deduce that $T(x) \in \mathbf{GL}(n, \mathrm{H}_0)$. Hence

$$G(x) = T^{-1}(x)A(x)T(x), \quad s = \eta(A, A) = \eta(G, G).$$

Let

$$G(x) = \sum_{j=0}^{\infty} G_j x^j, \quad G^{(k)} = \sum_{j=0}^{k} G_j x^j, \quad k = 0, 1, \ldots$$

We claim that $G^{(s)} \overset{a}{\approx} G(x)$. First note that

$$(I_n \otimes G(x) - G(x)^\top \otimes I_n) - (I_n \otimes G^{(s)}(x) - G^{(s)}(x)^\top \otimes I_n) = x^{s+1}O(1).$$

Lemma 1.15.2 implies that the matrices $(I_n \otimes G(x) - G(x)^\top \otimes I_n)$, $(I_n \otimes G^{(s)}(x) - G^{(s)}(x)^\top \otimes I_n)$ have the same local invariant polynomial up to the degree s. So $r(G, G) \le r(G^{(s)}, G^{(s)})$ which is equivalent to

$$\nu(G^{(s)}, G^{(s)}) \le \nu(G, G). \tag{3.9.5}$$

Let

$$\mathbf{Y}_k = \{\mathbf{y} = (y_1, \ldots, y_n)^\top : \quad y_j = 0 \text{ for } j > \psi_k\}, \quad k = 0, \ldots, p.$$

Clearly if $g_{ij} = 0$ then (i, j)th entry of $G^{(s)}$ is also equal to zero. By the definition $g_{ij}(x) = 0$ for $i > \psi_k$ and $\psi_k < j < \psi_{k+1}$. So

$G^{(s)}(x)\mathbf{Y}_{k+1} \subset \mathbf{Y}_k$ for $k = 0, \ldots, p-1$. Theorem 2.11.2 implies

$$\nu(G(x_0), G(x_0)) \leq \nu(G^{(s)}(x_0), G^{(s)}(x_0)) \qquad (3.9.6)$$

for all x_0 in the neighborhood of the origin. Hence, $\nu(G, G) \leq \nu(G^{(s)}, G^{(s)})$. This establishes equality in (3.9.5), which in return implies equality in (3.9.6) for $0 < |x_0| < \rho$. Theorem 2.11.2 yields that $G(x_0) \approx G^{(s)}(x_0)$ for $0 < |x_0| < \rho$. From Theorem 3.6.2 we deduce that $G \overset{r}{\approx} G^{(s)}$. As $G(x)I - IG^{(s)} = x^{s+1}O(1)$ Theorem 3.9.2 implies that $G \overset{a}{\approx} G^{(s)}$. This establishes the theorem in case that $A(x)$ is a nilpotent matrix.

We now consider the general case where $A(x)$ is an upper triangular matrix. Without loss of generality we may assume that $A(x)$ is of the form

$$A(x) = [A_{ij}]_1^\ell, \quad A_{ij} \in \mathrm{H}_0^{n_i \times n_j},$$

$$A_{ij}(x) = 0 \text{ for } j < i, \quad (A_{ii}(x) - \lambda_i(x)I_{n_i})^{n_i} = 0, \qquad (3.9.7)$$

$$\lambda_i \not\equiv \lambda_j(x), \text{ for } i \neq j, \quad i, j = 1, \ldots, \ell.$$

We already showed that

$$A_{ii}(x) = T_i(x)^{-1}F_{ii}(x)T_i(x), \quad T_i \in \mathbf{GL}(n, \mathrm{H}_0),$$

and each $F_{ii}(x) - \lambda_i(x)I_{n_i}$ is a nilpotent upper triangular matrix with polynomial entries of the form described above. Let

$$T(x) = \sum_{i=1}^\ell T_i(x), \quad G(x) = [G_{ij}(x)]_1^\ell = T(x)^{-1}A(x)T(x).$$

As $\lambda_i(x) \not\equiv \lambda_j(x)$ for $i \neq j$ Problem 3 implies $\nu(G, G) = \sum_{i=1}^\ell \nu(G_{ii}, G_{ii})$. Let $G^{(k)}(x) = [G_{ij}^{(k)}]$ be defined as above. Theorem 2.10.2 implies

$$\nu(G^{(k)}, G^{(k)}) \geq \sum_{i=1}^\ell \nu(G_{ii}^{(k)}, G_{ii}^{(k)}).$$

Using Theorem 2.11.2 as above we obtain $\nu(G_{ii}, G_{ii}) \leq \nu(G_{ii}^{(k)}, G_{ii}^{(k)})$. Combine the above inequalities we obtain $\nu(G, G) \leq \nu(G^{(s)}, G^{(s)})$.

Compare this inequality with the inequality (3.9.5) to deduce equality in (3.9.5). Hence

$$\nu(G_{ii}^{(s)}, G_{ii}^{(s)}) = \nu(G_{ii}, G_{ii}), \quad i = 1, \ldots, \ell. \tag{3.9.8}$$

Let

$$D_i(x) = \lambda_i(x) I_{n_i} = \sum_{j=0}^{\infty} D_{ij} x^j, \quad D_i^{(k)} = \sum_{j=0}^{k} D_{ij} x^j, \tag{3.9.9}$$

$$D(x) = \oplus_{i=1}^{\ell} D_i(x), \quad D^{(k)}(x) = \oplus_{i=1}^{\ell} D_i^{(k)}(x).$$

Then (3.9.8) is equivalent to

$$\nu(G_{ii}^{(s)} - D_i^{(s)}, G_{ii}^{(s)} - D_i^{(s)}) = \nu(G_{ii} - D_{ii}, G_{ii} - D_{ii}), \quad i = 1, \ldots, \ell.$$

As above Theorem 2.11.2 yields that $G_{ii}^{(s)} - D_i^{(s)} \overset{r}{\approx} G_{ii} - D_i \Rightarrow G_{ii}^{(s)} - D_i^{(s)} + D_i \overset{r}{\approx} G_{ii}$. Since $\lambda_i(x) \not\equiv \lambda_j(x)$ for $i \neq j$ we finally deduce that $G \overset{r}{\approx} G^{(s)} - D^{(s)} + D$. Also $GI - I(G^{(s)} - D^{(s)} + D) = x^{s+1} O(1)$. Theorem 3.9.2 yields $G \overset{a}{\approx} G^{(s)} - D^{(s)} + D$. The proof of the theorem is completed.   □

Theorem 3.9.4 *Let $P(x)$ and $Q(x)$ be matrices of the form* (3.9.4)

$$P(x) = \oplus_{i=1}^{p} P_i(x), \ P_i(x) \in H_0^{m_i \times m_i},$$

$$(\alpha_i I_{m_i} - P_i(0))^{m_i} = 0, \ \alpha_i \neq \alpha_j \text{ for } i \neq j, \ i, j = 1, \ldots, p,$$

$$Q(x) = \oplus_{j=1}^{q} Q_j(x), \ Q_j(x) \in H_0^{n_j \times n_j}, \tag{3.9.10}$$

$$(\beta_j I_{n_j} - Q_j(0))^{n_j} = 0, \ \beta_i \neq \beta_j \text{ for } i \neq j, \ i, j = 1, \ldots, q.$$

Assume furthermore that

$$\alpha_i = \beta_i, \ i = 1, \ldots, t, \ \alpha_j \neq \beta_j, \tag{3.9.11}$$

$$i = t+1, \ldots, p, \ j = t+1, \ldots, q, \ 0 \leq t \leq \min(p, q).$$

Then the nonconstant local invariant polynomials of $I \otimes P(x) - Q(x)^\top \otimes I$ are the nonconstant local invariant polynomials of

$I \otimes P_i(x) - Q_i(x)^\top \otimes I$ for $i = 1, \ldots, t$. *That is*

$$\kappa_p(P, Q) = \sum_{i=1}^{t} \kappa_p(P_i, Q_i), \quad p = 1, \ldots \qquad (3.9.12)$$

In particular, if $C(x)$ is of the form (3.9.4) *then*

$$\eta(C, C) = \max_{1 \leq i \leq \ell} \eta(C_i, C_i). \qquad (3.9.13)$$

Proof. Theorem 1.15.3 implies $\kappa_p(P, Q) = \dim \mathbf{W}_{p-1} - \dim \mathbf{W}_p$, where $\mathbf{W}_p \subset \mathbb{C}^{n \times n}$ is the subspace of $n \times n$ matrices X_0 such that

$$\sum_{j=0}^{k} P_{k-j} X_j - X_j Q_{k-j} = 0, \quad k = 0, \ldots, p. \qquad (3.9.14)$$

Here

$$P(x) = \sum_{j=0}^{\infty} P_j x^j, \ P_i(x) = \sum_{j=1}^{\infty} P_j^{(i)} x^j, \ P_j = \oplus_{i=1}^{p} P_j^{(i)},$$

$$Q(x) = \sum_{j=0}^{\infty} Q_j x^j, \ Q_i(x) = \sum_{j=1}^{\infty} Q_j^{(i)} x^j, \ Q_j = \oplus_{i=1}^{q} Q_j^{(i)}.$$

Partition X_j to $[X_{\alpha\beta}^{(j)}]$, $X_{\alpha\beta}^{(j)} \in \mathbb{C}^{m_\alpha \otimes n_\beta}$, $\alpha = 1, \ldots, p$, $\beta = 1, \ldots, q$. We claim that $X_{\alpha\beta}^{(j)} = 0$ if either $\alpha > t + 1$, or $\beta > t + 1$, or $\alpha \neq \beta$. Indeed in view of Lemma 2.8.1 the equation $P_0^{(\alpha)} Y - Y Q_0^{(\beta)} = 0$ has only the trivial solution for α, β satisfying the above conditions. Then the claim that $X_{\alpha\beta}^{(j)} = 0$ follows by induction. Thus (3.9.14) splits to the system

$$\sum_{j=0}^{k} P_{k-j}^{(i)} X_{ii}^{(j)} - X_{ii}^{(j)} Q_{k-j}^{(i)} = 0, \quad i = 1, \ldots, t.$$

Apply the characterizations of $\kappa_p(P, Q)$ and $\kappa_p(P_i, Q_i)$ for $i = 1, \ldots, t$ to deduce (3.9.12). Clearly (3.9.12) implies (3.9.13). $\qquad \square$

We conclude this section by remarking that main assumptions of Theorem 3.9.3, the splitting of the characteristic polynomial of $A(x)$ in H_0, is not a heavy restriction in view of the Weierstrass preparation theorem (Theorem 1.8.4). That is the eigenvalues of $A(y^m)$ split in H_0 for some value of m. Recall that m can be always be chosen $n!$, i.e. the minimal m divides $n!$ Problem 1 claims $A(x) \overset{a}{\approx} B(x) \iff A(y^m) \overset{a}{\approx} B(y^m)$. In view of Theorem 3.9.3 the classification problem of analytic similarity classes reduces to the description of the polynomial entries which are above the diagonal (in the matrix C in Theorem 3.9.3). Thus given the rational canonical form of $A(x)$ and the index $\eta(A, A)$ the set of all possible analytic similarity classes which correspond to A is a certain finite dimensional variety.

The case $n = 2$ is classified completely (Problem 2). In this case to a given rational canonical form there are at most countable number of analytic similarity classes. For $n = 3$ we have an example in which to a given rational canonical form there the family of distinct similarity classes correspond to a finite dimensional variety (Problem 3).

Problems

1. Let $A(x), B(x) \in H_0^{n \times n}$ and let m be a positive integer. Assume that $A(y^m)T(y) = T(y)A(y^m)$ where $T(y) \in H_0^{n \times n}$. Show

$$A(x)Q(x) = Q(x)B(x),$$

$$Q(y^m) = \frac{1}{m} \sum_{k=1}^{m} T(y e^{\frac{2\pi\sqrt{-1}k}{m}}), \quad Q(x) \in H_0^{n \times n}.$$

Prove $A(x) \overset{a}{\approx} B(x) \iff A(y^m) \overset{a}{\approx} B(y^m)$.

2. Let $A(x) \in H_0^{2 \times 2}$ and assume that

$$\det(\lambda I - A(x)) = (\lambda - \lambda_1(x))(\lambda - \lambda_2(x),$$

$$\lambda_i(x) = \sum_{j=0}^{\infty} \lambda_j^{(i)} x^j \in H_0, \quad i = 1, 2,$$

$$\lambda_j^{(1)} = \lambda_j^{(2)}, \ j = 0, \ldots, p, \quad \lambda_{p+1}^{(1)} \neq \lambda_{p+1}^{(2)}, \quad -1 \leq p \leq \infty,$$
$$p = \infty \iff \lambda_1(x) = \lambda_2(x).$$

Show that $A(x)$ is analytically similar either to a diagonal matrix or to

$$B(x) = \begin{bmatrix} \lambda_1(x) & x^k \\ 0 & \lambda_2(x) \end{bmatrix}, \quad k = 0, \ldots, p \ (p \geq 0).$$

Furthermore if $A(x) \overset{a}{\approx} B(x)$ then $\eta(A, A) = k$. (*Hint*: Use a similarity transformation of the form DAD^{-1}, where D is a diagonal matrix.)

3. Let $A(x) \in H_0^{3 \times 3}$. Assume that

$$A(x) \overset{r}{\approx} C(p), \quad p(\lambda, x) = \lambda(\lambda - x^{2m})(\lambda - x^{4m}), \quad m \geq 1.$$

Show that $A(x)$ is analytically similar to a matrix

$$B(x, a) = \begin{bmatrix} 0 & x^{k_1} & a(x) \\ 0 & x^{2m} & x^{k_2} \\ 0 & 0 & x^{4m} \end{bmatrix}, \quad 0 \leq k_1, k_2 \leq \infty \ (x^\infty = 0),$$

where $a(x)$ is a polynomial of degree $4m - 1$ at most. (Use Problem 2.) Assume that $k_1 = k_2 = m$. Show that $B(x, a) \overset{a}{\approx} B(x, b)$ if and only if

(1) if $a(0) \neq 1$ then $b - a$ is divisible by x^m.

(2) if $a(0) = 1$ and $\frac{d^i a}{dx^i} = 0$, $i = 1, \ldots, k - 1$, $\frac{d^k a}{dx^k} \neq 0$ for $1 \leq k < m$ then $b - a$ is divisible by x^{m+k}.

(3) if $a(0) = 1$ and $\frac{d^i a}{dx^i} = 0$, $i = 1, \ldots, m$ then $b - a$ is divisible by x^{2m}.

Then for $k_1 = k_2 = m$ and $a(0) \in \mathbb{C} \backslash \{1\}$ we can assume that $a(x)$ is a polynomial of degree less than m. Furthermore the similarity classes of $A(x)$ is uniquely determined by such $a(x)$. These similarity classes are parameterized by $\mathbb{C} \backslash \{1\} \times \mathbb{C}^{m-1}$ (the Taylor coefficients of $a(x)$).

4. Let P and Q satisfy the assumptions of Theorem 3.9.4. Show that P and Q are analytically similar if and only if

$$p = q = t, \quad m_i = n_i, \ P_i(x) \overset{a}{\approx} Q_i(x), \quad i = 1, \ldots, t.$$

3.10 Strict Similarity of Matrix Polynomials

Definition 3.10.1 *Let $A(x), B(x) \in \mathbb{C}[x]^{n \times n}$. Then $A(x)$ and $B(x)$ are called strictly similar $(A \overset{s}{\approx} B)$ if there exists $P \in \mathbf{GL}(n, \mathbb{C})$ such that $B(x) = PA(x)P^{-1}$.*

Definition 3.10.2 *Let ℓ be a positive integer and $(A_0, A_1, \ldots, A_\ell), (B_0, \ldots, B_\ell) \in (\mathbb{C}^{n \times n})^{\ell+1}$. Then $(A_0, A_1, \ldots, A_\ell)$ and $(B_0, \ldots, B_\ell)$ are called simultaneously similar $(A_0, A_1, \ldots, A_\ell) \approx (B_0, \ldots, B_\ell)$ if there exists $P \in \mathbf{GL}(n, \mathbb{C})$ such that $B_i = PA_iP^{-1}, i = 0, \ldots, \ell$, i.e. $(B_0, B_1, \ldots, B_\ell) = P(A_0, A_1, \ldots, A_\ell)P^{-1}$.*

Clearly

Proposition 3.10.3 *Let*

$$A(x) = \sum_{i=0}^{\ell} A_i x^i, \ B(x) = \sum_{i=0}^{\ell} B_i x^i \in \mathbb{C}[x]^{n \times n}. \tag{3.10.1}$$

Then $(A \overset{s}{\approx} B)$ if and only if $(A_0, A_1, \ldots, A_\ell) \approx (B_0, \ldots, B_\ell)$.

The problem of simultaneous similarity of matrices, i.e. to describe the similarity class of a given $m \ (\geq 2)$ tuple of matrices or to decide when a given two tuples of matrices are simultaneously similar, is a hard problem. See [Fri83]. There are some cases where this problem has a relatively simple solution.

Theorem 3.10.4 *Let $\ell \geq 1$ and $(A_0, \ldots, A_\ell) \in (\mathbb{C}^{n \times n})^{\ell+1}$. Then $(A_0, \ldots, A_\ell)$ is simultaneously similar to a diagonal tuple $(B_0, \ldots, B_\ell) \in \mathbb{C}^{n \times n \, \ell+1}$, i.e. each B_i is a diagonal matrix, if and only if $A_0, \ldots, A_\ell$ are $\ell + 1$ commuting diagonable matrices:*

$$A_i A_j = A_j A_i, \quad i, j = 0, \ldots, \ell. \tag{3.10.2}$$

Proof. Clearly if $(A_0, \ldots, A_\ell)$ is simultaneously similar to a diagonal tuple then $A_0, \ldots, A_\ell$ a set of commuting diagonal matrices. Assume that $A_0, \ldots, A_\ell$ a set of commuting diagonal matrices. We show that $(A_0, \ldots, A_\ell)$ is simultaneously similar to a diagonal tuple by the double induction on n and ℓ. It is convenient to let $\ell \geq 0$. For $n = 1$ the theorem trivially holds for any $\ell \geq 0$. For $\ell = 0$ the theorem trivially holds for any $n \geq 1$. Assume now that $p > 1$, $q \geq 1$ and assume that the theorem holds for $n \leq p - 1$ and all ℓ and for $n = p$ and $\ell \leq q - 1$. Assume that $A_0, \ldots, A_q \in \mathbb{C}^{p \times p}$ are $q + 1$ commuting diagonable matrices. Suppose first that $A_0 = aI_p$. The induction hypothesis yields that $(B_1, \ldots, B_q) = P(A_1, \ldots, A_q)P^{-1}$ is a diagonal q-tuple for some $P \in \mathbf{GL}(n, \mathbb{C})$. As $PA_0P^{-1} = A_0 = aI_p$ we deduce that $(A_0, B_1, \ldots, B_\ell) = P(A_0, A_1, \ldots, A_\ell)P^{-1}$.

Assume that A_0 is not a scalar matrix, i.e. $A_0 \neq \frac{1}{p} \operatorname{tr} A \, I_p$. Let

$$\tilde{A}_0 = QA_0Q^{-1} = \oplus_{i=1}^k a_iI_{p_i},$$

$$1 \leq p_i, \ a_i \neq a_j \text{ for } i \neq j, \ i, j = 1, \ldots, k, \ \sum_{i=1}^k n_i = p.$$

Then the $q + 1$ tuple $(\tilde{A}_0, \ldots, \tilde{A}_q) = Q(A_0, \ldots, A_q)Q^{-1}$ is a $q + 1$ tuple of diagonable commuting matrices. The specific form of $\tilde{A}_0$ and the assumption that $\tilde{A}_0$ and $\tilde{A}_j$ commute implies

$$\tilde{A}_j = \oplus_{i=1}^k \tilde{A}_{j,i}, \quad \tilde{A}_{j,i} \in \mathbb{C}^{p_i \times p_i}, \ i = 1, \ldots, k, \ j = 1, \ldots, q.$$

The assumption that $(\tilde{A}_0, \ldots, \tilde{A}_q)$ is a $q + 1$ tuple of diagonable commuting matrices implies that each i the tuple $(a_iI_{p_i}, \tilde{A}_{1,i} \ldots, \tilde{A}_{q,i})$ is $q + 1$ tuple of diagonable commuting matrices. Hence, the induction hypothesis yields that $(a_iI_{p_i}, \tilde{A}_{1,i} \ldots, \tilde{A}_{q,i})$ is similar to a $q + 1$ diagonal tuple for $i = 1, \ldots, k$. It follows straightforward that $(\tilde{A}_0, \tilde{A}_1 \ldots, \tilde{A}_q)$ is simultaneously similar to a diagonal $q + 1$ tuple. $\square$

The problem when $A(x) \in \mathbb{C}[x]^{n \times n}$ is strictly similar to an upper triangular matrix $B(x) \in \mathbb{C}[x]^{n \times n}$ is equivalent to the problem when an $\ell + 1$ tuple $(A_0, \ldots, A_\ell) \in (\mathbb{C}^{n \times n})^{\ell+1}$ is simultaneously an upper triangular tuple $(B_0, \ldots, B_\ell)$, i.e. each B_i is an upper triangular matrix, is solved in [DDG51]. We bring their result without a proof.

Definition 3.10.5 *Let $\mathbb{D}$ be a domain, let n, m be positive integers and let $C_1, \ldots, C_m \in \mathbb{D}^{n \times n}$. Then $\mathcal{A}(C_1, \ldots, C_m) \subset \mathbb{D}^{n \times n}$ denotes the minimal algebra in $\mathbb{D}^{n \times n}$ containing I_n and $C_1, \ldots, C_m$. That is every matrix $F \in \mathcal{A}(C_1, \ldots, C_m)$ is a noncommutative polynomial in $C_1, \ldots, C_m$.*

Theorem 3.10.6 *Let m, ℓ be positive integers and let $A_0, \ldots, A_\ell \in (\mathbb{C}^{n \times n})^{\ell+1}$. TFAE:*

(a) *$(A_0, \ldots, A_\ell)$ is simultaneously similar to an upper triangular tuple $(B_0, \ldots, B_\ell) \in \mathrm{M}_n(\mathbb{C})^{\ell+1}$.*

(b) *For any $0 \le i < j \le \ell$ and $F \in \mathcal{A}(A_0, \ldots, A_\ell)$ the matrix $(A_i A_j - A_j A_i)F$ is nilpotent.*

The implication $(a) \Rightarrow (b)$ is trivial. (See Problem 2.) The verification of condition (b) can be done quite efficiently. (See Problem 3.)

Corollary 3.10.7 *Let m, ℓ be positive integers and assume that $A_0, \ldots, A_\ell \in \mathbb{C}^{n \times n}$ are commuting matrices. Then $(A_0, \ldots, A_\ell)$ is simultaneously similar to an upper triangular tuple $(B_0, \ldots, B_\ell)$.*

See Problem 4.

Problems

1. Let $\mathbb{F}$ be a field. View $\mathbb{F}^{n \times n}$ as an n^2 dimensional vector space over $\mathbb{F}$. Note that any $A \in \mathbb{F}^{n \times n}$ acts as a linear transformation on $\mathbb{F}^{n \times n}$ by left multiplication: $B \mapsto AB$, $B \in \mathbb{C}^{n \times n}$. Let $A_0, \ldots, A_\ell \in \mathbb{F}^{n \times n}$. Let $\mathbf{W}_0 = \mathrm{span}\,(I_n)$ and define

$$\mathbf{W}_k = \mathbf{W}_{k-1} + \sum_{j=0}^{\ell} A_j \mathbf{W}_{k-1}, \quad k = 1, \ldots$$

Show that $\mathbf{W}_{k-1} \subset \mathbf{W}_k$ for each $k \ge 1$. Let p be the minimal non-negative integer for which the equality $\mathbf{W}_k = \mathbf{W}_{k+1}$ holds. Show that $\mathcal{A}(A_0, \ldots, A_\ell) = \mathbf{W}_p$. In particular $\mathcal{A}(A_0, \ldots, A_\ell)$ is a finite dimensional subspace of $\mathbb{F}^{n \times n}$.

2. Show the implication $(a) \Rightarrow (b)$ in Theorem 3.10.6.

3. Let the assumptions of Problem 1 hold. Let $\mathbf{X}_0 = \mathcal{A}(A_0, \ldots, A_\ell)$ and define recursively

$$\mathbf{X}_k = \sum_{0 \leq i < j \leq \ell} (A_i A_j - A_j A_i) \mathbf{X}_{k-1} \subset \mathbb{F}^{n \times n}, \quad k = 1, \ldots$$

Show that the condition (a) of Theorem 3.10.6 to the following two conditions:

(c) $A_i \mathbf{X}_k \subset \mathbf{X}_k$, $\quad i = 0, \ldots, \ell$, $k = 0, \ldots$.

(d) There exists $q \geq 1$ such that $\mathbf{X}_q = \{0\}$ and $\mathbf{X}_k$ is a strict subspace of $\mathbf{X}_{k-1}$ for $k = 1, \ldots, q$.

4. Let $A_0, \ldots, A_\ell \in \mathbb{F}^{n \times n}$. Assume that $\mathbf{0} \neq \mathbf{x} \in \mathbb{F}^n$ and $A_0 \mathbf{x} = \lambda_0 \mathbf{x}$. Suppose that $A_0 A_i = A_i A_0$, $i = 1, \ldots, \ell$.

   (a) Show that any nonzero vector in $\mathcal{A}(A_1, \ldots, A_\ell)$span $(\mathbf{x})(\supset$ span $(\mathbf{x}))$ is an eigenvector of A_0 corresponding λ_0.

   (b) Assume in addition that $A_1, \ldots, A_\ell$ are commuting matrices whose characteristic polynomials split in $\mathbb{F}$ to linear factors. Show by induction that there exists $\mathbf{0} \neq \mathbf{y} \in \mathcal{A}(A_1, \ldots, A_\ell)$ span $(\mathbf{x})$ such $A_i \mathbf{y} = \lambda_i \mathbf{y}$, $i = 0, \ldots, \ell$.

   (c) Show that if $A_0, \ldots, A_\ell \in \mathbb{F}^{n \times n}$ are commuting matrices whose characteristic polynomials split in $\mathbb{F}$ to linear factors then $(A_0, \ldots, A_\ell)$ is simultaneously similar over $\mathbf{GL}(n, \mathbb{F})$ to an upper triangular $\ell + 1$ tuple.

3.11   Similarity to Diagonal Matrices

Theorem 3.11.1 *Let $A(x) \in H_0^{n \times n}$ and assume that the characteristic polynomial of $A(x)$ splits in H_0 as in (3.9.3). Let*

$$B(x) = \operatorname{diag}(\lambda_1(x), \ldots, \lambda_n(x)). \tag{3.11.1}$$

Then $A(x)$ and $B(x)$ are not analytically similar if and only if there exists a non-negative integer p such that

$$\kappa_p(A, A) + \kappa_p(B, B) < 2\kappa_p(A, B),$$
$$\kappa_j(A, A) + \kappa_j(B, B) = 2\kappa_j(A, B), \ j = 0, \ldots, p - 1, \quad \text{if } p \geq 1.$$
$$(3.11.2)$$

In particular $A(x) \overset{a}{\approx} B(x)$ if and only if the three matrices given in (2.9.4) are equivalent over H_0.

Proof. Suppose first that (3.11.2) holds. Then the three matrices in (2.9.4) are not equivalent. Hence, $A(x) \overset{a}{\not\approx} B(x)$. Assume now that $A(x) \overset{a}{\not\approx} B(x)$. Without a loss in generality we may assume that $A(x) = C(x)$ where $C(x)$ is given in (3.9.4). Let

$$B(x) = \oplus_{j=1}^{\ell} B_j(x), \ B_j(0) = \alpha_j I_{n_j}, \ j = 1, \ldots, \ell.$$

We prove (3.11.2) by induction on n. For $n = 1$ (3.11.2) is obvious. Assume that the (3.11.2) holds for $n \leq N - 1$. Let $n = N$. If $A(0) \not\approx B(0)$ then Theorem 2.9.2 implies the inequality (3.11.2) for $p = 0$. Suppose now $A(0) \approx B(0)$. That is $A_j(0) = B_j(0) = \alpha_j I_{n_j}, \ j = 1, \ldots, \ell$. Suppose first that $\ell > 1$. Theorem 3.9.4 yields

$$\kappa_p(A, A) = \sum_{j=1}^{\ell} \kappa_p(A_j, A_j), \quad \kappa_p(A, B) = \sum_{j=1}^{\ell} \kappa_p(A_j, B_j),$$

$$\kappa_p(B, B) = \sum_{j=1}^{\ell} \kappa_p(B_j, B_j).$$

Problem 4 implies that $A(x) \overset{a}{\not\approx} B(x) \iff A_j(x) \overset{a}{\not\approx} B_j(x)$ for some j. Use the induction hypothesis to deduce (3.11.2). It is left to consider the case

$$A(0) = B(0) = \alpha_0, \ \kappa_0(A, A) = \kappa_0(A, B) = \kappa_0(B, B) = 0.$$

Let

$$A^{(1)}(x) = \frac{A(x) - \alpha_0 I}{x}, \quad B^{(1)}(x) = \frac{B(x) - \alpha_0 I}{x}.$$

Clearly

$$\kappa_p(A, A) = \kappa_{p-1}(A^{(1)}, A^{(1)}), \ \kappa_p(A, B) = \kappa_{p-1}(A^{(1)}, B^{(1)}),$$
$$\kappa_p(B, B) = \kappa_{p-1}(B^{(1)}, B^{(1)}).$$

Furthermore $A(x) \overset{a}{\approx} B(x) \iff A^{(1)}(x) \overset{a}{\approx} B^{(1)}(x)$. Continue this process. If at some (first) stage k either $A^{(k)}(0) \overset{}{\not\approx} B^{(k)}(0)$ or $A^{(k)}(0)$ has at least two distinct eigenvalues we conclude (3.11.2) as above. Suppose finally that such k does not exist. Then $A(x) = B(x) = \lambda(x)I$, which contradicts the assumption $A(x) \overset{a}{\not\approx} B(x)$. $\qquad\square$

Let $A(x) \in \mathrm{H}_0^{n \times n}$. The Weierstrass preparation theorem (Theorem 1.8.4) implies that the eigenvalues of $A(y^s)$ are analytic in y for some $s|n!$. That is the eigenvalues $\lambda_1(x), \ldots, \lambda_n(x)$ are multivalued analytic functions in x which have the expansion

$$\lambda_j(x) = \sum_{k=0}^{\infty} \lambda_{jk} x^{\frac{k}{s}}, \quad j = 1, \ldots, n.$$

In particular each λ_j has s_j branches, where $s_j|m$. For more properties of the eigenvalues $\lambda_1(x), \ldots, \lambda_n(x)$ see for example [Kat80, Chapter 2].

Let $A(x) \in \mathbb{C}[x]^{n \times n}$. Then

$$A(x) = \sum_{k=0}^{\ell} A_k x^k, \quad A_k \in \mathbb{C}^{n \times n}, \ k = 0, \ldots, \ell. \tag{3.11.3}$$

The eigenvalues of $A(x)$ satisfy the equation

$$\det\left(\lambda I - A(x)\right) = \lambda^n + \sum_{j=1}^{n} a_j(x)\lambda^{n-j}, \ a_j(x) \in \mathbb{C}[x], \ j = 1, \ldots, n.$$

$$\tag{3.11.4}$$

Thus the eigenvalues $\lambda_1(x), \ldots, \lambda_n(x)$ are algebraic functions of x. (See for example [GuR65].) For each $\zeta \in \mathbb{C}$ we apply the Weierstrass preparation theorem in H_ζ to obtain the *Puiseux* expansion of $\lambda_j(x)$

around $x = \zeta$:

$$\lambda_j(x) = \sum_{k=0}^{\infty} \lambda_{jk}(\zeta)(x - \zeta)^{\frac{k}{s}}, \quad j = 1, \ldots, n. \tag{3.11.5}$$

For simplicity of notation we choose $s \le n!$ for which the above expansion holds for each $\zeta \in \mathbb{C}$. (For example, $s = n!$ is always a valid choice.) Since $A(x)$ is a polynomial matrix each $\lambda_j(x)$ has Puiseux expansion at ∞. Let

$$A(x) = x^{\ell} B \left(\frac{1}{x} \right), \quad B(y) = \sum_{k=0}^{\ell} A_k y^{\ell-k}.$$

Then the Puiseux expansion of the eigenvalues of $B(y)$ at $y = 0$ yields

$$\lambda_j(x) = x^{\ell} \sum_{k=0}^{\infty} \lambda_{jk}(\infty) x^{-\frac{k}{s}}, \quad j = 1, \ldots, n. \tag{3.11.6}$$

Equivalently, we view the eigenvalues $\lambda_j(x)$ as multivalued analytic functions over the Riemann sphere $\mathbb{P} = \mathbb{C} \cup \infty$. To view $A(x)$ as a matrix function over $\mathbb{P}$ we need to homogenize as in §2.1.

Definition 3.11.2 *Let* $A(x)$ *be given by* (3.11.3). *Denote by* $A(x_0, x_1)$ *the corresponding homogeneous matrix*

$$A(x_0, x_1) = \sum_{k=0}^{\ell'} A_k x_0^{\ell'-k} x_1^k \in \mathbb{C}[x_0, x_1]^{n \times n}, \tag{3.11.7}$$

where $\ell' = -1$ *if* $A(x) = 0$ *and* $A_{\ell'} \ne 0$ *and* $A_j = 0$ *for* $\ell' < j \le \ell$ *if* $A(x) \ne 0$.

Let $A(x), B(x) \in \mathbb{C}[x]^{n \times n}$. Then $A(x)$ and $B(x)$ are similar over $\mathbb{C}[x]$, denoted by $A(x) \approx B(x)$, if $B(x) = P(x)A(x)P^{-1}(x)$ for some $P(x) \in \mathbf{GL}(n, \mathbb{C}[x])$. Lemma 2.9.4 implies that if $A(x) \approx B(x)$ then the three matrices in (3.6.2) are equivalent over $\mathbb{C}[x]$. Assume a stronger condition $A \overset{s}{\approx} B$. Clearly if $B(x) = PA(x)P^{-1}$ then

$B(x_0, x_1) = PA(x_0, x_1)P^{-1}$. According to Lemma 2.9.4 the matrices

$$I \otimes A(x_0, x_1) - A(x_0, x_1)^\top \otimes I, \tag{3.11.8}$$

$$I \otimes A(x_0, x_1) - B(x_0, x_1)^\top \otimes I, \quad I \otimes B(x_0, x_1) - B(x_0, x_1)^\top \otimes I,$$

are equivalent over $\mathbb{C}[x_0, x_1]$. Lemma 1.12.3 yields.

Lemma 3.11.3 *Let $A(x), B(x) \in \mathbb{C}[x]^{n \times n}$. Assume that $A(x) \overset{s}{\approx} B(x)$. Then the three matrices in (3.11.8) have the same invariant polynomials over $\mathbb{C}[x_0, x_1]$.*

Definition 3.11.4 *Let $A(x), B(x) \in \mathbb{C}[x]^{n \times n}$. Let $A(x_0, x_1)$, $B(x_0, x_1)$ be the homogeneous matrices corresponding to $A(x), B(x)$ respectively. Denote by $i_k(A, B, x_0, x_1)$, $k = 1, \ldots, r(A, B)$ the invariant factors of $I \otimes A(x_0, x_1) - B(x_0, x_1)^\top \otimes I$.*

The arguments of the proof of Lemma 2.1.2 imply that $i_k(A, B, x_0, x_1)$ is a homogeneous polynomial for $k = 1, \ldots, r(A, B)$. Moreover $i_k(A, B, 1, x)$ are the invariants factors of $I \otimes A(x) - B(x)^\top \otimes I$. (See Problems 5 and 6.)

Theorem 3.11.5 *Let $A(x) \in \mathbb{C}[x]^{n \times n}$. Assume that the characteristic polynomial of $A(x)$ splits to linear factors over $\mathbb{C}[x]$. Let $B(x)$ be the diagonal matrix of the form (3.11.1). Then $A(x) \approx B(x)$ if and only if the three matrices in (3.6.2) are equivalent over $\mathbb{C}[x]$. Furthermore $A(x) \overset{s}{\approx} B(x)$ if and only if the three matrices in (1.34.8) have the same invariant factors over $\mathbb{C}[x_0, x_1]$.*

Proof. Clearly if $A(x) \approx B(x)$ then the three matrices in (3.6.2) are equivalent over $\mathbb{C}[x]$. Similarly if $A(x) \overset{s}{\approx} B(x)$ then the three matrices in (1.34.8) have the same invariant factors over $\mathbb{C}[x_0, x_1]$. We now show the opposite implications.

Without loss of generality we may assume that $B(x)$ is of the form

$$B(x) = \oplus_{i=1}^{m}\lambda_i(x)I_{n_i} \in \mathbb{C}[x]^{n\times n},$$

$$\lambda_i(x) \neq \lambda_j(x),\ i \neq j,\ i,j = 1,\ldots,m. \qquad (3.11.9)$$

Thus for all but a finite number of points $\zeta \in \mathbb{C}$ we have that

$$\lambda_i(\zeta) \neq \lambda_j(\zeta) \text{ for } i \neq j, \quad i,j = 1,\ldots,m. \qquad (3.11.10)$$

Assume first that $A(x) \approx B(x)$. Let $P_j(A)$ be the projection of $A(x)$ on $\lambda_j(x)$ for $j = 1,\ldots,m$. Suppose that (3.11.10) is satisfied at ζ. Problem 3.4.10 yields that each $P_j(x)$ is analytic in the neighborhood of ζ. Assume that (3.11.10) does not hold for $\zeta \in \mathbb{C}$. The assumptions that the three matrices in (3.11.8) have the same invariant polynomials imply that the matrices in (3.6.2) are equivalent over H_ζ. Now use Theorem 3.11.1 to get that $A(x) = Q(x)B(x)Q(x)^{-1}$, $Q \in \mathbf{GL}(n, H_\zeta)$. Clearly $P_j(B)$, the projection of $B(x)$ on $\lambda_j(x)$, is $0 \oplus I_{n_j} \oplus 0$. In particular $P_j(B)$ is analytic in the neighborhood of any $\zeta \in \mathbb{C}$ and its rank is always equal to n_j. Problem 3.4.11 yields that $P_j(A)(x) = Q(x)P_j(B)(x)Q(x)^{-1} \in H_\zeta^{n\times n}$. Hence, rank $P_j(A)(\zeta) = n_j$ for all $\zeta \in \mathbb{C}$. Furthermore, $P_j(A)(x) \in H_{\mathbb{C}}^{n\times n}$, i.e. each entry of $P_j(A)$ is an entire function (analytic function on $\mathbb{C}$). Problem 3.4.14 yields that

$$P_j(A)(\zeta) = \prod_{k=1,k\neq j}^{n} \frac{A(\zeta) - \lambda_k(\zeta)I}{\lambda_j(\zeta) - \lambda_k(\zeta)}, \quad j = 1,\ldots,n. \qquad (3.11.11)$$

Hence, each entry of $P_j(A)(\zeta)$ is a rational function of ζ on $\mathbb{C}$. Since $P_i(A)(x)$ is analytic in the neighborhood of each $\zeta \in \mathbb{C}$ it follows that $P_i(x) \in \mathbb{C}[x]^{n\times n}$. We also showed that its rank is locally constant, hence rank $P_i(x) = n_i$, $i = 1,\ldots,m$. Therefore, the Smith normal form of $P_i(x)$ over $\mathbb{C}[x]$ is $P_i(x) = U_i(x)(I_{n_i} \oplus 0)V_i(x)$, $U_i, V_i \in \mathbf{GL}(n, \mathbb{C}[x])$. Let $\mathbf{u}_{1,i}(x),\ldots,\mathbf{u}_{n_i,i}(x)$ be the first n_i columns of $U_i(x)$. Then $P_i(x)\mathbb{C}^n = \text{span}\,(\mathbf{u}_{1,i}(x),\ldots,\mathbf{u}_{n_i,i}(x))$. Recall that $P_1(x)+\cdots+P_m(x) = I_n$. Hence, $\mathbf{u}_{1,1}(x),\ldots,\mathbf{u}_{n_1,1}(x),\ldots,\mathbf{u}_{1,m}(x),\ldots,\mathbf{u}_{n_m,m}(x)$ is a basis for $\mathbb{C}^n$ for each $x \in \mathbb{C}$. Let $S(x)$ be the matrix with the columns

$$\mathbf{u}_{1,1}(x),\dots,\mathbf{u}_{n_1,1}(x),\dots,\mathbf{u}_{1,m}(x),\dots,\mathbf{u}_{n_m,m}(x).$$

Then $S(x) \in \mathbf{GL}(n,\mathbb{C}[x])$. Let $D(x) = S^{-1}(x)A(x)S(x) \in \mathbb{C}[x]^{n\times n}$. Since $A(x)$ is pointwise diagonable $D(\zeta) = B(\zeta)$, where ζ satisfies (3.11.10) and $B(x)$ is of the form (3.11.9). Since only finite number of points $\zeta \in \mathbb{C}$ do not satisfy the condition (3.11.9) it follows that $D(x) = B(x)$. This proves the first part of the theorem.

Assume now that the three matrices in (1.34.8) have the same invariant factors over $\mathbb{C}[x_0, x_1]$. The same arguments imply that $A(x_0,1)\overset{a}{\approx}B(x_0,1)$ over the ring H_0. That is $P_j(A)$ is also analytic at the neighborhood $\zeta = \infty$. So $P_j(A)$ is analytic on $\mathbb{P}$ hence bounded, i.e. each entry of $P_j(A)$ is bounded. Hence, $P_j(A)$ is a constant matrix. Therefore, $S(x)$ is a constant invertible matrix, i.e. $A(x)\overset{s}{\approx}B(x)$. □

Let $A(x) \in \mathbb{C}[x]^{n\times n}$ be of the form (3.11.3) with $\ell \geq 1$ and $A_\ell \neq 0$. Assume that $A(x)$ is strictly similar to a diagonal matrix $B(x)$. Then $A(x)$ is pointwise *diagonable*, i.e. $A(x)$ is similar to a diagonal matrix for each $x \in \mathbb{C}$, and $A_\ell \neq 0$ is diagonable. Equivalently, consider the homogeneous polynomial matrix $A(x_0, x_1)$. Then $A(x_0, x_1)$ is pointwise diagonable (in $\mathbb{C}^2$). However, the assumption that any $A(x_0, x_1)$ is pointwise diagonable does not imply that $A(x)$ is strictly equivalent to a diagonal matrix. Consider for example

$$A(x) = \begin{bmatrix} x^2 & x \\ 0 & 1+x^2 \end{bmatrix} \Rightarrow A(x_0,x_1) = \begin{bmatrix} x_1^2 & x_0x_1 \\ 0 & x_0^2 + x_1^2 \end{bmatrix}. \tag{3.11.12}$$

(See Problem 2.)

Definition 3.11.6 *Let $A(x) \in \mathbb{C}[x]^{n\times n}$ be of the form (3.11.3) with $\ell \geq 1$ and $A_\ell \neq 0$. Let $\lambda_p(x)$ and $\lambda_q(x)$ be two distinct eigenvalues of $A(x)$. ($\lambda_p(x)$ and $\lambda_q(x)$ have distinct Puiseux expansion for any $\zeta \in \mathbb{P}$.) The eigenvalues $\lambda_p(x)$ and $\lambda_q(x)$ are said to be tangent* *at $\zeta \in \mathbb{P}$ if their Puiseux expansion at ζ satisfy*

$$\lambda_{pk}(\zeta) = \lambda_{qk}(\zeta), \quad k = 0, \ldots, s. \tag{3.11.13}$$

(Note that two distinct eigenvalues are tangent at ∞ if the corresponding eigenvalues of $A(x, 1)$ are tangent at 0.)

Note that for $A(x)$ given in (3.11.12) the two eigenvalues of $A(x)$ x^2 and $1 + x^2$ are tangent at one point $\zeta = \infty$. (The eigenvalues of $A(x, 1)$ are 1 and $1 + x^2$.)

Theorem 3.11.7 *Let $A(x) \in \mathbb{C}[x]^{n \times n}$ be of the form (3.11.3) with $\ell \geq 1$ and $A_\ell \neq 0$. Assume that $B(x) \in \mathbb{C}[x]^{n \times n}$ is a diagonal matrix of the form $\sum_{i=1}^{m} \lambda_i(x) I_{k_i}$, where $k_1, \ldots, k_m \geq 1$. Furthermore $\lambda_1(x), \ldots, \lambda_m(x)$ are m distinct polynomials satisfying the following conditions:*

(a) *$\ell \geq \deg \lambda_i(x)$, $i = 1, \ldots, m$.*

(b) *The polynomial $\lambda_i(x) - \lambda_j(x)$ has only simple roots in $\mathbb{C}$ for $i \neq j$. ($\lambda_i(\zeta) = \lambda_j(\zeta) \Rightarrow \lambda_i'(\zeta) \neq \lambda_j'(\zeta)$).*

Then one of the following conditions imply that $A(x) = S(x)B(x) S^{-1}(x)$, where $S(x) \in \mathbf{GL}(n, \mathbb{C}[x])$.

I. *The characteristic polynomial of $A(x)$ splits in $\mathbb{C}[x]$, i.e. all the eigenvalues of $A(x)$ are polynomials. $A(x)$ is pointwise diagonable in $\mathbb{C}$ and no two distinct eigenvalues are tangent at any $\zeta \in \mathbb{C}$.*

II. *$A(x)$ is pointwise diagonable in $\mathbb{C}$ and A_ℓ is diagonable. No two distinct eigenvalues are tangent at any point $\zeta \in \mathbb{C} \cup \{\infty\}$. Then $A(x)$ is strictly similar to $B(x)$, i.e. $S(x)$ can be chosen in $\mathbf{GL}(n, \mathbb{C})$. Furthermore $\lambda_1(x), \ldots, \lambda_m(x)$ satisfy the additional condition:*

(c) *$\deg \lambda_1(x) = l$. Furthermore, for $i \neq j$ either $\frac{d^\ell \lambda_i}{d^\ell x}(0) \neq \frac{d^\ell \lambda_j}{d^\ell x}(0)$ or $\frac{d^\ell \lambda_i}{d^\ell x}(0) = \frac{d^\ell \lambda_j}{d^\ell x}(0)$ and $\frac{d^{\ell-1} \lambda_i}{d^{\ell-1} x}(0) \neq \frac{d^{\ell-1} \lambda_j}{d^{\ell-1} x}(0)$.*

Proof. View $A(x)$ as matrix in $\mathcal{M}^{n \times n}$, where $\mathcal{M}$ is field of rational functions. Let $\mathbb{K}$ be a finite extension of $\mathcal{M}$ such that det $(\lambda I - A(x))$ splits to linear factors over $\mathbb{K}$. Then $A(x)$ has m distinct eigenvalues $\lambda_1, \ldots, \lambda_m \in \mathbb{K}$ of multiplicities $n_1, \ldots, n_m$ respectively. We view these eigenvalues as multivalued functions $\lambda_1(x), \ldots, \lambda_m(x)$. Thus for all but a finite number of points ζ (3.11.10) holds. Assume that ζ satisfies (3.11.10). Denote by $P_j(\zeta)$ the projection of $A(\zeta)$ on $\lambda_j(\zeta)$. Problem 10 implies that $P_j(x)$ is a multivalued analytic in the neighborhood of ζ and rank $P_j(\zeta) = n_j$. Problem 14 yields (3.11.11). We claim that in the neighborhood of any $\zeta \in \mathbb{C}$ each λ_j and P_j is multivalued analytic and rank $P_j(x) = n_j$. Let $\zeta \in \mathbb{C}$ for which (3.11.10) is violated. For simplicity of notation we consider $\lambda_1(x)$ and $P_1(x)$. Let

$$\lambda_1(\zeta) = \cdots = \lambda_r(\zeta) \neq \lambda_k(\zeta), \quad k = r+1, \ldots, m.$$

Theorem 3.5.7 implies the existence of $Q(x) \in \mathbf{GL}(n, H_\zeta)$ such that

$$Q^{-1}(x)A(x)Q(x) = C_1(x) \oplus C_2(x),$$

$$C_j(x) \in H_\zeta^{m_j \times m_j}, \ j = 1, 2, \quad m_1 = \sum_{i=1}^{r} n_i, \ m_2 = n - m_1.$$

The eigenvalues of $C_1(x)$ and $C_2(x)$ are $\lambda_1(x), \ldots, \lambda_r(x)$ and $\lambda_{r+1}(x), \ldots, \lambda_m(x)$ respectively in some neighborhood of ζ. Since $C(x)$ is pointwise diagonable in H_ζ it follows that $C_1(x)$ and $C_2(x)$ are pointwise diagonable in H_ζ. We claim that $\lambda_i(x) \in H_\zeta$, the projection $\hat{P}_i(x)$ of $C_1(x)$ on $\lambda_i(x)$ is in $H_\zeta^{m_1 \times m_1}$ and rank $\hat{P}_i(\zeta) = n_i$ for $i = 1, \ldots, r$. If $r = 1$ $\lambda_1(x) = \frac{1}{n_1} \operatorname{tr} C_1(x) \in H_\zeta$ and $\hat{P}_1(x) = I_{n_1}$. Assume that $r > 1$. Since $C_1(\zeta)$ is diagonable and has one eigenvalue $\lambda_1(\zeta)$ of multiplicity m_1 it follows that $C_1(\zeta) = \lambda_1(\zeta)I_{m_1}$. Hence

$$C_1(x) = \lambda_1(\zeta)I_{m_1} + (x - \zeta)\hat{C}_1(x), \quad \hat{C}_1(x) \in H_\zeta^{m_1 \times m_1}.$$

Clearly $\hat{C}_1(x)$ has r distinct eigenvalues $\hat{\lambda}_1(x), \ldots, \hat{\lambda}_r(x)$ such that

$$\lambda_i(x) = \lambda_1(\zeta) + (x - \zeta)\hat{\lambda}_i(x), \quad i = 1, \ldots, r.$$

Each $\hat{\lambda}_i(x)$ has Puiseux expansion (3.11.5). The above equality shows that for $1 \leq i < j \leq r$ $\lambda_i(x)$ and $\lambda_j(x)$ are not tangent if and only if $\hat{\lambda}_i(\zeta) \neq \hat{\lambda}_j(\eta)$. By the assumption of theorem no two different eigenvalues of $A(x)$ are tangent in $\mathbb{C}$. Hence, $\hat{\lambda}_i(\zeta) \neq \hat{\lambda}_j(\eta)$ for all $i \neq j \leq r$. That is $\hat{C}_1(\zeta)$ has r distinct eigenvalues. Apply Theorem 3.5.7 to $\hat{C}_1(\zeta)$ to deduce that $\hat{C}_1(\zeta)$ is analytically similar $\tilde{C}_1 \oplus \cdots \oplus \tilde{C}_r$ such that $\tilde{C}_i$ has a unique eigenvalues $\hat{\lambda}_i(x)$ of multiplicity n_i for $i = 1, \ldots, r$. Hence, $\hat{\lambda}_i(x) = \frac{1}{n_i} \operatorname{tr} \tilde{C}_i(x) \in \mathrm{H}_\zeta \Rightarrow \lambda_i(x) \in \mathrm{H}_\zeta$. Clearly the projection of $\tilde{C}_i(x)$ on $\hat{\lambda}_i(x)$ is I_{n_i}. Hence, $\hat{P}_i(x)$ is analytically similar to the projection to $0 \oplus \cdots \oplus I_{n_i} \cdots \oplus 0$. So $\hat{P}_i(x) \in \mathrm{H}_\zeta^{m_1 \times m_1}$, rank $P_i(x) = n_i$ for $i = 1, \ldots, r$. Hence, $P_1(x) \in \mathrm{H}_\zeta^{n \times n}$, rank $P_1(x) = n_1$ as we claimed.

Assume now that $\lambda_1(x), \ldots, \lambda_n(x)$ are polynomials. Then $P_i(x)$ are entire functions on $\mathbb{C}$. (See for example [Rud74].) Since $\lim_{|x| \to \infty} \frac{A(x)}{x^\ell} = A_l$ it follows that $\limsup_{|x| \to \infty} \frac{|\lambda_i(x)|}{|x|^\ell} \leq \rho(A)$, where $\rho(A)$ is the spectral radius of A_ℓ. Hence, each $\lambda_j(x)$ is polynomial of degree ℓ at most. Since $A_\ell \neq 0$ it follows that at least one of $\lambda_j(x)$ is a polynomial of degree ℓ exactly. We may assume that $\deg \lambda_1(x) = \ell$. This proves the condition (a) of the theorem. The condition (b) is equivalent to the statement that no two distinct eigenvalues of $A(x)$ are tangent in $\mathbb{C}$.

Define $P_i(\zeta)$ by (3.11.11). As in the proof of Theorem 3.11.5 it follows that $P_i(x) \in \mathbb{C}[x]^{n \times n}$ and rank $P_i(x) = n_i$, $i = 1, \ldots, m$. Furthermore we define $S(x) \in \mathbf{GL}(n, \mathbb{C}[x])$ as in the proof of Theorem 3.11.5 such that $B(x) = S^{-1}(x)A(x)S(x)$. This proves the first part of the theorem.

To prove the second part of the theorem observe that in view of our definition of tangency at ∞ the condition (c) is equivalent to the condition that no two distinct eigenvalues of A are tangent at infinity. Assume now that A_l is diagonable and no two distinct eigenvalues are tangent at ∞. Then the above arguments show that each $P_i(x)$ is also multivalued analytic at ∞. By considering $x^{-l}A(x)$ it follows that $P_i(x)$ is bounded at the neighborhood of ∞. Hence, $P_i(x) = P_i(0)$ for $i = 1, \ldots, m$. Thus $S \in \mathbf{GL}(n, \mathbb{C})$. So $A(x)$ is diagonable by a constant matrix. In particular all the eigenvalues of $A(x)$ are polynomials. Since no two distinct eigenvalues are tangent at ∞ we deduce the condition (c) holds. $\square$

Problems

1. Let $A(x) \in \mathbb{C}[x]^{n \times n}$. Assume that there exists an infinite sequence of distinct points $\{\zeta_k\}_1^\infty$ such that $A(\zeta_k)$ is diagonable for $k = 1, \ldots$ Show that $A(x)$ is diagonable for all but a finite number of points. (*Hint*: Consider the rational canonical form of $A(x)$ over the field of rational functions $\mathbb{C}(x)$.)

2. Consider the matrix $A(x)$ given in (3.11.12). Show

   (a) $A(x)$ and $A(x_0, x_1)$ are pointwise similar to diagonal matrices in $\mathbb{C}$ and $\mathbb{C}^2$ respectively.

   (b) The eigenvalues of $A(x)$ are not tangent at any point in $\mathbb{C}$.

   (c) Find $S(x) \in \mathbf{GL}(2, \mathbb{C}[x])$ such that $S^{-1}(x)A(x)S(x) = \mathrm{diag}(x^2, 1 + x^2)$.

   (d) Show that $A(x)$ is not strictly similar to $\mathrm{diag}(x^2, 1 + x^2)$.

   (c) Show that the eigenvalues of $A(x)$ are tangent at $\zeta = \infty$.

3.12   Property L

In this section and the next one we assume that all pencils $A(x) = A_0 + A_1 x$ are square pencils, i.e. $A(x) \in \mathbb{C}[x]^{n \times n}$, and $A_1 \neq 0$ unless stated otherwise. Then $A(x_0, x_1) = A_0 x_0 + A_1 x_1$.

Definition 3.12.1 *A pencil* $A(x) \in \mathbb{C}[x]^{n \times n}$ *has property L if all the eigenvalues of* $A(x_0, x_1)$ *are linear functions. That is* $\lambda_i(x_0, x_1) = \alpha_i x_0 + \beta_i x_1$ *is an eigenvalue of* $A(x_0, x_1)$ *of multiplicity* n_i *for* $i = 1, \ldots, m$, *where*

$$n = \sum_{i=1}^m n_i, \quad (\alpha_i, \beta_i) \neq (\alpha_j, \beta_j), \text{ for } 1 \leq i < j \leq m.$$

The proofs of the following propositions is left to the reader. (See Problems 1 and 2.)

Proposition 3.12.2 *Let* $A(x) = A_0 + xA_1$ *be a pencil in* $\mathbb{C}[x]^{n \times n}$. *TFAE:*

(a) $A(x)$ *has property L.*

(b) *The eigenvalues of* $A(x)$ *are polynomials of degree 1 at most.*

(c) *The characteristic polynomial of* $A(x)$ *splits to linear factors over* $\mathbb{C}[x]$.

(d) *There is an ordering of the eigenvalues of* A_0 *and* A_1, $a_1, \ldots, a_n$ *and* $b_1, \ldots, b_n$, *respectively, such that the eigenvalues of* $A_0 x_0 + A_1 x_1$ *are* $a_1 x_0 + b_1 x_1, \ldots, a_n x_0 + b_n x_1$.

Proposition 3.12.3 *Let* $A(x)$ *be a pencil in* $\mathbb{C}[x]^{n \times n}$. *Then* $A(x)$ *has property L if one of the following conditions hold:*

(a) $A(x)$ *is similar over* $\mathbb{C}(x)$ *to an upper triangular matrix* $U(x) \in \mathbb{C}(x)^{n \times n}$.

(b) $A(x)$ *is strictly similar to an upper triangular pencil* $U(x) = U_0 + U_1 x$, *i.e.* U_0, U_1 *are upper triangular.*

(c) $A(x)$ *is similar over* $\mathbb{C}[x]$ *to a diagonal matrix* $B(x) \in \mathbb{C}[x]^{n \times n}$.

(d) $A(x)$ *is strictly similar to diagonal pencil.*

Note that for pencils with property L any two distinct eigenvalues are not tangent at any point of $\mathbb{P}$. For pencils one can significantly improve Theorem 3.11.7.

Theorem 3.12.4 *Let* $A(x) = A_0 + A_1 x \in \mathbb{C}[x]^{n \times n}$ *be a non-constant pencil* $(A_1 \neq 0)$. *Assume that* $A(x)$ *is pointwise diagonable on* $\mathbb{C}$. *Then* $A(x)$ *has property L. Furthermore* $A(x)$ *is similar over* $\mathbb{C}[x]$ *to a diagonal pencil* $B(x) = B_0 + B_1 x$. *Suppose furthermore that* A_1 *is diagonable, i.e.* $A(x_0, x_1)$ *is pointwise diagonable on* $\mathbb{C}^2$. *Then* $A(x)$ *is strictly similar to the diagonal pencil* $B(x)$, *i.e.* A_0 *and* A_1 *are commuting diagonable matrices.*

Proof. We follow the proof of Theorem 3.11.7. Let $\lambda_1(x), \ldots,$ $\lambda_m(x)$ be the eigenvalues of $A(x)$ of multiplicities $n_1, \ldots, n_m$ respectively, where each $\lambda_j(x)$ is viewed as multivalued function of x. More precisely, there exists an irreducible polynomial

$$\phi(x, \lambda) = \lambda^p + \sum_{q=1}^{p} \phi_q(x) \lambda^{p-q} \in \mathbb{C}[x, \lambda],$$
(3.12.1)

$$\phi(x, \lambda) | \det (\lambda I - A(x)),$$

such that $\lambda_j(x)$ satisfies the algebraic equation

$$\phi(x, \lambda) = 0.$$
(3.12.2)

Moreover, all branches generated by $\lambda_j(x)$ on $\mathbb{C}$ will generate all the solutions $\lambda(x)$ of (3.12.2). Equivalently all pairs (x, λ) satisfying (3.12.2) form an affine algebraic variety $V_0 \subset \mathbb{C}^2$. If we compactify V_0 to a projective variety $V \subset \mathbb{P}^2$ then V is a compact Riemann surface. $V \backslash V_0$ consists of a finite number of points, the points of V_0 at infinity. The compactification of V_0 is equivalent to considering $\lambda_j(x)$ as a multivalued function on $\mathbb{P}$. See for example [GuR65]. Note that any local solution of (3.12.2) is some eigenvalue $\lambda_i(x)$ of $A(x)$. Since $A(\zeta)$ is diagonable at $\zeta \in \mathbb{C}$ Theorem 3.8.1 implies that the Puiseux expansion of $\lambda_j(x)$ around ζ in (3.11.5) is of the form

$$\lambda_j(x) = \lambda_j(\zeta) + \sum_{k=s}^{\infty} \lambda_{jk}(\zeta)(x - \zeta)^{\frac{k}{s}}.$$

Then

$$\frac{d\lambda_j(x)}{dx} = \sum_{k=s}^{\infty} \frac{k}{s} \lambda_{jk}(\zeta)(x - \zeta)^{\frac{k-s}{s}}.$$

So $\frac{d\lambda_j(x)}{dx}$ is a multivalued locally bounded function on $\mathbb{C}$. Equivalently, using the fact that $\lambda_j(x)$ satisfy (3.12.2) we deduce

$$\frac{d\lambda_j(x)}{dx} = -\frac{\frac{\partial \phi(x, \lambda)}{\partial x}}{\frac{\partial \phi(x, \lambda)}{\partial y}}.$$
(3.12.3)

Hence, $\frac{d\lambda_j(x)}{dx}$ is a rational function on V, which is analytic on V_0 in view of the assumption that $A(x)$ is pointwise diagonable in $\mathbb{C}$. The Puiseux expansion of $\lambda_j(x)$ it ∞ (3.11.6) is

$$\lambda_j(x) = x \sum_{k=0}^{\infty} \lambda_{jk}(\infty) x^{-\frac{k}{s}}.$$

Hence

$$\frac{d\lambda_j(x)}{dx} = \lambda_{j0}(\infty) + \sum_{k=1}^{\infty} \frac{s-k}{s} \lambda_{jk}(\infty) x^{-\frac{k}{s}}.$$

That is the multivalued function $\frac{d\lambda_j(x)}{dx}$ is bounded at the neighborhood of ∞. Equivalently the rational function in (3.12.3) is bounded at all points of $V \backslash V_0$. Thus the rational function in (3.12.3) is bounded on a compact Riemann surface (3.12.2). Hence it must be constant, i.e. $\frac{d\lambda_j(x)}{dx} = b_j \Rightarrow \lambda_j(x) = a_j + b_j x$. So we have property L by part (b) of Proposition 3.12.2. In particular two distinct eigenvalues of $A(x)$ are not tangent at any $\zeta \in \mathbb{P}$. The first part of Theorem 3.11.7 implies that $A(x)$ is similar to $B(x) = \sum_{j=1}^{m} \oplus (a_j + b_j x) I_{n_j}$ over $\mathbb{C}[x]$.

Assume now that A_1 is diagonable. Then the second part of Theorem 3.11.7 yields that $A(x)$ is strictly similar to $B(x)$, which is equivalent to the assumption that A_0, A_1 are commuting diagonable matrices (Theorem 3.10.4). $\square$

Theorem 3.12.5 *Let* $A(x) = A_0 + A_1 x \in \mathbb{C}[x]^{n \times n}$. *Assume that* A_1 *and* A_2 *are diagonable and* $A_0 A_1 \neq A_1 A_0$. *Then exactly one of the following conditions hold:*

(a) $A(x)$ *is not diagonable exactly at the points* $\zeta_1, \ldots, \zeta_p$, *where* $1 \leq p \leq n(n-1)$.

(b) $A(x)$ *is diagonable exactly at the points* $\zeta_1 = 0, \ldots, \zeta_q$ *for some* $q \geq 1$.

Proof. Combine the assumptions of the theorem with Theorem 3.12.4 to deduce the existence of $0 \neq \zeta \in \mathbb{C}$ such that $A(\zeta)$

is not diagonable. Consider the homogenized pencil $A(x_0, x_1) = A_0 x_0 + A_1 x_1$. Let

$$C(p_1, \ldots, p_k)(x_0, x_1) = \oplus_{j=1}^k C(p_j) \in \mathbb{C}[x_0, x_1]^{n \times n},$$

$$\prod_{i=1}^k p_i(x_0, x_1, \lambda) = \det(\lambda I - A(x_0, x_1)),$$

$$p_i(x_0, x_1, \lambda)$$

$$= \lambda^{m_i} + \sum_{j=1}^{m_i} \lambda^{m_i - j} p_{ij}(x) \in \mathbb{C}[x_0, x_1][\lambda], 1 \le m_i \quad i = 1, \ldots, k,$$

$$p_1 | p_2 | \ldots | p_k,$$

be the rational canonical form $A(x_0, x_1)$ over $\mathbb{C}(x_0, x_1)$. (See §2.3.) That is each $p_i(x_0, x_1, \lambda)$ is a nontrivial invariant polynomial of $\lambda I - A(x_0, x_1)$. Hence, each $p_i(x_0, x_1, \lambda)$ is a homogeneous polynomial of degree m_i in x_0, x_1, λ. Furthermore $A(x_0, x_1) = S(x_0, x_1) C(p_1, \ldots, p_k)(x_0, x_1) S(x_0, x_1)^{-1}$ for some $S(x_0, x_1) \in \mathbb{C}[x_0, x_1]^{n \times n} \cap \mathbf{GL}(n, \mathbb{C}(x_0, x_1))$. Choose $x_0 = \tau \neq 0$ such that $\det S(\tau, x_1)$ is not identically zero in x_1. Then $A(x) = A(1, x)$ is pointwise similar to $\frac{1}{\tau} C(p_1, \ldots, p_k)(\tau, \tau x)$ at all point for which $\det S(\tau, \tau x) \neq 0$, i.e. at all but a finite number of points in $\mathbb{C}$.

Since $\mathbb{C}[x_0, x_1, \lambda]$ is $\mathbb{D}_u$ then $p_k(x_0, x_1, \lambda) = \prod_{i=1}^r \phi_i(x_0, x_1, \lambda)^{\ell_i}$, where each ϕ_i is a nonconstant irreducible (homogeneous) polynomial and ϕ_i is coprime with ϕ_j for $i \neq j$. Assume first that some $\ell_i > 1$. Then $C(p_k)(\tau, t)$ has a multiple eigenvalue for any $t \in \mathbb{C}$, hence it is not diagonable. That is the condition (b) of the theorem holds.

Assume now that $\ell_1 = \cdots = \ell_r = 1$. This is equivalent to the assumption that $p_k(x_0, x_1, \lambda) = 0$ does not have multiple roots for some (x_0, x_1). We claim that it is possible to choose $\tau \neq 0$ such that $p_k(\tau, x_1, \lambda)$ has m_k pairwise distinct roots (in λ) except in the points $\zeta_1, \ldots, \zeta_q$. Consider the discriminant $D(x_0, x_1)$ of $p_k(x_0, x_1, \lambda) \in \mathbb{C}[x_0, x_1][\lambda]$. See §1.9. Since p_{ki} is a homogeneous polynomial of degree i for $i = 1, \ldots, m_k$ it follows that $D(x_0, x_1)$ is homogeneous polynomial of degree $m_k(m_k - 1) \le n(n - 1)$. Since

$p_k(x_0, x_1, \lambda) = 0$ does not have multiple roots for some (x_0, x_1) it follows that $D(x_0, x_1)$ is not a zero polynomial, and $p_k(x_0, x_1, \lambda) = 0$ has a multiple root if and only if $D(x_0, x_1) = 0$. Choose $\tau \neq 0$ such that $D(\tau, x_1)$ is not a zero polynomial. Let $\zeta_1, \ldots, \zeta_q$ be the distinct roots of $D(\tau, \tau x) = 0$. Since the degree of $D(x_0, x_1)$ is at most $n(n-1)$ it follows that the degree of $D(\tau, x)$ is at most $n(n-1)$. Hence, $0 \leq q \leq n(n-1)$. By the definition of the invariant polynomials it follows that $p_k(x_0, x_1, A(x_0, x_1)) = 0$. Hence, $p_k(\tau, \tau t, A(\tau, \tau t)) = 0$. Let $t \in X = \mathbb{C} \backslash \{\zeta_1, \ldots, \zeta_q\}$. Since $p_k(\tau, \tau t, \lambda)$ has m_k distinct roots, which are all eigenvalues of $A(\tau, \tau t)$ it follows that $A(\tau, \tau t) = \tau A(t)$ is a diagonable matrix. $\square$

For $n = 2$ the case (b) in Theorem 3.12.5 does not arise. See Problem 4. We do not know if the case (b) of Theorem 3.12.5 arises.

Problems

1. Prove Proposition 3.12.2.

2. (a) Show that property L is equivalent to the condition (a) of Proposition 3.12.3.

   (b) Prove the other conditions of Proposition 3.12.3.

3. Show that a pencil $A(x) = A_0 + A_1 x \in \mathbb{C}[x]^{2 \times 2}$ have property L if and only if $A(x)$ is strictly similar to an upper triangular pencil.

4. Let $A(x_0, x_1) = A_0 x_0 + A_1 x_1 \in \mathbb{C}[x_0, x_1]^{2 \times 2}$. Then exactly one the following conditions hold.

   (a) $A(x_0, x_1)$ is strictly similar to a diagonal pencil. (Property L holds.)

   (b) $A(x_0, x_1)$ is not diagonable except exactly for the points $(x_0, x_1) \neq (0, 0)$ lying on a line $a x_0 + b x_1 = 0$. (Property L holds, $A_0 A_1 = A_1 A_0$ but A_i is not diagonable for some $i \in \{1, 2\}$, $A(x_0, x_1)$ has a double eigenvalue.)

   (c) $A(x_0, x_1)$ is diagonable except exactly for the points $(x_0, x_1) \neq (0, 0)$ lying on a line $a x_0 + b x_1 = 0$. (Property L holds.)

(d) $A(x_0, x_1)$ is diagonable except exactly the points $(x_0, x_1) \neq (0,0)$ which lie on two distinct lines in $\mathbb{C}^2$. (Property L does not hold.)

5. Let

$$A_0 = \begin{bmatrix} 0 & 1 & 0 \\ 0 & 0 & 1 \\ 0 & 0 & 0 \end{bmatrix}, \quad A_1 = \begin{bmatrix} 1 & 1 & 2 \\ 1 & 1 & 2 \\ -1 & -1 & -2 \end{bmatrix}.$$

(a) Show that A_0, A_1 are nilpotent while $A_0 + A_1$ is nonsingular.

(b) Show that $A(x) = A_0 + A_1 x$ does not have property L.

(c) Show that $A(x)$ is diagonable for all $x \neq 0$.

3.13 Strict Similarity of Pencils and Analytic Similarity

Let $A(x) = A_0 + A_1 x, B(x) = B_0 + B_1 x \in \mathbb{C}[x]^{n \times n}$. Recall the notion of strict equivalence $A(x) \overset{s}{\sim} B(x)$ (2.1) and strict similarity $A(x) \overset{s}{\approx} B(x)$ (3.10). Clearly $A(x) \overset{s}{\approx} B(x) \Rightarrow A(x) \overset{s}{\sim} B(x)$ (2.9.3) yields.

Proposition 3.13.1 *Let* $A(x), B(x) \in \mathbb{C}[x]^{n \times n}$ *be two strictly similar pencils. Then the three pencils in* (3.6.2) *are strictly equivalent.*

Using Kronecker's result (Theorem 2.1.7) we can determine if the three pencils in (3.6.2) are strictly equivalent. We now study the implications of Proposition 3.13.1.

Lemma 3.13.2 *Let* $A(x) = A_0 + A_1 x, B(x) = B_0 + B_1 x \in \mathbb{C}[x]^{n \times n}$ *be two pencils such that*

$$I \otimes A(x) - A(x)^\top \otimes I \overset{s}{\sim} I \otimes A(x) - B(x)^\top \otimes I. \qquad (3.13.1)$$

Then there exists two nonzero $U, V \in \mathbb{C}^{n \times n}$ such that

$$A(x)U - UB(x) = 0, \quad VA(x) - B(x)V = 0. \tag{3.13.2}$$

In particular

$$A_0 \mathrm{Ker}\, V, A_1 \mathrm{Ker}\, V \subset \mathrm{Ker}\, V, \quad B_0 \mathrm{Ker}\, U, B_1 \mathrm{Ker}\, U \subset \mathrm{Ker}\, U. \tag{3.13.3}$$

Proof. As $A(x)I - IA(x) = 0$ it follows that the kernels of $I \otimes A(x) - A(x)^\top \otimes I \in \mathbb{C}[x]^{n^2 \times n^2}$ and its transpose contain a nonzero vector $\hat{I}_n \in \mathbb{C}^{n^2}$ which is induced by I_n. (See §2.8.) Hence, the kernel of $I \otimes A(x) - B(x)^\top \otimes I$ contain nonzero constant vectors. This is equivalent to (3.13.2).

Assume that (3.13.2) holds. Let $\mathbf{x} \in \mathrm{Ker}\, V$. Multiply the second equality in (3.13.2) from the right by $\mathbf{x}$ to deduce the first part (3.13.3). The second part of (3.13.3) is obtained similarly. $\square$

Definition 3.13.3 $A_0, A_1 \in \mathbb{C}^{n \times n}$ *have a common invariant subspace if there exist a subspace* $\mathbf{U} \subset \mathbb{C}^n$, $1 \le \dim \mathbf{U} \le n - 1$ *such that* $A_0 \mathbf{U}, A_1 \mathbf{U} \subset \mathbf{U}$.

The following claims are left to the reader (see Problems 1 and 2).

Proposition 3.13.4 *Let* $A(x) = A_0 + xA_1 \in \mathbb{C}[x]^{n \times n}$. *Then* $A(x)$ *is strictly similar to a block upper triangular pencil*

$$B(x) = \begin{bmatrix} B_{11}(x) & B_{12}(x) \\ 0 & B_{22}(x) \end{bmatrix},$$
$B_{11}(x) \in \mathbb{C}[x]^{n_1 \times n_1}$, $B_{12}(x) \in \mathbb{C}[x]^{n_1 \times n_2}$, $B_{22}(x) \in \mathbb{C}[x]^{n_2 \times n_2}$,

$1 \le n_1, n_2, \; n_1 + n_2 = n,$

$$\tag{3.13.4}$$

if and only if A_0, A_1 have a common invariant subspace.

Proposition 3.13.5 *Assume that $A(x) \in \mathbb{C}[x]^{n \times n}$ is similar over $\mathbb{C}(x)$ to a block upper triangular matrix $B(x)$ of the form (3.13.4). Then* $\det(\lambda I - A(x)) \in \mathbb{C}[x, \lambda]$ *is reducible.*

Theorem 3.13.6 *Let $A(x) = A_0 + A_1 x, B(x) = B_0 + B_1 x \in$ $\mathbb{C}[x]^{n \times n}$. Assume that either $\det(\lambda I - A(x))$ or $\det(\lambda I - B(x))$ is irreducible over $\mathbb{C}[x, \lambda]$. Then $A(x) \overset{s}{\approx} B(x)$ if and only if (3.13.1) holds.*

Proof. Assume that (3.13.1) holds. Suppose that $\det(\lambda I - A(x))$ is irreducible. Propositions 3.13.4–3.13.5 imply that A_0, A_1 do not have a common invariant subspace. Lemma 3.13.2 implies that the matrix V in (3.13.2) is invertible, i.e. $B(x) = VA(x)V^{-1}$. Similarly if $\det(\lambda I - B(x))$ is irreducible then $B(x) = U^{-1}A(x)U$. $\qquad \square$

Definition 3.13.7 *Let $\mathcal{I}_n \subset (\mathbb{C}^{n \times n})^2$ be the set of all pairs (A_0, A_1) such that $\det(\lambda I - (A_0 + A_1 x))$ is irreducible.*

We will show later that $\mathcal{I}_n = (\mathbb{C}^{n \times n})^2 \backslash X_n$ where X_n is a strict sub-variety of $(\mathbb{C}^{n \times n})^2$. That is, for most of the pencils $A(x), (A_0, A_1) \in$ $(\mathbb{C}^{n \times n})^2 \det(\lambda I - A(x))$ is irreducible. Clearly if $(A_0, A_1) \overset{s}{\approx} (B_0, B_1)$ then either $(A_0, A_1), (B_0, B_1) \in \mathcal{I}_n$ or $(A_0, A_1), (B_0, B_1) \notin \mathcal{I}_n$.

Corollary 3.13.8 *Let $(A_0, A_1), (B_0, B_1) \in \mathcal{I}_n$. Then $A(x) = A_0 + A_1 x$ is strictly similar to $B(x) = B_0 + B_1 x$ if and only if (3.13.1) holds.*

We now discuss the connection between the notion of analytic similarity of matrices over H_0 and strict similarity of pencils. Let $A(x), B(x) \in H_0^{n \times n}$ and assume that $\eta(A, A) = 1$. Suppose that $A \overset{r}{\approx} B(x)$. Theorem 3.9.2 claims that $A(x) \approx B(x)$ if and only if there exists two matrices $T_0 \in \mathbf{GL}(n, \mathbb{C}), T_1 \in \mathbb{C}^{n \times n}$ such that

$$A_0 T_0 = T_0 B_0, \quad A_1 T_0 + A_0 T_1 = T_0 B_1 + T_1 B_0.$$

Let

$$F(A_0, A_1) = \begin{bmatrix} A_0 & A_1 \\ 0 & A_0 \end{bmatrix} \in \mathbb{C}^{2n \times 2n}. \qquad (3.13.5)$$

Then (3.9.2) is equivalent in this case to

$$F(A_0, A_1)F(T_0, T_1) = F(T_0, T_1)F(B_0, B_1). \qquad (3.13.6)$$

As det $F(T_0, T_1) = (\det T_0)^2$ it follows that T_0 is invertible if and only if $F(T_0, T_1)$ is invertible.

Definition 3.13.9 *Let $A_0, A_1, B_0, B_1 \in \mathbb{C}^{n \times n}$. Then $F(A_0, A_1)$ and $F(B_0, B_1)$ are called strongly similar ($F(A_0, A_1) \cong F(A_0, A_1)$) if there exists $F(T_0, T_1) \in \mathbf{GL}(2n, \mathbb{C})$ such that (3.13.6) holds.*

Clearly $F(A_0, A_1) \cong F(A_0, A_1) \Rightarrow F(A_0, A_1) \approx \cdot F(B_0, B_1)$. It can be shown that the notion of strong similarity is stronger that the notion of similarity (Problem 10).

Proposition 3.13.10 *The matrices $F(A_0, A_1)$ and $F(B_0, B_1)$ are strongly similar if and only if the pencils*

$$A(x) = F(0, I) + F(A_0, A_1)x, \quad B(x) = F(0, I) + F(B_0, B_1)x$$

are strictly similar.

Proof. Let $[P_{ij}]_1^2 \in \mathbb{C}^{2n \times 2n}$. Then $F(0, I)P = PF(0, I)$ if and only if $P_{11} = P_{22}$, $P_{21} = 0$. That is $P = F(P_{11}, P_{12})$ and the proposition follows.   □

Clearly $F(A_0, A_1) \cong F(A_0, B_1) \Rightarrow A_0 \approx B_0$. Without loss of generality we may assume that $A_0 = B_0$. (See Problem 5.) Consider all matrices T_0, T_1 satisfying (3.13.6). For $A_0 = B_0$ (3.13.6) reduces 1to

$$A_0 T_0 = T_0 A_0, \quad A_0 T_1 - T_1 A_0 = T_0 B_1 - A_1 T_0.$$

Theorem 2.10.1 yields that the set of matrices T_0 which satisfies the above conditions is of the form

$$\mathcal{P}(A_1, B_1) = \{T_0 \in C(A_0) : \quad (3.13.7)$$

$$\mathrm{tr}(V(T_0 B_1 - A_1 T_0)) = 0, \text{ for all } V \in C(A_0)\}.$$

Hence

Proposition 3.13.11 *Suppose that $F(A_0, A_1) \cong F(A_0, B_1)$. Then*

$$\dim \mathcal{P}(A_1, A_1) = \dim \mathcal{P}(A_1, B_1) = \dim \mathcal{P}(B_1, B_1). \quad (3.13.8)$$

As in Theorem 2.9.5 for a fixed A_0, A_1 there exists a neighborhood $D(A_1, \rho)$ such that the first two equalities in (3.11.9) imply that $F(A_0, A_1) \cong F(A_0, B_1)$ for all $B_1 \in D(A_1, \rho)$ (Problem 4).

We now considering a splitting result analogous to Theorem 3.5.7.

Theorem 3.13.12 *Assume that*

$$A_0 = \begin{bmatrix} A_{11}^{(0)} & 0 \\ 0 & A_{22}^{(0)} \end{bmatrix}, \quad A_{ii}^{(0)} \in \mathbb{C}^{n_i \times n_i}, \ i = 1, 2, \tag{3.13.9}$$

where $A_{11}^{(0)}$ and $A_{22}^{(0)}$ do not have a common eigenvalue. Let

$$A_1 = \begin{bmatrix} A_{11}^{(1)} & A_{12}^{(1)} \\ A_{21}^{(1)} & A_{22}^{(1)} \end{bmatrix}, \quad B_1 = \begin{bmatrix} B_{11}^{(1)} & B_{12}^{(1)} \\ B_{21}^{(1)} & B_{22}^{(1)} \end{bmatrix}$$

be the block partition of A_1, B_1 as the block partition of A_0. Then

$$\mathcal{P}(A_1, B_1) = \mathcal{P}(A_{11}^{(1)}, B_{11}^{(1)}) \oplus \mathcal{P}(A_{22}^{(1)}, B_{22}^{(1)}). \tag{3.13.10}$$

Moreover

$$F(A_0, A_1) \cong F(A_0, B_1) \iff F(A_{ii}^{(0)}, A_{ii}^{(1)})$$

$$\cong F(A_{ii}^{(0)}, B_{ii}^{(1)}) \text{ for } i = 1, 2.$$

Proof. According to Problem 4 $C(A_0) = C(A_{11}^{(0)}) \oplus C(A_{22}^{(0)})$. Then the trace condition in (3.13.7) reduces to

$$\mathrm{tr}(V_1(T_1^{(0)} B_{11}^{(1)} - A_{11}^{(1)} T_1^{(0)}) + V_2(T_2^{(0)} B_{22}^{(1)} - A_{22}^{(1)} T_2^{(0)})) = 0,$$

where

$$V = V_1 \oplus V_2, \ T_0^{(0)} = T_1^{(0)} \oplus T_2^{(0)} \in C(A_{11}^{(0)}) \oplus C(A_{22}^{(0)}).$$

Choosing either $V_0 = 0$ or $V_1 = 0$ we obtain (3.13.10). The right implication of the last claim of the theorem is straightforward. As $\det T_0 = \det T_1^{(0)} \det T_2^{(0)}$ it follows that $T_0 \in \mathbf{GL}(n, C) \iff T_i^{(0)} \in \mathbf{GL}(n_i, \mathbb{C})$, $i = 1, 2$. This establishes the left implication of the last claim of the theorem. $\qquad \square$

Thus, the classification of strong similarity classes for matrices $F(A_0, A_1)$ reduces to the case where A_0 is nilpotent (Problem 6). In the case $A_0 = 0$ $F(0, A_1) \cong F(0, B_1) \iff A_1 \approx B_1$. In the case $A_0 = H_n$ the strong similarity classes of $F(H_n, A_1)$ classified completely (Problem 9). This case corresponds to the case discussed in Theorem 3.5.5. The case $A_0 = H_m \oplus H_m$ can be classified completely using the results of Problem 2 (Problem 3.13.11).

Problems

1. Prove Proposition 3.13.4.

2. Prove Proposition 3.13.5.

3. Let $A(x) \in \mathbb{C}[x]^{n \times n}$ and assume that $A(x)\mathbf{U} \subset \mathbf{U} \subset \mathbb{C}^n$ is a nontrivial invariant subspace of $A(x)$, i.e. $1 \leq \dim \mathbf{U} \leq n-1$. Let $p(x, \lambda) \in \mathbb{C}[x, \lambda]$ be the minimal polynomial of $A(x)|\mathbf{U}$. Thus $1 \leq \deg_\lambda p(x, \lambda) \leq n-1$. Show that $p(x, \lambda)|\det(\lambda I - A(x))$. Hence, $\det(\lambda I - A(x))$ is reducible over $\mathbb{C}[x, \lambda]$.

4. Modify the proof of Theorem 2.9.5 to show that for a fixed $A_0, A_1 \in \mathbb{C}^{n \times n}$ there exists $\rho > 0$ such that the first two equalities in (3.13.8) for $B_1 \in D(A_1, \rho)$ imply that $F(A_0, A_1) \cong F(A_0, B_1)$.

5. Show that for any $P \in \mathbf{GL}(n, \mathbb{C})$

$$F(A_0, A_1)$$
$$\cong F(B_0, B_1) \iff F(A_0, A_1) \cong F(PB_0P^{-1}, PB_1P^{-1}).$$

Assume that $F(A_0, A_1) \cong F(B_0, B_1)$. Show that there exists $P \in \mathbf{GL}(n, \mathbb{C})$ such that $A_0 = PB_0P^{-1}$.

6. Show that for any $\lambda \in \mathbb{C}$

$$F(A_0, A_1) \cong F(B_0, B_1) \iff F(A_0 - \lambda I, A_1) \cong F(B_0 - \lambda I, B_1).$$

7. Let $A_i \in \mathbb{C}^{n \times n}$, $i = 0, \dots, s-1$. Define

$$F(A_0, \dots, A_{s-1}) = \begin{bmatrix} A_0 & A_1 & A_2 & \dots & A_{s-1} \\ 0 & A_0 & A_1 & \dots & A_{s-2} \\ \vdots & \vdots & \vdots & \ddots & \vdots \\ 0 & 0 & 0 & \dots & A_0 \end{bmatrix} \in \mathbb{C}^{sn \times sn}.$$

$F(A_0, \dots, A_{s-1})$ and $F(B_0, \dots, B_{s-1})$ are called strongly similar $(F(A_0, \dots, A_{s-1}) \cong F(B_0, \dots, B_{s-1}))$

$$F(A_0, \dots, A_{s-1}) = F(T_0, \dots, T_{s-1}) F(B_0, \dots, B_{s-1})$$
$$F(T_0, \dots, T_{s-1})^{-1}, \quad F(T_0, \dots, T_{s-1}) \in \mathbf{GL}(sn, \mathbb{C}).$$

Show that $F(A_0, \dots, A_{s-1}) \cong F(B_0, \dots, B_{s-1})$ if and only if the equalities (3.9.2) hold for $k = 0, \dots, s-1$ and $T_0 \in \mathbf{GL}(n, \mathbb{C})$.

8. Let

$$Z = H_n \oplus \dots \oplus H_n, \quad X = [X_{pq}]_1^s, \ [Y_{pq}]_1^s \in \mathbb{C}^{sn \times sn},$$
$$X_{pq} = [x_{ij}^{(pq)}]_1^n, \ Y_{pq} = [y_{ij}^{(pq)}]_1^n \in \mathbb{C}^{n \times n}, \quad p, q = 1, \dots, s.$$

Show that if each X_{pq} is an upper triangular matrix then

$$\det X = \prod_{r=1}^n \det [x_{rr}^{(pq)}]_{p,q=1}^s.$$

(Expand the determinant of X by the rows $n, 2n, \dots, sn$ and use the induction.) Define

$$A_r = \left[a_{pq}^{(r)} \right]_1^s, \ B_r = \left[b_{pq}^{(r)} \right]_1^s \in \mathbb{C}^{s \times s},$$

$$a_{pq}^{(r)} = \sum_{i=1}^{r+1} x_{(n-r+i-1)i}^{(pq)},$$

$$b_{pq}^{(r)} = \sum_{i=1}^{r+1} y_{(n-r+i-1)i}^{(pq)}, \quad r = 0, \dots, n-1.$$

Using Theorem 2.8.3 shows

$$F(Z, X) \cong F(Z, Y) \iff F(A_0, \dots, A_{n-1}) \cong F(B_0, \dots, B_{n-1}).$$

9. Let $X = [x_{ij}]_1^n$, $Y = [y_{ij}]_1^n \in \mathbb{C}^{n \times n}$. Using Problems 7–8 show that $F(H_n, X) \cong F(H_n, Y)$ if and only if

$$\sum_{i=1}^{r} x_{(n-r+i)i} = \sum_{i=1}^{r} y_{(n-r+i)i}, \quad \text{for } r = 1, \ldots, n.$$

10. Let $X = [x_{ij}]_1^2 \in \mathbb{C}^{2 \times 2}$. Show that if $x_{21} \neq 0$ then $F(H_2, X) \cong H_4$. Combine this result with Problem 9 to show the existence of $Y \in \mathbb{C}^{2 \times 2}$ such that $F(H_2, X)$ is similar to $F(H_2, Y)$ but $F(H_2, X)$ is not strongly similar to $F(H_2, Y)$.

$$(3.13.11)$$

Assume in Problem 8 $s = 2$. Let

$$A(x) = \sum_{i=0}^{n-1} A_i x^i, \ B(x) = \sum_{i=0}^{n-1} B_i x^i \in H_0^{2 \times 2}.$$

Use the results of Problems 7 and 8, (3.9.2) and Problem 2 to show that $F(Z, X) \cong F(Z, Y)$ if and only if the three matrices in (3.6.2) have the same local invariant polynomials up to degree $n - 1$.

3.14   Historical Remarks

The exposition of §3.1 is close to [Gan59]. The results of §3.2 were inspired by [Rot81]. The notion of local indices (Problem 3.2.3) can be found in [FrS80]. The content of §3.3 are standard. Theorem 3.3.2 can be found in [Wie67] and Problem §3.3.1 in [Gan59]. The use of Cauchy integration formula to study the properties of analytic functions of operators and matrices as in §3.4 is now common, e.g. [Kat80] and [Kat82]. Theorem 3.4.6 is standard. Theorem 3.4.9 is a part of the Kreiss matrix stability theorem [Kre62]. The inequality (3.4.16) is due to [Tad81]. The results of Problem 3.4.7 are from [Fri81a]. The results of §3.5 influenced by Arnold [Arn71], in particular Theorem 3.5.3 is from [Arn71]. See also [Was77]. The subject of §3.6 and its applications in theory of differential equations in neighborhood of singularities was emphasized in works of Wasow [Was63], [Was77] and [Was78]. Theorem 3.6.4 for one complex variable appears in [Fri80b]. Corollary 3.6.5 is due to [Was63]. Theorem 3.7.1 for simply connected domain is due to [Gin78]. See [Was78] for the extension of Theorem 3.7.1 to certain domains $\Omega \subset \mathbb{C}^p$. It is shown there that Theorem 3.7.1 fails even for simply connected domains in $\mathbb{C}^3$.

Theorem 3.8.1 can be found in [Kat80] or [Fri78]. The results of §3.9 were taken from [Fri80b]. It is worthwhile to mention that the conjecture stated in [Fri80b] that $A(x)$ and $B(x)$ are analytically similar over H_0 if the three matrices in (3.6.2) are equivalent over H_0 is false [Gur81, §6]. The contents of §3.10 are known to the experts. The nontrivial part of this section (Theorem 3.10.6) is due to [DDG51]. Theorem 3.11.1 is stated in [Fri80b]. Some other results in §3.11, in particular Theorem 3.11.5, seem to be new. Property L of §3.12 was introduced by Motzkin–Taussky [MoT52] and [MoT55]. Theorem 3.12.4 is a slight improvement of [MoT55]. Our proof of property L in Theorem 3.12.4 follows [Kat80]. Many results in §3.13 are taken from [Fri80a] and [Fri80b]. It connects the analytic similarity of matrices with simultaneous similarity of certain pairs of matrices. Simultaneous similarity of matrices is discussed in [Fri83].

Chapter 4

Inner Product Spaces

4.1   Inner Product

Definition 4.1.1 *Let* $\mathbb{F} = \mathbb{R}, \mathbb{C}$ *and let* $\mathbf{V}$ *be a vector space over* $\mathbb{F}$. *Then* $\langle \cdot, \cdot \rangle : \mathbf{V} \times \mathbf{V} \to \mathbb{F}$ *is called an inner product if the following conditions hold:*

(a) $\langle a\mathbf{x} + b\mathbf{y}, \mathbf{z} \rangle = a\langle \mathbf{x}, \mathbf{z} \rangle + b\langle \mathbf{y}, \mathbf{z} \rangle$, *for all* $a, b \in \mathbb{F}$, $\mathbf{x}, \mathbf{y}, \mathbf{z} \in \mathbf{V}$,

(br)   *for* $\mathbb{F} = \mathbb{R}$ $\langle \mathbf{y}, \mathbf{x} \rangle = \langle \mathbf{x}, \mathbf{y} \rangle$, *for all* $\mathbf{x}, \mathbf{y} \in \mathbf{V}$;

(bc)   *for* $\mathbb{F} = \mathbb{C}$ $\langle \mathbf{y}, \mathbf{x} \rangle = \overline{\langle \mathbf{x}, \mathbf{y} \rangle}$, *for all* $\mathbf{x}, \mathbf{y} \in \mathbf{V}$;

(c) $\langle \mathbf{x}, \mathbf{x} \rangle > 0$ *for all* $\mathbf{x} \in \mathbf{V} \backslash \{\mathbf{0}\}$.

$\|\mathbf{x}\| := \sqrt{\langle \mathbf{x}, \mathbf{x} \rangle}$ *is called the norm (length) of* $\mathbf{x} \in \mathbf{V}$.

Other standard properties of inner products are mentioned in Problems 1 and 2. We will use the abbreviation IPS for inner product space. In this chapter, we assume that $\mathbb{F} = \mathbb{R}, \mathbb{C}$ unless stated otherwise.

Proposition 4.1.2 *Let* $\mathbf{V}$ *be a vector space over* $\mathbb{R}$. *Identify* $\mathbf{V}_{\mathbb{C}}$, *called the* **complexification** *of* $\mathbf{V}$, *with the set of pairs* $(\mathbf{x}, \mathbf{y})$,

195

$\mathbf{x}, \mathbf{y} \in \mathbf{V}$. *Then* $\mathbf{V}_{\mathbb{C}}$ *is a vector space over* $\mathbb{C}$ *with*

$$(a + \sqrt{-1}b)(\mathbf{x}, \mathbf{y}) := a(\mathbf{x}, \mathbf{y}) + b(-\mathbf{y}, \mathbf{x}), \quad \text{for all } a, b \in \mathbb{R},$$

$$\mathbf{x}, \mathbf{y} \in \mathbf{V}.$$

If $\mathbf{V}$ *has a basis* $\mathbf{e}_1, \ldots, \mathbf{e}_n$ *over* $\mathbb{R}$ *then* $(\mathbf{e}_1, \mathbf{0}), \ldots, (\mathbf{e}_n, \mathbf{0})$ *is a basis of* $\mathbf{V}_{\mathbb{C}}$ *over* $\mathbb{C}$. *Any inner product* $\langle \cdot, \cdot \rangle$ *on* $\mathbf{V}$ *over* $\mathbb{F}$ *induces the following inner product on* $\mathbf{V}_{\mathbb{C}}$:

$$\langle (\mathbf{x}, \mathbf{y}), (\mathbf{u}, \mathbf{v}) \rangle = \langle \mathbf{x}, \mathbf{u} \rangle + \langle \mathbf{y}, \mathbf{v} \rangle + \sqrt{-1}(\langle \mathbf{y}, \mathbf{u} \rangle - \langle \mathbf{x}, \mathbf{v} \rangle),$$

$$\mathbf{x}, \mathbf{y}, \mathbf{u}, \mathbf{v} \in \mathbf{V}.$$

We leave the proof of this proposition to the reader (Problem 3).

Definition 4.1.3 *Let* $\mathbf{V}$ *be an IPS. Then*

(a) $\mathbf{x}, \mathbf{y} \in \mathbf{V}$ *are called orthogonal if* $\langle \mathbf{x}, \mathbf{y} \rangle = 0$.

(b) $S, T \subset \mathbf{V}$ *are called orthogonal if* $\langle \mathbf{x}, \mathbf{y} \rangle = 0$ *for any* $\mathbf{x} \in S$, $\mathbf{y} \in T$.

(c) *For any* $S \subset \mathbf{V}$, $S^{\perp} \subset \mathbf{V}$ *is the maximal orthogonal set to* S.

(d) $\mathbf{x}_1, \ldots, \mathbf{x}_m$ *is called an orthonormal set if*

$$\langle \mathbf{x}_i, \mathbf{x}_j \rangle = \delta_{ij}, \text{ for all } i, j = 1, \ldots, m.$$

(e) $\mathbf{x}_1, \ldots, \mathbf{x}_n$ *is called an orthonormal basis if it is an orthonormal set which is a basis in* $\mathbf{V}$.

Definition 4.1.4 (Gram–Schmidt algorithm) *Let* $\mathbf{V}$ *be an IPS and* $S = \{\mathbf{x}_1, \ldots, \mathbf{x}_m\} \subset \mathbf{V}$ *a finite (possibly empty) set* $(m \geq 0)$. *Then* $\tilde{S} = \{\mathbf{e}_1, \ldots, \mathbf{e}_p\}$ *is the orthonormal set* $(p \geq 1)$ *or the empty set* $(p = 0)$ *obtained from* S *using the following recursive steps:*

(a) *If* $\mathbf{x}_1 = \mathbf{0}$ *remove it from* S. *Otherwise replace* $\mathbf{x}_1$ *by* $\|\mathbf{x}_1\|^{-1}\mathbf{x}_1$.

(b) *Assume that* $\mathbf{x}_1, \ldots, \mathbf{x}_k$ *is an orthonormal set and* $1 \leq k < m$. *Let* $\mathbf{y}_{k+1} = \mathbf{x}_{k+1} - \sum_{i=1}^{k} \langle \mathbf{x}_{k+1}, \mathbf{x}_i \rangle \mathbf{x}_i$. *If* $\mathbf{y}_{k+1} = \mathbf{0}$ *remove* $\mathbf{x}_{k+1}$ *from* S. *Otherwise replace* $\mathbf{x}_{k+1}$ *by* $\|\mathbf{y}_{k+1}\|^{-1}\mathbf{y}_{k+1}$.

Corollary 4.1.5 *Let* $\mathbf{V}$ *be an IPS and* $S = \{\mathbf{x}_1, \ldots, \mathbf{x}_n\} \subset \mathbf{V}$ *be* n *linearly independent vectors. Then the Gram–Schmidt algorithm on* S *is given as follows:*

$$\mathbf{y}_1 := \mathbf{x}_1, \quad r_{11} := \|\mathbf{y}_1\|, \quad \mathbf{e}_1 := \frac{1}{r_{11}}\mathbf{y}_1,$$

$$r_{ji} := \langle \mathbf{x}_i, \mathbf{e}_j \rangle, \quad j = 1, \ldots, i-1, \tag{4.1.1}$$

$$\mathbf{y}_i := \mathbf{x}_i - \sum_{j=1}^{i-1} r_{ji}\mathbf{e}_j, \quad r_{ii} := \|\mathbf{y}_i\|, \quad \mathbf{e}_i := \frac{1}{r_{ii}}\mathbf{y}_i, \quad i = 2, \ldots, n.$$

In particular, $\mathbf{e}_i \in S_i$ *and* $\|\mathbf{y}_i\| = \mathrm{dist}(\mathbf{x}_i, S_{i-1})$, *where* $S_i = \mathrm{span}\,(\mathbf{x}_1, \ldots, \mathbf{x}_i)$ *for* $i = 1, \ldots, n$ *and* $S_0 = \{\mathbf{0}\}$. *(See Problem 4 below for the definition of* $\mathrm{dist}(\mathbf{x}_i, S_{i-1})$.)

Corollary 4.1.6 *Any (ordered) basis in a finite dimensional IPS* $\mathbf{V}$ *induces an orthonormal basis by the Gram–Schmidt algorithm.*

See Problem 4 for some known properties related to the above notions.

Problems

1. Let $\mathbf{V}$ be an IPS over $\mathbb{F}$. Show

$$\langle \mathbf{0}, \mathbf{x} \rangle = \langle \mathbf{x}, \mathbf{0} \rangle = 0,$$

for $\mathbb{F} = \mathbb{R}$ $\langle \mathbf{z}, a\mathbf{x} + b\mathbf{y} \rangle$
$$= a\langle \mathbf{z}, \mathbf{x} \rangle + b\langle \mathbf{z}, \mathbf{y} \rangle, \text{ for all } a, b \in \mathbb{R}, \mathbf{x}, \mathbf{y}, \mathbf{z} \in \mathbf{V},$$

for $\mathbb{F} = \mathbb{C}$ $\langle \mathbf{z}, a\mathbf{x} + b\mathbf{y} \rangle$
$$= \bar{a}\langle \mathbf{z}, \mathbf{x} \rangle + \bar{b}\langle \mathbf{z}, \mathbf{y} \rangle, \quad \text{for all } a, b \in \mathbb{C}, \mathbf{x}, \mathbf{y}, \mathbf{z} \in \mathbf{V}.$$

2. Let $\mathbf{V}$ be an IPS. Show

(a) $\|a\mathbf{x}\| = |a|\,\|\mathbf{x}\|$ for $a \in \mathbb{F}$ and $\mathbf{x} \in \mathbf{V}$.

(b) The Cauchy–Schwarz inequality:

$$|\langle \mathbf{x}, \mathbf{y} \rangle| \le \|\mathbf{x}\| \, \|\mathbf{y}\|,$$

and equality holds if and only if $\mathbf{x}, \mathbf{y}$ are linearly dependent (collinear).

(c) The triangle inequality

$$\|\mathbf{x} + \mathbf{y}\| \le \|\mathbf{x}\| + \|\mathbf{y}\|,$$

and equality holds if either $\mathbf{x} = 0$ or $\mathbf{y} = a\mathbf{x}$ for $a \in \mathbb{R}_+$.

3. Prove Proposition 4.1.2.

4. Let $\mathbf{V}$ be a finite dimensional IPS of dimension n. Assume that $S \subset \mathbf{V}$. Show

(a) If $\mathbf{x}_1, \ldots, \mathbf{x}_m$ is an orthonormal set then $\mathbf{x}_1, \ldots, \mathbf{x}_m$ are linearly independent.

(b) Assume that $\mathbf{e}_1, \ldots, \mathbf{e}_n$ is an orthonormal basis in $\mathbf{V}$. Show that for any $\mathbf{x} \in \mathbf{V}$ the orthonormal expansion holds

$$\mathbf{x} = \sum_{i=1}^{n} \langle \mathbf{x}, \mathbf{e}_i \rangle \mathbf{e}_i. \qquad (4.1.2)$$

Furthermore, for any $\mathbf{x}, \mathbf{y} \in \mathbf{V}$

$$\langle \mathbf{x}, \mathbf{y} \rangle = \sum_{i=1}^{n} \langle \mathbf{x}, \mathbf{e}_i \rangle \overline{\langle \mathbf{y}, \mathbf{e}_i \rangle}. \qquad (4.1.3)$$

(c) Assume that S is a finite set. Let $\tilde{S}$ be the set obtained by the Gram–Schmidt process. Show that $\tilde{S} = \emptyset \iff$ span $S = \{\mathbf{0}\}$. Show that if $\tilde{S} \ne \emptyset$ then $\mathbf{e}_1, \ldots, \mathbf{e}_p$ is an orthonormal basis in span S.

(d) There exists an orthonormal basis $\mathbf{e}_1, \ldots, \mathbf{e}_n$ in $\mathbf{V}$ and $0 \leq m \leq n$ such that

$$\mathbf{e}_1, \ldots, \mathbf{e}_m \in S, \quad \text{span } S = \text{span } (\mathbf{e}_1, \ldots, \mathbf{e}_m),$$

$$S^\perp = \text{span } (\mathbf{e}_{m+1}, \ldots, \mathbf{e}_n),$$

$$(S^\perp)^\perp = \text{span } S.$$

(e) Assume from here to the end of the problem that S is a subspace. Show $\mathbf{V} = S \oplus S^\perp$.

(f) Let $\mathbf{x} \in \mathbf{V}$ and let $\mathbf{x} = \mathbf{u} + \mathbf{v}$ for unique $\mathbf{u} \in S$, $\mathbf{v} \in S^\perp$. Let $P(\mathbf{x}) := \mathbf{u}$ be the projection of $\mathbf{x}$ on S. Show that $P : \mathbf{V} \to \mathbf{V}$ is a linear transformation satisfying

$$P^2 = P, \quad \text{Range } P = S, \quad \text{Ker } P = S^\perp.$$

(g) Show

$$\text{dist}(\mathbf{x}, S) := \|\mathbf{x} - P\mathbf{x}\| \leq \|\mathbf{x} - \mathbf{w}\| \text{ for any } \mathbf{w} \in S$$

$$\text{and equality} \iff \mathbf{w} = P\mathbf{x}. \tag{4.1.4}$$

(h) Show that $\text{dist}(\mathbf{x}, S) = \|\mathbf{x} - \mathbf{w}\|$ for some $\mathbf{w} \in S$ if and only if $\mathbf{x} - \mathbf{w}$ is orthogonal to S.

5. Let $X \in \mathbb{C}^{m \times n}$ and assume that $m \geq n$ and rank $X = n$. Let $\mathbf{x}_1, \ldots, \mathbf{x}_n \in \mathbb{C}^m$ be the columns of X, i.e. $X = [\mathbf{x}_1 \ldots \mathbf{x}_n]$. Assume that $\mathbb{C}^m$ is an IPS with the standard inner product $\langle \mathbf{x}, \mathbf{y} \rangle = \mathbf{y}^*\mathbf{x}$. Perform the Gram–Schmidt algorithm (4.1.5) to obtain the matrix $Q = [\mathbf{e}_1 \ldots \mathbf{e}_n] \in \mathbb{C}^{m \times n}$. Let $R = [r_{ji}]_1^n \in \mathbb{C}^{n \times n}$ be the upper triangular matrix with r_{ji}, $j \leq i$ given by (4.1.1). Show that $\bar{Q}^T Q = I_n$ and $X = QR$. (This is the QR decomposition.) Show that if in addition $X \in \mathbb{R}^{m \times n}$, then Q and R are real valued matrices.

6. Let $C \in \mathbb{C}^{n \times n}$ and assume that $\{\lambda_1, \ldots, \lambda_n\}$ are n eigenvalues of C counted with their multiplicities. View C as an operator

$C : \mathbb{C}^n \to \mathbb{C}^n$. View $\mathbb{C}^n$ as $2n$-dimensional vector space over $\mathbb{R}^{2n}$. Let $C = A + \sqrt{-1}B$, $A, B \in \mathbb{R}^{n \times n}$.

a. Then $\tilde{C} := \begin{bmatrix} A & -B \\ B & A \end{bmatrix} \in \mathbb{R}^{(2n) \times (2n)}$ represents the operator $C : \mathbb{C}^n \to \mathbb{C}^n$ as an operator over $\mathbb{R}$ in suitably chosen basis.

b. Show that $\{\lambda_1, \bar{\lambda}_1, \ldots, \lambda_n, \bar{\lambda}_n\}$ are the $2n$ eigenvalues of $\tilde{C}$ counting with multiplicities.

c. Show that the Jordan canonical form of $\tilde{C}$, is obtained by replacing each Jordan block $\lambda I + H$ in C by two Jordan blocks $\lambda I + H$ and $\bar{\lambda} I + H$.

4.2   Special Transformations in IPS

Proposition 4.2.1 *Let* $\mathbf{V}$ *be an IPS and* $T : \mathbf{V} \to \mathbf{V}$ *a linear transformation. Then there exists a unique linear transformation* $T^* : \mathbf{V} \to \mathbf{V}$ *such that* $\langle T\mathbf{x}, \mathbf{y} \rangle = \langle \mathbf{x}, T^*\mathbf{y} \rangle$ *for all* $\mathbf{x}, \mathbf{y} \in \mathbf{V}$.

See Problems 1 and 2.

Definition 4.2.2 *Let* $\mathbf{V}$ *be an IPS and let* $T : \mathbf{V} \to \mathbf{V}$ *be a linear transformation. Then*

(a) *T is called self-adjoint if* $T^* = T$;

(b) *T is called anti self-adjoint if* $T^* = -T$;

(c) *T is called unitary if* $T^*T = TT^* = I$;

(d) *T is called normal if* $T^*T = TT^*$.

Denote by $\mathbf{S(V)}$, $\mathbf{AS(V)}$, $\mathbf{U(V)}$, $\mathbf{N(V)}$ *the sets of self-adjoint, anti self-adjoint, unitary and normal operators on* $\mathbf{V}$ *respectively.*

Proposition 4.2.3 *Let* **V** *be an IPS over* $\mathbb{F} = \mathbb{R}, \mathbb{C}$ *with an orthonormal basis* $E = \{\mathbf{e}_1, \ldots, \mathbf{e}_n\}$. *Let* $T : \mathbf{V} \to \mathbf{V}$ *be a linear transformation. Let* $A = [a_{ij}] \in \mathbb{F}^{n \times n}$ *be the representation matrix of* T *in the basis* E:

$$a_{ij} = \langle T\mathbf{e}_j, \mathbf{e}_i \rangle, \quad i, j = 1, \ldots, n. \tag{4.2.1}$$

Then for $\mathbb{F} = \mathbb{R}$:

(a) T^* *is represented by* $A^\top$,

(b) T *is self-adjoint* $\iff A = A^\top$,

(c) T *is anti self-adjoint* $\iff A = -A^\top$,

(d) T *is unitary* $\iff A$ *is orthogonal* $\iff AA^\top = A^\top A = I$,

(e) T *is normal* $\iff A$ *is normal* $\iff AA^\top = A^\top A$,

and for $\mathbb{F} = \mathbb{C}$:

(a) T^* *is represented by* $A^* \, (:= \bar{A}^\top)$,

(b) T *is self-adjoint* $\iff A$ *is hermitian* $\iff A = A^*$,

(c) T *is anti self-adjoint* $\iff A$ *is anti hermitian*

$$\iff A = -A^*,$$

(d) T *is unitary* $\iff A$ *is unitary* $\iff AA^* = A^*A = I$,

(e) T *is normal* $\iff A$ *is normal* $\iff AA^* = A^*A$.

See Problem 3.

Proposition 4.2.4 *Let* **V** *be an IPS over* $\mathbb{R}$, *and let* $T \in$ Hom $(\mathbf{V})$. *Let* $\mathbf{V}_\mathbb{C}$ *be the complexification of* **V**. *Then there exists a unique* $T_\mathbb{C} \in$ Hom $(\mathbf{V}_\mathbb{C})$ *such that* $T_\mathbb{C}|\mathbf{V} = T$. *Furthermore* T *is self-adjoint, unitary or normal if and only if* $T_\mathbb{C}$ *is self-adjoint, unitary or normal respectively.*

See Problem 4.

Definition 4.2.5 *For a domain* $\mathbb{D}$ *with identity* 1 *let*

$$\mathbf{S}(n, \mathbb{D}) := \{A \in \mathbb{D}^{n \times n} : \quad A = A^\top\},$$

$$\mathbf{AS}(n, \mathbb{D}) := \{A \in \mathbb{D}^{n \times n} : \quad A = -A^\top\},$$

$$\mathbf{O}(n, \mathbb{D}) := \{A \in \mathbb{D}^{n \times n} : \quad AA^\top = A^\top A = I\},$$

$$\mathbf{SO}(n, \mathbb{D}) := \{A \in \mathbf{O}(n, \mathbb{D}) : \quad \det A = 1\},$$

$$\mathbf{DO}(n, \mathbb{D}) := \mathbf{D}(n, \mathbb{D}) \cap \mathbf{O}(n, \mathbb{D}),$$

$$\mathbf{N}(n, \mathbb{R}) := \{A \in \mathbb{R}^{n \times n} : \quad AA^\top = A^\top A\},$$

$$\mathbf{N}(n, \mathbb{C}) := \{A \in \mathbb{C}^{n \times n} : \quad AA^* = A^* A\},$$

$$\mathbf{H}_n := \{A \in \mathbb{C}^{n \times n} : \quad A = A^*\},$$

$$\mathbf{AH}_n := \{A \in \mathbb{C}^{n \times n} : \quad A = -A^*\},$$

$$\mathbf{U}_n := \{A \in \mathbb{C}^{n \times n} : \quad AA^* = A^* A = I\},$$

$$\mathbf{SU}_n := \{A \in \mathbf{U}_n : \quad \det A = 1\},$$

$$\mathbf{DU}_n := \mathbf{D}(n, \mathbb{C}) \cap \mathbf{U}_n.$$

See Problem 5 for relations between these classes.

Theorem 4.2.6 *Let* $\mathbf{V}$ *be an IPS over* $\mathbb{C}$ *of dimension* n. *Then a linear transformation* $T : \mathbf{V} \to \mathbf{V}$ *is normal if and only if* $\mathbf{V}$ *has an orthonormal basis consisting of eigenvectors of* T.

Proof. Suppose first that $\mathbf{V}$ has an orthonormal basis $\mathbf{e}_1, \ldots, \mathbf{e}_n$ such that $T\mathbf{e}_i = \lambda_i \mathbf{e}_i$, $i = 1, \ldots, n$. From the definition of T^* it follows that $T^* \mathbf{e}_i = \bar{\lambda}_i \mathbf{e}_i$, $i = 1, \ldots, n$. Hence, $TT^* = T^* T$.

Assume now T is normal. Since $\mathbb{C}$ is algebraically closed T has an eigenvalue λ_1. Let $\mathbf{V}_1$ be the subspace of $\mathbf{V}$ spanned by all eigenvectors of T corresponding to the eigenvalue λ_1. Clearly $T\mathbf{V}_1 \subset \mathbf{V}_1$. Let $\mathbf{x} \in \mathbf{V}_1$. Then $T\mathbf{x} = \lambda_1 \mathbf{x}$. Thus

$$T(T^*\mathbf{x}) = (TT^*)\mathbf{x} = (T^*T)\mathbf{x} = T^*(T\mathbf{x}) = \lambda_1 T^*\mathbf{x} \Rightarrow T^*\mathbf{V}_1 \subset \mathbf{V}_1.$$

Hence, $T\mathbf{V}_1^\perp, T^*\mathbf{V}_1^\perp \subset \mathbf{V}_1^\perp$. Since $\mathbf{V} = \mathbf{V}_1 \oplus \mathbf{V}_1^\perp$ it is enough to prove the theorem for $T|\mathbf{V}_1$ and $T|\mathbf{V}_1^\perp$.

As $T|\mathbf{V}_1 = \lambda_1 I_{\mathbf{V}_1}$ it is straightforward to show $T^*|\mathbf{V}_1 = \bar{\lambda}_1 I_{\mathbf{V}_1}$ (see Problem 2). Hence, for $T|\mathbf{V}_1$, the theorem trivially holds. For $T|\mathbf{V}_1^\perp$ the theorem follows by induction. $\qquad\square$

The proof of Theorem 4.2.6 yields:

Corollary 4.2.7 *Let* $\mathbf{V}$ *be an IPS over* $\mathbb{R}$ *of dimension* n. *Then the linear transformation* $T : \mathbf{V} \to \mathbf{V}$ *with a real spectrum is normal if and only if* $\mathbf{V}$ *has an orthonormal basis consisting of eigenvectors of* T.

Proposition 4.2.8 *Let* $\mathbf{V}$ *be an IPS over* $\mathbb{C}$. *Let* $T \in \mathbf{N}(\mathbf{V})$. *Then*

T *is self-adjoint* $\iff$ spec $(T) \subset \mathbb{R}$,

T *is unitary* $\iff$ spec $(T) \subset S^1 = \{z \in \mathbb{C} : \quad |z| = 1\}$.

Proof. Since T is normal there exists an orthonormal basis $\mathbf{e}_1, \ldots, \mathbf{e}_n$ such that $T\mathbf{e}_i = \lambda_i \mathbf{e}_i$, $i = 1, \ldots, n$. Hence, $T^*\mathbf{e}_i = \bar{\lambda}_i \mathbf{e}_i$. Then

$$T = T^* \iff \lambda_i = \bar{\lambda}_i, \ i = 1, \ldots, n,$$
$$TT^* = T^*T = I \iff |\lambda_i| = 1, \ i = 1, \ldots, n.$$

$\qquad\square$

Combine Proposition 4.2.4 and Corollary 4.2.7 with the above proposition to deduce:

Corollary 4.2.9 *Let* $\mathbf{V}$ *be an IPS over* $\mathbb{R}$ *and let* $T \in \mathbf{S}(\mathbf{V})$. *Then* spec $(T) \subset \mathbb{R}$ *and* $\mathbf{V}$ *has an orthonormal basis consisting of the eigenvectors of* T.

Proposition 4.2.10 *Let* $\mathbf{V}$ *be an IPS over* $\mathbb{R}$ *and let* $T \in \mathbf{U}(\mathbf{V})$. *Then* $\mathbf{V} = \oplus_{i \in \{-1,1,2,\ldots,k\}} \mathbf{V}_i$, *where* $k \geq 1$, $\mathbf{V}_i$ *and* $\mathbf{V}_j$ *are orthogonal for* $i \neq j$, *such that*

(a) $T|\mathbf{V}_{-1} = -I_{\mathbf{V}_{-1}}$ dim $\mathbf{V}_{-1} \geq 0$,

(b) $T|\mathbf{V}_1 = I_{\mathbf{V}_1}$ dim $\mathbf{V}_1 \geq 0$,

(c) $TV_i = V_i$, dim $V_i = 2$,   and   spec $(T|V_i) \subset S^1 \setminus \{-1, 1\}$ for $i = 2, \ldots, k$.

See Problem 7.

Proposition 4.2.11 *Let* V *be an IPS over* $\mathbb{R}$ *and let* $T \in$ **AS(V)**. *Then* $V = \oplus_{i \in \{1, 2, \ldots, k\}} V_i$, *where* $k \geq 1$, *and* V_i *and* V_j *are orthogonal for* $i \neq j$, *such that*

(a) $T|V_1 = 0_{V_1}$ dim $V_0 \geq 0$,

(b) $TV_i = V_i$, dim $V_i = 2$, spec $(T|V_i) \subset \sqrt{-1}\mathbb{R} \setminus \{0\}$ *for* $i = 2, \ldots, k$.

See Problem 8.

Theorem 4.2.12 *Let* V *be an IPS over* $\mathbb{C}$ *of dimension* n. *Let* $T \in \mathrm{Hom}\,(V)$. *Let* $\lambda_1, \ldots, \lambda_n \in \mathbb{C}$ *be* n *eigenvalues of* T *counted with their multiplicities. Then there exists a orthonormal basis* $g_1, \ldots, g_n$ *of* V *with the following properties:*

$$T\mathrm{span}\,(g_1, \ldots, g_i) \subset \mathrm{span}\,(g_1, \ldots, g_i),$$
$$\langle Tg_i, g_i \rangle = \lambda_i, \ i = 1, \ldots, n. \qquad (4.2.2)$$

Next, let V *be an IPS over* $\mathbb{R}$ *of dimension* n. *Let* $T \in \mathrm{Hom}\,(V)$ *and assume that* spec $(T) \subset \mathbb{R}$. *Let* $\lambda_1, \ldots, \lambda_n \in \mathbb{R}$ *be* n *eigenvalues of* T *counted with their multiplicities. Then there exists an orthonormal basis* $g_1, \ldots, g_n$ *of* V *such that* (4.2.2) *holds.*

Proof. Assume first that V is an IPS over $\mathbb{C}$ of dimension n. The proof is by induction on n. For $n = 1$ the theorem is trivial. Assume that $n > 1$. Since $\lambda_1 \in$ spec (T) it follows that there exists $g_1 \in V$, $\langle g_1, g_1 \rangle = 1$ such that $Tg_1 = \lambda_1 g_1$. Let $U := \mathrm{span}\,(g_1)^{\perp}$. Let P be the orthogonal projection on U. Let $T_1 := PT|_U$. Then $T_1 \in \mathrm{Hom}\,(U)$. Let $\tilde{\lambda}_2, \ldots, \tilde{\lambda}_n$ be the eigenvalues of T_1 counted with their multiplicities. The induction hypothesis yields the existence of an orthonormal basis $\mathbf{g}_2, \ldots, \mathbf{g}_n$ of $\mathbf{U}$ such that

$$T_1 \text{span} \, (\mathbf{g}_2, \ldots, \mathbf{g}_i) \subset \text{span} \, (\mathbf{g}_2, \ldots, \mathbf{g}_i),$$

$$\langle T_1 \mathbf{g}_i, \mathbf{g}_i \rangle = \tilde{\lambda}_i, \; i = 1, \ldots, n.$$

It is straightforward to show that $T \text{span} \, (\mathbf{g}_1, \ldots, \mathbf{g}_i) \subset \text{span} \, (\mathbf{g}_1, \ldots, \mathbf{g}_i)$ for $i = 1, \ldots, n$. Hence, in the orthonormal basis $\mathbf{g}_1, \ldots, \mathbf{g}_n$ T is presented by an upper diagonal matrix $B = [b_{ij}]_1^n$, with $b_{11} = \lambda_1$ and $b_{ii} = \tilde{\lambda}_i, \; i = 2, \ldots, n$. Hence, $\lambda_1, \tilde{\lambda}_2, \ldots, \tilde{\lambda}_n$ are the eigenvalues of T counted with their multiplicities. This establishes the theorem in this case. The real case is treated similarly.

$\square$

Combine the above results with Problems 6 and 12 to deduce:

Corollary 4.2.13 *Let $A \in \mathbb{C}^{n \times n}$. Let $\lambda_1, \ldots, \lambda_n \in \mathbb{C}$ be n eigenvalues of A counted with their multiplicities. Then there exists an upper triangular matrix $B = [b_{ij}]_1^n \in \mathbb{C}^{n \times n}$, such that $b_{ii} = \lambda_i, \; i = 1, \ldots, n$, and a unitary matrix $U \in \mathbf{U}_n$ such that $A = UBU^{-1}$. If $A \in \mathbf{N}(n, \mathbb{C})$ then B is a diagonal matrix.*

Next, let $A \in \mathbb{R}^{n \times n}$ and assume that $\mathrm{spec} \, (T) \subset \mathbb{R}$. Then $A = UBU^{-1}$ where U can be chosen to be a real orthogonal matrix and B is a real upper triangular matrix. If $A \in \mathbf{N}(n, \mathbb{R})$ and $\mathrm{spec} \, (A) \subset \mathbb{R}$ then B is a diagonal matrix.

It is easy to show that U in the above Corollary can be chosen in $\mathbf{SU}_n$ or $\mathbf{SO}(n, \mathbb{R})$ respectively (see Problem 11).

Definition 4.2.14 *Let $\mathbf{V}$ be a vector space and assume that $T : \mathbf{V} \to \mathbf{V}$ is a linear operator. Let $0 \neq \mathbf{v} \in \mathbf{V}$. Then $\mathbf{W} = \text{span} \, (\mathbf{v}, T\mathbf{v}, T^2\mathbf{v}, \ldots)$ is called a cyclic invariant subspace of T generated by $\mathbf{v}$. (It is also referred as a Krylov subspace of T generated by $\mathbf{v}$.) Sometimes we will refer to $\mathbf{W}$ as simply a cyclic subspace, or a Krylov subspace.*

Theorem 4.2.15 *Let* **V** *be a finite dimensional IPS. Let* T : **V** $\rightarrow$ **V** *be a linear operator. For* $0 \neq \mathbf{v} \in \mathbf{V}$ *let* **W** $=$ span $(\mathbf{v}, T\mathbf{v}, \ldots, T^{r-1}\mathbf{v})$ *be a cyclic* T-*invariant subspace of dimension* r *generated by* **v**. *Let* $\mathbf{u}_1, \ldots, \mathbf{u}_r$ *be an orthonormal basis of* **W** *obtained by the Gram–Schmidt process from the basis* $[\mathbf{v}, T\mathbf{V}, \ldots, T^{r-1}\mathbf{v}]$ *of* **W**. *Then* $\langle T\mathbf{u}_i, \mathbf{u}_j \rangle = 0$ *for* $1 \leq i \leq j - 2$, *i.e. the representation matrix of* $T|\mathbf{W}$ *in the basis* $[\mathbf{u}_1, \ldots, \mathbf{u}_r]$ *is upper Hessenberg. If* T *is self-adjoint then the representation matrix of* $T|\mathbf{W}$ *in the basis* $[\mathbf{u}_1, \ldots, \mathbf{u}_r]$ *is a tridiagonal hermitian matrix.*

Proof. Let $\mathbf{W}_j = \mathrm{span}\,(\mathbf{v}, \ldots, T^{j-1}\mathbf{v})$ for $j = 1, \ldots, r + 1$. Clearly $T\mathbf{W}_j \subset \mathbf{W}_{j+1}$ for $j = 1, \ldots, r$. The assumption that **W** is a T-invariant subspace yields that $\mathbf{W} = \mathbf{W}_r = \mathbf{W}_{r+1}$. Since dim $\mathbf{W} = r$, it follows that $\mathbf{v}, \ldots, T^{r-1}\mathbf{v}$ are linearly independent. Hence, $[\mathbf{v}, \ldots, T^{r-1}\mathbf{v}]$ is a basis for **W**. Recall that span $(\mathbf{u}_1, \ldots, \mathbf{u}_j) = \mathbf{W}_j$ for $j = 1, \ldots, r$. Let $r \geq j \geq i + 2$. Then $T\mathbf{u}_i \in T\mathbf{W}_i \subset \mathbf{W}_{i+1}$. As $\mathbf{u}_j \perp \mathbf{W}_{i+1}$ it follows that $\langle T\mathbf{u}_i, \mathbf{u}_j \rangle = 0$. Assume that $T^* = T$. Let $r \geq i \geq j + 2$. Then $\langle T\mathbf{u}_i, \mathbf{u}_j \rangle = \langle \mathbf{u}_i, T\mathbf{u}_j \rangle = 0$. Hence, the representation matrix of $T|\mathbf{W}$ in the basis $[\mathbf{u}_1, \ldots, \mathbf{u}_r]$ is a tridiagonal hermitian matrix. $\qquad\square$

Problems

1. Prove Proposition 4.2.1.

2. Let $P, Q \in \mathrm{Hom}\,(\mathbf{V})$, and $\mathbf{a}, b \in \mathbb{F}$. Show that $(aP + bQ)^* = \bar{a}P^* + \bar{b}Q^*$.

3. Prove Proposition 4.2.3.

4. Prove Proposition 4.2.4 for finite dimensional **V**. (*Hint*: Choose an orthonormal basis in **V**.)

5. Show the following

$$\mathbf{SO}(n, \mathbb{D}) \subset \mathbf{O}(n, \mathbb{D}) \subset \mathbf{GL}(n, \mathbb{D}),$$

$$\mathbf{S}(n, \mathbb{R}) \subset \mathbf{H}_n \subset \mathbf{N}(n, \mathbb{C}),$$

$$\mathbf{AS}(n, \mathbb{R}) \subset \mathbf{AH}_n \subset \mathbf{N}(n, \mathbb{C}),$$

$\mathbf{S}(n, \mathbb{R}), \mathbf{AS}(n, \mathbb{R}) \subset \mathbf{N}(n, \mathbb{R}) \subset \mathbf{N}(n, \mathbb{C})$,

$\mathbf{O}(n, \mathbb{R}) \subset \mathbf{U}_n \subset \mathbf{N}(n, \mathbb{C})$,

$\mathbf{SO}(n, \mathbb{D}), \mathbf{O}(n, \mathbb{D}), \mathbf{SU}_n, \mathbf{U}_n$ are groups,

$\mathbf{S}(n, \mathbb{D})$ is a $\mathbb{D}$-module of dimension $\binom{n+1}{2}$,

$\mathbf{AS}(n, \mathbb{D})$ is a $\mathbb{D}$-module of dimension $\binom{n}{2}$,

$\mathbf{H}_n$ is an $\mathbb{R}$- vector space of dimension n^2,

$\mathbf{AH}_n = \sqrt{-1}\,\mathbf{H}_n$.

6. Let $E = \{e_1, \ldots, e_n\}$ be an orthonormal basis in an IPS $\mathbf{V}$ over $\mathbb{F}$. Let $G = \{g_1, \ldots, g_n\}$ be another basis in $\mathbf{V}$. Show that G is an orthonormal basis if and only if the transfer matrix either from E to G or from G to E is a unitary matrix.

7. Prove Proposition 4.2.10.

8. Prove Proposition 4.2.11.

9. (a) Show that $A \in \mathbf{SO}(2, \mathbb{R})$ is of the form $A = \begin{bmatrix} \cos\theta & \sin\theta \\ -\sin\theta & \cos\theta \end{bmatrix}, \theta \in \mathbb{R}$.

   (b) Show that $\mathbf{SO}(2, \mathbb{R}) = e^{\mathbf{AS}(2,\mathbb{R})}$. That is, for any $B \in \mathbf{AS}(2, \mathbb{R})$, $e^B \in \mathbf{SO}(2, \mathbb{R})$, and any $A \in \mathbf{SO}(n, \mathbb{R})$ is e^B for some $B \in \mathbf{AS}(2, \mathbb{R})$. (*Hint*: Consider the power series for e^B, $B = \begin{bmatrix} 0 & \theta \\ -\theta & 0 \end{bmatrix}$.)

   (c) Show that $\mathbf{SO}(n, \mathbb{R}) = e^{\mathbf{AS}(n,\mathbb{R})}$. (*Hint*: Use Propositions 4.2.10 and 4.2.11 and part b.)

   (d) Show that $\mathbf{SO}(n, \mathbb{R})$ is a path connected space. (See part e.)

   (e) Let $\mathbf{V}$ be an $n(>1)$-dimensional IPS over $\mathbb{F} = \mathbb{R}$. Let $p \in [n-1]$. Assume that $x_1, \ldots, x_p$ and $y_1, \ldots, y_p$ are two orthonormal systems in $\mathbf{V}$. Show that these two o.n.s. are path connected. That is, there are p continuous mappings $\mathbf{z}_i(t) : [0,1] \to \mathbf{V}$, $i = 1, \ldots, p$ such that for each $t \in [0,1]$ $\mathbf{z}_1(t), \ldots, \mathbf{z}_p(t)$ is an o.n.s. and $\mathbf{z}_i(0) = \mathbf{x}_i, \mathbf{z}_i(1) = \mathbf{y}_i, i = 1, \ldots, p$.

10. (a) Show that $\mathbf{U}_n = e^{\mathbf{AH}_n}$. (*Hint:* Use Proposition 4.2.8 and its proof.)

    (b) Show that $\mathbf{U}_n$ is path connected.

    (c) Prove Problem 9(e) for $\mathbb{F} = \mathbb{C}$.

11. Show

    (a) $D_1 D D_1^* = D$ for any $D \in \mathbf{D}(n, \mathbb{C})$, $D_1 \in \mathbf{DU}_n$.

    (b) $A \in \mathbf{N}(n, \mathbb{C}) \iff A = UDU^*$, for some $U \in \mathbf{SU}_n$, $D \in \mathbf{D}(n, \mathbb{C})$.

    (c) $A \in \mathbf{N}(n, \mathbb{R})$, $\sigma(A) \subset \mathbb{R} \iff A = UDU^\top$, for some $U \in \mathbf{SO}_n$, $D \in \mathbf{D}(n, \mathbb{R})$.

12. Show that an upper triangular or a lower triangular matrix $B \in \mathbb{C}^{n \times n}$ is normal if and only if B is diagonal. (*Hint:* Consider the equality $(BB^*)_{11} = (B^*B)_{11}$.)

13. Let the assumptions of Theorem 4.2.15 hold. Show that instead of performing the Gram–Schmidt process on $\mathbf{v}, T\mathbf{v}, \ldots, T^{r-1}\mathbf{v}$ one can perform the following process. Let $\mathbf{w}_1 := \frac{1}{\|\mathbf{v}\|}\mathbf{v}$. Assume that one already obtained i orthonormal vectors $\mathbf{w}_1, \ldots, \mathbf{w}_i$. Let $\tilde{\mathbf{w}}_{i+1} := T\mathbf{w}_i - \sum_{j=1}^{i}\langle T\mathbf{w}_i, \mathbf{w}_j\rangle\mathbf{w}_j$. If $\tilde{\mathbf{w}}_{i+1} = 0$ then stop the process, i.e. one is left with i orthonormal vectors. If $\mathbf{w}_{i+1} \neq 0$ then $\mathbf{w}_{i+1} := \frac{1}{\|\tilde{\mathbf{w}}_{i+1}\|}\tilde{\mathbf{w}}_{i+1}$ and continue the process. Show that the process ends after obtaining r orthonormal vectors $\mathbf{w}_1, \ldots, \mathbf{w}_r$ and $\mathbf{u}_i = \mathbf{w}_i$ for $i = 1, \ldots, r$. (This is a version of *Lanczos tridiagonalization* process.)

4.3 Symmetric Bilinear and Hermitian Forms

Definition 4.3.1 *Let* $\mathbf{V}$ *be a module over* $\mathbb{D}$ *and* $Q : \mathbf{V} \times \mathbf{V} \to \mathbb{D}$. *Q is called a symmetric bilinear form (on* $\mathbf{V}$*) if the following conditions are satisfied:*

(a) $Q(\mathbf{x}, \mathbf{y}) = Q(\mathbf{y}, \mathbf{x})$ *for all* $\mathbf{x}, \mathbf{y} \in \mathbf{V}$ *(symmetricity);*

(b) $Q(a\mathbf{x} + b\mathbf{z}, y) = aQ(\mathbf{x}, \mathbf{y}) + bQ(\mathbf{z}, \mathbf{y})$ *for all* $a, b \in \mathbb{D}$ *and* $\mathbf{x}, \mathbf{y}, \mathbf{z} \in \mathbf{V}$ *(bilinearity).*

For $\mathbb{D} = \mathbb{C}$, *Q is called a hermitian form (on* $\mathbf{V}$*) if Q satisfies the conditions* (a') *and* (b) *where*
(a') $Q(\mathbf{x}, \mathbf{y}) = \overline{Q(\mathbf{y}, \mathbf{x})}$ *for all* $\mathbf{x}, \mathbf{y} \in \mathbf{V}$ *(barsymmetricity).*

The following results are elementary (see Problems 1 and 2):

Proposition 4.3.2 *Let* $\mathbf{V}$ *be a module over* $\mathbb{D}$ *with a basis* $E = \{\mathbf{e}_1, \ldots, \mathbf{e}_n\}$. *Then there is a* $1 - 1$ *correspondence between a symmetric bilinear form* Q *on* $\mathbf{V}$ *and* $A \in \mathbf{S}(n, \mathbb{D})$:

$$Q(\mathbf{x}, \mathbf{y}) = \eta^{\top} A \xi, \; \text{where}$$

$$\mathbf{x} = \sum_{i=1}^{n} \xi_i \mathbf{e}_i, \; \mathbf{y} = \sum_{i=1}^{n} \eta_i \mathbf{e}_i,$$

$$\xi = (\xi_1, \ldots, \xi_n)^{\top}, \eta = (\eta_1, \ldots, \eta_n)^{\top} \in \mathbb{D}^n.$$

Let $\mathbf{V}$ *be a vector space over* $\mathbb{C}$ *with a basis* $E = \{\mathbf{e}_1, \ldots, \mathbf{e}_n\}$. *Then there is a* $1 - 1$ *correspondence between a hermitian form* Q *on* $\mathbf{V}$ *and* $A \in \mathbf{H}_n$:

$$Q(\mathbf{x}, \mathbf{y}) = \eta^* A \xi, \quad \text{where}$$

$$\mathbf{x} = \sum_{i=1}^{n} \xi_i \mathbf{e}_i, \; \mathbf{y} = \sum_{i=1}^{n} \eta_i \mathbf{e}_i,$$

$$\xi = (\xi_1, \ldots, \xi_n)^{\top}, \eta = (\eta_1, \ldots, \eta_n)^{\top} \in \mathbb{C}^n.$$

Definition 4.3.3 *Let the assumptions of Proposition 4.3.2 hold. Then A is called the representation matrix of Q in the basis E.*

Proposition 4.3.4 *Let the assumptions of Proposition 4.3.2 hold. Let $F = \{\mathbf{f_1}, \ldots, \mathbf{f_n}\}$ be another basis of the $\mathbb{D}$ module $\mathbf{V}$. Then the symmetric bilinear form Q is represented by $B \in \mathbf{S}(n, \mathbb{D})$ in the basis F, where B is congruent to A:*

$$B = U^\top A U, \quad U \in \mathbf{GL}(n, \mathbb{D})$$

and U is the matrix corresponding to the change of basis from F to E. For $\mathbb{D} = \mathbb{C}$ the hermitian form Q is presented by $B \in \mathbf{H}_n$ in the basis F, where B is hermicongruent to A:

$$B = U^* A U, \quad U \in \mathbf{GL}(n, \mathbb{C})$$

and U is the matrix corresponding to the change of basis from F to E.

In what follows we assume that $\mathbb{D} = \mathbb{F} = \mathbb{R}, \mathbb{C}$.

Proposition 4.3.5 *Let $\mathbf{V}$ be an n-dimensional vector space over $\mathbb{R}$. Let $Q : \mathbf{V} \times \mathbf{V} \to \mathbb{R}$ be a symmetric bilinear form. Let $A \in \mathbf{S}(n, \mathbb{R})$ be the representation matrix of Q with respect to a basis E in $\mathbf{V}$. Let $\mathbf{V}_\mathbb{C}$ be the complexification of $\mathbf{V}$ over $\mathbb{C}$. Then there exists a unique hermitian form $Q_\mathbb{C} : \mathbf{V}_\mathbb{C} \times \mathbf{V}_\mathbb{C} \to \mathbb{C}$ such that $Q_\mathbb{C}|_{\mathbf{V} \times \mathbf{V}} = Q$ and $Q_\mathbb{C}$ is presented by A with respect to the basis E in $\mathbf{V}_\mathbb{C}$.*

See Problem 3.

Normalization 4.3.6 *Let $\mathbf{V}$ be a finite dimensional IPS over $\mathbb{F}$. Let $Q : \mathbf{V} \times \mathbf{V} \to \mathbb{F}$ be either a symmetric bilinear form for $\mathbb{F} = \mathbb{R}$ or a hermitian form for $\mathbb{F} = \mathbb{C}$. Then a representation matrix A of Q is chosen with respect to an orthonormal basis E.*

The following proposition is straightforward (see Problem 4).

Proposition 4.3.7 *Let $\mathbf{V}$ be an n-dimensional IPS over $\mathbb{F}$. Let $Q : \mathbf{V} \times \mathbf{V} \to \mathbb{F}$ be either a symmetric bilinear form for $\mathbb{F} = \mathbb{R}$ or a hermitian form for $\mathbb{F} = \mathbb{C}$. Then there exists a unique $T \in \mathbf{S}(\mathbf{V})$ such* *that $Q(\mathbf{x}, \mathbf{y}) = \langle T\mathbf{x}, \mathbf{y} \rangle$ for any $\mathbf{x}, \mathbf{y} \in \mathbf{V}$. In any orthonormal basis of $\mathbf{V}$, Q and T are represented by the same matrix A. In particular, the characteristic polynomial $p(\lambda)$ of T is called the characteristic polynomial of Q. Q has only real roots:*

$$\lambda_1(Q) \geq \cdots \geq \lambda_n(Q),$$

which are called the eigenvalues of Q. Furthermore, there exists an orthonormal basis $F = \{\mathbf{f}_1, \ldots, \mathbf{f}_n\}$ in $\mathbf{V}$ such that $D = \operatorname{diag}(\lambda_1(Q), \ldots, \lambda_n(Q))$ is the representation matrix of Q in F.

Vice versa, for any $T \in \mathbf{S}(\mathbf{V})$ and any subspace $\mathbf{U} \subset \mathbf{V}$ the form $Q(T, \mathbf{U})$ defined by

$$Q(T, \mathbf{U})(\mathbf{x}, \mathbf{y}) := \langle T\mathbf{x}, \mathbf{y} \rangle \quad \text{for } \mathbf{x}, \mathbf{y} \in \mathbf{U}$$

is either a symmetric bilinear form for $\mathbb{F} = \mathbb{R}$ or a hermitian form for $\mathbb{F} = \mathbb{C}$.

In the rest of the book, we use the following normalization unless stated otherwise.

Normalization 4.3.8 *Let $\mathbf{V}$ be an n-dimensional IPS over $\mathbb{F}$. Assume that $T \in \mathbf{S}(\mathbf{V})$. Then arrange the eigenvalues of T counted with their multiplicities in the decreasing order*

$$\lambda_1(T) \geq \cdots \geq \lambda_n(T).$$

Note that the same normalization applies to real symmetric matrices and complex hermitian matrices.

Problems

1. Prove Proposition 4.3.2.

2. Prove Proposition 4.3.4.

3. Prove Proposition 4.3.5.

4. Prove Proposition 4.3.7.

4.4   Max–Min Characterizations of Eigenvalues

Definition 4.4.1 *Let* **V** *be a finite dimensional vector space over the field* $\mathbb{F}$. *Denote by* $\mathrm{Gr}(m, \mathbf{V})$ *the space of all* m-*dimensional subspaces in* **V** *where* $m \in \{0, \dots, n\}$, *and* n *is the dimension of* **V**.

Theorem 4.4.2 (*The convoy principle*) *Let* **V** *be an* n-*dimensional IPS. Let* $T \in \mathbf{S}(\mathbf{V})$. *Then*

$$\lambda_k(T) = \max_{\mathbf{U} \in \mathrm{Gr}(k, \mathbf{V})} \min_{0 \neq \mathbf{x} \in \mathbf{U}} \frac{\langle T\mathbf{x}, \mathbf{x} \rangle}{\langle \mathbf{x}, \mathbf{x} \rangle} \tag{4.4.1}$$

$$= \max_{\mathbf{U} \in \mathrm{Gr}(k, \mathbf{V})} \lambda_k(Q(T, \mathbf{U})), \quad k = 1, \dots, n,$$

where the quadratic form $Q(T, \mathbf{U})$ *is defined in Proposition 4.3.7. For* $k \in \{1, \dots, n\}$ *let* **U** *be an invariant subspace of* T *spanned by eigenvectors* $\mathbf{e}_1, \dots, \mathbf{e}_k$ *corresponding to the eigenvalues* $\lambda_1(T), \dots, \lambda_k(T)$. *Then* $\lambda_k(T) = \lambda_k(Q(T, \mathbf{U}))$. *Let* $\mathbf{U} \in \mathrm{Gr}(k, \mathbf{V})$ *and assume that* $\lambda_k(T) = \lambda_k(Q(T, \mathbf{U}))$. *Then* **U** *contains an eigenvector of* T *corresponding to* $\lambda_k(T)$.

In particular

$$\lambda_1(T) = \max_{0 \neq \mathbf{x} \in \mathbf{V}} \frac{\langle T\mathbf{x}, \mathbf{x} \rangle}{\langle \mathbf{x}, \mathbf{x} \rangle}, \quad \lambda_n(T) = \min_{0 \neq \mathbf{x} \in \mathbf{V}} \frac{\langle T\mathbf{x}, \mathbf{x} \rangle}{\langle \mathbf{x}, \mathbf{x} \rangle}. \tag{4.4.2}$$

Moreover, for any $\mathbf{x} \neq \mathbf{0}$

$$\lambda_1(T) = \frac{\langle T\mathbf{x}, \mathbf{x} \rangle}{\langle \mathbf{x}, \mathbf{x} \rangle} \iff T\mathbf{x} = \lambda_1(T)\mathbf{x},$$

$$\lambda_n(T) = \frac{\langle T\mathbf{x}, \mathbf{x} \rangle}{\langle \mathbf{x}, \mathbf{x} \rangle} \iff T\mathbf{x} = \lambda_n(T)\mathbf{x}.$$

The quotient $\frac{\langle T\mathbf{x}, \mathbf{x} \rangle}{\langle \mathbf{x}, \mathbf{x} \rangle}$, $\mathbf{0} \neq \mathbf{x} \in \mathbf{V}$ is called the *Rayleigh quotient*. The characterization (4.4.2) is called the *convoy principle*.

Proof. Choose an orthonormal basis $E = \{e_1, \ldots, e_n\}$ such that

$$Te_i = \lambda_i(T)e_i, \quad \langle e_i, e_j \rangle = \delta_{ij} \quad i, j = 1, \ldots, n. \qquad (4.4.3)$$

Then

$$\frac{\langle Tx, x \rangle}{\langle x, x \rangle} = \frac{\sum_{i=1}^{n} \lambda_i(T)|x_i|^2}{\sum_{i=1}^{n} |x_i|^2}, \quad x = \sum_{i=1}^{n} x_i e_i \neq 0. \qquad (4.4.4)$$

The above equality yields straightforward (4.4.2) and the equality cases in these characterizations. Let $U \in \mathrm{Gr}(k, V)$. Then the minimal characterization of $\lambda_k(Q(T, U))$ yields the equality

$$\lambda_k(Q(T, U)) = \min_{0 \neq x \in U} \frac{\langle Tx, x \rangle}{\langle x, x \rangle} \quad \text{for any } U \in \mathrm{Gr}(k, U). \qquad (4.4.5)$$

Next there exists $0 \neq x \in U$ such that $\langle x, e_i \rangle = 0$ for $i = 1, \ldots, k-1$. (For $k = 1$ this condition is void.) Hence

$$\frac{\langle Tx, x \rangle}{\langle x, x \rangle} = \frac{\sum_{i=k}^{n} \lambda_i(T)|x_i|^2}{\sum_{i=k}^{n} |x_i|^2} \leq \lambda_k(T) \Rightarrow \lambda_k(T) \geq \lambda_k(Q(T, U)).$$

Let

$$\lambda_1(T) = \cdots = \lambda_{n_1}(T) > \lambda(T)_{n_1+1}(T) = \cdots = \lambda_{n_2}(T) > \cdots >$$

$$\lambda_{n_{r-1}+1}(T) = \cdots = \lambda_{n_r}(T) = \lambda_n(T), \quad n_0 = 0 < n_1 < \cdots < n_r = n.$$
$$(4.4.6)$$

Assume that $n_{j-1} < k \leq n_j$. Suppose that $\lambda_k(Q(T, U)) = \lambda_k(T)$. Then for $x \in U$ such that $\langle x, e_i \rangle = 0$ we have the equality $\lambda_k(Q(T, U)) = \lambda_k(T)$ if and only if $x = \sum_{i=k}^{n_j} x_i e_i$. Thus $Tx = \lambda_k(T)x$.

Let $U_k = \mathrm{span}\,(e_1, \ldots, e_k)$. Let $0 \neq x = \sum_{i=1}^{k} \in U_k$. Then

$$\frac{\langle Tx, x \rangle}{\langle x, x \rangle} = \frac{\sum_{i=1}^{k} \lambda_i(T)|x_i|^2}{\sum_{i=1}^{k} |x_i|^2} \geq \lambda_k(T) \Rightarrow \lambda_k(Q(T, U_k)) \geq \lambda_k(T).$$

Hence, $\lambda_k(Q(T, U_k)) = \lambda_k(T)$. $\qquad \square$

It can be shown that for $k > 1$ and $\lambda_1(T) > \lambda_k(T)$ there exists $\mathbf{U} \in \mathrm{Gr}(k, \mathbf{V})$ such that $\lambda_k(T) = \lambda_k(T, \mathbf{U})$ and $\mathbf{U}$ is not an invariant subspace of T, in particular $\mathbf{U}$ does not contain all $\mathbf{e}_1, \ldots, \mathbf{e}_k$ satisfying (4.4.3). (See Problem 1.)

Corollary 4.4.3 *Let the assumptions of Theorem 4.4.2 hold. Let $1 \le \ell \le n$. Then*

$$\lambda_k(T) = \max_{\mathbf{W} \in \mathrm{Gr}(\ell, \mathbf{V})} \lambda_k(Q(T, \mathbf{W})), \quad k = 1, \ldots, \ell. \qquad (4.4.7)$$

Proof. For $k \le \ell$ apply Theorem 4.4.2 to $\lambda_k(Q(T, \mathbf{W}))$ to deduce that $\lambda_k(Q(T, \mathbf{W})) \le \lambda_k(T)$. Let $\mathbf{U}_\ell = \mathrm{span}\,(\mathbf{e}_1, \ldots, \mathbf{e}_\ell)$. Then

$$\lambda_k(Q(T, \mathbf{U}_\ell)) = \lambda_k(T), \quad k = 1, \ldots, \ell.$$

$\square$

Theorem 4.4.4 (*Courant–Fischer principle*) *Let $\mathbf{V}$ be an n-dimensional IPS and $T \in \mathbf{S}(\mathbf{V})$. Then*

$$\lambda_k(T) = \min_{\mathbf{W} \in \mathrm{Gr}(k-1, \mathbf{V})} \max_{0 \ne \mathbf{x} \in \mathbf{W}^\perp} \frac{\langle T\mathbf{x}, \mathbf{x} \rangle}{\langle \mathbf{x}, \mathbf{x} \rangle}, \quad k = 1, \ldots, n.$$

See Problem 2 for the proof of the theorem and the following corollary.

Corollary 4.4.5 *Let $\mathbf{V}$ be an n-dimensional IPS and $T \in \mathbf{S}(\mathbf{V})$. Let $k, \ell \in [n]$ be integers satisfying $k \le l$. Then*

$$\lambda_{n-\ell+k}(T) \le \lambda_k(Q(T, \mathbf{W})) \le \lambda_k(T), \quad \text{for any } \mathbf{W} \in \mathrm{Gr}(\ell, \mathbf{V}).$$

Theorem 4.4.6 *Let $\mathbf{V}$ be an n-dimensional IPS and $S, T \in \mathbf{S}(\mathbf{V})$. Then for any $i, j \in \mathbb{N}, i + j - 1 \le n$ the inequality $\lambda_{i+j-1}(S + T) \le \lambda_i(S) + \lambda_j(T)$ holds.*

Proof. Let $\mathbf{U}_{i-1}, \mathbf{V}_{j-1} \subset \mathbf{V}$ be eigenspaces of S, T spanned by the first $i - 1, j - 1$ eigenvectors of S, T respectively. So

$$\langle S\mathbf{x}, \mathbf{x} \rangle \le \lambda_i(S) \langle \mathbf{x}, \mathbf{x} \rangle,$$

$$\langle T\mathbf{y}, \mathbf{y} \rangle \le \lambda_j(T) \langle \mathbf{y}, \mathbf{y} \rangle \text{ for all } \mathbf{x} \in \mathbf{U}_{i-1}^\perp, \mathbf{y} \in \mathbf{V}_{j-1}^\perp.$$

Note that dim $\mathbf{U}_{i-1} = i - 1, \dim \mathbf{V}_{j-1} = j - 1$. Let $\mathbf{W} = \mathbf{U}_{i-1} + \mathbf{V}_{j-1}$. Then dim $\mathbf{W} = l-1 \leq i+j-2$. Assume that $\mathbf{z} \in \mathbf{W}^{\perp}$. Then $\langle (S+T)\mathbf{z}, \mathbf{z} \rangle = \langle S\mathbf{z}, \mathbf{z} \rangle + \langle T\mathbf{z}, \mathbf{z} \rangle \leq (\lambda_i(S)+\lambda_j(T))\langle \mathbf{z}, \mathbf{z} \rangle$. Hence, $\max_{0 \neq \mathbf{z} \in \mathbf{W}^{\perp}} \frac{\langle (S+T)\mathbf{z}, \mathbf{z} \rangle}{\langle \mathbf{z}, \mathbf{z} \rangle} \leq \lambda_i(S)+\lambda_j(T)$. Use Theorem 4.4.4 to deduce that $\lambda_{i+j-1}(S+T) \leq \lambda_l(S+T) \leq \lambda_i(S) + \lambda_j(T)$. $\qquad\square$

Definition 4.4.7 *Let* $\mathbf{V}$ *be an* n-*dimensional IPS. Fix an integer* $k \in [n]$. *Then* $F_k = \{\mathbf{f}_1, \ldots, \mathbf{f}_k\}$ *is called an orthonormal* k-*frame if* $\langle \mathbf{f}_i, \mathbf{f}_j \rangle = \delta_{ij}$ *for* $i, j = 1, \ldots, k$. *Denote by* $\mathrm{Fr}(k, \mathbf{V})$ *the set of all orthonormal* k-*frames in* $\mathbf{V}$.

Note that each $F_k \in \mathrm{Fr}(k, \mathbf{V})$ induces $\mathbf{U} = \mathrm{span}\, F_k \in \mathrm{Gr}(k, \mathbf{V})$. Vice versa, any $\mathbf{U} \in \mathrm{Gr}(k, \mathbf{V})$ induces the set $\mathrm{Fr}(k, \mathbf{U})$ of all orthonormal k-frames which span $\mathbf{U}$.

Theorem 4.4.8 *Let* $\mathbf{V}$ *be an* n-*dimensional IPS and* $T \in \mathbf{S}(\mathbf{V})$. *Then for any integer* $k \in [n]$

$$\sum_{i=1}^{k} \lambda_i(T) = \max_{\{\mathbf{f}_1, \ldots, \mathbf{f}_k\} \in \mathrm{Fr}(k, \mathbf{V})} \sum_{i=1}^{k} \langle T\mathbf{f}_i, \mathbf{f}_i \rangle.$$

Furthermore

$$\sum_{i=1}^{k} \lambda_i(T) = \sum_{i=1}^{k} \langle T\mathbf{f}_i, \mathbf{f}_i \rangle$$

for some k-*orthonormal frame* $F_k = \{\mathbf{f}_1, \ldots, \mathbf{f}_k\}$ *if and only if* $\mathrm{span}\, F_k$ *is spanned by* $\mathbf{e}_1, \ldots, \mathbf{e}_k$ *satisfying* (4.4.3).

Proof. Define

$$\mathrm{tr}\, Q(T, \mathbf{U}) := \sum_{i=1}^{k} \lambda_i(Q(T, \mathbf{U})) \quad \text{for } \mathbf{U} \in \mathrm{Gr}(k, \mathbf{V}),$$

(4.4.8)

$$\mathrm{tr}_k\, T := \sum_{i=1}^{k} \lambda_i(T).$$

Let $F_k = \{\mathbf{f}_1, \ldots, \mathbf{f}_k\} \in \mathrm{Fr}(k, \mathbf{V})$. Set $\mathbf{U} = \mathrm{span}\, F_k$. Then in view of Corollary 4.4.3

$$\sum_{i=1}^{k} \langle T\mathbf{f}_i, \mathbf{f}_i \rangle = \mathrm{tr}\, Q(T, \mathbf{U}) \leq \sum_{i=1}^{k} \lambda_i(T).$$

Let $E_k := \{\mathbf{e}_1, \ldots, \mathbf{e}_k\}$ where $\mathbf{e}_1, \ldots, \mathbf{e}_n$ are given by (4.4.3). Clearly $\mathrm{tr}_k\, T = \mathrm{tr}\, Q(T, \mathrm{span}\, E_k)$. This shows the maximal characterization of $\mathrm{tr}_k\, T$.

Let $\mathbf{U} \in \mathrm{Gr}(k, \mathbf{V})$ and assume that $\mathrm{tr}_k\, T = \mathrm{tr}\, Q(T, \mathbf{U})$. Hence, $\lambda_i(T) = \lambda_i(Q(T, \mathbf{U}))$ for $i = 1, \ldots, k$. Then there exists $G_k = \{\mathbf{g}_1, \ldots, \mathbf{g}_k\} \in \mathrm{Fr}(k, \mathbf{U}))$ such that

$$\min_{0 \neq \mathbf{x} \in \mathrm{span}\,(\mathbf{g}_1, \ldots, \mathbf{g}_i)} \frac{\langle T\mathbf{x}, \mathbf{x} \rangle}{\langle \mathbf{x}, \mathbf{x} \rangle} = \lambda_i(Q(T, \mathbf{U})) = \lambda_i(T), \ i = 1, \ldots, k.$$

Use Theorem 4.4.2 to deduce that $T\mathbf{g}_i = \lambda_i(T)\mathbf{g}_i$ for $i = 1, \ldots, k$. $\square$

Theorem 4.4.9 *Let* $\mathbf{V}$ *be an* n-*dimensional IPS and* $T \in \mathbf{S}(\mathbf{V})$. *Then for any integer* $k, l \in [n]$, *such that* $k + l \leq n$, *we have that*

$$\sum_{i=l+1}^{l+k} \lambda_i(T) = \min_{\mathbf{W} \in \mathrm{Gr}(l, \mathbf{V})} \ \max_{\{\mathbf{f}_1, \ldots, \mathbf{f}_k\} \in \mathrm{Fr}(k, \mathbf{V} \cap \mathbf{W}^\perp)} \sum_{i=1}^{k} \langle T\mathbf{f}_i, \mathbf{f}_i \rangle.$$

Proof. Let $\mathbf{W}_j := \mathrm{span}\,(\mathbf{e}_1, \ldots, \mathbf{e}_j), j = 1, \ldots, n$, where $\mathbf{e}_1, \ldots, \mathbf{e}_n$ are given by (4.4.3). Then $\mathbf{V}_1 := \mathbf{V} \cap \mathbf{W}_l$ is an invariant subspace of T. Let $T_1 := T|\mathbf{V}_1$. Then $\lambda_i(T_1) = \lambda_{l+i}(T)$ for $i = 1, \ldots, n - l$. Theorem 4.4.8 for T_1 yields

$$\max_{\{\mathbf{f}_1, \ldots, \mathbf{f}_k\} \in \mathrm{Fr}(k, \mathbf{V} \cap \mathbf{W}_l^\perp)} \sum_{i=1}^{k} \langle T\mathbf{f}_i, \mathbf{f}_i \rangle = \sum_{i=l+1}^{l+k} \lambda_i(T).$$

Let $T_2 := T|\mathbf{W}_{l+k}$ and $\mathbf{W} \in \mathrm{Gr}(l, \mathbf{V})$. Set $\mathbf{U} := \mathbf{W}_{l+k} \cap \mathbf{W}^\perp$. Then $\dim \mathbf{U} \geq k$. Apply Theorem 4.4.8 to $-T_2$ to deduce

$$\sum_{i=1}^{k} \lambda_i(-T_2) \geq \sum_{i=1}^{k} \langle -T\mathbf{f}_i, \mathbf{f}_i \rangle \text{ for } \{\mathbf{f}_1, \ldots, \mathbf{f}_k\} \in \mathrm{Fr}(k, \mathbf{U}).$$

The above inequality is equal to the inequality

$$\sum_{i=l+1}^{l+k} \lambda_i(T) \le \sum_{i=1}^{k} \langle T\mathbf{f}_i, \mathbf{f}_i \rangle \text{ for } \{\mathbf{f}_1, \ldots, \mathbf{f}_k\} \in \text{Fr}(k, \mathbf{U})$$

$$\le \max_{\{\mathbf{f}_1, \ldots, \mathbf{f}_k\} \in \text{Fr}(k, \mathbf{V} \cap \mathbf{W}^\perp)} \sum_{i=1}^{k} \langle T\mathbf{f}_i, \mathbf{f}_i \rangle.$$

The above inequalities yield the theorem. □

Definition 4.4.10 *Let* $\mathbf{V}$ *be an n-dimensional IPS. Assume that* $k \in [n]$ *and* $1 \le i_1 < \cdots < i_k \le n$ *are integers. Let* $U_0 := \{0\}$ *and* $\mathbf{U}_j \in \text{Gr}(i_j, \mathbf{V})$ *for* $j = 1, \ldots, k$. $(\mathbf{U}_1, \ldots, \mathbf{U}_k)$ *is called a* flag *in* $\mathbf{V}$ *if* $\mathbf{U}_1 \subset \mathbf{U}_2 \subset \cdots \subset \mathbf{U}_k$. *The set of all such flags is denoted by* $\mathcal{F}(i_1, \ldots, i_k, \mathbf{V})$.

Theorem 4.4.11 *Let* $\mathbf{V}$ *be an n-dimensional IPS. Assume that* $k \in [n]$ *and* $1 \le i_1 < \cdots < i_k \le n$ *are integers. Then for each* $T \in \mathbf{S}(\mathbf{V})$ *the following characterization holds:*

$$\sum_{j=1}^{k} \lambda_{i_j}(T) = \max_{(\mathbf{U}_1, \ldots, \mathbf{U}_k) \in \mathcal{F}(i_1, \ldots, i_k)}$$

$$\min_{\mathbf{x}_1, \ldots, \mathbf{x}_k, \mathbf{x}_p \in \mathbf{U}_p, \langle \mathbf{x}_p, \mathbf{x}_q \rangle = \delta_{pq}, p, q \in [k]} \sum_{j=1}^{k} \langle T\mathbf{x}_j, \mathbf{x}_j \rangle.$$

Proof. We first prove by induction on the dimension $n = \mathbf{S}(\mathbf{V})$ the inequality

$$\sum_{j=1}^{k} \lambda_{i_j}(T) \ge \min_{\mathbf{x}_1, \ldots, \mathbf{x}_k, \mathbf{x}_p \in \mathbf{U}_p, \langle \mathbf{x}_p, \mathbf{x}_q \rangle = \delta_{pq}, p, q \in [k]} \sum_{j=1}^{k} \langle T\mathbf{x}_j, \mathbf{x}_j \rangle, \quad (4.4.9)$$

for each $k \in [n]$ and each flag $(\mathbf{U}_1, \ldots, \mathbf{U}_k)$.

Clearly, for $n = 1$ equality holds. Assume that the above inequality holds for $n = N$. Let $n = N + 1$. Suppose first that $k = N + 1$. Then $\text{tr}\, T = \sum_{j=1}^{N} \langle T\mathbf{x}_j, \mathbf{x}_j \rangle$ and equality holds in the above inequality.

Assume that $k \in [N]$. Suppose first that $i_k \leq N$. Let $\mathbf{W} \in$ $\mathrm{Gr}(N, \mathbf{V})$ such that $\mathbf{U}_k \subset \mathbf{W}$. Then $(\mathbf{U}_1, \ldots, \mathbf{U}_k)$ is a flag in $\mathbf{W}$. The induction hypothesis yields that the right-hand side of (4.4.9) is bounded above by $\sum_{j=1}^{k} \lambda_{i_j}(Q(T, \mathbf{W}))$. Corollary 4.4.3 yields that

$$\sum_{j=1}^{k} \lambda_{i_j} T) \geq \sum_{j=1}^{k} \lambda_{i_j}(Q(T, \mathbf{W})).$$

Hence, (4.4.9) holds.

Assume now that $i_j = N + j - k + 1$ for $j = k, \ldots, l$ and $i_{l-1} < N + l - k$ if $l > 1$. Let $\mathbf{y}_1, \ldots, \mathbf{y}_n$ be an orthonormal set of eigenvectors of T corresponding to the eigenvalues $\lambda_1(T), \ldots, \lambda_n(T)$ respectively. Let $\mathbf{W}$ be N dimensional subspace of $\mathbf{V}$ which contains $\mathbf{U}_{l-1}$ and span $(\mathbf{x}_{N+1}, \ldots, \mathbf{x}_{N+1+l-k})$. Clearly, $i_j - 1 \leq \dim \mathbf{U}_j \cap \mathbf{W} \leq i_j$ for $j \in [k]$. Hence,

$$\mathbf{U}_{l-1} \subset \mathbf{U}_l \cap \mathbf{W} \subseteq \cdots \subseteq \mathbf{U}_k \cap \mathbf{W}.$$

Therefore, there exists a flag $(\mathbf{U}'_1, \ldots, \mathbf{U}'_k)$ in $\mathbf{W}$ with the following properties: First, $\mathbf{U}'_j = \mathbf{U}_j$ for $j = 1, \ldots, l - 1$. Second $i'_j := \dim \mathbf{U}'_j = i_j - 1$ for $j = l, \ldots, k$. Third, $\mathbf{U}'_j \subset \mathbf{U}_j$ for $j = l, \ldots, k$. (Note that if $i_j - 1 = \dim \mathbf{U}_j \cap \mathbf{W}$ then $\mathbf{U}'_j = \mathbf{U}_j \cap \mathbf{W}$.) Let $i'_j = i_j$ for $j = 1, \ldots, l - 1$ if $l > 1$.

The induction hypothesis yields

$$\sum_{j=1}^{k} \lambda_{i_j}(Q(T, \mathbf{W})) \geq \min_{\mathbf{x}_1, \ldots, \mathbf{x}_k, \mathbf{x}_p \in \mathbf{U}'_p, \langle \mathbf{x}_p, \mathbf{x}_q \rangle = \delta_{pq}, p, q \in [k]} \sum_{j=1}^{k} \langle T\mathbf{x}_j, \mathbf{x}_j \rangle.$$

Clearly, the right-hand side of the above inequality is not less than the right-hand side of (4.4.9). It is left to show that the left-hand side of the above inequality is not more then the left-hand side of (4.4.9). Since span $(\mathbf{x}_{N+1}, \ldots, \mathbf{x}_{N+1+l-k}) \subseteq \mathbf{W}$ it follows that $\lambda_{i_j-1}(Q(T, \mathbf{W})) = \lambda_{i_j}(T)$ for $j = k, \ldots, l$. Corollary 4.4.3 yields that $\lambda_i(T) \geq \lambda_i(Q(T, \mathbf{W}))$ for $i \in [N]$. Hence, the right-hand side of (4.4.9) is not less than the right-hand side of the above inequality. This establishes (4.4.9).

Let $\mathbf{U}_{i_j} = \operatorname{span}(\mathbf{y}_1, \ldots, \mathbf{y}_{i_j})$ for $j = 1, \ldots, k$. Recall that $\langle \mathbf{x}_j, \mathbf{x}_j \rangle \geq \lambda_{i_j}(T)$ for each $\mathbf{x}_j \in \mathbf{U}_j$ satisfying $\langle \mathbf{x}_j, \mathbf{x}_j \rangle = 1$. Hence, the right-hand side of (4.4.9) is at least $\sum_{j=1}^{k} \lambda_{i_j}(T)$. This establishes the theorem. $\qquad\square$

Problems

1. Let $\mathbf{V}$ be a three-dimensional IPS and $T \in \operatorname{Hom}(\mathbf{V})$ be self-adjoint. Assume that

$$\lambda_1(T) > \lambda_2(T) > \lambda_3(T), \quad T\mathbf{e}_i = \lambda_i(T)\mathbf{e}_i, \ i = 1, 2, 3.$$

Let $\mathbf{W} = \operatorname{span}(\mathbf{e}_1, \mathbf{e}_3)$.

   (a) Show that for each $t \in [\lambda_3(T), \lambda_1(T)]$ there exists a unique $\mathbf{W}(t) \in \operatorname{Gr}(1, \mathbf{W})$ such that $\lambda_1(Q(T, \mathbf{W}(t))) = t$.

   (b) Let $t \in [\lambda_2(T), \lambda_1(T)]$. Let $\mathbf{U}(t) = \operatorname{span}(\mathbf{W}(t), \mathbf{e}_2) \in \operatorname{Gr}(2, \mathbf{V})$. Show that $\lambda_2(T) = \lambda_2(Q(T, \mathbf{U}(t)))$.

2. (a) Let the assumptions of Theorem 4.4.4 hold. Let $\mathbf{W} \in \operatorname{Gr}(k-1, \mathbf{V})$. Show that there exists $\mathbf{0} \neq \mathbf{x} \in \mathbf{W}^\perp$ such that $\langle \mathbf{x}, \mathbf{e}_i \rangle = 0$ for $k+1, \ldots, n$, where $\mathbf{e}_1, \ldots, \mathbf{e}_n$ satisfy (4.4.3). Conclude that $\lambda_1(Q(T, \mathbf{W}^\perp)) \geq \frac{\langle T\mathbf{x}, \mathbf{x} \rangle}{\langle \mathbf{x}, \mathbf{x} \rangle} \geq \lambda_k(T)$.

   (b) Let $\mathbf{U}_\ell = \operatorname{span}(\mathbf{e}_1, \ldots, \mathbf{e}_\ell)$. Show that $\lambda_1(Q(T, \mathbf{U}_\ell^\perp)) = \lambda_{\ell+1}(T)$ for $\ell = 1, \ldots, n-1$.

   (c) Prove Theorem 4.4.4.

   (d) Prove Corollary 4.4.5. (*Hint*: Choose $\mathbf{U} \in \operatorname{Gr}(k, \mathbf{W})$ such that $\mathbf{U} \subset \mathbf{W} \cap \operatorname{span}(\mathbf{e}_{n-\ell+k+1}, \ldots, \mathbf{e}_n)^\perp$. Then $\lambda_{n-\ell+k}(T) \leq \lambda_k(Q(T, \mathbf{U})) \leq \lambda_k(Q(T, \mathbf{W}))$.)

3. Let $B = [b_{ij}]_{i,j=1}^n \in \mathbf{H}_n$ and denote by $A \in \mathbf{H}_{n-1}$ the matrix obtained from B by deleting the jth row and column.

   (a) Show the Cauchy interlacing inequalities

$$\lambda_i(B) \geq \lambda_i(A) \geq \lambda_{i+1}(B), \text{ for } i = 1, \ldots, n-1.$$

(b) Show the inequality $\lambda_1(B) + \lambda_n(B) \le \lambda_1(A) + b_{ii}$.

Hint: Express the traces of B and A respectively in terms of eigenvalues to obtain

$$\lambda_1(B) + \lambda_n(B) = b_{ii} + \lambda_1(A) + \sum_{i=2}^{n-1}(\lambda_i(A) - \lambda_i(B)).$$

Then use the Cauchy interlacing inequalities.

4. Show the following generalization of Problem 3(b) ([Big96, p. 56]). Let $B \in \mathbf{H}_n$ be the following 2×2 block matrix $B = \begin{bmatrix} B_{11} & B_{12} \\ B_{12}^* & B_{22} \end{bmatrix}$. Show that

$$\lambda_1(B) + \lambda_n(B) \le \lambda_1(B_{11}) + \lambda_1(B_{22}).$$

Hint: Assume that $B\mathbf{x} = \lambda_1(B)\mathbf{x}, \mathbf{x}^\top = (\mathbf{x}_1^\top, \mathbf{x}_2^\top)$, partitioned as B. Consider $\mathbf{U} = \mathrm{span}\,((\mathbf{x}_1^\top, \mathbf{0})^\top, (\mathbf{0}, \mathbf{x}_2^\top)^\top)$. Analyze $\lambda_1(Q(T, \mathbf{U})) + \lambda_2(Q(T, \mathbf{U}))$.

5. Let $B = [b_{ij}]_1^n \in \mathbf{H}_n$. Show that all eigenvalues of B are positive if and only if $\det [b_{ij}]_1^k > 0$ for $k = 1, \ldots, n$.

6. Let $T \in \mathbf{S}(\mathbf{V})$. Denote by $\iota_+(T), \iota_0(T), \iota_-(T)$ the number of positive, negative and zero eigenvalues among $\lambda_1(T) \ge \cdots \ge \lambda_n(T)$. The triple $\iota(T) := (\iota_+(T), \iota_0(T), \iota_-(T))$ is called the inertia of T. For $B \in \mathbf{H}_n$ let $\iota(B) := (\iota_+(B), \iota_0(B), \iota_-(B))$ be the inertia of B, where $\iota_+(B), \iota_0(B), \iota_-(B)$ is the number of positive, negative and zero eigenvalues of B respectively. Let $\mathbf{U} \in \mathrm{Gr}(k, \mathbf{V})$. Show

(a) Assume that $\lambda_k(Q(T, \mathbf{U})) > 0$, i.e. $Q(T, \mathbf{U}) > 0$. Then $k \le \iota_+(T)$. If $k = \iota_+(T)$ then $\mathbf{U}$ can be the invariant subspace of $\mathbf{V}$ spanned by the eigenvectors of T corresponding to positive eigenvalues of T.

(b) Assume that $\lambda_k(Q(T, \mathbf{U})) \ge 0$, i.e. $Q(T, \mathbf{U}) \ge 0$. Then $k \le \iota_+(T) + \iota_0(T)$. If $k = \iota_+(T) + \iota_0(T)$ then $\mathbf{U}$ can be the invariant subspace of $\mathbf{V}$ spanned by the eigenvectors of T corresponding to non-negative eigenvalues of T.

(c) Assume that $\lambda_1(Q(T, \mathbf{U})) < 0$, i.e. $Q(T, \mathbf{U}) < 0$. Then $k \leq \iota_-(T)$. If $k = \iota_-(T)$ then $\mathbf{U}$ can be the invariant subspace of $\mathbf{V}$ spanned by the eigenvectors of T corresponding to negative eigenvalues of T.

(d) Assume that $\lambda_1(Q(T, \mathbf{U})) \leq 0$, i.e. $Q(T, \mathbf{U}) \leq 0$. Then $k \leq \iota_-(T) + \iota_0(T)$. If $k = \iota_-(T) + \iota_0(T)$ then $\mathbf{U}$ can be the invariant subspace of $\mathbf{V}$ spanned by the eigenvectors of T corresponding to nonpositive eigenvalues of T.

7. Let $B \in \mathbf{H}_n$ and assume that $A = PBP^*$ for some $P \in \mathbf{GL}(n, \mathbb{C})$. Then $\iota(A) = \iota(B)$.

4.5 Positive Definite Operators and Matrices

Definition 4.5.1 *Let $\mathbf{V}$ be a finite dimensional IPS over $\mathbb{F} = \mathbb{C}, \mathbb{R}$. Let $S, T \in \mathbf{S}(\mathbf{V})$. Then $T \succ S$, $(T \succeq S)$ if $\langle T\mathbf{x}, \mathbf{x} \rangle > \langle S\mathbf{x}, \mathbf{x} \rangle$, $(\langle T\mathbf{x}, \mathbf{x} \rangle \geq \langle S\mathbf{x}, \mathbf{x} \rangle)$ for all $\mathbf{0} \neq \mathbf{x} \in \mathbf{V}$. T is called positive (non-negative) definite if $T \succ 0$ $(T \succeq 0)$, where 0 is the zero operator in $\mathrm{Hom}\,(\mathbf{V})$.*

Denote by $\mathbf{S}_+(\mathbf{V})^o \subset \mathbf{S}_+(\mathbf{V}) \subset \mathbf{S}(\mathbf{V})$ the open set of positive definite self-adjoint operators and the closed set of non-negative self-adjoint operators respectively. Denote by $\mathbf{S}_{+,1}(\mathbf{V})$ the closed set of non-negative definite self-adjoint operators of trace one.

Let P, Q be either quadratic forms, if $\mathbb{F} = \mathbb{R}$, or hermitian forms, if $\mathbb{F} = \mathbb{C}$. Then we write $Q \succ P$, $(Q \succeq P)$ if $Q(\mathbf{x}, \mathbf{x}) > P(\mathbf{x}, \mathbf{x})$, $(Q(\mathbf{x}, \mathbf{x}) \geq P(\mathbf{x}, \mathbf{x}))$ for all $\mathbf{0} \neq \mathbf{x} \in \mathbf{V}$. Q is called positive (non-negative) definite if $Q \succ 0$ $(Q \succeq 0)$, where 0 is the zero operator in $\mathrm{Hom}\,(\mathbf{V})$.

For $A, B \in \mathbf{H}_n$, we write $B \succ A$, $(B \succeq A)$ if $\mathbf{x}^* B\mathbf{x} > \mathbf{x}^* A\mathbf{x}$ $(\mathbf{x}^* B\mathbf{x} \geq \mathbf{x}^* A\mathbf{x})$ for all $\mathbf{0} \neq \mathbf{x} \in \mathbb{C}^n$. $B \in \mathbf{H}_n$ is called is called positive (non-negative) definite if $B \succ 0$, $(B \succeq 0)$. Denote by $\mathbf{H}_{n,+}^o \subset \mathbf{H}_{n,+} \subset \mathbf{H}_n$ the open set of positive definite $n \times n$ hermitian matrices and the closed set of $n \times n$ non-negative hermitian matrices respectively. Denote by $\mathbf{H}_{n,+,1}$ the closed set of $n \times n$ non-negative definite

hermitian matrices of trace one. Let $\mathbf{S}_+(n, \mathbb{R}) := \mathbf{S}(n, \mathbb{R}) \cap \mathbf{H}_{n,+},$
$\mathbf{S}_+(n, \mathbb{R})^o := \mathbf{S}(n, \mathbb{R}) \cap \mathbf{H}_{n,+}^o,$ $\mathbf{S}_{+,1}(n, \mathbb{R}) := \mathbf{S}(n, \mathbb{R}) \cap \mathbf{H}_{n,+,1}.$

Use (4.4.1) to deduce:

Corollary 4.5.2 *Let* $\mathbf{V}$ *be an* n-*dimensional IPS. Let* $T \in \mathbf{S}(\mathbf{V})$.
Then $T \succ 0, (T \succeq 0)$ *if and only if* $\lambda_n(T) > 0, (\lambda_n(T) \geq 0)$.
Let $S \in \mathbf{S}(\mathbf{V})$ *and assume that* $T \succ S, (T \succeq S)$. *Then* $\lambda_i(T) >$
$\lambda_i(S), (\lambda_i(T) \geq \lambda_i(S))$ *for* $i = 1, \ldots, n$.

Proposition 4.5.3 *Let* $\mathbf{V}$ *be a finite dimensional IPS. Assume
that* $T \in \mathbf{S}(\mathbf{V})$. *Then* $T \succeq 0$ *if and only if there exists* $S \in \mathbf{S}(\mathbf{V})$
such that $T = S^2$. *Furthermore* $T \succ 0$ *if and only if* S *is invertible.
For* $0 \preceq T \in \mathbf{S}(\mathbf{V})$ *there exists a unique* $0 \preceq S \in \mathbf{S}(\mathbf{V})$ *such that*
$T = S^2$. *This* S *is called the square root of* T *and is denoted by* $T^{\frac{1}{2}}$.

Proof. Assume first that $T \succeq 0$. Let $\mathbf{e}_1, \ldots, \mathbf{e}_n$ be an orthonormal basis consisting of eigenvectors of T as in (4.4.3).   Since
$\lambda_i(T) \geq 0$, $i = 1, \ldots, n$ we can define $P \in \text{Hom}(\mathbf{V})$ as follows

$$P\mathbf{e}_i = \sqrt{\lambda_i(T)}\mathbf{e}_i, \quad i = 1, \ldots, n.$$

Clearly P is self-adjoint non-negative and $T = P^2$.

Suppose now that $T = S^2$ for some $S \in \mathbf{S}(\mathbf{V})$. Then $T \in \mathbf{S}(\mathbf{V})$
and $\langle T\mathbf{x}, \mathbf{x} \rangle = \langle S\mathbf{x}, S\mathbf{x} \rangle \geq 0$. Hence, $T \succeq 0$. Clearly $\langle T\mathbf{x}, \mathbf{x} \rangle =$
$0 \iff S\mathbf{x} = 0$. Hence, $T \succ 0 \iff S \in \mathbf{GL}(\mathbf{V})$. Suppose that
$S \succeq 0$. Then $\lambda_i(S) = \sqrt{\lambda_i(T)}$, $i = 1, \ldots, n$. Furthermore each
eigenvector of S is an eigenvector of T. It is straightforward to show
that $S = P$, where P is defined above. Clearly $T \succ 0$ if and only if
$\sqrt{\lambda_n(T)} > 0$, i.e. if and only if S is invertible. $\square$

Corollary 4.5.4 *Let* $B \in \mathbf{H}_n, (\mathbf{S}(n, \mathbb{R}))$. *Then* $B \succeq 0$ *if and
only there exists* $A \in \mathbf{H}_n$ $(\mathbf{S}(n, \mathbb{R}))$ *such that* $B = A^2$. *Furthermore
$B \succ 0$ if and only if* A *is invertible. For* $B \succeq 0$ *there exists a unique
$A \succeq 0$ such that* $B = A^2$. *This* A *is denoted by* $B^{\frac{1}{2}}$.

Theorem 4.5.5 *Let* **V** *be an IPS over* $\mathbb{F} = \mathbb{C}, \mathbb{R}$. *Let* $\mathbf{x}_1, \ldots, \mathbf{x}_n \in \mathbf{V}$. *Then the grammian matrix* $G(\mathbf{x}_1, \ldots, \mathbf{x}_n) :=$ $[\langle \mathbf{x}_i, \mathbf{x}_j \rangle]_1^n$ *is a hermitian non-negative definite matrix.* (*If* $\mathbb{F} = \mathbb{R}$ *then* $G(\mathbf{x}_1, \ldots, \mathbf{x}_n)$ *is real symmetric non-negative definite.*) $G(\mathbf{x}_1, \ldots, \mathbf{x}_n) \succ 0$ *if and only* $\mathbf{x}_1, \ldots, \mathbf{x}_n$ *are linearly independent.* *Furthermore, for any integer* $k \in [n-1]$

$$\det G(\mathbf{x}_1, \ldots, \mathbf{x}_n) \leq \det G(\mathbf{x}_1, \ldots, \mathbf{x}_k) \det G(\mathbf{x}_{k+1}, \ldots, \mathbf{x}_n).$$
$$(4.5.1)$$

Equality holds if and only if either $\det G(\mathbf{x}_1, \ldots, \mathbf{x}_k) \det G(\mathbf{x}_{k+1}, \ldots, \mathbf{x}_n) = 0$ *or* $\langle \mathbf{x}_i, \mathbf{x}_j \rangle = 0$ *for* $i = 1, \ldots, k$ *and* $j = k+1, \ldots, n$.

Proof. Clearly $G(\mathbf{x}_1, \ldots, \mathbf{x}_n) \in \mathbf{H}_n$. If **V** is an IPS over $\mathbb{R}$ then $G(\mathbf{x}_1, \ldots, \mathbf{x}_n) \in \mathbf{S}(n, \mathbb{R})$. Let $\mathbf{a} = (a_1, \ldots, a_n)^\top \in \mathbb{F}^n$. Then

$$\mathbf{a}^* G(\mathbf{x}_1, \ldots, \mathbf{x}_n) \mathbf{a} = \left\langle \sum_{i=1}^n a_i \mathbf{x}_i, \sum_{j=1}^n a_j \mathbf{x}_j \right\rangle \geq 0.$$

Equality holds if and only if $\sum_{i=1}^n a_i \mathbf{x}_i = 0$. Hence, $G(\mathbf{x}_1, \ldots, \mathbf{x}_n) \geq 0$ and $G(\mathbf{x}_1, \ldots, \mathbf{x}_n) > 0$ if and only if $\mathbf{x}_1, \ldots, \mathbf{x}_n$ are linearly independent. In particular, $\det G(\mathbf{x}_1, \ldots, \mathbf{x}_n) \geq 0$ and $\det G(\mathbf{x}_1, \ldots, \mathbf{x}_n) > 0$ if and only if $\mathbf{x}_1, \ldots, \mathbf{x}_n$ are linearly independent.

We now prove the inequality (4.5.1). Assume first that the right-hand side of (4.5.1) is zero. Then either $\mathbf{x}_1, \ldots, \mathbf{x}_k$ or $\mathbf{x}_{k+1}, \ldots, \mathbf{x}_n$ are linearly dependent. Hence, $\mathbf{x}_1, \ldots, \mathbf{x}_n$ are linearly dependent and $\det G = 0$.

Assume now that the right-hand side of (4.5.1) is positive. Hence, $\mathbf{x}_1, \ldots, \mathbf{x}_k$ and $\mathbf{x}_{k+1}, \ldots, \mathbf{x}_n$ are linearly independent. If $\mathbf{x}_1, \ldots, \mathbf{x}_n$ are linearly dependent then $\det G = 0$ and strict inequality holds in (4.5.1). It is left to show the inequality (4.5.1) and the equality case when $\mathbf{x}_1, \ldots, \mathbf{x}_n$ are linearly independent. Perform the Gram–Schmidt algorithm on $\mathbf{x}_1, \ldots, \mathbf{x}_n$ as given in (4.1.1). Let $S_j = \text{span}(\mathbf{x}_1, \ldots, \mathbf{x}_j)$ for $j = 1, \ldots, n$. Corollary 4.1.1 yields that $\text{span}(\mathbf{e}_1, \ldots, \mathbf{e}_{n-1}) = S_{n-1}$. Hence, $\mathbf{y}_n = \mathbf{x}_n - \sum_{j=1}^{n-1} b_j \mathbf{x}_j$ for some $b_1, \ldots, b_{n-1} \in \mathbb{F}$. Let G' be the matrix obtained from $G(\mathbf{x}_1, \ldots, \mathbf{x}_n)$

by subtracting from the nth row b_j times the jth row. Thus the last row of G' is $(\langle \mathbf{y}_n, \mathbf{x}_1 \rangle, \ldots, \langle \mathbf{y}_n, \mathbf{x}_n \rangle) = (0, \ldots, 0, \|\mathbf{y}_n\|^2)$. Clearly $\det G(\mathbf{x}_1, \ldots, \mathbf{x}_n) = \det G'$. Expand $\det G'$ by the last row to deduce

$$\det G(\mathbf{x}_1, \ldots, \mathbf{x}_n) = \det G(\mathbf{x}_i, \ldots, \mathbf{x}_{n-1}) \|\mathbf{y}_n\|^2 = \cdots$$

$$= \det G(\mathbf{x}_1, \ldots, \mathbf{x}_k) \prod_{i=k+1}^{n} \|\mathbf{y}_i\|^2,$$

$$= \det G(\mathbf{x}_1, \ldots, \mathbf{x}_k) \prod_{i=k+1}^{n} \mathrm{dist}(\mathbf{x}_i, S_{i-1})^2,$$

$$k = n - 1, \ldots, 1. \tag{4.5.2}$$

Perform the Gram–Schmidt process on $\mathbf{x}_{k+1}, \ldots, \mathbf{x}_n$ to obtain the orthogonal set of vectors $\hat{\mathbf{y}}_{k+1}, \ldots, \hat{\mathbf{y}}_n$ such that

$$\hat{S}_j := \mathrm{span}\,(\mathbf{x}_{k+1}, \ldots, \mathbf{x}_j)$$

$$= \mathrm{span}\,(\hat{\mathbf{y}}_{k+1}, \ldots, \hat{\mathbf{y}}_j),\ \mathrm{dist}(\mathbf{x}_j, \hat{S}_{j-1}) = \|\hat{\mathbf{y}}_j\|,$$

for $j = k+1, \ldots, n$, where $\hat{S}_k = \{\mathbf{0}\}$. Use (4.5.2) to deduce that $\det G(\mathbf{x}_{k+1}, \ldots, \mathbf{x}_n) = \prod_{j=k+1}^{n} \|\hat{\mathbf{y}}_j\|^2$. As $\hat{S}_{j-1} \subset S_{j-1}$ for $j > k$ it follows that

$$\|\mathbf{y}_j\| = \mathrm{dist}(\mathbf{x}_j, S_{j-1}) \le \mathrm{dist}(\mathbf{x}_j, \hat{S}_{j-1}) = \|\hat{\mathbf{y}}_j\|, \quad j = k+1, \ldots, n.$$

This shows (4.5.1). Assume now equality holds in (4.5.1). Then $\|\mathbf{y}_j\| = \|\hat{\mathbf{y}}_j\|$ for $j = k+1, \ldots, n$. Since $\hat{S}_{j-1} \subset S_{j-1}$ and $\hat{\mathbf{y}}_j - \mathbf{x}_j \in \hat{S}_{j-1} \subset S_{j-1}$, it follows that $\mathrm{dist}(\mathbf{x}_j, S_{j-1}) = \mathrm{dist}(\hat{\mathbf{y}}_j, S_{j-1}) = \|\mathbf{y}_j\|$. Hence, $\|\hat{\mathbf{y}}_j\| = \mathrm{dist}(\hat{\mathbf{y}}_j, S_{j-1})$. Part (h) of Problem 4.1.4 yields that $\hat{\mathbf{y}}_j$ is orthogonal on S_{j-1}. In particular, each $\hat{\mathbf{y}}_j$ is orthogonal to S_k for $j = k+1, \ldots, n$. Hence, $\mathbf{x}_j \perp S_k$ for $j = k+1, \ldots, n$, i.e. $\langle \mathbf{x}_j, \mathbf{x}_i \rangle = 0$ for $j > k$ and $i \le k$. Clearly, if the last condition holds then $\det G(\mathbf{x}_1, \ldots, \mathbf{x}_n) = \det G(\mathbf{x}_1, \ldots, \mathbf{x}_k) \det G(\mathbf{x}_{k+1}, \ldots, \mathbf{x}_n)$. $\qquad\square$

$\det G(\mathbf{x}_1, \ldots, \mathbf{x}_n)$ has the following geometric meaning. Consider a parallelepiped Π in $\mathbf{V}$ spanned by $\mathbf{x}_1, \ldots, \mathbf{x}_n$ starting from the origin $\mathbf{0}$. Then Π is a convex hull spanned by the vectors

$\mathbf{0}$ and $\sum_{i \in S} \mathbf{x}_i$ for all nonempty subsets $S \subset \{1, \ldots, n\}$ and $\sqrt{\det G(\mathbf{x}_1, \ldots, \mathbf{x}_n)}$ is the n-volume of Π. The inequality (4.5.1) and equalities (4.5.2) are "obvious" from this geometrical point of view.

Corollary 4.5.6 *Let* $0 \leq B = (b_{ij})_1^n \in \mathbf{H}_{n,+}$. *Then*

$$\det B \leq \det [b_{ij}]_1^k \det [b_{ij}]_{k+1}^n, \quad \text{for } k = 1, \ldots, n-1.$$

For a fixed k equality holds if and only if either the right-hand side of the above inequality is zero or $b_{ij} = 0$ for $i = 1, \ldots, k$ and $j = k+1, \ldots, n$.

Proof. From Corollary 4.5.4 it follows that $B = X^2$ for some $X \in \mathbf{H}_n$. Let $\mathbf{x}_1, \ldots, \mathbf{x}_n \in \mathbb{C}^n$ be the n-columns of $X^T = [\mathbf{x}_1 \ldots \mathbf{x}_n]$. Let $\langle \mathbf{x}, \mathbf{y} \rangle = \mathbf{y}^* \mathbf{x}$. Since $X \in \mathbf{H}_n$ we deduce that $B = G(\mathbf{x}_1, \ldots, \mathbf{x}_n)$. $\square$

Theorem 4.5.7 *Let* $\mathbf{V}$ *be an n-dimensional IPS. Let* $T \in \mathbf{S}$. *TFAE:*

(a) $T \succ 0$.

(b) *Let* $\mathbf{g}_1, \ldots, \mathbf{g}_n$ *be a basis of* $\mathbf{V}$. *Then* $\det (\langle T\mathbf{g}_i, \mathbf{g}_j \rangle)_{i,j=1}^k > 0$, $k = 1, \ldots, n$.

Proof. (a) $\Rightarrow$ (b). According to Proposition 4.5.3 $T = S^2$ for some $S \in \mathbf{S}(\mathbf{V}) \cap \mathbf{GL}(\mathbf{V})$. Then $\langle T\mathbf{g}_i, \mathbf{g}_j \rangle = \langle S\mathbf{g}_i, S\mathbf{g}_j \rangle$. Hence, $\det (\langle T\mathbf{g}_i, \mathbf{g}_j \rangle)_{i,j=1}^k = \det G(S\mathbf{g}_1, \ldots, S\mathbf{g}_k)$. Since S is invertible and $\mathbf{g}_1, \ldots, \mathbf{g}_k$ are linearly independent it follows that $S\mathbf{g}_1, \ldots, S\mathbf{g}_k$ are linearly independent. Theorem 4.5.1 implies that $\det G(S\mathbf{g}_1, \ldots, S\mathbf{g}_k) > 0$ for $k = 1, \ldots, n$. (b) $\Rightarrow$ (a). The proof is by induction on n. For $n = 1$ (a) is obvious. Assume that (a) holds for $n = m - 1$. Let $\mathbf{U} := \text{span}(\mathbf{g}_1, \ldots, \mathbf{g}_{n-1})$ and $Q := Q(T, \mathbf{U})$. Then there exists $P \in \mathbf{S}(\mathbf{U})$ such that $\langle P\mathbf{x}, \mathbf{y} \rangle = Q(\mathbf{x}, \mathbf{y}) = \langle T\mathbf{x}, \mathbf{y} \rangle$ for any $\mathbf{x}, \mathbf{y} \in \mathbf{U}$. By induction $P \succ 0$. Corollary 4.4.3 yields that $\lambda_{n-1}(T) \geq \lambda_{n-1}(P) > 0$. Hence, T has at least $n - 1$ positive eigenvalues. Let $\mathbf{e}_1, \ldots, \mathbf{e}_n$ be given by (4.4.3). Then $\det (\langle T\mathbf{e}_i, \mathbf{e}_j \rangle)_{i,j=1}^n =$

$\prod_{i=1}^n \lambda_i(T) > 0$. Let $A = (a_{pq})_1^n \in \mathbf{GL}(n, \mathbb{C})$ be the transformation matrix from the basis $\mathbf{g}_1, \ldots, \mathbf{g}_n$ to $\mathbf{e}_1, \ldots, \mathbf{e}_n$, i.e.

$$\mathbf{g}_i = \sum_{p=1}^n a_{pi} \mathbf{e}_p, \ i = 1, \ldots, n.$$

It is straightforward to show that

$$(\langle T\mathbf{g}_i, \mathbf{g}_j \rangle)_1^n = A^T (\langle T\mathbf{e}_p, \mathbf{e}_q \rangle) \bar{A} \Rightarrow$$

$$\tag{4.5.3}$$

$$\det (\langle T\mathbf{g}_i, \mathbf{g}_j \rangle)_1^n = \det (\langle T\mathbf{e}_i, \mathbf{e}_j \rangle)_1^n |\det A|^2 = |\det A|^2 \prod_{i=1}^n \lambda_i(T).$$

Since $\det (\langle T\mathbf{g}_i, \mathbf{g}_j \rangle)_1^n > 0$ and $\lambda_1(T) \geq \cdots \geq \lambda_{n-1}(T) > 0$ it follows that $\lambda_n(T) > 0$. $\square$

Corollary 4.5.8 *Let* $B = [b_{ij}]_1^n \in \mathbf{H}_n$. *Then* $B \succ 0$ *if and only if* $\det [b_{ij}]_1^k > 0$ *for* $k = 1, \ldots, n$.

The following result is straightforward (see Problem 1):

Proposition 4.5.9 *Let* $\mathbf{V}$ *be a finite dimensional IPS over* $\mathbb{F} = \mathbb{R}, \mathbb{C}$ *with the inner product* $\langle \cdot, \cdot \rangle$. *Assume that* $T \in \mathbf{S}(\mathbf{V})$. *Then* $T \succ 0$ *if and only if* $(\mathbf{x}, \mathbf{y}) := \langle T\mathbf{x}, \mathbf{y} \rangle$ *is an inner product on* $\mathbf{V}$. *Vice versa any inner product* $(\cdot, \cdot) : \mathbf{V} \times \mathbf{V} \to \mathbb{R}$ *is of the form* $(\mathbf{x}, \mathbf{y}) = \langle T\mathbf{x}, \mathbf{y} \rangle$ *for a unique self-adjoint positive definite operator* $T \in \mathrm{Hom}\,(\mathbf{V})$.

Example 4.5.10 *Each* $0 \prec B \in \mathbf{H}_n$ *induces an inner product on* $\mathbb{C}^n$: $(\mathbf{x}, \mathbf{y}) = \mathbf{y}^* B\mathbf{x}$. *Each* $0 \prec B \in \mathbf{S}(n, \mathbb{R})$ *induces an inner product on* $\mathbb{R}^n$: $(\mathbf{x}, \mathbf{y}) = \mathbf{y}^T B\mathbf{x}$. *Furthermore, any inner product on* $\mathbb{C}^n$ *or* $\mathbb{R}^n$ *is of the above form. In particular, the standard inner products on* $\mathbb{C}^n$ *and* $\mathbb{R}^n$ *are induced by the identity matrix* I.

Definition 4.5.11 *Let* $\mathbf{V}$ *be a finite dimensional IPS with the inner product* $\langle \cdot, \cdot \rangle$. *Let* $S \in \mathrm{Hom}\,(V)$. *Then* S *is called symmetrizable if there exists an inner product* $(\cdot, \cdot)$ *on* $\mathbf{V}$ *such that* S *is self-adjoint with respect to* $(\cdot, \cdot)$.

Problems

1. Show Proposition 4.5.9.

2. Recall the Hölder inequality

$$\sum_{l=1}^{n} x_l y_l a_l \le \left(\sum_{l=1}^{n} x_l^p a_l \right)^{\frac{1}{p}} \left(\sum_{l=1}^{n} y_l^q a_l \right)^{\frac{1}{q}} \qquad (4.5.4)$$

for any $\mathbf{x} = (x_1, \ldots, x_n)^\top, \mathbf{y} = (y_1, \ldots, y_n)^\top, \mathbf{a} = (a_1, \ldots, a_n) \in \mathbb{R}_+^n$ and $p, q \in (1, \infty)$ such that $\frac{1}{p} + \frac{1}{q} = 1$. Show

   (a) Let $A \in \mathbf{H}_{n,+}, \mathbf{x} \in \mathbb{C}^n$ and $0 \le i < j < k$ be three integers. Then

$$\mathbf{x}^* A^j \mathbf{x} \le (\mathbf{x}^* A^i \mathbf{x})^{\frac{k-j}{k-i}} (\mathbf{x}^* A^k \mathbf{x})^{\frac{j-i}{k-i}}. \qquad (4.5.5)$$

   Hint: Diagonalize A.

   (b) Assume that $A = e^B$ for some $B \in \mathbf{H}_n$. Show that (4.5.5) holds for any three real numbers $i < j < k$.

4.6 Convexity

Definition 4.6.1 *Let* $\mathbf{V}$ *be a finite dimensional vector space over* $\mathbb{R}$.

1. *For any set* $\mathcal{T} \subset \mathbf{V}$ *we let* $\mathrm{Cl}\,\mathcal{T}$ *be the closure of* $\mathcal{T}$ *in the standard topology in* $\mathbf{V}$ *(which is identified with the standard topology of* $\mathbb{R}^{\dim_\mathbb{R} \mathbf{V}}$*).*

2. *For any two points* $\mathbf{x}, \mathbf{y} \in \mathbf{V}$ *denote by* $(\mathbf{x}, \mathbf{y})$ *and* $[\mathbf{x}, \mathbf{y}]$*, the open and the closed interval spanned by* $\mathbf{x}, \mathbf{y}$ *respectively. That is, the set of points of the form* $t\mathbf{x} + (1-t)\mathbf{y}$*, where* $t \in (0, 1)$ *and* $[0, 1]$*, respectively.*

3. *Set* $\mathrm{C} \subseteq \mathbf{V}$ *is called convex if for any* $\mathbf{x}, \mathbf{y} \in \mathrm{C}$ *the open interval* $(\mathbf{x}, \mathbf{y})$ *is in* C. *(Note that a convex set is connected.)*

4. *Assume that* $C \subseteq \mathbf{V}$ *is a nonempty convex set and let* $\mathbf{x} \in C$. *Denote by* $C - \mathbf{x}$ *the set* $\{\mathbf{z} : \mathbf{z} = \mathbf{y} - \mathbf{x}, \mathbf{y} \in C\}$. *Let* $\mathbf{U} = \operatorname{span}(C - \mathbf{x})$. *Then the dimension of* C, *denoted by* $\dim C$, *is the dimension of the vector space* $\mathbf{U}$. *The interior of* $C - \mathbf{x}$ *as a subset of* $\mathbf{U}$ *is called the* relative interior *and denoted by* $\operatorname{ri}(C - \mathbf{x})$. *Then the relative interior of* C *is defined as* $\operatorname{ri}(C - \mathbf{x}) + \mathbf{x}$.

5. $H \subset \mathbf{V}$ *is called a hyperplane if* $H = \mathbf{U} + \mathbf{x}$, *where* $\mathbf{x} \in \mathbf{V}$ *and* $\mathbf{U}$ *is a subspace of* $\mathbf{V}$ *of codimension one, i.e.* $\dim \mathbf{U} = \dim \mathbf{V} - 1$.

6. *Assume that* $C \subset \mathbf{V}$ *is convex. A point* $\mathbf{e} \in C$ *is called an* extremal point *if for any* $\mathbf{x}, \mathbf{y} \in C$ *such that* $\mathbf{e} \in [\mathbf{x}, \mathbf{y}]$ *the equality* $\mathbf{x} = \mathbf{y} = \mathbf{e}$ *holds. For a convex set* C *denote by* $\operatorname{ext} C$ *the set of the extremal points of* C.

7. *Denote by* $\Pi_n \subset \mathbb{R}_+^n$ *the set of probability vectors, i.e. all vectors with non-negative entries that add up to one.*

8. *Let* $S \subset \mathbf{V}$. *The convex hull of* S, *denoted by* $\operatorname{conv} S$, *is the minimal convex set containing* S. (*We also call* $\operatorname{conv} S$ *the convex set generated by* S.) *For each* $j \in \mathbb{N}$ *let* $\operatorname{conv}_{j-1} S$ *be the set of vectors* $\mathbf{z} \in \mathbf{V}$ *such that* $\mathbf{z} = \sum_{i=1}^{j} p_i \mathbf{x}_i$ *for all* $\mathbf{p} = (p_1, \ldots, p_j)^\top \in \Pi_j$ *and* $\mathbf{x}_1, \ldots, \mathbf{x}_j \in S$.

9. *Let* C *be convex.* $F \subset C$ *is called a face of* C *if* F *is a convex set satisfying the following property: Let* $\mathbf{x}, \mathbf{y} \in C$ *and asume that* $\frac{1}{2}\mathbf{x} + \frac{1}{2}\mathbf{y} \in F$. *Then* $\mathbf{x}, \mathbf{y} \in \mathbb{F}$. F *is called* $\dim F$-*face of* C.

10. *Let* C *be a convex set. For* $\mathbf{f} \in \mathbf{V}' \setminus \{\mathbf{0}\}$ *and* $\mathbf{x} \in \mathbf{V}$ *denote*

$$H_0(\mathbf{f}, \mathbf{x}) := \{\mathbf{y} \in \mathbf{V}, \mathbf{f}(\mathbf{y}) = \mathbf{f}(\mathbf{x})\},$$

$$H_+(\mathbf{f}, \mathbf{x}) := \{\mathbf{y} \in \mathbf{V}, \mathbf{f}(\mathbf{y}) \geq \mathbf{f}(\mathbf{x})\},$$

$$H_-(\mathbf{f}, \mathbf{x}) := \{\mathbf{y} \in \mathbf{V}, \mathbf{f}(\mathbf{y}) \leq \mathbf{f}(\mathbf{x})\}.$$

$H_+(\mathbf{f}, \mathbf{x})$ *and* $H_-(\mathbf{f}, \mathbf{x})$ *are called the upper and the lower half spaces respectively, or simply the half spaces.*

It is straightforward to show that dim C, ri C do not depend on the choice of $\mathbf{x} \in$ C. Furthermore, ri C is convex. See Problem 3 or [Roc70]. It is known that ext (conv S) $\subset$ S [Roc70] or see Problem 1. Suppose that S is a finite set of cardinality $N \in \mathbb{N}$. Then conv S $=$ conv$_{N-1}$S and conv S is called a *finitely generated convex set.* The following result is well known [Roc70]. (See Problem 8 for finitely generated convex sets.)

Theorem 4.6.2 *Let* **V** *be a real vector space of finite dimension. Let* C $\subset$ **V** *be a nonempty compact convex set. Then* conv(ext C) $=$ C. *Let* $d :=$ dim C. *Then* conv$_d$(ext C) $=$ C. *More general, for any* S $\subset$ **V** *let* $d =$ dim (conv S). *Then* conv$_d$ S $=$ conv S. *That is, every vector in* conv S *is a convex combination of some* $d+1$ *extreme points* (*Carathéodory's theorem*).

In many case we shall identify a finite dimensional vector space **V** over $\mathbb{R}$ with $\mathbb{R}^d$. Assume that the C $\subset$ **V** is a nonempty compact convex set of dimension d. Then the following facts are known. If $d = 2$ then ext C is a closed set. For $d \geq 3$ there exist C such that ext C is not closed.

Assume that **V** is a complex finite dimensional subspace, of dimension n. Then **V** can be viewed as a real vector space $\mathbf{V}_{\mathbb{R}}$ of dimension $2n$. A convex set C $\subset$ **V** is a convex set $C_{\mathbb{R}} \subset \mathbf{V}_{\mathbb{R}}$. However, as we see later, sometimes it is natural to consider convex sets as subsets of complex vector space **V**, rather then subsets of $\mathbf{V}_{\mathbb{R}}$.

Assume that $\mathbf{f} \in \mathbf{V}' \setminus \{\mathbf{0}\}$. Then we associate with $\mathbf{f}$ a real valued functional $\mathbf{f}_{\mathbb{R}} : \mathbf{V} \to \mathbb{R}$ by letting $\mathbf{f}_{\mathbb{R}}(\mathbf{x}) = \Re f(\mathbf{x})$. Then

$$H_0(\mathbf{f}, \mathbf{x}) := H_0(\mathbf{f}_{\mathbb{R}}, \mathbf{x}), \quad H_+(\mathbf{f}, \mathbf{x}) := H_+(\mathbf{f}_{\mathbb{R}}, \mathbf{x}), \quad H_-(\mathbf{f}, \mathbf{x}) ;= H_-(\mathbf{f}_{\mathbb{R}}, \mathbf{x})$$

are the real hyperplane and the real half spaces in **V**. Note also that $H_-(\mathbf{f}, \mathbf{x}) = H_+(-\mathbf{f}, \mathbf{x})$.

In what follows we assume that **V** is a finite dimensinal vector space over $\mathbb{F} = \mathbb{R}, \mathbb{C}$, which is treated as a real vector space.

Definition 4.6.3 *An intersection of a finite number of half spaces* $\cap_{i=1}^m H_+(\mathbf{f}_i, \mathbf{x}_i)$ *is called a polyhedron. A nonempty compact polyhedron is called polytope.*

Clearly, a polyhedron is a closed convex set. Given a polyhedron C, it is a natural problem to find if this polyhedron is empty or not empty. The complexity of finding out if this polyhedron is empty or not depends polynomially on: the dimension of $\mathbf{V}$, and the complexity of all the half spaces in the characterizing C. This is not a trivial fact, which is obtained using an ellipsoid method. See [Kha79, Kar84, Lov86].

It is well known that any polytope has a finite number of extreme points, and is equal to the convex hull of its extreme points [Roc70, p. 12]. Moreover, if S is a finite set then conv S is a polytope.

In general it is a difficult problem to find explicitly all the extreme points of the given compact convex set, or even of a given polyhedron. The following example is a classic in matrix theory.

Theorem 4.6.4 *Let* $\mathbf{H}_{n,+,1} \subset \mathbb{C}^{n \times n}$ *be the convex set of non-negative definite hermitian matrices with trace* 1. *Then*

$$\text{ext} \left(\mathbf{H}_{n,+,1} \right) = \{ \mathbf{x}\mathbf{x}^*, \mathbf{x} \in \mathbb{C}^n, \mathbf{x}^*\mathbf{x} = 1 \}, \qquad (4.6.1)$$

$$\text{ext} \left(\mathbf{H}_{n,+,1} \cap S(n, \mathbb{R}) \right) = \{ \mathbf{x}\mathbf{x}^\top, \mathbf{x} \in \mathbb{R}^n, \mathbf{x}^\top\mathbf{x} = 1 \}. \qquad (4.6.2)$$

Each matrix in $\mathbf{H}_{n,+,1}$ *or* $\mathbf{H}_{n,+,1} \cap S(n, \mathbb{R})$ *is a convex combination of at most* n *extreme points.*

Proof. Let $A = \mathbf{x}\mathbf{x}^*, \mathbf{x} \in \mathbb{C}^n, \mathbf{x}^*\mathbf{x} = 1$. Clearly $A \in \mathbf{H}_{n,+,1}$. Suppose that $A = aB + (1 - a)C$ for some $B, C \in \mathbf{H}_{n,+,1}$ and $a \in (0, 1)$. Hence, $A \succcurlyeq aB \succcurlyeq 0$. Since $\mathbf{y}^*A\mathbf{y} \geq a\mathbf{y}^*B\mathbf{y} \geq 0$ it follows that $\mathbf{y}^*B\mathbf{y} = 0$ for $\mathbf{y}^*\mathbf{x} = 0$. Hence, $B\mathbf{y} = \mathbf{0}$ for $\mathbf{y}^*\mathbf{x} = 0$. Thus B is a rank one non-negative definite matrix of the form $t\mathbf{x}\mathbf{x}^*$ where $t > 0$. Since $\text{tr}\,B = 1$ we deduce that $t = 1$ and $B = A$. Similarly $C = A$. Hence, A is an extremal point.

Let $F \in \mathbf{H}_{n,+,1}$. Then the spectral decomposition of F yields that $F = \sum_{i=1}^n \lambda_i \mathbf{x}_i \mathbf{x}_i^*$, where $\mathbf{x}_i^* \mathbf{x}_j = \delta_{ij}, i, j = 1, \ldots, n$. Furthermore, since F is non-negative definite of trace 1, $\lambda_1, \ldots, \lambda_n$, the eigenvalues of F, are non-negative and sum to 1. So $F \in \text{conv}\{\mathbf{x}_1\mathbf{x}_1^*, \ldots, \mathbf{x}_n\mathbf{x}_n^*\}$. Similar arguments apply to non-negative real symmetric matrices of rank 1. $\qquad \square$

Definition 4.6.5 *Let* $C_1, C_2 \subset \mathbf{V}$, *where* $\mathbf{V}$ *is a finite dimensional vector space over* $\mathbb{F} = \mathbb{R}, \mathbb{C}$. C_1, C_2 *are called hyperplane separated if there exists* $\mathbf{f} \in \mathbf{V}' \setminus \{0\}$ *and* $\mathbf{x} \in \mathbf{V}$ *such that* $C_1 \subset H_+(\mathbf{f}, \mathbf{x}), C_2 \subset H_-(\mathbf{f}, \mathbf{x})$. $H_0(\mathbf{f}, \mathbf{x})$ *is called the separating (real) hyperplane.* $H_0(\mathbf{f}, \mathbf{x})$ *is said to separate* C_1 *and* C_2 *properly if* $H_0(\mathbf{f}, \mathbf{x})$ *separates* C_1 *and* C_2 *and* $H_0(\mathbf{f}, \mathbf{x})$ *does contain* C_1 *and* C_2.

The following result is well known [Roc70, Theorem 11.3].

Theorem 4.6.6 *Let* C_1, C_2 *be nonempty convex sets in a finite dimensional vector space* $\mathbf{V}$. *Then there exists a hyperplane separating* C_1 *and* C_2 *properly if and only* ri $C_1 \cap$ ri $C_2 = \emptyset$.

Corollary 4.6.7 *Let* C_1 *be a compact convex set in a finite dimensional vector space* $\mathbf{V}$ *over* $\mathbb{F} = \mathbb{R}, \mathbb{C}$. *Assume that* C_1 *contains more than one point. Let* $\mathbf{x}$ *be a an extreme point of* C. *Then there exists a hyperplane which supports properly* C_1 *at* $\mathbf{x}$. *That is, there exists* $f \in \mathbf{V}' \setminus \{0\}$, *such that* $\Re f(\mathbf{x}) \leq \Re f(\mathbf{y})$ *for each* $\mathbf{y} \in C$. *Furthermore, there exists* $\mathbf{y} \in C$ *such that* $\Re f(\mathbf{x}) < \Re f(\mathbf{y})$.

Proof. Let $C_2 = \{\mathbf{x}\}$. So C_2 is a convex set. Problem 4 yields that ri $C_1 \cap$ ri $C_2 = \emptyset$. Use Theorem 4.6.4 to deduce the corollary. $\qquad\qquad\square$

Definition 4.6.8 *A point* $\mathbf{x}$ *of a convex set* C *in a finite dimensional vector space* $\mathbf{V}$ *is called exposed, if there there exist a linear functional* $\mathbf{f} \in \mathbf{V}' \setminus \{0\}$ *such that* $\Re f(\mathbf{x}) > \Re f(\mathbf{y})$ *for any* $\mathbf{y} \in C \setminus \{\mathbf{x}\}$.

Clearly, an exposed point of C is an extreme point (Problem 5). There exist compact convex sets with extreme points which are not exposed (see Problem 6). In what follows we need Straszewiz [Str35].

Theorem 4.6.9 *Let* C *be a closed convex set. Then the set of exposed points of* C *is a dense subset of extreme points of* C. *Thus every extreme point is the limit of some sequence of exposed points.*

Corollary 4.6.10 *Let* C *be a closed convex set. Let* $\mathbf{x} \in$ C *be an isolated extreme point. (That is, there is a neighborhood of* $\mathbf{x}$, *where* $\mathbf{x}$ *is the only extreme point of* C.) *Then* $\mathbf{x}$ *is an exposed point.*

Definition 4.6.11 *Let* D $\subseteq \mathbb{R}^n$. D *is called a regular set if the interior of* D, *denoted by* $D^o \subset \mathbb{R}^n$, *is a nonempty set, and* D *is a subset of* Cl D^o. D *is called a domain if* D *is open and connected.*

Assume that D *is a regular set. A function* $f :$ D $\to \mathbb{R}$ *is in the class* $C^k(D)$, *i.e.* f *has* k *continuous derivatives, if* $f \in C^k(D^o)$ *and* f *and any of its derivatives of order not greater than* k *have a continuous extension to* D.

Assume that $f \in C^1(D)$. *Then for each* $\mathbf{x} = (x_1, \ldots, x_n)^\top \in D^o$ *the gradient of* f *at* $\mathbf{x}$ *is given as* $\nabla f(\mathbf{x}) := ((\frac{\partial f}{\partial x_1}(\mathbf{x}), \ldots, \frac{\partial f}{\partial x_m}(\mathbf{x}))^\top$. $\mathbf{x} \in D^o$ *is called critical if* $\nabla f(\mathbf{x}) = \mathbf{0}$.

Definition 4.6.12 *Let* C $\subseteq$ **V** *be a convex set. Assume that* $f :$ C $\to \mathbb{R}$. f *is called convex if*

$$f(t\mathbf{x} + (1-t)\mathbf{y})) \leq tf(\mathbf{x}) + (1-t)f(\mathbf{y}) \quad \text{for all } t \in (0,1), \mathbf{x}, \mathbf{y} \in C.$$
$$(4.6.3)$$

f *is called log-convex if* $f \geq 0$ *on* C *and*

$$f(t\mathbf{x} + (1-t\mathbf{y})) \leq (f(\mathbf{x}))^t f(\mathbf{y})^{(1-t)} \quad \text{for all } t \in (0,1), \mathbf{x}, \mathbf{y} \in C.$$
$$(4.6.4)$$

f *is called strictly convex or strictly log-convex if strict inequality holds in* (4.6.3) *and* (4.6.4), *respectively.*

A function $g :$ C $\to \mathbb{R}$ *is called (strictly) concave if the function* $-g$ *is (strictly) convex.*

Proposition 4.6.13 *Let* C $\subset$ **V** *be convex. Then* Cl C *is convex. Assume that* C *is a regular set and* $f \in C^0(\text{Cl} \, C)$. *Then* f *is convex in* Cl C *if and only if* f *is convex in* C.

See Problem 12.

The following result is well known. (See Problems 13.)

Theorem 4.6.14 *Let* $D \subset \mathbb{R}^d$ *be a regular convex set. Assume that* $f \in C^2(D)$. *Then* f *is convex if and only if the symmetric matrix* $H(f) := [\frac{\partial^2 f}{\partial x_i x_j}]^d_{i,j=1}$ *is non-negative definite for each* $\mathbf{y} \in D$. *Furthermore, if* $H(f)$ *is positive definite for each* $\mathbf{y} \in D$, *then* f *is strictly convex.*

Definition 4.6.15 *Denote by* $\overline{\mathbb{R}} := \mathbb{R} \cup \{-\infty, \infty\}$ *the extended real line. Then* $a + \infty = \infty + a = \infty$ *for* $a \in \mathbb{R} \cup \{\infty\}$, $a - \infty = -\infty + a = -\infty$ *for* $a \in \mathbb{R} \cup \{-\infty\}$ *and* $\infty - \infty$, $-\infty + \infty$ *are not defined. For* $a > 0$ *we let* $a\infty = \infty a = \infty$, $a(-\infty) = (-\infty)a = -\infty$ *and* $0\infty = \infty 0 = 0$, $0(-\infty) = (-\infty)0 = 0$. *Clearly for any* $a \in \mathbb{R}$ $-\infty < a < \infty$. *Let* C *be a convex set. Then* $f : C \to \overline{\mathbb{R}}$ *is called an extended convex function if* (4.6.3) *holds.*

Let $f : C \to \mathbb{R}$ be a convex function. Then f has the following continuity and differentiability properties:

In the one-dimensional case, where $C = (a, b) \subset \mathbb{R}$, f is continuous on C and f has a derivative $f'(x)$ at all but a countable set of points. $f'(x)$ is a nondecreasing function (where defined). In particular, f has left and right derivatives at each x, which are given as the left and the right limits of $f'(x)$ (where defined).

In the general case where $C \subset \mathbf{V}$, f is a continuous function in ri C, f has a differential Df in a dense set C_1 of ri C, the complement of C_1 in ri C has a zero measure, and Df is continuous in C_1. Furthermore at each $\mathbf{x} \in$ ri C f has a subdifferential $\phi \in \text{Hom}(\mathbf{V}, \mathbb{R})$ such that

$$f(\mathbf{y}) \geq f(\mathbf{x}) + \phi(\mathbf{y} - \mathbf{x}) \quad \text{for all } \mathbf{y} \in C. \qquad (4.6.5)$$

See for example [Roc70].

The following result is well known (see Problems 14 and 15):

Proposition 4.6.16 (*The maximal principle*) *Let* C *be a convex set and let* $f_\phi : C \to \overline{\mathbb{R}}$ *be an extended convex function for each* ϕ *in a set* Φ. *Then*

$$f(\mathbf{x}) := \sup_{\phi \in \Phi} f_\phi(\mathbf{x}), \quad \text{for each } \mathbf{x} \in \mathbf{V},$$

is an extended convex function on C.

Proposition 4.6.17 *Let* $S \subset V$ *and assume that* $f : \operatorname{conv} S \to \overline{\mathbb{R}}$ *is a convex function. Then*

$$\sup_{\mathbf{x} \in \operatorname{conv} S} f(\mathbf{x}) = \sup_{\mathbf{y} \in S} f(\mathbf{y}).$$

If in addition S *is compact and* f *is continuous on* $\operatorname{conv} S$ *then one can replace* sup *by* max.

See Problem 16 for a generalization of this proposition.

Corollary 4.6.18 *Let* $\mathbf{V}$ *be a finite dimensional IPS over* $\mathbb{F} = \mathbb{R}, \mathbb{C}$. *Let* $S \subset \mathbf{S}(\mathbf{V})$. *Let* $f : \operatorname{conv} S \to \overline{\mathbb{R}}$ *be a convex function. Then*

$$\sup_{A \in \operatorname{conv} S} f(A) = \sup_{B \in S} f(B).$$

Definition 4.6.19 *For* $\mathbf{x} = (x_1, \ldots, x_n)^\top, \mathbf{y} = (y_1, \ldots, y_n)^\top \in \mathbb{R}^n$ *let* $\mathbf{x} \le \mathbf{y} \iff x_i \le y_i, \ i = 1, \ldots, n$. *Let* $D \subset \mathbb{R}^n$ *and* $f : D \to \overline{\mathbb{R}}$. f *is called a nondecreasing function on* D *if for any* $\mathbf{x}, \mathbf{y} \in D$ *one has the implication* $\mathbf{x} \le \mathbf{y} \Rightarrow f(\mathbf{x}) \le f(\mathbf{y})$.

Problems

1. Show

   (a) For any nonempty subset S of a finite dimensional vector space $\mathbf{V}$ over $\mathbb{F}$, $\operatorname{conv} S$ is a convex set.

   (b) Furthermore, if S is compact, then $C := \operatorname{conv} S$ is compact and $\operatorname{ext}(C) \subset S$.

2. Let C be a convex set in a finite dimensional subspace, with the set of extreme points $\operatorname{ext}(C)$. Let $E_1 \subset \operatorname{ext}(C)$ and $C_1 = \operatorname{conv} E_1$. Show that $\operatorname{ext}(C_1) = E_1$.

3. Let $\mathbf{V}$ be a finite dimensional space and $C \subset \mathbf{V}$ a nonempty convex set. Let $\mathbf{x} \in C$. Show

   (a) The subspace $\mathbf{U} := \operatorname{span}(C - \mathbf{x})$ does not depend on $\mathbf{x} \in C$.

(b) $C - \mathbf{x}$ has a nonempty convex interior in $\mathbf{U}$ and the definition of ri C does not depend on $\mathbf{x} \in C$.

4. Let C be a convex set in a finite dimensional vector space $\mathbf{V}$. Assume that C contains at least two distinct points. Show

   (a) Show that $\dim C \geq 1$.
   (b) Show that ri $C \cap \operatorname{ext}(C) = \emptyset$.

5. Let $\mathbf{x} \in C$ be an exposed point. Show that $\mathbf{x}$ is an extreme point of C.

6. Consider the convex set $C \in \mathbb{R}^2$, which is a union of the three convex sets:

$$C_1 := \{(x, y)^\top, |x| \leq 1, |y| \leq 1\},$$
$$C_2 = \{(x, y)^\top, (x - 1)^2 + y^2 \leq 1\},$$
$$C_3 = \{(x, y)^\top, (x + 1)^2 + y^2 \leq 1\}.$$

Show that C has exactly four extreme points $(\pm 1, \pm 1)^\top$ which are not exposed points.

7. Show that $\Pi_n := \operatorname{conv}\{\mathbf{e}_1, \ldots, \mathbf{e}_n\} \subset \mathbb{R}^n$, where $\mathbf{e}_1, \ldots, \mathbf{e}_n$ is the standard basis in $\mathbb{R}^n$, is the set of *probability* vectors in $\mathbb{R}^n$. Furthermore, $\operatorname{ext} \Pi_n = \{\mathbf{e}_1, \ldots, \mathbf{e}_n\}$.

8. Let $S \subset \mathbb{R}^n$ be a nonempty finite set. Show

   (a) Let $S = \{\mathbf{x}_1, \ldots, \mathbf{x}_N\}$. Then $\operatorname{conv} S = \operatorname{conv}_{N-1}(S)$.
   (b) Any finitely generated convex set is compact.
   (c) $S \subset \operatorname{ext} \operatorname{conv} S$.
   (d) Let $f_1, \ldots, f_m : \mathbb{R}^n \to \mathbb{R}$ be linear functions and $a_1, \ldots, a_m \in \mathbb{R}^m$. Denote by A the affine space $\{\mathbf{x} \in \mathbb{R}^n : f_i(\mathbf{x}) = a_i, \ i = 1, \ldots, m\}$. Assume that $C := \operatorname{conv} S \cap A \neq \emptyset$. Then C is a finitely generated convex set such that $\operatorname{ext} C \subset \operatorname{conv}_m S$. (*Hint*: Describe C by $m + 1$ equations with $\#S$ variables as in part a. Use the fact that any homogeneous system in $m + 1$ equations and $l > m + 1$ variables has a nontrivial solution.)

(e) Prove Theorem 4.6.2 for a finitely generated convex set C and a finite S.

9. Let C be a convex set of dimension d with a nonempty ext C. Let $C' = \text{conv ext } C$. Show that ext $C' = \text{ext } C$.

10. Let C be a convex set in a finite dimensional space, with the set of extreme points ext (C). Let $E_1 \subset \text{ext } (C)$ and $C_1 = \text{conv } E_1$. Show that ext $(C_1) = E_1$.

11. Let $C \subset \mathbf{V}$ be convex set and assume that $f : C \to \mathbb{R}$ be a convex function. Show that for any $k \geq 3$ one has the inequality

$$f\left(\sum_{j=1}^{k} p_j \mathbf{u}_j\right) \leq \sum_{j=1}^{k} p_j f(\mathbf{u}_j), \qquad (4.6.6)$$

for any $\mathbf{u}_1, \ldots, \mathbf{u}_k \in C, \mathbf{p} := (p_1, \ldots, p_k)^\top \in \Pi_k$. Assume in addition that f is strictly convex, $\mathbf{p} > \mathbf{0}$ and not all $\mathbf{u}_1, \ldots, \mathbf{u}_k$ are equal. Then strict inequality holds in (4.6.6).

12. Prove Proposition 4.6.13.

13. (a) Let $f \in C^1((a, b))$. Show that f is convex on (a, b) if and only if $f'(x)$ is nondecreasing on (a, b). Show that if $f'(x)$ is increasing on (a, b) then f is strictly convex on (a, b).

(b) Let $f \in C([a, b]) \cap C^1((a, b))$. Show that f is convex in $[a, b]$ if and only if f is convex in (a, b). Show that if $f'(x)$ is increasing on (a, b) then f is strictly convex on $[a, b]$.

(c) Let $f \in C^2((a, b))$. Show that f is convex on (a, b) if and only if f'' is a non-negative function on (a, b). Show that if $f''(x) > 0$ for each $x \in (a, b)$ then f is strictly convex on (a, b).

(d) Prove Theorem 4.6.14.

14. Prove Proposition 4.6.16.

15. Prove Proposition 4.6.17.

16. Assume the assumptions of Proposition 4.6.17. Assume in addition that $f(\operatorname{conv} S) \subset [0, \infty]$ and $g : \operatorname{conv} S \to (0, \infty)$ is concave. Then

$$\sup_{\mathbf{x} \in \operatorname{conv} S} \frac{f(\mathbf{x})}{g(\mathbf{x})} = \sup_{\mathbf{y} \in S} \frac{f(\mathbf{y})}{g(\mathbf{y})}.$$

If in addition S is compact and f, g are continuous on $\operatorname{conv} S$ then one can replace sup by max.

17. Show Proposition 4.7.2.

18. Prove Lemma 4.7.5.

19. Let $\mathbf{x}, \mathbf{y} \in \mathbb{R}^n$. Show that $\mathbf{x} \prec \mathbf{y} \iff -\mathbf{y} \prec -\mathbf{x}$.

20. Prove Corollary 4.7.12.

4.7 Majorization

Definition 4.7.1 *Let*

$$\mathbb{R}^n_\searrow := \{\mathbf{x} = (x_1, \ldots, x_n)^\top \in \mathbb{R}^n : \quad x_1 \geq x_2 \geq \cdots \geq x_n\}.$$

For $\mathbf{x} = (x_1, \ldots, x_n)^\top \in \mathbb{R}^n$ *let* $\underline{\mathbf{x}} = (\underline{x}_1, \ldots, \underline{x}_n)^\top \in \mathbb{R}^n_\searrow$ *be the unique rearrangement of the coordinates of* $\mathbf{x}$ *in a decreasing order. That is, there exists a permutation* π *on* $\{1, \ldots, n\}$ *such that* $\underline{x}_i = x_{\pi(i)}$, $i = 1, \ldots, n$.

Let $\mathbf{x} = (x_1, \ldots, x_n)^\top, \mathbf{y} = (y_1, \ldots, y_n)^\top \in \mathbb{R}^n$. *Then* $\mathbf{x}$ *is weakly majorized by* $\mathbf{y}$ *(or* $\mathbf{y}$ *weakly majorizes* $\mathbf{x}$), *denoted by* $\mathbf{x} \preceq \mathbf{y}$, *if*

$$\sum_{i=1}^k \underline{x}_i \leq \sum_{i=1}^k \underline{y}_i, \quad k = 1, \ldots, n. \tag{4.7.1}$$

$\mathbf{x}$ *is majorized by* $\mathbf{y}$ *(or* $\mathbf{y}$ *majorizes* $\mathbf{x}$), *denoted by* $\mathbf{x} \prec \mathbf{y}$, *if* $\mathbf{x} \preceq \mathbf{y}$ *and* $\sum_{i=1}^n x_i = \sum_{i=1}^n y_i$.

Proposition 4.7.2 *Let* $\mathbf{y} = (y_1, \ldots, y_n)^\top \in \mathbb{R}^n_\searrow$. *Let*

$$M(\mathbf{y}) := \{\mathbf{x} \in \mathbb{R}^n_\searrow : \quad \mathbf{x} \prec \mathbf{y}\}.$$

Then $M(\mathbf{y})$ *is a closed convex set.*

See Problem 1.

Definition 4.7.3 $A \in \mathbb{R}_+^{n \times n}$ *is called a doubly stochastic matrix if the sum of each row and column of A is equal to 1. Denote by $\Omega_n \subset \mathbb{R}_+^{n \times n}$ the set of doubly stochastic matrices. Denote by $\frac{1}{n} J_n$ the $n \times n$ doubly stochastic matrix in which every entry equals $\frac{1}{n}$, i.e. $J_n \in \mathbb{R}_+^{n \times n}$ is the matrix in which every entry is 1.*

Definition 4.7.4 $P \in \mathbb{R}_+^{n \times n}$ *is called a permutation matrix if each row and column of P contains exactly one nonzero element which is equal to 1. Denote by $\mathcal{P}_n$ the set of $n \times n$ permutation matrices.*

Lemma 4.7.5 *The following properties hold:*

1. $A \in \mathbb{R}_+^{n \times n}$ *is doubly stochastic if and only if $A\mathbf{1} = A^\top \mathbf{1} = \mathbf{1}$, where $\mathbf{1} = (1, \ldots, 1)^\top \in \mathbb{R}^n$.*

2. $\Omega_1 = \{1\}$.

3. Ω_n *is a convex set.*

4. $A, B \in \Omega_n \Rightarrow AB \in \Omega_n$.

5. $\mathcal{P}_n \subset \Omega_n$.

6. $\mathcal{P}_n$ *is a group with respect to the multiplication of matrices, with I_n the identity and $P^{-1} = P^\top$.*

7. $A \in \Omega_l$, $B \in \Omega_m \rightarrow A \oplus B \in \Omega_{l+m}$.

See Problem 2.

Theorem 4.7.6 Ω_n *is a convex set generated by $\mathcal{P}_n$. That is $\Omega_n = \operatorname{conv} \mathcal{P}_n$. Furthermore, $\mathcal{P}_n$ is the set of the extreme points* *of Ω_n. That is, $A \in \mathbb{R}_+^{n \times n}$ is doubly stochastic if and only*

$$A = \sum_{P \in \mathcal{P}_n} a_P P \text{ for some } a_P \geq 0, \quad \text{where } P \in \mathcal{P}_n, \quad \text{and}$$

$$\sum_{P \in \mathcal{P}_n} a_P = 1. \tag{4.7.2}$$

Furthermore, if $A \in \mathcal{P}_n$ then for any decomposition of A of the above form one has the equality $a_P = 0$ for $P \neq A$.

Proof. In view of properties 3 and 5 of Lemma 4.7.5 it follows that any A of the form (4.7.2) is doubly stochastic. We now show by induction on n that any $A \in \Omega_n$ is of the form (4.7.2). For $n = 1$ the result trivially holds. Assume that the result holds for $n = m - 1$ and assume that $n = m$.

Assume that $A = (a_{ij}) \in \Omega_n$. Let $l(A)$ be the number of nonzero entries of A. Since each row sum of A is 1 it follows that $l(A) \geq n$. Suppose first $l(A) \leq 2n - 1$. Then there exists a row i of A which has exactly one nonzero element, which must be 1. Hence, there exists $i, j \in [n]$ such that $a_{ij} = 1$. Then all other elements of A on the row i and column j are zero. Denote by $A_{ij} \in \mathbb{R}_+^{(n-1) \times (n-1)}$ the matrix obtained from A by deleting the row and column j. Clearly $A_{ij} \in \Omega_{n-1}$. Use the induction hypothesis on A_{ij} to deduce (4.7.2), where $a_P = 0$ if the entry (i, j) of P is not 1.

We now show by induction on $l(A) \geq 2n - 1$ that A is of the form (4.7.2). Suppose that any $A \in \Omega_n$ such that $l(A) \leq l - 1, l \geq 2n$ is of the form (4.7.2). Assume that $l(A) = l$. Let $\mathcal{S} \subset [n] \times [n]$ be the set of all indices $(i, j) \in [n] \times [n]$ where $a_{ij} > 0$. Note $\#\mathcal{S} = l(A) \geq 2n$. Consider the following system of equations in n^2 variables, which are the entries $X = [x_{ij}]_{i,j=1}^n \in \mathbb{R}^{n \times n}$:

$$\sum_{j=1}^n x_{ij} = \sum_{j=1}^n x_{ji} = 0, \quad i = 1, \ldots, n.$$

Since the sum of all rows of X is equal to the sum of all columns of X we deduce that the above system has at most $2n - 1$ linear independent equations. Assume furthermore the conditions $x_{ij} = 0$

for $(i,j) \notin S$. Since we have at least $2n$ variables it follows that there exist $X \neq 0_{n \times n}$ satisfying the above conditions. Note that X has a zero entry in the places where A has a zero entry. Furthermore, X has at least one positive and one negative entry. Therefore, there exist $b, c > 0$ such that $A - bX, A + cX \in \Omega_n$ and $l(A - bX), l(A + cX) < l$. So $A - bX, A + cX$ are of the form (4.7.2). As $A = \frac{c}{b+c}(A - bX) + \frac{b}{b+c}(A + cX)$ we deduce that A is of the form (4.7.2).

Assume that $A \in \mathcal{P}_n$. Consider the decomposition (4.7.2). Suppose that $a_P > 0$. So $A - a_P P \geq 0$. Therefore, if the (i,j) entry of A is zero then the (i,j) entry of P is zero. Since A is a permutation matrix it follows that the nonzero entries of A and P must coincide. Hence, $A = P$. $\qquad \square$

Theorem 4.7.7 *Let $\Omega_{n,s} \subset \Omega_n$ be the set of symmetric doubly stochastic matrices. Denote*

$$\mathcal{P}_{n,s} = \{A. \; A = \frac{1}{2}(P + P^\top) \text{ for some } P \in \mathcal{P}_n\}. \qquad (4.7.3)$$

Then $A \in \mathcal{P}_{n,s}$ if and only if there exists a permutation matrix matrix $P \in \mathcal{P}_n$ such that $A = PBP^\top$, where $B = \mathrm{diag}(B_1, \ldots, B_t)$ and each B_j is a doubly stochastic symmetric matrix of the following form:

1. 1×1 *matrix* $[1]$.

2. 2×2 *matrix* $\begin{bmatrix} 0 & 1 \\ 1 & 0 \end{bmatrix}$.

3. $n \times n$ *matrix* $\begin{bmatrix} 0 & \frac{1}{2} & 0 & \cdots & 0 & \frac{1}{2} \\ \frac{1}{2} & 0 & \frac{1}{2} & \cdots & 0 & 0 \\ \vdots & \vdots & \vdots & \cdots & \vdots & \vdots \\ \frac{1}{2} & 0 & 0 & \cdots & \frac{1}{2} & 0 \end{bmatrix}$.

Furthermore, $\Omega_{n,s} = \text{conv}\, \mathcal{P}_{n,s}$ *and* $\mathcal{P}_{n,s} = \text{ext}\, \Omega_{n,s}$. *That is,* $A \in \Omega_{n,s}$ *if and only if*

$$A = \sum_{R \in \mathcal{P}_{n,s}} b_R R \text{ for some } b_R \geq 0,$$

$$\text{where } R \in \mathcal{P}_{n,s}, \quad \text{and} \quad \sum_{R \in \mathcal{P}_{n,s}} b_R = 1. \tag{4.7.4}$$

Moreover, if $A \in \mathcal{P}_{n,s}$ *then for any decomposition of* A *of the above form* $b_R = 0$ *unless* $R = A$.

Proof. Assume that $A \in \Omega_{n,s}$. As A is doubly stochastic (4.7.2) holds. Clearly, $A = A^\top = \sum_{P \in \mathcal{P}_n} a_P P^\top$. Hence

$$A = \frac{1}{2}(A + A^\top) = \sum_{P \in \mathcal{P}_n} a_P \left(\frac{1}{2}(P + P^\top) \right).$$

This establishes (4.7.4).

Let $Q \in \mathcal{P}_n$. So Q represents a permutation (bijection) $\sigma : [n] \to [n]$. Namely $Q = [q_{ij}]_{i,j=1}^n$ where $q_{ij} = \delta_{\sigma(i)j}$ for $i, j \in [n]$. Fix $i \in [n]$ and consider the orbit of i under the iterations of σ: $\sigma^k(i)$ for $k \in \mathbb{N}$. Then σ decomposes to t-cycles. Namely, we have a decomposition of $[n]$ to a disjoint union $\cup_{j \in [t]} C_j$. On each $C_j = \{c_{1,j}, \ldots, c_{l_j,j}\}$ σ acts as a cyclic permutation $c_{1,j} \to c_{2,j} \to \cdots \to c_{l_j-1,j} \to c_{l_j,j} \to c_{1,j}$ Rename the integers in $[n]$ such that: Each C_j consists of consecutive integers; $c_{p,j} = c_{1,j} + p - 1$ for $p = 1, \ldots, l_j$; $1 \leq l_1 \leq \ldots \leq l_t$. Equivalently, there exists a permutation P such that $PQP^{-1} = PQP^\top$ has the following block diagonal form: $\text{diag}(Q_1, \ldots, Q_t)$. Q_j is [1] if and only if $l_j = 1$. For $l_j \geq 2$ $Q_j = [\delta_{(i+1)j}]_{i,j=1}^n$, where $n + 1 \equiv 1$. It now follows that $\frac{1}{2}P(Q + Q^\top)P^\top$ has one of the forms 1,2,3.

Suppose that $A \in \mathcal{P}_{n,s}$ has a decomposition (4.7.4). Assume that $b_R > 0$. Then $A - b_R R \geq 0$. If the (i,j) entry of A is zero then the (i,j) entry of R is also zero. Without loss of generality we may assume that $A = \text{diag}(B_1, \ldots, B_t)$ here each B_j has one of the forms 1,2,3. So $R = \text{diag}(R_1, \ldots, R_t)$. Recall that $R = \frac{1}{2}(Q + Q^\top)$ for some $Q \in \mathcal{P}_n$. Hence, $Q = \text{diag}(Q_1, \ldots, Q_t)$, where each Q_j is a permutation matrix of corresponding order. Clearly, each row and column of R_j has at most two nonzero entries. If B_j has the form 1 or 2 then $R_j = Q_j = B_j$. Suppose that B_j is of the form 3. So $B_j = \frac{1}{2}(F_j + F_j^\top)$ for a corresponding cyclic permutation and $R_j = \frac{1}{2}(Q_j + Q_j^\top)$. Then a straightforward argument show that either $Q_j = F_j$ or $Q_j = F_j^\top$. □

Theorem 4.7.8 *Let* $\mathbf{x}, \mathbf{y} \in \mathbb{R}^n$. *Then* $\mathbf{x} \prec \mathbf{y}$ *if and only if there exists* $A \in \Omega_n$ *such that* $\mathbf{x} = A\mathbf{y}$.

Proof. Assume first that $\mathbf{x} = P\mathbf{y}$ for some $P \in \mathcal{P}_n$. Then it is straightforward to see that $\mathbf{x} \prec \mathbf{y}$. Assume that $\mathbf{x} = A\mathbf{y}$ for some $A \in \Omega_n$. Use Theorem 4.7.6 to deduce that $\mathbf{x} \prec \mathbf{y}$.

Assume now that $\mathbf{x}, \mathbf{y} \in \mathbb{R}^n$ and $\mathbf{x} \prec \mathbf{y}$. Since $\mathbf{x} = P\underline{\mathbf{x}}, \mathbf{y} = Q\underline{\mathbf{y}}$ for some $P, Q \in \mathcal{P}_n$, it follows that $\underline{\mathbf{x}} \prec \underline{\mathbf{y}}$. In view of Lemma 4.7.5 it is enough to show that $\underline{\mathbf{x}} = B\underline{\mathbf{y}}$ some $B \in \Omega_n$. We prove this claim by induction on n. For $n = 1$ this claim is trivial. Assume that if $\underline{\mathbf{x}} \prec \underline{\mathbf{y}} \in \mathbb{R}^l$ then $\underline{\mathbf{x}} = B\underline{\mathbf{y}}$ for some $B \in \Omega_l$ for all $l \le m - 1$. Assume that $n = m$ and $\underline{\mathbf{x}} \prec \underline{\mathbf{y}}$. Suppose first that for some $1 \le k \le n - 1$ we have the equality $\sum_{i=1}^k \underline{x}_i = \sum_{i=1}^k \underline{y}_i$. Let

$$\mathbf{x}_1 = (\underline{x}_1, \ldots, \underline{x}_k)^\top, \mathbf{y}_1 = (\underline{y}_1, \ldots, \underline{y}_k)^\top \in \mathbb{R}^k,$$

$$\mathbf{x}_2 = (\underline{x}_{k+1}, \ldots, \underline{x}_n)^\top, \mathbf{y}_2 = (\underline{y}_{k+1}, \ldots, \underline{y}_n)^\top \in \mathbb{R}^{n-k}.$$

Then $\mathbf{x}_1 \prec \mathbf{y}_1, \mathbf{x}_2 \prec \mathbf{y}_2$. Use the induction hypothesis that $\mathbf{x}_i = B_i \mathbf{y}_i, i = 1, 2$ where $B_1 \in \Omega_k, B_2 \in \Omega_{n-k}$. Hence, $\underline{\mathbf{x}} = (B_1 \oplus B_2)\mathbf{y}$ and $B_1 \oplus B_2 \in \Omega_n$.

It is left to consider the case where strict inequalities hold in (4.7.1) for $k = 1, \ldots, n-1$. We now define a finite number of vectors

$$\underline{\mathbf{y}} = \mathbf{z}_1 \succ \mathbf{z}_2 = \underline{\mathbf{z}}_2 \succ \ldots \succ \mathbf{z}_N = \underline{\mathbf{z}}_N \succ \mathbf{x},$$

where $N \ge 2$, such that

1. $\mathbf{z}_{i+1} = B_i \mathbf{z}_i$ for $i = 1, \ldots, N - 1$.

2. $\sum_{i=1}^k \underline{x}_i = \sum_{i=1}^k w_i$ for some $k \in [n-1]$, where $\mathbf{z}_N = \mathbf{w} = (w_1, \ldots, w_n)^\top$.

Observe first that we cannot have $\underline{y}_1 = \cdots = \underline{y}_n$. Otherwise $\mathbf{x} = \mathbf{y}$ and we have equalities in (4.7.1) for all $k \in [n]$, which contradicts our assumptions. Assume that we defined

$$\mathbf{y} = \mathbf{z}_1 \succ \mathbf{z}_2 = \mathbf{z}_2 \succ \ldots \succ \mathbf{z}_r = \mathbf{z}_r = (u_1, \ldots, u_n)^\top \succ \mathbf{x},$$

for $1 \leq r$ such that $\sum_{i=1}^k x_i < \sum_{i=1}^k u_i$ for $k = 1, \ldots; n-1$. Assume that $u_1 = \cdots = u_p > u_{p+1} = \cdots = u_{p+q}$, where $u_{p+q} > u_{p+q+1}$ if $p + q < n$. Let $C(t) = ((1-t)I_{p+q} + \frac{t}{p+q}J_{p+q}) \oplus I_{n-(p+q)})$ for $t \in [0,1]$ and define $\mathbf{u}(t) = C(t)\mathbf{z}_r$. We vary t continuously from $t = 0$ to $t = 1$. Note that $\mathbf{u}(t) = \underline{\mathbf{u}}(t)$ for all $t \in [0,1]$. We have two possibilities. First there exists $t_0 \in (0,1]$ such that $\mathbf{u}(t) \succ \mathbf{x}$ for all $t \in [0, t_0]$. Furthermore for $\mathbf{w} = \mathbf{u}(t_0) = (w_1, \ldots, w_n)^\top$ we have the equality $\sum_{i=1}^k x_i = \sum_{i=1}^k w_i$ for some $k \in \langle n-1 \rangle$. In that case $r = N - 1$ and $\mathbf{z}_N = \mathbf{u}(t_0)$.

Otherwise let $\mathbf{z}_{r+1} = \mathbf{u}(1) = (v_1, \ldots, v_n)^\top$, where $v_1 = \cdots = v_{p+q} > v_{p+q+1}$. Repeat this process for $\mathbf{z}_{r+1}$ and so on until we deduce the conditions 1 and 2. So $\mathbf{x} = B_N \mathbf{z}_N = B_N B_{N-1} \mathbf{z}_{N-1} = B_N \ldots B_1 \mathbf{y}$. In view of 4 of Lemma 4.7.5 we deduce that $\mathbf{x} = A\mathbf{y}$ for some $A \in \Omega_n$. $\qquad\square$

Combine Theorems 4.7.8 and 4.7.6 to deduce:

Corollary 4.7.9 *Let* $\mathbf{x}, \mathbf{y} \in \mathbb{R}^n$. *Then* $\mathbf{x} \prec \mathbf{y}$ *if and only if*

$$\mathbf{x} = \sum_{P \in \mathcal{P}_n} a_P P\mathbf{y} \quad \text{for some } a_P \geq 0, \quad \text{where} \quad \sum_{P \in \mathcal{P}_n} a_P = 1.$$

$$(4.7.5)$$

Furthermore, if $\mathbf{x} \prec \mathbf{y}$ *and* $\mathbf{x} \neq P\mathbf{y}$ *for all* $P \in \mathcal{P}_n$ *then in* (4.7.5) *each* $a_P < 1$.

Theorem 4.7.10 *Let* $\mathbf{x} = (x_1, \ldots, x_n)^\top \prec \mathbf{y} = (y_1, \ldots, y_n)^\top$. *Let* $\phi : [\underline{y}_n, \underline{y}_1] \to \mathbb{R}$ *be a convex function. Then*

$$\sum_{i=1}^n \phi(x_i) \leq \sum_{i=1}^n \phi(y_i). \qquad (4.7.6)$$

If ϕ *is strictly convex on* $[\underline{y}_n, \underline{y}_1]$ *and* $P\mathbf{x} \neq \mathbf{y}$ *for all* $P \in \mathcal{P}_n$ *then strict inequality holds in* (4.7.6).

Proof. Problem 3 implies that if $\mathbf{x} = (x_1, \ldots, x_n)^\top \prec \mathbf{y} = (y_1, \ldots, y_n)^\top$ then $x_i \in [\underline{y}_n, \underline{y}_1]$ for $i = 1, \ldots, n$. Use Corollary 4.7.9 and the convexity of ϕ, (see Problem 11), to deduce:

$$\phi(x_i) \leq \sum_{P \in \mathcal{P}_n} a_P \phi((P\mathbf{y})_i), \quad i = 1, \ldots, n.$$

Observe next that $\sum_{i=1}^n \phi(y_i) = \sum_{i=1}^n \phi((P\mathbf{y})_i)$ for all $P \in \mathcal{P}_n$. Sum up the above inequalities to deduce (4.7.6).

Assume now that ϕ is strictly convex and $\mathbf{x} \neq P\mathbf{y}$ for all $P \in \mathcal{P}_n$. Then Corollary 4.7.9 and the strict convexity of ϕ implies that at least in one the above ith inequality one has strict inequality. Hence, strict inequality holds in (4.7.6). □

Definition 4.7.11 *Let $\mathbf{V}$ be an n-dimensional IPS, and $T \in \mathbf{S}(\mathbf{V})$. Define the eigenvalue vector of T to be $\boldsymbol{\lambda}(T) := (\lambda_1(T), \ldots, \lambda_n(T))^\top \in \mathbb{R}_{\searrow}^n$.*

Corollary 4.7.12 *Let $\mathbf{V}$ be an n-dimensional IPS. Let $T \in \mathbf{S}(\mathbf{V})$, and $F_n = \{\mathbf{f}_1, \ldots, \mathbf{f}_n\} \in \mathrm{Fr}(n, \mathbf{V})$. Then*

$$(\langle T\mathbf{f}_1, \mathbf{f}_1 \rangle, \ldots, \langle T\mathbf{f}_n, \mathbf{f}_n \rangle)^\top \prec \boldsymbol{\lambda}(T).$$

Furthermore, if $\phi : [\lambda_n(T), \lambda_1(T)] \to \mathbb{R}$ is a convex function, then

$$\sum_{i=1}^n \phi(\lambda_i(T)) = \max_{\{\mathbf{f}_1, \ldots \mathbf{f}_n\} \in \mathrm{Fr}(n, \mathbf{V})} \sum_{i=1}^n \phi(\langle T\mathbf{f}_i, \mathbf{f}_i \rangle).$$

Finally, if ϕ is strictly convex, then $\sum_{i=1}^n \phi(\lambda_i(T)) = \sum_{i=1}^n \phi(\langle T\mathbf{f}_i, \mathbf{f}_i \rangle)$ if and only if $\mathbf{f}_1, \ldots, \mathbf{f}_n$ is a set of n orthonormal eigenvectors of T.

See Problem 4.

Problems

1. Show Proposition 4.7.2.

2. Prove Lemma 4.7.5.

3. Let $\mathbf{x}, \mathbf{y} \in \mathbb{R}^n$. Show that $\mathbf{x} \prec \mathbf{y} \iff -\mathbf{y} \prec -\mathbf{x}$.

4. Prove Corollary 4.7.12.

4.8 Spectral Functions

Definition 4.8.1

1. D *is called a* Schur set *if* $D \subset \mathbb{R}^n_{\searrow}$ *and the following property holds: Assume that* $\mathbf{y} \in D$. *Then each* $\mathbf{x} \in \mathbb{R}^n_{\searrow}$ *which satisfies* $\mathbf{x} \prec \mathbf{y}$ *belongs to* D.

2. *A function* $h : D \to \mathbb{R}$ *is called* Schur's order preserving *if*

$$h(\mathbf{x}) \le h(\mathbf{y}) \quad \text{for any } \mathbf{x}, \mathbf{y} \in D \text{ such } \mathbf{x} \prec \mathbf{y}.$$

h is called strict Schur's order preserving *if a strict inequality holds in the above inequality whenever* $\mathbf{x} \ne \mathbf{y}$. *h is called* strong Schur's order preserving *if*

$$h(\mathbf{x}) \le h(\mathbf{y}) \quad \text{for any } \mathbf{x}, \mathbf{y} \in D \text{ such } \mathbf{x} \preceq \mathbf{y}.$$

3. *Let* $\mathcal{T} \subset \mathbf{S}(\mathbf{V})$, *where* $\mathbf{V}$ *is an n-dimensional IPS over* $\mathbb{F} = \mathbb{R}, \mathbb{C}$. *Let* $\boldsymbol{\lambda}(\mathcal{T}) := \{\boldsymbol{\lambda}(T) \in \mathbb{R}^n_{\searrow} : T \in \mathcal{T}\}$. *A function* $f : \mathcal{T} \to \mathbb{R}$ *is called a* spectral function *if there exists a set* $D \subset \mathbb{R}^n_{\searrow}$ *and* $h : D \to \mathbb{R}$ *such that* $\boldsymbol{\lambda}(\mathcal{T}) \subset D$ *and* $f(T) = h(\boldsymbol{\lambda}(T))$.

Note that if $h((x_1, \ldots, x_n)) := \sum_{i=1}^n g(x_i)$ for some convex function $g : \mathbb{R} \to \mathbb{R}$ then Corollary 4.7.12 implies that $h : \mathbb{R}^n_{\searrow} \to \mathbb{R}$ is Schur's order preserving. The results of §4.4 yield:

Proposition 4.8.2 *Let* $\mathbf{V}$ *be an n-dimensional IPS,* $D \subset \mathbb{R}^n_{\searrow}$ *be a Schur set and* $h : D \to R$ *be a Schur's order preserving function. Let* $T \in \mathbf{S}(\mathbf{V})$ *and assume that* $\lambda(T) \in D$. *Then*

$$h(\lambda(T)) = \max_{\mathbf{x} \in D, \mathbf{x} \prec \lambda(T)} h(\mathbf{x}).$$

Theorem 4.8.3 *Let* $D \subset \mathbb{R}^n_\searrow$ *be a regular Schur set in* $\mathbb{R}^n$. *Let* $F \in C^1(D)$. *Then* F *is Schur's order preserving if and only if*

$$\frac{\partial F}{\partial x_1}(\mathbf{x}) \geq \cdots \geq \frac{\partial F}{\partial x_n}(\mathbf{x}), \text{ for each } \mathbf{x} = (x_1, \ldots, x_n)^\top \in D. \quad (4.8.1)$$

If for any point $\mathbf{x} = (x_1, \ldots, x_n)^\top \in D$ *such that* $x_i > x_{i+1}$ *the inequality* $\frac{\partial F}{\partial x_i}(\mathbf{x}) > \frac{\partial F}{\partial x_{i+1}}(\mathbf{x})$ *holds then* F *is strict Schur's order preserving.*

Proof. Assume that $F \in C^1(D)$ and F is Schur's order preserving. Let $\mathbf{x} = (x_1, \ldots, x_n)^\top \in D^o$. Hence, $x_1 > \cdots > x_n$. Let $\mathbf{e}_i = (\delta_{i1}, \ldots, \delta_{in})^\top, i = 1, \ldots, n$. For $i \in [1, n-1] \cap \mathbb{Z}_+$ let $\mathbf{x}(t) := \mathbf{x} + t(\mathbf{e}_i - \mathbf{e}_{i+1})$. Then

$$\mathbf{x}(t) \in R^n_\searrow \text{ for } |t| \leq \tau := \frac{\min_{j \in [1, n-1] \cap \mathbb{Z}_+} x_j - x_{j+1}}{2}, \quad (4.8.2)$$

and

$$\mathbf{x}(t_1) \prec \mathbf{x}(t_2) \text{ for } -\tau \leq t_1 \leq t_2 \leq \tau.$$

See Problem 1. Since D^o is open there exists $\epsilon > 0$ such that $\mathbf{x}(t) \in D^o$ for $t \in [-\epsilon, \epsilon]$. Then $f(t) := F(\mathbf{x}(t))$ is an increasing function on $[-\epsilon, \epsilon]$. Hence, $f'(0) = \frac{\partial F}{\partial x_i}(\mathbf{x}) - \frac{\partial F}{\partial x_{i+1}}(\mathbf{x}) \geq 0$. This proves (4.8.1) in D^o. The continuity argument yields (4.8.1) in D.

Assume now that (4.8.1) holds. Let $\mathbf{y} = (y_1, \ldots, y_n)^\top, \mathbf{z} = (z_1, \ldots, z_n)^\top \in D$ and define

$$\mathbf{y}(t) := (1-t)\mathbf{y} + t\mathbf{z}, \quad g(t) := F((1-t)\mathbf{y} + t\mathbf{z}), \quad \text{for } t \in [0, 1].$$

Suppose that $\mathbf{y} \prec \mathbf{z}$. Then $\mathbf{y}(t_1) \prec \mathbf{y}(t_2)$ for $0 \leq t_1 \leq t_2 \leq 1$. Since D is a Schur set, $[\mathbf{y}, \mathbf{z}] \subset D$. Then

$$g'(t) = \sum_{i=1}^{n-1} \left(\frac{\partial F(\mathbf{y}(t))}{\partial x_i} - \frac{\partial F(\mathbf{y}(t))}{\partial x_{i+1}} \right) \sum_{j=1}^{i} (z_j - y_j). \quad (4.8.3)$$

See Problem 2. Hence $g'(t) \geq 0$, i.e. $g(t)$ is a nondecreasing function on $[0, 1]$. Thus $F(\mathbf{y}) = g(0) \leq g(1) = F(\mathbf{z})$. Assume that for any point $\mathbf{x} = (x_1, \ldots, x_n)^\top \in D$ such that $x_i > x_{i+1}$ the inequality

$\frac{\partial F}{\partial x_i}(\mathbf{x}) > \frac{\partial F}{\partial x_{i+1}}(\mathbf{x})$ holds. Suppose that $\mathbf{y} \neq \mathbf{z}$. Then $g'(t) > 0$ and $F(\mathbf{y}) = g(0) < g(1) = F(\mathbf{z})$. $\qquad\qquad\qquad\qquad\qquad\qquad\square$

Theorem 4.8.4 *Let* $\mathrm{D} \subset \mathbb{R}^n$ *be a regular Schur set in* $\mathbb{R}^n$. *Let* $F \in \mathrm{C}^1(\mathrm{D})$. *If F is strong Schur's order preserving then*

$$\frac{\partial F}{\partial x_1}(\mathbf{x}) \geq \cdots \geq \frac{\partial F}{\partial x_n}(\mathbf{x}) \geq 0, \quad \text{for each } \mathbf{x} = (x_1, \ldots, x_n)^\top \in \mathrm{D}.$$

$$(4.8.4)$$

Suppose that in addition to the above assumptions D *is convex. If F satisfies the above inequalities then F is strong Schur's order preserving. If for any point* $\mathbf{x} = (x_1, \ldots, x_n)^\top \in \mathrm{D}$ $\frac{\partial F}{\partial x_n}(\mathbf{x}) > 0$ *and* $\frac{\partial F}{\partial x_i}(\mathbf{x}) > \frac{\partial F}{\partial x_{i+1}}(\mathbf{x})$ *whenever $x_i > x_{i+1}$ holds, then F is strict strong Schur's order preserving.*

Proof. Assume that F is strong Schur's order preserving. Since F is Schur's order preserving (4.8.1) holds. Let $\mathbf{x} = (x_1, \ldots, x_n)^\top \in \mathrm{D}^o$. Define $\mathbf{w}(t) = \mathbf{x} + t\mathbf{e}_n$. Then there exists $\epsilon > 0$ such that $\mathbf{w}(t) \in \mathrm{D}^o$ for $t \in [-\epsilon, \epsilon]$. Clearly $\mathbf{w}(t_1) \preceq \mathbf{w}(t_2)$ for $-\epsilon \leq t_1 \leq t_2 \leq \epsilon$. Hence, the function $h(t) := F(\mathbf{w}(t))$ is not decreasing on the interval $[-\epsilon, \epsilon]$. Thus $\frac{\partial F}{\partial x_n}(\mathbf{x}) = h'(0) \geq 0$. Use the continuity argument to deduce that $\frac{\partial F}{\partial x_n}(\mathbf{x}) \geq 0$ for any $\mathbf{x} \in \mathrm{D}$.

Assume that D is convex and (4.8.4) holds. Let $\mathbf{y}, \mathbf{z} \in \mathrm{D}$ and define $\mathbf{y}(t)$ and $g(t)$ as in the proof of Theorem 4.8.3. Then

$$g'(t) = \sum_{i=1}^{n-1} \left(\frac{\partial F(\mathbf{y}(t))}{\partial x_i} - \frac{\partial F(\mathbf{y}(t))}{\partial x_{i+1}} \right) \sum_{j=1}^{i}(z_j - y_j)$$

$$+ \frac{\partial F(\mathbf{y}(t))}{\partial x_n} \sum_{j=1}^{n}(z_j - y_j). \tag{4.8.5}$$

See Problem 2. Assume that $\mathbf{y} \preceq \mathbf{z}$. Then $g'(t) \geq 0$. Hence, $F(\mathbf{y}) \leq F(\mathbf{z})$.

Assume now that for any point $\mathbf{x} = (x_1, \ldots, x_n)^\top \in \mathrm{D}$ $\frac{\partial F}{\partial x_n}(\mathbf{x}) > 0$ and $\frac{\partial F}{\partial x_i}(\mathbf{x}) > \frac{\partial F}{\partial x_{i+1}}(\mathbf{x})$ whenever $x_i > x_{i+1}$. Let $\mathbf{y}, \mathbf{z} \in \mathrm{D}$ and assume that $\mathbf{y} \preceq \mathbf{z}$ and $\mathbf{y} \neq \mathbf{z}$. Define $g(t)$ on $[0, 1]$ as above. Use (4.8.5) to deduce that $g'(t) > 0$ on $[0, 1]$. Hence, $F(\mathbf{y}) < F(\mathbf{z})$.   □

Theorem 4.8.5 *Let* $\mathbf{V}$ *be an* n-*dimensional IPS over* $\mathbb{F} = \mathbb{R}, \mathbb{C}$. *Then the function* $\operatorname{tr}_i : \mathbf{S}(\mathbf{V}) \to \mathbb{R}$, *where* $\operatorname{tr}_i(T) = \sum_{j=1}^{i} \lambda_i(T)$, *is a continuous homogeneous convex function for* $i = 1, \dots, n - 1$. $\operatorname{tr}_n(T) = \operatorname{tr} T$ *is a linear function on* $\mathbf{S}(\mathbf{V})$.

Proof. Clearly $\operatorname{tr}_i(aT) = a \operatorname{tr}_i(T)$ for $a \in [0, \infty)$. Hence, tr_i is a homogeneous function. Since the eigenvalues of T are continuous it follows that tr_i is a continuous function. Clearly tr_n is a linear function on the vector space $\mathbf{S}(\mathbf{V})$. Combine Theorem 4.4.8 with Proposition 4.6.16 to deduce that tr_i is convex.   □

Corollary 4.8.6 *Let* $\mathbf{V}$ *be a finite dimensional IPS over* $\mathbb{F} = \mathbb{R}, \mathbb{C}$. *Then*

$$\boldsymbol{\lambda}(\alpha A + (1 - \alpha)B) \prec \alpha \boldsymbol{\lambda}(A) + (1 - \alpha)\boldsymbol{\lambda}(B),$$
$$\text{for any } A, B \in \mathbf{S}(\mathbf{V}), \ \alpha \in [0, 1].$$

For $\alpha \in (0, 1)$ *equality holds if and only if there exists an orthonormal basis* $[\mathbf{v}_1, \dots, \mathbf{v}_n]$ *in* $\mathbf{V}$ *such that*

$$A\mathbf{v}_i = \lambda_i(A)\mathbf{u}_i, \ B\mathbf{v}_i = \lambda_i(B)\mathbf{v}_i, \quad i = 1, \dots, n.$$

See Problem 4.

Proposition 4.8.7 *Let* $\mathbf{V}$ *be an* n-*dimensional IPS over* $\mathbb{F} = \mathbb{R}, \mathbb{C}$. *For* $\mathrm{D} \subset \mathbb{R}^n$ *let*

$$\boldsymbol{\lambda}^{-1}(\mathrm{D}) := \{T \in \mathbf{S}(\mathbf{V}) : \boldsymbol{\lambda}(T) \in \mathrm{D}\}. \tag{4.8.6}$$

If $\mathrm{D} \subset \mathbb{R}^n$ *is a regular convex Schur set then* $\boldsymbol{\lambda}^{-1}(\mathrm{D})$ *is regular convex set in the vector space* $\mathbf{S}(\mathbf{V})$.

Proof. The continuity of the function $\boldsymbol{\lambda} : \mathbf{S}(\mathbf{V}) \to \mathbb{R}^n_{\searrow}$ implies that $\boldsymbol{\lambda}^{-1}(\mathrm{D})$ is a regular set in $\mathbf{S}(\mathbf{V})$. Suppose that $A, B \in \boldsymbol{\lambda}(\mathrm{D})^{-1}$ and $\alpha \in [0, 1]$. Since D is convex, $\alpha \boldsymbol{\lambda}(A) + (1 - \alpha)\boldsymbol{\lambda}(B) \in \mathrm{D}$. Since D is a Schur set Corollary 4.8.6 yields that $\boldsymbol{\lambda}(\alpha A + (1 - \alpha)B) \in \mathrm{D}$. Hence, $\alpha A + (1 - \alpha)B \in \boldsymbol{\lambda}^{-1}(\mathrm{D})$.   □

Definition 4.8.8 *For* $\mathbf{x} = (x_1, \ldots, x_n)^\top \in \mathbb{D}^n$ *denote* $D(\mathbf{x}) :=$ $\mathrm{diag}(x_1, \ldots, x_n)$.

Theorem 4.8.9 *Let* $\mathrm{D} \subset \mathbb{R}^n$ *be a regular convex Schur set and let* $h : \mathrm{D} \to \mathbb{R}$. *Let* $\mathbf{V}$ *be an n-dimensional IPS over* $\mathbb{F} = \mathbb{R}, \mathbb{C}$. *Let* $f : \boldsymbol{\lambda}^{-1}(\mathrm{D}) \to \mathbb{R}$ *be the spectral function given by* $f(A) := h(\boldsymbol{\lambda}(A))$. *Then the following are equivalent:*

(a) f *is (strictly) convex on* $\boldsymbol{\lambda}^{-1}(\mathrm{D})$.

(b) h *is (strictly) convex and (strictly) Schur's order preserving on* D.

Proof. Choose a fixed orthonormal basis $[\mathbf{u}_1, \ldots, \mathbf{u}_n]$. For simplicity of the argument we assume that $\mathbb{F} = \mathbb{C}$. Identify $\mathbf{S}(\mathbf{V})$ with $\mathbf{H}_n$. Thus we view $\mathcal{T} := \boldsymbol{\lambda}^{-1}(\mathrm{D})$ as a subset of $\mathbf{H}_n$. Since D is a regular convex Schur set, Proposition 4.8.7 yields that $\mathcal{T}$ is a regular convex set. Let $\mathbf{x} = (x_1, \ldots, x_n)^\top \in \mathbb{R}^n$. Then $\boldsymbol{\lambda}(D(\mathbf{x})) = \underline{\mathbf{x}}$. Thus $D(\mathbf{x}) \in \mathcal{T} \iff \underline{\mathbf{x}} \in \mathrm{D}$ and $f(D(\mathbf{x})) = h(\underline{\mathbf{x}})$ for $\underline{\mathbf{x}} \in \mathrm{D}$.

(a) $\Rightarrow$ (b). Assume that f is convex on $\mathcal{T}$. By restricting f to $D(\mathbf{x}), \mathbf{x} \in \mathrm{D}$, we deduce that h is convex on D. If f is strictly convex on $\mathcal{T}$ we deduce that h is strictly convex on D.

Let $\mathbf{x}, \mathbf{y} \in \mathrm{D}$ and assume that $\mathbf{x} \prec \mathbf{y}$. Then (4.7.5) holds. Hence

$$D(\mathbf{x}) = \sum_{P \in \mathcal{P}_n} a_P P D(\mathbf{y}) P^\top.$$

Clearly $\boldsymbol{\lambda}(PD(\mathbf{y})P^\top) = \boldsymbol{\lambda}(D(\mathbf{y})) = \mathbf{y}$. The convexity of f yields

$$h(\mathbf{x}) = f(D(\mathbf{x})) \le \sum_{P \in \mathcal{P}_n} a_P f(PD(\mathbf{y})P^\top) = f(D(\mathbf{y})) = h(\mathbf{y}).$$

See Problem 4.7.11. Hence, h is Schur's order preserving. If f is strictly convex on $\mathcal{T}$ then in the above inequality one has a strict inequality if $\mathbf{x} \ne \mathbf{y}$. Hence, h is strictly Schur's order preserving.

(b) $\Rightarrow$ (a). Assume that h is convex. Then for $A, B \in \mathcal{T}$

$$\alpha f(A) + (1 - \alpha)f(B) = \alpha h(\boldsymbol{\lambda}(A)) + (1 - \alpha)h(\boldsymbol{\lambda}(B))$$
$$\geq h(\alpha\boldsymbol{\lambda}(A) + (1 - \alpha)\boldsymbol{\lambda}(B)).$$

Use Corollary 4.8.6 and the assumption that h is Schur's order preserving to deduce the convexity of f. Suppose that h is strictly convex and strictly Schur's order preserving. Assume that $f(\alpha A + (1 - \alpha)B) = \alpha f(A) + (1 - \alpha)f(B)$ for some $A, B \in \mathcal{T}$ and $\alpha \in (0, 1)$. Hence, $\boldsymbol{\lambda}(A) = \boldsymbol{\lambda}(B)$ and $\boldsymbol{\lambda}(\alpha A + (1 - \alpha)B) = \alpha\boldsymbol{\lambda}(A) + (1 - \alpha)\boldsymbol{\lambda}(B)$. Use Corollary 4.8.6 to deduce that $A = B$. Hence, f is strictly convex. $\qquad\square$

Theorem 4.8.10 *Let* $\mathrm{D} \subset \mathbb{R}^n_{\searrow}$ *be a regular convex Schur set and let* $h \in \mathrm{C}^1(\mathrm{D})$. *Let* $\mathbf{V}$ *be an n-dimensional IPS over* $\mathbb{F} = \mathbb{R}, \mathbb{C}$. *Let* $f : \boldsymbol{\lambda}^{-1}(\mathrm{D}) \to \mathbb{R}$ *be the spectral function given by* $f(A) := h(\boldsymbol{\lambda}(A))$. *Then the following are equivalent:*

(a) *f is convex on $\boldsymbol{\lambda}^{-1}(\mathrm{D})$ and $f(A) \leq f(B)$ for any $A, B \in \boldsymbol{\lambda}^{-1}(\mathrm{D})$ such that $A \preceq B$.*

(b) *h is convex and strongly Schur's order preserving on* D.

Proof. We repeat the proof of Theorem 4.8.9 with the following modifications.

(b) $\Rightarrow$ (a). Since h is convex and Schur's order preserving, Theorem 4.8.9 yields that f is convex on $\mathcal{T}$. Let $A, B \in \mathcal{T}$ and assume that $A \preceq B$. Then $\boldsymbol{\lambda}(A) \preceq \boldsymbol{\lambda}(B)$. As h is strongly Schur's order preserving $h(\boldsymbol{\lambda}(A)) \leq h(\boldsymbol{\lambda}(B)) \Rightarrow f(A) \leq f(B)$.

(a) $\Rightarrow$ (b). Since f is convex on $\mathcal{T}$, Theorem 4.8.9 implies that h is convex and Schur's order preserving. Since $h \in \mathrm{C}^1(\mathrm{D})$, Theorem 4.8.3 yields that h satisfies the inequalities (4.8.1). Let $\mathbf{x} \in \mathrm{D}^o$ and define $\mathbf{x}(t) := \mathbf{x} + t\mathbf{e}_n$. Then for a small $a > 0$, $\mathbf{x}(t) \in \mathrm{D}^o$ for $t \in (-a, a)$. Clearly $D(\mathbf{x}(t_1)) \leq D(\mathbf{x}(t_2))$ for $t_1 \leq t_2$. Hence, $g(t) := f(D(\mathbf{x}(t))) = h(\mathbf{x}(t))$ is a nondecreasing function on $(-a, a)$. Hence, $\frac{\partial h}{\partial x_n}(\mathbf{x}) = g'(0) \geq 0$. Use the continuity hypothesis to deduce that h satisfies (4.8.4). Theorem 4.8.4 yields that h is strong Schur's order preserving. $\square$

Theorem 4.8.11 *Assume that $n < N$ are two positive integers. Let $\mathbf{V}$ be an N-dimensional IPS over $\mathbb{F} = \mathbb{R}, \mathbb{C}$. Let $\boldsymbol{\lambda}_{(n)} : \mathbf{S(V)} \to \mathbb{R}^n_{\searrow}$ be the map $A \mapsto \boldsymbol{\lambda}_{(n)}(A) := (\lambda_1(A), \ldots, \lambda_n(A))^\top$. Assume that $\mathrm{D} \subset \mathbb{R}^n_{\searrow}$ is a regular convex Schur set and let $T := \boldsymbol{\lambda}_{(n)}^{-1}(\mathrm{D}) \subset \mathbf{S(V)}$. Let $f : T \to \mathbb{R}$ be the spectral function given by $f(A) := h(\boldsymbol{\lambda}_{(n)}(A))$. Then the following are equivalent:*

(a) *f is convex on T.*

(b) *h is convex and strongly Schur's order preserving on D.*

(c) *f is convex on T and $f(A) \leq f(B)$ for any $A, B \in T$ such that $A \preceq B$.*

Proof. Let $\pi : \mathbb{R}^N_{\searrow} \to \mathbb{R}^n_{\searrow}$ be the projection on the first n coordinates. Let $\mathrm{D}_1 := \pi^{-1}(\mathrm{D}) \subset \mathbb{R}^N_{\searrow}$. It is straightforward to show that D_1 is a regular convex set. Let $h_1 := h \circ \pi : \mathrm{D}_1 \to \mathbb{R}$. Then $\frac{\partial h_1}{\partial x_i} = 0$ for $i = n+1, \ldots, N$.

(a) $\Rightarrow$ (b). Suppose that f is convex on T. Then Theorem 4.8.9 yields that h_1 is convex and Schur's order preserving. Theorem 4.8.3 yields the inequalities (4.8.1). Hence, $\frac{\partial h_1}{\partial x_n}(\mathbf{y}) \geq \frac{\partial h_1}{\partial x_{n+1}}(\mathbf{y}) = 0$ for any $\mathbf{y} \in \mathrm{D}_1$. Clearly h is convex and

$$\frac{\partial h}{\partial x_i}(\mathbf{x}) = \frac{\partial h_1}{\partial x_i}(\mathbf{y}), \quad i = 1, \ldots, n, \text{ where } \mathbf{x} \in \mathrm{D}, \ \mathbf{y} \in \mathrm{D}_1, \text{ and } \pi(\mathbf{y}) = \mathbf{x}.$$

Thus h satisfies (4.8.4). Theorem 4.8.4 yields that h is strongly Schur's order preserving.

Other nontrivial implications follow as in the proof of Theorem 4.8.10.

Problems

1. Let $\mathbf{x} = (x_1, \ldots, x_n)^\top \in \mathbb{R}^n$ and assume $x_1 > \cdots > x_n$. Let $\mathbf{x}(t)$ be defined as in the proof of Theorem 4.8.3. Prove (4.8.2).

2. Let $D \subset \mathbb{R}^n$ be a regular set and assume that $[\mathbf{y}, \mathbf{z}] \subset D, \mathbf{y} = (y_1, \ldots, y_n)^\top, \mathbf{z} = (z_1, \ldots, z_n)^\top$. Let $F \in C^1(D)$ and assume that $g(t)$ is defined as in the proof of Theorem 4.8.3. Show the equality (4.8.5). Suppose furthermore that $\sum_{i=1}^n y_i = \sum_{i=1}^n z_i$. Show the equality (4.8.3).

3. (*This problem offers an alternative proof of Theorem 4.7.10.*) Let $a < b$ and $n \in \mathbb{N}$. Denote

$$[a, b]_\searrow^n := \{(x_1, \ldots, x_n) \in \mathbb{R}_\searrow^n : \quad x_i \in [a, b], \ i = 1, \ldots, n\}.$$

   (a) Show that $[a, b]_\searrow^n$ is a regular convex Schur domain.

   (b) Let $f \in C^1([a, b])$ be a convex function. Let $F : [a, b]^n \to \mathbb{R}$ be defined by $F((x_1, \ldots, x_n)^\top) := \sum_{i=1}^n f(x_i)$. Show that F satisfies the condition (4.8.1) on $[a, b]_\searrow^n$. Hence, Theorem 4.7.10 holds for any $\mathbf{x}, \mathbf{y} \in [a, b]^n$ such that $\mathbf{x} \prec \mathbf{y}$.

   (c) Assume that any convex $f \in C([a, b])$ can be uniformly approximated by as sequence of convex $f_k \in C^1([a, b]), k = 1, \ldots$ Show that Theorem 4.7.10 holds for $\mathbf{x}, \mathbf{y} \in [a, b]^n$ such that $\mathbf{x} \prec \mathbf{y}$.

4. Use Theorem 4.4.8 to show the equality case in Corollary 4.8.6

5. For $p \in [1, \infty)$ let $\|\mathbf{x}\|_{p,\mathbf{w}} := \left(\sum_{i=1}^n w_i |x_i|^p\right)^{\frac{1}{p}}$, where $\mathbf{x} = (x_1, \ldots, x_n)^\top \in \mathbb{R}^n$ and $\mathbf{w} = (w_1, \ldots, w_n)^\top \in \mathbb{R}_+^n$.

   (a) Show that $\| \cdot \|_{p,\mathbf{w}} : \mathbb{R}^n \to \mathbb{R}$ is a homogeneous convex function. Furthermore this function is strictly convex if and only if $p > 1$ and $w_i > 0$ for $i = 1, \ldots, n$. (*Hint:* First prove the case $\mathbf{w} = (1, \ldots, 1)$.)

   (b) For $q > 1$ show that $\| \cdot \|_{p,\mathbf{w}}^q : \mathbb{R}^n \to \mathbb{R}$ is a convex function. Furthermore, this function is strictly convex if and only if $w_i > 0$ for $i = 1, \ldots, n$. (*Hint:* Use the fact that $f(x) = x^q$ is strictly convex on $[0, \infty)$.)

   (c) Show that for $q > 0$ the function $\| \cdot \|_{p,\mathbf{w}}^q : \mathbb{R}_{+,\searrow}^n \to \mathbb{R}$ is strong Schur's order preserving if and only if $w_1 \geq \cdots \geq$

$w_n \geq 0$. Furthermore, this function is strictly strong Schur's order preserving if and only if $w_1 \geq \cdots \geq w_n > 0$.

(d) Let $\mathbf{V}$ be an n-dimensional IPS over $\mathbb{F} = \mathbb{R}, \mathbb{C}$. Show that for $q \geq 1$, $w_1 \geq \cdots \geq w_n \geq 0$ the spectral function $T \to \|\boldsymbol{\lambda}(T)\|_{p,\mathbf{w}}^q$ is a convex function on $\mathbf{S}(\mathbf{V})_+$ (the positive self-adjoint operators on $\mathbf{V}$.) If in addition $w_n > 0$ and $\max(p, q) > 1$ then the above function is strictly convex on $\mathbf{S}(\mathbf{V})_+$.

6. Use the differentiability properties of convex functions to show that Theorems 4.8.10 and 4.8.11 hold under the lesser assumption $h \in C(D)$.

7. Show that on $\mathbf{H}_{n,+}^o$ the function $\log \det A$ is a strictly concave function, i.e. $\det(\alpha A + (1 - \alpha)B) \geq (\det A)^\alpha (\det B)^{1-\alpha}$. (*Hint:* Observe that $-\log x$ is a strictly convex function on $(0, \infty)$.)

4.9 Inequalities for Traces

Let $\mathbf{V}$ be a finite dimensional IPS over $\mathbb{F} = \mathbb{R}, \mathbb{C}$. Let $T : \mathbf{V} \to \mathbf{V}$ be a linear operator. Then $\operatorname{tr} T$ is the trace of the representation matrix A with respect to any orthonormal basis of $\mathbf{V}$. See Problem 1.

Theorem 4.9.1 *Let $\mathbf{V}$ be an n-dimensional IPS over $\mathbb{F} = \mathbb{R}, \mathbb{C}$. Assume that $S, T \in \mathbf{S}(\mathbf{V})$. Then $\operatorname{tr} ST$ is bounded below and above by*

$$\sum_{i=1}^n \lambda_i(S)\lambda_{n-i+1}(T) \leq \operatorname{tr} ST \leq \sum_{i=1}^n \lambda_i(S)\lambda_i(T). \tag{4.9.1}$$

Equality for the upper bound holds if and only if $ST = TS$ and there exists an orthonormal basis $\mathbf{x}_1, \ldots, \mathbf{x}_n \in \mathbf{V}$ such that

$$S\mathbf{x}_i = \lambda_i(S)\mathbf{x}_i, \quad T\mathbf{x}_i = \lambda_i(T)\mathbf{x}_i, \quad i = 1, \ldots, n. \tag{4.9.2}$$

Equality for the lower bound holds if and only if $ST = TS$ and there exists an orthonormal basis $\mathbf{x}_1, \ldots, \mathbf{x}_n \in \mathbf{V}$ such that

$$S\mathbf{x}_i = \lambda_i(S)\mathbf{x}_i, \quad T\mathbf{x}_i = \lambda_{n-i+1}(T)\mathbf{x}_i, \quad i = 1, \ldots, n. \tag{4.9.3}$$

Proof. Let $\mathbf{y}_1, \ldots, \mathbf{y}_n$ be an orthonormal basis of $\mathbf{V}$ such that

$$T\mathbf{y}_i = \lambda_i(T)\mathbf{y}_i, \quad i = 1, \ldots, n,$$

$$\lambda_1(T) = \cdots = \lambda_{i_1}(T) > \lambda_{i_1+1}(T) = \cdots = \lambda_{i_2}(T) > \cdots >$$

$$\lambda_{i_{k-1}+1}(T) = \cdots = \lambda_{i_k}(T) = \lambda_n(T), \quad 1 \leq i_1 < \cdots < i_k = n.$$

If $k = 1 \iff i_1 = n$ it follows that $T = \lambda_1 I$ and the theorem is trivial in this case. Assume that $k > 1$. Then

$$\operatorname{tr} ST = \sum_{i=1}^{n} \lambda_i(T)\langle S\mathbf{y}_i, \mathbf{y}_i \rangle$$

$$= \sum_{i=1}^{n-1}(\lambda_i(T) - \lambda_{i+1}(T))\left(\sum_{l=1}^{i}\langle S\mathbf{y}_l, \mathbf{y}_l \rangle\right) + \lambda_n(T)\left(\sum_{l=1}^{n}\langle S\mathbf{y}_l, \mathbf{y}_l \rangle\right)$$

$$= \sum_{j=1}^{k-1}(\lambda_{i_j}(T) - \lambda_{i_j+1}(T))\sum_{l=1}^{i_j}\langle S\mathbf{y}_l, \mathbf{y}_l \rangle + \lambda_n(T)\operatorname{tr} S.$$

Theorem 4.4.8 yields that $\sum_{l=1}^{i_j}\langle S\mathbf{y}_l, \mathbf{y}_l \rangle \leq \sum_{l=1}^{i_j}\lambda_l(S)$. Substitute these inequalities for $j = 1, \ldots, k - 1$ in the above identity to deduce the upper bound in (4.9.1). Clearly the condition (4.9.2) implies that $\operatorname{tr} ST$ is equal to the upper bound in (4.9.1). Assume now that $\operatorname{tr} ST$ is equal to the upper bound in (4.9.1). Then $\sum_{l=1}^{i_j}\langle S\mathbf{y}_l, \mathbf{y}_l \rangle = \sum_{l=1}^{i_j}\lambda_l(S)$ for $j = 1, \ldots, k - 1$. Theorem 4.4.8 yields that span $(\mathbf{y}_1, \ldots, \mathbf{y}_{i_j})$ is spanned by some i_j eigenvectors of S corresponding to the first i_j eigenvalues of S for $j = 1, \ldots, k - 1$. Let $\mathbf{x}_1, \ldots, \mathbf{x}_{i_1}$ be an orthonormal basis of span $(\mathbf{y}_1, \ldots, \mathbf{y}_{i_1})$ consisting of the eigenvectors of S corresponding to the eigenvalues of $\lambda_1(S), \ldots, \lambda_{i_1}(S)$. Since any $0 \neq \mathbf{x} \in$ span $(\mathbf{y}_1, \ldots, \mathbf{y}_{i_1})$ is an eigenvector of T corresponding to the eigenvalue $\lambda_{i_1}(T)$ it follows that (4.9.2) holds for $i = 1, \ldots, i_1$. Consider span $(\mathbf{y}_1, \ldots, \mathbf{y}_{i_2})$. The above arguments imply that this subspace contains i_2 eigenvectors of S and T corresponding to the first i_2 eigenvalues of S and T. Hence $\mathbf{U}_2$, the orthogonal complement of span $(\mathbf{x}_1, \ldots, \mathbf{x}_{i_1})$ in span $(\mathbf{y}_1, \ldots, \mathbf{y}_{i_2})$, spanned by $\mathbf{x}_{i_1+1}, \ldots, \mathbf{x}_{i_2}$, which are $i_2 - i_1$ orthonormal eigenvectors of S corresponding to the eigenvalues $\lambda_{i_1+}(S), \ldots, \lambda_{i_2}(S)$. Since any nonzero vector in $\mathbf{U}_2$ is an eigenvector of T corresponding to the eigenvalue $\lambda_{i_2}(T)$ we deduce that (4.9.2) holds for $i = 1, \ldots, i_2$. Continuing in the same manner we obtain (4.9.2).

To prove the equality case in the lower bound consider the equality in the upper bound for $\operatorname{tr} S(-T)$. $\qquad\square$

Corollary 4.9.2 *Let* $\mathbf{V}$ *be an* n-*dimensional IPS over* $\mathbb{F} = \mathbb{R}, \mathbb{C}$. *Assume that* $S, T \in \mathbf{S}(\mathbf{V})$. *Then*

$$\sum_{i=1}^{n} (\lambda_i(S) - \lambda_i(T))^2 \leq \operatorname{tr}(S - T)^2. \qquad (4.9.4)$$

Equality holds if and only if $ST = TS$ *and* $\mathbf{V}$ *has an orthonormal basis* $\mathbf{x}_1, \ldots, \mathbf{x}_n$ *satisfying* (4.9.2).

Proof. Note

$$\sum_{i=1}^{n} (\lambda_i(S) - \lambda_i(T))^2 = \operatorname{tr} S^2 + \operatorname{tr} T^2 - 2 \sum_{i=1}^{n} \lambda_i(S)\lambda_i(T).$$

$\qquad\square$

Corollary 4.9.3 *Let* $S, T \in \mathbf{H}_n$. *Then the inequalities* (4.9.1) *and* (4.9.4) *hold. Equalities in the upper bounds hold if and only if there exists* $U \in \mathbf{U}_n$ *such that* $S = U \operatorname{diag} \lambda(S) U^*, T = U \operatorname{diag} \lambda(T) U^*$. *Equality in the lower bound of* (4.9.1) *if and only if there exists* $V \in \mathbf{U}_n$ *such that* $S = V \operatorname{diag} \lambda(S) V^*, -T = V \operatorname{diag} \lambda(-T) V^*$.

Problems

1. Let $\mathbf{V}$ be a n-dimensional IPS over $\mathbb{F} = \mathbb{R}, \mathbb{C}$.

(a) Assume that $T : \mathbf{V} \to \mathbf{V}$ is a linear transformation. Show that for any orthonormal basis $\mathbf{x}_1, \ldots, \mathbf{x}_n$

$$\operatorname{tr} T = \sum_{i=1}^{n} \langle T\mathbf{x}_i, \mathbf{x}_i \rangle.$$

Furthermore, if $\mathbb{F} = \mathbb{C}$ then $\operatorname{tr} T$ is the sum of the n eigenvalues of T.

(b) Let $S, T \in \mathbf{S}(\mathbf{V})$. Show that $\operatorname{tr} ST = \operatorname{tr} TS \in \mathbb{R}$.

2. (a) Let $S, T \in \mathbf{S}(\mathbf{V})$. Show that $\operatorname{tr} ST \in \mathbb{R}$. Furthermore, $\langle S, T \rangle := \operatorname{tr} ST$ is an inner product on $\mathbf{S}(\mathbf{V})$ over $\mathbb{R}$.

(b) Let $\phi : \mathbf{S}(\mathbf{V}) \to \mathbb{R}$ be a linear functional. Show that there exists $F \in \mathbf{S}(\mathbf{V})$ such that $\phi(X) = \operatorname{tr} XF$ for each $X \in \mathbf{S}(\mathbf{V})$.

3. Assume that $S, T \in \mathbf{S}_+(\mathbf{V})$.

(a) Show that $\operatorname{tr} ST \geq 0$. Furthermore $\operatorname{tr} ST = 0$ if and only if $ST = TS = 0$.

(b) Suppose furthermore that $T \in \mathbf{S}_+(\mathbf{V})^\circ$. Then $\operatorname{tr} ST = 0$ if and only if $S = 0$.

(c) Show that $P \in \mathbf{S}(\mathbf{V})$ is non-negative definite if and only if $\operatorname{tr} PS \geq 0$ for each $S \in \mathbf{S}_+(\mathbf{V})$. Furthermore P is positive definite if $\operatorname{tr} PS > 0$ for each $S \in \mathbf{S}_+(\mathbf{V}) \setminus \{0\}$.

4.10 Singular Value Decomposition (SVD)

Let $\mathbf{U}, \mathbf{V}$ be finite dimensional IPS over $\mathbb{F} = \mathbb{R}, \mathbb{C}$, with the inner products $\langle \cdot, \cdot \rangle_{\mathbf{U}}, \langle \cdot, \cdot \rangle_{\mathbf{V}}$ respectively. Let $\mathbf{u}_1, \ldots, \mathbf{u}_m$ and $\mathbf{v}_1, \ldots, \mathbf{v}_n$ be bases in $\mathbf{U}$ and $\mathbf{V}$ respectively. Let $T : \mathbf{V} \to \mathbf{U}$ be a linear operator. In these bases T is represented by a matrix $A \in \mathbb{F}^{m \times n}$ as given by (1.11.2). Let $T^* : \mathbf{U}^* = \mathbf{U} \to \mathbf{V}^* = \mathbf{V}$. Then $T^*T : \mathbf{V} \to \mathbf{V}$

and $TT^* : \mathbf{U} \to \mathbf{U}$ are self-adjoint operators. As

$$\langle T^*T\mathbf{v}, \mathbf{v}\rangle_{\mathbf{V}} = \langle T\mathbf{v}, T\mathbf{v}\rangle_{\mathbf{V}} \geq 0, \quad \langle TT^*\mathbf{u}, \mathbf{u}\rangle_{\mathbf{U}} = \langle T^*\mathbf{u}, T^*\mathbf{u}\rangle_{\mathbf{U}} \geq 0$$

it follows that $T^*T \geq 0, TT^* \geq 0$. Let

$$T^*T\mathbf{c}_i = \lambda_i(T^*T)\mathbf{c}_i, \quad \langle \mathbf{c}_i, \mathbf{c}_k\rangle_{\mathbf{V}} = \delta_{ik}, \quad i, k = 1, \dots, n,$$

$$\lambda_1(T^*T) \geq \cdots \geq \lambda_n(T^*T) \geq 0, \tag{4.10.1}$$

$$TT^*\mathbf{d}_j = \lambda_j(TT^*)\mathbf{d}_j, \quad \langle \mathbf{d}_j, \mathbf{d}_l\rangle_{\mathbf{U}} = \delta_{jl}, \; j, l = 1, \dots, m,$$

$$\lambda_1(TT^*) \geq \cdots \geq \lambda_m(TT^*) \geq 0. \tag{4.10.2}$$

Proposition 4.10.1 *Let* $\mathbf{U}, \mathbf{V}$ *be finite dimensional IPS over* $\mathbb{F} = \mathbb{R}, \mathbb{C}$. *Let* $T : \mathbf{V} \to \mathbf{U}$. *Then* rank T = rank T* = rank T*T = rank TT* = r. *Furthermore the self-adjoint non-negative definite operators* T^*T *and* TT^* *have exactly* r *positive eigenvalues, and*

$$\lambda_i(T^*T) = \lambda_i(TT^*) > 0, \quad i = 1, \dots, \text{rank T}. \tag{4.10.3}$$

Moreover, for $i \in [1, r]$ $T\mathbf{c}_i$ *and* $T^*\mathbf{d}_i$ *are eigenvectors of* TT^* *and* T^*T *corresponding to the eigenvalue* $\lambda_i(TT^*) = \lambda_i(T^*T)$ *respectively. Furthermore if* $\mathbf{c}_1, \dots, \mathbf{c}_r$ *satisfy* (4.10.1) *then* $\tilde{\mathbf{d}}_i := \frac{T\mathbf{c}_i}{\|T\mathbf{c}_i\|}, i = 1, \dots, r$ *satisfy* (4.10.2) *for* $i = 1, \dots, r$. *A similar result holds for* $\mathbf{d}_1, \dots, \mathbf{d}_r$.

Proof. Clearly $T\mathbf{x} = \mathbf{0} \iff \langle T\mathbf{x}, T\mathbf{x}\rangle = 0 \iff T^*T\mathbf{x} = \mathbf{0}$. Hence

$$\text{rank T}^*\text{T} = \text{rank T} = \text{rank T}^* = \text{rank TT}^* = \text{r}.$$

Thus T^*T and TT^* have exactly r positive eigenvalues. Let $i \in [r]$. Then $T^*T\mathbf{c}_i \neq \mathbf{0}$. Hence, $T\mathbf{c}_i \neq \mathbf{0}$. (4.10.1) yields that $TT^*(T\mathbf{c}_i) = \lambda_i(T^*T)(T\mathbf{c}_i)$. Similarly $T^*T(T^*\mathbf{d}_i) = \lambda_i(TT^*)(T^*\mathbf{d}_i) \neq 0$. Hence, (4.10.3) holds. Assume that $\mathbf{c}_1, \dots, \mathbf{c}_r$ satisfy (4.10.1). Let $\tilde{\mathbf{d}}_1, \dots, \tilde{\mathbf{d}}_r$ be defined as above. By the definition $\|\tilde{\mathbf{d}}_i\| = 1, i = 1, \dots, r$. Let

$1 \leq i < j \leq r$. Then

$$0 = \langle \mathbf{c}_i, \mathbf{c}_j \rangle = \lambda_i(T^*T)\langle \mathbf{c}_i, \mathbf{c}_j \rangle$$
$$= \langle T^*T\mathbf{c}_i, \mathbf{c}_j \rangle = \langle T\mathbf{c}_i, T\mathbf{c}_j \rangle \Rightarrow \langle \tilde{\mathbf{d}}_i, \tilde{\mathbf{d}}_j \rangle = 0.$$

Hence, $\tilde{\mathbf{d}}_1, \ldots, \tilde{\mathbf{d}}_r$ is an orthonormal system. $\qquad\square$

Let

$$\sigma_i(T) = \sqrt{\lambda_i(T^*T)} \text{ for } i = 1, \ldots, r, \quad \sigma_i(T) = 0 \text{ for } i > r,$$

$$\tag{4.10.4}$$

$$\boldsymbol{\sigma}_{(p)}(T) := (\sigma_1(T), \ldots, \sigma_p(T))^\top \in \mathbb{R}^p_{\searrow}, \quad p \in \mathbb{N}.$$

Then $\sigma_i(T) = \sigma_i(T^*), i = 1, \ldots, \min(m, n)$ are called the singular values of T and T^* respectively. Note that the singular values are arranged in a decreasing order. The positive singular values are called principal singular values of T and T^* respectively. Note that

$$\|T\mathbf{c}_i\|^2 = \langle T\mathbf{c}_i, T\mathbf{c}_i \rangle = \langle T^*T\mathbf{c}_i, \mathbf{c}_i \rangle = \lambda_i(T^*T) = \sigma_i^2 \Rightarrow$$

$$\|T\mathbf{c}_i\| = \sigma_i, \ i = 1, \ldots, n, \quad \text{and}$$

$$\|T^*\mathbf{d}_j\|^2 = \langle T^*\mathbf{d}_j, T^*\mathbf{d}_j \rangle = \langle TT^*\mathbf{d}_j, \mathbf{d}_i \rangle = \lambda_i(TT^*) = \sigma_j^2 \Rightarrow$$

$$\|T\mathbf{d}_j\| = \sigma_j, \ j = 1, \ldots, m.$$

Let $\mathbf{c}_1, \ldots, \mathbf{c}_n$, be an orthonormal basis of $\mathbf{V}$ satisfying (4.10.1). Choose an orthonormal basis $\mathbf{d}_1, \ldots, \mathbf{d}_m$ as follows. Set $\mathbf{d}_i := \frac{T\mathbf{c}_i}{\sigma_i}, i = 1, \ldots, r$. Then complete the orthonormal set $\{\mathbf{d}_1, \ldots, \mathbf{d}_r\}$ to an orthonormal basis of $\mathbf{U}$. Since span $(\mathbf{d}_1, \ldots, \mathbf{d}_r)$ is spanned by all eigenvectors of TT^* corresponding to nonzero eigenvalues of TT^* it follows that ker $T^* = $ span $(\mathbf{d}_{r+1}, \ldots, \mathbf{d}_m)$. Hence, (4.10.2) holds. In these orthonormal bases of $\mathbf{U}$ and $\mathbf{V}$ the operators T and T^* are represented quite simply:

$$T\mathbf{c}_i = \sigma_i(T)\mathbf{d}_i, \ i = 1, \ldots, n, \quad \text{where } \mathbf{d}_i = \mathbf{0} \quad \text{for } i > m,$$

$$\tag{4.10.5}$$

$$T^*\mathbf{d}_j = \sigma_j(T)\mathbf{c}_j, \ j = 1, \ldots, m, \quad \text{where } \mathbf{c}_j = \mathbf{0} \quad \text{for } j > n.$$

Let

$$\Sigma = [s_{ij}]_{i,j=1}^{m,n}, \quad s_{ij} = 0 \quad \text{for } i \neq j, \quad s_{ii} = \sigma_i \quad \text{for}$$

$$i = 1, \ldots, \min(m, n). \tag{4.10.6}$$

In the case $m \neq n$ we call Σ a diagonal matrix with the diagonal $\sigma_1, \ldots, \sigma_{\min(m,n)}$. Then in the bases $[\mathbf{d}_1, \ldots, \mathbf{d}_m]$ and $[\mathbf{c}_1, \ldots, \mathbf{c}_n]$ T and T^* represented by the matrices Σ and $\Sigma^\top$ respectively.

Lemma 4.10.2 *Let* $[\mathbf{u}_1, \ldots, \mathbf{u}_m], [\mathbf{v}_1, \ldots, \mathbf{v}_n]$ *be orthonormal bases in the vector spaces* $\mathbf{U}, \mathbf{V}$ *over* $\mathbb{F} = \mathbb{R}, \mathbb{C}$ *respectively. Then* T *and* T^* *are represented by the matrices* $A \in \mathbb{F}^{m \times n}$ *and* $A^* \in \mathbb{F}^{n \times m}$ *respectively. Let* $U \in \mathbf{U}_m$ *and* $V \in \mathbf{U}_n$ *be the unitary matrices representing the change of bases* $[\mathbf{d}_1, \ldots, \mathbf{d}_m]$ *to* $[\mathbf{u}_1, \ldots, \mathbf{u}_m]$ *and* $[\mathbf{c}_1, \ldots, \mathbf{c}_n]$ *to* $[\mathbf{v}_1, \ldots, \mathbf{v}_n]$ *respectively. (If* $\mathbb{F} = \mathbb{R}$ *then* U *and* V *are orthogonal matrices.) Then*

$$A = U\Sigma V^* \in \mathbb{F}^{m \times n}, \quad U \in \mathbf{U}_m, \quad V \in \mathbf{U}_n. \tag{4.10.7}$$

Proof. By definition $T\mathbf{v}_j = \sum_{i=1}^{m} a_{ij}\mathbf{u}_i$. Let $U = (u_{ip})_{i,p=1}^{m}, V = (v_{jq})_{j,q=1}^{n}$. Then

$$T\mathbf{c}_q = \sum_{j=1}^{n} v_{jq} T\mathbf{v}_j = \sum_{j=1}^{n} v_{jq} \sum_{i=1}^{m} a_{ij}\mathbf{u}_i = \sum_{j=1}^{n} v_{jq} \sum_{i=1}^{m} a_{ij} \sum_{p=1}^{m} \bar{u}_{ip}\mathbf{d}_p.$$

Use the first equality of (4.10.5) to deduce that $U^*AV = \Sigma$. $\qquad\square$

Definition 4.10.3 (4.10.7) *is called SVD of* A.

Proposition 4.10.4 *Let* $\mathbb{F} = \mathbb{R}, \mathbb{C}$ *and denote by* $\mathcal{R}_{m,n,k}(\mathbb{F}) \subset \mathbb{F}^{m \times n}$ *the set of all matrices of rank at most* $k \in [\min(m, n)]$. *Then* $A \in \mathcal{R}_{m,n,k}(\mathbb{F})$ *if and only if* A *can be expressed as a sum of at most* k *matrices of rank 1. Furthermore* $\mathcal{R}_{m,n,k}(\mathbb{F})$ *is a variety in* $\mathbb{F}^{m \times n}$ *given by the polynomial condition that each* $(k + 1) \times (k + 1)$ *minor of* A *is equal to zero.*

For the proof see Problem 2.

Definition 4.10.5 *Let $A \in \mathbb{C}^{m \times n}$ and assume that A has the SVD given by (4.10.7), where $U = [\mathbf{u}_1, \ldots, \mathbf{u}_m], V = [\mathbf{v}_1, \ldots, \mathbf{v}_n]$. Denote by $A_k := \sum_{i=1}^{k} \sigma_i \mathbf{u}_i \mathbf{v}_i^* \in \mathbb{C}^{m \times n}$ for $k = 1, \ldots, \operatorname{rank} A$. For $k > \operatorname{rank} A$ we define $A_k := A \ (= A_{\operatorname{rank} A})$.*

Note that for $1 \le k < \operatorname{rank} A$, the matrix A_k is uniquely defined if and only if $\sigma_k > \sigma_{k+1}$. (See Problem 1.)

Theorem 4.10.6 *For $\mathbb{F} = \mathbb{R}, \mathbb{C}$ and $A = [a_{ij}] \in \mathbb{F}^{m \times n}$ the following conditions hold:*

$$\|A\|_F := \sqrt{\operatorname{tr} A^* A} = \sqrt{\operatorname{tr} AA^*} = \sqrt{\sum_{i=1}^{\operatorname{rank} A} \sigma_i(A)^2}. \qquad (4.10.8)$$

$$\|A\|_2 := \max_{\mathbf{x} \in \mathbb{F}^n, \|\mathbf{x}\|_2 = 1} \|A\mathbf{x}\|_2 = \sigma_1(A). \qquad (4.10.9)$$

$$\min_{B \in \mathcal{R}_{m,n,k}(\mathbb{F})} \|A - B\|_2 = \|A - A_k\| = \sigma_{k+1}(A), \quad k = 1, \ldots, \operatorname{rank} A - 1. \qquad (4.10.10)$$

$$\sigma_i(A) \ge \sigma_i([a_{i_p j_q}]_{p=1, q=1}^{m', n'}) \ge \sigma_{i + (m - m') + (n - n')}(A),$$

$$(4.10.11)$$

$$m' \in [m], \ n' \in [n], \ 1 \le i_1 < \cdots < i_{m'} \le m, \ 1 \le j_1 < \cdots < j_{n'} \le n.$$

Proof. The proof of (4.10.8) is left as Problem 7. We now show the equality in (4.10.9). View A as an operator $A : \mathbb{C}^n \to \mathbb{C}^m$. From the definition of $\|A\|_2$ it follows

$$\|A\|_2^2 = \max_{0 \ne \mathbf{x} \in \mathbb{R}^n} \frac{\mathbf{x}^* A^* A \mathbf{x}}{\mathbf{x}^* \mathbf{x}} = \lambda_1(A^* A) = \sigma_1(A)^2,$$

which proves (4.10.9).

We now prove (4.10.10). In the SVD decomposition of A (4.10.7) assume that $U = [\mathbf{u}_1 \ldots \mathbf{u}_m]$ and $V = [\mathbf{v}_1 \ldots \mathbf{v}_n]$. Then (4.10.7) is equivalent to the following representation of A:

$$A = \sum_{i=1}^{r} \sigma_i \mathbf{u}_i \mathbf{v}_i^*, \quad \mathbf{u}_1, \ldots, \mathbf{u}_r \in \mathbb{R}^m, \quad \mathbf{v}_1, \ldots, \mathbf{v}_r \in \mathbb{R}^n,$$

$$\mathbf{u}_i^* \mathbf{u}_j = \mathbf{v}_i^* \mathbf{v}_j = \delta_{ij}, \quad i, j = 1, \ldots, r, \tag{4.10.12}$$

where $r = \text{rank } A$. Let $B = \sum_{i=1}^{k} \sigma_i \mathbf{u}_i \mathbf{v}_i^* \in \mathcal{R}_{m,n,k}(\mathbb{F})$. Then in view of (4.10.9)

$$\|A - B\|_2 = \|\sum_{k+1}^{r} \sigma_i \mathbf{u}_i \mathbf{v}_i^*\|_2 = \sigma_{k+1}.$$

Let $B \in \mathcal{R}_{m,n,k}(\mathbb{F})$. To show (4.10.10) it is enough to show that $\|A - B\|_2 \geq \sigma_{k+1}$. Let

$$\mathbf{W} := \{\mathbf{x} \in \mathbb{R}^n : \quad B\mathbf{x} = \mathbf{0}\}.$$

Then codim $\mathbf{W} \geq k$. Furthermore

$$\|A - B\|_2^2 \geq \max_{\|\mathbf{x}\|_2=1, \mathbf{x} \in \mathbf{W}} \|(A - B)\mathbf{x}\|^2$$

$$= \max_{\|\mathbf{x}\|_2=1, \mathbf{x} \in \mathbf{W}} \mathbf{x}^* A^* A \mathbf{x} \geq \lambda_{k+1}(A^* A) = \sigma_{k+1}^2,$$

where the last inequality follows from the min–max characterization of $\lambda_{k+1}(A^* A)$.

Let $C = [a_{i j_q}]_{i,q=1}^{m,n'}$. Then $C^* C$ is a principal submatrix of $A^* A$ of dimension n'. The interlacing inequalities between the eigenvalues of $A^* A$ and $C^* C$ yields (4.10.11) for $m' = m$. Let $D = [a_{i_p j_q}]_{p,q=1}^{m',n'}$. Then DD^* is a principle submatrix of CC^*. Use the interlacing properties of the eigenvalues of CC^* and DD^* to deduce (4.10.11). $\square$

We now restate the above results for linear operators.

Definition 4.10.7 *Let* $\mathbf{U}, \mathbf{V}$ *be finite dimensional vector spaces over* $\mathbb{F} = \mathbb{R}, \mathbb{C}$. *For* $k \in \mathbb{Z}_+$ *denote* $L_k(\mathbf{V}, \mathbf{U}) := \{T \in L(\mathbf{V}, \mathbf{U}) : \text{rank } T \leq k\}$. *Assume furthermore that* $\mathbf{U}, \mathbf{V}$ *are IPS. Let* $T \in L(\mathbf{V}, \mathbf{U})$ *and assume that the orthonormal bases of*

$[\mathbf{d}_1, \ldots, \mathbf{d}_m], [\mathbf{c}_1, \ldots, \mathbf{c}_n]$ *of* $\mathbf{U}, \mathbf{V}$ *respectively satisfy* (4.10.5). *Define* $T_0 := \mathbf{0}$ *and* $T_k := T$ *for an integer* $k \geq \mathrm{rank}\, T$. *Let* $k \in [\mathrm{rank}\, T - 1]$. *Define* $T_k \in \mathrm{L}(\mathbf{V}, \mathbf{U})$ *by the equality* $T_k(\mathbf{v}) = \sum_{i=1}^{k} \sigma_i(T)\langle \mathbf{v}, \mathbf{c}_i \rangle \mathbf{d}_i$ *for any* $\mathbf{v} \in \mathbf{V}$.

It is straightforward to show that $T_k \in \mathrm{L}_k(\mathbf{V}, \mathbf{U})$ and T_k is unique if and only if $\sigma_k(T) > \sigma_{k+1}(T)$. See Problem 8. Theorem 4.10.6 yields:

Corollary 4.10.8 *Let* $\mathbf{U}$ *and* $\mathbf{V}$ *be finite dimensional IPS over* $\mathbb{F} = \mathbb{R}, \mathbb{C}$. *Let* $T : \mathbf{V} \to \mathbf{U}$ *be a linear operator. Then*

$$||T||_F := \sqrt{\mathrm{tr}\, T^*T} = \sqrt{\mathrm{tr}\, TT^*} = \sqrt{\sum_{i=1}^{\mathrm{rank}\, T} \sigma_i(T)^2}. \quad (4.10.13)$$

$$||T||_2 := \max_{\mathbf{x} \in \mathbf{V}, ||\mathbf{x}||_2 = 1} ||T\mathbf{x}||_2 = \sigma_1(T). \quad (4.10.14)$$

$$\min_{Q \in \mathrm{L}_k(\mathbf{V}, \mathbf{U})} ||T - Q||_2 = \sigma_{k+1}(T),$$

$$k = 1, \ldots, \mathrm{rank}\, T - 1. \quad (4.10.15)$$

Problems

1. Let $\mathbf{U}, \mathbf{V}$ be finite dimensional inner product spaces. Assume that $T \in \mathrm{L}(\mathbf{U}, \mathbf{V})$. Show that for any complex number $t \in \mathbb{C}$, $\sigma_i(tT) = |t|\sigma_i(T)$ for all i.

2. Prove Proposition 4.10.4. (Use SVD to prove the nontrivial part of the proposition.)

3. Let $A \in \mathbb{C}^{m \times n}$ and assume that $U \in \mathbf{U}_m, V \in \mathbf{U}_n$. Show that $\sigma_i(UAV) = \sigma_i(A)$ for all i.

4. Let $A \in \mathbf{GL}(n, \mathbb{C})$. Show that $\sigma_1(A^{-1}) = \sigma_n(A)^{-1}$.

5. Let $\mathbf{U}, \mathbf{V}$ be inner product spaces of dimensions m and n respectively. Assume that

$$\mathbf{U} = \mathbf{U}_1 \oplus \mathbf{U}_2, \quad \dim \mathbf{U}_1 = m_1, \quad \dim \mathbf{U}_2 = m_2,$$

$$\mathbf{V} = \mathbf{V}_1 \oplus \mathbf{V}_2, \quad \dim \mathbf{V}_1 = n_1, \quad \dim \mathbf{V}_2 = n_2.$$

Assume that $T \in \mathrm{L}(\mathbf{V}, \mathbf{U})$. Suppose furthermore that $T\mathbf{V}_1 \subseteq \mathbf{U}_1$, $T\mathbf{V}_2 \subseteq \mathbf{U}_2$. Let $T_i \in \mathrm{L}(\mathbf{V}_i, \mathbf{U}_i)$ be the restriction of T to $\mathbf{V}_i$ for $i = 1, 2$. Then $\mathrm{rank}\, T = \mathrm{rank}\, T_1 + \mathrm{rank}\, T_2$ and $\{\sigma_1(T), \ldots, \sigma_{\mathrm{rank}\, T}(T)\} = \{\sigma_1(T_1), \ldots, \sigma_{\mathrm{rank}\, T_1}(T_1)\} \cup \{\sigma_1(T_2), \ldots, \sigma_{\mathrm{rank}\, T_2}(T_2)\}$.

6. Let the assumptions of the Definition 4.10.5 hold. Show that for $1 \le k < \mathrm{rank}\, A$, A_k is uniquely defined if and only if $\sigma_k > \sigma_{k+1}$.

7. Prove the equalities in (4.10.8).

8. Let the assumptions of Definition 4.10.7 hold. Show that for $k \in [\mathrm{rank}\, T - 1]$ $\mathrm{rank}\, T_k = k$ and T_k is unique if and only if $\sigma_k(T) > \sigma_{k+1}(T)$.

9. Let $\mathbf{V}$ be an n-dimensional IPS. Assume that $T \in \mathrm{L}(\mathbf{V})$ is a normal operator. Let $\lambda_1(T), \ldots, \lambda_n(T)$ be the eigenvalues of T arranged in the order $|\lambda_1(T)| \ge \cdots \ge |\lambda_n(T)|$. Show that $\sigma_i(T) = |\lambda_i(T)|$ for $i = 1, \ldots, n$.

4.11 Characterizations of Singular Values

Theorem 4.11.1 *Let* $\mathbb{F} = \mathbb{R}, \mathbb{C}$ *and assume that* $A \in \mathbb{F}^{m \times n}$. *Define*

$$H(A) = \begin{bmatrix} 0 & A \\ A^* & 0 \end{bmatrix} \in \mathrm{H}_{m+n}. \tag{4.11.1}$$

Then

$$\lambda_i(H(A)) = \sigma_i(A), \ \lambda_{m+n+1-i}(H(A)) = -\sigma_i(A), \ i = 1, \dots, \text{rank } A,$$

$$(4.11.2)$$

$$\lambda_j(H(A)) = 0, \ j = \text{rank } A + 1, \dots, n + m - \text{rank } A.$$

View A as an operator $A : \mathbb{F}^n \to \mathbb{F}^m$. Choose orthonormal bases $[\mathbf{d}_1, \dots, \mathbf{d}_m], \ [\mathbf{c}_1, \dots, \mathbf{c}_n]$ in $\mathbb{F}^m, \mathbb{F}^n$ respectively satisfying (4.10.5). Then

$$\begin{bmatrix} 0 & A \\ A^* & 0 \end{bmatrix} \begin{bmatrix} \mathbf{d}_i \\ \mathbf{c}_i \end{bmatrix} = \sigma_i(A) \begin{bmatrix} \mathbf{d}_i \\ \mathbf{c}_i \end{bmatrix}, \quad \begin{bmatrix} 0 & A \\ A^* & 0 \end{bmatrix} \begin{bmatrix} \mathbf{d}_i \\ -\mathbf{c}_i \end{bmatrix} = -\sigma_i(A) \begin{bmatrix} \mathbf{d}_i \\ -\mathbf{c}_i \end{bmatrix},$$

$$i = 1, \dots, r = \text{rank } A, \qquad (4.11.3)$$

$$\text{Ker } H(A) = \text{span} \left(\begin{bmatrix} \mathbf{d}_{r+1} \\ 0 \end{bmatrix}, \dots, \begin{bmatrix} \mathbf{d}_m \\ 0 \end{bmatrix}, \begin{bmatrix} 0 \\ \mathbf{c}_{r+1} \end{bmatrix}, \dots, \begin{bmatrix} 0 \\ \mathbf{c}_m \end{bmatrix} \right).$$

Proof. It is straightforward to show the equalities (4.11.3). Since all the eigenvectors appearing in (4.11.3) are linearly independent we deduce (4.11.2). $\qquad \square$

Corollary 4.11.2 *Let $\mathbb{F} = \mathbb{R}, \mathbb{C}$ and assume that $A \in \mathbb{F}^{m \times n}$. Let $\hat{A} := A[\alpha, \beta] \in \mathbb{F}^{p \times q}$ be a submatrix of A, formed by the set of rows and columns $\alpha \in Q_{p,m}, \beta \in Q_{q,n}$ respectively. Then*

$$\sigma_i(\hat{A}) \le \sigma_i(A) \quad \text{for } i = 1, \dots. \qquad (4.11.4)$$

For $l \in [\text{rank } A]$ the equalities $\sigma_i(\hat{A}) = \sigma_i(A), i = 1, \dots, l$ hold if and only if there exist two orthonormal systems of l right and left singular vectors $\mathbf{c}_1, \dots, \mathbf{c}_l \in \mathbb{F}^n, \ \mathbf{d}_1, \dots, \mathbf{d}_l \in \mathbb{F}^n$, satisfying (4.11.3) for $i = 1, \dots, l$ such that the nonzero coordinate vectors $\mathbf{c}_1, \dots, \mathbf{c}_l$ and $\mathbf{d}_1, \dots, \mathbf{d}_l$ are located at the indices β, α respectively.

See Problem 1.

Corollary 4.11.3 *Let $\mathbf{V}, \mathbf{U}$ be inner product spaces over $\mathbb{F} = \mathbb{R}, \mathbb{C}$. Assume that $\mathbf{W}$ is a subspace of $\mathbf{V}$. Let $T \in \text{L}(\mathbf{V}, \mathbf{U})$ and* *denote by* $\hat{T} \in \mathrm{L}(\mathbf{W}, \mathbf{U})$ *the restriction of* T *to* $\mathbf{W}$. *Then* $\sigma_i(\hat{T}) \leq \sigma_i(T)$ *for any* $i \in \mathbb{N}$. *Furthermore* $\sigma_i(\hat{T}) = \sigma_i(T)$ *for* $i = 1, \ldots, l \leq$ *rank* T *if and only if* $\mathbf{U}$ *contains a subspace spanned by the first* l *right singular vectors of* T.

See Problem 2.

Define by $\mathbb{R}^n_{+,\searcalow} := \mathbb{R}^n_\searcalow \cap \mathbb{R}^n_+$. Then $\mathrm{D} \subset \mathbb{R}^n_{+,\searcalow}$ is called a strong Schur set if for any $\mathbf{x}, \mathbf{y} \in \mathbb{R}^n_{+,\searcalow}, \mathbf{x} \preceq \mathbf{y}$ we have the implication $\mathbf{y} \in D \Rightarrow \mathbf{x} \in D$.

Theorem 4.11.4 *Let* $p \in \mathbb{N}$ *and* $\mathrm{D} \subset \mathbb{R}^p_{+,\searcalow}$ *be a regular convex strong Schur domain. Fix* $m, n \in \mathbb{N}$ *and let* $\boldsymbol{\sigma}_{(p)}(\mathrm{D}) := \{A \in \mathbb{F}^{m \times n} : \boldsymbol{\sigma}_{(p)}(A) \in \mathrm{D}\}$. *Let* $h : \mathrm{D} \to \mathbb{R}$ *be a convex and strongly Schur's order preserving function on* D. *Let* $f : \boldsymbol{\sigma}_{(p)} :\to \mathbb{R}$ *be given as* $h \circ \boldsymbol{\sigma}_{(p)}$. *Then* f *is a convex function.*

See Problem 3.

Corollary 4.11.5 *Let* $\mathbb{F} = \mathbb{R}, \mathbb{C}$, $m, n, p \in \mathbb{N}$, $q \in [1, \infty)$ *and* $w_1 \geq w_2 \geq \cdots \geq w_p > 0$. *Then the following function*

$$f : \mathbb{F}^{m \times n} \to \mathbb{R}, \quad where\ f(A) := \left(\sum_{i=1}^p w_i \sigma_i(A)^q \right)^{\frac{1}{q}}, \ A \in \mathbb{F}^{m \times n}$$

is a convex function.

See Problem 4.

We now translate Theorem 4.11.1 to the operator setting.

Lemma 4.11.6 *Let* $\mathbf{U}, \mathbf{V}$ *be finite dimensional inner product spaces with the inner products* $\langle \cdot, \cdot \rangle_\mathbf{U}, \langle \cdot, \cdot \rangle_\mathbf{V}$ *respectively. Define* $\mathbf{W} := \mathbf{V} \oplus \mathbf{U}$ *to be the induced IPS with*

$$\langle (\mathbf{y}, \mathbf{x}), (\mathbf{v}, \mathbf{u}) \rangle_\mathbf{W} := \langle \mathbf{y}, \mathbf{v} \rangle_\mathbf{V} + \langle \mathbf{x}, \mathbf{u} \rangle_\mathbf{U}.$$

Let $T : \mathbf{V} \to \mathbf{U}$ be a linear operator, and $T^* : \mathbf{U} \to \mathbf{V}$ be the adjoint of T. Define the operator

$$\hat{T} : \mathbf{W} \to \mathbf{W}, \quad \hat{T}(\mathbf{y}, \mathbf{x}) := (T^*\mathbf{x}, T\mathbf{y}). \qquad (4.11.5)$$

Then $\hat{T}$ is self-adjoint operator and $\hat{T}^2 = T^*T \oplus TT^*$. Hence, the spectrum of $\hat{T}$ is symmetric with respect to the origin and $\hat{T}$ has exactly 2 rank T nonzero eigenvalues. More precisely, if $\dim \mathbf{U} = m, \dim \mathbf{V} = n$ then:

$$\lambda_i(\hat{T}) = -\lambda_{m+n-i+1}(\hat{T}) = \sigma_i(T), \quad \text{for } i = 1, \ldots, \text{rank } T,$$

$$(4.11.6)$$

$$\lambda_j(\hat{T}) = 0, \quad \text{for } j = \text{rank } T + 1, \ldots, n + m - \text{rank } T.$$

Let $\{\mathbf{d}_1, \ldots, \mathbf{d}_{\min(m,n)}\} \in \mathrm{Fr}(\min(m,n), \mathbf{U})$, $\{\mathbf{c}_1, \ldots, \mathbf{c}_{\min(m,n)}\} \in \mathrm{Fr}(\min(m,n), \mathbf{V})$ be the set of vectors satisfying (4.10.5). Define

$$\mathbf{z}_i := \frac{1}{\sqrt{2}}(\mathbf{c}_i, \mathbf{d}_i), \quad \mathbf{z}_{m+n-i+1} := \frac{1}{\sqrt{2}}(\mathbf{c}_i, -\mathbf{d}_i), \quad (4.11.7)$$

$$i = 1, \ldots, \min(m, n).$$

Then $\{\mathbf{z}_1, \mathbf{z}_{m+n}, \ldots, \mathbf{z}_{\min(m,n)}, \mathbf{z}_{m+n-\min(m,n)+1}\} \in \mathrm{Fr}(2\min(m,n), \mathbf{W})$. Furthermore $\hat{T}\mathbf{z}_i = \sigma_i(T)\mathbf{z}_i, \hat{T}\mathbf{z}_{m+n-i+1} = -\sigma_i(T)\mathbf{z}_{m+n-i+1}$ for $i = 1, \ldots, \min(m, n)$.

See Problem 5.

Theorem 4.11.7 *Let $\mathbf{U}, \mathbf{V}$ be m and n-dimensional inner product spaces over $\mathbb{C}$ respectively. Let $T : \mathbf{V} \to \mathbf{U}$ be a linear operator. Then for each $k \in [\min(m, n)]$*

$$\sum_{i=1}^{k} \sigma_i(T) = \max_{\{\mathbf{f}_1, \ldots, \mathbf{f}_k\} \in \mathrm{Fr}(k, \mathbf{U}), \{\mathbf{g}_1, \ldots, \mathbf{g}_k\} \in \mathrm{Fr}(k, \mathbf{V})} \sum_{i=1}^{k} \Re \langle T\mathbf{g}_i, \mathbf{f}_i \rangle_{\mathbf{U}}$$

$$= \max_{\{\mathbf{f}_1, \ldots, \mathbf{f}_k\} \in \mathrm{Fr}(k, \mathbf{U}), \{\mathbf{g}_1, \ldots, \mathbf{g}_k\} \in \mathrm{Fr}(k, \mathbf{V})} \sum_{i=1}^{k} |\langle T\mathbf{g}_i, \mathbf{f}_i \rangle_{\mathbf{U}}|.$$

$$(4.11.8)$$

' *Furthermore* $\sum_{i=1}^{k} \sigma_i(T) = \sum_{i=1}^{k} \Re\langle T\mathbf{g}_i, \mathbf{f}_i\rangle_\mathbf{U}$ *for some two* k-*orthonormal frames* $F_k = \{\mathbf{f}_1, \ldots, \mathbf{f}_k\}, G_k = \{\mathbf{g}_1, \ldots, \mathbf{g}_k\}$ *if and only* span $((\mathbf{g}_1, \mathbf{f}_1), \ldots, (\mathbf{g}_k, \mathbf{f}_k))$ *is spanned by* k *eigenvectors of* $\hat{T}$ *corresponding to the first* k *eigenvalues of* $\hat{T}$.

Proof. Assume that $\{\mathbf{f}_1, \ldots, \mathbf{f}_k\} \in \mathrm{Fr}(k, \mathbf{U}), \{\mathbf{g}_1, \ldots, \mathbf{g}_k\} \in \mathrm{Fr}(k, \mathbf{V})$. Let $\mathbf{w}_i := \frac{1}{\sqrt{2}}(\mathbf{g}_i, \mathbf{f}_i), i = 1, \ldots, k$. Then $\{\mathbf{w}_1, \ldots, \mathbf{w}_k\} \in \mathrm{Fr}(k, \mathbf{W})$. A straightforward calculation shows $\sum_{i=1}^{k}\langle \hat{T}\mathbf{w}_i, \mathbf{w}_i\rangle_\mathbf{W} = \sum_{i=1}^{k} \Re\langle T\mathbf{g}_i, \mathbf{f}_i\rangle_\mathbf{U}$. The maximal characterization of $\sum_{i=1}^{k} \lambda_i(\hat{T})$, (Theorem 4.4.8), and (4.11.6) yield the inequality $\sum_{i=1}^{k} \sigma_i(\hat{T}) \geq \sum_{i=1}^{k} \Re\langle T\mathbf{g}_i, \mathbf{f}_i\rangle_\mathbf{U}$ for $k \in [\min(m,n)]$. Let $\mathbf{c}_1, \ldots, \mathbf{c}_{\min(m,n)}$, $\mathbf{d}_1, \ldots, \mathbf{d}_{\min(m,n)}$ satisfy (4.10.5). Then Lemma 4.11.6 yields that $\sum_{i=1}^{k} \sigma_i(\hat{T}) = \sum_{i=1}^{k} \Re\langle T\mathbf{c}_i, \mathbf{d}_i\rangle_\mathbf{U}$ for $k \in [\min(m,n)]$. This proves the first equality of (4.11.8). The second equality of (4.11.8) is straightforward. (See Problem 6.)

Assume now that $\sum_{i=1}^{k} \sigma_i(T) = \sum_{i=1}^{k} \Re\langle T\mathbf{g}_i, \mathbf{f}_i\rangle_\mathbf{U}$ for some two k-orthonormal frames $F_k = \{\mathbf{f}_1, \ldots, \mathbf{f}_k\}, G_k = \{\mathbf{g}_1, \ldots, \mathbf{g}_k\}$. Define $\mathbf{w}_1, \ldots, \mathbf{w}_k$ as above. The above arguments yield that $\sum_{i=1}^{k}\langle \hat{T}\mathbf{w}_i, \mathbf{w}_i\rangle_\mathbf{W} = \sum_{i=1}^{k} \lambda_i(\hat{T})$. Theorem 4.4.8 yields that span $((\mathbf{g}_1, \mathbf{f}_1), \ldots, (\mathbf{g}_k, \mathbf{f}_k))$ is spanned by k eigenvectors of $\hat{T}$ corresponding to the first k eigenvalues of $\hat{T}$. Vice versa, assume that $\{\mathbf{f}_1, \ldots, \mathbf{f}_k\} \in \mathrm{Fr}(k, \mathbf{U}), \{\mathbf{g}_1, \ldots, \mathbf{g}_k\} \in \mathrm{Fr}(k, \mathbf{V})$ and span $((\mathbf{g}_1, \mathbf{f}_1), \ldots, (\mathbf{g}_k, \mathbf{f}_k))$ is spanned by k eigenvectors of $\hat{T}$ corresponding to the first k eigenvalues of $\hat{T}$. Define $\{\mathbf{w}_1, \ldots, \mathbf{w}_k\} \in \mathrm{Fr}(\mathbf{W})$ as above. Then span $(\mathbf{w}_1, \ldots, \mathbf{w}_k)$ contains k linearly independent eigenvectors corresponding to the the first k eigenvalues of $\hat{T}$. Theorem 4.4.8 and Lemma 4.11.6 yield that $\sigma_i(T) = \sum_{i=1}^{k}\langle \hat{T}\mathbf{w}_i, \mathbf{w}_i\rangle_\mathbf{W} = \sum_{i=1}^{k} \Re\langle T\mathbf{g}_i, \mathbf{f}_i\rangle_\mathbf{U}$. $\qquad\square$

Theorem 4.11.8 $\mathbf{U}, \mathbf{V}$ *be* m *and* n *dimensional inner product spaces. Let* $S, T : \mathbf{V} \to \mathbf{U}$ *be linear operators. Then*

$$\Re\,\mathrm{tr}(S^*T) \leq \sum_{i=1}^{\min(m,n)} \sigma_i(S)\sigma_i(T). \qquad (4.11.9)$$

Equality holds if and only if there exists two orthonormal sets $\{\mathbf{d}_1, \ldots, \mathbf{d}_{\min(m,n)}\} \in \mathrm{Fr}(\min(m,n), \mathbf{U})$, $\{\mathbf{c}_1, \ldots, \mathbf{c}_{\min(m,n)}\} \in \mathrm{Fr}(\min$

$(m, n), \mathbf{V})$, *such that*

$$S\mathbf{c}_i = \sigma_i(S)\mathbf{d}_i, \quad T\mathbf{c}_i = \sigma_i(T)\mathbf{d}_i, \qquad (4.11.10)$$

$$S^*\mathbf{d}_i = \sigma_i(S)\mathbf{c}_i, \quad T^*\mathbf{d}_i = \sigma_i(T)\mathbf{c}_i, \quad i = 1, \dots, \min(m, n).$$

Proof. Let $A, B \in \mathbb{C}^{n \times m}$. Then $\operatorname{tr} B^*A = \overline{\operatorname{tr} AB^*}$. Hence $2\Re \operatorname{tr} AB^* = \operatorname{tr} H(A)H(B)$. Therefore $2\Re \operatorname{tr} S^*T = \operatorname{tr} \hat{S}\hat{T}$. Use Theorem 4.9.1 for $\hat{S}, \hat{T}$ and Lemma 4.11.6 to deduce (4.11.9). Equality in (4.11.9) if and only if $\operatorname{tr} \hat{S}\hat{T} = \sum_{i=1}^{m+n} \lambda_i(\hat{S})\lambda_i(\hat{T})$.

Clearly, the assumptions that $\{\mathbf{d}_1, \dots, \mathbf{d}_{\min(m,n)}\} \in \operatorname{Fr}(\min(m, n), \mathbf{U})$, $\{\mathbf{c}_1, \dots, \mathbf{c}_{\min(m,n)}\} \in \operatorname{Fr}(\min(m, n), \mathbf{V})$, and the equalities (4.11.10) imply equality in (4.11.9).

Assume equality in (4.11.9). Theorem 4.9.1 and the definitions of $\hat{S}, \hat{T}$ yield the existence $\{\mathbf{d}_1, \dots, \mathbf{d}_{\min(m,n)}\} \in \operatorname{Fr}(\min(m, n), \mathbf{U})$, $\{\mathbf{c}_1, \dots, \mathbf{c}_{\min(m,n)}\} \in \operatorname{Fr}(\min(m, n), \mathbf{V})$, such that (4.11.10) hold. $\square$

Theorem 4.11.9 *Let* $\mathbf{U}$ *and* $\mathbf{V}$ *be finite dimensional inner product spaces over* $\mathbb{F} = \mathbb{R}, \mathbb{C}$. *Let* $T : \mathbf{V} \to \mathbf{U}$ *be a linear operator. Then*

$$\min_{Q \in \mathrm{L}_k(\mathbf{V}, \mathbf{U})} \|T - Q\|_F = \sqrt{\sum_{i=k+1}^{\operatorname{rank} T} \sigma_i^2(T)}, \quad k = 1, \dots, \operatorname{rank} T - 1.$$

$$(4.11.11)$$

Furthermore $\|T - Q\|_F = \sqrt{\sum_{i=k+1}^{\operatorname{rank} T} \sigma_i^2(T)}$ *for some* $Q \in \mathrm{L}_k(\mathbf{V}, \mathbf{U})$, $k < \operatorname{rank} T$, *if and only there* $Q = T_k$, *where* T_k *is defined in Definition 4.10.7.*

Proof. Use Theorem 4.11.8 to deduce that for any $Q \in \mathrm{L}(\mathbf{V}, \mathbf{U})$ one has

$$\|T - Q\|_F^2 = \operatorname{tr} T^*T - 2\Re \operatorname{tr} Q^*T + \operatorname{tr} Q^*Q$$

$$\geq \sum_{i=1}^{\operatorname{rank} T} \sigma_i^2(T) - 2\sum_{i=1}^{k} \sigma_i(T)\sigma_i(Q) + \sum_{i=1}^{k} \sigma_i^2(Q)$$

$$= \sum_{i=1}^{k} (\sigma_i(T) - \sigma_i(Q))^2 + \sum_{i=k+1}^{\operatorname{rank} T} \sigma_i^2(T) \geq \sum_{i=k+1}^{\operatorname{rank} T} \sigma_i^2(T).$$

Clearly $||T - T_k||_F^2 = \sum_{i=k+1}^{\text{rank } T} \sigma_i^2(T)$. Hence, (4.11.11) holds. Vice versa if $Q \in L_k(\mathbf{V}, \mathbf{U})$ and $||T - Q||_F^2 = \sum_{i=k+1}^{\text{rank } T} \sigma_i^2(T)$ then the equality case in Theorem 4.11.8 yields that $Q = T_k$. $\qquad\square$

Corollary 4.11.10 *Let* $F = \mathbb{R}, \mathbb{C}$ *and* $A \in \mathbb{F}^{m \times n}$. *Then*

$$\min_{B \in \mathcal{R}_{m,n,k}(\mathbb{F})} ||A - B||_F = \sqrt{\sum_{i=k+1}^{\text{rank } A} \sigma_i^2(A)}, \quad k = 1, \ldots, \text{rank } A - 1.$$

$$(4.11.12)$$

Furthermore $||A - B||_F = \sqrt{\sum_{i=k+1}^{\text{rank } A} \sigma_i^2(A)}$ *for some* $B \in \mathcal{R}_{m,n,k}(\mathbb{F}), k < \text{rank } A$, *if and only if* $B = A_k$, *where* A_k *is defined in Definition* 4.10.5.

Theorem 4.11.11 *Let* $\mathbb{F} = \mathbb{R}, \mathbb{C}$ *and* $A \in \mathbb{F}^{m \times n}$. *Then*

$$\min_{B \in \mathcal{R}_{m,n,k}(\mathbb{F})} \sum_{i=1}^{j} \sigma_i(A - B) = \sum_{i=k+1}^{k+j} \sigma_i(A), \qquad (4.11.13)$$

$$j = 1, \ldots, \min(m, n) - k, \ k = 1, \ldots, \min(m, n) - 1.$$

Proof. Clearly, for $B = A_k$ we have the equality $\sum_{i=1}^{j} \sigma_i(A - B) = \sum_{i=k+1}^{k+j} \sigma_i(A)$. Let $B \in \mathcal{R}_{m,n,k}(\mathbb{F})$. Let $\mathbf{X} \in \text{Gr}(k, \mathbb{C}^n)$ be a subspace which contains the columns of B. Let $\mathbf{W} = \{(\mathbf{0}^\top, \mathbf{x}^\top)^\top \in \mathbb{F}^{m+n}, \mathbf{x} \in \mathbf{X}\}$. Observe that for any $\mathbf{z} \in \mathbf{W}^\perp$ one has the equality $\mathbf{z}^* H((A - B))\mathbf{z} = \mathbf{z}^* H(A)\mathbf{z}$. Combine Theorems 4.4.9 and 4.11.1 to deduce $\sum_{i=1}^{j} \sigma_i(B - A) \geq \sum_{i=k+1}^{k+j} \sigma_i(A)$. $\qquad\square$

Theorem 4.11.12 *Let* $\mathbf{V}$ *be an* n-*dimensional IPS over* $\mathbb{C}$. *Let* $T : \mathbf{V} \to \mathbf{V}$ *be a linear operator. Assume the* n *eigenvalues of* $T, \lambda_1(T), \ldots, \lambda_n(T)$, *are arranged in the order* $|\lambda_1(T)| \geq \cdots \geq |\lambda_n(T)|$. *Let* $\boldsymbol{\lambda}_a(T) := (|\lambda_1(T)|, \ldots, |\lambda_n(T)|)$, $\boldsymbol{\sigma}(T) := (\sigma_1(T), \ldots, \sigma_n(T))$. *Then* $\boldsymbol{\lambda}_a(T) \preceq \boldsymbol{\sigma}(T)$. *That is*

$$\sum_{i=1}^{k} |\lambda_i(T)| \leq \sum_{i=1}^{k} \sigma_i(T), \quad i = 1, \ldots, n. \qquad (4.11.14)$$

Furthermore, $\sum_{i=1}^{k} |\lambda_i(T)| = \sum_{i=1}^{k} \sigma_i(T)$ for some $k \in [n]$ if and only if the following conditions hold: There exists an orthonormal basis $\mathbf{x}_1, \ldots, \mathbf{x}_n$ of $\mathbf{V}$ such that

1. *$T\mathbf{x}_i = \lambda_i(T)\mathbf{x}_i, T^*\mathbf{x}_i = \overline{\lambda_i(T)}\mathbf{x}_i$ for $i = 1, \ldots, k$.*

2. *Denote by $S : \mathbf{U} \to \mathbf{U}$ the restriction of T to the invariant subspace $\mathbf{U} = \mathrm{span}\,(\mathbf{x}_{k+1}, \ldots, \mathbf{x}_n)$. Then $||S||_2 \le |\lambda_k(T)|$.*

Proof. Use Theorem 4.2.12 to choose an orthonormal basis $\mathbf{g}_1, \ldots, \mathbf{g}_n$ of $\mathbf{V}$, such that T is represented by an upper diagonal matrix $A = [a_{ij}] \in \mathbb{C}^{n \times n}$ where $a_{ii} = \lambda_i(T), i = 1, \ldots, n$. Let $\epsilon_i \in \mathbb{C}, |\epsilon_i| = 1$ such that $\bar{\epsilon}_i \lambda_i(T) = |\lambda_i(T)|$ for $i = 1, \ldots, n$. Let $S \in \mathrm{L}(\mathbf{V})$ be presented in the basis $\mathbf{g}_1, \ldots, \mathbf{g}_n$ by a diagonal matrix $\mathrm{diag}(\epsilon_1, \ldots, \epsilon_k, 0, \ldots, 0)$. Clearly, $\sigma_i(S) = 1$ for $i = 1, \ldots, k$ and $\sigma_i(S) = 0$ for $i = k+1, \ldots, n$. Furthermore, $\Re \,\mathrm{tr}\, S^*C = \sum_{i=1}^{k} |\lambda_i(T)|$. Hence, Theorem 4.11.8 yields (4.11.14).

Assume now that $\sum_{i=1}^{k} |\lambda_i(T)| = \sum_{i=1}^{k} \sigma_i(T)$. Then equality holds in (4.11.9). Hence, there exists two orthonormal bases $\{\mathbf{c}_1, \ldots, \mathbf{c}_n\}, \{\mathbf{d}_1, \ldots, \mathbf{d}_n\}$ in $\mathbf{V}$ such that (4.11.10) holds. It easily follows that $\{\mathbf{c}_1, \ldots, \mathbf{c}_k\}, \{\mathbf{d}_1, \ldots, \mathbf{d}_k\}$ are orthonormal bases of $\mathbf{W} := \mathrm{span}\,(\mathbf{g}_1, \ldots, \mathbf{g}_k)$. Hence, $\mathbf{W}$ is an invariant subspace of T and T^*. Hence, $A = A_1 \oplus A_2$, i.e. A is a block diagonal matrix. Thus $A_1 = a_{ij}\rfloor_{i,j=1}^{k} \in \mathbb{C}^{k \times k}, A_2 = [a_{ij}]_{i,j=k+1}^{n} \in \mathbb{C}^{(n-k) \times (n-k)}$ represent the restriction of T to $\mathbf{W}, \mathbf{U} := \mathbf{W}^{\perp}$, denoted by T_1 and T_2 respectively. Hence, $\sigma_i(T_1) = \sigma_i(T)$ for $i = 1, \ldots, k$. Note that the restriction of S to $\mathbf{W}$, denoted by S_1 is given by the diagonal matrix $D_1 := \mathrm{diag}(\epsilon_1, \ldots, \epsilon_k) \in \mathbf{U}(k)$. (4.11.10) yields that $S_1^{-1}T_1\mathbf{c}_i = \sigma_i(T)\mathbf{c}_i$ for $i = 1, \ldots, k$, i.e. $\sigma_1(T), \ldots, \sigma_k(T)$ are the eigenvalues of $S_1^{-1}T_1$. Clearly $S_1^{-1}T_1$ is presented in the basis $[\mathbf{g}_1, \ldots, \mathbf{g}_k]$ by the matrix $D_1^{-1}A_1$, which is a diagonal matrix with $|\lambda_1(T)|, \ldots, |\lambda_k(T)|$ on the main diagonal. That is $S_1^{-1}T_1$ has eigenvalues $|\lambda_1(T)|, \ldots, |\lambda_k(T)|$. Therefore, $\sigma_i(T) = |\lambda_i(T)|$ for $i = 1, \ldots, k$. Theorem 4.10.6 yields that

$$\mathrm{tr}\, A_1^* A_1 = \sum_{i,j=1}^{k} |a_{ij}|^2 = \sum_{i=1}^{k} \sigma_i^2(A_1) = \sum_{i=1}^{k} \sigma_i^2(T_1) = \sum_{i=1}^{k} |\lambda_i(T)|^2.$$

As $\lambda_1(T), \ldots, \lambda_k(T)$ are the diagonal elements of A_1 is follows from the above equality that A_1 is a diagonal matrix. Hence, we can choose $\mathbf{x}_i = \mathbf{g}_i$ for $i = 1, \ldots, n$ to obtain Part 1 of the equality case.

Let $T\mathbf{x} = \lambda\mathbf{x}$ where $\|\mathbf{x}\| = 1$ and $\rho(T) = |\lambda|$. Recall $\|T\|_2 = \sigma_1(T)$, where $\sigma_1(T)^2 = \lambda_1(T^*T)$ is the maximal eigenvalue of the self-adjoint operator T^*T. The maximum characterization of $\lambda_1(T^*T)$ yields that $|\lambda|^2 = \langle T\mathbf{x}, T\mathbf{x} \rangle = \langle T^*T\mathbf{x}, \mathbf{x} \rangle \le \lambda_1(T^*T) = \|T\|_2^2$. Hence, $\rho(T) \le \|T\|_2$.

Assume now that $\rho(T) = \|T\|_2$. $\rho(T) = 0$ then $\|T\|_2 = 0 \Rightarrow T = 0$, and theorem holds trivially in this case. Assume that $\rho(T) > 0$. Hence, the eigenvector $\mathbf{x}_1 := \mathbf{x}$ is also the eigenvector of T^*T corresponding to $\lambda_1(T^*T) = |\lambda|^2$. Hence, $|\lambda|^2\mathbf{x} = T^*T\mathbf{x} = T^*(\lambda\mathbf{x})$, which implies that $T^*\mathbf{x} = \bar{\lambda}\mathbf{x}$. Let $\mathbf{U} = \text{span}(\mathbf{x})^\perp$ be the orthogonal complement of span $(\mathbf{x})$. Since $T\text{span}(\mathbf{x}) = \text{span}(\mathbf{x})$ it follows that $T^*\mathbf{U} \subseteq \mathbf{U}$. Similarly, since $T^*\text{span}(\mathbf{x}) = \text{span}(\mathbf{x})$ $T\mathbf{U} \subseteq \mathbf{U}$. Thus $\mathbf{V} = \text{span}(\mathbf{x}) \oplus \mathbf{U}$ and span $(\mathbf{x}), \mathbf{U}$ are invariant subspaces of T and T^*. Hence, span $(\mathbf{x}), \mathbf{U}$ are invariant subspaces of T^*T and TT^*. Let T_1 be the restriction of T to $\mathbf{U}$. Then $T_1^*T_1$ is the restriction of T^*T. Therefore, $\|T_1\|_2^2 = \lambda_1(T_1 * T_1) \ge \lambda_1(T^*T) = \|T\|_2^2$. This establishes the second part of theorem, labeled 1 and 2.

The above result implies that the conditions 1 and 2 of the theorem yield the equality $\rho(T) = \|T\|_2$. $\qquad\square$

Corollary 4.11.13 *Let $\mathbf{U}$ be an n-dimensional IPS over $\mathbb{C}$. Let $T : \mathbf{U} \to \mathbf{U}$ be a linear operator. Then $|\boldsymbol{\lambda}_a(T)| = \boldsymbol{\sigma}(T)$ if and only if T is a normal operator.*

Problems

1. Let the assumptions of Corollary 4.11.2 hold. Denote by $\#\alpha, \#\beta$ the cardinalities of the sets α, β.

   (a) Since rank $\hat{A} \le$ rank A show that the inequalities (4.11.4) reduce to $\sigma_i(\hat{A}) = \sigma_i(A) = 0$ for $i >$ rank A.

   (b) Noting that $H(\hat{A})$ is a submatrix of $H(A)$, use the Cauchy interlacing principle to deduce the inequalities (4.11.4) for

$i = 1, \ldots, \operatorname{rank} A$. Furthermore, if $p' := m - \#\alpha, q' = n - \#\beta$ then the Cauchy interlacing principle gives the complementary inequalities $\sigma_i(\hat{A}) \geq \sigma_{i+p'+q'}(A)$ for any $i \in \mathbb{N}$.

(c) Assume that $\sigma_i(\hat{A}) = \sigma_i(A)$ for $i = 1, \ldots, l \leq \operatorname{rank} A$. Compare the maximal characterization of the sum of the first k eigenvalues of $H(\hat{A})$ and $H(A)$ given by Theorem 4.4.8 for $k = 1, \ldots, l$ to deduce the last part of Corollary (4.11.2).

2. Prove Corollary 4.11.3 by choosing any orthonormal basis in $\mathbf{U}$, an orthonormal basis in $\mathbf{V}$ whose first dim $\mathbf{W}$ elements span $\mathbf{W}$, and using Problem 1.

3. Combine Theorems 4.8.11 and 4.11.1 to deduce Theorem 4.11.4.

4. (a) Prove Corollary 4.11.5.

   (b) Recall the definition of a norm on a vector space over $\mathbb{F} = \mathbb{R}, \mathbb{C}$ 7.1.1. Show that the function f defined in Corollary 4.11.5 is a norm. For $p = \min(m, n)$ and $w_1 = \cdots = w_p = 1$ this norm is called the *q-Schatten* norm.

5. Prove Lemma 4.11.6.

6. Under the assumptions of Theorem 4.11.7 show.

   (a)
   $$\max_{\{\mathbf{f}_1,\ldots,\mathbf{f}_k\}\in\mathrm{Fr}(k,\mathbf{U}),\{\mathbf{g}_1,\ldots,\mathbf{g}_k\}\in\mathrm{Fr}(k,\mathbf{V})} \sum_{i=1}^{k} \Re\langle T\mathbf{g}_i, \mathbf{f}_i\rangle_{\mathbf{U}}$$

   $$= \max_{\{\mathbf{f}_1,\ldots,\mathbf{f}_k\}\in\mathrm{Fr}(k,\mathbf{U}),\{\mathbf{g}_1,\ldots,\mathbf{g}_k\}\in\mathrm{Fr}(k,\mathbf{V})} \sum_{i=1}^{k} |\langle T\mathbf{g}_i, \mathbf{f}_i\rangle_{\mathbf{U}}|.$$

   (b) For $w_1 \geq \cdots \geq w_k \geq 0$

   $$\sum_{i=1}^{k} w_i \sigma_i(T) = \max_{\{\mathbf{f}_1,\ldots,\mathbf{f}_k\}\in\mathrm{Fr}(k,\mathbf{U}),\{\mathbf{g}_1,\ldots,\mathbf{g}_k\}\in\mathrm{Fr}(k,\mathbf{V})}$$

   $$\times \sum_{i=1}^{k} w_i \Re\langle T\mathbf{g}_i, \mathbf{f}_i\rangle_{\mathbf{U}}.$$

7. Let the assumptions of Theorem 4.11.7 hold. Show

$$\sum_{i=1}^{k} \sigma_i(T) \leq \max_{\{\mathbf{f}_1,\ldots,\mathbf{f}_k\}\in\mathrm{Fr}(k,\mathbf{U})} \sum_{i=1}^{k} \|T\mathbf{f}_i\|_{\mathbf{V}}.$$

Furthermore, equality holds if and only if $\sigma_1(T) = \cdots = \sigma_k(T)$.

Hint: First study the case rank $T = 2$.

8. Let $\mathbf{U}, \mathbf{V}$ be finite dimensional inner product spaces. Assume that $P, T \in \mathrm{L}(\mathbf{U}, \mathbf{V})$. Show that $\Re\,\mathrm{tr}(P^*T) \geq -\sum_{i=1}^{\min(m,n)} \sigma_i(S)\sigma_i(T)$ and that equality holds if and only if $S = -P$ and T satisfy the conditions of Theorem 4.11.8.

4.12   Moore–Penrose Generalized Inverse

Let $A \in \mathbb{C}^{m\times n}$. Then (4.10.12) is called the *reduced* SVD of A. It can be written as

$$A = U_r \Sigma_r V_r^*, \quad r = \mathrm{rank}\,A,$$

$$\Sigma_r := \mathrm{diag}(\sigma_1(A),\ldots,\sigma_r(A)) \in S_r(\mathbb{R}),$$

$$U_r = [\mathbf{u}_1,\ldots,\mathbf{u}_r] \in \mathbb{C}^{m\times r},$$

$$V_r = [\mathbf{v}_1,\ldots,\mathbf{v}_r] \in \mathbb{C}^{n\times r}, U_r^*U_r = V_r^*V_r = I_r. \tag{4.12.1}$$

Recall that

$$AA^*\mathbf{u}_i = \sigma_i(A)^2\mathbf{u}_i, A^*A\mathbf{v}_i = \sigma_i(A)^2\mathbf{v}_i,$$

$$\mathbf{v}_i = \frac{1}{\sigma_i(A)}A^*\mathbf{u}_i, \quad \mathbf{u}_i = \frac{1}{\sigma_i(A)}A\mathbf{v}_i, i = 1,\ldots,r.$$

Then

$$A^\dagger := V_r\Sigma_r^{-1}U_r^* \in \mathbb{C}^{n\times m} \tag{4.12.2}$$

is the *Moore–Penrose* generalized inverse of A. If $A \in \mathbb{R}^{m\times n}$ then we assume that $U \in \mathbb{R}^{m\times r}$ and $V \in R^{n\times r}$, i.e. U, V are real valued matrices.

Theorem 4.12.1 *Let* $A \in \mathbb{C}^{m \times n}$. *Then the Moore–Penrose generalized inverse* $A^\dagger \in \mathbb{C}^{n \times m}$ *satisfies the following properties.*

1. rank A = rank $A^\dagger$.

2. $A^\dagger A A^\dagger = A^\dagger$, $AA^\dagger A = A$, $A^* A A^\dagger = A^\dagger A A^* = A^*$.

3. $A^\dagger A$ *and* $AA^\dagger$ *are Hermitian non-negative definite idempotent matrices, i.e.* $(A^\dagger A)^2 = A^\dagger A$ *and* $(AA^\dagger)^2 = AA^\dagger$, *having the same rank as* A.

4. *The least square solution of* $A\mathbf{x} = \mathbf{b}$, *i.e. the solution of the system* $A^* A\mathbf{x} = A^*\mathbf{b}$, *has a solution* $\mathbf{y} = A^\dagger \mathbf{b}$. *This solution has the minimal norm* $\|\mathbf{y}\|$, *for all possible solutions of* $A^* A\mathbf{x} = A^*\mathbf{b}$.

5. *If* rank $A = n$ *then* $A^\dagger = (A^* A)^{-1} A^*$. *In particular, if* $A \in \mathbb{C}^{n \times n}$ *is invertible then* $A^\dagger = A^{-1}$.

To prove the above theorem we need the following proposition.

Proposition 4.12.2 *Let* $E \in \mathbb{C}^{l \times m}, G \in \mathbb{C}^{m \times n}$. *Then* rank $EG \le \min(\text{rank } E, \text{rank } G)$. *If* $l = m$ *and* E *is invertible then* rank EG = rank G. *If* $m = n$ *and* G *is invertible then* rank EG = rank E.

Proof. Let $\mathbf{e}_1, \ldots, \mathbf{e}_m \in \mathbb{C}^l, \mathbf{g}_1, \ldots, \mathbf{g}_n \in \mathbb{C}^m$ be the columns of E and G respectively. Then rank E = dim span $(\mathbf{e}_1, \ldots, \mathbf{e}_l)$. Observe that $EG = [E\mathbf{g}_1, \ldots, E\mathbf{g}_n] \in \mathbb{C}^{l \times n}$. Clearly $E\mathbf{g}_i$ is a linear combination of the columns of E. Hence, $E\mathbf{g}_i \in$ span $(\mathbf{e}_1, \ldots, \mathbf{e}_l)$. Therefore span $(E\mathbf{g}_1, \ldots, E\mathbf{g}_n) \subseteq$ span $(\mathbf{e}_1, \ldots, \mathbf{e}_l)$, which implies that rank $EG \le$ rank E. Note that $(EG)^T = G^T E^T$. Hence, rank EG = rank $(EG)^T \le$ rank G^T = rank G. Thus rank $EG \le$ $\min(\text{rank } E, \text{rank } G)$. Suppose E is invertible. Then rank $EG \le$ rank G = rank $E^{-1}(EG) \le$ rank EG. Hence, rank EG = rank G. Similarly rank EG = rank E if G is invertible. ☐

Proof of Theorem 4.12.1.

1. Proposition 4.12.2 yields that rank $A^\dagger$ = rank $V_r \Sigma_r^{-1} U_r^* \leq$ rank $\Sigma_r^{-1} U_r^* \leq$ rank Σ_r^{-1} = r = rank A. Since $\Sigma_r = V_r^* A^\dagger U_r$, Proposition 4.12.2 yields that rank $A^\dagger \geq$ rank Σ_r^{-1} = r. Hence, rank A = rank $A^\dagger$.

2. $AA^\dagger = (U_r \Sigma_r V_r^*)(V_r \Sigma_r^{-1} U_r^*) = U_r \Sigma_r \Sigma_r^{-1} U_r^* = U_r U_r^*$. Hence

$$AA^\dagger A = (U_r U_r^*)(U_r \Sigma_r V_r^*) = U_r \Sigma V_r^* = A.$$

Hence, $A^* AA^\dagger = (V_r \Sigma_r U_r^*)(U_r U_r^*) = A^*$. Similarly $A^\dagger A = V_r V_r^*$ and $A^\dagger AA^\dagger = A^\dagger$, $A^\dagger AA^* = A^*$.

3. Since $AA^\dagger = U_r U_r^*$ we deduce that $(AA^\dagger)^* = (U_r U_r^*)^* = (U_r^*)^* U_r^* = AA^\dagger$, i.e. $AA^\dagger$ is Hermitian. Next $(AA^\dagger)^2 = (U_r U_r^*)^2 = (U_r U_r^*)(U_r U_r^*) = (U_r U_r^*) = AA^\dagger$, i.e. $AA^\dagger$ is idempotent. Hence, $AA^\dagger$ is non-negative definite. As $AA^\dagger = U_r I_r U_r^*$, the arguments in Part 1 yield that rank $AA^\dagger = r$. Similar arguments apply to $A^\dagger A = V_r V_r^*$.

4. Since $A^* AA^\dagger = A^*$ it follows that $A^* A(A^\dagger \mathbf{b}) = A^* \mathbf{b}$, i.e. $\mathbf{y} = A^\dagger \mathbf{b}$ is a least square solution. It is left to show that if $A^* A\mathbf{x} = A^* \mathbf{b}$ then $||\mathbf{x}|| \geq ||A^\dagger \mathbf{b}||$ and equality holds if and only if $\mathbf{x} = A^\dagger \mathbf{b}$.

We now consider the system $A^* A\mathbf{x} = A^* \mathbf{b}$. To analyze this system we use the full form of SVD given in (4.10.7). It is equivalent to $(V \Sigma^T U^*)(U \Sigma V^*)\mathbf{x} = V \Sigma^T U^* \mathbf{b}$. Multiplying by V^* we obtain the system $\Sigma^T \Sigma(V^* \mathbf{x}) = \Sigma^T (U^* \mathbf{b})$. Let $\mathbf{z} = (z_1, \ldots, z_n)^T :=$ $V^* \mathbf{x}$, $\mathbf{c} = (c_1, \ldots, c_m)^T := U^* \mathbf{b}$. Note that $\mathbf{z}^* \mathbf{z} = \mathbf{x}^* V V \mathbf{x} = \mathbf{x}^* \mathbf{x}$, i.e. $||\mathbf{z}|| = ||\mathbf{x}||$. After these substitutions the least square system in $z_1, \ldots, z_n$ variables is given in the form $\sigma_i(A)^2 z_i = \sigma_i(A)c_i$ for $i = 1, \ldots, n$. Since $\sigma_i(A) = 0$ for $i > r$ we obtain that $z_i = \frac{1}{\sigma_i(A)} c_i$ for $i = 1, \ldots, r$ while $z_{r+1}, \ldots, z_n$ are free variables. Thus $||\mathbf{z}||^2 = \sum_{i=1}^r \frac{1}{\sigma_i(A)^2} + \sum_{i=r+1}^n |z_i|^2$. Hence, the least square solution with the minimal length $||\mathbf{z}||$ is the solution with $z_i = 0$ for $i = r + 1, \ldots, n$. This solution corresponds to $\mathbf{x} = A^\dagger \mathbf{b}$.

5. Since rank $A^*A = \text{rank } A = n$ it follows that A^*A is an invertible matrix. Hence, the least square solution is unique and is given by $\mathbf{x} = (A^*A)^{-1}A^*\mathbf{b}$. Thus for each $\mathbf{b}$ one has $(A^*A)^{-1}A^*\mathbf{b} = A^\dagger\mathbf{b}$, hence $A^\dagger = (A^*A)^{-1}A^*$.

If A is an $n \times n$ matrix and is invertible it follows that $(A^*A)^{-1}A^* = A^{-1}(A^*)^{-1}A^* = A^{-1}$. □

Problems

1. $P \in \mathbb{C}^{n \times n}$ is called a *projection* if $P^2 = P$. Show that P is a projection if and only if the following two conditions are satisfied:

   - Each eigenvalue of P is either 0 or 1.

   - P is a diagonable matrix.

2. $P \in \mathbb{R}^{n \times n}$ is called an *orthogonal* projection if P is a projection and a symmetric matrix. Let $\mathbb{V} \subseteq \mathbb{R}^n$ be the subspace spanned by the columns of P. Show that for any $\mathbf{a} \in \mathbb{R}^n, \mathbf{b} \in \mathbb{V}, \|\mathbf{a}-\mathbf{b}\| \geq \|\mathbf{a} - P\mathbf{a}\|$ and equality holds if and only if $\mathbf{b} = P\mathbf{a}$. That is, $P\mathbf{a}$ is the orthogonal projection of $\mathbf{a}$ on the column space of P.

3. Let $A \in \mathbb{R}^{m \times n}$ and assume that the SVD of A is given by (4.10.7), where $U \in \mathbf{O}(m, \mathbb{R}), V \in \mathbf{O}(n, \mathbb{R})$.

   (a) What is the SVD of A^T?

   (b) Show that $(A^T)^\dagger = (A^\dagger)^T$.

   (c) Suppose that $B \in \mathbb{R}^{l \times m}$. Is it true that $(BA)^\dagger = A^\dagger B^\dagger$?

4.13   Approximation by Low Rank Matrices

We now restate Theorem 4.11.8 in matrix terms. That is we view $A, B \in \mathbb{C}^{m \times n}$ as linear operators $A, B : \mathbb{C}^n \to \mathbb{C}^m$, where $\mathbb{C}^m, \mathbb{C}^n$ are inner product spaces equipped with the standard inner product.

Theorem 4.13.1 *Let $A, B \in \mathbb{C}^{m \times n}$, and assume that $\sigma_1(A) \geq \sigma_2(A) \geq \cdots \geq 0, \sigma_1(B) \geq \sigma_2(B) \geq \cdots \geq 0$, where $\sigma_i(A) = 0$ and $\sigma_j(B) = 0$ for $i > \text{rank } A$ and $j > \text{rank } B$ respectively. Then*

$$- \sum_{i=1}^{m} \sigma_i(A)\sigma_i(B) \leq \Re \text{tr } AB^* \leq \sum_{i=1}^{m} \sigma_i(A)\sigma_i(B). \qquad (4.13.1)$$

Equality in the right-hand side holds if and only if $\mathbb{C}^n, \mathbb{C}^m$ have two orthonormal bases $[\mathbf{c}_1, \ldots, \mathbf{c}_n], [\mathbf{d}_1, \ldots, \mathbf{d}_m]$ such that (4.11.10) is satisfied for $T = A$ and $S = B$. Equality for the left-hand side holds if and only if $\mathbb{C}^n, \mathbb{C}^m$ have two orthonormal bases $[\mathbf{c}_1, \ldots, \mathbf{c}_n], [\mathbf{d}_1, \ldots, \mathbf{d}_m]$ such that (4.11.10) is satisfied for $T = A$ and $S = -B$.

Theorem 4.11.9 yields:

Corollary 4.13.2 *For $A \in \mathbb{C}^{m \times n}$ let A_k be defined as in Definition 4.10.5. Then $\min_{B \in \mathcal{R}_{m,n,k}(\mathbb{F})} \|A - B\|_F^2 = \|A - A_k\|^2 = \sum_{i=k+1}^{m} \sigma_i(A)^2$. A_k is the unique solution to this minimal problem if and only if $1 \leq k < \text{rank } A$ and $\sigma_k(A) \geq \sigma_{k+1}(A)$.*

We now give a generalization of Corollary 4.11.9. Let $A \in \mathbb{C}^{m \times n}$ and assume that $A = U_A \Sigma_A V_A^*$ is the SVD of A given in (4.10.7). Let $U_A = [\mathbf{u}_1 \ \mathbf{u}_2 \ \ldots \mathbf{u}_m], V_A = [\mathbf{v}_1 \ \mathbf{v}_2 \ \ldots \mathbf{v}_n]$ be the representations of U, V in terms of their m, n columns respectively. Then

$$P_{A,\text{left}} := \sum_{i=1}^{\text{rank } A} \mathbf{u}_i \mathbf{u}_i^* \in \mathbb{C}^{m \times m}, \quad P_{A,\text{right}} := \sum_{i=1}^{\text{rank } A} \mathbf{v}_i \mathbf{v}_i^* \in \mathbb{C}^{n \times n},$$

$$(4.13.2)$$

are the orthogonal projections on the range of A and A^*, respectively.

Theorem 4.13.3 *Let $A \in \mathbb{C}^{m \times n}, C \in \mathbb{C}^{m \times p}, R \in \mathbb{C}^{q \times n}$ be given. Then $X = C^\dagger (P_{C,\text{left}} A P_{R,\text{right}})_k R^\dagger$ is a solution to the minimal* *problem*

$$\min_{X \in \mathcal{R}_{p,q,k}(\mathbb{C})} \|A - CXR\|_F, \tag{4.13.3}$$

having the minimal $\|X\|_F$. *This solution is unique if and only if either* $k \geq \operatorname{rank} P_{C,\text{left}} A P_{R,\text{right}}$ *or* $1 \leq k < \operatorname{rank} P_{C,\text{left}} A P_{R,\text{right}}$ *and* $\sigma_k(P_{C,\text{left}} A P_{R,\text{right}}) > \sigma_{k+1}(P_{C,\text{left}} A P_{R,\text{right}})$.

Proof. Assume that $C = U_C \Sigma_C V_C^*, R = U_R \Sigma_R V_R^*$ are the SVD decompositions of C and R, respectively. Recall that the Frobenius norm is invariant under the multiplication from the left and the right by the corresponding unitary matrices. Hence, $\|A - BXC\|_F = \|\tilde{A} - \Sigma_C \tilde{X} \Sigma_R\|$, where $\tilde{A} := U_C^* A V_R, \tilde{X} := V_C^* X U_R$. Clearly, X and $\tilde{X}$ have the same rank and the same Frobenius norm. Thus it is enough to consider the minimal problem $\min_{\tilde{X} \in \mathcal{R}_{p,q,k}(\mathbb{C})} \|\tilde{A} - \Sigma_C \tilde{X} \Sigma_R\|_F$. Let $s = \operatorname{rank} C, t = \operatorname{rank} R$. Clearly if C or R is a zero matrix, then $X = 0_{p \times q}$ is the solution to the minimal problem (4.13.3). In this case either $P_{C,\text{left}}$ or $P_{R,\text{right}}$ are zero matrices, and the theorem holds trivially in this case.

It is left to consider the case $1 \leq s, 1 \leq t$. Define

$$C_1 := \operatorname{diag}(\sigma_1(C), \ldots, \sigma_s(C)) \in \mathbb{C}^{s \times s},$$

$$R_1 := \operatorname{diag}(\sigma_1(R), \ldots, \sigma_t(R)) \in \mathbb{C}^{t \times t}.$$

Partition $\tilde{A}$ and $\tilde{X}$ to 2×2 block matrices $\tilde{A} = [A_{ij}]_{i,j=1}^2$ and $\tilde{X} = [X_{ij}]_{i,j=1}^2$, where $A_{11}, X_{11} \in \mathbb{C}^{s \times t}$. (For certain values of s and t, we may have to partition $\tilde{A}$ or $\tilde{X}$ to less than 2×2 block matrices.) Observe next that $Z := \Sigma_C \tilde{X} \Sigma_R = [Z_{ij}]_{i,j=1}^2$, where $Z_{11} = C_1 X_{11} R_1$ and all other blocks Z_{ij} are zero matrices. Hence

$$\|\tilde{A} - Z\|_F^2 = \|A_{11} - Z_{11}\|_F^2 + \sum_{2 < i+j \leq 4} \|A_{ij}\|_F^2$$

$$\geq \|A_{11} - (A_{11})_k\|_F^2 + \sum_{2 < i+j \leq 4} \|A_{ij}\|_F^2.$$

Thus $\hat{X} = [X_{ij}]_{i,j=1}^2$, where $X_{11} = C_1^{-1}(A_{11})_k R_1^{-1}$ and $X_{ij} = 0$ for all $(i,j) \neq (1,1)$ is a solution $\min_{\tilde{X} \in \mathcal{R}_{p,q,k}(\mathbb{C})} \|\tilde{A} - \Sigma_C \tilde{X} \Sigma_R\|_F$

with the minimal Frobenius form. This solution is unique if and only if the solution $Z_{11} = (A_{11})_k$ is the unique solution to $\min_{Z_{11}\in\mathcal{R}_{s,t,k}(\mathbb{C})} \|A_{11} - Z_{11}\|_F$. This happens if either $k \geq \text{rank } A_{11}$ or $1 \leq k < \text{rank } A_{11}$ and $\sigma_k(A_{11}) > \sigma_{k+1}(A_{11})$. A straightforward calculation shows that $\hat{X} = \Sigma_C^\dagger (P_{\Sigma_C,\text{left}}\tilde{A}P_{\Sigma_R,\text{right}})_k\Sigma_R^\dagger$. This shows that $X = C^\dagger(P_{C,\text{left}}AP_{R,\text{right}})_kR^\dagger$ is a solution of (4.13.3) with the minimal Frobenius norm. This solution is unique if and only if either $k \geq \text{rank } P_{C,\text{left}}AP_{R,\text{right}}$ or $1 \leq k < \text{rank } P_{C,\text{left}}AP_{R,\text{right}}$ and $\sigma_k(P_{C,\text{left}}AP_{R,\text{right}}) > \sigma_{k+1}(P_{C,\text{left}}AP_{R,\text{right}})$. $\qquad\square$

Corollary 4.13.4 *Let the assumptions of Theorem 4.13.3 hold. Then $X = C^\dagger AR^\dagger$ is the unique solution to the minimal problem $\min_{X\in\mathbb{C}^{p\times q}} \|A - CXR\|_F$ with the minimal Frobenius norm.*

We now give a version of Theorem 4.13.3 for the operator norm $\|A\|_2 = \sigma_1(A)$.

Theorem 4.13.5 *Let $A \in \mathbb{C}^{m\times n}, C \in \mathbb{C}^{m\times p}, R \in \mathbb{C}^{q\times n}$ be given. Consider the minimum problem $\min_{X\in\mathcal{R}_{p,q,k}(\mathbb{C})} \|A - CXR\|_2$. Use SVDs of C and R, as in proof of Theorem 4.13.3, to replace this minimum problem with*

$$\mu := \min_{Z_{11}\in\mathcal{R}_{s,t,k}(\mathbb{C})} \left\| \begin{bmatrix} A_{11} & A_{12} \\ A_{21} & A_{22} \end{bmatrix} - \begin{bmatrix} Z_{11} & 0 \\ 0 & 0 \end{bmatrix} \right\|_2. \qquad (4.13.4)$$

Then

$$\max(\sigma_{k+1}(A_{11}), \|[A_{21}\ A_{22}]\|_2, \|[A_{12}^*\ A_{22}^*]\|_2) \leq \mu \qquad (4.13.5)$$

$$\leq 3\max(\sigma_{k+1}(A_{11})\,\|[A_{21}\ A_{22}]\|_2, \|[A_{12}^*\ A_{22}^*]\|_2).$$

Proof. (4.10.10) yields that $\sigma_{k+1}(A_{11}) \leq \|A_{11} - Z_{11}\|_2$. Let $F := \begin{bmatrix} A_{11} - Z_{11} & A_{12} \\ A_{21} & A_{22} \end{bmatrix}$. Corollary 4.11.2 yields that the operator norm of a submatrix of F is not greater than $\|F\|_2$. Combine these two results, and the fact that the positive singular values of B and B^*

coincide, to deduce the lower bound for μ in (4.13.5). The triangle inequality yields

$$\|F\|_2 \leq \left\|\begin{bmatrix} A_{11} - Z_{11} & 0 \\ 0 & 0 \end{bmatrix}\right\|_2 + \left\|\begin{bmatrix} 0 & 0 \\ A_{21} & A_{22} \end{bmatrix}\right\|_2 + \left\|\begin{bmatrix} 0 & A_{12} \\ 0 & 0 \end{bmatrix}\right\|_2 .$$

Clearly, the positive singular values of $\begin{bmatrix} 0 & 0 \\ A_{21} & A_{22} \end{bmatrix}$ are equal to the positive singular values of $[A_{21} \ A_{22}]$. Choose Z_{11} such that $\sigma_{k+1}(A_{11}) = \|A_{11} - Z_{11}\|_2$. Use the above inequality and Corollary 4.11.2 to deduce the upper bound for μ in (4.13.5). $\square$

Theorem 4.13.6 *Let* $\mathbf{a}_1, \ldots, \mathbf{a}_n \in \mathbb{C}^m$ *and* $k \in [m-1]$ *be given. Let* $A = [\mathbf{a}_1 \ldots \mathbf{a}_n] \in \mathbb{C}^{m \times n}$. *Denote by* $L_k \in \mathrm{Gr}(k, \mathbb{C}^m)$ *a* k-*dimensional subspace spanned by the first* k *left singular vectors of* A. *Then*

$$\min_{L \in \mathrm{Gr}(k,\mathbb{C}^m)} \sum_{i=1}^{n} \min_{\mathbf{b}_i \in L} \|\mathbf{a}_i - \mathbf{b}_i\|_2^2 = \sum_{i=1}^{n} \min_{\mathbf{b}_i \in L_k} \|\mathbf{a}_i - \mathbf{b}_i\|_2^2. \qquad (4.13.6)$$

Proof. Let $L \in \mathrm{Gr}(k, \mathbb{C}^m)$ and $\mathbf{b}_1, \ldots, \mathbf{b}_n \in L$. Then $B := [\mathbf{b}_1 \ldots \mathbf{b}_n] \in \mathcal{R}_{m,n,k}(\mathbb{C})$. Vice versa, given $B \in \mathcal{R}_{m,n,k}(\mathbb{C})$, the column space of B is contained in some $L \in \mathrm{Gr}(k, \mathbb{C}^m)$. Hence, $\sum_{i=1}^{n} \|\mathbf{a}_i - \mathbf{b}_i\|_2^2 = \|A - B\|_2^2$. Corollary 4.13.2 implies that the minimum stated in the left-hand side of (4.13.6) is achieved by the n columns of A_k. Clearly, the column space of A is equal to L_k. (Note that L_k is not unique. See Problem 3.) $\square$

Problems

1. Let $A \in \mathbf{S}(n, \mathbb{R})$ and assume the $A = Q^T \Lambda Q$, where $Q \in \mathbf{O}(n, \mathbb{R})$ and $\Lambda = \mathrm{diag}(\alpha_1, \ldots, \alpha_n)$ is a diagonal matrix, where $|\alpha_1| \geq \cdots \geq |\alpha_n| \geq 0$.

   (a) Find the SVD of A.

(b) Show that $\sigma_1(A) = \max(\lambda_1(A), |\lambda_n(A)|)$, where $\lambda_1(A) \geq \cdots \geq \lambda_n(A)$ are the n eigenvalues of A arranged in a decreasing order.

2. Let k, m, n be positive integers such that $k \leq \min(m, n)$. Show that the function $f : \mathbb{R}^{m \times n} \to [0, \infty)$ given by $f(A) = \sum_{i=1}^{k} \sigma_i(A)$ is a convex function on $\mathbb{R}^{m \times n}$.

3. Show that the minimal subspace for the problem (4.13.6) is unique if and only if $\sigma_k(A) > \sigma_{k+1}(A)$.

4. Prove Corollary 4.13.4.

4.14 CUR-Approximations

Let $A = [a_{ij}]_{i,j=1}^{m,n} \in \mathbb{C}^{m \times n}$, where m, n are big, e.g. $m, n \geq 10^6$. Then the low rank approximation of A given by its SVD has prohibitively high computational complexity and storage requirements. In this section we discuss a low rank approximation of A of the form CUR, where $C \in \mathbb{C}^{m \times p}, R \in \mathbb{C}^{q \times n}$ are obtained from A by reading p, q columns and rows of A, respectively. If one chooses U as the best least squares approximation given by Corollary 4.13.4 then $U = C^\dagger A R^\dagger$. Again, for very large m, n this U has too high computational complexity. In this section, we give different ways to compute U of a relatively low computational complexity.

Let

$$I = \{1 \leq \alpha_1 < \cdots < \alpha_q \leq m\} \subset [m],$$
$$J = \{1 < \beta_1 < \cdots < \beta_p \leq n\} \subset [n]$$

be two nonempty sets of cardinality q, p respectively. Using the indices in I, J, we consider the submatrices

$$A_{IJ} = [a_{\alpha_k \beta_l}]_{k,l=1}^{q,p} \in \mathbb{C}^{q \times p},$$

$$R = A_{I[n]} = [a_{\alpha_k j}]_{k,j=1}^{q,n} \in \mathbb{C}^{q \times n}, \tag{4.14.1}$$

$$C = A_{[m]J} = [a_{i \beta_l}]_{i,l=1}^{m,p} \in \mathbb{C}^{m \times p}.$$

Thus, $C = A_{[m]J}$ and $R = A_{I[n]}$ are composed of the columns in J and the rows I of A, respectively. The read entries of A are in the index set

$$\mathcal{S} := [m] \times [n] \setminus (([m] \setminus I) \times ([n] \setminus J)), \quad \#\mathcal{S} = mp + qn - pq.$$
$$(4.14.2)$$

We look for a matrix $F = CUR \in \mathbb{C}^{m \times n}$, with $U \in \mathbb{C}^{p \times q}$ still to be determined. We determine U_{opt} as a solution to the least square problem of minimizing $\sum_{(i,j) \in \mathcal{S}} |a_{ij} - (CUR)_{ij}|^2$, i.e.

$$U_{\mathrm{opt}} = \arg \min_{U \in \mathbb{C}^{p \times q}} \sum_{(i,j) \in \mathcal{S}} |a_{ij} - (CUR)_{ij}|^2. \qquad (4.14.3)$$

It is straightforward to see that the above least squares is the least squares solution of the following overdetermined system

$$T\hat{U} = \hat{A}, \quad T = [t_{(i,j)(k,l)}] \in \mathbb{C}^{(mp+qn-pq) \times pq}, \quad t_{(i,j)(k,l)} = a_{ik}a_{lj},$$
$$(4.14.4)$$
$$\hat{U} = [u_{(k,l)}] \in \mathbb{C}^{pq}, \quad \hat{A} = [a_{(i,j)}] \in \mathbb{C}^{mp+qn-pq},$$
$$(i,j) \in \mathcal{S}, \quad (k,l) \in [p] \times [q].$$

Here $\hat{U}, \hat{A}$ is viewed as a vector whose coordinates are the entries of U and the entries of A which are either in C or R. Note that T is a corresponding submatrix of $A \otimes A$.

Theorem 4.14.1 *Let $A \in \mathbb{C}^{m \times n}$, and let $I \subset [m], J \subset [n]$ have cardinality q and p, respectively. Let $C = A_{[m]J} \in \mathbb{C}^{m \times p}$, and $R = A_{I[n]} \in \mathbb{C}^{p \times n}$ be as in (4.14.1) and suppose that A_{IJ} is invertible. Then the overdetermined system (4.14.4) has a unique solution $U = A_{IJ}^{-1}$, i.e. the rows in I and the columns in J of the matrix $CA_{IJ}^{-1}R$ are equal to the corresponding rows and columns of A, respectively.*

Proof. For any $I \subset [m], J \subset [n]$, with $\#I = q, \#J = p$, and $U \in \mathbb{C}^{m \times n}$ we have the identity

$$(A_{[m]J}UA_{I[n]})_{IJ} = A_{IJ}UA_{IJ}. \qquad (4.14.5)$$

Hence, the part of the system (4.14.4) corresponding to $(CUR)_{IJ} = A_{IJ}$ reduces to the equation

$$A_{IJ}UA_{IJ} = A_{IJ}. \tag{4.14.6}$$

If A_{IJ} is a square matrix and invertible, then the unique solution to this matrix equation is $U = A_{IJ}^{-1}$. Furthermore

$$(A_{[m]J}A_{IJ}^{-1}A_{I[n]})_{I[n]} = A_{IJ}A_{IJ}^{-1}A_{I[n]} = A_{I[n]},$$

$$(A_{[m]J}A_{IJ}^{-1}A_{I[n]})_{[m]J} = A_{[m]J}A_{IJ}^{-1}A_{IJ} = A_{[m]J}.$$

$$\square$$

This results extends to the general nonsquare case.

Theorem 4.14.2 *Let* $A \in \mathbb{C}^{m \times n}$, *and let* $I \subset [m], J \subset [n]$ *have cardinality* q *and* p, *respectively. Let* $C = A_{[m]J} \in \mathbb{C}^{m \times p}$, *and* $R = A_{I[n]} \in \mathbb{C}^{p \times n}$ *be as in* (4.14.1). *Then* $U = A_{IJ}^\dagger$ *is the minimal solution* (*with respect to the Frobenius norm*) *of* (4.14.3).

Proof. Consider the SVD decomposition of A_{IJ}

$$A_{IJ} = W\Sigma V^*, \ W \in \mathbb{C}^{q \times q}, \ V \in \mathbb{C}^{p \times p},$$

$$\Sigma = \text{diag}(\sigma_1, \ldots, \sigma_r, 0, \ldots, 0) \in \mathbb{R}_+^{q \times p},$$

where W, V are unitary matrices and $\sigma_1, \ldots, \sigma_r$ are the positive singular values of A_{IJ}. In view of Theorem 4.14.1 it is enough to assume that $\max(p, q) > r$. W.l.o.g. we may assume that $I = [q], J = [p]$. Let

$$W_1 = \begin{bmatrix} W & 0_{q \times (m-q)} \\ 0_{(m-q) \times q} & I_{m-q} \end{bmatrix} \in \mathbb{C}^{m \times m},$$

$$V_1 = \begin{bmatrix} V & 0_{p \times (n-p)} \\ 0_{(n-p) \times p} & I_{n-p} \end{bmatrix} \in \mathbb{C}^{n \times n}.$$

Replace A by $A_1 = W_1 A V_1^*$. It is easy to see that it is enough to prove the theorem for A_1. For simplicity of the notation we assume that $A_1 = A$. That is, we assume that $A_{IJ} = \Sigma_r \oplus 0_{(q-r) \times (p-r)}$, where

$\Sigma_r = \operatorname{diag}(\sigma_1, \ldots, \sigma_r)$ and $r = \operatorname{rank} A_{IJ}$. For $U \in \mathbb{C}^{p \times q}$ denote by $U_r \in \mathbb{C}^{p \times q}$ the matrix obtained from U by replacing the last $p - r$ rows and $q - r$ columns by rows and columns of zeroes, respectively. Note that then $CUR = CU_rR$ and $\|U_r\|_F \leq \|U\|_F$, and equality holds if and only if $U = U_r$. Hence, the minimal Frobenius norm least squares solution U of is given by $U = U_r$. Using the fact that the rows $r + 1, \ldots, q$ and columns $r + 1, \ldots, p$ of CUR are zero it follows that the minimum in (4.14.3) is reduced to the minimum on $\mathcal{S}' = [m] \times [r] \cup [r] \times [n]$. Then, by Theorem 4.14.1 the solution to the minimal Frobenius norm least square problem is given by $\Sigma^\dagger$. $\square$

For a matrix A define the *entrywise maximal norm*

$$\|A\|_{\infty,e} := \max_{i \in [m], j \in [n]} |a_{ij}|, \quad A = [a_{ij}] \in \mathbb{C}^{m \times n}. \qquad (4.14.7)$$

Theorem 4.14.3 *Let* $A \in \mathbb{C}^{m \times n}, p \in [\operatorname{rank} A]$. *Define*

$$\mu_p := \max_{I \subset [m], J \subset [n], \#I = \#J = p} |\det A_{IJ}| > 0. \qquad (4.14.8)$$

Suppose that

$$|\det A_{IJ}| \geq \delta \mu_p, \delta \in (0, 1], I \subset [m], J \subset [n], \#I = \#J = p. \quad (4.14.9)$$

Then for C, R *defined by* (4.14.1) *we have*

$$\|A - CA_{IJ}^{-1}R\|_{\infty,e} \leq \frac{p+1}{\delta}\sigma_{p+1}(A). \qquad (4.14.10)$$

Proof. We now estimate $|a_{ij} - (CA_{IJ}^{-1}R)_{ij}|$ from above. In the case $p = \operatorname{rank} A$, i.e. $\sigma_{p+1}(A) = 0$, we deduce from Problem 1 that $a_{ij} - (CA_{IJ}^{-1}R)_{ij} = 0$. Assume $\sigma_{p+1}(A) > 0$. By Theorem 4.14.1 $a_{ij} - (CA_{IJ}^{-1}R)_{ij} = 0$ if either $i \in I$ or $j \in J$. It is left to consider the case $i \in [m]\backslash I, j \in [n]\backslash J$. Let $K = I\cup\{i\}, L = J\cup\{j\}$. Let $B = A_{KL}$. If rank $B = p$ then Problem 1 yields that $B = B_{KJ}A_{IJ}^{-1}B_{IK}$. Hence, $a_{ij} - (CA_{IJ}^{-1}R)_{ij} = 0$. Assume that $\det B \neq 0$. We claim that

$$a_{ij} - (CA_{IJ}^{-1}R)_{ij} = \pm\frac{\det B}{\det A_{IJ}}. \qquad (4.14.11)$$

It is enough to consider the case where $I = J = [p], i = j = p + 1K = L = [p + 1]$. In view of Theorem 4.14.1 $B - B_{KJ}A_{IJ}^{-1}B_{JL} =$

$\mathrm{diag}(0,\ldots,0,t)$, where t is equal to the left-hand side of (4.14.11). Multiply this matrix equality from the left by $B^{-1} = [b_{st,-1}]_{s,t=1}^{p+1}$. Note that the last row of $B^{-1}B_{KI}$ is zero. Hence, we deduce that $b_{(p+1)(p+1),-1}t = 1$, i.e $t = b_{(p+1)(p+1),-1}^{-1}$. Use the identity $B^{-1} = (\det B)^{-1}\mathrm{adj}\, B$ to deduce the equality (4.14.11).

We now estimate $\sigma_1(B^{-1})$ from above. Note that each entry of $B^{-1} = (\det B)^{-1}\mathrm{adj}\, B$ is bounded above by $|\det B|^{-1}\mu_p$. Hence, $\sigma_1(B^{-1}) \leq \frac{(p+1)\mu_p}{|\det B|}$. Recall that $\sigma_1(B^{-1}) = \sigma_{p+1}(B)^{-1}$. Thus

$$\frac{|\det B|}{\mu_p} \leq (p+1)\sigma_{p+1}(B) \Rightarrow \frac{|\det B|}{|\det A_{IJ}|} \leq \frac{(p+1)\sigma_{p+1}(B)}{\delta}.$$

Since B is a submatrix of A we deduce $\sigma_{p+1}(B) \leq \sigma_{p+1}(A)$. $\qquad \square$

Problems

1. Let $A \in \mathbb{C}^{m \times n}$, $\mathrm{rank}\, A = r$. Assume that $I \subset [m], J \subset [n], \#I = \#J = r$. Assume that $\det A_{IJ} \neq 0$. Show that $A = CA_{IJ}^{-1}R$.

4.15 Some Special Maximal Spectral Problems

Theorem 4.15.1 *Let* $\mathbf{V}$ *be an n-dimensional IPS over $\mathbb{R}$. Let* $p \in [n]$ *and* $D \subset \mathbb{R}_{\searrow}^n$ *be a convex Schur set. Let D_p be the projection of D on the first p coordinates. Let $h : D_p \to \mathbb{R}$ and assume that* $f : \boldsymbol{\lambda}^{-1}(D) \to \mathbb{R}$ *is the spectral function given by $A \mapsto h(\boldsymbol{\lambda}_{(p)}(A))$, where $\boldsymbol{\lambda}_{(p)}(A) = (\lambda_1(A),\ldots,\lambda_p(A))^\top$. Let $\mathrm{S} \subset \boldsymbol{\lambda}^{-1}(D)$. Assume that h is nondecreasing on D_p. Then*

$$\sup_{A \in \mathrm{conv}\, \mathrm{S}} f(A) = \sup_{B \in \mathrm{conv}\,_{\binom{p+1}{2}-1}\mathrm{S}} f(B), \qquad (4.15.1)$$

and this result is sharp.

Proof. Since $\dim \mathrm{conv}\, \mathrm{S} \leq \dim \mathbf{S}(\mathbf{V}) = \binom{n+1}{2}$, Theorem 4.6.2 implies that it is enough to prove the theorem in the case $\mathrm{S} = \mathrm{T} := \{A_1,\ldots,A_N\}$, where $N \leq \binom{n+1}{2} + 1$. Observe next that since D is a convex Schur set and $S \subset \boldsymbol{\lambda}^{-1}(D)$ it follows that $\operatorname{conv} S \subset \boldsymbol{\lambda}^{-1}(D)$ (Problem 2).

Let $A \in \operatorname{conv} T$. Assume that $\mathbf{x}_1, \ldots, \mathbf{x}_p$ are p-orthonormal eigenvectors of A corresponding to the eigenvalues $\lambda_1(A), \ldots, \lambda_p(A)$. For any $B \in \mathbf{S}(\mathbf{V})$ let $B(\mathbf{x}_1, \ldots, \mathbf{x}_p) := (\langle B\mathbf{x}_i, \mathbf{x}_j \rangle)_{i,j=1}^{p} \in \mathbf{S}_p(\mathbb{R})$. We view $\mathbf{S}_p(\mathbb{R})$ as a real vector space of dimension $\binom{p+1}{2}$. Let $T' := \{A_1(\mathbf{x}_1, \ldots, \mathbf{x}_p), \ldots, A_N(\mathbf{x}_1, \ldots, \mathbf{x}_p)\} \subset \mathbf{S}_p(\mathbb{R})$. It is straightforward to show that for any $B \in \operatorname{conv} T$ one has $B(\mathbf{x}_1, \ldots, \mathbf{x}_p) \in \operatorname{conv} T'$. Let $\tilde{T}$ be the restriction of $\operatorname{conv} T'$ to the line in $\mathbf{S}_p(\mathbb{R})$

$$\{X = (x_{ij}) \in \mathbf{S}_p(\mathbb{R}) : \quad x_{ij} = \lambda_i(A)\delta_{ij}, \quad \text{for } i + j > 2\}.$$

Clearly $A(\mathbf{x}_1, \ldots, \mathbf{x}_p) \in \tilde{T}$. Hence $\tilde{T} = [C(\mathbf{x}_1, \ldots, \mathbf{x}_p), D(\mathbf{x}_1, \ldots, \mathbf{x}_p)]$ for some $C, D \in \operatorname{conv} T$. It is straightforward to show that $C, D \in \operatorname{conv}_{\binom{p+1}{2}-1} T$. (See Problem 8.) Hence $\max_{X \in \tilde{T}} x_{11} = \max(\langle C\mathbf{x}_1, \mathbf{x}_1 \rangle, \langle D\mathbf{x}_1, \mathbf{x}_1 \rangle)$. Without loss of generality we may assume that the above maximum is achieved for the matrix C. Hence, $C(\mathbf{x}_1, \ldots, \mathbf{x}_p)$ is a diagonal matrix such that $\lambda_1(C(\mathbf{x}_1, \ldots, \mathbf{x}_p) \geq \lambda_1(A)$ and $\lambda_i(C(\mathbf{x}_1, \ldots, \mathbf{x}_p))) = \lambda_i(A)$ for $i = 2, \ldots, p$. Let $\mathbf{U} = \operatorname{span}(\mathbf{x}_1, \ldots, \mathbf{x}_p)$. Since $\mathbf{x}_1, \ldots, \mathbf{x}_p$ are orthonormal it follows that $\lambda_i(Q(C, \mathbf{U})) = \lambda_i(C(\mathbf{x}_1, \ldots, \mathbf{x}_p))$ for $i = 1, \ldots, p$. Corollary 4.4.7 yields that $\boldsymbol{\lambda}_{(p)}(C) \geq \boldsymbol{\lambda}_{(p)}(A)$. Since h is increasing on D we get $h(\boldsymbol{\lambda}_{(p)}(C)) \geq h(\boldsymbol{\lambda}_{(p)}(A))$. See Problem 3 which shows that (4.15.4) is sharp. $\qquad \square$

Theorem 4.15.2 *Let* $\mathbf{V}$ *be an* n-*dimensional IPS over* $\mathbb{C}$. *Let* $p \in [n]$ *and* $D \subset \mathbb{R}^n_{\searrow}$ *be a convex Schur set. Let* D_p *be the projection of* D *on the first* p *coordinates. Let* $h : D_p \to \mathbb{R}$ *and assume that* $f : \boldsymbol{\lambda}^{-1}(D) \to \mathbb{R}$ *is the spectral function given by* $A \mapsto h(\boldsymbol{\lambda}_{(p)}(A))$. *Let* $S \subset \boldsymbol{\lambda}^{-1}(D)$. *Assume that* h *is nondecreasing on* D_p. *Then*

$$\sup_{A \in \operatorname{conv} S} f(A) = \sup_{B \in \operatorname{conv}_{p^2-1} S} f(B), \tag{4.15.2}$$

and this result is sharp.

See Problem 4 for the proof of the theorem.

It is possible to improve Theorems 4.15.1 and 4.15.2 in special interesting cases for $p > 1$.

Definition 4.15.3 *Let* $\mathbf{V}$ *be an n-dimensional IPS over* $\mathbb{F} = \mathbb{R}, \mathbb{C}$ *and* $p \in [n]$. *Let* $A \in \mathbf{S}(\mathbf{V})$. *Then the p-upper multiplicity of* $\lambda_p(A)$, *denoted by* upmul(A, p), *is a natural number in* $[p]$ *such that*

$$\lambda_{p-\mathrm{upmul}(A,p)}(A) > \lambda_{p-\mathrm{upmul}(A,p)+1}(A)$$
$$= \cdots = \lambda_p(A), \quad \text{where } \lambda_0(A) = \infty.$$

For any $\mathrm{C} \subset \mathbf{S}(\mathbf{V})$ *let* upmul$(\mathrm{C}, p) := \max_{A \in \mathrm{C}} \mathrm{upmul}(A, p)$.

See Problem 10 for sets satisfying upmul$(\mathrm{C}, p) \leq k$ for any $k \in \mathbb{N}$.

Theorem 4.15.4 *Let* $\mathbf{V}$ *be an n-dimensional IPS over* $\mathbb{R}$. *Let* $p \in [n]$ *and denote* $\mu(p) := \mathrm{upmul}(\mathrm{conv}\, S, p)$. *Then*

$$\sup_{A \in \mathrm{conv}\, S} \lambda_p(A) = \sup_{B \in \mathrm{conv}\, \frac{\mu(p)(2p-\mu(p)+1)}{2} - 1 S} \lambda_p(B). \qquad (4.15.3)$$

Proof. For $\mu(p) = p$ (4.15.4) follows from Theorem 4.15.1. Thus, it is enough to consider the case $p > 1$ and $\mu(p) < p$. As in the proof of Theorem 4.15.1 we may assume that $S = \{A_1, \ldots, A_N\}$ where $N \leq \binom{n+1}{2} + 1$. Let $\mathcal{M} := \{B \in \mathrm{conv}\, S : \lambda_p(B) = \max_{A \in \mathrm{conv}\, S} \lambda_p(A)\}$. Since $\lambda_p(A)$ is a continuous function on $\mathbf{S}(\mathbf{V})$ and conv S is a compact set it follows that $\mathcal{M}$ is a nonempty compact set of conv S. Let $\nu := \frac{\mu(p)(2p-\mu(p)+1)}{2} - 1$. Assume to the contrary that the theorem does not hold, i.e. $\mathcal{M} \cap \mathrm{conv}_\nu S = \emptyset$. Let $\mathcal{M}' := \{\mathbf{p} = (p_1, \ldots, p_N)^\top \in \mathcal{P}_N : \sum_{i=1}^n A_i \in \mathcal{M}\}$. Then $\mathcal{M}'$ is a nonempty compact set of $\mathcal{P}_N$ and any $\mathbf{p} \in \mathcal{M}'$ has at least $\nu + 2$ positive coordinates. Introduce the following complete order on $\mathcal{P}_N$. Let $\mathbf{x} = (x_1, \ldots, x_N)^\top, \mathbf{y} = (y_1, \ldots, y_N)^\top \in \mathbb{R}^N$. As in Definition 4.7.1 let $\bar{\mathbf{x}} = (\bar{x}_1, \ldots, \bar{x}_N)^\top, \bar{\mathbf{y}} = (\bar{y}_1, \ldots, \bar{y}_N)^\top \in \mathbb{R}^N$ be the rearrangements of the coordinates of the vectors $\mathbf{x}$ and $\mathbf{y}$ in the nonincreasing order. Then $\mathbf{x} \ll \mathbf{y}$ if either $\bar{\mathbf{x}} = \bar{\mathbf{y}}$ or $\bar{x}_i = \bar{y}_i$ for $i = 0, \ldots, m-1$ and $\bar{x}_m < \bar{y}_m$ for some $m \in [n]$. (We assume that $\mathbf{x}_0 = \mathbf{y}_0 = \infty$.) Since $\mathcal{M}'$ is compact there exists a maximal element $\mathbf{p} = (p_1, \ldots, p_N)^\top \in \mathcal{M}'$, i.e. $\mathbf{q} \in \mathcal{M}' \Rightarrow \mathbf{q} \ll \mathbf{p}$. Let

$\mathcal{I} := \{i \in \langle N \rangle : p_i > 0\}$. Then $\#\mathcal{I} \geq \nu + 2$. Let $B = \sum_{i=1}^{N} p_i A_i \in$ conv S be the corresponding matrix with the maximal λ_p on conv S. Assume that $\mathbf{x}_1, \ldots, \mathbf{x}_n \in \mathbf{V}$ is an orthonormal basis of $\mathbf{V}$, consisting of the eigenvectors of B corresponding to the eigenvalues $\lambda_1(B), \ldots, \lambda_n(B)$ respectively. Let $m := \mathrm{upmul}(B, p) \leq \mu(p)$. Consider the following systems of $\frac{m(2p-m+1)}{2}$ equations in $\#\mathcal{I}$ unknowns $q_i \in \mathbb{R}, i \in \mathcal{I}$:

$$q_i = 0, \text{ for } i \in \langle N \rangle \backslash \mathcal{I}, \quad \sum_{i \in \mathcal{I}} q_i = 0, \quad \sum_{i \in \mathcal{I}} q_i \langle A_i \mathbf{x}_j, \mathbf{x}_k \rangle = 0,$$

$$j = 1, \ldots, k - 1, \ k = p, \ldots, p - m + 1, \quad \sum_{i \in \mathcal{I}} q_i \langle A_i \mathbf{x}_j, \mathbf{x}_j \rangle$$

$$= \sum_{i \in \mathcal{I}} q_i \langle A_i \mathbf{x}_p, \mathbf{x}_p \rangle j = p - 1, \ldots, p - m + 1 \text{ if } m > 1.$$

Since $\#\mathcal{I} \geq \nu + 2 = \frac{\mu(p)(2p-\mu(p)+1)}{2} + 1 > \frac{m(2p-m+1)}{2}$ it follows that there exists $0 \neq \mathbf{q} = (q_1, \ldots, q_N)^\top \in \mathbb{R}^N$ whose coordinates satisfy the above equations. Let $B(t) := B + tC$, $C := \sum_{i=1}^{N} q_i A_i, t \in \mathbb{R}$. Then there exists $a > 0$ such that for $t \in [-a, a]$ $\mathbf{p}(t) := \mathbf{p} + t\mathbf{q} \in \mathcal{P}_N \Rightarrow B(t) \in$ conv S. As in the proof of Theorem 4.15.1 consider the matrix $B(t)(\mathbf{x}_1, \ldots, \mathbf{x}_p) \in \mathbf{S}_p(\mathbb{R})$. Since $B(0)(\mathbf{x}_1, \ldots, \mathbf{x}_p) = B(\mathbf{x}_1, \ldots, \mathbf{x}_p)$ is the diagonal matrix $\mathrm{diag}(\lambda_1(B), \ldots, \lambda_p(B))$ the conditions on the coordinates of $\mathbf{q}$ imply that $B(t)(\mathbf{x}_1, \ldots, \mathbf{x}_p)$ is of the form $(\mathrm{diag}(\lambda_1(B), \ldots, \lambda_{p-m}(B)) + tC_1) \oplus (\lambda_p + tb)I_m$ for a corresponding $C_1 \in \mathbf{S}_{p-m}(\mathbb{R})$. Since $\lambda_{p-m}(B) > \lambda_p(B)$ it follows that there exists $a' \in (0, a]$ such that

$$\lambda_{p-m}(B(t)(\mathbf{x}_1, \ldots, \mathbf{x}_p))$$

$$= \lambda_{p-m}(\mathrm{diag}(\lambda_1(B), \ldots, \lambda_{p-m}(B)) + tC_1)$$

$$> \lambda_p(B) + |tb|, \ \lambda_p(B(t)) = \lambda_p(B) + tb, \quad \text{for } |t| \leq a'.$$

Hence, $\lambda_p(B(t)) \geq \lambda_p(B) + tb$ for $|t| \leq a'$. As $B(t) \in$ conv S for $|t| \leq a'$ and $\lambda_p(B) \geq \lambda_p(B(t))$ for $|t| \leq a'$ it follows that $b = 0$ and $\lambda_p(B(t)) = \lambda_p(B)$ for $|t| \leq a'$. Hence, $\mathbf{p} + t\mathbf{q} \in \mathcal{M}'$ for $|t| \leq a'$. Since

$\mathbf{q} \neq 0$, it is impossible to have the inequalities $\mathbf{p} - a'\mathbf{q} \ll \mathbf{p}$ and $\mathbf{p} + a'\mathbf{q} \ll \mathbf{p}$. This contradiction proves the theorem. □

It is possible to show that the above theorem is sharp in the case $\mu(p) = 1$, see Problem 9 (d2). Similarly one can show that: (See Problem 6.)

Theorem 4.15.5 *Let* $\mathbf{V}$ *be an* n-*dimensional IPS over* $\mathbb{C}$. *Let* $p \in [n]$ *and denote* $\mu(p) := \mathrm{upmul}(\mathrm{conv}\, S, p)$. *Then*

$$\sup_{A \in \mathrm{conv}\, S} \lambda_p(A) = \sup_{B \in \mathrm{conv}_{\mu(p)(2p-\mu(p))-1} S} \lambda_p(B). \qquad (4.15.4)$$

Problems

1. (a) Let $\mathbf{x}, \mathbf{y} \in \mathbb{R}^n$. Show the implication $\mathbf{x} \leq \mathbf{y} \Rightarrow \mathbf{x} \preceq \mathbf{y}$.

   (b) Let $D \subset \mathbb{R}^n_{\searrow}$ and assume that $f : D \to \mathbb{R}$ is strong Schur's order preserving. Show that f is nondecreasing on D.

   (c) Let $i \in [2, n] \cap \mathbb{N}$ and f be the following function on $\mathbb{R}^n$: $(x_1, \ldots, x_n)^\top \mapsto x_i$. Show that f is nondecreasing on $\mathbb{R}^n$ but not Schur's order preserving on $\mathbb{R}^n_{\searrow}$.

2. Let $D \subset \mathbb{R}^n_{\searrow}$ be a convex Schur set. Let $\mathbf{V}$ be an n-dimensional IPS over $\mathbb{F} = \mathbb{R}, \mathbb{C}$. Let $S \subset \mathbf{S}(\mathbf{V})$ be a finite set such that $S \subset \boldsymbol{\lambda}^{-1}(D)$. Show that $\mathrm{conv}\, S \subset \boldsymbol{\lambda}^{-1}(D)$.

3. (a) Let $A \in \mathbf{H}_n$ and assume that $\mathrm{tr}\, A = 1$. Show that $\lambda_n(A) \leq \frac{1}{n}$ and equality holds if and only if $A = \frac{1}{n} I_n$.

   (b) Let $E_{kl} := [\frac{\delta_{(k,l)(i,j)} + \delta_{(l,k)(i,j)}}{2}]_{i,j=1}^p \in \mathbf{S}(p, \mathbb{R})$ for $1 \leq k \leq l \leq p$ be the symmetric matrices which have at most two nonzero equal entries at the locations (k, l) and (l, k) which sum to

1. Let $Q_1, \ldots, Q_{\binom{p+1}{2}} \in \mathbf{S}(p, \mathbb{R})$ be defined as follows:

$$Q_1 := E_{11} + E_{12}, \quad Q_2 := E_{11} - E_{12} + E_{13}, \ldots,$$

$$Q_p := E_{11} - E_{1p} + E_{23}, \ldots,$$

$$Q_{2p-3} := E_{11} - E_{2(p-1)} + E_{2p}, \ldots, Q_{\binom{p}{2}}$$

$$:= E_{11} - E_{(p-2)p} + E_{(p-1)p},$$

$$Q_{\binom{p}{2}+1} = E_{11} - E_{(p-1)p}, \quad Q_{\binom{p}{2}+i} = E_{ii}, \text{ for } i = 2, \ldots, p.$$

Let $S = \{Q_1, \ldots, Q_{\binom{p+1}{2}}\}$. Show that $\frac{1}{p} I_p \in \text{conv } S = \text{conv}_{\binom{p+1}{2}-1} S$ and $\frac{1}{p} I_p \notin \text{conv}_{\binom{p+1}{2}-2} S$.

(c) Let $S \subset \mathbf{S}(p, \mathbb{R})$ be defined as in (b). Show that $\text{tr } A = 1$ for each $A \in \text{conv } S$. Hence

$$\max_{A \in \text{conv } S} \lambda_p(A) = \lambda_p\left(\frac{1}{p} I_p\right) = \frac{1}{p} > \max_{B \in \text{conv}_{\binom{p+1}{2}-2} S} \lambda_p(B).$$

(d) Assume that $n > p$ and let $R_i := Q_i \oplus 0 \in \mathbf{S}(n, \mathbb{R})$, where Q_i is defined in b, for $i = 1, \ldots, \binom{p+1}{2}$. Let $S = \{R_1, \ldots, R_{\binom{p+1}{2}}\}$. Show that

$$\max_{A \in \text{conv } S} \lambda_p(A) = \lambda_p\left(\frac{1}{p} I_p \oplus 0\right) = \frac{1}{p} > \max_{B \in \text{conv}_{\binom{p+1}{2}-2} S} \lambda_p(B).$$

4. (a) Prove Theorem 4.15.2 repeating the arguments of Theorem 4.15.1. (*Hint:* Note that the condition $\langle B\mathbf{x}_i, \mathbf{x}_j \rangle = 0$ for two distinct orthonormal vectors $\mathbf{x}_i, \mathbf{x}_j \in \mathbf{V}$ is equivalent to two real conditions, while the condition $\langle B\mathbf{x}_i, \mathbf{x}_i \rangle = \lambda_i(A)$ is one real conditions for $B \in \mathbf{S}(\mathbf{V})$.)

(b) Modify the example in Problem 3 to show that Theorem 4.15.2 is sharp.

5. Let $C = A + \sqrt{-1}B \in \mathbb{C}^{n \times n}, A, B \in \mathbb{R}^{n \times n}$.

(a) Show $C \in \mathbf{H}_n$ if and only if A is symmetric and B is anti-symmetric: $B^\top = -B$.

(b) Assume that $C \in \mathbf{H}_n$ and let $\hat{C} \in \mathbb{R}^{(2n) \times (2n)}$ be defined as in Problem 6. Show that $\hat{C} \in \mathbf{S}(2n, \mathbb{R})$ and $\lambda_{2i-1}(\hat{C}) = \lambda_{2i}(\hat{C}) = \lambda_i(C)$ for $i = 1, \ldots, n$.

(c) Use the results of (b) to obtain a weaker version of Theorem 4.15.2 directly from Theorem 4.15.1.

6. Prove Theorem 4.15.5.

7. Let $\mathbb{F}$ be a field and $k \in \mathbb{Z}_+$. $A = [a_{ij}] \in \mathbb{F}^{n \times n}$ is called a $2k + 1$-diagonal matrix if $a_{ij} = 0$ if $|i - j| > k$. (1-diagonal are diagonal and 3-diagonal are called tridiagonal.) Then the entries $a_{1(k+1)}, \ldots, a_{(n-k)n}$ are called the k-upper diagonal.

(a) Assume that $A \in \mathbb{F}^{n \times n}$, $n > k$ and A is $2k + 1$-diagonal. Suppose furthermore that the k-upper diagonal of A does not have zero elements. Show that rank $A \geq n - k$.

(b) Suppose in addition to the assumptions in (a) that $A \in \mathbf{H}_n$. Show that $\mathrm{upmul}(A, p) \leq k$ for any $p \in [n]$.

8. Let $\mathbf{V}$ be an n-dimensional IPS over $\mathbb{F} = \mathbb{R}, \mathbb{C}$. Let $S \subset \mathbf{S}(\mathbf{V})$ and $p \in [n]$. Define the *weak* p-upper multiplicity denoted by $\mathrm{wupmul}(\mathrm{conv}\, S, p)$ as follows. It is the smallest positive integer $m \leq p$ such that for any $N = \binom{n+1}{2} + 1$ operators $A_1, \ldots, A_N \in S$ there exists a sequence $A_{j,k} \in \mathbf{S}(\mathbf{V}), j \in [N], k \in \mathbb{N}$, such that $\lim_{k \to \infty} A_{j,k} = A_j, j \in [N]$ and $\mathrm{upmul}(\mathrm{conv}\{A_{1,k}, \ldots, A_{N,k}\}, p) \leq m$ for $k \in \mathbb{N}$.

(a) Show that $\mathrm{wupmul}(\mathrm{conv}\, S, p) \leq \mathrm{upmul}(\mathrm{conv}\, S, p)$.

(b) Show that in Theorems 4.15.4 and 4.15.5 one can replace $\mathrm{upmul}(\mathrm{conv}\, S, p)$ by $\mathrm{wupmul}(\mathrm{conv}\, S, p)$.

9. (a) Show that for any set $S \subset \mathbf{D}(n, \mathbb{R})$ and $p \in [n]$ $\mathrm{wupmul}(\mathrm{conv}\, S, p) = 1$. (*Hint*: Use Problem 7.)

(b) Let $D_i = \mathrm{diag}(\delta_{i1}, \ldots, \delta_{in}), i = 1, \ldots, n$. Let S := $\{D_1, \ldots, D_n\}$. Show that for $p \in [2, n] \cap \mathbb{N}$

$$\max_{D \in \mathrm{conv\,S}} \lambda_p(D) = \max_{D \in \mathrm{conv}_{p-1}\mathrm{S}} \lambda_p(D)$$

$$= \frac{1}{p} > \max_{D \in \mathrm{conv}_{p-2}\mathrm{S}} \lambda_p(D) = 0.$$

(c) Show that the variation of Theorem 4.15.4 as in Problem 8(b) for wupmul(conv S, p) = 1 is sharp.

(d) Let $A \in \mathbf{S}(n, \mathbb{R})$ be a tridiagonal matrix with nonzero elements on the first upper diagonal as in 7(b). Let $t \in \mathbb{R}$ and define $D_i(t) = D_i + tA$, where D_i is defined as in b, for $i = 1, \ldots, n$. Let S(t) = $\{D_1(t), \ldots, D_n(t)\}$. Show

d1. For $t \neq 0$ upmul(conv S(t), p) = 1 for $p \in [2, n] \cap \mathbb{Z}$.

d2. There exists $\epsilon > 0$ such that for any $|t| \leq \epsilon$

$$\max_{A \in \mathrm{conv\,S}(t)} \lambda_p(A) = \max_{B \in \mathrm{conv}_{p-1}\mathrm{S}(t)} \lambda_p(B)$$

$$> \max_{C \in \mathrm{conv}_{p-2}\mathrm{S}(t)} \lambda_p(C).$$

Hence, Theorem 4.15.4 is sharp in the case upmul (conv S, p) = 1.

10. (a) Let $S \subset H_n$ be a set of $2k+1$-diagonal matrices. Assume that either each k-upper diagonal of any $A \in S$ consists of positive elements, or all k-upper diagonals of $A \in S$ are equal and consist of nonzero elements. Show that upmul(conv S, p) $\leq k$.

(b) Let $S \subset H_n$ be a set of $2k + 1$-diagonal matrices. Show that wupmul(conv S, p) $\leq k + 1$.

4.16 Multiplicity Index of a Subspace of S(V)

Definition 4.16.1 *Let* **V** *be a finite dimensional IPS over* $\mathbb{F} = \mathbb{R}, \mathbb{C}$. *Let* **U** *be a nontrivial subspace of* $\mathbf{S}(V)$. *Then the* *multiplicity index of* **U** *is defined*

$$\text{mulind } \mathbf{U} := \{\max p \in \mathbb{N} : \exists A \in \mathbf{U} \backslash \{0\} \text{ such that } \lambda_1(A)$$
$$= \cdots = \lambda_p(A)\}.$$

Clearly for any nontrivial **U** mulind $\mathbf{U} \in [1, \dim \check{\mathbf{V}}]$. Also mulind $\mathbf{U} = \dim \mathbf{V} \iff I \in \mathbf{V}$. Let

$$\kappa(r, n, \mathbb{R}) := \frac{(r-1)(2n-r+2)}{2},$$

$$\kappa(r, n, \mathbb{C}) := (r-1)(2n-r+1). \tag{4.16.1}$$

The aim of this section to prove the following theorem.

Theorem 4.16.2 *Let* **V** *be an IPS over* $\mathbb{F} = \mathbb{R}, \mathbb{C}$ *of dimension* $n \geq 3$. *Assume that* $r \in [n-1] \backslash \{1\}$. *Let* **U** *be a subspace of* $\mathbf{S}(\mathbf{V})$. *Then* mulind $\mathbf{U} \geq r$ *if* dim $\mathbf{U} \geq \kappa(r, n, \mathbb{F})$ *and this result is sharp.*

Proof. Assume first that $\mathbb{F} = \mathbb{R}$. Fix the value of $n \geq 3$ and denote $\kappa(r) := \kappa(r, n, \mathbb{F})$. Suppose to the contrary that the theorem is wrong. Let $n \geq 3$ be the minimal positive integer for which the theorem is false. Let $r \in [n-1] \backslash \{1\}$ and $\mathbf{U} \in \mathrm{Gr}(\kappa(r), \mathbf{S}(\mathbf{V}))$ be the minimal r and a corresponding subspace $\mathbf{U} = \mathrm{span}\,(B_1, \ldots, B_{\kappa(r)})$ for which the theorem is false. The minimality of r yields that there exists $A \in \mathbf{U}$ such that $\lambda_1(A) = \cdots = \lambda_{r-1}(A) > \lambda_r(A)$. Assume that

$$A\mathbf{x}_i = \lambda_i(A)\mathbf{x}_i, \quad \mathbf{x}_i \in \mathbf{V}, \quad \langle \mathbf{x}_i, \mathbf{x}_j \rangle = \delta_{ij}, \ i, j = 1, \ldots, n.$$

For $r - 1$ orthonormal vectors $\mathbf{y}_1, \ldots, \mathbf{y}_{r-1} \in \mathbf{V}$ denote

$$\mathcal{B}(\mathbf{y}_1, \ldots \mathbf{y}_{r-1}) := \{B \in \mathbf{U}, B = \sum_{j=1}^{\kappa(r)} t_j B_j, \ \sum_{j=1}^{\kappa(r)} t_j B_j \mathbf{y}_i - t_0 \mathbf{y}_i = \mathbf{0},$$

$$i = 1, \ldots, r-1, \text{ for some } t_0, \ldots, t_{\kappa(r)} \in \mathbb{R}\}.$$

We claim that dim $\mathcal{B}(\mathbf{y}_1, \ldots, \mathbf{y}_{r-1}) \geq 1$. Indeed, by representing $\mathbf{S}(\mathbf{V})$ as $S(n, \mathbb{R})$ with respect to the orthonormal basis $\mathbf{y}_1, \ldots, \mathbf{y}_n$, we may assume that **U** is a $\kappa(r)$ dimensional subspace of $\mathbf{S}(n, \mathbb{R})$

and $\mathbf{y}_1, \ldots, \mathbf{y}_n$ is a standard basis in $\mathbb{R}^n$. Then the number of variables in the equations defining $\mathcal{B}(\mathbf{y}_1, \ldots, \mathbf{y}_{r-1})$ is $\kappa(r) + 1$, while the number of equations is $\kappa(r)$. Note also that if in these equations $t_1 = \cdots = t_{\kappa(r)} = 0$ then $t_0 = 0$. Hence, there exists a nontrivial $B \in \mathcal{B}(\mathbf{y}_1, \ldots, \mathbf{y}_{r-1})$.

Consider next $\mathcal{B}(\mathbf{x}_1, \ldots, \mathbf{x}_{r-1})$. Clearly, $A \in \mathcal{B}(\mathbf{x}_1, \ldots, \mathbf{x}_{r-1})$. Suppose that $\dim \mathcal{B}(\mathbf{x}_1, \ldots, \mathbf{x}_{r-1}) > 1$. So there exists $B \in \mathcal{B}(\mathbf{x}_1, \ldots, \mathbf{x}_{r-1})$, $B \notin \mathrm{span}\,(A)$ such that $B\mathbf{x}_i = s\mathbf{x}_i$ for $i = 1, \ldots, r-1$. By considering $-B$ instead of B if necessary, we may assume that $\lambda_1(B) > 0$. Consider $A(b) := A + bB, b \in \mathbb{R}$. Clearly $A(b)\mathbf{x}_i = (\lambda_1(A) + bs)\mathbf{x}_i$ for $i = 1, \ldots, r-1$. Furthermore, $\lambda_i(A(0)) = \lambda_1(A) > \lambda_r(A) = \lambda_r(A(0))$ for $i = 1, \ldots, r-1$. Hence, there exists a positive ϵ so that $\lambda_1(A(b)) = (\lambda_1(A) + bs)$ for $|b| < \epsilon$.

Suppose first that $s = 0$. Increase continuously b from the value 0. Since we assumed that our theorem fails for $\mathbf{U}$ it follows that for each $b \geq 0$ $\lambda_1(A(b)) = \lambda_1(A)$. One the other hand the Weyl's inequality yields that $\lambda_1(A(b)) \geq b\lambda_1(B) + \lambda_n(A)$. For $b \gg 1$ we obtain a contradiction.

Hence, $s \neq 0$. By considering $A(b)$ for $|b| < \epsilon$ we deduce that

$$\lambda_1(A(b)) = \cdots = \lambda_{r-1}(A(b)) = \lambda_1(A) + bs > \lambda_r(A(b)).$$

Replacing A by $A(b)$ if needed, we can assume that $\lambda_1(A) \neq 0$. By considering $B' := B - s'A$ we can assume that $B'\mathbf{x}_i = \mathbf{0}$ for $i = 1, \ldots, r-1$. This gives us a contradiction as above. Hence, we conclude that $\dim \mathcal{B}(\mathbf{x}_1, \ldots, \mathbf{x}_r) = 1$.

Next we claim that $\dim \mathcal{B}(\mathbf{y}_1, \ldots, \mathbf{y}_{r-1}) = 1$ for any system of $r-1$ orthonormal vectors $\mathbf{y}_1, \ldots, \mathbf{y}_{r-1}$ in $\mathbf{V}$. Furthermore, one can choose $B(\mathbf{y}_1, \ldots, \mathbf{y}_{r-1}) \in \mathcal{B}(\mathbf{y}_1, \ldots, \mathbf{y}_{r-1}) \setminus \{0\}$ such that

$$B(\mathbf{y}_1, \ldots, \mathbf{y}_{r-1})\mathbf{y}_i = \lambda_1 \mathbf{y}_i, \quad i = 1, \ldots, r-1. \tag{4.16.2}$$

Since $r < n$ it is known that there exists a continuous family of $r-1$ orthonormal vectors $\mathbf{x}_1(t), \ldots, \mathbf{x}_{r-1}(t)$ for $t \in [0,1]$ such that

$$\mathbf{x}_i(0) = \mathbf{x}_i, \quad \mathbf{x}_i(1) = \mathbf{y}_i, \quad i = 1, \ldots, r-1. \tag{4.16.3}$$

See Problem 4.2.9(e). Since $\dim \mathcal{B}(\mathbf{x}_1, \ldots, \mathbf{x}_{r-1}) = 1$ it follows that there exists $\epsilon > 0$ such that for $t \in [0, \epsilon)$ $\dim \mathcal{B}(\mathbf{x}_1(t), \ldots, \mathbf{x}_{r-1}(t)) = 1$. Let $s \in (0, 1]$ and assume that for each $t \in [0, s]$ $\dim \mathcal{B}(\mathbf{x}_1(t), \ldots, \mathbf{x}_{r-1})(t) = 1$. Then we can choose a unique $B(t) \in \mathcal{B}(\mathbf{x}_1(t), \ldots, \mathbf{x}_{r-1}(t)), t \in [0, s]$ with the following properties. First, $B(0) = A$. Second, $\operatorname{tr} B(t)^2 = \operatorname{tr} A^2$. Third, $B(t)$ depends continuously on t. Since $\mathbf{U}$ does not satisfy the theorem it follows that $\lambda_1(B(t)) = \cdots = \lambda_{r-1}(B(t)) > \lambda_r(B(t))$. Furthermore, $\mathbf{x}_1(t), \ldots, \mathbf{x}_{r-1}(t)$ are $r-1$ orthonormal eigenvectors of $B(t)$ corresponding to $\lambda_1(B(t))$. Suppose first that $s = 1$. Then our claim holds for $\mathcal{B}(\mathbf{y}_1, \ldots, \mathbf{y}_{r-1})$. Assume now that $s_0 \in (0, 1]$ is the minimal $t \in (0, 1]$ satisfying $\dim \mathcal{B}(\mathbf{x}_1(t), \ldots, \mathbf{x}_{r-1}(t)) > 1$. Let $B(t)$ be the unique matrix in $\mathcal{B}(\mathbf{x}_1(t), \ldots, \mathbf{x}_{r-1}(t))$ defined as above for $t < s_0$. Let $s_j, j \in \mathbb{N}$ be an increasing sequence converging to s_0. By taking a subsequence of $s_j, j \in \mathbb{N}$ we can assume that $\lim_{j \to \infty} B(s_j) = C$. Then $C \in \mathcal{B}(\mathbf{x}_1(s_0), \ldots, \mathbf{x}_{r-1}(s_0))$. Furthermore, $\lambda_1(C) = \cdots = \lambda_{r-1}(C)$ and $\mathbf{x}_1(s_0), \ldots, \mathbf{x}_{r-1}(s_0)$ are $r-1$ orthonormal vectors corresponding to $\lambda_1(C)$. The above arguments show that the assumption $\dim \mathcal{B}(\mathbf{x}_1(s_0), \ldots, \mathbf{x}_{r-1}(s_0)) > 1$ contradicts that $\mathbf{U}$ violates our theorem. Hence, $\dim \mathcal{B}(\mathbf{x}_1(t), \ldots, \mathbf{x}_{r-1})(t) = 1$ for each $t \in [0, 1]$ and our claim is proved.

We finally obtain a contradiction to the assumption that $\mathbf{U}$ violates our theorem by constructing a nonzero matrix $C \in \mathbf{U}$ satisfying

$$C\mathbf{y}_i = \lambda_{i+1}(C)\mathbf{y}_i, \quad i = 1, \ldots, r-1, \quad \lambda_1(C) > \lambda_2(C) > \lambda_n(C),$$

for some $r-1$ orthonormal vectors $\mathbf{y}_1, \ldots, \mathbf{y}_{r-1}$.

Suppose first that $r = 2$. Since $\mathbf{U}$ violates our theorem it follows that any $C \in \mathbf{U} \setminus \{0\}$ satisfies the above condition. Assume now that $r > 2$. Consider a $\kappa(r) - 1$ dimensional subspace $\mathbf{U}'$ of $\mathbf{U}$ which does not contain A. Let $\mathbf{x}_1, \ldots, \mathbf{x}_n$ be an orthonormal set of eigenvectors of A as defined above. In $\mathbf{U}'$ consider the subspace of all matrices B satisfying $B\mathbf{x}_i = \mathbf{0}$ for $i = 2, \ldots, r-1$. By assuming $\mathbf{x}_1, \ldots, \mathbf{x}_n$ is a standard basis in $\mathbb{R}^n$ we deduce that one has exactly $\kappa(r-1)$ linear conditions on B viewed as real symmetric matrices. Since $\kappa(r-1) < \kappa(r) - 1$ one has a nonzero $B \in \mathbf{U}'$

satisfying the above conditions. If $\lambda_2(B) = \cdots = \lambda_{n-1}(B) = 0$ then $C = B$. Otherwise we can assume that $\lambda_1(B) \geq \lambda_2(B) > 0$. Let $B(t) = A + tB$. So $\lambda_1(A)$ is an eigenvalue of $B(t)$ of multiplicity $r - 2$ at least. Clearly, Then there exists $\epsilon > 0$ such that for $|t| < \epsilon$ $\lambda_r(B(t)) < \lambda_1(B(0)) = \lambda_1(A)$. Thus, for $|t| \leq \epsilon$ $\lambda_2(B(t)) \leq \lambda_1(A)$. For $t \gg 1$ $\lambda_1(B(t)) \geq \lambda_2(B(t)) > \lambda_1(A)$. Let $T := \{t > 0, \ \lambda_2(B(t)) > \lambda_1(A)\}$. So each $t \in T$ satisfies $t \geq \epsilon$. Let $t_0 := \inf\{t, \ t \in T\}$. We claim that $\lambda_1(B(t_0)) > \lambda_2(B(t_0)) = \cdots = \lambda_r(B(t_0)) = \lambda_1(A)$. In view of definition t_0 we must have equality $\lambda_2(B(t_0)) = \lambda_1(A)$. On the other hand for $t > t_0$ $\lambda_2(B(t)) > \lambda_1(A)$. Hence, $\lambda_2(B(t_0))$ has to be an eigenvalue of multiplicity $r - 1$ at least. Since $\mathbf{U}$ violates our theorem we must have the inequality $\lambda_1(B(t_0)) > \lambda_2(B(t_0))$. Also if $\lambda_2(B(t_0)) = \lambda_n(B(t_0))$ then $\lambda_1(-B(t_0)) = \cdots = \lambda_{n-1}(-B(t_0))$ which contradicts our assumption that $\mathbf{U}$ violates out theorem. Hence, $C = B(t_0) \in \mathbf{U}$ is whose existence was claimed above.

We now show that such C contradicts our previous results. Assume that $C\mathbf{y}_i = \lambda_2(C)\mathbf{y}_i$ for $i = 2, \ldots, r$, where $\mathbf{y}_2, \ldots, \mathbf{y}_r$ is an orthonormal system. Consider the set $B(\mathbf{y}_2, \ldots, \mathbf{y}_r)$. Clearly, $C \in B(\mathbf{y}_2, \ldots, \mathbf{y}_r)$. We showed that $\dim B(\mathbf{y}_2, \ldots, \mathbf{y}_r) = 1$. Hence, span $(C) = B(\mathbf{y}_2, \ldots, \mathbf{y}_r)$. We also showed that the maximal eigenvalue of either C or $-C$ has multiplicity $r - 1$. This contradicts our results on C. Hence, for each subspace $\mathbf{U}$ of $\mathbf{S}(\mathbf{V})$ over $\mathbb{R}$ of dimension $\kappa(r)$ at least mulind $\mathbf{U} \geq r$.

We now show that our result is sharp. Let $\mathbf{U} \subset \mathbf{S}(n, \mathbb{R})$ be a subspace of matrices $A = [a_{ij}]_{i,j=1}^n$ satifying

$$a_{ij} = 0, \ i, j = 1, \ldots, n-r+1, \quad \mathrm{tr}\, A = 0.$$

Clearly, $\dim \mathbf{U} = \kappa(r) - 1$. We claim that there exists a nonzero matrix A in $\mathbf{U}$ such that $\lambda_1(A) = \cdots = \lambda_r(A)$. Assume to the contrary that such nonzero A exists. As $\mathrm{tr}\, A = 0$ we must have that $\lambda_1(A) > 0$. Consider the matrix $B = \lambda_1(A)I_n - A$. So rank $B \leq n-r$. One the other hand the $(n-r+1) \times (n-r+1)$ submatrix based on the first $n-r+1$ rows and columns if $\lambda_1(A)I_{n-r+1}$. So rank $B \geq n-r+1$ which contradicts the previous observation on the rank of B.

The proof of the theorem for $\mathbb{F} = \mathbb{C}$ is similar, and is left as a Problem 1. □

Denote

$$f(r, \mathbb{R}) := \frac{r(r+1)}{2}, \quad f(r, \mathbb{C}) = r^2. \tag{4.16.4}$$

Theorem 4.16.3 *Let* $\mathbf{V}$ *be an IPS over* $\mathbb{F} = \mathbb{R}, \mathbb{C}$ *of dimension* $n \geq 2$. *Assume that* $r \in [n-1]$. *Let* $\mathbf{W}$ *be a subspace of* $\mathbf{S}(\mathbf{V})$. *Assume that* $\mathbf{W}$ *has the following property: If* $\mathbf{x}_1, \ldots, \mathbf{x}_r \in \mathbf{V}$ *satisfy the equalities*

$$\sum_{i=1}^{r} \langle A\mathbf{x}_i, \mathbf{x}_i \rangle = 0 \ \text{for all } A \in \mathbf{W} \tag{4.16.5}$$

then $\mathbf{x}_1 = \cdots = \mathbf{x}_r = \mathbf{0}$. *Suppose that*

$$\dim \mathbf{W} < f(r+1, \mathbb{F}) - \left\lfloor \frac{r+1}{n} \right\rfloor. \tag{4.16.6}$$

Then $\mathbf{W}$ *contains a positive definite operator.*

To prove the theorem we need the following results. Observe first that $\mathbf{S}(\mathbf{V})$ is a real vector space of dimension $f(\dim \mathbf{V}, \mathbb{F})$. $\mathbf{S}(\mathbf{V})$ has a natural inner product $\langle S, T \rangle = \text{tr}(ST)$. Furthermore, any linear functional $\phi : \mathbf{S}(\mathbf{V}) \to \mathbb{R}$ is of the form $\phi(X) = \text{tr}(XF)$ for some $F \in \mathbf{S}(\mathbf{V})$. See Problem 4.9.2. For a subspace $\mathbf{W} \subset \mathbf{S}(\mathbf{V})$ denote by $\mathbf{W}^\perp \subset \mathbf{S}(\mathbf{V})$ the orthogonal complement of $\mathbf{W}$. Recall that $\mathbf{S}_+(\mathbf{V})$ and $\mathbf{S}_+(\mathbf{V})^o$ are the closed cone of non-negative definite operators and the positive definite operators respectively in $\mathbf{S}(\mathbf{V})$.

Lemma 4.16.4 *Let* $\mathbf{W}$ *be a nontrivial subspace of* $\mathbf{S}(\mathbf{V})$. *Then*

$$\mathbf{W} \cap \mathbf{S}_+(\mathbf{V}) = \{0\} \iff \mathbf{W}^\perp \cap \mathbf{S}_+(\mathbf{V})^o \neq \emptyset. \tag{4.16.7}$$

Proof. Suppose that the right-hand side of (4.16.7) holds. Problem 4.9.2 implies the left-hand side of (4.16.7).

Assume now that the left-hand side of (4.16.7) holds. Let $\mathbf{S}_{+,1}(\mathbf{V})$ be the set of all non-negative definite operators with trace one. Clearly, $\mathbf{S}_{+,1}(\mathbf{V})$ is a compact convex set. The left-hand of

(4.16.7) is equivalent to $\mathbf{W} \cap \mathbf{S}_{+,1}(\mathbf{V}) = \emptyset$. Since $\mathbf{W}$ is a closed convex set, there is a linear functional $\phi : \mathbf{S}(\mathbf{V}) \to \mathbb{R}$ which separates $\mathbf{W}$ and $\mathbf{S}_{+,1}(\mathbf{V})$ [Roc70]. That is, there exists $a \in \mathbb{R}$ such that $\phi(X) \le a$ for each $X \in \mathbf{W}$ and $\phi(S) > a$ for each $S \in \mathbf{S}_{+,1}(\mathbf{V})$. Since $\mathbf{W}$ is a nontrivial subspace, it follows that ϕ vanishes on $\mathbf{W}$. So $a \ge 0$. Assume that $\phi(X) = \operatorname{tr} XT$ for all $X \in \mathbf{S}(\mathbf{V})$. So $\operatorname{tr} ST > 0$ for each $S \in \mathbf{S}_{+,1}(\mathbf{V})$. Problem 4.9.3 yields that $T \in \mathbf{S}_{+}(\mathbf{V})^{o}$. As $\operatorname{tr} XT = 0$ for $X \in \mathbf{W}$ we deduce that $T \in \mathbf{W}^{\perp}$. $\qquad\square$

Proof of Theorem 4.16.3. Let $\mathbf{U} = \mathbf{W}^{\perp}$. According to Lemma 4.16.4 one needs to show that $\mathbf{U} \cap \mathbf{S}_{+}(\mathbf{V}) = \{0\}$. Assume to the contrary that $\mathbf{U} \cap \mathbf{S}_{+}(\mathbf{V})$ contains a nonzero non-negative definite S. Let T be a nonzero non-negative matrix in $\mathbf{U}$ with a minimal rank m. The assumption that the condition (4.16.5) yields that $\mathbf{x}_1 = \cdots \mathbf{x}_r = 0$ implies that $m > r$.

Observe that

$$\dim \mathbf{U} > f(n, \mathbb{F}) - f(r+1, \mathbb{F}) + \left\lfloor \frac{r+1}{n} \right\rfloor$$

$$= \kappa(n - r, n, \mathbb{F}) + \left\lfloor \frac{r+1}{n} \right\rfloor.$$

Assume first that $m = n$. So $T = P^2$ for some positive definite P. Let $\mathbf{U}_1 \subset \mathbf{U}$ be a subspace of codimension one such that $P \notin \mathbf{U}_1$. Clearly, $\dim \mathbf{U}_1 \ge \kappa(n - r, n, \mathbb{F}) + \lfloor \frac{r+1}{n} \rfloor \ge 1$ for $r \in [n - 1]$. (Note that for $r = n - 1$ one has $\kappa(1, n, \mathbb{F}) = 0, \lfloor \frac{r+1}{n} \rfloor = 1$.) Consider the subspaces $\mathbf{U}_1' := P^{-1} \mathbf{U}_1 P^{-1} \subset \mathbf{U}' := P^{-1} \mathbf{U} P^{-1} \subset \mathbf{S}(\mathbf{V})$. So $I \in \mathbf{U}'$ and $I \notin \mathbf{U}_1'$. Suppose first that $r \le n - 2$, i.e. $n - r \in [n - 1] \setminus \{1\}$. Since $\dim \mathbf{U}_1' \ge \kappa(n - r)$, Theorem (4.16.2) yields the existence of a nonzero $A \in \mathbf{U}_1'$ such that $\lambda_1 = \cdots = \lambda_{n-r}(A)$. Hence, $B := \lambda_1(A)I - A$ is a nonzero non-negative definite operator at most of rank r.

Suppose now that $r = n - 1$. Then $\mathbf{U}_1'$ contains a nonzero matrix A. So $B := \lambda_1(A)I - A$ a nonzero non-negative definite operator of rank $r = n - 1$ at most. Therefore, PBP is a nonzero non-negative definite operator of at most rank r in $\mathbf{U}$. This contradicts the minimality of the rank of T.

It is left to consider the case where rank $T = m < n$. For $r = n-1$ such T does not exists. Hence, $\mathbf{U} \cap \mathbf{S}_+(\mathbf{V}) = \{0\}$ as claimed.

Assume now that $r \in [n-2]$. Suppose furthermore that $\mathbf{S}(\mathbf{V})$ is either $\mathbf{S}(n, \mathbb{R})$ or $\mathbf{H}_n$. By considering the subspace $P^{-1}\mathbf{U}P^{-1}$ we may assume without loss of generality that $T = A_1 = \mathrm{diag}(I_m, 0)$. Let $A_1, \ldots, A_d$ be a basis of $\mathbf{U}$. So $d \geq \kappa(n - r, n, \mathbb{F}) + 1$ Partition $A_j = [A_{j,1} \ A_{j,2}]$, where each $A_{j,1}$ is $n \times m$ matrix. We claim that $A_{2,2}, \ldots, A_{2,d}$ are linearly dependent over $\mathbb{R}$. Assume that $\sum_{j=2}^{m} a_j A_{j,2} = 0$. The number of variables $a_2, \ldots, a_d$ is at least $\kappa(n - r, n, \mathbb{F})$. The number of real equations is $f(n, \mathbb{F}) - f(m, \mathbb{F}) = \kappa(n - m + 1, \mathbb{F})$. Thus, if $m > r + 1$ we have always a nontrivial solution in $a_2, \ldots, a_d$.

So assume to the contrary that $m = r + 1, d = \kappa(n - r) + 1$ and $A_{2,2}, \ldots, A_{2,d}$ are linearly independent. Partition each $A_{j,2} = \begin{bmatrix} E_{j,2} \\ F_{j,2} \end{bmatrix}$ for $j = 2, \ldots, d$. Hence, span $(A_2, \ldots, A_d)$ contains a block diagonal matrix $\mathrm{diag}(C, F)$, where F is an arbitrary diagonal matrix. In particular $A = \mathrm{diag}(C, I_{n-r-1}) \in \mathbf{U}$. Hence, $a_1 A_1 + A$ is positive definite for $a_1 \gg 1$. This fact contradicts our assumption that $\mathbf{U}$ does not contain a positive definite matrix. Hence $A_{2,2}, \ldots, A_{2,d}$ are linearly dependent over $\mathbb{R}$. So there exists a nonzero matrix in span $(A_2, \ldots, A_d)$ of the form $A = \mathrm{diag}(F, 0)$. As A_1 and A are linearly independent it follows that $F \neq \lambda_1(F)I_m$. Hence $\lambda_1(F)A_1 - A$ is a nonzero non-negative definite matrix of rank at most $m - 1$. This contradicts the minimality of the rank of T. $\qquad\square$

Problems

1. Prove Theorem 4.16.2 for $\mathbb{F} = \mathbb{C}$.

2. (Calabi's theorem.) Let $S_1, S_2 \in \mathbf{S}(n, \mathbb{R})$. Assume that if $\mathbf{x} \in \mathbb{R}^n$ satisfies $\mathbf{x}^\top S_1 \mathbf{x} = \mathbf{x}^\top S_2 \mathbf{x} = 0$ then $\mathbf{x} = \mathbf{0}$.

   (a) Show that if $n \geq 3$ then there exists $a_1, a_2 \in \mathbb{R}$ such that $a_1 S_1 + a_2 S_2$ is positive definite.

(b) Assume that $n = 2$. Give an example of $S_1, S_2 \in \mathbf{S}(2, \mathbb{R})$ satisfying the above conditions, such that span $(S_1, S_2) \cap \mathbf{S}_+(2, \mathbb{R}) = \{0\}$.

3. Show that Theorem 4.16.3 implies Theorem 4.16.2.

4.17   Rellich's Theorem

Definition 4.17.1

1. *For $S \subset \mathbb{C}$ denote $\bar{S} := \{z, \ \bar{z} \in S\}$.*

2. *A domain $\Omega \subset \mathbb{C}$ is called $\mathbb{R}$-symmetric if $\bar{\Omega} = \Omega$.*

3. *Assume that a domain $\Omega \subset \mathbb{C}$ is $\mathbb{R}$-symmetric. Let $A(z), U(z) \in \mathrm{H}(\Omega)^{n \times n}$. Then $A(z)$ and $U(z)$ are called hermitian and unitary analytic respectively if*

$$A(z) = A(\bar{z})^*, \quad U(z)^{-1} = U(\bar{z})^*, \quad \textit{for all } z \in \Omega. \quad (4.17.1)$$

Theorem 4.17.2 (*Rellich's theorem*) *Let Ω be an $\mathbb{R}$-symmetric domain. Assume that $A \in \mathrm{H}(\Omega)^{n \times n}$ is hermitian analytic in Ω. Suppose that J is a real open interval in Ω. Then there exists an $\mathbb{R}$-symmetric domain Ω_1, where $J \subset \Omega_1 \subsetneq \Omega$, such that the following properties hold:*

1. *The characteristic polynomial of $A(z)$ splits in $\mathrm{H}(\Omega_1)$. That is, $\det(\alpha I_n - A(z)) = \prod_{i=1}^{n}(\alpha - \alpha_i(z))$, where $\alpha_1(z), \ldots, \alpha_n(z) \in \mathrm{H}(\Omega_1)$.*

2. *There exists a unitary analytic $U \in \mathrm{H}(\Omega_1)^{n \times n}$ such that $A(z) = U(z) \operatorname{diag}(\alpha_1(z), \ldots, \alpha_n(z)) U(\bar{z})^*$.*

To prove the above theorem we start with the following result.

Lemma 4.17.3 *Let $\Omega \subset \mathbb{C}$ be an $\mathbb{R}$-symmetric domain. Assume that $A \in \mathrm{H}(\Omega)^{n \times n}$ is hermitian analytic in Ω. Let $C(p_1, \ldots, p_k) \in \mathrm{H}(\Omega)^{n \times n}$ be the rational canonical form of A over the field of rational functions $\mathcal{M}(\Omega)$. $p_1(\alpha, z), \ldots, p_k(\alpha, z) \in \mathrm{H}(\Omega)[\alpha]$ are nontrivial* *invariant polynomials of $\alpha I_n - A(z)$, monic in α, such that $p_j | p_{j+1}$ for $j = 1, \ldots, k-1$ and $\prod_{j=1}^{k} p_k(\alpha, z) = \det(\alpha I_n - A(z))$. Then there exists a nonzero analytic function $f \in \mathrm{H}(\Omega)$ such that the following conditions hold:*

1. *Denote by $Z(f)$ the zero set of f in Ω. (So $Z(f)$ is a countable set whose accumulation points are on the boundary of Ω.)*

2. *Let $\Omega_2 := \Omega \setminus Z(f)$. Then A is similar to $C(p_1, \ldots, p_k)$ over $\mathrm{H}(\Omega_2)$.*

3. *For each $\zeta \in \Omega_2$ the polynomial $p_k(\alpha, \zeta)$ has simple roots.*

4. *For each $\zeta \in \Omega_2$ the matrix $A(\zeta)$ is diagonable.*

Proof. Theorem 2.3.6 yields that $C(p_1, \ldots, p_k) \in \mathrm{H}(\Omega)^{n \times n}$. As A is similar to $C(p_1, \ldots, p_k)$ over $\mathcal{M}(\Omega)$ it follows that $A = XC(p_1, \ldots, p_k)X^{-1}$ for some $X \in \mathbf{GL}(n, \mathcal{M}(\Omega))$. Let g be the product of the denominators of the entries of X. Then $Y := gX \in \mathrm{H}(\Omega)^{n \times n} \cap \mathbf{GL}(n, \mathcal{M}(\Omega))$ and $A = YC(p_1, \ldots, p_k)Y^{-1}$. Let $f_1 := \det Y$. Then $Z(f_1)$ is a countable set in Ω whose accumulation points are on the boundary of Ω. So for each $\zeta \in \Omega_3 := \Omega \setminus Z(f_1)$ $A(\zeta)$ is similar to $C(p_1, \ldots, p_k)(\zeta)$. Let $f_2 \in \mathrm{H}(\Omega)$ be the discriminant of $p_k(\alpha, z)$. We claim that f_2 is not identically zero. Assume to the contrary that $f_2 = 0$. Then for each $\zeta \in \Omega$ the polynomial $p_k(\alpha, \zeta)$ has a mulitple root. Hence, $C(p_k)(\zeta)$ and $C(p_1, \ldots, p_k)(\zeta)$ are not diagonable. Since Ω is $\mathbb{R}$-symmetric it contains a real open interval J. (See Problem 3.4.17.2.) Clearly, $J \setminus Z(f_1)$ is a countable union of open intervals. Let $\zeta \in J \setminus Z(f_1)$. As $A(\zeta)$ is a hermitian matrix, $A(\zeta)$ is diagonable. This contradicts our results that $A(\zeta)$ is similar to $C(p_1, \ldots, p_k)(\zeta)$, which is not diagonable. Hence, $f_2 \neq 0$. Let $f = f_1 f_2$ and $\Omega_2 := \Omega \setminus Z(f) \subset \Omega_3$. Let $\zeta \in \Omega_2$. Then the polynomial $p_k(\alpha, \zeta)$ has simple roots. Hence, each $p_j(\alpha, \zeta)$ has simple roots. Therefore, $C(p_1, \ldots, p_k)(\zeta)$ is similar to a diagonal matrix, which implies that $A(\zeta)$ is similar to a diagonal matrix. $\qquad\square$

Proof of Theorem 4.17.2 We first show that for $\zeta \in J$ the characteristic polynomial of $A(z)$ viewed as a matrix in $\mathrm{H}_\zeta^{n \times n}$ splits in H_ζ. Since each p_j divides p_k, it is enough to show that $p_k(\alpha, z)$ splits in H_ζ. Assume that the degree of $p_k(\alpha, z)$ with respect to α is m. Let $Z(f)$ and Ω_2 be defined as Lemma 4.17.3. So $J \cap Z(f)$ consists of a countable number of points who can accumulate only to the end points of J. Furthermore, $J \cap \Omega_2 = J \setminus Z(f)$ consists of a countable number of intervals. Assume that $\zeta \in J \setminus Z(f)$. Since all the roots of $p_k(\alpha, \zeta)$ are distinct, the implicit function theorem yields that $p_k(\alpha, z)$ splits in H_ζ. It is left to consider the case when $\zeta \in J \cap Z(f)$. So $p_k(\alpha, z)$ has m simple roots in some punctured disk $0 < |z - \zeta| < r$. As $p_k(\alpha, z) \in H_\zeta[\alpha]$ is a monic polynomial in α of degree m, each root $\beta_j(z)$ satisfying $p_k(\beta_j(z), z) = 0$ is a multivalued function in $0 < |z - \zeta| < r$. That is, when we take one branch $\beta_j(z)$ on a closed circle $z(\theta) = \zeta + r_1 e^{\theta \sqrt{-1}}, r_1 \in (0, r)$ for $\theta \in [0, 2\pi]$ then $\beta_j(\zeta + e^{2\pi \sqrt{-1}} r_1)$ may give another branch $\beta_{j'}(\zeta + r_1)$. Theorem 1.8.4 claims that if we replace $z - \zeta$ by w^s then each $\beta_j(w^s)$ is locally analytic in w. Since $A(\xi)$ is a hermitian matrix for each $\xi \in \mathbb{R}$ it follows that all the eigenvalues of $A(\xi)$ are real. So $\beta_j(w^s)$ is real value for $w \geq 0$. Hence the Taylor coefficients of each $\beta_j(w^s)$ are real. Equivalently, $\beta_j(z)$ has the Puiseux expansion

$$\beta_j(z) = \sum_{l=0}^{\infty} \beta_{j,l}(z - \zeta)^{\frac{l}{s}}, \tag{4.17.2}$$

where $\beta_{j,l} \in \mathbb{R}$, $l = 0, \ldots$, $j = 1, \ldots, d$. By considering the branch $\beta_j(z)$ on a closed circle $z = \zeta + r_1 e^{\theta \sqrt{-1}}$ for $\theta \in [0, 2\pi]$ we get the the the Puiseux expansion of $\beta_{j'}(z)$:

$$\beta_{j'}(z) = \sum_{l=0}^{\infty} \beta_{j,l} e^{\frac{2l\pi \sqrt{-1}}{s}} (z - \zeta)^{\frac{l}{s}}. \tag{4.17.3}$$

Hence, each coefficient $\beta_{j,l} e^{\frac{2l\pi \sqrt{-1}}{s}}$ is real. Therefore, $\beta_{j,l} = 0$ if s does not divide l. Thus we showed that the Puiseux expansion of each $\beta_j(z)$ is a Taylor expansion. This shows that $p_k(\alpha, z)$ splits in H_ζ.

For each $\zeta \in J$ let $r(\zeta) > 0$ be the largest r such that the disk $|z - \zeta| < r(\zeta)$ is contained in Ω and each $\beta_j(z)$ is analytic in this disk. Let $\Omega_1' := \cup_{\zeta \in J}\{z, |z - \zeta| < r(\zeta)\}$. So Ω_1' is a simply connected $\mathbb{R}$-symmetric domain contained in Ω. Hence, each $\beta_j(z)$ is analytic in Ω_1'. This proves part (1) of the theorem for Ω_1'.

Since $p_{j-1}|p_j$ for $j = 1, \ldots, k-1$ it follows that $\beta_1(z), \ldots, \beta_m(z) \in H(\Omega_1')$ are the distinct roots of the characteristic polynomial of $A(z)$. So

$$p_k(\alpha, z) = \prod_{i=1}^m (\alpha - \beta_i(z)), \quad \det(\alpha I_n - A(z)) = \prod_{i=1}^m (\alpha - \beta_i(z))^{n_i},$$

$$(4.17.4)$$

where $n_j \in \mathbb{N}$, $j = 1, \ldots, m$, $n_1 + \cdots + n_m = n$. Hence, for each $\zeta \in \Omega_1' \setminus Z(f)$ the multiplicity of the eigenvalue $\beta_j(\zeta)$ is exactly n_j.

Since Ω_1' is EDD domain it follows that the null space of $A - \beta_j I_n$ has a basis consisting of $n_j' = \operatorname{nul}(A - \beta_j I_n)$. Clearly, n_j' is also the nullity of $A - \beta_j I_n$ in $\Omega_1' \setminus Z(f)$. Hence, $n_j' = n_j$ for $j = 1, \ldots, m$.

Assume that $1 \le j < l \le m$. Suppose that $\mathbf{u} \in \operatorname{nul}(A - \beta_j I_n), \mathbf{v} \in \operatorname{nul}(A - \beta_l I_n)$. We claim that $h(z) := \mathbf{u}(\bar{z})^* \mathbf{v}(z) \in H(\Omega_1')$ is identically zero in Ω_1'. Indeed, let $\zeta \in J \setminus Z(f)$. As $\mathbf{u}(\zeta)$ and $\mathbf{v}(\zeta)$ are eigenvectors of $A(\zeta)$ corresponding to two different eigenvalues it follows that $h(\zeta) = 0$. Since $Z(f)$ is countable set, the continuity of $h(z)$ yields that $h(z)$ is zero on J. Hence, h is identically zero in Ω_1'.

We now show that we can choose an orthogonal basis in each $\operatorname{nul}(A - \mu_j I_n)$. Let $\mathbf{u}_{1,j}, \ldots, \mathbf{u}_{n_j,j}$ be a basis in $\operatorname{nul}(A - \mu_j I_n) \subset H(\Omega_1')$. Hence, $\mathbf{u}_{q,j}(\zeta) \ne \mathbf{0}$ for each $\zeta \in H(\Omega_1')$. Let $h_{1,j}(z) := \mathbf{u}_{1,j}(\bar{z})^* \mathbf{u}_{1,j}(z) \in H(\Omega_1')$. So $h_{1,j}(\zeta) > 0$ for each $\zeta \in J$. Hence, $\{\mathbf{v}, \mathbf{v} \in \operatorname{nul}(A - \mu_j I_n), \mathbf{u}_{1,j}^* \mathbf{v} = 0\}$ is a subspace of $\operatorname{nul}(A - \mu_j I_n)$ of dimension $n_j - 1$. Continuing this process we obtain an orthogonal basis of each $\operatorname{nul}(A - \mu_j I_n)$. That is, we can assume that $\mathbf{u}_{1,j}, \ldots, \mathbf{u}_{n_j,j}$ is a basis in $\operatorname{nul}(A - \mu_j I_n)$ satisfying $\mathbf{u}_{i,j}^* \mathbf{u}_{q,j} = 0$ for $i \ne q$. In view of the orthogonality conditions of $\operatorname{nul}(A - \mu_j I_n)$ and $\operatorname{nul}(A - \mu_l I_n)$ for $j \ne l$ we conclude that we obtained an orthogonal basis $\mathbf{v}_1, \ldots, \mathbf{v}_n$ of $H(\Omega_1)^n$ which consists of eigenvectors of A. That is, $\mathbf{v}_i^* \mathbf{v}_q = 0$ for $i \ne q$. Let $g_i := \mathbf{v}_i^* \mathbf{v}_i \in H(\Omega_1')$ for $i = 1, \ldots, n$. The arguments above imply that $g_i(\zeta) > 0$ for each $\zeta \in J$. Let $r'(\zeta)$ be the maximal r such that each $g_i(z) \ne 0$ in the disk $|z - \zeta| < r$ contained in Ω_1'. Define $\Omega_1 := \cup_{\zeta \in J}\{z, |z - \zeta| < r'(\zeta)\}$. So Ω_1 is a simply connected $\mathbb{R}$-symmetric domain in $\mathbb{C}$. Let $\sqrt{g_i} \in H(\Omega_1)$ be the unique square root of g_i which is positive on J. Define $\mathbf{w}_i := \frac{1}{\sqrt{g_i}}\mathbf{v}_i \in H(\Omega_1)^n$ for $i = 1, \ldots, n$. Let $U = [\mathbf{w}_1 \ldots \mathbf{w}_n] \in H(\Omega_1)^{n \times n}$. Then U is unitary analytic and $A = U \operatorname{diag}(\alpha_1, \ldots, \alpha_n)U^*$ in Ω_1. $\qquad\square$

Problems

1. Let $\Omega \subset \mathbb{C}$ be a domain. Assume that $J \subset \mathbb{R}$ is an interval contained in Ω. Show that there exists an $\mathbb{R}$-symmetric domain Ω_1 such that $J \subset \Omega_1 \subseteq \Omega$.

2. Assume that $\Omega \subset\subset \mathbb{C}$ is an $\mathbb{R}$-symmetric domain and $A(z), U(z) \in H(\Omega)^{n \times n}$.

   (a) For each $z \in \Omega$ let $A_1(z) := A(\bar{z})^*, U_1(z) := U(\bar{z})^*$. Show that $A_1, U_1 \in H(\Omega)^{n \times n}$.

   (b) Show that Ω contains an open interval $J \subset \mathbb{R}$.

   (c) Assume that $x_0 \in \mathbb{R} \cap \Omega$. Let

   $$A(z) = \sum_{j=0}^{\infty}(z - x_0)^j A_j, \quad U(z) = \sum_{j=0}^{\infty}(z - x_0)^j U_j. \quad (4.17.5)$$

   Show that $A(z), U(z)$ are hermitian and unitary analytic respectively if and only if the following conditions hold:

   $$A_j^* = A_j, \quad \sum_{k=0}^{j} U_k U_{j-k}^* = \delta_{0j}I_n, \quad j = 0, \ldots. \quad (4.17.6)$$

   Hint: To show that the above equalities imply that $A(z)$, $U(z)$ are hermitian and unitary analytic, use the analytic continuation principle for $A(z) - A_1(z)$ and $U(z)U_1(z) - I_n$ respectively in Ω.

4.18   Hermitian Pencils

Definition 4.18.1 $A(z) := A_0 + zA_1$ *is called a hermitian pencil if A_0 and A_1 are hermitian matrices of the same order.*

In this section we assume that A_0, A_1 are hermitian matrices of order n. Observe that the pencil $A(z) = A_0 + zA_1$ is hermitan analytic in $\mathbb{C}$, i.e. $A(\bar{z})^* = A(z)$ for $z \in \mathbb{C}$. The precise version of Lemma 4.17.3 and Theorem 4.17.2 for $A(z)$ is as follows:

Theorem 4.18.2 *Let $A(z) = A_0 + zA_1 \in \mathbb{C}[z]^{n \times n}$ be a hermitian pencil. Then $C(p_1, \ldots, p_k) \in \mathbb{C}[z]^{n \times n}$ is the rational canonical form of $A(z)$ over $\mathbb{C}(z)$. $p_1(\alpha, z), \ldots, p_k(\alpha, z) \in \mathbb{C}[z][\alpha]$ are the nontrivial invariant polynomials of $\alpha I_n - A(z)$, monic in α, such that $p_j | p_{j+1}$ for $j = 1, \ldots, k-1$ and $\prod_{j=1}^{k} p_k(\alpha, z) = \det(\alpha I_n - A(z))$. Furthermore, the following conditions hold:*

1. *$p_j(\alpha, z) = \alpha_j^{m_j} + \sum_{i=1}^{m_j} p_{j,i}(z) \alpha^{m_j - i}$, where $\deg p_{j,i}(z) \leq i$ for $i = 1, \ldots, m_j$ and $j = 1, \ldots, k$.*

2. *The discriminant of $p_k(\alpha, z)$ with respect to α is a nonzero polynomial $D(p_k)(z)$ of degree $m(m-1)$ at most, where $m = m_k$.*

3. *Let $Z(D(p_k)) \subset \mathbb{C}$ be the zero set of $D(p_k)$. Then for each $\zeta \in \mathbb{C} \setminus Z(D(p_k))$ $A(\zeta)$ is similar to a diagonal matrix.*

4. *The roots of $p_k(\alpha, z) = 0$, $\beta_1(z), \ldots, \beta_m(z)$, are finite multivalued analytic functions in $\mathbb{C} \setminus D(p_k)$. The set $\{\beta_1(z), \ldots, \beta_m(z)\}$ is the spectrum of $A(z)$ for $z \in \mathbb{C} \setminus D(p_k)$. Furthermore, the following conditions are satisfied:*

   (a) *Let $\zeta \in D(p_k)$. Then each $\beta_j(z)$ has Puiseux expansion (4.17.2) in a disk $D(\zeta, r(\zeta)) := \{z \in \mathbb{C}, 0 < |z - \zeta| < r(\zeta)\}$, where $r(\zeta)$ is the biggest r such that $\{z, 0 < |z - \zeta| < r\} \cap D(p_k) = \emptyset$. A closed circle $z(\theta) := \zeta + r_1 e^{\theta \sqrt{-1}}, r_1 \in (0, r(\zeta))$ for $\theta \in [0, 2\pi]$ in $D(\zeta, r(\zeta))$ induces a permutation $\sigma(\zeta) : [m] \to [m]$. That is, $\beta_j(z(2\pi)) = \beta_{\sigma(\zeta)(j)}(\zeta + r_1)$ for $j = 1, \ldots, m$. Each cycle of this permutation of length q consists of q branches of eigenvalues $\beta_1(z), \ldots, \beta_m(z)$, which are cyclically permuted by $\sigma(\zeta)$ after completing one closed circle $z(\theta)$.*

   (b) *Each eigenvalue $\beta_j(z)$ has a fixed multiplicity n_j in $\mathbb{C} \setminus D(p_k)$ for $j = 1, \ldots, m$. The multiplicities of each $\beta_i(z)$ are* *invariant under the action of $\sigma(\zeta)$ for $\zeta \in D(p_k)$. That is (4.17.4) holds.*

(c) *For each $\zeta \in D(p_k)$ the polynomial $p_k(\alpha, \zeta)$ has at least one multiple root. That is, there exists $1 \leq i < j \leq m$ such that $\beta_i(\zeta) = \beta_j(\zeta)$.*

5. *There exists a simply connected $\mathbb{R}$-symmetric domain $\Omega_1 \subset \mathbb{C}$ containing the real line $\mathbb{R}$ such that $\beta_1(z), \ldots, \beta_m(z)$, are analytic functions in Ω_1. Furthermore $\mathbf{x}_1, \ldots, \mathbf{x}_n \in \mathrm{H}(\Omega_1)^n$ are the corresponding normalized eigenfunctions:*

$$A(z)\mathbf{x}_i(z) = \alpha_i(z)\mathbf{x}_i(z),$$
$$\mathbf{x}_i(\bar{z})^*\mathbf{x}_j(z) = \delta_{ij}, \quad i, j = 1, \ldots, n, \quad z \in \Omega_1, \quad (4.18.1)$$

where $\alpha_1, \ldots, \alpha_n \in \mathrm{H}(\Omega_1)$ are the n eigenvalues of $A(z)$ for $z \in \Omega_1$.

6. *TFAE*

(a) $A_0 A_1 = A_1 A_0$.

(b) *There exists an orthonormal basis $\mathbf{x}_1, \ldots, \mathbf{x}_n$ in $\mathbb{C}^n$ such that*

$$A_i \mathbf{x}_j = a_{i,j}\mathbf{x}_j, \quad j = 1, \ldots, n, \quad i = 0, 1. \quad (4.18.2)$$

(c) *For each ζ in $\mathbb{C}$ $A(\zeta)$ is diagonable.*

(d) *$A(z)$ has property L.*

Proof. Consider the homogeneous pencil $A(z, w) = wA_0 + zA_1 \in \mathbb{C}[z, w]$. Clearly, each minor of the matrix $\alpha I_n - A(z, w)$ of order l is a homogeneous polynomial in α, z, w of total degree l. Hence, the nontrivial invariant polynomials of $\alpha I_n - A(z, w)$ over the field $\mathbb{C}(z, w)$ are homogeneous polynomials $\tilde{p}_j(\alpha, z, w), j = 1, \ldots, k$, which are monic in α and of total degrees $m_1, \ldots, m_k$ respectively. So $\tilde{p}_{j-1}|\tilde{p}_j$ for $j = 1, \ldots, k - 1$. Furthermore, $\prod_{j=1}^{k} \tilde{p}_j = \det(\alpha I_n - A(z, w))$. Clearly, $\tilde{p}_j(\alpha, z, w) = \alpha^{m_j} + \sum_{i=1}^{m_j} \tilde{p}_{j,i}(z, w)\alpha^{m_j - i}$ and $\deg \tilde{p}_{j,i} = m_j - i$ for $i = 1, \ldots, m_j$ and $j = 1, \ldots, k$. Let $p_j(\alpha, z) := \tilde{p}_j(\alpha, z, 1)$

for $j = 1, \ldots, k$. Then $p_1(\alpha, z), \ldots, p_k(\alpha, z)$ are the nontrivial invariant polynomials of $\alpha I_n - A(z)$ over $\mathbb{C}(z)$. Furthermore, $\prod_{j=1}^{k} p_j = \det(\alpha I_n - A(z))$.

1. Clearly, $p_{j,i}(z) = \tilde{p}_{j,i}(z, 1)$. Hence, $\deg p_{j,i} \leq m_j - i$ for $i = 1, \ldots, m_j$ and $j = 1, \ldots, k$.

2. Let $D(\tilde{p}_k)(z, w)$ be the discriminant of $\tilde{p}_k(\alpha, z, w)$ with respect to α. Then $D(p_k)(z) = D(\tilde{p}_k)(z, 1)$. The proof of Lemma 4.17.3 yields that $D(p_k)(z)$ is a nonzero polynomial. Hence, $D(\tilde{p}_k)(z, w)$ is a nonzero homogeneous polynomial of total degree $m(m - 1)$. Thus $D(p_k)(z)$ is of degree at most $m(m - 1)$.

3. This claim follows from part 4 of Lemma 4.17.3.

4. This claim follows from the proof of Lemma 4.17.3.

   (a) The Puiseux expansion (4.17.2) follows from Theorem 1.8.4. Let $z(\theta) := \zeta + r_1 e^{\theta\sqrt{-1}}, r_1 \in (0, r(\zeta))$ for $\theta \in [0, 2\pi]$ be a closed circle in $D(\zeta, r(\zeta))$. Then this closed circle induces the following permutation permutation $\sigma(\zeta) : [n] \to [n]$. Namely, the Puiseux expansion of $\alpha_{\sigma(\zeta)(j)}$ is given by (4.17.3). Clearly, the cycle $\sigma(\zeta)^{l-1}(j) \mapsto \sigma(\zeta)^l(j)$ for $j = 1, \ldots, s$ gives all branches of $\alpha_j(z)$ in $D(\zeta, r(\zeta))$.

   (b) (4.17.4) implies that the multiplicity of $\beta_j(z)$ is n_j for $z \in \mathbb{C} \setminus Z(D(p_k))$. The action of $\sigma(\zeta)$ for $\zeta \in Z(D(p_k))$ yields that $n_{\sigma(\zeta)(j)} = n_j$.

   (c) Since $p_k(\alpha, z)$ is a monic polynomial with respect to α, $p_k(\alpha, \zeta)$ has a multiple root if and only if $\zeta \in Z(D(p_k))$.

5. This claim follows from Rellich's theorem 4.17.2.

6. The equivalence of (a) and (b) follows from Problem 1. The implication (b) and (c) is trivial. Theorem 3.12.4 yields the implication (c) and (d).

Assume that (d) holds. So $\alpha_j(z) = a_{0,j} + a_{1,j} z$ for $j = 1, \ldots, n$. Thus $a_{i,1}, \ldots, a_{i,n}$ are the n eigenvalues of A_i for $i = 0, 1$. Without loss of generality we can assume that $a_{1,1} \geq \cdots \geq a_{1,n}$. We show that A_0, A_1 commute by induction on n. Clearly, for $n = 1$ A_0, A_1 commute. Suppose this claim is true for each $n < N$. Assume that $n = N$.

Suppose first that $a_{1,1} = \cdots = a_{1,n}$. Then $A_1 = a_{1,1} I_n$ and A_0, A_1 commute. Suppose now that $a_{1,1} = \cdots a_{1,q} > a_{1,q+1}$ for some $q \in [n-1]$. Assume that

$$\lambda_1(t) \geq \cdots \geq \lambda_n(t) \text{ are eigenvalues of } A(t), \ t \in \mathbb{R}. \qquad (4.18.3)$$

Clearly, there exists $t_0 > 0$ such that $\lambda_j(t) = a_{0,j} + a_{1,j} t$ for $t \geq t_0$ and $j = 1, \ldots, q$. Choose $t > t_0$. Let $\mathbf{x}_1, \ldots, \mathbf{x}_q$ be the orthonormal eigenvectors of $A(t)$ corresponding $\lambda_1(t), \ldots, \lambda_q(t)$ respectively. Hence

$$\sum_{j=1}^{q} (a_{0,j} + a_{1,j} t) = \sum_{j=1}^{q} \lambda_j(t)$$

$$= \sum_{j=1}^{t} \mathbf{x}_j^* A(t) \mathbf{x}_j = \sum_{j=1}^{t} \mathbf{x}_j^* A(t_0) \mathbf{x}_j + \sum_{j=1}^{t} \mathbf{x}_j^*(t - t_0) A_1 \mathbf{x}_j.$$

Theorem 4.4.9 yields:

$$\sum_{j=1}^{t} \mathbf{x}_j^* A(t_0) \mathbf{x}_j \leq \sum_{j=1}^{q} \lambda_j(t_0), \quad \sum_{j=1}^{t} \mathbf{x}_j^*(t - t_0) A_1 \mathbf{x}_j$$

$$\leq \sum_{j=1}^{q} \lambda_j((t - t_0) A_1).$$

Recall that

$$\sum_{j=1}^{q} \lambda_j(t_0) = \sum_{j=1}^{q} (a_{0,j} + a_{1,j} t_0),$$

$$\sum_{j=1}^{q} \lambda_j((t - t_0) A_1) = (t - t_0) \sum_{j=1}^{q} a_{1,j}.$$

Hence, in the above two inequalities we have equalities. Let $\mathbf{U} = \mathrm{span}\,(\mathbf{x}_1, \ldots, \mathbf{x}_q)$. The equality case in Theorem 4.4.9 yields that $\mathbf{U}$ is spanned by the q eigenvectors of $A(t_0)$ and A_1 corresponding to $\lambda_1(t_0), \ldots, \lambda_q(t_0)$ and A_1 respectively. In particular, $\mathbf{U}$ is a common invariant subspace of $A(t_0)$ and A_1, hence $\mathbf{U}$ is an invariant subspace of A_0 and A_1. Therefore, $\mathbf{U}^\perp$ is a common invariant subspace of A_0 and A_1. So the restrictions of $A(z)$ to $\mathbf{U}$ and $\mathbf{U}^\perp$ are hermitian pencils which have property L. The induction assumption yields that the restrictions of A_0, A_1 to $\mathbf{U}, \mathbf{U}^\perp$ commute. Hence, A_0, A_1 commute. $\qquad\qquad\qquad\qquad\qquad\qquad\qquad\qquad\qquad\qquad\quad\square$

Definition 4.18.3 *Let $A(z)$ be a hermitian pencil. Assume that $\zeta \in \mathbb{C}$.*

1. *ζ is called a crossing point of the pencil $A(z)$ if $\beta_i(\zeta) = \beta_j(\zeta)$ for $i \neq j$.*

2. *ζ is called a regular point of $\beta_i(z)$ if $\beta_i(z) \in \mathrm{H}_\zeta$. (See Problem 3.)*

3. *ζ is called a resonance point of $A(z)$ if $A(\zeta)$ is not diagonable.*

Theorem 4.18.2 yields:

Corollary 4.18.4 *Let $A(z) = A_0 + zA_1 \in \mathbb{C}^{n \times n}$ be a hermitian pencil. Then*

1. *ζ is a crossing point if and only if $\zeta \in Z(D(p_k))$.*

2. *The number of crossing points is at most $n(n-1)$.*

3. *Every resonance point of $A(z)$ is a crossing point.*

4. *$A(z)$ has a resonance point if and only if $A_0 A_1 \neq A_1 A_0$.*

5. *Assume that $A_0 A_1 \neq A_1 A_0$. Then there exists $2q$ distinct points $\zeta_1, \bar\zeta_1, \ldots, \zeta_q, \bar\zeta_q \in \mathbb{C} \setminus \mathbb{R}$, where $1 \leq q \leq \frac{n(n-1)}{2}$, such that $A(z)$ is not diagonable if and only if $z \in \{\zeta_1, \bar\zeta_1, \ldots, \zeta_q, \bar\zeta_q\}$.*

Let $A(z) = A_1 + zA_1$ be a hermitian pencil. Rellich's theorem yields that the eigenvalues and orthonormal eigenvectors of $A(z)$ can be chosen to be analytic on the real line. That is, (4.18.1) holds. Fix $t \in \mathbb{R}$. Expand each $\alpha_j(z), \mathbf{x}_j(z)$ in Taylor series at t:

$$\alpha_j(z) = \sum_{l=0}^{\infty} \alpha_{j,l}(z-t)^l, \quad \mathbf{x}_j(z) = \sum_{l=0}^{\infty}(z-t)^l \mathbf{x}_{j,l}. \qquad (4.18.4)$$

The orthonormality conditions in (4.18.1) yield:

$$\mathbf{x}_{j,0}^* \mathbf{x}_{r,0} = \delta_{jr}, \quad \sum_{l=0}^{s} \mathbf{x}_{j,l}^* \mathbf{x}_{r,s-l} = 0, \quad j,r \in [n], \ q \in \mathbb{N}. \qquad (4.18.5)$$

The eigenvalue condition $A(z)\mathbf{x}_j(z) = \alpha_j(z)\mathbf{x}_j(z)$ is equivalent to:

$$A(t)\mathbf{x}_{j,0} = \alpha_{j,0}\mathbf{x}_{j,0}, \quad A(t)\mathbf{x}_{j,s} + A_1\mathbf{x}_{j,s-1} = \sum_{l=0}^{s} \alpha_{j,l}\mathbf{x}_{j,s-l},$$

$$j \in [n], s \in \mathbb{N}. \qquad (4.18.6)$$

In some situations, as in the proof of Theorem 4.18.2, we need to arrange the eigenvalues of $A(t)$ in a nonincreasing order as in (4.18.3). Usually, $\lambda_i(t)$ is not analytic on $\mathbb{R}$. However, if t is not a crossing point then each $\lambda_i(t)$ is analytic in some disk $|z - t| < r(t)$. Assume that $\zeta \in \mathbb{R}$. We call ζ a noncrossing point of $\lambda_i(\zeta)$ if there is an open interval $I = \{t, |t - \zeta| < r\}$ such that $\lambda_i(t)$ has a constant multiplicity as an eigenvalue of $A(t)$. Otherwise ζ is called a crossing point of $\lambda_i(\zeta)$. We call ζ a point of analyticity of λ_i if $\lambda_i(t) = \beta_j(t)$ on I and ζ is a regular point of $\beta_j(z)$. See Problem 3.

Lemma 4.18.5 *Let the assumptions and the definitions of Theorem 4.18.2 hold. Assume (4.18.3). Then $\mathbb{R} \setminus Z(D(p_k))$ is the union of q disjoint open intervals $I_i = (a_{i-1}, a_i)$ for $i = 1, \ldots, q$, where $a_0 = -\infty < a_1 < \cdots < a_q = \infty$.*

1. *For each interval I_i there is a permutation $\sigma_i : [n] \to [n]$ such that $\lambda_j(t) = \alpha_{\sigma_i(j)}(t)$ for $t \in [a_{i-1}, a_i]$ and $j \in [n]$.*

2. *Let* $t \in I_i$. *Assume that the multiplicity of* $\lambda_j(t)$ *is* n_j. *Let* $\mathbf{u}_1, \ldots, \mathbf{u}_{n_j}$ *be an orthonormal basis for eigenspace of* $A(t)$ *corresponding to* $\lambda_j(t)$. *Then there exists* $\beta_l(z) \in \mathrm{H}(\Omega_1)$ *such that* $\lambda_j(t) = \beta_l(t)$ *for* $t \in I_i$. *Furthermore, there exist* $\mathbf{z}_1(z), \ldots, \mathbf{z}_{n_j}(z) \in \mathrm{H}(\Omega_1)^n$, *which form an orthonormal basis of the eigenspace of* $A(z)$ *corresponding to* $\beta_l(z)$, *such that* $\mathbf{z}_r(t) = \mathbf{u}_r$ *for* $r = 1, \ldots, n_j$.

3. *Let* $t \in I_i$. *Then* $\lambda_j'(t) = \mathbf{y}_j^* A_1 \mathbf{y}_j$ *for any eigenvector* $A(t)\mathbf{y}_j = \lambda_j(t)\mathbf{y}_j, \mathbf{y}_j^* \mathbf{y}_j^* = 1$.

4. *Assume that* $\zeta \in \mathbb{R} \cap Z(D(p_k))$. *That is,* ζ *is one of the end points of the two adjacent intervals* $I_i = (a_{i-1}, \zeta)$ *and* $I_{i+1} = (\zeta, a_{i+1})$.

   (a) *Suppose that* ζ *is a noncrossing point of* $\lambda_j(t)$. *Then there exists* $\beta_l(z) \in \mathrm{H}(\Omega_1)$ *such that* $\lambda_j(t) = \beta_l(t)$ *for* $t \in (a_{i-1}, a_{i+1})$. *Furthermore,* $\lambda_j'(\zeta) = \mathbf{y}_j^* A_1 \mathbf{y}_j$ *for any eigenvector* $A(\zeta)\mathbf{y}_j = \lambda_j(\zeta)\mathbf{y}_j$, $\mathbf{y}_j^* \mathbf{y}_j^* = 1$. $\lambda_j(\zeta)$ *has multiplicity* n_j, *which is the multiplicity of* $\lambda_j(t)$ *for* $t \in I_i$. *Let* $\mathbf{u}_1, \ldots, \mathbf{u}_{n_j}$ *be an orthonormal basis of for eigenspace of* $A(\zeta)$ *corresponding to* $\lambda_j(\zeta)$. *Then there exists* $\mathbf{z}_1(z), \ldots, \mathbf{z}_{n_j}(z) \in \mathrm{H}(\Omega_1)^n$ *orthonormal eigenvectors of* $A(z)$ *corresponding to* $\beta_l(z)$ *such that* $\mathbf{z}_r(\zeta) = \mathbf{u}_r$ *for* $r = 1, \ldots, n_j$.

   (b) *Assume that* ζ *is a crossing point of* $\lambda_j(\zeta)$. *Suppose that* $\lambda_j(t) = \beta_l(t)$ *for* $t \in I_i$ *and* $\lambda_j(t) = \beta_{l'}(t)$ *for in* I_{i+1}. *Let* $n_l, n_{l'}$ *be the multiplicities of* $\beta_l(z), \beta_{l'}(z) \in \mathrm{H}(\Omega_1)$ *respectively. Assume furthermore that* $\mathbf{u}_1(z), \ldots, \mathbf{u}_{n_l}(z), \mathbf{v}_1(z), \ldots, \mathbf{v}_{n_{l'}}(z) \in \mathrm{H}(\Omega_1)^n$ *are the orthonormal eigenvectors of* $A(z)$ *corresponding to* $\beta_l(z), \beta_{l'}(z)$ *respectively. Let* $\mathbf{U} = \mathrm{span}\,(\mathbf{u}_1(\zeta), \ldots, \mathbf{u}_{n_l}(\zeta))$, $\mathbf{V} = \mathrm{span}\,(\mathbf{v}_1(\zeta), \ldots, \mathbf{v}_{n_{l'}}(\zeta))$. *Then the left and the right derivatives of* $\lambda_j(t)$ *at* ζ *are given by:*

   $$\lambda_j'(\zeta-) = \mathbf{u}^* A_1 \mathbf{u}, \quad \lambda_j'(\zeta+) = \mathbf{v}^* A_1 \mathbf{v}, \tag{4.18.7}$$

   for any $\mathbf{u} \in \mathbf{U}, \mathbf{v} \in \mathbf{V}, \mathbf{u}^* \mathbf{u} = \mathbf{v}\mathbf{v}^* = 1$.

Proof. Since $D(p_k)$ is a nonzero polynomial it follows that $Z(D(p_k))$ is a finite (possibly empty) set in $\mathbb{C}$. Hence, $\mathbb{R} \setminus Z(D(p_k))$ is a finite union of disjoint intervals, whose closure is $\mathbb{R}$.

1. As the interval I_i does not contain a crossing point $\beta_j(t) \neq \beta_l(t)$ for $1 \leq j < l \leq m$ and $t \in I_i$. So each $\beta_l(t)$ is an eigenvalue of $A(t)$ of multiplicity n_l for $l = 1, \ldots, m$. Hence, there exists a permutation $\sigma_i : [n] \to [n]$ such that $\lambda_j(t) = \alpha_{\sigma_i(j)}(t)$ for $j = 1, \ldots, n$. Since each $\lambda_j(t)$ is a continuous function n $\mathbb{R}$ it follows that the equality $\lambda_j(t) = \alpha_{\sigma_i(j)}(t)$ is valid for the closed interval $[a_{i-1}, a_i]$.

2. Recall that $\alpha_j(z) = \beta_l(z)$ for some $l \in [m]$. $\beta_l(z)$ is of multiplicity n_l. Hence the multiplicity n_j of $\lambda_j(t)$ is n_l for $t \in I_i$. Let $\mathbf{v}_1(z), \ldots, \mathbf{v}_{n_j}(z) \in H(\Omega_1)^n$ be a set of orthonormal eigenvectors of $A(z)$ corresponding to $\beta_l(z)$. Clearly, $\mathbf{u}_1, \ldots, \mathbf{u}_{n_j}$ for a basis in span $(\mathbf{v}_1(t), \ldots, \mathbf{v}_{n_j}(t))$. Hence, there exists a unitary matrix $O \in \mathbf{U}(n_j)$ such that $O\mathbf{v}_r(t) = \mathbf{u}_r$ for $r = 1, \ldots, n_j$. Let $\mathbf{z}_r(z) = O\mathbf{v}_r(z)$ for $r = 1, \ldots, n_j$. Then $\mathbf{z}_1(z), \ldots, \mathbf{z}_{n_j}(z)$ is an orthonormal basis of the eigenspace of $A(z)$ in $H(\Omega_1)^n$ corresponding to $\beta_l(z)$.

3. Rename the analytic eigenvalues of $A(z)$ in $H(\Omega_1)$ and the corresponding orthonormal eigenvectors so that $\lambda_j(t) = \alpha_j(t)$ for $t \in I_i$ for $j = 1, \ldots, n$. Let $\mathbf{y}_j$ be an eigenvector of $A(t)$ of unit length corresponding to $\lambda_j(t)$. The arguments of the proof of part 2 imply that we can assume that $\mathbf{x}_j(z) = \mathbf{z}_1(z)$ and $\mathbf{x}_j(t) = \mathbf{y}_j$.

   Recall that the eigenvalue condition $A(z)\mathbf{x}_j(z) = \alpha_j(z)\mathbf{x}_j(z)$ is equivalent to (4.18.6). Furthermore

   $$\alpha_{j,0} = \lambda_j(t), \quad \alpha_{j,1} = \lambda_j'(t), \quad \mathbf{x}_j(t) = \mathbf{x}_{j,0}, \quad \mathbf{x}_{j,0}^* \mathbf{x}_{j,0} = 1.$$

   The first condition and the second condition of (4.18.6) for $s = 1$ are

   $$A(t)\mathbf{x}_{j,0} = \alpha_{j,0}\mathbf{x}_{j,0}, \quad A(t)\mathbf{x}_{1,0} + A_1\mathbf{x}_{j,0} = \alpha_{j,0}\mathbf{x}_{1,0} + \alpha_{j,1}\mathbf{x}_{j,0}.$$

Multiply the second equality by $\mathbf{x}_{j,0}^*$ to deduce that $\alpha_{j,1} = \mathbf{x}_{j,0}^*$ $A_1\mathbf{x}_{j,0}$. Hence, the equality $\alpha_{j,1} = \mathbf{x}_{j,0}^* A_1\mathbf{x}_{j,0}$ yields the equality $\lambda_j'(t) = \mathbf{y}_j^* A_1\mathbf{y}_j$.

4. Assume that $\zeta \in \mathbb{R} \cap Z(D(p_k))$. So $\zeta = a_i$ for $i \in [q-1]$. Hence, ζ is a crossing point for some $\beta_r(z)$.

(a) Suppose that ζ is a noncrossing point of $\lambda_j(\zeta)$. Then there exists $\beta_l(z)$ such that $\lambda_j(\zeta) = \beta_j(\zeta)$. Hence, ζ is a regular point of $\beta_l(z)$. Therefore, $\lambda_j(t) = \beta_i(t)$ for $t \in (a_{i-1}, a_{i+1})$. Hence, the results of 2–3 hold for each $t \in (a_{i-1}, a_{i+1})$, in particular for $t = \zeta$.

(b) Assume that ζ is a crossing point of $\lambda_j(\zeta)$. So $\lambda_j(t) = \beta_l(t)$ for $t \in I_i$ and $\lambda_j(t) = \beta_{l'}(t)$ for $t \in I_{i+1}$. (We do not exclude the possibility that $l = l'$.) Recall that $\lambda_j(\zeta) = \beta_l(\zeta)$. Hence, $\lambda_j'(\zeta-) = \beta_l'(\zeta)$. Similarly, $\lambda_j'(\zeta+) = \beta_{l'}'(\zeta)$. The equalities (4.18.7) follow from part 3.

$\square$

Theorem 4.18.6 *Let* $A_0, A_1 \in \mathbf{H}_n$ *and* $A(t) = A_0 + tA_1$ *for* $t \in \mathbb{R}$. *Assume* (4.18.3). *Suppose that* $k \in [n-1]$ *and* $1 \le i_1 < \cdots < i_k \le n$ *are integers. Then the following inequality holds:*

$$\sum_{j=1}^{k} \lambda_{i_j}(A_0 + A_1) \le \sum_{j=1}^{k} \lambda_{i_j}(A_0) + \sum_{j=1}^{k} \lambda_j(A_1). \qquad (4.18.8)$$

Equality holds if and only if the following conditions hods: There exists r *invariant subspaces* $\mathbf{U}_1, \ldots, \mathbf{U}_r \subset \mathbb{C}^n$ *of* A_0 *and* A_1 *such that each* $\mathbf{U}_l$ *is spanned by* k-*orthonormal vectors of* A_1 *corresponding to* $\lambda_1(A_1), \ldots, \lambda_k(A_1)$. *Let* $\mu_{1,l}(t) \ge \cdots \ge \mu_{k,l}(t)$ *be the eigenvalues of the restriction of* $A(t)$ *to* $\mathbf{U}_l$ *for* $l = 1, \ldots, r$. *Then there exists* $b_0 = 0 < b_1 < \cdots < b_{r-1} < b_r = 1$ *with the following properties: For each* $l \in [r]$ *and* $t \in [b_{l-1}, b_l]$ $\mu_{j,l}(t) = \lambda_{i_j}(t)$ *for* $j = 1, \ldots, k$.

Proof. We use the notations and the results of Lemma 4.18.5. Recall that each $\lambda_j(t)$ is continuous on $\mathbb{R}$ and analytic in each I_i for $i = 1, \ldots, q$. That is, each $\lambda_j(t)$ is piecewise smooth. Fix I_i.

Then there exist n analytic orthonormal vectors $\mathbf{x}_1(t), \ldots, \mathbf{x}_n(t)$ in I_i corresponding to $\lambda_1(t), \ldots, \lambda_n(t)$ respectively. Assume that $t \in I_i$. Part 3 of Lemma 4.18.5 yields that $\lambda'_j(t) = \mathbf{x}_j(t)^* A_1 \mathbf{x}_j(t)$ for $j \in [n]$. Let $f(t) = \sum_{j=1}^k \lambda_{i_j}(t)$. Then $f'(t) = \sum_{j=1}^k \mathbf{x}_{i_j}(t)^* A_1 \mathbf{x}_{i_j}(t)$. Theorem 4.4.9 yields that $f'(t) \leq \sum_{j=1}^k \lambda_j(A_1)$. Hence

$$f(1) - f(0) = \int_0^1 f'(t)dt \leq \int_0^1 \left(\sum_{j=1}^k \lambda_j(A_1) \right) dt = \sum_{j=1}^k \lambda_j(A_1).$$

The above inequality is equivalent to (4.18.8).

We now discuss the equality case in (4.18.8). Suppose first that equality holds in (4.18.8). Let $0 < b_1 < \cdots < b_{r-1} < 1$ be the intersection points of $(0,1) \cap Z(D(p_k))$, i.e. $(0,1) \cap Z(D(p_k)) = \cup_{l=1}^{r-1}\{b_l\}$. Let $b_0 = 0, b_r = 1$. Since $f(t)$ is analytic in each (b_{l-1}, b_l) our proof of inequality (4.18.8) and the equality assumption implies

$$f'(t) = \sum_{j=1}^k \mathbf{x}_{i_j,l}(t) A_1 \mathbf{x}_{i_j,l}(t) = \sum_{j=1}^k \lambda_j(A_1).$$

Here, $\mathbf{x}_{1,l}(z), \ldots, \mathbf{x}_{n,l}(z)$ is a set of orthonormal eigenvectors of $A(z)$ in $H(\Omega_1)^n$ such that the eigenvector $\mathbf{x}_{j,l}(t)$ corresponds to to $\lambda_j(t)$ for $t \in (b_{l-1}, b_l)$. Theorem 4.4.9 implies that $\mathbf{U}(t) := \text{span}(\mathbf{x}_{i_1}(t), \ldots, \mathbf{x}_{i_k}(t))$ is spanned by k orthonormal eigenvectors of A_1 corresponding to eigenvalues $\lambda_1(A_1), \ldots, \lambda_k(A_1)$. Hence, $A_1\mathbf{U}(t), A(t)\mathbf{U}(t) \subset \mathbf{U}(t)$. Therefore, $A_0\mathbf{U}(t) \subset \mathbf{U}(t)$. Fix $t_0 \in (b_{l-1}, b_l)$. Let $\mathbf{U}_l = \mathbf{U}(t_0)$. So $\mathbf{U}_l$ is an invariant subspace of A_0, A_1. Let $A(z, \mathbf{U}_l)$ be the restriction of $A(z)$ to $\mathbf{U}_l$. So $A(z, \mathbf{U}_i)$ can be viewed as a hermitian pencil in $\mathbb{C}[z]^{k \times k}$. So its eigenvalues $\gamma_1(z), \ldots, \gamma_k(z) \in H(\Omega_1)$ are k analytic eigenvalues of $A(z, \mathbf{U}_l)$. The proof of Theorem 4.17.2 yields that there exists an $\mathbb{R}$-symmetric domain $\hat{\Omega}_1$ such that $\mathbf{y}_1(z), \ldots, \mathbf{y}_k(z) \in H(\hat{\Omega}_1)^n \cap \mathbf{U}_i$ are orthonormal vectors of $A(z, \mathbf{U}_l)$ corresponding to $\gamma_1(z), \ldots, \gamma_k(z)$. Furthermore $\mu_{j,i}(t) = \gamma_j(t)$ for $j = 1, \ldots, k$ for $t \in (b_{l-1}, b_l)$. By the construction $\gamma_j(t_0) = \lambda_{i_j}(t_0)$ for $j = 1, \ldots, k$. Hence $\mu_{j,l}(t) = \gamma_j(t) = \lambda_{i_j}(t)$ for

$t \in (b_{l-1}, b_l)$. Since all these function are continuous on $[b_{l-1}, b_l]$ we deduce the equalities hold for $t \in [b_{l-1}, b_l]$. This proves our claim on $\mathbf{U}_1, \ldots, \mathbf{U}_r$.

Assume now that there are r k-dimensional subspaces $\mathbf{U}_1, \ldots, \mathbf{U}_r$ which are invariant for A_0 and A_1. Let $A_0(\mathbf{U}_l), A_1(\mathbf{U}_l)$ be the restrictions of A_0 and A_1 to $\mathbf{U}_l$ respectively. Denote by $\mu_{1,l}(t) \geq \cdots \geq \mu_{k,l}(t)$ the k eigenvalues of $A(t, \mathbf{U}_l)$ for $l = 1, \ldots, r$. We also assume that there exists $b_0 < b_1 < \cdots < b_{r-1} < b_r$ such that $\mu_{j,l}(t) = \lambda_{i_j}(t)$ for $t \in [b_{l-1}, b_l]$ for $l = 1, \ldots, r$. Observe first that

$$\sum_{j=1}^{k} \lambda_{i_j}(b_l) - \sum_{j=1}^{k} \lambda_{i_j}(b_{l-1})$$

$$= \sum_{j+1}^{k} \mu_{j,l}(b_l) - \sum_{j=1}^{k} \mu_{j,l}(b_{l-1})$$

$$= \operatorname{tr} A(b_l, \mathbf{U}_l) - \operatorname{tr} A(b_{l-1}, \mathbf{U}_l)$$

$$= (b_l - b_{l-1}) \operatorname{tr} A_1(\mathbf{U}_l) = (b_l - b_{l-1}) \sum_{j=1}^{k} \lambda_j(A_1).$$

Add the above equalities for $l = 1, \ldots, r$ to deduce (4.18.8). $\qquad\square$

Problems

1. Let $A_0, A_1 \in \mathbb{C}^{n \times n}$ be normal matrices. Show that $A_0 A_1 = A_1 A_0$ if and only if (4.18.2) holds. *Hint*: Use Corollary 3.10.7 to show that commuting A_0, A_1 have a common eigenvector.

2. Let $A(z) = A_1 + z A_2$ be a hermitian pencil. Show that $A(z)$ does not have a crossing point if and only if $A_1 = a_1 I_n$. *Hint*: Show that if $A(z)$ does not have a crossing point then $p_k(\alpha, z)$ splits in $\mathbb{C}[z]$. Hence, $A(z)$ has property L.

3. Let $A(z) = A_1 + z A_2$ be a hermitian pencil. Let $p_k(\alpha, z) \in \mathbb{C}[\alpha, z]$ be defined as in Theorem 4.18.2.

(a) Let ζ be a crossing point of $A(z)$. Let $\beta_i(z)$ be the multi-valued root of $p_k(\alpha, z)$ in $0 < |z - \zeta| < r(\zeta)$. Show that $\lim_{z \to \zeta} \beta_i(z) = \beta_i(\zeta)$, where $\beta_i(\zeta)$ is a corresponding root of $p_k(\alpha, \zeta)$.

(b) Suppose $\beta_i(\zeta)$ is a simple root of $p_k(\alpha, \zeta)$. Show that $\beta_i(z)$ is analytic in $|z - \zeta| < r(\zeta)$.

(c) Give an example of a hermitian pencil with property L, where each $\lambda_i(t)$ is not analytic at least one point $\zeta_i \in \mathbb{R}$.

4. Prove the inequality (4.18.8) using Theorems 4.4.11 and 4.4.9.

4.19 Eigenvalues of Sum of Hermitian Matrices

Let $\mathbf{V}$ be an n-dimensional IPS over $\mathbb{F} = \mathbb{R}, \mathbb{C}$. Recall that $\boldsymbol{\lambda}(C) = (\lambda_1(C), \ldots, \lambda_n(C))^\top$ for $C \in \mathbf{S}(\mathbf{V})$. Clearly, for $t \geq 0$ we have the equality $\lambda_i(tC) = t\lambda_i(C)$ for $i = 1, \ldots, n$, i.e. each $\lambda_i(C)$ is homogeneous for $t \geq 0$. Assume that $A, B \in \mathbf{S}(\mathbf{V})$. An interesting and important question is the relations between the eigenvalues of A, B and $A + B$. Some of these relations we discussed in §4.4 and §4.8. Corollary (4.8.6) for $\alpha = \frac{1}{2}$ yields

$$\boldsymbol{\lambda}(A + B) \prec \boldsymbol{\lambda}(A) + \boldsymbol{\lambda}(B). \tag{4.19.1}$$

Equality holds if and only if the conditions of Corollary 4.8.6 hold.

Theorem 4.19.1 *Let $\mathbf{V}$ be an IPS over $\mathbb{F} = \mathbb{R}, \mathbb{C}$ of dimension n. Assume that $A, B \in \mathbf{S}(\mathbf{V})$. Then TFAE:*

1. $\boldsymbol{\lambda}(A + B) - \boldsymbol{\lambda}(A) \prec \boldsymbol{\lambda}(B)$.

2. $\boldsymbol{\lambda}(A + B)$ *lies in the convex hull of the set* $\{\boldsymbol{\lambda}(A) + P\boldsymbol{\lambda}(B), P \in \mathcal{P}_n\}$, *where* $\mathcal{P}_n$ *is the group of $n \times n$ permutation matrices.*

3. *Let $k \in [n-1]$. Then for each k distinct integers $i_1, \ldots, i_k$ in $[n]$ the following inequality holds:*

$$\sum_{j=1}^{k} \lambda_{i_j}(A+B) \leq \sum_{j=1}^{k} \lambda_{i_j}(A)) + \sum_{j=1}^{k} \lambda_j(B). \qquad (4.19.2)$$

Furthermore, all the above statements hold.

Proof. 1⇒2. Theorem 4.7.8 yields that $\boldsymbol{\lambda}(A+B) - \boldsymbol{\lambda}(A) = F\boldsymbol{\lambda}(B)$ for some doubly stochastic matrix F. Theorem 4.7.6 implies that F is a convex combination of permutation matrices. Hence, $\boldsymbol{\lambda}(A+B)$ is in the claimed convex set.

2⇒3. *2* is equivalent to the statement that $\boldsymbol{\lambda}(A+B) - \boldsymbol{\lambda}(A)$ is in the convex hull of the set spanned by $P\boldsymbol{\lambda}(B), P \in \mathcal{P}_n$. Since the coordinates of $\boldsymbol{\lambda}(B)$ are arranged in a nonincreasing order it follows that $\sum_{j=1}^{k}(P\boldsymbol{\lambda}(B))_{i_j} \leq \sum_{j=1}^{k} \lambda_j(B)$ for each $P \in \mathcal{P}_n$ for any k distinct integers $i_1, \ldots, i_k$ in $[n]$. Hence (4.19.2) holds.

3⇒1. Clearly, $\operatorname{tr}(A+B) = \operatorname{tr} A + \operatorname{tr} B$. The assumption that (4.19.2) for each k integers $i_1, \ldots, i_k$ in $[n]$ and for each $k \in [n-1]$ is equivalent to $\boldsymbol{\lambda}(A+B) - \boldsymbol{\lambda}(A) \prec \boldsymbol{\lambda}(B)$.

Theorem 4.18.6 yields (4.19.2). Hence, conditions 1–3 hold. □

Recall the definition of the p-norm in $\mathbb{R}^n$: $\|\mathbf{x}\|_p := \left(\sum_{j=1}^{n} |x_j|^p \right)^{\frac{1}{p}}$, where $\mathbf{x} = (x_1, \ldots, x_n)^\top$ and $p \in [1, \infty]$. Problem 1 and Theorem 4.7.10 yield:

Corollary 4.19.2 *Let $\mathbf{V}$ be an IPS over $\mathbb{F} = \mathbb{R}, \mathbb{C}$ of dimension n. Assume that $A, B \in \mathbf{S}(\mathbf{V})$. Let $f : [\lambda_n(A-B), \lambda_1(A-B)]$ be a convex function. Then*

$$\sum_{j=1}^{n} f(\lambda_j(A) - \lambda_j(B)) \leq \sum_{j=1}^{n} f(\lambda_j(A-B)). \qquad (4.19.3)$$

In particular,

$$\sum_{j=1}^{n} |\lambda_j(A) - \lambda_j(B)|^p \leq \sum_{j=1}^{n} |\lambda_j(A - B)|^p \quad \text{for each } p \geq 1,$$

(4.19.4)

$$\max_{j \in [n]} |\lambda_j(A) - \lambda_j(B)| \leq \max_{j \in [n]} |\lambda_j(A - B)|.$$

(4.19.5)

Equivalently:

$$\|\boldsymbol{\lambda}(A) - \boldsymbol{\lambda}(B)\|_p \leq \|\boldsymbol{\lambda}(A - B)\|_p \quad \text{for each } p \in [1, \infty].$$

(4.19.6)

For $A \in \mathbb{C}^{m \times n}$ denote by $\boldsymbol{\sigma}(A) := (\sigma_1(A), \ldots, \sigma_{\min(m,n)})^{\top}$.

Theorem 4.19.3 *Let $A, B \in \mathbb{C}^{m \times n}$. Then*

$$\boldsymbol{\sigma}(A) - \boldsymbol{\sigma}(B) \preceq \boldsymbol{\sigma}(A - B),$$

(4.19.7)

$$\|\boldsymbol{\sigma}(A) - \boldsymbol{\sigma}(B)\|_p \leq \|\boldsymbol{\sigma}(A - B)\|_p \quad \text{for each } p \in [1, \infty].$$

(4.19.8)

Proof. Let $l = \min(m, n)$ and $H(A) \in \mathbf{H}_{n+m}$ be defined by (4.11.1). Recall that $\boldsymbol{\lambda}(H(A)) = (\sigma_1(A), \ldots, \sigma_l(A), 0, \ldots, 0, -\sigma_l(A), \ldots, -\sigma_1(A))^{\top}$. (Note that if $m = n$ then there are no zero coordinates in $\boldsymbol{\lambda}(H(A))$.) Theorem 4.19.1 for $H(A), H(B)$ yields $\boldsymbol{\lambda}(H(A)) - \boldsymbol{\lambda}(H(B)) \prec \boldsymbol{\lambda}(H(A - B))$. This relation yields (4.19.7). Observe next:

$$\|\boldsymbol{\lambda}(H(A)) - \boldsymbol{\lambda}(H(B))\|_p = 2^{\frac{1}{p}} \|\boldsymbol{\sigma}(A) - \boldsymbol{\sigma}(B)\|_p,$$

$$\|\boldsymbol{\lambda}(H(A - B))\|_p = 2^{\frac{1}{p}} \|\boldsymbol{\sigma}(A - B)\|_p.$$

Use Corollary 4.19.2 for $H(A), H(B)$ to deduce (4.19.8). $\square$

Problems

1. Let $\mathbf{V}$ be an IPS over $\mathbb{F} = \mathbb{R}, \mathbb{C}$ of dimension n. Assume that $A, B \in \mathbf{S}(\mathbf{V})$. Show

   (a) $\boldsymbol{\lambda}(A) - \boldsymbol{\lambda}(B) \prec \boldsymbol{\lambda}(A - B)$.

(b) $\lambda_j(A) - \lambda_j(B) \in [\lambda_n(A-B), \lambda_1(A-B)]$ for $j = 1, \ldots, n$.

(c) Prove (4.19.5). *Hint:* Take the $\frac{1}{p}$ power if the inequality (4.19.4) and let $p \to \infty$.

2. Assume that $A, B \in \mathbb{C}^{m \times n}$. Show that (4.19.8) for $p = \infty$ is equivalent to $|\sigma_i(A) - \sigma_i(B)| \le \sigma_1(A-B)$ for $i = 1, \ldots, \min(m, n)$.

4.20 Perturbation Formulas for Eigenvalues and Eigenvectors of Hermitian Pencils

Let $B \in \mathbf{H}_n$. Denote by $B^\dagger \in \mathbf{H}_n$ the Moore–Penrose inverse of B. (See §4.12.) That is, B is uniquely characterized by the condition that $B^\dagger B = BB^\dagger$ is the projection on the subspace spanned by all eigenvectors of B corresponding to the nonzero eigenvalues of B.

Theorem 4.20.1 *Let $A(z) = A_0 + zA_1 \in \mathbb{C}[z]^{n \times n}$ be a hermitian pencil. Assume that the eigenvalue a_0 is a simple eigenvalue for A_0. Suppose furthermore that $A_0 x_0 = a_0 x_0$, $x_0^* x_0 = 1$. Let $\alpha(z) \in H(\Omega_1)$ and $\mathbf{x}(z) \in H(\Omega_1)^n$ be the analytic eigenvalue and the the corresponding analytic normalized eigenvector satisfying $\alpha(0) = a_0$ as described in part 5 of Theorem 4.18.2. Then $\alpha(z)$ and $\mathbf{x}(z)$ has power series:*

$$\alpha(z) = \sum_{j=0}^{\infty} a_j z^j, \quad \mathbf{x}(z) = \sum_{j=0}^{\infty} z^j \mathbf{x}_j. \qquad (4.20.1)$$

It is possible to choose $\mathbf{x}(z)$ such that

$$a_1 = x_0^* A_1 x_0,$$

$$x_1 = (a_0 I - A_0)^\dagger A_1 x_0,$$

$$a_2 = x_0^* A_1 x_1' = x_0^* A_1 (a_0 I - A_0)^\dagger A_1 x_0, \qquad (4.20.2)$$

$$x_2 = (a_0 I - A_0)^\dagger (A_1 - a_1 I) x_1 - \left(\frac{1}{2} x_1^* x_1\right) x_0,$$

$$a_3 = x_0^* A_1 ((a_0 I - A_0)^\dagger)^2 A_1 x_0.$$

Proof. By replacing A_0, A_1 with U^*A_0U, U^*A_1U, where $U \in \mathbf{U}_n$, we may assume that A_0 is a diagonal matrix $\mathrm{diag}(d_1, \ldots, d_n)$, where $d_1 = a_0$ and $d_i \neq a_0$ for $i > 1$. Moreover, we can assume that $\mathbf{x}_0 = (1, 0, \ldots, 0)^\top$. Note that $(a_0I - A_0)^\dagger = \mathrm{diag}(0, (a_0 - d_1)^{-1}, \ldots, (a_0 - d_n)^{-1})$. (We are not going to use explicitly these assumptions, but the reader can see more transparently our arguments using these assumptions.)

The orthogonality condition $\mathbf{x}(\bar{z})^*\mathbf{x}(z) = 1$ yields:

$$\sum_{j=0}^{k} \mathbf{x}_j^*\mathbf{x}_{k-j} = 0, \quad k \in \mathbb{N}. \tag{4.20.3}$$

The equality $A(z)\mathbf{x}(z) = \alpha(z)\mathbf{x}(z)$ yields

$$A_0\mathbf{x}_k + A_1\mathbf{x}_{k-1} = \sum_{j=0}^{k} a_j\mathbf{x}_{k-j}, \quad k \in \mathbb{N}. \tag{4.20.4}$$

Since $\alpha(s)$ is real for a real s we deduce that $a_j \in \mathbb{R}$. Consider the equality (4.20.4) for $k = 1$. Multiply it by $\mathbf{x}_0^*$ and use the equality $\mathbf{x}_0^*A_0 = a_0\mathbf{x}_0^*$ to deduce the well-known equality $a_1 = \mathbf{x}_0^*A_0\mathbf{x}_0$, which is the first equality of (4.20.2). (See part 3 of Lemma 4.18.5.) The equality (4.20.4) for $k = 1$ is equivalent to $(a_0I - A_0)\mathbf{x}_1 = A_1\mathbf{x}_0 - a_1\mathbf{x}_0$. Hence, $\mathbf{x}_1$ is of the form

$$\mathbf{x}_1 = (a_0I - A_0)^\dagger(A_1\mathbf{x}_0 - a_1\mathbf{x}_0) + b_1\mathbf{x}_0 = (a_0I - A)^\dagger A_1\mathbf{x}_0 + b_1\mathbf{x}_0, \tag{4.20.5}$$

for some b_1. The orthogonality condition $\Re\mathbf{x}_0^*\mathbf{x}_1 = 0$ implies that $\Re b_1 = 0$. So $b_1 = \sqrt{-1}c_1$ for some $c_1 \in \mathbb{R}$. Replace the eigenvector $\mathbf{x}(z)$ by

$$\mathbf{y}(z) := e^{\sqrt{-1}c_1z}\mathbf{x}(z) = \sum_{j=0}^{\infty} z^j\mathbf{y}_j,$$

$$\mathbf{y}_0 = \mathbf{x}_0, \mathbf{y}_1 = \mathbf{x}_1 - \sqrt{-1}c_1\mathbf{x}_0 = (a_0I - A)^\dagger(A_1\mathbf{x}_0 - a_1\mathbf{x}_0).$$

Note that $\mathbf{y}(\bar{z})^*\mathbf{y}(z) = 0$. Hence, we can assume the second equality of (4.20.2).

Multiply the equality (4.20.4) for $k = 2$ by $\mathbf{x}_0^*$ and use $\mathbf{x}_0^* A_0 = a_0 \mathbf{x}_0^*, \mathbf{x}_0^* \mathbf{x}_1 = 0$ to obtain $a_2 = \mathbf{x}_0^* A_1 \mathbf{x}_1$. This establishes the third equality of (4.20.2). Rewrite the equality (4.20.4) for $k = 2$ as $(a_0 I - A_0) \mathbf{x}_2 = A_1 \mathbf{x}_1 - a_1 \mathbf{x}_1 - a_2 \mathbf{x}_0$. Hence

$$\mathbf{x}_2 = (a_0 I - A_0)^\dagger (A_1 - a_1 I) \mathbf{x}_1 + b_2 \mathbf{x}_0. \tag{4.20.6}$$

Let $c_2 = \Im b_2$. Replace $\mathbf{x}(z)$ by $\mathbf{y}(z) = e^{-\sqrt{-1} c_2 z^2} \mathbf{x}(z)$ to deduce that we can assume that $b_2 \in \mathbb{R}$. Multiply the above equality by $\mathbf{x}_0^*$ to deduce that $\mathbf{x}_0^* \mathbf{x}_2 = b_2$. (4.20.3) for $k = 2$ yields $2 \Re \mathbf{x}_0^* \mathbf{x}_2 + \mathbf{x}_1^* \mathbf{x}_1 = 0$. Hence, $b_2 = -\frac{1}{2} \mathbf{x}_1^\top \mathbf{x}_1$. This establishes the fourth equality of (4.20.2).

Multiply the equality (4.20.4) for $k = 3$ by $\mathbf{x}_0^*$ to deduce

$$a_3 = \mathbf{x}_0^* A_1 \mathbf{x}_2 - a_1 \mathbf{x}_0^* \mathbf{x}_2 = \mathbf{x}_0^* (A_1 - a_1 I) \mathbf{x}_2.$$

Observe next that from the first equality in (4.20.2), $\mathbf{x}_0^* (A_1 - a_1 I) \mathbf{x}_0 = 0$. Also $\mathbf{x}_0^* (a_0 I - A_0)^\dagger = \mathbf{0}^*$. This establishes the last equality of (4.20.2). $\qquad\square$

Note that if A_0, A_1 are real symmetric then $\mathbf{x}(t)$ can be chosen to be in $\mathbb{R}^n$ for all $t \in \mathbb{R}$. In this case $\mathbf{x}(t)$ is determined up to ± 1. Hence, if $\mathbf{x}_0 \in \mathbb{R}^n$ is an eigenvector of A_0 satisfying $\mathbf{x}_0^\top \mathbf{x}_0 = 1$ then $\mathbf{x}(t)$ is determined uniquely. In particular, $\mathbf{x}_1, \mathbf{x}_2$ are determined uniquely by (4.20.2).

Theorem 4.20.2 *Let $A_0 \in \mathbf{H}_n, n \geq 2$. Assume that a_0 is a simple eigenvalue of A_0, with the corresponding eigenvector $A\mathbf{x}_0 = a_0 \mathbf{x}_0, \mathbf{x}_0^* \mathbf{x}_0 = 1$. Suppose furthermore that $|\lambda - a_0| \geq r > 0$ for any other eigenvalue λ of A_0. Let $A_1 \in \mathbf{H}_n, A_1 \neq 0$, and denote by $\|A_1\|$ the l_2 norm of A_1, i.e. the maximal absolute value of the eigenvalues of A_1. Let $\alpha(z)$ be the eigenvalue of $A(z) = A_0 + z A_1$, which is analytic in the neighborhood of $\mathbb{R}$ and satisfying the condition $\alpha(0) = a_0$. Let a_1, a_2 be given by (4.20.2). Fix $0 < c < \frac{r}{2\|A_1\|}$. Then*

$$|\alpha(s) - (a_0 + a_1 s + a_2 s^2)| \leq \frac{4\|A_1\|^3 |s|^3}{(r - 2c\|A_1\|)^2} \text{ for all } s \in [-c, c].$$

$$\tag{4.20.7}$$

Proof. Let $\lambda_1(s) \geq \cdots \geq \lambda_n(s)$ be the eigenvalues of $A(s), s \in \mathbb{R}$. Note that $\lambda_1(0), \ldots, \lambda_n(0)$ are the eigenvalues of A_0. Assume that $\lambda_i(0) = a_1$. Let $\rho(s) = \min(\lambda_{i-1}(s) - \lambda_i(s), \lambda_i(s) - \lambda_{i+1}(s))$, where $\lambda_0(s) = \infty, \lambda_{n+1}(s) = -\infty$. Thus $r \leq \rho(0)$. Let $\beta_1 \geq \cdots \geq \beta_n$ be the eigenvalues of A_1. Then $||A_1|| = \max(|\beta_1|, |\beta_n|)$. (4.19.5) for $A = A(s)$ and $B = A_0$ yields

$$|\lambda_j(s) - \lambda_j(0)| \leq |s| \, ||A_1||, \quad j = 1, \ldots, n.$$

Hence

$$\rho(s) \geq \rho(0) - 2|s| ||A_1|| > 0 \text{ for } s \in \left(-\frac{\rho(0)}{2||A_1||}, \frac{\rho(0)}{2||A_1||} \right). \quad (4.20.8)$$

In particular, $\lambda_i(s)$ is a simple eigenvalue of $A(s)$ in the above interval. Assume that s is in the interval given in (4.20.8). It is straightforward to show that

$$||(\alpha(s)I - A(s))^\dagger|| = \frac{1}{\rho(s)} \leq \frac{1}{\rho(0) - 2|s| \, ||A_1||}. \quad (4.20.9)$$

(One can assume that $A(s)$ is a diagonal matrix.)

Use the Taylor theorem with remainder to obtain the equality

$$\alpha(s) - (a_0 + s a_1 + s^2 a_2) = \frac{1}{6} \alpha^{(3)}(t) s^3 \text{ for some } t, \quad |t| < |s|. \quad (4.20.10)$$

Use Theorem 4.20.1 to deduce that

$$\frac{1}{6}\alpha^{(3)}(t) = \mathbf{x}_i(t)^* ((A_1 - \alpha_i'(t)I)(\alpha_i(t)I - A(t))^\dagger)^2 A_1 \mathbf{x}_i(t),$$

where $\mathbf{x}_i(s)$ is an eigenvector of $A(s)$ of length one corresponding to $\alpha_i(s)$. As $\alpha'(t) = \mathbf{x}_i(t)^* A_1 \mathbf{x}_i(t)$ we deduce that $|\alpha_i'(t)| \leq ||A_1||$. Hence $||A_1 - \alpha_i'(t)I|| \leq 2||A_1||$. Therefore

$$\left| \frac{1}{6}\alpha^{(3)}(t) \right| \leq ||((A_1 - \alpha_i'(t)I)(\alpha_i(t)I - A(t))^\dagger)^2 A_1|| \leq \frac{4||A_1||^3}{\rho(t)^2}.$$

Use the inequality (4.20.9) and the inequality $r \leq \rho(0)$ to deduce the theorem. □

4.21 Historical Remarks

§4.1 and §4.2 are standard. Corollary 4.2.13 is Schur's unitary triangulation theorem [Schu09]. Theorem 4.2.15 is Lanczos method [Lanc50]. §4.3 is standard. A hermitian form is also called a *sesquilinear* form (see also [GolV96]). §4.4 is well known. The maximal and minimal characterizations (4.4.2) are called Rayleigh's principle [Ray73]. The convoy principle, Theorem 4.4.2, is stated in the paper by Pólya and Schiffer [PS54]. It is a precise version of Poincaré's min–max characterization [Poi90]. The Courant–Fischer principle, Theorem 4.4.4, is due to Fischer [Fis05] and Courant [Cou20]. Inequalities of Theorem 4.4.6 are called Weyl's inequalities. Theorem 4.4.8 is Ky Fan's inequality [Fan49]. Theorem 4.4.11 is Wielandt's characterization [Wie56, Wey12]. §4.5 is standard. The results of §4.6 on convex sets and functions are well known. §4.7 is well known. The notion of *majorization* was in introduced formally by Hardy–Littlewood–Pólya [HLP52]. See also [MOA11]. Theorem 4.7.6 is called Birkhoff's theorem [Bir46]. Sometime it is referred as Birkhoff–von Neumann theorem [vNe53]. Theorem 4.7.7 is due to Katz [Katz70]. Theorems 4.7.8 and 4.7.10 are due to Hardy–Littlewood–Pólya [HLP52]. The results of §4.8 are close to the results in [Fri84]. See also [Dav57]. Theorem 4.9.1 goes back to von Neumann [vNe37]. Corollary 4.9.3 is a special case of Hoffman–Wielandt theorem [HW53]. §4.10 is a classical subject. See [Ste93] for a historical survey. Most of the results of §4.11 are well known to the experts. Theorem 4.11.8 goes back to von Neumann [vNe37]. The notion of Moore–Penrose generalized inverse in §4.12 was introduced in [Moo20] and [Pen55]. Most of the results in §4.13 are well known. Theorem 4.13.3 is due to Friedland–Torokhti [FriT07]. The notion of CUR decomposition and some results about it discussed in §4.14 appear in Goreinov–Tyrtyshnikov–Zamarashkin [GorTZ95]. The results of §4.15 are taken from [Fri73]. The results of §4.16 [GorTZ97, GorT01] are taken from Friedland–Loewy [FrL76] (see also [Fri78]). Theorem 4.16.3 is due to H.F. Bohnenblust (unpublished). It is a generalization of [Cal64]. Most of the results of §4.17 are due to Rellich [Rel37, Rel69].

Our exposition follows some results in [Kat80] and [Wim86]. In §4.18, we combine Rellich's theorem with our results in Chapter 3, in particular §3.11. Corollary 4.18.3 is due to Moiseyev–Friedland [MoF80]. Theorem 4.18.6 states the inequality (4.18.8) due to Wielandt [Wie56]. Furthermore, it gives necessary and sufficient conditions for the equality case in Wielandt's inequality. The proof of Theorem 4.18.6 follows [Fri15]. Other results in this section should be known to the experts. The results of §4.19 are known to the experts. Theorem 4.19.1 is due to Lidskii and Wielandt [Lid50, Wie56]. See also [Smi68]. Corollary 4.19.2 is due to Kato [Kat80]. A more general problem is to characterize the eigenvalues of $A + B$, where A, B are hermitian matrices with given eigenvalues. The characterization of such a set was conjectured by Horn [Hor62]. This conjecture was settled by the works of Klyachko [Kly98] and Knutson–Tao [KT99]. See Friedland [Fri00] for generalization of Horn's characterization for hermitian matrices and compact operators, and improved conditions by Fulton [Ful00]. The contents of §4.20 should be known to the experts.

Chapter 5

Elements of Multilinear Algebra

5.1 Tensor Product of Two Free Modules

Let $\mathbb{D}$ be a domain. Recall that $\mathbf{N}$ is called a *free* finite dimensional module over $\mathbb{D}$ if $\mathbf{N}$ has a finite basis $\mathbf{e}_1, \ldots, \mathbf{e}_n$, i.e. $\dim \mathbf{N} = n$. Then $\mathbf{N}' := \mathrm{Hom}\,(\mathbf{N}, \mathbb{D})$ is a free n-dimensional module. Furthermore we can identify $\mathrm{Hom}\,(\mathbf{N}', \mathbb{D})$ with $\mathbf{N}$. (See Problem 1.)

Definition 5.1.1 *Let* $\mathbf{M}, \mathbf{N}$ *be two free finite dimensional modules over an integral domain* $\mathbb{D}$. *Then the tensor product* $\mathbf{M} \otimes \mathbf{N}$ *is identified with* $\mathrm{Hom}\,(\mathbf{N}', \mathbf{M})$. *Moreover, for each* $\mathbf{m} \in \mathbf{M}, \mathbf{n} \in \mathbf{N}$ *we identify* $\mathbf{m} \otimes \mathbf{n} \in \mathbf{M} \otimes_{\mathbb{D}} \mathbf{N}$ *with the linear transformation* $\overline{\mathbf{m} \otimes \mathbf{n}}$: $\mathbf{N}' \to \mathbf{M}$ *given by* $\mathbf{f} \mapsto \mathbf{f}(\mathbf{n})\mathbf{m}$ *for any* $\mathbf{f} \in \mathbf{N}'$.

Proposition 5.1.2 *Let* $\mathbf{M}, \mathbf{N}$ *be free modules over a domain* $\mathbb{D}$ *with bases* $[\mathbf{d}_1, \ldots, \mathbf{d}_m], [\mathbf{e}_1, \ldots, \mathbf{e}_n]$ *respectively. Then* $\mathbf{M} \otimes_{\mathbb{D}} \mathbf{N}$ *is a free module with the basis* $\mathbf{d}_i \otimes \mathbf{e}_j, i = 1, \ldots, m, j = 1, \ldots, n$. *In particular*

$$\dim \mathbf{M} \otimes \mathbf{N} = \dim \mathbf{M} \dim \mathbf{N}. \qquad (5.1.1)$$

(See Problem 3.) For an abstract definition of $\mathbf{M} \otimes_{\mathbb{D}} \mathbf{N}$ for any two $\mathbb{D}$-modules see Problem 16.

Intuitively, one views $\mathbf{M} \otimes \mathbf{N}$ as a linear span of all elements of the form $\mathbf{m} \otimes \mathbf{n}$, where $\mathbf{m} \in \mathbf{M}, \mathbf{n} \in \mathbf{N}$ satisfy the following *natural* properties:

1. $a(\mathbf{m} \otimes \mathbf{n}) = (a\mathbf{m}) \otimes \mathbf{n} = \mathbf{m} \otimes (a\mathbf{n})$ for all $a \in \mathbb{D}$.

2. $(a_1\mathbf{m}_1 + a_2\mathbf{m}_2) \otimes \mathbf{n} = a_1(\mathbf{m}_1 \otimes \mathbf{n}) + a_2(\mathbf{m}_2 \otimes \mathbf{n})$ for all $a_1, a_2 \in \mathbb{D}$.

   (Linearity in the first variable.)

3. $\mathbf{m} \otimes (a_1\mathbf{n}_1 + a_2\mathbf{n}_2) = a_1(\mathbf{m} \otimes \mathbf{n}_1) + a_2(\mathbf{u} \otimes \mathbf{n}_2)$ for all $a_1, a_2 \in \mathbb{D}$.

   (Linearity in the second variable.)

The element $\mathbf{m} \otimes \mathbf{n}$ is called a *decomposable tensor*, or *decomposable element (vector)*, or *rank one tensor*.

Proposition 5.1.3 *Let* $\mathbf{M}, \mathbf{N}$ *be free modules over a domain* $\mathbb{D}$ *with bases* $[\mathbf{d}_1, \ldots, \mathbf{d}_m]$, $[\mathbf{e}_1, \ldots, \mathbf{e}_n]$ *respectively. Then any* $\tau \in \mathbf{M} \otimes_{\mathbb{D}} \mathbf{N}$ *is given by*

$$\tau = \sum_{i=j=1}^{i=m, j=n} a_{ij}\mathbf{d}_i \otimes \mathbf{e}_j, \quad A = [a_{ij}] \in \mathbb{D}^{m \times n}. \tag{5.1.2}$$

Let $[\mathbf{u}_1, \ldots, \mathbf{u}_m]$, $[\mathbf{v}_1, \ldots, \mathbf{v}_n]$ *be different bases of* $\mathbf{M}, \mathbf{N}$ *respectively. Assume that* $\tau = \sum_{i,j=1}^{m,n} b_{ij}\mathbf{u}_i \otimes \mathbf{v}_j$ *and let* $B = [b_{ij}] \in \mathbb{D}^{m \times n}$. *Then* $B = PAQ^T$, *where* P *and* Q *are the transition matrices from the bases* $[\mathbf{d}_1, \ldots, \mathbf{d}_m]$ *to* $[\mathbf{u}_1, \ldots \mathbf{u}_m]$ *and* $[\mathbf{e}_1, \ldots, \mathbf{e}_n]$ *to* $[\mathbf{v}_1, \ldots, \mathbf{v}_n]$. *(That is,* $[\mathbf{d}_1, \ldots, \mathbf{d}_m] = [\mathbf{u}_1, \ldots \mathbf{u}_m]P$, $[\mathbf{e}_1, \ldots, \mathbf{e}_n] = [\mathbf{v}_1, \ldots, \mathbf{v}_n]Q$.*)*

See Problem 6.

Definition 5.1.4 *Let* $\mathbf{M}, \mathbf{N}$ *be free finite dimensional modules over a domain* $\mathbb{D}$. *Let* $\tau \in \mathbf{M} \otimes_{\mathbb{D}} \mathbf{N}$ *be given by* (5.1.2). *The rank of* τ, *denoted by* rank τ, *is the rank of the representation matrix* A, *i.e.* rank τ = rank A. *The tensor rank of* τ, *denoted by* Rank τ, *is the minimal* k *such that* $\tau = \sum_{l=1}^{k} \mathbf{m}_l \otimes \mathbf{n}_l$ *for some* $\mathbf{m}_l \in \mathbf{M}, \mathbf{n}_l \in \mathbf{N}, l = 1, \ldots, k$.

The rank τ is independent of the choice of bases in $\mathbf{M}$ and $\mathbf{N}$. (See Problem 7.) Since $\mathbf{M} \otimes_{\mathbb{D}} \mathbf{N}$ has a basis consisting of decomposable tensors it follows that

$$\text{Rank } \tau \le \min(\dim \mathbf{M}, \dim \mathbf{N}) \text{ for any } \tau \in \mathbf{M} \otimes_{\mathbb{D}} \mathbf{N}. \qquad (5.1.3)$$

See Problem 8.

Proposition 5.1.5 *Let* $\mathbf{M}, \mathbf{N}$ *be free finite dimensional modules over a domain* $\mathbb{D}$. *Let* $\tau \in \mathbf{M} \otimes_{\mathbb{D}} \mathbf{N}$. *Then* $\text{rank } \tau \le \text{Rank } \tau$. *If* $\mathbb{D}$ *is a Bezout domain then* $\text{rank } \tau = \text{Rank } \tau$.

Proof. Assume that $\mathbf{M}, \mathbf{N}$ have bases as in Proposition 5.1.3. Suppose that (5.1.2) holds. Let $\tau = \sum_{l=1}^{k} \mathbf{m}_l \otimes \mathbf{n}_l$. Clearly, each $\mathbf{m}_l \otimes \mathbf{n}_l = \sum_{i,j=1}^{m,n} a_{ij,l} \mathbf{d}_i \otimes \mathbf{e}_j$, where $A_l := [a_{ij,l}]_{i,j=1}^{m,n} \in \mathbb{D}^{m \times n}$ is rank one matrix. Then $A = \sum_{l=1}^{k} A_l$. It is straightforward to show that $\text{rank } A \le k$. This shows that $\text{rank } \tau \le \text{Rank } \tau$.

Assume that $\mathbb{D}$ is a Bezout domain. Let $P \in \mathbf{GL}(m, \mathbb{D})$ such that $PA = [b_{ij}] \in \mathbb{D}^{m \times n}$ is a Hermite normal form of A. In particular, the first $r := \text{rank } A$ rows of B are nonzero rows, and all other rows of B are zero rows. Let $[\mathbf{u}_1, \ldots, \mathbf{u}_m] := [\mathbf{d}_1, \ldots, \mathbf{d}_m] P^{-1}$ be a basis in $\mathbf{M}$. Proposition 5.1.3 yields that $\tau = \sum_{i,j=1}^{m,n} b_{ij} \mathbf{u}_i \otimes \mathbf{e}_j$. Define $\mathbf{n}_l = \sum_{j=1}^{n} b_{lj} \mathbf{e}_j, l = 1, \ldots, r$. Then $\tau = \sum_{l=1}^{r} \mathbf{u}_l \otimes \mathbf{n}_l$. Hence $r \ge \text{Rank } \tau$, which implies that $\text{rank } \tau = \text{Rank } \tau$. $\square$

Proposition 5.1.6 *Let* $\mathbf{M}_i, \mathbf{N}_i$ *be free finite dimensional modules over* $\mathbb{D}$. *Let* $T_i : \mathbf{M}_i \to \mathbf{N}_i$ *be homomorphisms. Then there exists a unique homomorphism on* $T : \mathbf{M}_1 \otimes \mathbf{M}_2 \to \mathbf{N}_1 \otimes \mathbf{N}_2$ *such that* $T(\mathbf{m}_1 \otimes \mathbf{m}_2) = (T_1 \mathbf{m}_1) \otimes (T_2 \mathbf{m}_2)$ *for all* $\mathbf{m}_1 \in \mathbf{M}_1, \mathbf{m}_2 \in \mathbf{M}_2$. *This homomorphism is denoted by* $T_1 \otimes T_2$.

Suppose furthermore that $\mathbf{W}_1, \mathbf{W}_2$ *are free finite dimensional* $\mathbb{D}$-*modules, and* $P_i : \mathbf{N}_i \to \mathbf{W}_i, i = 1, 2$ *are homomorphisms. Then* $(P_1 \otimes P_2)(T_1 \otimes T_2) = (P_1 T_1) \otimes (P_2 T_2)$.

See Problem 9.

Since each homomorphism $T_i : \mathbf{M}_i \to \mathbf{N}_i, i = 1, 2$ is represented by a matrix, one can reduce the definition of $T_1 \otimes T_2$ to the notion of *tensor* product of two matrices $A_1 \in \mathbb{D}^{n_1 \times m_1}, A_2 \in \mathbb{D}^{n_2 \times m_2}$. This tensor product is called the *Kronecker* product.

Definition 5.1.7 *Let* $A = [a_{ij}]_{i,j=1}^{m,n} \in \mathbb{D}^{m \times n}, B = [b_{ij}]_{i,j=1}^{p,q} \in \mathbb{D}^{p \times q}$. *Then* $A \otimes B \in \mathbb{D}^{mp \times nq}$ *is the following block matrix:*

$$A \otimes B := \begin{bmatrix} a_{11}B & a_{12}B & ... & a_{1n}B \\ a_{21}B & a_{22}B & ... & a_{2n}B \\ \vdots & \vdots & \vdots & \vdots \\ a_{m1}B & a_{m_1 2}B & ... & a_{mn}B \end{bmatrix}. \tag{5.1.4}$$

In the rest of the section, we discuss the symmetric and skew symmetric tensor products of $\mathbf{M} \otimes \mathbf{M}$.

Definition 5.1.8 *Let* $\mathbf{M}$ *be a free finite dimensional module over* $\mathbb{D}$. *Denote* $\mathbf{M}^{\otimes 2} := \mathbf{M} \otimes \mathbf{M}$. *The submodule* $\mathrm{Sym}^2 \mathbf{M} \subset \mathbf{M}^{\otimes 2}$, *called a 2-symmetric power of* $\mathbf{M}$, *is spanned by tensors of the form* $\mathrm{sym}^2(\mathbf{m}, \mathbf{n}) := \mathbf{m} \otimes \mathbf{n} + \mathbf{n} \otimes \mathbf{m}$ *for all* $\mathbf{m}, \mathbf{n} \in \mathbf{M}$. $\mathrm{sym}^2(\mathbf{m}, \mathbf{n}) = \mathrm{sym}^2(\mathbf{n}, \mathbf{m})$ *is called a 2-symmetric product of* $\mathbf{m}$ *and* $\mathbf{n}$, *or simply a symmetric product. Any vector* $\tau \in \mathrm{Sym}^2 \mathbf{M}$ *is a called a 2-symmetric tensor, or simply a symmetric tensor. The subspace* $\bigwedge^2 \mathbf{M} \subset \mathbf{M}^{\otimes 2}$, *called 2-exterior power of* $\mathbf{M}$, *is spanned by all tensors of the form* $\mathbf{m} \wedge \mathbf{n} := \mathbf{m} \otimes \mathbf{n} - \mathbf{n} \otimes \mathbf{m}$, *for all* $\mathbf{m}, \mathbf{n} \in \mathbf{M}$. $\mathbf{m} \wedge \mathbf{n} = -\mathbf{n} \wedge \mathbf{m}$ *is called the* wedge product *of* $\mathbf{m}$ *and* $\mathbf{n}$. *Any vector* $\tau \in \bigwedge^2 \mathbf{M}$ *is called a 2-skew symmetric tensor, or simply a skew symmetric tensor.*

Since $\mathbf{M}^{\otimes 2}$ can be identified with $\mathbb{D}^{m \times m}$ it follows that $\mathrm{Sym}^2(\mathbf{M})$ and $\bigwedge^2 \mathbf{M}$ can be identified with the submodules of symmetric and skew symmetric matrices respectively. See Problem 12. Observe next that $2\mathbf{m} \otimes \mathbf{n} = \mathrm{sym}^2(\mathbf{m}, \mathbf{n}) + \mathbf{m} \wedge \mathbf{n}$. Assume that 2 is a unit in $\mathbb{D}$. Then $\mathbf{M}^{\otimes 2} = \mathrm{Sym}^2(\mathbf{M}) \oplus \bigwedge^2 \mathbf{M}$. Hence any tensor $\tau \in \mathbf{M}^{\otimes 2}$ can be decomposed uniquely to a sum $\tau = \tau_s + \tau_a$ where $\tau_s, \tau_a \in \mathbf{M}^{\otimes 2}$ are symmetric and skew symmetric tensors respectively. (See Problem 12.)

Proposition 5.1.9 *Let* $\mathbf{M}, \mathbf{N}$ *be a finite dimensional module over* $\mathbb{D}$. *Let* $T : \mathrm{Hom}\,(\mathbf{M}, \mathbf{N})$. *Then*

$$T \otimes T : \mathrm{Sym}^2\mathbf{M} \to \mathrm{Sym}^2\mathbf{N}, \quad T \otimes T : \overset{2}{\bigwedge}\mathbf{M} \to \overset{2}{\bigwedge}\mathbf{N}.$$

See Problem 13.

Definition 5.1.10 *Let* $\mathbf{M}, \mathbf{N}$ *be finite dimensional modules over* $\mathbb{D}$. *Let* $T : \mathrm{Hom}\,(\mathbf{M}, \mathbf{N})$. *Then* $T \wedge T \in \mathrm{Hom}\,(\bigwedge^2 \mathbf{M}, \bigwedge^2 \mathbf{N})$ *is defined as the restriction of* $T \otimes T$ *to* $\bigwedge^2 \mathbf{M}$.

Proposition 5.1.11 *Let* $\mathbf{M}, \mathbf{N}$ *be a finite dimensional module over* $\mathbb{D}$. *Let* $T : \mathrm{Hom}\,(\mathbf{M}, \mathbf{N})$. *Then*

1. *Assume that* $[\mathbf{d}_1, \dots, \mathbf{d}_m]$ *is a basis of* $\mathbf{M}$. *Then* $\mathbf{d}_i \wedge \mathbf{d}_j, 1 \leq i < j \leq m$ *is a basis of* $\bigwedge^2 \mathbf{M}$.

2. *Assume that* $S : \mathrm{Hom}\,(\mathbf{L}, \mathbf{M})$. *Then* $ST \wedge ST = (S \wedge S)(T \wedge T)$.

See Problem 14.

Problems

1. Let $\mathbf{N}$ be a free module with a basis $[\mathbf{e}_1, \dots, \mathbf{e}_n]$. Show

   (a) $\mathbf{N}' := \mathrm{Hom}\,(\mathbf{N}, \mathbb{D})$ is a free module with a basis $[\mathbf{f}_1, \dots, \mathbf{f}_n]$, where $\mathbf{f}_i(\mathbf{e}_j) = \delta_{ij}, i, j = 1, \dots, n$.

   (b) Show that $(\mathbf{N}')'$ can be identified with $\mathbf{N}$ as follows. To each $\mathbf{n} \in \mathbf{N}$ associate the following functional $\widehat{\mathbf{n}} : \mathbf{N}' \to \mathbb{D}$ defined by $\widehat{\mathbf{n}}(\mathbf{f}) = \mathbf{f}(\mathbf{n})$ for each $\mathbf{f} \in \mathbf{N}'$. Show that $\widehat{\mathbf{n}}$ is a linear functional on $\mathbf{N}'$ and any $\tau \in (\mathbf{N}')'$ is equal to a unique $\widehat{\mathbf{n}}$.

2. Let $\mathbb{F}$ be a field and $\mathbf{V}$ be an n-dimensional subspace of $\mathbf{V}$. Then $\mathbf{V}' := \mathrm{Hom}\,(\mathbf{V}, \mathbb{F})$ is called the *dual space* of $\mathbf{V}$. Show

   (a) $(\mathbf{V}')'$ can be identified with $\mathbf{V}$. That is, for each $\mathbf{v} \in \mathbf{V}$, let $\hat{\mathbf{v}} : \mathbf{V}' \to \mathbb{F}$ be the linear functional given by $\hat{\mathbf{v}}(\mathbf{f}) = \mathbf{f}(\mathbf{v})$. Then any $\psi \in (\mathbf{V}')'$ is of the form $\hat{\mathbf{v}}$ for some $\mathbf{v} \in \mathbf{V}$.

(b) For $X \subseteq \mathbf{V}$, $F \subseteq \mathbf{V}'$ denote by $X^{\perp} := \{\mathbf{f} \in \mathbf{V}' : \mathbf{f}(\mathbf{x}) = 0,$ $\forall \mathbf{x} \in X\}, F^{\perp} := \{\mathbf{v} \in \mathbf{V} : \mathbf{f}(\mathbf{v}) = 0, \forall \mathbf{f} \in F\}$. Then $X^{\perp}, F^{\perp}$ are subspaces of $\mathbf{V}', \mathbf{V}$ respectively satisfying

$$(X^{\perp})^{\perp} = \text{span}\,(X), \quad \dim X^{\perp} = n - \dim \text{span}\,(X),$$
$$(F^{\perp})^{\perp} = \text{span}\,(F), \quad \dim F^{\perp} = n - \dim \text{span}\,(F).$$

(c) Let $\mathbf{U}_1, \ldots, \mathbf{U}_k$ be k-subspaces of either $\mathbf{V}$ or $\mathbf{V}'$. Then

$$(\cap_{i=1}^{k} \mathbf{U}_i)^{\perp} = \sum_{i=1}^{k} \mathbf{U}_i^{\perp}, \quad \left(\sum_{i=1}^{k} \mathbf{U}_i\right)^{\perp} = \cap_{i=1}^{k} \mathbf{U}_i^{\perp}.$$

(d) For each bases $\{\mathbf{v}_1, \mathbf{v}_2, \ldots, \mathbf{v}_n\}, \{\mathbf{f}_1, \ldots, \mathbf{f}_n\}$ in $\mathbf{V}, \mathbf{V}'$ respectively there exists unique dual bases $\{\mathbf{g}_1, \mathbf{g}_2, \ldots, \mathbf{g}_n\}$, $\{\mathbf{u}_1, \ldots, \mathbf{u}_n\}$ in $\mathbf{V}', \mathbf{V}$ respectively such that $\mathbf{g}_i(\mathbf{v}_j) = \mathbf{f}_i(\mathbf{u}_j) = \delta_{ij}, i, j = 1, \ldots, n$.

(e) Let $\mathbf{U} \subset \mathbf{V}, \mathbf{W} \subset \mathbf{V}'$ two m-dimensional subspaces. TFAE

    i. $\mathbf{U} \cap \mathbf{W}^{\perp} = \{\mathbf{0}\}$.

    ii. $\mathbf{U}^{\perp} \cap \mathbf{W} = \{\mathbf{0}\}$.

    iii. There exists bases $\{\mathbf{u}_1, \ldots, \mathbf{u}_m\}, \{\mathbf{f}_1, \ldots, \mathbf{f}_m\}$ in $\mathbf{U}, \mathbf{W}$ respectively such that $\mathbf{f}_j(\mathbf{u}_i) = \delta_{ij}, i, j = 1, \ldots, m$.

3. Show Proposition 5.1.2.

4. Let $\mathbf{U}$ be the space of all polynomials in variable x of degree less than m: $p(x) = \sum_{i=0}^{m-1} a_i x^i$ with coefficients in $\mathbb{F}$. Let $\mathbf{V}$ be the space of all polynomials in variable y of degree less than n: $q(y) = \sum_{j=0}^{n-1} b_j y^j$ with coefficients in $\mathbb{F}$. Then $\mathbf{U} \otimes \mathbf{V}$ is identified with the vector space of all polynomials in two variables x, y of the form $f(x,y) = \sum_{i=j=0}^{m-1,n-1} c_{ij} x^i y^j$ with the coefficients in $\mathbb{F}$. The decomposable elements are $p(x)q(y), p \in \mathbf{U}, q \in \mathbf{V}$. (The tensor products of this kind are basic tools for solving partial differential equations (PDE), using *separation of variables*, i.e. Fourier series.)

5. Let $\mathbf{M} = \mathbb{D}^m, \mathbf{N} = \mathbb{D}^n$. Show

   (a) $\mathbf{M} \otimes \mathbf{N}$ can be identified with the space of $m \times n$ matrices $\mathbb{D}^{m \times n}$. More precisely each $A \in \mathbb{D}^{m \times n}$ is viewed as a homomorphism $A : \mathbb{D}^n \to \mathbb{D}^m$, where $\mathbb{D}^m$ is identified with $\mathbf{M}'$.

   (b) The decomposable tensor $\mathbf{m} \otimes \mathbf{n}$ is identified with $\mathbf{mn}^T$. (*Note*: $\mathbf{mn}^T$ is indeed rank one matrix.)

6. Prove Proposition 5.1.3.

7. Show that rank τ defined in Definition 5.1.4 is independent of the choice of bases in $\mathbf{M}$ and $\mathbf{N}$.

8. Let the assumptions of Proposition 5.1.3 holds. Show that the equalities

$$\tau = \sum_{i=1}^{m} \mathbf{d}_i \otimes \left(\sum_{j=1}^{n} b_{ij} \mathbf{e}_j \right) = \sum_{j=1}^{n} \left(\sum_{i=1}^{m} b_{ij} \mathbf{d}_i \right) \otimes \mathbf{e}_j$$

   yield (5.1.3).

9. Prove Proposition 5.1.6.

10. Let the assumptions of Proposition 5.1.2 hold. Arrange the basis of $\mathbf{M} \otimes_{\mathbb{D}} \mathbf{N}$ in the lexicographical order: $\mathbf{d}_1 \otimes \mathbf{e}_1, \ldots, \mathbf{d}_1 \otimes \mathbf{e}_n, \mathbf{d}_2 \otimes \mathbf{e}_1, \ldots, \mathbf{d}_2 \otimes \mathbf{e}_n, \ldots, \mathbf{d}_m \otimes \mathbf{e}_1, \ldots, \mathbf{d}_m \otimes \mathbf{e}_n$. We denote this basis by $[\mathbf{d}_1, \ldots, \mathbf{d}_m] \otimes [\mathbf{e}_1, \ldots, \mathbf{e}_n]$.

    Let $\mathbf{M}_l, \mathbf{N}_l$ be free modules with the bases $[\mathbf{d}_{1,l}, \ldots, \mathbf{d}_{m_l,l}]$, $[\mathbf{e}_{1,l}, \ldots, \mathbf{e}_{n_l,l}]$ for $l = 1, 2$. Let $T_l : \mathbf{M}_l \to \mathbf{N}_l$ be a homomorphism represented by $A_l \in \mathbb{D}^{n_l \times m_l}$ in the above bases for $l = 1, 2$. Show that $T_1 \otimes T_2$ is represented by the matrices $A_1 \otimes A_2$ with respect to the bases $[\mathbf{d}_{1,1}, \ldots, \mathbf{d}_{m_1,1}] \otimes [\mathbf{e}_{1,1}, \ldots, \mathbf{e}_{n_1,1}]$ and $[\mathbf{d}_{1,2}, \ldots, \mathbf{d}_{m_2,2}] \otimes [\mathbf{e}_{1,2}, \ldots, \mathbf{e}_{n_2,2}]$.

11. Let $A \in \mathbb{D}^{m \times n}, B \in \mathbb{D}^{p \times q}$. Show

    (a) If $m = n$ and A is upper triangular, then $A \otimes B$ is block upper triangular.

(b) If $m = n, p = q$ and A and B are upper triangular, then $A \otimes B$ is upper triangular.

(c) If A and B are diagonal then $A \otimes B$ is diagonal. In particular $I_m \otimes I_p = I_{mp}$.

(d) Let $C \in \mathbb{D}^{l \times m}, D \in \mathbb{D}^{r \times p}$. Then $(C \otimes D)(A \otimes B) = (CA) \otimes (DB)$.

(e) $A \in \mathbf{GL}(m, \mathbb{D}), B \in \mathbf{GL}(p, \mathbb{D})$ then $A \otimes B \in \mathbf{GL}(mp, \mathbb{D})$ and $(A \otimes B)^{-1} = A^{-1} \otimes B^{-1}$.

(f) rank $A \otimes B$ = rank A rank B. (Use the fact that over the quotient field $\mathbb{F}$ of $\mathbb{D}$, A and B are equivalent to diagonal matrices.)

(g) Let $m = n, p = q$. Show that det $A \otimes B$ = det A det B.

12. Let $\mathbf{M}$ be a free module with a basis $[\mathbf{d}_1, \ldots, \mathbf{d}_m]$. Identify $\mathbf{M}^{\otimes 2}$ with $\mathbb{D}^{m \times m}$. Show that $\mathrm{Sym}^2 \mathbf{M}$ is identified with $S_m(\mathbb{D}) \subset \mathbb{D}^{m \times m}$, the module of $m \times m$ symmetric matrices: $A^T = A$, and $\bigwedge^2 \mathbf{M}$ is identified with $\mathbf{AS}(m, \mathbb{D})$, the module of $m \times m$ skew symmetric matrices: $A^T = -A$.

    Assume that 2 is a unit in $\mathbb{D}$. Show the decomposition $\tau \in \mathbf{M}^{\otimes 2}$ as sum of symmetric and skew symmetric tensor is equivalent to the following fact: Any matrix $A \in \mathbb{D}^{m \times m}$ is of the form $A = 2^{-1}(A + A^T) + 2^{-1}(A - A^T)$, which is the unique decomposition of a sum of symmetric and skew symmetric matrices.

13. (a) Prove Proposition 5.1.9.

    (b) Show that $(\mathrm{Sym}^2 \mathbf{M}, \mathrm{Sym}^2 \mathbf{N})$ and $(\bigwedge^2 \mathbf{M}, \bigwedge^2 \mathbf{N})$ are the only invariant pairs of submodules of $T^{\otimes 2}$ for all choices of $T \in \mathrm{Hom} \, (\mathbf{M}, \mathbf{N})$.

14. Prove Proposition 5.1.11.

15. Let $\mathbf{M}$ be a module over the domain $\mathbb{D}$. Let $X \subseteq \mathbf{M}$ be a subset of $\mathbf{M}$. Then span X is the set of all finite linear combinations of the elements from X.

    (a) Show that span X is a submodule of $\mathbf{M}$.

(b) span X is called the submodule generated by X.

16. Let X be a nonempty set. For a given domain $\mathbb{D}$ denote by $\mathbf{M}_{\mathbb{D}}(X)$ the free $\mathbb{D}$-module generated by X. That is, $\mathbf{M}_{\mathbb{D}}(X)$ has a set of elements $\mathbf{e}(x), x \in X$ with the following properties:

(a) For each finite nonempty subset $Y \subseteq X$, the set of vectors $\mathbf{e}(y), y \in Y$ are linearly independent.

(b) $M_{\mathbb{D}}(X)$ is generated by $\{\mathbf{e}(x), x \in X\}$.

Let $\mathbf{M}, \mathbf{N}$ be two modules over an integral domain $\mathbb{D}$. Let $\mathbf{P}$ be the free module generated by $\mathbf{M} \times \mathbf{N} := \{(\mathbf{m}, \mathbf{n}) : \mathbf{m} \in \mathbf{M}, \mathbf{n} \in \mathbf{N}\}$. Let $\mathbf{Q} \subseteq \mathbf{P}$ generated by the elements of the form

$$\mathbf{e}((a\mathbf{m}_1 + b\mathbf{m}_2, c\mathbf{n}_1 + d\mathbf{n}_2)) - ac\mathbf{e}((\mathbf{m}_1, \mathbf{n}_1))$$
$$-ad\mathbf{e}((\mathbf{m}_1, \mathbf{n}_2)) - bc\mathbf{e}((\mathbf{m}_2, \mathbf{n}_1) - bd\mathbf{e}(\mathbf{m}_2, \mathbf{n}_2)),$$

for all $a, b, c, d \in \mathbb{D}$ and $\mathbf{m}_1, \mathbf{m}_2 \in \mathbf{M}, \mathbf{n}_1, \mathbf{n}_2 \in \mathbf{N}$. Then $\mathbf{M} \otimes_{\mathbb{D}} \mathbf{N} := \mathbf{P}/\mathbf{Q}$ is called the tensor product of $\mathbf{M}$ and $\mathbf{N}$ over $\mathbb{D}$.

Show that if $\mathbf{M}, \mathbf{N}$ are two free finite dimensional modules then the above definition of $\mathbf{M} \otimes_{\mathbb{D}} \mathbf{N}$ is isomorphic to Definition 5.1.1.

5.2   Tensor Product of Several Free Modules

Definition 5.2.1 *Let* $\mathbf{M}_i$ *be free finite dimensional modules over a domain* $\mathbb{D}$ *for* $i = 1, \ldots, k$, *where* $k \geq 2$. *Then* $\mathbf{M} := \otimes_{i=1}^{k} \mathbf{M}_i = \mathbf{M}_1 \otimes \mathbf{M}_2 \otimes \cdots \otimes \mathbf{M}_k$ *is the* tensor product space *of* $\mathbf{M}_1, \ldots, \mathbf{M}_k$, *and is defined as follows. For* $k = 2$ $\mathbf{M}_1 \otimes \mathbf{M}_2$ *is defined in Definition* 5.1.1. *For* $k \geq 3$ $\otimes_{i=1}^{k} \mathbf{M}_i$ *is defined recursively as* $(\otimes_{i=1}^{k-1} \mathbf{M}_i) \otimes \mathbf{M}_k$.

Note that from now on we suppress in our notation the dependence on $\mathbb{D}$. When we need to emphasize $\mathbb{D}$ we use the notation

$\mathbf{M}_1 \otimes_{\mathbb{D}} \cdots \otimes_{\mathbb{D}} \mathbf{M}_k$. $\mathbf{M}$ is spanned by the *decomposable tensors*

$$\otimes_{i=1}^k \mathbf{m}_i := \mathbf{m}_1 \otimes \mathbf{m}_2 \otimes \cdots \otimes \mathbf{m}_k, \quad \mathbf{m}_i \in \mathbf{M}_i, i = 1, \ldots, k,$$

also called *rank one tensors*. One has the basic identity:

$$a(\mathbf{m}_1 \otimes \mathbf{m}_2 \otimes \cdots \otimes \mathbf{m}_k) = (a\mathbf{m}_1) \otimes \mathbf{m}_2 \otimes \cdots \otimes \mathbf{m}_k$$
$$= \mathbf{m}_1 \otimes (a\mathbf{m}_2) \otimes \cdots \otimes \mathbf{m}_k = \cdots$$
$$= \mathbf{m}_1 \otimes \mathbf{m}_2 \otimes \cdots \otimes (a\mathbf{m}_k).$$

Furthermore, the above decomposable tensor is multilinear in each variable. Clearly

$$\otimes_{i=1}^k \mathbf{m}_{j_i,i}, \, j_i = 1, \ldots, m_i, i = 1, \ldots, k \text{ is a basis of } \otimes_{i=1}^k \mathbf{M}_i \tag{5.2.1}$$

if $\mathbf{m}_{1,i}, \ldots, \mathbf{m}_{m_i,i}$ is a basis of $\mathbf{M}_i$ for $i = 1, \ldots, k$.

Hence

$$\dim \otimes_{i=1}^k \mathbf{M}_i = \prod_{i=1}^k \dim \mathbf{M}_i. \tag{5.2.2}$$

Thus

$$\alpha = \sum_{j_1 = j_2 = \cdots = j_k = 1}^{m_1, m_2, \ldots, m_k} a_{j_1 j_2 \ldots j_k} \otimes_{i=1}^k \mathbf{m}_{j_i,i}, \text{ for any } \alpha \in \otimes_{i=1}^k \mathbf{M}_i. \tag{5.2.3}$$

Denote

$$\mathbb{D}^{m_1 \times \cdots \times m_k} := \otimes_{i=1}^k \mathbb{D}^{m_i}, \text{ for } k \in \mathbb{N} \text{ and } m_i \in \mathbb{N}, i = 1, \ldots, k. \tag{5.2.4}$$

$\mathcal{A} \in \mathbb{D}^{m_1 \times \cdots \times m_k}$ is given as $\mathcal{A} := [a_{j_1 \ldots j_k}]_{j_1 = \cdots = j_k = 1}^{m_1, \ldots, m_k}$, where $a_{j_1 \ldots j_k} \in \mathbb{D}, j_i = 1, \ldots, m_i, i = 1, \ldots, k$. $\mathcal{A}$ is called a *k-tensor*. So 1-tensor is a vector and 2-tensor is a matrix.

In particular $\otimes_{i=1}^k \mathbf{M}_i$ is isomorphic to $\mathbb{D}^{m_1 \times \cdots \times m_k}$. Furthermore, after choosing a basis of $\otimes_{i=1}^k \mathbf{M}_i$ of the form (5.2.1) we correspond to each $\tau \in \otimes_{i=1}^k \mathbf{M}_i$ of the form (5.2.3) the tensor $\mathcal{A} = [a_{j_1 \ldots j_k}]_{j_1 = \cdots = j_k = 1}^{m_1, \ldots, m_k} \in \mathbb{D}^{m_1 \times \cdots \times m_k}$.

Proposition 5.2.2 *Let $\mathbf{M}_i, \mathbf{N}_i, i = 1, \ldots, k$ be free finite dimensional modules over $\mathbb{D}$. Let $T_i : \mathbf{M}_i \to \mathbf{N}_i, i = 1, \ldots, k$ be homomorphisms. Then there exists a unique homomorphism on $T : \otimes_{i=1}^k \mathbf{M}_i \to \otimes_{i=1}^k \mathbf{N}_i$ such that $T(\otimes_{i=1}^k \mathbf{m}_i) = \otimes_{i=1}^k (T_i \mathbf{m}_i)$ for all $\mathbf{m}_i \in \mathbf{M}_i, i = 1, \ldots, k$. This homomorphism is denoted by $\otimes_{i=1}^k T_i$.*

Suppose furthermore that $\mathbf{W}_i, i = 1, \ldots, k$ are free finite dimensional $\mathbb{D}$-modules, and $P_i : \mathbf{N}_i \to \mathbf{W}_i, i = 1, \ldots, k$ are homomorphisms. Then $(\otimes_{i=1}^k P_i)(\otimes_{i=1}^k T_i) = \otimes_{i=1}^k (P_i T_i)$.

See Problem 3.

Since each homomorphism $T_i : \mathbf{M}_i \to \mathbf{N}_i, i = 1, \ldots, k$ is represented by a matrix, one can reduce the definition of $\otimes_{i=1}^k T_i$ to the notion of the *tensor* product of k matrices.

Definition 5.2.3 *Let $A_i = [a_{lj,i}]_{l,j=1}^{m_i,n_i} \in \mathbb{D}^{m_i \times n_i}, i = 1, \ldots, k.$ Then the Kronecker product $A := \otimes_{i=1}^k A_i \in \mathbb{D}^{m_1 \cdots m_k \times n_1 \cdots n_k}$ is the matrix with the entries*

$$A = [a_{(l_1, \ldots, l_k)(j_1, \ldots, j_k)}], \quad a_{(l_1, \ldots, l_k)(j_1, \ldots, j_k)} := \prod_{i=1}^k a_{l_i j_i, i},$$

$$\text{for } l_i = 1, \ldots, m_i, \ j_i = 1, \ldots, n_i, \ i = 1, \ldots, k,$$

where the indices $(l_1, \ldots, l_k), l_i = 1, \ldots, m_i, i = 1, \ldots, k,$ and the indices $(j_1, \ldots, j_k), j_i = 1, \ldots, n_i, i = 1, \ldots, k$ are arranged in lexicographical order.

It is straightforward to show that the above tensor product of matrices can be recursively defined by the Kronecker product of two matrices as defined in Definition 5.1.7. See Problem 4. The tensor products of k matrices have similar properties as in the case $k = 2$. See Problem 5.

We now consider the k-symmetric and k-exterior products of a free finite dimensional module $\mathbf{M}$. In view of the previous section we may assume that $k \geq 3$. Recall that Σ_k the permutation group of k elements of $[k]$ and sign $\sigma \in \{1, -1\}$ is the sign of $\sigma \in \Sigma_k$.

Definition 5.2.4 *Let* $\mathbf{M}$ *be a free finite dimensional module over* $\mathbb{D}$ *and* $2 \leq k \in \mathbb{N}$. *Denote* $\mathbf{M}^{\otimes k} := \otimes_{i=1}^{k} \mathbf{M}_i$, *where* $\mathbf{M}_i = \mathbf{M}$ *for* $i = 1, \ldots, k$. *The submodule* $\mathrm{Sym}^k \mathbf{M} \subset \mathbf{M}^{\otimes k}$, *called a* k-*symmetric power of* $\mathbf{M}$, *is spanned by tensors of the form*

$$\mathrm{sym}^k(\mathbf{m}_1, \ldots, \mathbf{m}_k) := \sum_{\sigma \in \Sigma_k} \otimes_{i=1}^{k} \mathbf{m}_{\sigma(i)}, \qquad (5.2.5)$$

for all $\mathbf{m}_i \in \mathbf{M}, i = 1, \ldots, k$. $\mathrm{sym}^k(\mathbf{m}_1, \ldots, \mathbf{m}_k)$ *is called a* k-*symmetric product of* $\mathbf{m}_1, \ldots, \mathbf{m}_k$, *or simply a symmetric product. Any tensor* $\tau \in \mathrm{Sym}^k \mathbf{M}$ *is called a* k-*symmetric tensor, or simply a symmetric tensor. The subspace* $\bigwedge^k \mathbf{M} \subset \mathbf{M}^{\otimes k}$, *called the* k-*exterior power of* $\mathbf{M}$, *is spanned by all tensors of the form*

$$\wedge_{i=1}^{k} \mathbf{m}_i = \mathbf{m}_1 \wedge \cdots \wedge \mathbf{m}_k := \sum_{\sigma \in \Sigma_k} \mathrm{sign}\, \sigma \otimes_{i=1}^{k} \mathbf{m}_{\sigma(i)} \qquad (5.2.6)$$

for all $\mathbf{m}_i \in \mathbf{M}, i = 1, \ldots, k$. $\wedge_{i=1}^{k} \mathbf{m}_i$ *is called the* k-*wedge product of* $\mathbf{m}_1, \ldots, \mathbf{m}_k$. *Any vector* $\tau \in \bigwedge^k \mathbf{M}$ *is called a* k-*skew symmetric tensor, or simply a skew symmetric tensor.*

Proposition 5.2.5 *Let* $\mathbf{M}, \mathbf{N}$ *be free finite dimensional modules over* $\mathbb{D}$. *Let* $T \in \mathrm{Hom}\,(\mathbf{M}, \mathbf{N})$. *For* $k \in \mathbb{N}$ *let* $T^{\otimes k} : \mathbf{M}^{\otimes k} \to \mathbf{N}^{\otimes k}$ *be* $\underbrace{T \otimes \cdots \otimes T}_{k}$. *Then*

$$T^{\otimes k} : \mathrm{Sym}^k \mathbf{M} \to \mathrm{Sym}^k \mathbf{N}, \quad T^{\otimes k} : \bigwedge^k \mathbf{M} \to \bigwedge^k \mathbf{N}.$$

See Problem 6.

Definition 5.2.6 *Let* $\mathbf{M}, \mathbf{N}$ *be free finite dimensional modules over* $\mathbb{D}$. *Let* $T \in \mathrm{Hom}\,(\mathbf{M}, \mathbf{N})$. *Then* $\wedge^k T \in \mathrm{Hom}\,(\bigwedge^k \mathbf{M}, \bigwedge^k \mathbf{N})$ *is defined as the restriction of* $T^{\otimes k}$ *to* $\bigwedge^k \mathbf{M}$.

Proposition 5.2.7 *Let* $\mathbf{M}, \mathbf{N}$ *be free finite dimensional modules over* $\mathbb{D}$. *Let* $T \in \mathrm{Hom}\,(\mathbf{M}, \mathbf{N})$. *Then*

1. Let $[\mathbf{d}_1,\ldots,\mathbf{d}_m], [\mathbf{e}_1,\ldots,\mathbf{e}_n]$ be bases in $\mathbf{M},\mathbf{N}$ respectively. Assume that T is represented by the matrix $A = [a_{ij}] \in \mathbb{D}^{n \times m}$ in these bases. Then $\wedge^k T$ is represented in the bases

$$\wedge_{i=1}^k \mathbf{d}_{j_i}, \ 1 \leq j_1 < \cdots < j_k \leq m,$$

$$\wedge_{i=1}^k \mathbf{e}_{l_i}, \ 1 \leq l_1 < \cdots < l_k \leq n,$$

by the matrix $\wedge^k A \in \mathbb{D}^{\binom{n}{k} \times \binom{m}{k}}$, where the entry $((l_1,\ldots,l_k),$ $(j_1,\ldots,j_k))$ of $\wedge^k A$ is the $k \times k$ minor of A obtained by deleting all rows and columns of A except the $(l_1,\ldots,l_k)$ rows and $(j_1,\ldots,j_k)$ columns.

2. Let $\mathbf{L}$ be a free finite dimensional module and assume that $S :$ Hom $(\mathbf{L},\mathbf{M})$. Then $\wedge^k(TS) = (\wedge^k T)(\wedge^k S)$.

See Problem 6.

Remark 5.2.8 *In the classical matrix books such as* [Gan59] *and* [MaM64] *the matrix* $\wedge^k A$ *is called the kth compound matrix or kth adjugate of* A.

The following proposition is proven straightforward:

Proposition 5.2.9 *Let* $\mathbf{M}_1,\ldots,\mathbf{M}_k, \mathbf{M} := \otimes_{i=1}^k \mathbf{M}_i$ *be free finite dimensional modules over* $\mathbb{D}$ *with bases given in* (5.2.1). *Let* $[\mathbf{n}_{1,i},\ldots,\mathbf{n}_{m_i,i}] = [\mathbf{m}_{1,i},\ldots,\mathbf{m}_{m_i,i}]T_i^{-1}$, $T_i = [t_{lj,i}] \in \mathbf{GL}(m_i,\mathbb{D})$ *be another basis of* $\mathbf{M}_i$ *for* $i = 1,\ldots,m_i$. *Let* $\alpha \in \mathbf{M}$ *be given by* (5.2.3). *Then*

$$\alpha = \sum_{l_1=\cdots=l_k=1}^{m_1,\ldots,m_k} b_{l_1\ldots l_k} \otimes_{i=1}^k \mathbf{n}_{l_i,i}, \ where \tag{5.2.7}$$

$$b_{l_1,\ldots,l_k} = \sum_{j_1,\ldots,j_k=1}^{m_1,\ldots,m_k} \left(\prod_{i=1}^k t_{l_i j_i,i}\right) a_{j_1\ldots j_k} \ for \ l_i = 1,\ldots,m_i, i = 1,\ldots,k.$$

That is if $\mathcal{A} := [a_{j_1\ldots j_k}], \mathcal{B} := [b_{l_1\ldots l_k}]$ *then* $\mathcal{B} = (\otimes_{i=1}^k T_i)\mathcal{A}$.

Definition 5.2.10 *Let* $\mathbf{M}_1, \ldots, \mathbf{M}_k$ *be free finite dimensional modules over a domain* $\mathbb{D}$. *Let* $\tau \in \otimes_{i=1}^k \mathbf{M}_i$. *The tensor rank of* τ, *denoted by* Rank τ, *is the minimal* R *such that* $\tau = \sum_{l=1}^R \otimes_{i=1}^k \mathbf{m}_{l,i}$ *for some* $\mathbf{m}_{l,i} \in \mathbf{M}_i, l = 1, \ldots, R, i = 1, \ldots, k.$

We shall see that for $k \geq 3$ it is hard to determine the tensor rank of a general k-tensor, even in the case $\mathbb{D} = \mathbb{C}$.

Let $\mathbf{M}$ be a $\mathbb{D}$-module, and let $\mathbf{M}' = \mathrm{Hom}(\mathbf{M}, \mathbb{D})$ be the *dual* module of $\mathbf{M}$. For $\mathbf{m} \in \mathbf{M}, \mathbf{g} \in \mathbf{M}'$ we denote $\langle \mathbf{m}, \mathbf{g} \rangle := \mathbf{g}(\mathbf{m})$. Let

$$\mathbf{m}_1, \ldots, \mathbf{m}_k \in \mathbf{M}, \quad \mathbf{g}_1, \ldots, \mathbf{g}_k \in \mathbf{M}'.$$

It is straightforward to show

$$\langle \mathbf{m}_1 \wedge \cdots \wedge \mathbf{m}_k, \mathbf{g}_1 \wedge \cdots \wedge \mathbf{g}_k \rangle = k! \langle \otimes_{i=1}^k \mathbf{m}_i, \mathbf{g}_1 \wedge \cdots \wedge \mathbf{g}_k \rangle$$

$$= k! \det \begin{bmatrix} \langle \mathbf{m}_1, \mathbf{g}_1 \rangle & \cdots & \langle \mathbf{m}_1, \mathbf{g}_k \rangle \\ \vdots & \ddots & \vdots \\ \langle \mathbf{m}_k, \mathbf{g}_1 \rangle & \cdots & \langle \mathbf{m}_k, \mathbf{g}_k \rangle \end{bmatrix}.$$

$$(5.2.8)$$

See Problem 9(b).

Assume that $\mathbf{M}$ is an m-dimensional free module over $\mathbb{D}$, with the basis $\mathbf{d}_1, \ldots, \mathbf{d}_m$. Recall that $\mathbf{M}'$ is an m-dimensional free module with the *dual basis* $\mathbf{f}_1, \ldots, \mathbf{f}_m$:

$$\langle \mathbf{d}_i, \mathbf{f}_j \rangle = \mathbf{f}_j(\mathbf{d}_i) = \delta_{ij}, \quad i, j = 1, \ldots, m. \tag{5.2.9}$$

Let $\mathbf{M}_1, \ldots, \mathbf{M}_k, \mathbf{M} := \otimes_{i=1}^k \mathbf{M}_i$ be free finite dimensional modules over $\mathbb{D}$ with bases given in (5.2.1). Let $\mathbf{f}_{1,i}, \ldots, \mathbf{f}_{m_i,i}$ be the dual basis of $\mathbf{M}_i'$ for $i = 1, \ldots, k$. Then $\mathbf{M}'$ is isomorphic to $\otimes_{i=1}^k \mathbf{M}_i'$, where we assume that

$$\langle \otimes_{i=1}^k \mathbf{m}_i, \otimes_{i=1}^k \mathbf{g}_i \rangle := \prod_{i=1}^k \langle \mathbf{m}_i, \mathbf{g}_i \rangle, \ \mathbf{m}_i \in \mathbf{M}_i, \mathbf{g}_i \in \mathbf{M}', \ i = 1, \ldots, k.$$

$$(5.2.10)$$

In particular, $\mathbf{M}'$ has the dual basis $\otimes_{i=1}^k \mathbf{f}_{j_i,i}, j_i = 1, \ldots, m_i, i = 1, \ldots, k.$

Assume that $\mathbf{d}_1, \ldots, \mathbf{d}_m$ is a basis of $\mathbf{M}$ and $\mathbf{f}_1, \ldots, \mathbf{f}_m$ is the dual basis of $\mathbf{M}'$. Note that $\bigwedge^k \mathbf{M}'$ is a submodule of $(\bigwedge^k \mathbf{M})'$. See Problem 9(c). Note that if $\mathbb{Q} \subseteq \mathbb{D}$ then $\bigwedge^k \mathbf{M}' = (\bigwedge^k \mathbf{M})'$.

Let $\mathbf{N}$ be a module over $\mathbb{D}$ of dimension n, as defined in Problem 1.6.2. Assume that $\mathbf{M} \subseteq \mathbf{N}$ is a submodule of dimension $m \leq n$. For any $k \in \mathbb{N}$ we view $\bigwedge^k \mathbf{M}$ as a submodule of $\bigwedge^k \mathbf{N}$. $\bigwedge^0 \mathbf{M} := 1$, $\bigwedge^m \mathbf{M}$ is a one dimensional module, while for $k > m$ it is agreed that $\bigwedge^k \mathbf{M}$ is a trivial subspace consisting of zero vector. (See Problem 11.)

Let $\mathbf{O} \subseteq \mathbf{N}$ be another submodule of $\mathbf{N}$. Then $(\bigwedge^p \mathbf{M}) \bigwedge (\bigwedge^q \mathbf{O})$ is a submodule of $\bigwedge^{p+q} (\mathbf{M} + \mathbf{O})$ of $\bigwedge^{p+q} \mathbf{N}$, spanned by $(\mathbf{m}_1 \wedge \cdots \wedge \mathbf{m}_p) \wedge (\mathbf{o}_1 \wedge \cdots \wedge \mathbf{o}_q)$, where $\mathbf{m}_1, \ldots, \mathbf{m}_p \in \mathbf{U}, \mathbf{o}_1, \ldots, \mathbf{o}_q \in \mathbf{O}$ for $p, q \geq 1$. If $p = 0$ or $q = 0$ then $(\bigwedge^p \mathbf{M}) \bigwedge (\bigwedge^q \mathbf{O})$ is equal to $\bigwedge^q \mathbf{O}$ or $\bigwedge^p \mathbf{M}$ respectively.

In the next sections we need the following lemma.

Lemma 5.2.11 *Let $\mathbf{V}$ be an n-dimensional vector space over $\mathbb{F}$. Assume that $0 \leq p_1, p_2$, $1 \leq q_1, q_2$, $k := p_1 + q_1 = p_2 + q_2 \leq n$. Suppose that $\mathbf{U}_1, \mathbf{U}_2, \mathbf{W}_1, \mathbf{W}_2$ are subspaces of $\mathbf{V}$ such that $\dim \mathbf{U}_i = p_i$, $\dim \mathbf{W}_i \geq q_i$ for $i = 1, 2$ and $\mathbf{U}_1 \cap \mathbf{W}_1 = \mathbf{U}_2 \cap \mathbf{W}_2 = \{\mathbf{0}\}$. Then*

$$\left(\bigwedge^{p_1} \mathbf{U}_1 \right) \bigwedge \left(\bigwedge^{q_1} \mathbf{W}_1 \right) \cap \left(\bigwedge^{p_2} \mathbf{U}_2 \right) \bigwedge \left(\bigwedge^{q_2} \mathbf{W}_1 \right) \neq \{\mathbf{0}\} \quad (5.2.11)$$

if and only if the following condition holds. There exists a subspace $\mathbf{V}_1 \subseteq \mathbf{V}$ of dimension k at such that

$$\mathbf{U}_1 \subset \mathbf{V}_1, \quad \mathbf{U}_2 \subset \mathbf{V}_1, \quad \mathbf{V}_1 \subseteq (\mathbf{U}_1 + \mathbf{W}_1), \quad \mathbf{V}_1 \subseteq (\mathbf{U}_2 + \mathbf{W}_2).$$

$$(5.2.12)$$

Proof. Assume first that (5.2.11) holds. Note that

$$\left(\bigwedge^{p_1} \mathbf{U}_1 \right) \bigwedge \left(\bigwedge^{q_1} \mathbf{W}_1 \right) \subseteq \bigwedge^{k} (\mathbf{U}_1 + \mathbf{W}_1), \quad \left(\bigwedge^{p_2} \mathbf{U}_2 \right) \bigwedge \left(\bigwedge^{q_2} \mathbf{W}_2 \right)$$

$$\subseteq \bigwedge^{k} (\mathbf{U}_2 + \mathbf{W}_2).$$

Let $\mathbf{V}_2 := (\mathbf{U}_1 + \mathbf{W}_1) \cap (\mathbf{U}_2 + \mathbf{W}_2)$. Problem 11(a) below yields that

$$\left(\overset{p_1}{\bigwedge} \mathbf{U}_1 \right) \wedge \left(\overset{q_1}{\bigwedge} \mathbf{W}_1 \right) \cap \left(\overset{p_2}{\bigwedge} \mathbf{U}_2 \right) \wedge \left(\overset{q_2}{\bigwedge} \mathbf{W}_1 \right) \subseteq \overset{k}{\bigwedge} \mathbf{V}_2.$$

$$(5.2.13)$$

The assumption (5.2.11) implies that $\dim \mathbf{V}_2 \geq k$. We now show that $\mathbf{U}_1 \subset \mathbf{V}_2$. Assume to the contrary that $\dim \mathbf{U}_1 \cap \mathbf{V}_2 = i < p_1$. Choose a basis $\mathbf{v}_1, \ldots, \mathbf{v}_n$ such that in $\mathbf{V}$ such that $\mathbf{v}_1, \ldots, \mathbf{v}_{p_1}$ and $\mathbf{v}_1, \ldots, \mathbf{v}_i, \mathbf{v}_{p_1+1}, \ldots, \mathbf{v}_r$ are bases of $\mathbf{U}_1$ and $\mathbf{V}_2$ respectively. Observe that the span of vectors $\mathbf{v}_1 \wedge \cdots \wedge \mathbf{v}_{p_1} \wedge \mathbf{v}_{i_1} \ldots \mathbf{v}_{i_{q_1}}$ for $p_1 < i_1 < \cdots < i_{q_1} \leq n$ contain the subspace $(\bigwedge^{p_1} \mathbf{U}_1) \bigwedge (\bigwedge^{q_1} \mathbf{W}_1)$. On the other hand the subspace $\bigwedge^k \mathbf{V}_2$ has a basis formed by the exterior products of k vectors out of $\mathbf{v}_1, \ldots, \mathbf{v}_i, \mathbf{v}_{p_1+1}, \ldots, \mathbf{v}_r$. Hence, $((\bigwedge^{p_1} \mathbf{U}_1) \bigwedge (\bigwedge^{q_1} \mathbf{W}_1)) \cap \bigwedge^k \mathbf{V}_2 = \{\mathbf{0}\}$, which contradicts (5.2.11)–(5.2.13). So $\mathbf{U}_1 \subset \mathbf{V}_2$. Similarly $\mathbf{U}_2 \subset \mathbf{V}_2$.

Next we claim that $\dim (\mathbf{U}_1 + \mathbf{U}_2) \leq k$. Assume to the contrary that $\dim (\mathbf{U}_1 + \mathbf{U}_2) = j > k$. Let $\mathbf{u}_1, \ldots, \mathbf{u}_n$ be a basis of $\mathbf{V}$, such that

$$\mathbf{u}_1, \ldots, \mathbf{u}_{p_1} \quad \text{and} \quad \mathbf{u}_1, \ldots, \mathbf{u}_{p_1+p_2-j}, \mathbf{u}_{p_1+1}, \ldots, \mathbf{u}_j$$

are bases of $\mathbf{U}_1$ and $\mathbf{U}_2$ respectively. Then $(\bigwedge^{p_1} \mathbf{U}_1) \bigwedge (\bigwedge^{q_1} \mathbf{W}_1)$ is spanned by $\binom{n-p_1}{q_1}$ linearly independent vectors $\mathbf{u}_{i_1} \wedge \cdots \mathbf{u}_{i_k}$, where $1 \leq i_1 < \cdots < i_k \leq n$ and $\{1, \ldots, p_1\} \subset \{i_1, \ldots, i_k\}$. Similarly, $(\bigwedge^{p_2} \mathbf{U}_2) \bigwedge (\bigwedge^{q_2} \mathbf{W}_2)$ is spanned by $\binom{n-p_2}{q_2}$ linearly independent vectors $\mathbf{u}_{j_1} \wedge \cdots \mathbf{u}_{j_k}$, where $1 \leq j_1 < \cdots < j_k \leq n$ and $\{1, \ldots, p_1 + p_2 - j, p_1 + 1, \ldots, j\} \subset \{i_1, \ldots, i_k\}$. Since $j > k$ it follows that these two subset of vectors of the full set of the basis of $\bigwedge^k V$ do not have any common vector, which contradicts (5.2.11). So $\dim (\mathbf{U}_1 + \mathbf{U}_2) \leq k$. Choose $\mathbf{V}_1$ any k-dimensional subspace of $\mathbf{V}_2$ which contains $\mathbf{U}_1 + \mathbf{U}_2$.

Vice versa, suppose that $\mathbf{V}_1$ is a k-dimensional subspace of $\mathbf{V}$ satisfying (5.2.12). So $\bigwedge^k \mathbf{V}_1$ is a one-dimensional subspace which is contained in $(\bigwedge^{p_i} \mathbf{U}_i) \bigwedge (\bigwedge^{q_i} \mathbf{W}_i)$ for $i = 1, 2$. Hence, (5.2.11) holds. $\qquad \square$

Problems

1. Let $\mathbf{M}_1, \ldots, \mathbf{M}_k, \mathbf{N}$ be free finite dimensional modules over $\mathbb{D}$. A map $f : \mathbf{M}_1 \times \cdots \times \mathbf{M}_k \to \mathbf{N}$ is called multilinear, if by fixing all variables $\mathbf{m}_1, \ldots, \mathbf{m}_{i-1}, \mathbf{m}_{i+1}, \ldots, \mathbf{m}_k$, the map f is linear on $\mathbf{M}_i$ for $i = 1, \ldots, k$. Show that $\otimes_{i=1}^k \mathbf{M}_i$ is determined uniquely by the following universal lifting property: There exists a unique multilinear $F : \mathbf{M}_1 \times \cdots \times \mathbf{M}_k \to \otimes_{i=1}^k \mathbf{M}_i$ such that each multilinear map $f : \mathbf{M}_1 \times \cdots \times \mathbf{M}_k \to \mathbf{N}$ can be lifted to a linear map $\tilde{f} : \otimes_{i=1}^k \mathbf{M}_i \to \mathbf{N}$ satisfying $f = \tilde{f} \circ F$.

2. Let $\mathbf{M}_1, \ldots, \mathbf{M}_k$ be free finite dimensional modules over $\mathbb{D}$. Show that for any $\sigma \in \Sigma_k$ $\otimes_{i=1}^k \mathbf{M}_{\sigma(i)}$ is isomorphic to $\otimes_{i=1}^k \mathbf{M}_i$.

3. Prove Proposition 5.2.2.

4. Show

   (a) Let $A \in \mathbb{D}^{m \times n}$ and $B \in \mathbb{D}^{p \times q}$. Then the definitions of $A \otimes B$ given by Definitions 5.1.7 and 5.2.3 coincide.

   (b) Let the assumptions of Definition 5.2.3 hold. Assume that $k \geq 3$. Then the recursive definition of $\otimes_{i=1}^k A_i := (\otimes_{i=1}^{k-1} A_i) \otimes A_k$ coincides with the definition of $\otimes_{i=1}^k A_i$ given in Definition 5.2.3.

5. Let $A_i \in \mathbb{D}^{m_i \times n_i}, i = 1, \ldots, k \geq 3$. Show

   (a) $\otimes_{i=1}^k (a_i A_i) = (\prod_{i=1}^k a_i) \otimes_{i=1}^k A_i$.

   (b) $(\otimes_{i=1}^k A_i)^T = \otimes_{i=1}^k A_i^T$.

   (c) If $m_i = n_i$ and A_i is upper triangular for $i = 1, \ldots, k$ then $\otimes_{i=1}^k A_i$ is upper triangular.

   (d) If $A_1, \ldots, A_k$ are diagonal matrices then $\otimes_{i=1}^k A_i$ is a diagonal matrix. In particular $\otimes_{i=1}^k I_{m_i} = I_{m_1 \ldots m_k}$.

   (e) Let $B_i \in \mathbb{D}^{l_i \times m_i}, i = 1, \ldots, k$. Then $(\otimes_{i=1}^k B_i)(\otimes_{i=1}^k A_i) = \otimes_{i=1}^k (B_i A_i)$.

   (f) $A_i \in \mathbf{GL}(m_i, \mathbb{D}), i = 1, \ldots, k$ then $\otimes_{i=1}^k A_i \in \mathbf{GL}(m_1 \ldots m_k, \mathbb{D})$ and $(\otimes_{i=1}^k A_i)^{-1} = \otimes_{i=1}^k A_i^{-1}$.

(g) rank $\otimes_{i=1}^{k} A = \prod_{i=1}^{k}$ rank A_i.

(h) For $m_i = n_i, i = 1, \ldots, k,$ det $\otimes_{i=1}^{k} A_i = \prod_{i=1}^{k}$ $(\det A_i)^{\frac{\prod_{j=1}^{k} m_j}{m_i}}$.

6. Prove Proposition 5.2.7.

7. (a) Let $A \in \mathbb{D}^{m \times n}, B \in \mathbb{D}^{n \times p}$. Show that $\wedge^k AB = \wedge^k A \wedge^k B$ for any $k \in [1, \min(m, n, p)] \cap \mathbb{N}$.

(b) Let $A \in \mathbb{D}^{n \times n}$. Then $\wedge^k A$ is upper triangular, lower triangular, or diagonal if A is upper triangular, lower triangular, or diagonal respectively.

(c) $\wedge^k I_n = I_{\binom{n}{k}}$.

(d) If $A \in \mathbf{GL}(n, \mathbb{D})$ then $\wedge^k A \in \mathbf{GL}(\binom{n}{k}, \mathbb{D})$ and $(\wedge^k A)^{-1} = \wedge^k A^{-1}$.

8. Let $\mathbb{F}$ be an algebraically closed field. Recall that over an algebraically closed $A \in \mathbb{F}^{n \times n}$ is similar to an upper triangular matrix.

(a) Let $A_i \in \mathbb{F}^{n_i \times n_i}$ for $i = 1, \ldots, k$. Show that there exists $T_i \in \mathbf{GL}(n_i, \mathbb{F})$ such that $(\otimes_{i=1}^{k} T_i)(\otimes_{i=1}^{k} A_i)(\otimes_{i=1}^{k} T_i)^{-1}$ is an un upper triangular matrix. Furthermore, let $\lambda_{1,i}, \ldots, \lambda_{n_i,i}$ be the eigenvalues of A_i, counted with their multiplicities. Then $\prod_{i=1}^{k} \lambda_{j_i,i}$ for $j_i = 1, \ldots, n_i, i = 1, \ldots, k$ are the eigenvalues of $\otimes_{i=1}^{k} A_i$ counted with their multiplicities.

(b) Let $A \in \mathbb{F}^{n \times n}$ and assume that $\lambda_1, \ldots, \lambda_n$ are the eigenvalues of A counted with their multiplicities. Show that $\prod_{i=1}^{k} \lambda_{j_i}$ for $1 \le j_1 < \cdots < j_k \le n$ are all the eigenvalues of $\wedge^k A$ counted with their multiplicites.

9. Let $\mathbf{M}$ be a finitely generated module over $\mathbb{D}$.

(a) Let $\mathbf{m}_1, \ldots, \mathbf{m}_k \in \mathbf{M}$. Show that for any $\sigma \in \Sigma_k$ $\mathbf{m}_{\sigma(1)} \wedge \cdots \wedge \mathbf{m}_{\sigma(k)} = \text{sign } \sigma \mathbf{m}_1 \wedge \cdots \wedge \mathbf{m}_k$. In particular, if $\mathbf{m}_i = \sum_{j \ne i} a_j \mathbf{m}_j$ then $\mathbf{m}_1 \wedge \cdots \wedge \mathbf{m}_k = \mathbf{0}$.

(b) Prove the equality (5.2.8).

(c) Assume that $\mathbf{d}_1, \ldots, \mathbf{d}_m$ is a basis of $\mathbf{M}$ and $\mathbf{f}_1, \ldots, \mathbf{f}_m$ is a dual basis of $\mathbf{M}'$. Show that $\frac{1}{k!}\mathbf{f}_{i_1} \wedge \cdots \wedge \mathbf{f}_{i_k}, 1 \leq i_1 < \cdots < i_k \leq m$ can be viewed as a basis for $(\bigwedge^k \mathbf{M})'$ for $k \in [1, m]$.

10. Let $\mathbf{M}$ be an m-dimensional module over $\mathbb{D}$ as defined in Problem 1.6.2. Show

   (a) $\bigwedge^m \mathbf{M}$ is a one-dimensional module over $\mathbb{D}$.

   (b) $\bigwedge^k \mathbf{V}$ is a zero module over $\mathbb{D}$ for $k > m$.

11. (a) Let $\mathbf{V}$ be an finite dimensional vector space over $\mathbb{F}$ and assume that $\mathbf{U}, \mathbf{W}$ are subspaces of $\mathbf{V}$. Show that $\bigwedge^k \mathbf{U} \cap \bigwedge^k \mathbf{W} = \bigwedge^k (\mathbf{U} \cap \mathbf{W})$.

   Hint: Choose a basis $\mathbf{v}_1, \ldots, \mathbf{v}_n$ in $\mathbf{V}$ satisfying the following property. $\mathbf{v}_1, \ldots, \mathbf{v}_m$ and $\mathbf{v}_1, \ldots, \mathbf{v}_l, \mathbf{v}_{m+1}, \ldots \mathbf{v}_{m+p-l}$ are bases for $\mathbf{U}$ and $\mathbf{W}$ respectively. Recall that $\mathbf{v}_{i_1} \wedge \cdots \wedge \mathbf{v}_{i_k}, 1 \leq i_1 < \cdots < i_k \leq n$ form a basis in $\bigwedge^k \mathbf{V}$. Observe next that bases of $\mathbf{U}$ and $\mathbf{W}$ are of the form of exterior, (wedge), products of k vectors from $\mathbf{v}_1, \ldots, \mathbf{v}_m$ and $\mathbf{v}_1, \ldots, \mathbf{v}_l, \mathbf{v}_{m+1}, \ldots \mathbf{v}_{m+p-l}$ respectively.

   (b) Assume that $\mathbf{V}$ is an n-dimensional module of $\mathbb{D}_b$. Suppose furthermore that $\mathbf{U}, \mathbf{W}$ are finitely generated submodules of $\mathbf{V}$. Show that $\bigwedge^k \mathbf{U} \cap \bigwedge^k \mathbf{W} = \bigwedge^k (\mathbf{U} \cap \mathbf{W})$.

12. Let $\mathbf{V}$ be an n-dimensional vector space over $\mathbb{F}$ and $\mathbf{U} \subset \mathbf{V}$, $\mathbf{W} \subset \mathbf{V}'$ be m-dimensional subspaces. Show

   (a) Let $\{\mathbf{u}_1, \ldots, \mathbf{u}_m\}, \{\mathbf{f}_1, \ldots, \mathbf{f}_m\}$ be bases of $\mathbf{U}, \mathbf{W}$ respectively. Then the vanishing of the determinant $\det [\langle \mathbf{u}_i, \mathbf{f}_j \rangle]_{i,j=1}^m$ is independent of the choice of bases in $\mathbf{U}, \mathbf{W}$.

   (b) Let $\mathbb{F}$ be a field of infinite characteristic. TFAE

      i. $\dim \mathbf{U}^\perp \cap \mathbf{W} > 0$.

      ii. $\dim \mathbf{U} \cap \mathbf{W}^\perp > 0$.

      iii. $\bigwedge^{m-1} \mathbf{U} \subset (\bigwedge^{m-1} \mathbf{W})^\perp$.

      iv. $\bigwedge^{m-1} \mathbf{W} \subset (\bigwedge^{m-1} \mathbf{U})^\perp$.

v. For any bases $\{\mathbf{u}_1, \ldots, \mathbf{u}_m\}, \{\mathbf{f}_1, \ldots, \mathbf{f}_m\}$ of $\mathbf{U}, \mathbf{W}$ respectively $\langle \mathbf{u}_1 \wedge \cdots \wedge \mathbf{u}_m, \mathbf{f}_1 \wedge \cdots \wedge \mathbf{f}_m \rangle = 0$.

Hint: If $\dim \mathbf{U}^\perp \cap \mathbf{W} = 0$ use Problem 2(e). If $\dim \mathbf{U}^\perp \cap \mathbf{W} > 0$ choose at least one vector of a basis in $\mathbf{W}$ to be in $\mathbf{U}^\perp \cap \mathbf{W}$ and use (5.2.8).

5.3   Sparse Bases of Subspaces

Definition 5.3.1   1. *For* $0 \neq \mathbf{x} \in \mathbb{F}^n$ *denote* $\operatorname{span}(\mathbf{x})^* := \operatorname{span}(\mathbf{x}) \backslash \{\mathbf{0}\}$.

2. *The* support *of* $\mathbf{x} = (x_1, \ldots, x_n)^\top \in \mathbb{F}^n$ *is defined as* $\operatorname{supp}(\mathbf{x}) = \{i \in \{1, \ldots, n\} : x_i \neq 0\}$.

3. *For a nonzero subspace* $\mathbf{U} \subseteq \mathbb{F}^n$, *a nonzero vector* $\mathbf{x} \in \mathbf{U}$ *is called* elementary *if for every* $0 \neq \mathbf{y} \in \mathbf{U}$ *the condition* $\operatorname{supp}(\mathbf{y}) \subseteq \operatorname{supp}(\mathbf{x})$ *implies* $\operatorname{supp}(\mathbf{y}) = \operatorname{supp}(\mathbf{x})$. $\operatorname{span}(\mathbf{x})^*$ *is called an* elementary class, *in* $\mathbf{U}$, *if* $\mathbf{x} \in \mathbf{U}$ *is elementary.*

4. *Denote by* $\mathcal{E}(\mathbf{U})$ *the union of all elementary classes in* $\mathbf{U}$.

5. *A basis in* $\{\mathbf{u}_1, \ldots, \mathbf{u}_m\}$ *in* $\mathbf{U}$ *is called* sparse *if* $\mathbf{u}_1, \ldots, \mathbf{u}_m$ *are elementary.*

Proposition 5.3.2 *Let* $\mathbf{U}$ *be a subspace of* $\mathbb{F}^n$ *of dimension* $m \in [n]$. *Then*

1. $\mathbf{x} \in \mathbf{U}$ *is elementary if and only if for each* $0 \neq \mathbf{y} \in \mathbf{U}$ *the condition* $\operatorname{supp}(\mathbf{y}) \subseteq \operatorname{supp}(\mathbf{x})$ *implies that* $\mathbf{y} \in \operatorname{span}(\mathbf{x})^*$.

2. $\mathcal{E}(\mathbf{U})$ *consists of a finite number of elementary classes.*

3. $\operatorname{span}(\mathcal{E}(\mathbf{U})) = \mathbf{U}$.

4. *For each subset* I *of* $\{1, \ldots, n\}$ *of cardinality* $m - 1$ *there exists an elementary* $\mathbf{x} \in \mathbf{U}$ *such that* $\operatorname{supp}(\mathbf{x})^c := \{1, \ldots, n\} \backslash \operatorname{supp}(\mathbf{x})$ *contains* I.

See Problem 1 for proof.

Definition 5.3.3 *Let* $\mathbb{F}$ *be a field of* 0 *characteristic.*

1. $A = [a_{ij}] \in \mathbb{F}^{k \times n}$ *is called* generic *if all the entries of* A *are algebraically independent over* $\mathbb{Q}$, *i.e. there is no nontrivial polynomial* p *in* kn *variable with integer coefficients such that* $p(a_{11}, \ldots, a_{kn}) = 0.$

2. A *is called* nondegenerate *if all* $\min(k, n)$ *minors of* A *are nonzero.*

3. *An* $1 \le m$-*dimensional subspace* $\mathbf{U} \subseteq \mathbb{F}^n$ *is called* nondegenerate *if for* $J \subset \{1, \ldots, n\}$ *of cardinality* $n - m + 1$ *there exists a unique elementary set* span $\mathbf{x}^*$ *such that* $J = \operatorname{supp}(\mathbf{x}).$

Lemma 5.3.4 *Let* $A \in \mathbb{F}^{k \times n}, 1 \le k < n$ *be of rank* k. *TFAE:*

1. A *is nondegenerate.*

2. *The row space of* A, *(viewed as a column space of* $A^\top$), *is nondegenerate.*

3. *The null space of* A *is nondegenerate.*

Proof. Consider first the column space of $A^\top$ denoted by $\mathbf{U} \subseteq \mathbb{F}^n$. Recall that any vector in $\mathbf{U}$ is of the form $\mathbf{x} = A^\top \mathbf{y}$ for some $\mathbf{y} \in \mathbb{F}^k$. Let $I \subset \{1, \ldots, n\}$ be a set of cardinality $k - 1$. Let $B = (A^\top)[I, :] \in \mathbb{F}^{k-1 \times k}$ be a submatrix of $A^\top$ with the rows indexed by the set I. The condition that $\operatorname{supp}(\mathbf{x}) \subseteq I^c$ is equivalent to the condition $B\mathbf{y} = \mathbf{0}$. Since rank $B \le k - 1$, there exists $\mathbf{0} \ne \mathbf{x} \in \mathbf{U}$ such that $\operatorname{supp}(\mathbf{x}) \subseteq I^c$. Let $\mathbf{d}$ be defined as in Problem 3.

Assume that rank $B < k - 1$. Then $\mathbf{d} = \mathbf{0}$, (see Problem 3(b)). Furthermore, it is straightforward to show that for each $j \in I^c$ there exists a nonzero $\mathbf{x} \in \mathbf{U}$ such that $\operatorname{supp}(\mathbf{x}) \subseteq (I \cup \{j\})^c$. So $\det A[:, I \cup \{j\}] = 0$ and A is not degenerate.

Suppose that rank $B = k - 1$, i.e. $\mathbf{d} \ne \mathbf{0}$. Then for any nonzero $\mathbf{x} \in \mathbf{U}, \operatorname{supp}(\mathbf{x}) \subset I^c$ is in span $(A^\top \mathbf{d})^*$. Let $j \in I^c$. Expand $\det A[:, I \cup \{j\}]$ by the column j to deduce that $(A^\top \mathbf{d})_j = \pm \det A[:, I \cup \{j\}]$. Thus $\operatorname{supp}(\mathbf{x}) = I^c$ if and only $\det A[:, I \cup \{j\}] \ne 0$ for each $j \in I^c$. These arguments show the equivalence of 1 and 2.

The equivalence of 1 and 3 are shown in a similar way and are discussed in Problem 4.   □

For a finite set J denote $\#J$ the cardinality of J.

Definition 5.3.5 *Let* $\mathcal{J} = \{J_1, \ldots, J_t\}$ *be* t *subsets of* $[n]$, *each of cardinality* $m - 1$. *Then* $\mathcal{J}$ *satisfies the* m-*intersection property provided that*

$$\# \cap_{i \in P} J_i \leq m - \#P \ \text{for all} \ \emptyset \neq P \subseteq [t]. \qquad (5.3.1)$$

It is known that for given a set $\mathcal{J}$ one can check efficiently, i.e. in polynomial time, whether $\mathcal{J}$ satisfies the m-intersection property. See Problems 5–7.

The aim of this section is to prove the following theorem.

Theorem 5.3.6 *Let* $\mathbb{F}$ *be a field of* 0 *characteristic and assume that* $A \in \mathbb{F}^{k \times n}$ *is generic over* $\mathbb{Q}$.

1. *Let* $\mathcal{I} = \{I_1, \ldots, I_s\}$ *denote the collection of* $s \leq k$ *subsets of* $[n]$ *each of cardinality* $n - k + 1$. *Then the elementary vectors* $\mathbf{x}(I_1), \ldots, \mathbf{x}(I_s)$ *in the row space of* A *with supports* $I_1, \ldots, I_s$ *are linearly independent if and only if* $\mathcal{I}' := \{I_1^c, \ldots, I_s^c\}$, *consisting of the complements of the supports, have the* k-*intersection property.*

2. *Let* $\mathcal{J} = \{J_1, \ldots, J_t\}$ *denote the collection of* $t \leq n - k$ *subsets of* $[n]$ *each of cardinality* $k + 1$. *Then the elementary vectors* $\mathbf{y}(J_1), \ldots, \mathbf{y}(J_t)$ *in the null space of* A *with supports* $J_1, \ldots, J_t$ *are linearly independent if and only if* $\mathcal{J}' := \{J_1^c, \ldots, J_t^c\}$, *consisting of the complements of the supports, have the* $n - k$-*intersection property.*

The proof of this theorem needs a number of auxiliary results.

Lemma 5.3.7 *Let* $A \in \mathbb{F}^{k \times n}$ *be nondegenerate.*

1. *Let* $\mathcal{I} = \{I_1, \ldots, I_s\}$ *denote the collection of* $s \leq k$ *subsets of* $[n]$ *each of cardinality* $n - k + 1$. *Then the elementary vectors* $\mathbf{x}(I_1), \ldots, \mathbf{x}(I_s)$ *in the row space of* A *with supports* $I_1, \ldots, I_s$ *are linearly independent if and only if the* $k \times s$ *submatrix of* $\wedge^{k-1} A$ *determined by its columns indexed by* $I_1^c, \ldots, I_s^c$ *has rank* s.

2. Let $\mathbf{b}_1, \ldots, \mathbf{b}_{n-k} \in \mathbb{R}^n$ be a basis in the null space of A and denote by $B^\top \in \mathbb{F}^{n \times (n-k)}$ the matrix whose columns are $\mathbf{b}_1, \ldots, \mathbf{b}_{n-k}$. Let $\mathcal{J} = \{J_1, \ldots, J_t\}$ denote the collection of $t \le n - k$ subsets of $[n]$ each of cardinality $k + 1$. Then the elementary vectors $\mathbf{y}(J_1), \ldots, \mathbf{y}(J_t)$ in the null space of A with supports $J_1, \ldots, J_t$ are linearly independent if and only if the $(n - k - 1) \times t$ matrix $\wedge^{n-k-1} B$ determined by its columns indexed by $J_1^c, \ldots, J_t^c$ has rank t.

See Problems 8–9 for the proof of the lemma.

Corollary 5.3.8 *Let $A \in \mathbb{F}^{k \times n}$ be nondegenerate.*

1. Let $\mathcal{I} = \{I_1, \ldots, I_k\}$ denote the collection of k subsets of $[n]$ each of cardinality $n - k + 1$. Then the elementary vectors $\mathbf{x}(I_1), \ldots, \mathbf{x}(I_s)$ in the row space of A with supports $I_1, \ldots, I_s$ are not linearly independent if and only if the determinant of the full row $k \times k$ submatrix of $\wedge^{k-1} X$ determined by its columns indexed by $I_1^c, \ldots, I_k^c$ is identically zero for any $X \in \mathbb{F}^{k \times n}$.

2. Let $\mathbf{b}_1, \ldots, \mathbf{b}_{n-k} \in \mathbb{R}^n$ be a basis in the null space of A and denote by $B^\top \in \mathbb{F}^{n \times (n-k)}$ the matrix whose columns are $\mathbf{b}_1, \ldots, \mathbf{b}_{n-k}$. Let $\mathcal{J} = \{J_1, \ldots, J_t\}$ denote the collection of $t \le n - k$ subsets of $[n]$ each of cardinality $k + 1$. Then the elementary vectors $\mathbf{y}(J_1), \ldots, \mathbf{y}(J_{n-k})$ in the null space of A with supports $J_1, \ldots, J_{n-k}$ are linearly independent if and only if the determinant of the full row $(n - k - 1) \times (n - k - 1)$ submatrix $\wedge^{n-k-1} Y$ determined by its columns indexed by $J_1^c, \ldots, J_{n-k}^c$ is identically zero for any $Y \in \mathbb{F}^{(n-k) \times n}$.

(One may use Problem 10 to show part 2 of the above corollary.)

Definition 5.3.9 *Let $\mathbf{V}$ be an n-dimensional vector space over $\mathbb{F}$. Let $\mathbf{U}_1, \ldots, \mathbf{U}_t \subset \mathbf{V}$ be t subspaces of dimension $m - 1$. Then $\{\mathbf{U}_1, \ldots, \mathbf{U}_t\}$ satisfies the dimension m-intersection property provided that*

$$\dim \cap_{i \in P} \mathbf{U}_i \le m - \#P \text{ for all } \emptyset \ne P \subseteq [t]. \tag{5.3.2}$$

Theorem 5.3.6 follows from the following theorem.

Theorem 5.3.10 *Let* $\mathbf{V}$ *be an n-dimensional vector space over a field $\mathbb{F}$ of 0 characteristic, and where $n \geq 2$. Let $2 \leq m \in [n]$ and assume that $\mathbf{U}_1, \ldots, \mathbf{U}_m \in \mathrm{Gr}(m-1, \mathbf{V}')$. Let $\mathcal{W}_m(\mathbf{U}_1, \ldots, \mathbf{U}_m) \subseteq \mathrm{Gr}(m, \mathbf{V})$ be the variety of all subspaces $\mathbf{X} \in \mathrm{Gr}(m, \mathbf{V})$ such that the one-dimensional subspace $\mathbf{Y} := \bigwedge^m(\bigwedge^{m-1}\mathbf{X}) \subset \otimes^{m(m-1)}\mathbf{V}$ is orthogonal on the subspace $\mathbf{W} := (\bigwedge^{m-1}\mathbf{U}_1)\bigwedge(\bigwedge^{m-1}\mathbf{U}_2)\bigwedge\cdots\bigwedge(\bigwedge^{m-1}\mathbf{U}_m) \subset \otimes^{m(m-1)}\mathbf{V}'$ of dimension one at most. Then $\mathcal{W}_m(\mathbf{U}_1, \ldots, \mathbf{U}_m)$ is a strict subvariety of $\mathrm{Gr}(m, \mathbf{V})$ if and only if $\mathbf{U}_1, \ldots, \mathbf{U}_m$ satisfy the dimension m-intersection property.*

Proof. Since each $\mathbf{U}_i$ is $m-1$ dimensional we assume that $\bigwedge^{m-1}\mathbf{U}_i = \mathrm{span}\,(\mathbf{w}_i)$ for some $\mathbf{w}_i \in \bigwedge^{m-1}\mathbf{U}_i$ for $i = 1, \ldots, m$. Then $\mathbf{W} = \mathrm{span}\,(\mathbf{w}_1 \wedge \cdots \wedge \mathbf{w}_m)$. Choose a basis $\mathbf{x}_1, \ldots, \mathbf{x}_m$ in $\mathbf{X}$. Let $\mathbf{y}_i$ be the wedge product of $m-1$ vectors from $\{\mathbf{x}_1, \ldots, \mathbf{x}_m\}\setminus\{\mathbf{x}_i\}$ for $i = 1, \ldots, m$. Then $\mathbf{y}_1, \ldots, \mathbf{y}_m$ are linearly independent and $\mathbf{Y} = \mathrm{span}\,(\mathbf{y}_1 \wedge \cdots \wedge \mathbf{y}_n)$. The condition that $\mathbf{Y} \perp \mathbf{W}$, i.e. $\mathbf{Y}^{\perp} \cap \mathbf{W}$ is a nontrivial subspace, is equivalent to the condition

$$\langle \mathbf{y}_1 \wedge \cdots \wedge \mathbf{y}_m, \mathbf{w}_1 \wedge \cdots \wedge \mathbf{w}_m \rangle = m!\det\left(\langle \mathbf{y}_i, \mathbf{w}_j \rangle\right)_{i,j=1}^{m} = 0.$$

$$(5.3.3)$$

See Problem 5.2.12. Since $\mathbb{F}$ has 0 characteristic, the condition (5.3.3) is equivalent to the vanishing of the determinant in the above formula. We will use the formula (5.2.8) for each $\langle \mathbf{y}_i, \mathbf{w}_j \rangle$.

Assume first that $\mathbf{U}_1, \ldots, \mathbf{U}_m$ do not satisfy the dimension intersection property. By interchanging the order of $\mathbf{U}_1, \ldots, \mathbf{U}_m$ if necessary, we may assume that there exists $2 \leq p \leq m$ such that $\mathbf{Z} := \cap_{j=1}^{p}\mathbf{U}_j$ has dimension $m-p+1$ at least. Let $\mathbf{Z}_1 \subseteq \mathbf{Z}$ be a subspace of dimension $m-p+1$. Then $\dim \mathbf{X} \cap \mathbf{Z}_1^{\perp} \geq m-(m-p+1) = p-1$. Let $\mathbf{F} \subseteq \mathbf{X} \cap \mathbf{Z}_1^{\perp}$ be a subspace of dimension $p-1$. Assume that $\mathbf{x}_1, \ldots, \mathbf{x}_m$ is a basis of $\mathbf{X}$ such that $\mathbf{x}_1, \ldots, \mathbf{x}_{p-1}$ is a basis of $\mathbf{F}$. So

$\mathbf{X}_i \subset \mathbf{F}$ for $i = p, \ldots, m$. Hence

$$\mathbf{X}_i \cap \mathbf{U}_j^\perp \supseteq \mathbf{X}_i \cap \mathbf{Z}^\perp \supseteq \mathbf{X}_i \cap \mathbf{Z}_1^\perp \supseteq \mathbf{F} \cap \mathbf{Z}_1^\perp \neq \{\mathbf{0}\}$$

$$\text{for } i = p, \ldots, m, \ j = 1, \ldots, p.$$

Thus $\langle \mathbf{y}_i, \mathbf{w}_j \rangle = 0$ for $i = p, \ldots, m$, $j = 1, \ldots, p$. See Problem 5.2.12. Hence any $p \times p$ submatrix $[\langle \mathbf{y}_i, \mathbf{w}_j \rangle]_{i,j=1}^m$, with the set of columns $\langle p \rangle$, must have a zero row. Expand $\det [\langle \mathbf{y}_i, \mathbf{w}_j \rangle]_{i,j=1}^m$ by the columns $\langle p \rangle$ to deduce that this determinant is zero. Hence, $\mathcal{W}_m(\mathbf{U}_1, \ldots, \mathbf{U}_m) = \mathrm{Gr}(m, \mathbf{V})$.

We now show by induction on m that if $\mathbf{U}_1, \ldots, \mathbf{U}_m \in \mathrm{Gr}(m-1, \mathbf{V}')$ satisfy the dimension m-intersection property then there exists $\mathbf{X} \in \mathrm{Gr}(m, \mathbf{V})$ such that $\dim \mathbf{Y}^\perp \cap \mathbf{W} = 0$, for each $n = m, m + 1, \ldots$ Assume that $m = 2$. As $\dim (\mathbf{U}_1 \cap \mathbf{U}_2) = 0$ we deduce that $\dim (\mathbf{U}_1 + \mathbf{U}_2) = 2$. Let $\mathbf{U}_i = \mathrm{span}\,(\mathbf{u}_i), i = 1, 2$. Then $\{\mathbf{u}_1, \mathbf{u}_2\}$ is a basis in $\mathbf{Z} = \mathrm{span}\,(\mathbf{u}_1, \mathbf{u}_2)$. Hence $\mathbf{Z}^\perp$ is a subspace of $\mathbf{V}$ of dimension $n-2$. Thus there exists a subspace $\mathbf{X} \in \mathrm{Gr}(2, \mathbf{V})$ such that $\dim \mathbf{X} \cap \mathbf{Z}^\perp = 0$. Note that $\bigwedge^{m-1} \mathbf{X} = \mathbf{X}, \bigwedge^{m-1} \mathbf{U}_i = \mathbf{U}_i, i = 1, 2$. Let $\mathbf{x}_1, \mathbf{x}_2$ be a basis in $\mathbf{X}$. The negation of the condition (5.3.3) is equivalent to $\langle \mathbf{x}_1 \wedge \mathbf{x}_2, \mathbf{u}_1 \wedge \mathbf{u}_2 \rangle \neq 0$. Use Problems 5.1.2(e) and 5.2.12 to deduce this negation.

Assume the induction hypothesis, that is, for $2 \leq l < n$ and any l dimensional subspaces $\hat{\mathbf{U}}_1, \ldots, \hat{\mathbf{U}}_l \subset \mathbf{V}'$ satisfying the l-dimensional intersection property there exists $\hat{\mathbf{X}} \in \mathrm{Gr}(l, \mathbf{V})$ such that $\dim \hat{\mathbf{Y}}^\perp \cap \hat{\mathbf{W}} = 0$. Let $m = l + 1$ and assume that $\mathbf{U}_1, \ldots, \mathbf{U}_m$ satisfy the m-dimensional intersection property. Let $\mathcal{P} := \{P \subseteq \langle m - 1 \rangle : \dim \cap_{i \in P} \mathbf{U}_i = m - \#P\}$. Note that $\{i\} \in \mathcal{P}$ for each $i \in \langle m-1 \rangle$. The m-intersection property yields that $\mathbf{U}_m \cap (\cap_{i \in P} \mathbf{U}_i)$ is a strict subspace of $\cap_{i \in P} \mathbf{U}_i$ for each $P \in \mathcal{P}$. That is, $\cap_{i \in P} \mathbf{U}_i \not\subseteq \mathbf{U}_m$ for each $P \in \mathcal{P}$. Equivalently $(\cap_{i \in P} \mathbf{U}_i)^\perp \not\supseteq \mathbf{U}_m^\perp$. Problem 12(d) yields that $\mathbf{U}_m^\perp \setminus \cup_{P \in \mathcal{P}} (\cap_{i \in P} \mathbf{U}_i)^\perp \neq \emptyset$. Let $\mathbf{x}_m \in \mathbf{U}_m^\perp \setminus \cup_{P \in \mathcal{P}} (\cap_{i \in P} \mathbf{U}_i)^\perp$. Define $\hat{\mathbf{U}}_i := \mathbf{U}_i \cap \{\mathbf{x}_m\}^\perp, i = 1, \ldots, l$. For $i \in \langle l \rangle$ we have that $\{i\} \in \mathcal{P}$, hence $\mathbf{x}_m \notin \mathbf{U}_i^\perp$. Thus $\hat{\mathbf{U}}_i \in \mathrm{Gr}(l-1, \mathbf{V}), i = 1, \ldots, l$. We claim that $\hat{\mathbf{U}}_1, \ldots, \hat{\mathbf{U}}_l$ satisfy the l-dimensional intersection property.

Assume to the contrary that the l-dimensional intersection property is violated. By renaming the indices in $\langle l \rangle$ we may assume that there is $2 \le k \in \langle l \rangle$ such that dim $\cap_{i \in \langle k \rangle} \hat{\mathbf{U}}_i > l - k = m - k - 1$. Since $\hat{\mathbf{U}}_i \subset \mathbf{U}_i, i \in \langle l \rangle$ we deduce that dim $\cap_{i \in \langle k \rangle} \mathbf{U}_i > m - k - 1$. The assumption that $\mathbf{U}_1, \ldots, \mathbf{U}_m$ satisfy the m-dimensional intersection property yields dim $\cap_{i \in \langle k \rangle} \mathbf{U}_i = m - k$, i.e. $\langle k \rangle \in \mathcal{P}$. Since $\mathbf{x}_m \notin (\cap_{i \in \langle k \rangle} \mathbf{U}_i)^\perp$ we deduce that dim $(\cap_{i \in \langle k \rangle} \mathbf{U}_i) \cap \{\mathbf{x}_m\}^\perp =$ dim $\cap_{i \in \langle k \rangle} \hat{\mathbf{U}}_i = m - k - 1$, contradicting our assumption. Hence, $\hat{\mathbf{U}}_1, \ldots, \hat{\mathbf{U}}_l$ satisfy the l-dimensional intersection property.

Let $\mathbf{v}_1, \ldots, \mathbf{v}_{n-1}, \mathbf{x}_m$ be a basis in $\mathbf{V}$. Let $\mathbf{f}_1, \ldots, \mathbf{f}_n$ be the dual basis in $\mathbf{V}'$. (See Problem 5.1.2(d).) Note that $\hat{\mathbf{U}}_i \subset$ span $(\mathbf{f}_1, \ldots, \mathbf{f}_{n-1})$. Let $\mathbf{V}_1 = $ span $(\mathbf{v}_1, \ldots, \mathbf{v}_{n-1})$. Then we can identify span $(\mathbf{f}_1, \ldots, \mathbf{f}_{n-1})$ with $\mathbf{V}_1'$. The induction hypothesis yields the existence of $\hat{\mathbf{X}} \in \mathrm{Gr}(l, \mathbf{V}_1)$ such that dim $\hat{\mathbf{Y}}^\perp \cap \hat{\mathbf{W}} = 0$. Assume that $\hat{\mathbf{X}}$ is the columns space of the matrix $X = [x_{ij}] \in \mathbb{F}^{n \times l}$. The existence of the above $\mathbf{X}$ is equivalent to the statement that the polynomial $p_{\hat{\mathbf{U}}_1, \ldots, \hat{\mathbf{U}}_l}(x_{11}, \ldots, x_{nl})$, defined in as in the Problem 15, is not identically zero. Recall that $\mathbf{U}_m \in$ span $(\mathbf{f}_1, \ldots, \mathbf{f}_{n-1})$. Problem 13 yields the existence of a nontrivial polynomial $p_{\mathbf{U}}(x_{11}, \ldots, x_{nl})$ such that $\hat{\mathbf{X}} \in \mathrm{Gr}(l, \mathbf{V}_1)$, equal to the column space of $X = [x_{ij}] \in \mathbb{F}^{n \times l}$, satisfies the condition dim $\hat{\mathbf{X}} \cap \mathbf{U}_m^\perp = 0 \iff p_{\mathbf{U}}(x_{11}, \ldots, x_{nl}) \ne 0$. As $p_{\mathbf{U}_m} p_{\hat{\mathbf{U}}_1, \ldots, \hat{\mathbf{U}}_l}$ is a nonzero polynomial we deduce the existence of $\hat{\mathbf{X}} \in \mathrm{Gr}(l, \mathbf{V}_1)$ such that dim $\hat{\mathbf{X}} \cap \mathbf{U}_m^\perp = 0$ and $\hat{\mathbf{X}} \notin \mathcal{W}_m(\hat{\mathbf{U}}_1, \ldots, \hat{\mathbf{U}}_l)$.

Assume that $\mathbf{x}_1, \ldots, \mathbf{x}_{m-1}$ is a basis of $\hat{\mathbf{X}}$. Let $\mathbf{X} :=$ span $(\mathbf{x}_1, \ldots, \mathbf{x}_m)$. We claim that $\mathbf{X} \notin \mathcal{W}_m(\mathbf{U}_1, \ldots, \mathbf{U}_m)$. Let $\mathbf{X}_i$ be the $m - 1$ dimensional subspace spanned by $\{\mathbf{x}_1, \ldots, \mathbf{x}_m\} \backslash \{\mathbf{x}_i\}$ for $i = 1, \ldots, m$. Then $\bigwedge^{m-1} \mathbf{X}_i = $ span $(\mathbf{y}_i), i = 1, \ldots, m$ and $\bigwedge^{m-1} \mathbf{X} = $ span $(\mathbf{y}_1, \ldots, \mathbf{y}_m)$. Let $\bigwedge^{m-1} \mathbf{U}_i = $ span $(\mathbf{w}_i), i = 1, \ldots, m$. Note that $\mathbf{x}_m \in \mathbf{X}_i \cap \mathbf{U}_m^\perp$ for $i = 1, \ldots, m - 1$. Problem 13 yields that $\langle \mathbf{y}_i, \mathbf{w}_m \rangle = 0$ for $i = 1, \ldots, m - 1$. Hence, det $[\langle \mathbf{y}_i, \mathbf{w}_j \rangle]_{i,j=1}^m = \langle \mathbf{y}_m, \mathbf{w}_m \rangle \det [\langle \mathbf{y}_i, \mathbf{w}_j \rangle]_{i,j=1}^{m-1}$. Since $\mathbf{X}_m = \hat{\mathbf{X}}$ we obtain that dim $\mathbf{X}_m \cap \mathbf{U}_m^\perp = 0$. Hence, $\langle \mathbf{y}_m, \mathbf{w}_m \rangle \ne 0$. It is left to show that det $[\langle \mathbf{y}_i, \mathbf{w}_j \rangle]_{i,j=1}^{m-1} \ne 0$. Let $\hat{\mathbf{X}}_i \subset \mathbf{X}_i$ be the subspace of dimension $l - 1 = m - 2$ spanned by $\{\mathbf{x}_1, \ldots, \mathbf{x}_{m-1}\} \backslash \{\mathbf{x}_i\}$ for $i = 1, \ldots, m - 1$. Note that $\hat{\mathbf{X}}_i \subset \mathbf{X}$. So $\bigwedge^{l-1} \hat{\mathbf{X}}_i = $ span $(\hat{\mathbf{y}}_i)$ and we can assume that $\mathbf{y}_i = \hat{\mathbf{y}}_i \wedge \mathbf{x}_m$ for $i = 1, \ldots, m-1$. Recall that $\hat{\mathbf{U}}_i = \{\mathbf{x}_m\}^\perp \cap \mathbf{U}_i$. As $\dim \hat{\mathbf{U}}_i = \dim \mathbf{U}_i - 1$ we deduce that there exists $\mathbf{u}_i \in \mathbf{U}_i$ such that $\langle \mathbf{x}_m, \mathbf{u}_i \rangle = 1$ for $i = 1, \ldots, m-1$. So $\mathbf{U}_i = \hat{\mathbf{U}}_i \oplus \mathrm{span}\,(\mathbf{u}_i)$. Let $\bigwedge^{l-1} \hat{\mathbf{U}}_i = \mathrm{span}\,(\hat{\mathbf{w}}_i)$. We can assume that $\mathbf{w}_i = \hat{\mathbf{w}}_i \wedge \mathbf{u}_i$ for $i = 1, \ldots, m-1$. Problem 14 yields that $\langle \hat{\mathbf{y}}_i \wedge \mathbf{x}_m, \hat{\mathbf{w}}_j \wedge \mathbf{u}_j \rangle = l \langle \hat{\mathbf{y}}_i, \hat{\mathbf{w}}_j \rangle$ for $i, j = 1, \ldots, m-1$. Hence, $\det\,[\langle \mathbf{y}_i, \mathbf{w}_j \rangle]_{i,j=1}^{m-1} = l^{m-1} \det\,[\langle \hat{\mathbf{y}}_i, \hat{\mathbf{w}}_j \rangle]_{i,j=1}^{m-1}$. Since $\hat{\mathbf{X}} \notin \mathcal{W}_m(\hat{\mathbf{U}}_1, \ldots, \hat{\mathbf{U}}_l)$ we deduce that $\det\,[\langle \hat{\mathbf{y}}_i, \hat{\mathbf{w}}_j \rangle]_{i,j=1}^{m-1} \neq 0$, i.e. $\mathbf{X} \notin \mathcal{W}_m(\mathbf{U}_1, \ldots, \mathbf{U}_m)$. $\qquad\square$

Lemma 5.3.11 *Let $J_1, \ldots, J_t$ be $t < m \le n$ subsets of $[n]$ each of cardinality $m-1$. Assume that $J_1, \ldots, J_t$ satisfy the m-intersection property. Then there exists $m-t$ subsets $J_{t+1}, \ldots, J_m$ of $[n]$ of cardinality $m-1$ such that the sets $J_1, \ldots, J_m$ satisfy the m-intersection property.*

Proof. It suffices to show that there is a subset $J_{t+1} \subset [n]$ of cardinality $m-1$ such that $J_1, \ldots, J_{t+1}$ that satisfies the m-intersection property. If $t = 1$ then choose $J_2 \neq J_1$. Assume that $t \ge 2$. Let $\mathcal{P} = \{P \subset [t] : \# \cap_{i \in P} J_i = m - \#P\}$. Note that $\{i\} \in \mathcal{P}$ for $i \in [t]$.

Let $P, Q \in \mathcal{P}$ and assume that $P \cap Q \neq \emptyset$. We claim that $P \cup Q \in \mathcal{P}$. Let $X := \cap_{i \in P} J_i, Y := \cap_{j \in Q} J_j$. Then $\#X = m - \#P, \#Y = m - \#Q$. Furthermore $\#(X \cap Y) = \#X + \#Y - \#(X \cup Y)$. Observe next $X \cup Y \subset \cap_{k \in P \cap Q} J_k$. Hence the m-intersection property of $J_1, \ldots, J_t$ yields $\#(X \cup Y) \le m - \#(P \cap Q)$. Combine the m-intersection property with all the above facts to deduce

$$m - \#\,(P \cup Q) \ge \# \cap_{l \in P \cup Q} J_l = \#(X \cap Y)$$
$$= m - \#P + m - \#Q - \#(X \cup Y)$$
$$\ge m - \#P + m - \#Q - (m - \#(P \cap Q)) = m - \#(P \cup Q).$$

It follows that there exists a partition $\{P_1, \ldots, P_l\}$ of $[t]$ into l sets such that equality in (5.3.1) holds for each P_i, and each $P \subset [t]$ satisfying equality in (5.3.1) is a subset of some P_i.

As $\# \cap_{i \in P_1} J_i = m - \#P_1 \ge m - t \ge 1$, we let $x \in \cap_{i \in P_1} J_i$. Choose J_{t+1} be any subset of cardinality $m-1$ such that $J_{t+1} \cap (\cap_{i \in P_1} J_i) =$

$(\cap_{i \in P_1} J_i) \backslash \{x\}$. Since $\#J_{t+1} = m - 1$ it follows that J_{t+1} contains exactly $\#P_1$ elements not in $\cap_{i \in P_1} J_i$.

We now show that $J_1, \ldots, J_{t+1}$ satisfy the m-intersection property. Let $Q \subseteq [t]$ and $P := Q \cup \{t + 1\}$. If $Q \notin \mathcal{P}$ then $\# \cap_{i \in P} J_i \leq m - \#Q - 1 = m - \#P$. Assume that $Q \in \mathcal{P}$. To show (5.3.1) we need to show that $\cap_{i \in Q} J_i \subsetneqq J_{t+1}$. Suppose first that $Q \subseteq P_1$. Then $x \in \cap_{i \in Q} J_i$ and $x \notin J_{t+1}$. Assume that $Q \subset P_j, j > 1$. So $P_1 \cap Q = \emptyset$ and $P_1 \cup Q \notin \mathcal{P}$. Hence

$$q := \#((\cap_{i \in P_1} J_i) \cap (\cap_{j \in Q} J_i))$$
$$= \# \cap_{k \in P_1 \cup Q} J_k \leq m - (\#P_1 + \#Q) - 1.$$

Thus $\#((\cap_{j \in Q} J_i) \backslash (\cap_{i \in P_1} J_i)) = m - \#Q - q \geq \#P_1 + 1$. We showed above that $\#(J_t \backslash (\cap_{i \in P_1} J_i)) = \#P_1$. Therefore, $\cap_{i \in Q} J_i \subsetneqq J_{t+1}$.   □

Proof of Theorem 5.3.6.   1.   Suppose first that $J_1 := I_1^c, \ldots, J_s := I_s^c$ do not satisfy the intersection k intersection property. Let $P \subset [s]$ for which $q := \#(\cap_{i \in P} J_i) \geq k - \#P + 1$. Note that $\#P \geq 2$. We can assume that $P = \langle p \rangle$ for some $2 \leq p \leq s$. We claim that $\mathbf{y}_i := \wedge^{k-1} A[;, J_i], i = 1, \ldots, p$ are linearly dependent. Let $J = \cap_{i=1}^p J_i$. By renaming the indices if necessary we may assume that $J = \langle q \rangle$. Suppose first that the columns $i = 1, \ldots, q$ are linearly dependent. Hence, any $k - 1$ columns in J_i are linearly dependent for $i = 1, \ldots, p$. Thus $\mathbf{y}_i = \mathbf{0}$ for $i = 1, \ldots, p$ and $\mathbf{y}_1, \ldots, \mathbf{y}_p$ are linearly dependent.

Assume now that the columns in $i = 1, \ldots, q$ are linearly independent. Let $C \in \mathbb{F}^{k \times k}$ be an invertible matrix. Then $\wedge^{k-1} C$ is also invertible. Thus $\mathbf{y}_1, \ldots, \mathbf{y}_p$ are linearly dependent if and only if $(\wedge^{k-1} C)\mathbf{y}_1, \ldots, (\wedge^{k-1} C)\mathbf{y}_p$ are linearly dependent. Thus we may replace A by $A_1 := CA$. Choose C such that $A_1 = \begin{bmatrix} I_q & X \\ O & F \end{bmatrix}$, where $O \in \mathbb{F}^{(k-q) \times q}$ is the zero matrix and $F \in \mathbb{F}^{(k-q) \times (n-q)}$.

Consider a $k - 1$ minor $A_1[\{i\}^c, K]$ for some $K \subset [n]$ of cardinality $k - 1$ containing set J. Expanding this minor by the first q columns we deduce that it is equal to zero, unless $i = q + 1, \ldots, k$.

Let $J_i' := J_i \backslash J, i = 1, \ldots, p$. Observe next that the submatrix of $\wedge^{k-1} A_1$ based on the rows $\{q+1\}^c, \ldots, \{k\}^c$ and columns $J_1, \ldots, J_p$ is equal to the matrix $\wedge^{k-q-1} F[;, \{J_1', \ldots, J_p'\}]$. Hence, rank $\wedge^{k-1} A_1[;, \{J_1, \ldots, J_p\}] = $ rank $\wedge^{k-q-1} F[;, \{J_1', \ldots, J_p'\}]$. Since $q \geq k - p + 1$ it follows that F has at most $k - (k - p + 1) = p - 1$ rows, which implies that $\wedge^{k-q-1} F$ has at most $p - 1$ rows. Hence, rank $\wedge^{k-q-1} F[;, \{J_1', \ldots, J_p'\}] \leq $ rank $\wedge^{k-q-1} F \leq p - 1$. Lemma 5.3.7 implies that $\mathbf{x}(I_1), \ldots, \mathbf{x}(I_p)$ are linearly dependent, which yield that $\mathbf{x}(I_1), \ldots, \mathbf{x}(I_s)$ are linearly dependent.

Assume now that $J_1 := I_1^c, \ldots, J_s := I_s^c$ satisfy the intersection k intersection property. By Lemma 5.3.11 we can extend these s sets to k subsets $J_1, \ldots, J_k \subset [n]$ of cardinality $k - 1$ which satisfy the intersection k intersection property. Let $\mathbf{V} := \mathbb{F}^n$ and identify $\mathbf{V}' := \mathbb{F}^n$, where $\langle \mathbf{v}, \mathbf{f} \rangle = \mathbf{f}^\top \mathbf{v}$. Let $\{\mathbf{f}_1, \ldots, \mathbf{f}_n\}$ be the standard basis in $\mathbb{F}^n$. Let $\mathbf{U}_i = \oplus_{j \in J_i} \text{span}(\mathbf{f}_j), j = 1, \ldots, k$. Then $\mathbf{U}_1, \ldots, \mathbf{U}_k$ have the k-dimensional intersection property. (See Problem 16). Theorem 5.3.10 yields that there exists a subspace $\mathbf{X} \in \text{Gr}(k, \mathbf{V})$ such that $\mathbf{X} \notin \mathcal{W}_k(\mathbf{U}_1, \ldots, \mathbf{U}_k)$. Assume that $\mathbf{X}$ is the column space of $B^\top \in \mathbb{F}^{n \times k}$, and that the columns of $B^\top$ are $\mathbf{b}_1, \ldots, \mathbf{b}_k$. As in the proof of Theorem 5.3.10 let $\mathbf{y}_i = \wedge_{j \in \langle k \rangle \backslash \{i\}} \mathbf{b}_j$, $\mathbf{w}_i = \wedge_{j \in J_i} \mathbf{f}_j, i = 1, \ldots, k$. Note that $\mathbf{y}_i$ is ith column of $\wedge^{k-1} B^\top$. Furthermore $\langle \mathbf{y}_i, \mathbf{w}_j \rangle = \wedge^{k-1} B[\{i\}^c, J_j]$. The choice of B is equivalent to the condition $\det [\langle \mathbf{y}_i, \mathbf{w}_j \rangle]_{i,j=1}^k \neq 0$. This is equivalent to the condition that the minor of $k \times k$ submatrix of $\wedge^{k-1} B$ based on the columns $J_1, \ldots, J_k$ is not equal to zero. Since A is generic, the corresponding minor of $\wedge^{k-1} A \neq 0$. (Otherwise the entries of A will satisfy some nontrivial polynomial equations with integer coefficients.) Hence, the k columns of $\wedge^{k-1} A$ corresponding to $J_1, \ldots, J_k$ are linearly independent. In particular the s columns of $\wedge^{k-1} A$ corresponding to $J_1, \ldots, J_s$ are linearly independent. Lemma 5.3.7 implies that $\mathbf{x}(I_1), \ldots, \mathbf{x}(I_s)$ are linearly independent.

2. For a generic A let $B \in \mathbb{F}^{(n-k) \times n}$ be such that the columns of $B^\top$ span the null space of A. So $AB^\top = 0$ and rank $B = n - k$. According to Problem 4(e) B is generic. Note that for any $J \subset [n]$ of cardinality $k + 1$ $\mathbf{x}(J, B) = \mathbf{y}(J, A)$.

Assume that $J_1^c, \ldots, J_t^c$ do not satisfy the the $n - k$ intersection property. The above arguments and 1 imply that $\mathbf{y}(J_1, A), \ldots, \mathbf{y}(J_t, A)$ are linearly dependent.

Suppose now that $J_1^c, \ldots, J_t^c$ satisfy the $n - k$ intersection property. Extend this set to the set $J_1, \ldots, J_{n-k}$, each a set of cardinality $k + 1$, such that $J_1^c, \ldots, J_{n-k}^c$ satisfy the $n - k$ intersection property. Let $B \in \mathbb{F}^{(n-k) \times n}$ be generic. 1 implies the $n - k$ vectors $\mathbf{x}(J_1, B), \ldots, \mathbf{x}(J_{n-k}, B)$ in the row space of B are linearly independent. Let $A \in \mathbb{F}^{k \times n}$ such that the columns of $A^\top$ span the null space of B. So rank $A = k$ and $BA^\top = 0$. According to Problem 4(e) A is generic. Hence, it follows that $\mathbf{x}(J_i, B) = \mathbf{y}(J_i, A), i = 1, \ldots, n - k$ are linearly independent vectors. Problems 3–4 yield that we can express the coordinate of each elementary vector in the null space of A in terms of corresponding $k \times k$ minors of A. Form the matrix $C = [\mathbf{y}(J_1, A) \ \cdots \ \mathbf{y}(J_{n-k}, A)] \in \mathbb{F}^{n \times (n-k)}$. Since rank $C = n - k$ it follows that some $(n - k)$ minor of C is in different form zero. Hence for any generic A the corresponding minor of C is different from zero. That is, the vectors $\mathbf{y}(J_1, A), \ldots, \mathbf{y}(J_{n-k}, A)$ are always linearly independent for a generic A. In particular, the vectors $\mathbf{y}(J_1, A), \ldots, \mathbf{y}(J_t, A)$ are linearly independent.   □

Problems

1. Prove Proposition 5.3.2.

2. Let $A \in \mathbb{F}^{k \times n}$ be generic. Show

   (a) All entries of A are nonzero.

   (b) A is nondegenerate.

3. Let $\mathbb{D}$ be a domain and $B \in \mathbb{D}^{(k-1) \times k}$, where $k \geq 1$. Let $B_i \in \mathbb{D}^{(k-1) \times (k-1)}$ be the matrix obtained by deleting the column i for $i = 1, \ldots, k$. Denote $\mathbf{d} = (d_1, -d_2, \ldots, (-1)^{k-1} d_k)^\top$. Show

   (a) $\mathbf{d} = \mathbf{0}$ if and only if rank $B < k - 1$.

   (b) $B\mathbf{d} = \mathbf{0}$.

(c) Assume that $\mathbf{x} \in \ker B$. If rank $B = k - 1$ then $\mathbf{x} = b\mathbf{d}$ for some b in the division field of $\mathbb{D}$.

4. Let $A \in \mathbb{F}^{k \times n}, 1 \le k \le n$. Assume that $1 \le \operatorname{rank} A = 1 \le k$. Show

   (a) For any $I \subset \{1, \ldots, n\}$ of cardinality $n - l - 1$ there exist $\mathbf{0} \ne \mathbf{x} \in \operatorname{nul} A$ such that supp $(\mathbf{x}) \subseteq I^c$.

   (b) Let $I \subset \{1, \ldots, n\}$ be of cardinality $n - k - 1$ and denote $B := A[:, I^c] \in \mathbb{R}^{k \times (k+1)}$. Then dim $\{\mathbf{x} \in \operatorname{nul} A : \text{ supp } (\mathbf{x}) \subseteq I^c\} = 1$ if and only if rank $B = k$.

   (c) Let $I \subset \{1, \ldots, n\}$ be of cardinality $n - k - 1$ and denote $B := A[:, I^c] \in \mathbb{R}^{k \times (k+1)}$. Then there exists an elementary vector $\mathbf{x} \in \operatorname{nul} A$ with supp $(\mathbf{x}) = I^c$ if and only if for each $j \in I^c$, det $A[:, I^c \backslash \{j\}] \ne 0$.

   (d) The conditions 1 and 3 of Lemma 5.3.4 are equivalent.

   (e) Let rank $A = 1 < k$. Then nul A is nondegenerate if all minors of A of order l are nonzero.

5. Let $\mathcal{J}$ be defined in Definition 5.3.5. Show

   (a) The condition (5.3.1) is equivalent to
   $$\#(\cup_{i \in P} J_i^c) \ge n - m + \#P \text{ for all } \emptyset \ne P \subseteq [t].$$

   (b) Assume that $\mathcal{J}$ satisfies (5.3.1). Let J_{t+1} be a subset of $[n]$ of cardinality $m - 1$. Then $\mathcal{J}' := \mathcal{J} \cup \{J_{t+1}\}$ satisfies the m-intersection property if and only if
   $$\# \cup_{i \in P} (J_i^c \cap J_{t+1}) \ge \#P \text{ for all } \emptyset \ne P \subseteq [t].$$

   In particular, if $\mathcal{J}'$ satisfies m-intersection property then each $J_i^c \cap J_{t+1}$ is nonempty for $i = 1, \ldots, t$. *Hint*: Observe that $J_{t+1}^c \cup (\cup_{i \in P} J_i^c)$ decomposes to union of two disjoint sets J_{t+1}^c and $\cup_{i \in P} (J_i^c \cap J_{t+1})$.

6. Let $S_1, \ldots, S_t$ be t nonempty subsets of a finite nonempty set S of cardinality at least t. $S_1, \ldots, S_t$ is said to have *a set of*

distinct representatives if there exists a subset $\{s_1, \ldots, s_t\} \subseteq \mathcal{S}$ of cardinality t such that $s_i \in S_i$ for $i = 1, \ldots, t$. Show that if $S_1, \ldots, S_t$ has a set of distinct representatives then

$$\# \cup_{i \in P} S_i \geq \# P \text{ for all } \emptyset \neq P \subseteq [t].$$

Hall's Theorem states that the above conditions are necessary and sufficient for existence of a set of distinct representatives [Hal35].

7. Let the assumptions of Problem 6 hold. Let G be a bipartite graph on a set of vertices $V = [t] \cup \mathcal{S}$ and the set of edges $E \subseteq [t] \times \mathcal{S}$ as follows. $(i, s) \in [t] \times \mathcal{S}$ if and only if $s \in S_i$. Show that $S_1, \ldots, S_t$ has a set of distinct representatives if and only if G has a match $M \subset E$, i.e. no two distinct edges in M have a common vertex, of cardinality t.

Remark: There exist effective algorithms in bipartite graphs $G = (V_1 \cup V_2, E), E \subseteq V_1 \times V_2$ to find a match of size $\min(\# V_1, \# V_2)$.

8. Let $A \in \mathbb{F}^{k \times n}$ be nondegenerate. Show

   (a) Let $I \subseteq [n]$ be of cardinality $n - k + 1$. Then there exists $\mathbf{x}(I) = (x_1, \ldots, x_n)$ in the row space of A, with $\operatorname{supp}(\mathbf{x}(I)) = I$, whose nonzero coordinates are given by $x_j = (-1)^{p_j+1} \det A[:, I^c \cup \{j\}]$ for any $j \in I$, where p_j is the number of integers in I^c less than j.

   (b) Let I and $\mathbf{x}(I)$ be defined as in (a). Show that there exists a unique $\mathbf{z}(I) = (z_1, \ldots, z_k) \in \mathbb{F}^k$ such that $\mathbf{x}(I) = \mathbf{z}(I)A$. Use the fact that $(\mathbf{z}A)_j = 0$ for any $j \in I^c$ and Cramer's rule to show that $z_i = (-1)^i \det A[\{i\}^c, I^c]$ for $i = 1, \ldots, k$.

   (c) Let $I_1, \ldots, I_s \subset [n]$ be sets of cardinality $n - k + 1$. Let $\mathbf{x}(I_1), \mathbf{z}(I_1), \ldots, \mathbf{x}(I_s), \mathbf{z}(I_s)$ be defined as above. Then

      i. $\mathbf{x}(I_1), \ldots, \mathbf{x}(I_s)$ are linearly independent if and only if $\mathbf{z}(I_1), \ldots, \mathbf{z}(I_s)$ are linearly independent.

      ii. Let $D = \operatorname{diag}(-1, 1, -1, \ldots) \in \mathbb{F}^{k \times k}$. Then the matrix $D[\mathbf{z}(I_1)^\top \quad \mathbf{z}(I_2) \quad \cdots \quad \mathbf{z}(I_s)] \in \mathbb{F}^{k \times s}$ is the submatrix $\wedge^{k-1} A[:, \{I_1^c, \ldots, I_s^c\}]$. Hence, $\mathbf{z}(I_1), \ldots, \mathbf{z}(I_s)$ are linearly independent if and only if the submatrix $\wedge^{k-1}A[;,\{I_1^c,\ldots,I_s^c\}]$ has rank s.

iii. The submatrix $\wedge^{k-1}A[;,\{I_1^c,\ldots,I_s^c\}]$ has rank s if and only if not all the determinants det $\wedge^{k-1} A[\{i_1\}^c, \ldots,\{i_s\}^c\},\{\{I_1^c,\ldots,I_s^c\}]$ for $1 \leq i_1 < i_2 < \cdots < i_s \leq k$ are equal to zero.

iv. $\mathbf{x}(I_1),\ldots,\mathbf{x}(I_k)$ is a basis in the row space of A if and only if the determinant of the full row submatrix of $\wedge^{k-1}A$ corresponding to the columns determined by $I_1^c,\ldots,I_k^c$ is not equal to zero.

9. Let $A \in \mathbb{F}^{k \times n}$ be nondegenerate. Let $\mathbf{b}_1,\ldots,\mathbf{b}_{n-k} \in \mathbb{R}^n$ be a basis in the null space of A and denote by $B^\top \in \mathbb{F}^{n \times (n-k)}$ the matrix whose columns are $\mathbf{b}_1,\ldots,\mathbf{b}_{n-k}$. Show

(a) Let $J \subseteq [n]$ be of cardinality $k + 1$. Then there exists $\mathbf{y}(J) = (y_1,\ldots,y_n)^\top$ in the column space of $B^\top$, with supp $(\mathbf{y}(J)) = J$, whose nonzero coordinates are given by $y_j = (-1)^{p_j+1}\det B[:, J^c \cup \{j\}]$ for any $j \in J$, where p_j is the number of integers in J^c less than j.

(b) Let J and $\mathbf{y}(J)$ be defined as in (a). Show that there exists a unique $\mathbf{u}(J) = (u_1,\ldots,u_{n-k})^\top \in \mathbb{F}^{n-k}$ such that $\mathbf{y}(J) = B^\top\mathbf{u}(J)$. Use the fact that $(B^\top\mathbf{u})_j = 0$ for any $j \in J^c$ and Cramer's rule to show that $u_i = (-1)^i\det B[\{i\}^c, J^c]$ for $i = 1,\ldots,n - k$.

(c) Let $J_1,\ldots,J_t \subset [n]$ be sets of cardinality $k + 1$. Let $\mathbf{y}(J_1), \mathbf{u}(J_1),\ldots,\mathbf{y}(J_t), \mathbf{u}(J_t)$ be defined as above. Then

i. $\mathbf{y}(J_1),\ldots,\mathbf{y}(J_t)$ are linearly independent if and only if $\mathbf{u}(J_1),\ldots,\mathbf{u}(J_t)$ are linearly independent.

ii. Let $D = \operatorname{diag}(-1,1,-1,\ldots) \in \mathbb{F}^{n-k \times n-k}$. Then the matrix $D[\mathbf{u}(J_1)\ \ \mathbf{u}(J_2)\ \ \ldots\ \ \mathbf{u}(J_t)] \in \mathbb{F}^{(n-k) \times t}$ is the submatrix $\wedge^{n-k-1}B[;,\{J_1^c,\ldots,J_t^c\}]$. Hence, $\mathbf{u}(J_1),\ldots,\mathbf{u}(J_t)$ are linearly independent if and only if the submatrix $\wedge^{n-k-1}B[;,\{J_1^c,\ldots,J_t^c\}]$ has rank t.

iii. The submatrix $\wedge^{n-k-1}B[;,\{J_1^c,\ldots,J_t^c\}]$ has rank t if and only if not all the determinants det $\wedge^{n-k-1}$

$B[\{i_1\}^c, \ldots, \{i_t\}^c\}, \{\{J_1^c, \ldots, J_t^c\}]$ for $1 \le i_1 < i_2 < \cdots$
$< i_t \le n - k$ are equal to zero.

    iv. $\mathbf{y}(J_1), \ldots, \mathbf{y}(J_{n-k})$ is a basis in the null space of A if and only if the determinant of the full row submatrix of $\wedge^{n-k-1} B$ corresponding to the columns determined by $J_1^c, \ldots, J_{n-k-1}^c$ is not equal to zero.

10. Let $C \in \mathbb{F}^{n \times (n-k)}$ be a matrix of rank $n - k$. Show that there exists $A \in \mathbb{F}^{k \times n}$ of rank k such that $AC = \mathbf{0}$.

11. Let $\mathbb{F}$ be a field of 0 characteristic. Let $p(x_1, \ldots, x_n) \in \mathbb{F}[x_1, \ldots, x_n]$. Show that $p(x_1, \ldots, x_n) = 0$ for all $\mathbf{x} = (x_1, \ldots, x_n)^\top \in \mathbb{F}^n$ if and only if p is the zero polynomial. *Hint*: Use induction.

12. Let $\mathbb{F}$ be a field of 0 characteristic. Assume that $\mathbf{V} = \mathbb{F}^n$. Identify $\mathbf{V}'$ with $\mathbb{F}^n$. So for $\mathbf{u} \in \mathbf{V}, \mathbf{f} \in \mathbf{V}'$ $\langle \mathbf{u}, \mathbf{f} \rangle = \mathbf{f}^\top \mathbf{v}$. Show

    (a) $\mathbf{U} \subset \mathbf{V}$ is a subspace of dimension of $n - 1$ if and only if there exists a nontrivial linear polynomial $l(\mathbf{x}) = a_1 x_1 + \cdots + a_n x_n$ such that $\mathbf{U}$ is the zero set of $l(\mathbf{x})$, i.e. $\mathbf{U} = Z(l)$.

    (b) Let $\mathbf{U}_1, \ldots, \mathbf{U}_k$ be k subspaces of $\mathbf{V}$ of dimension $n - 1$. Show that there exists a nontrivial polynomial $p = \prod_{i=1}^k l_i \in \mathbb{F}[x_1, \ldots, x_n]$, where each l_i is a nonzero linear polynomial, such that $\cup_{i=1} \mathbf{U}_i$ is $Z(p)$.

    (c) Show that if $\mathbf{U}_1, \ldots, \mathbf{U}_k$ are k strict subspaces of $\mathbf{V}$ then $\cup_{i=1}^k \mathbf{U}_i$ is a strict subset of $\mathbf{V}$. *Hint*: One can assume that $\dim \mathbf{U}_i = n - 1, i = 1, \ldots, k$ and use Problem 11.

    (d) Let $\mathbf{U}, \mathbf{U}_1, \ldots, \mathbf{U}_k$ be subspaces of $\mathbf{V}$. Assume that $\mathbf{U} \subseteq \cup_{i=1}^k \mathbf{U}_i$. Show that there exists a subspace $\mathbf{U}_i$ which contains $\mathbf{U}$. *Hint*: Observe $\mathbf{U} = \cup_{i=1}^k (\mathbf{U}_i \cap \mathbf{U})$.

13. Let the assumptions of Problem 12 hold. Let $X = [x_{ij}], U = [u_{ij}] \in \mathbb{F}^{n \times l}$. View the matrices $\wedge^l X, \wedge^l U$ as column vectors in $\mathbb{F}^{\binom{n}{l}}$. Let $p_U(x_{11}, \ldots, x_{nl}) := \det(X^\top U) = (\wedge^l X)^\top \wedge^l U$. View p_U as a polynomial in nl variables with coefficients in $\mathbb{F}$. Show

(a) p_U a homogeneous multilinear polynomial of degree l.

(b) p_U is a trivial polynomial if and only if rank $U < l$.

(c) Let $\mathbf{X} \in \mathrm{Gr}(l, \mathbf{V})$, $\mathbf{U} \in \mathrm{Gr}(l, \mathbf{V}')$ and assume that the column space of $X = [x_{ij}] = [\mathbf{x}_1, \ldots, \mathbf{x}_l]$, $U = [u_{ij}] = [\mathbf{u}_1, \ldots, \mathbf{u}_l] \in \mathbb{F}^{n \times l}$ are $\mathbf{X}, \mathbf{U}$ respectively. Then

$$p_U(x_{11}, \ldots, x_{nl}) = \det [\mathbf{u}_j^\top \mathbf{x}_i]_{i,j=1}^l,$$

$$\langle \mathbf{x}_1 \wedge \cdots \wedge \mathbf{x}_l, \mathbf{u}_1 \wedge \cdots \wedge \mathbf{u}_l \rangle = l! p_U(x_{11}, \ldots, x_{nl}).$$

In particular, $\dim \mathbf{X} \cap \mathbf{U}^\perp = 0 \iff p_U(x_{11}, \ldots, x_{nl}) \neq 0$.

14. Let $\mathbb{F}$ be a field of 0 characteristic. Assume that $\mathbf{V}$ is an n-dimensional vector space with $n \geq 2$. Let $\mathbf{X} \subset \mathbf{V}, \mathbf{U} \subset \mathbf{V}'$ be $m \geq 2$ dimensional subspaces. Assume that $\mathbf{X} \not\subseteq \mathbf{U}^\perp$. Let $\mathbf{x}_m \in \mathbf{X} \backslash \mathbf{U}^\perp$. Let $\hat{\mathbf{U}} = \{\mathbf{x}_m\}^\perp \cap \mathbf{U}$. Let $\hat{\mathbf{X}}$ be any $m - 1$ dimensional subspace of $\mathbf{X}$ which does not contain $\mathbf{x}_m$. Show

(a) $\dim \hat{\mathbf{U}} = m - 1, \mathbf{X} = \hat{\mathbf{X}} \oplus \mathrm{span}\,(\mathbf{x}_m)$.

(b) There exists $\mathbf{u}_m \in \mathbf{U}$ such that $\langle \mathbf{x}_m, \mathbf{u}_m \rangle = 1$. Furthermore $\mathbf{U} = \hat{\mathbf{U}} \oplus \mathrm{span}\,(\mathbf{u}_m)$.

(c) Let $\{\mathbf{x}_1, \ldots, \mathbf{x}_{m-1}\}, \{\mathbf{u}_1, \ldots, \mathbf{u}_{m-1}\}$ be bases of $\hat{\mathbf{X}}, \hat{\mathbf{Y}}$ respectively. Then

$$\langle \mathbf{x}_1 \wedge \cdots \wedge \mathbf{x}_m, \mathbf{u}_1 \wedge \mathbf{u}_1 \wedge \cdots \wedge \mathbf{u}_m \rangle$$

$$= m \langle \mathbf{x}_1 \wedge \cdots \wedge \mathbf{x}_{m-1}, \mathbf{u}_1 \wedge \mathbf{u}_1 \wedge \cdots \wedge \mathbf{u}_{m-1} \rangle,$$

where $\mathbf{u}_m$ is defined in (b). *Hint*: Use (5.2.8) and expand the determinant by the last row.

(d) Assume that $\bigwedge^{m-1} \hat{\mathbf{X}} = \mathrm{span}\,(\hat{\mathbf{y}}), \bigwedge^{m-1} \hat{\mathbf{U}} = \mathrm{span}\,(\hat{\mathbf{w}})$. Then $\langle \hat{\mathbf{y}} \wedge \mathbf{x}_m, \hat{\mathbf{w}} \wedge \mathbf{u}_n \rangle = m \langle \hat{\mathbf{y}}, \hat{\mathbf{w}} \rangle$.

15. Let the assumptions of Theorem 5.3.10 hold. View $\mathbf{V}$ and $\mathbf{V}'$ as $\mathbb{F}^n$. So for $\mathbf{u} \in \mathbf{V}, \mathbf{f} \in \mathbf{V}'$ $\langle \mathbf{u}, \mathbf{f} \rangle = \mathbf{f}^\top \mathbf{v}$. Let $\mathbf{X} \in \mathrm{Gr}(m, \mathbf{V})$ be the column space of $X = [x_{ij}] \in \mathbb{F}^{n \times m}$. Show that there exists a homogeneous polynomial $p_{\mathbf{U}_1, \ldots, \mathbf{U}_m}(x_{11}, \ldots, x_{nm})$ of degree $m(m - 1)$ such that $\mathbf{X} \in \mathcal{W}_m(\mathbf{U}_1, \ldots, \mathbf{U}_m)$ if and only if $p_{\mathbf{U}_1, \ldots, \mathbf{U}_m}(x_{11}, \ldots, x_{nm}) = 0$. *Hint*: Choose a basis of $\mathbf{X}$ to be

the columns of X. Then use the first paragraph of Proof of Theorem 5.3.10 and Problem 13.

16. Let $\mathbb{F}$ be a field and $\mathbf{V}$ an n-dimensional subspace over $\mathbb{F}$. Let $\mathbf{v}_1, \ldots, \mathbf{v}_n$ be a basis of $\mathbf{V}$. For $\emptyset \neq K \subset [n]$ let $\mathbf{U}_K = \oplus_{i \in K} \mathrm{span}\,(\mathbf{v}_i)$. Let $t \leq m$ and assume that $K_1, \ldots, K_t \subset [n]$ be sets of cardinality $m - 1$ for any $2 \leq m \in [n]$. Show that $\mathbf{U}_{K_1}, \ldots, \mathbf{U}_{K_t}$ satisfy the m-dimensional intersection property if and only if $K_1, \ldots, K_t$ satisfy the m-intersection property.

17. Let $\mathbf{V}$ be an n-dimensional vector space over $\mathbb{F}$ of characteristic 0. Show

    (a) Let $2 \leq m \leq n$. If $\mathbf{U}_1, \ldots, \mathbf{U}_m$ are $m - 1$ dimensional vector spaces satisfying the m-dimensional intersection property then $\dim \sum_{i=1}^{m} \bigwedge^{m-1} \mathbf{U}_i = m$.

    (b) For $m = 3, n = 4$ there exist $\mathbf{U}_1, \mathbf{U}_2, \mathbf{U}_3$, which do not satisfy the three-dimensional intersection property such that $dim \sum_{i=1}^{3} \bigwedge^2 \mathbf{U}_i = 3$. *Hint*: Choose a basis in $\mathbf{V}$ and assume that each $\mathbf{U}_i$ is spanned by some two vectors in the basis.

    (c) Show that for $2 \leq t \leq m = n$, $\mathbf{U}_1, \ldots, \mathbf{U}_t$ satisfy the n-intersection property if and only if $\dim \sum_{i=1}^{t} \bigwedge^{n-1} \mathbf{U}_i = t$.

5.4   Tensor Products of Inner Product Spaces

Let $\mathbb{F} = \mathbb{R}, \mathbb{C}$ and assume that $\mathbf{V}_i$ is a n_i-dimensional vector space with the inner product $\langle \cdot, \cdot \rangle_i$ for $i = 1, \ldots, k$. Then $\mathbf{Y} := \otimes_{i=1}^{k} \mathbf{V}_i$ has a unique inner product $\langle \cdot, \cdot \rangle$ satisfying the property

$$\langle \otimes_{i=1}^{k} \mathbf{x}_i, \otimes_{i=1}^{k} \mathbf{y}_j \rangle = \prod_{i=1}^{k} \langle \mathbf{x}_i, \mathbf{y}_i \rangle_i, \text{ for all } \mathbf{x}_i, \mathbf{y}_i \in \mathbf{V}_i, \ i = 1, \ldots, k.$$

$$(5.4.1)$$

(See Problem 1.) We will assume that $\mathbf{Y}$ has the above *canonical* inner product, unless stated otherwise.

Proposition 5.4.1 *Let* $\mathbf{U}_i, \mathbf{V}_i$ *be finite dimensional inner product spaces over* $\mathbb{F} := \mathbb{R}, \mathbb{C}$ *with the inner product* $\langle \cdot, \cdot \rangle_{\mathbf{U}_i}, \langle \cdot, \cdot \rangle_{\mathbf{V}_i}$ *for* $i = 1, \ldots, k$ *respectively. Let* $\mathbf{X} := \otimes_{i=1}^k \mathbf{U}_i, \mathbf{Y} := \otimes_{i=1}^k \mathbf{V}_i$ *be an IPS with the canonical inner products* $\langle \cdot, \cdot \rangle_{\mathbf{X}}, \langle \cdot, \cdot \rangle_{\mathbf{Y}}$ *respectively. Then the following claims hold.*

1. *Assume that* $T_i \in \mathrm{L}(\mathbf{V}_i, \mathbf{U}_i)$ *for* $i = 1, \ldots, k$. *Then* $\otimes_{i=1}^k T_i \in \mathrm{L}(\mathbf{Y}, \mathbf{X})$ *and* $(\otimes_{i=1}^k T_i)^* = \otimes_{i=1}^k T_i^* \in \mathrm{L}(\mathbf{X}, \mathbf{Y})$.

2. *Assume that* $T_i \in \mathrm{L}(\mathbf{V}_i)$ *is normal for* $i = 1, \ldots, k$. *Then* $\otimes_{i=1}^k T_i \in \mathrm{L}(\mathbf{Y})$ *is normal. Moreover,* $\otimes_{i=1}^k T_i$ *is hermitian or unitary, if each* T_i *is hermitian or unitary, respectively.*

3. *Assume that* $T_i \in \mathrm{L}(\mathbf{V}_i, \mathbf{U}_i)$ *for* $i = 1, \ldots, k$. *Let* $\sigma_1(T_i) \geq \cdots \geq \sigma_{\mathrm{rank}\, T_i}(T_i) > 0$, *where* $\sigma_j(T_i) = 0$, *for* $j > \mathrm{rank}\, T_i$, *be the singular values of* T_i. *Let* $\mathbf{c}_{1,i}, \ldots, \mathbf{c}_{n_i,i}$ *and* $\mathbf{d}_{1,i}, \ldots, \mathbf{d}_{m_i,i}$ *be orthonormal bases of* $\mathbf{V}_i$ *and* $\mathbf{U}_i$ *consisting of right and left singular vectors of* T_i *as described in* (4.10.5):

$$T_i \mathbf{c}_{j_i,i} = \sigma_{j_i}(T_i) \mathbf{d}_{j_i,i}, \quad j_i = 1, \ldots, \ i = 1, \ldots, k.$$

Then

$$(\otimes_{i=1}^k T_i) \otimes_{i=1}^k \mathbf{c}_{j_i,i} = \left(\prod_{i=1}^k \sigma_{j_i}(T_i) \right) \otimes_{i=1}^k \mathbf{d}_{j_i,i},$$

$$j_i = 1, \ldots, \ i = 1, \ldots, k.$$

In particular

$$\| \otimes_{i=1}^k T_i \| = \sigma_1(\otimes_{i=1}^k T_i) = \prod_{i=1}^k \|T_i\| = \prod_{i=1}^k \sigma_1(T_i), \quad (5.4.2)$$

$$\sigma_{\prod_{i=1}^k \mathrm{rank}\, T_i}(\otimes_{i=1}^k T_i) = \prod_{i=1}^k \sigma_{\mathrm{rank}\, T_i}(T_i).$$

We consider a fixed IPS vector space $\mathbf{V}$ of dimension n and its exterior products $\bigwedge^k \mathbf{V}$ for $k = 1, \ldots, n$. Since $\bigwedge^k \mathbf{V}$ is a subspace of $\mathbf{Y} := \otimes_{i=1}^k \mathbf{V}_i, \mathbf{V}_1 = \cdots = \mathbf{V}_k = \mathbf{V}$, it follows that $\bigwedge^k \mathbf{V}$ has a canonical inner product induced by $\langle \cdot, \cdot \rangle_{\mathbf{Y}}$. See Problem 3(a).

Proposition 5.4.2 *Let* $\mathbf{V}, \mathbf{U}$ *be inner product spaces of dimension* n *and* m *respectively. Assume that* $T \in \mathrm{L}(\mathbf{V}, \mathbf{U})$. *Suppose that* $\mathbf{c}_1, \ldots, \mathbf{c}_n$ *and* $\mathbf{d}_1, \ldots, \mathbf{d}_m$ *are orthonormal bases of* $\mathbf{V}$ *and* $\mathbf{U}$ *composed of the right and left singular eigenvectors of* T *respectively, as given in (4.10.5). Let* $k \in [\min(m, n)]$. *Then the orthonormal bases*

$$\frac{1}{\sqrt{k!}}\mathbf{c}_{i_1} \wedge \cdots \wedge \mathbf{c}_{i_k} \in \bigwedge^k \mathbf{V},\ 1 \leq i_1 < \cdots < i_k \leq n,$$

$$\frac{1}{\sqrt{k!}}\mathbf{d}_{j_1} \wedge \cdots \wedge \mathbf{d}_{j_k} \in \bigwedge^k \mathbf{U},\ 1 \leq j_1 < \cdots < j_k \leq m,$$

are the right and the left singular vectors of $\wedge^k T \in \mathrm{L}(\bigwedge^k \mathbf{V}, \bigwedge^k \mathbf{U})$, *with the corresponding singular values* $\prod_{l=1}^k \sigma_{i_l}(T)$ *and* $\prod_{l=1}^k \sigma_{j_l}(T)$ *respectively. In particular*

$$\wedge^k T \mathbf{c}_1 \wedge \cdots \wedge \mathbf{c}_k = \| \wedge^k T \| \mathbf{d}_1 \wedge \cdots \wedge \mathbf{d}_k, \| \wedge^k T \| = \sigma_1(\wedge^k T)$$

$$= \prod_{l=1}^k \sigma_i(T),\ \wedge^k T \mathbf{c}_{\mathrm{rank}\ T-k+1} \wedge \cdots \wedge \mathbf{c}_{\mathrm{rank}\ T}$$

$$= \prod_{l=1}^k \sigma_{\mathrm{rank}\ T-k+l}(T)\mathbf{d}_{\mathrm{rank}\ T-k+1} \wedge \cdots \wedge \mathbf{d}_{\mathrm{rank}\ T}$$

are the biggest and the smallest positive singular values of $\wedge^k T$ *for* $k \leq \mathrm{rank}\ \mathrm{T}$.

Corollary 5.4.3 *Suppose that* $\mathbf{V}$ *is an IPS of dimension* n. *Assume that* $T \in \mathrm{S}_+(\mathbf{V})$. *Let* $\lambda_1(T) \geq \cdots \geq \lambda_n(T) \geq 0$ *be the eigenvalues of* T *with the corresponding orthonormal eigenbasis* $\mathbf{c}_1, \ldots, \mathbf{c}_n$ *of* $\mathbf{V}$. *Then* $\wedge^k T \in \mathrm{S}_+(\bigwedge^k \mathbf{V})$. *Let* $k \in [n]$. *Then the orthonormal base* $\frac{1}{\sqrt{k!}}\mathbf{c}_{i_1} \wedge \cdots \wedge \mathbf{c}_{i_k}$, $1 \leq i_1 < \cdots < i_k \leq n$ *of* $\bigwedge^k \mathbf{V}$ *is an eigensystem of* $\wedge^k T$, *with the corresponding eigenvalues* $\prod_{l=1}^k \lambda_{i_l}(T)$. *In* *particular*

$$\wedge^k T \mathbf{c}_1 \wedge \cdots \wedge \mathbf{c}_k = \| \wedge^k T \| \mathbf{c}_1 \wedge \cdots \wedge \mathbf{c}_k, \quad \| \wedge^k T \| = \lambda_1(\wedge^k T)$$

$$= \prod_{l=1}^{k} \lambda_i(T), \quad \wedge^k T \mathbf{c}_{\text{rank } T-k+1} \wedge \cdots \wedge \mathbf{c}_{\text{rank } T}$$

$$= \prod_{l=1}^{k} \lambda_{\text{rank } T-k+l}(T) \mathbf{d}_{\text{rank } T-k+1} \wedge \cdots \wedge \mathbf{d}_{\text{rank } T}$$

are the biggest and the smallest positive eigenvalues of $\wedge^k T$ for $k \leq$ rank T.

See Problem 4.

Assume that the assumptions of Proposition 5.4.2 hold. Then $\mathbf{c}_1, \ldots, \mathbf{c}_k$ and $\mathbf{d}_1, \ldots, \mathbf{d}_k$ are called the first k-right and k-left singular vectors respectively.

Theorem 5.4.4 *Let $\mathbf{U}, \mathbf{V}, \mathbf{W}$ be finite dimensional inner product spaces. Assume that $P \in \mathrm{L}(\mathbf{U}, \mathbf{W})$, $T \in \mathrm{L}(\mathbf{V}, \mathbf{U})$. Then*

$$\prod_{i=1}^{k} \sigma_i(PT) \leq \prod_{i=1}^{k} \sigma_i(P)\sigma_i(T), \quad k = 1, \ldots \qquad (5.4.3)$$

For $k \leq \min(\text{rank } P, \text{rank } T)$, equality in (5.4.3) holds if and only if the following condition is satisfied: There exists a k-dimensional subspace $\mathbf{V}_k$ of $\mathbf{V}$ which is spanned by the first k-orthonormal right singular vectors of T, such that $T\mathbf{V}_k$ is a k-dimensional subspace of $\mathbf{U}$ which spanned the first k-orthonormal right singular vectors of P.

Proof. Suppose first that $k = 1$. Then $\|PT\| = \|PT\mathbf{v}\|$, where $\mathbf{v} \in \mathbf{V}, \|\mathbf{v}\| = 1$ is the right singular vector of PT. Clearly, $\|PT\mathbf{v}\| = \|P(T\mathbf{v})\| \leq \|P\| \|T\mathbf{v}\| \leq \|P\| \|T\|$, which implies the inequality (5.4.3) for $k = 1$. Assume that $\|P\| \|T\| > 0$. For the equality $\|PT\| = \|P\| \|T\|$ we must have that $T\mathbf{v}$ is the right singular vector corresponding to P and $\mathbf{v}$ the the right singular vector corresponding to T. This shows the equality case in the theorem for $k = 1$.

Assume that $k > 1$. If the right-hand side of (5.4.3) is zero then rank $PT \leq \min(\text{rank } P, \text{rank } Q) < k$ and $\sigma_k(PT) = 0$. Hence, (5.4.3) trivially holds. Assume that $k \leq \min(\text{rank } P, \text{rank } Q)$. Then the right-hand side of (5.4.3) is positive. Clearly $\min(\text{rank } P, \text{rank } T) \leq \min(\dim \mathbf{U}, \dim \mathbf{V}, \dim \mathbf{W})$. Observe that $\wedge^k T \in L(\bigwedge^k \mathbf{V}, \bigwedge^k \mathbf{U}), \wedge^k P \in L(\bigwedge^k \mathbf{U}, \bigwedge^k \mathbf{W})$. Hence, (5.4.3) for $k = 1$ applied to $\wedge^k PT = \wedge^k P \wedge^k T$ yields $\sigma_1(\wedge^k PT) \leq \sigma_1(\wedge^k P)\sigma_1(\wedge^k T)$. Use Proposition 5.4.2 to deduce (5.4.3). In order to have $\sigma_1(\wedge^k PT) = \sigma_1(\wedge^k P)\sigma_1(\wedge^k T)$, the operator $\wedge^k T$ must have a right first singular vector $\mathbf{x} \in \bigwedge^k \mathbf{V}$, such that $\mathbf{0} \neq \wedge^k T\mathbf{x}$ is a right singular vector of $\wedge^k P$ corresponding to $\sigma_1(\wedge^k P)$. It is left to show that $\mathbf{x}$ can be chosen as $\mathbf{c}_1 \wedge \cdots \wedge \mathbf{c}_k$, where $\mathbf{c}_1, \ldots, \mathbf{c}_k$ are the first k-right singular vectors of T. Suppose that

$$\sigma_1(T) = \cdots = \sigma_{l_1}(T) > \sigma_{l_1+1}(T) = \cdots$$
$$= \sigma_{l_2}(T) > \cdots > 0 = \sigma_j(T) \quad \text{for } j > l_p. \qquad (5.4.4)$$

Assume first that $k = l_i$ for some $i \leq p$. Then $\sigma_1(\wedge^k T) > \sigma_2(\wedge^k T)$ and $\mathbf{c}_1 \wedge \cdots \wedge \mathbf{c}_k$ is the right singular vector of $\wedge^k T$ corresponding to $\sigma_1(\wedge^k T)$. Then $\sigma_1(\wedge^k P \wedge^k T) = \sigma_1(\wedge^k P)\sigma_1(\wedge^k T)$ if and only if $(\wedge^k T)\mathbf{c}_1 \wedge \cdots \wedge \mathbf{c}_k = T\mathbf{c}_1 \wedge \cdots \wedge T\mathbf{c}_k$ is the right singular vector of $\wedge^k P$ corresponding to $\sigma_1(\wedge^k P)$.

Assume that

$$\sigma_1(P) = \ldots = \sigma_{m_1}(P) > \sigma_{m_1+1}(P) = \cdots$$
$$= \sigma_{m_2}(p) > \ldots > 0 = \sigma_j(P) \text{ for } j > m_q.$$

Suppose that $k = m_{j-1} + r$, where $1 \leq r \leq m_j - m_{j-1}$. (We assume here that $m_0 = 0$.) Let $\mathbf{U}_1$ be the subspace spanned by the m_{j-1} right singular vectors of P corresponding to the first m_{j-1} singular values of P and $\mathbf{W}_1$ be the subspace spanned by $m_j - m_{j-1}$ right singular vectors of P corresponding the $\sigma_{m_{j-1}+1}(P), \ldots, \sigma_{m_j}(P)$. Then any right singular vector of $\wedge^k P$ corresponding to $\sigma_1(\wedge^k P)$ is in the subspace $(\bigwedge^{m_{j-1}} \mathbf{U}_1) \bigwedge (\bigwedge^r \mathbf{W}_1)$. Let $\mathbf{V}_k = \text{span}(\mathbf{c}_1, \ldots, \mathbf{c}_k)$. So $\mathbf{c}_1 \wedge \cdots \wedge \mathbf{c}_k$ is a nonzero vector in $\bigwedge^k \mathbf{V}_k$ and $(\wedge^k T)\mathbf{c}_1 \wedge \cdots \wedge \mathbf{c}_k$ is a nonzero vector in $\bigwedge^k \mathbf{W}_2$, where $\mathbf{W}_2 := T\mathbf{V}_k$ and $\mathbf{U}_2 = \{\mathbf{0}\}$.

The equality in (5.4.3) yields that $\left(\wedge^{m_j-1} \mathbf{U}_1 \right) \wedge \left(\wedge^r \mathbf{W}_1 \right) \cap$ $\left(\wedge^o \mathbf{U}_2 \right) \wedge \left(\wedge^k \mathbf{W}_2 \right) \neq \{\mathbf{0}\}$. Lemma 5.2.11 yields that $\mathbf{U}_1 \subset T\mathbf{V}_k$ and $T\mathbf{V}_k \subseteq \mathbf{U}_1 + \mathbf{W}_1$. So $T\mathbf{V}_k$ is spanned by the first k right singular vectors of P.

Assume now that $k = l_{i-1} + s, 1 \leq s < l_i - l_{i-1}$. Then the subspace spanned by all right singular vectors of $\wedge^k T$ corresponding to $\sigma_1(\wedge^k T)$ is equal to $\left(\wedge^{l_{i-1}} \mathbf{U}_2 \right) \wedge \left(\wedge^s \mathbf{W}_2 \right)$, where $\mathbf{U}_3$ and $\mathbf{W}_3$ are the subspaces spanned by the right singular vectors of T corresponding to the first l_{i-1} and the next $l_i - l_{i-1}$ singular values of T respectively. Let $\mathbf{U}_2 := T\mathbf{U}_3, \mathbf{W}_2 := T\mathbf{W}_3$. The equality in (5.4.3) yields that $\left(\wedge^{m_j-1} \mathbf{U}_1 \right) \wedge \left(\wedge^r \mathbf{W}_1 \right) \cap \left(\wedge^{l_{i-1}} \mathbf{U}_2 \right) \wedge \left(\wedge^s \mathbf{W}_2 \right)$ contains a right singular vector of $\wedge^k P$ corresponding to $\sigma_1(\wedge^k P)$. Lemma 5.2.11 yields that there exists a k-dimensional subspace $\mathbf{V}_1'$ such that $\mathbf{V}_1' \supset \mathbf{U}_1 + \mathbf{U}_2$ and $\mathbf{V}_1' \subset (\mathbf{U}_1 + \mathbf{W}_1) \cap (\mathbf{U}_2 + \mathbf{W}_2)$. Hence, there exists a k-dimensional subspace $\mathbf{V}_k$ of $\mathbf{U}_3 + \mathbf{W}_3$ containing $\mathbf{U}_3$ such that $\mathbf{V}_1' = T\mathbf{V}_k$ contains $\mathbf{U}_1$ and is contained in $\mathbf{U}_1 + \mathbf{W}_1$. Hence, $T\mathbf{V}_k$ is spanned by the first k right singular vectors of P.

Assume now that $\prod_{i=1}^l \sigma_i(P)\sigma_i(T) > 0$. Then $0 < \sigma_i(P), 0 < \sigma_i(T)$ for $i = 1, \ldots, l$. Assume that for $k = 1, \ldots, l$ equality holds in (5.4.3). We prove the existence of orthonormal sets $\mathbf{c}_1, \ldots, \mathbf{c}_l$, $\mathbf{d}_1, \ldots, \mathbf{d}_l$ of right singular vectors of T and P respectively such that $\frac{1}{\sigma_k(T)}\mathbf{c}_k = \mathbf{d}_k, k = 1, \ldots, l$ by induction on l. For $l = 1$ the result is trivial. Assume that the result holds for $l = m$ and let $l = m + 1$. The equality in (5.4.3) for $k = m + 1$ yields the existence of $m + 1$ dimensional subspace $\mathbf{X} \subseteq \mathbf{U}$ such that $\mathbf{X}$ is spanned by the first $m + 1$ right singular vectors of T and $T\mathbf{X}$ is spanned by the first $m + 1$ right singular vectors of P. $\qquad\qquad \square$

Theorem 5.4.5 *Let the assumptions of Theorem 5.4.4 hold. Then equalities in (5.4.3) hold for $k = 1, \ldots, l$, where $l \leq \min(\operatorname{rank} P, \operatorname{rank} T)$, if and only if there exist first l-right singular vectors $\mathbf{c}_1, \ldots, \mathbf{c}_l$ of T, such that $\frac{1}{\sigma_1(T)}T\mathbf{c}_1, \ldots, \frac{1}{\sigma_l(T)}T\mathbf{c}_l$ are the first l-right singular vectors of P.*

Proof. We prove the theorem by induction on l. For $l = 1$ the theorem follows from Theorem 5.4.4. Suppose that the theorem holds for $l = j$. Let $l = j + 1$. Since we assumed that equality holds in (5.4.3) for $k = l$ Theorem 5.4.4 yields that there exists an l-dimensional subspace $\mathbf{V}_l$ of $\mathbf{V}$ which is spanned by the first l right singular vectors of T, and $T\mathbf{V}_l$ is spanned by the first l right singular vectors of P. Let $\hat{T} \in \mathrm{L}(\mathbf{V}_l, T\mathbf{V}_l), \hat{P} \in \mathrm{L}(T\mathbf{V}_l, PT\mathbf{V}_l)$ be the restrictions of T and P to the subspaces $\mathbf{V}_l, T\mathbf{V}_l$ respectively. Clearly

$$\sigma_i(T) = \sigma_i(\hat{T}) > 0, \ \sigma_i(P) = \sigma_i(\hat{P}) > 0, \ \text{for } i = 1, \ldots, l. \quad (5.4.5)$$

The equalities in (5.4.3) for $k = 1, \ldots, l$ imply that $\sigma_i(PT) = \sigma_i(P)\sigma_i(T)$ for $i = 1, \ldots, l$. Let $\hat{Q} := \hat{P}\hat{T} \in \mathrm{L}(\mathbf{V}_l, PT\mathbf{V}_l)$. Clearly $\hat{Q}$ is the restriction of $Q := PT$ to $\mathbf{V}_l$. Corollary 4.11.3 yields that $\sigma_i(\hat{Q}) \le \sigma_i(Q)$ for $i = 1, \ldots$ Since $\det \hat{Q} = \det \hat{P} \det \hat{T}$ we deduce that $\prod_{i=1}^l \sigma_i(\hat{Q}) = \prod_{i=1}^l \sigma_i(\hat{P}) \prod_{i=1}^l \sigma_i(\hat{T})$. The above arguments show that $\prod_{i=1}^l \sigma_i(\hat{Q}) = \prod_{i=1}^l \sigma_i(Q) > 0$. Corollary 4.11.3 yields that $\sigma_i(\hat{Q}) = \sigma_i(Q)$. Hence, we have equalities $\prod_{i=1}^k \sigma_i(\hat{P}\hat{Q}) = \prod_{i=1}^k \sigma_i(\hat{P})\sigma_i(\hat{T})$ for $i = 1, \ldots, l$. The induction hypothesis yields that there exist first $l - 1$-right singular vectors of $\hat{T}$ $\mathbf{c}_1, \ldots, \mathbf{c}_{l-1}$, such that $\frac{1}{\sigma_1(\hat{T})}\hat{T}\mathbf{c}_1, \ldots, \frac{1}{\sigma_l(\hat{T})}\hat{T}\mathbf{c}_{l-1}$ are the first l-right singular vectors of $\hat{P}$. Complete $\mathbf{c}_1, \ldots, \mathbf{c}_{l-1}$ to an orthonormal basis $\mathbf{c}_1, \ldots, \mathbf{c}_l$ of $\mathbf{V}_l$. Then $\mathbf{c}_l$ is a right singular vector of $\hat{T}$ corresponding $\sigma_l(\hat{T})$. Since $\hat{T}\mathbf{c}_l$ is orthogonal to $\hat{T}\mathbf{c}_1, \ldots, \hat{T}\mathbf{c}_{l-1}$, which are right singular vectors of $\hat{P}$ it follows that $\frac{1}{\sigma_l(\hat{T})}\hat{T}\mathbf{c}_l$ is a right singular vector of $\hat{P}$ corresponding to $\sigma_l(\hat{T})$. Use (5.4.5) and the fact that $\hat{T}$ and $\hat{P}$ are the corresponding restrictions of T and P respectively to deduce the theorem. $\qquad \square$

In what follows we need to consider the half closed infinite interval $[-\infty, \infty)$ which is a subset of $\bar{\mathbb{R}}$, see §4.6. Denote by $[-\infty, \infty)^n_{\searrow} \subset [-\infty, \infty)^n$ the set of $\mathbf{x} = (x_1, \ldots, x_n)$ where $x_1 \ge \cdots \ge x_n \ge -\infty$.

We now extend the notions of majorization, Schur set, Schur order preserving function to subsets of $[-\infty, \infty)^n_{\searrow}$. Let $\mathbf{x} = (x_1, \ldots, x_n), \mathbf{y} = (y_1, \ldots, y_n) \in [-\infty, \infty)^n_{\searrow}$. Then $\mathbf{x} \preceq \mathbf{y}$, i.e. $\mathbf{x}$ is weakly majorized by $\mathbf{y}$, if the inequalities $\sum_{i=1}^{k} x_i \leq \sum_{i=1}^{k} y_i$ hold for $i = 1, \ldots, n$. $\mathbf{x} \prec \mathbf{y}$, i.e. $\mathbf{x}$ majorized by $\mathbf{y}$, if $\mathbf{x} \preceq \mathbf{y}$ and $\sum_{i=1}^{n} x_i = \sum_{i=1}^{n} y_i$. A set $D \subseteq [-\infty, \infty)_{\searrow}^{n}$ is called a *Schur set* if for any $\mathbf{y} \in D$ and any $\mathbf{x} \prec \mathbf{y}$ $\mathbf{x} \in D$.

Let $I \subseteq [-\infty, \infty)$ be an interval, which may be open, closed or half closed. Denote by I_o the interior of I. $f : I \to \mathbb{R}$ is called continuous if the following conditions hold: First, $f|I_0$ is continuous. Second, if $a \in [-\infty, \infty) \cap I$ is an end point of I then f is continuous at a from the left or right respectively. Suppose that $-\infty \in I$. Then $f : I \to \mathbb{R}$ is called convex on I if f is continuous on I and a nondecreasing convex function on I_0. (See Problem 6.) f is called strictly convex on I if f continuous on I and strictly convex on I_0. If $-\infty \in I$ then f is continuous on I, and is also an increasingly strictly convex function on I_0. Note that the function e^x is a strictly convex function on $[-\infty, \infty)$.

Let $D \subseteq [-\infty, \infty)^n$. Then $f : D \to \mathbb{R}$ is continuous, if for any $\mathbf{x} \in D$ and any sequence of points $\mathbf{x}_k \in D$, with $k \in \mathbb{N}$, the equality $\lim_{k \to \infty} f(\mathbf{x}_k) = f(\mathbf{x})$ holds when $\lim_{k \to \infty} \mathbf{x}_k = \mathbf{x}$. D is convex if for any $\mathbf{x}, \mathbf{y} \in D$ the point $t\mathbf{x} + (1 - t)\mathbf{y}$ is in D for any $t \in (0, 1)$. For a convex D, $f : D \to \mathbb{R}$ is convex if f is continuous and $f(t\mathbf{x} + (1 - t)\mathbf{y}) \leq tf(\mathbf{x}) + (1 - t)f(\mathbf{y})$ for any $t \in (0, 1)$. For a Schur set $D \subseteq [-\infty, \infty)^n$, $f : D \to \mathbb{R}$ is called Schur order preserving, strict Schur order preserving, strong Schur order preserving, strict strong Schur order preserving if f is a continuous function satisfying the properties described in the beginning of §4.8. It is straightforward to generalize the results on Schur order preserving functions established in §4.8 using Problem 7.

Let the assumptions of Theorem 5.4.4 hold. For any $k \in \mathbb{N}$ let

$$\boldsymbol{\sigma}_k(T) := (\sigma_1(T), \ldots, \sigma_k(T)) \in \mathbb{R}_{+,\searrow}^{k}, \qquad (5.4.6)$$

$$\log \boldsymbol{\sigma}_k := (\log \sigma_1(T), \ldots, \log \sigma_k(T)) \in [-\infty, \infty)^k.$$

Theorem 5.4.4 yields

$$\log \boldsymbol{\sigma}_k(PT) \preceq \log \boldsymbol{\sigma}_k(P) + \log \boldsymbol{\sigma}_k(T)$$

$$\text{for any } k \in [1, \max(\text{rank } P, \text{rank } T)],$$

$$\log \sigma_k(PT) \prec \log \sigma_k(P) + \log \sigma_k(T)$$

$$\text{for } k > \max(\text{rank } P, \text{rank } T),$$

$$\log \sigma_k(PT) \prec \log \sigma_k(P) + \log \sigma_k(T)$$

$$\text{if } k = \text{rank } P = \text{rank } T = \text{rank } PT, \qquad (5.4.7)$$

See Problem 8.

Theorem 5.4.6 *Let* $\mathbf{U}, \mathbf{V}, \mathbf{W}$ *be inner product spaces. Assume that* $T \in L(\mathbf{V}, \mathbf{U}), P \in L(\mathbf{U}, \mathbf{W})$ *and* $l \in \mathbb{N}$.

1. *Assume that* $D \subset [-\infty, \infty)^l_{\searrow}$ *is a strong Schur set containing* $\log \sigma_l(PT), \log \sigma_l(P) + \log \sigma_l(T)$. *Let* $h : D \to \mathbb{R}$ *be a strong Schur order preserving function. Then* $h(\log \sigma_l(PT)) \le h(\log \sigma_l(PT) + \log \sigma_l(PT))$. *Suppose furthermore that* h *is strict strong Schur order preserving. Then equality holds in the above inequality if and only if equality holds in (5.4.3) for* $k = 1, \ldots, l$.

2. *Assume that* $\log \sigma_l(PT) \prec \log \sigma_l(P) + \log \sigma_l(T)$, *and* $D \subset [-\infty, \infty)^l_{\searrow}$ *is a Schur set containing* $\log \sigma_l(PT), \log \sigma_l(P) + \log \sigma_l(T)$. *Let* $h : D \to \mathbb{R}$ *be a Schur order preserving function. Then* $h(\log \sigma_l(PT)) \le h(\log \sigma_l(PT) + \log \sigma_l(PT))$. *Suppose furthermore that* h *is strict Schur order preserving. Then equality holds in the above inequality if and only if equality holds in (5.4.3) for* $k = 1, \ldots, l - 1$.

See Problem 9.

Corollary 5.4.7 *Let* $\mathbf{U}, \mathbf{V}, \mathbf{W}$ *be inner product spaces. Assume that* $T \in L(\mathbf{V}, \mathbf{U}), P \in L(\mathbf{U}, \mathbf{W})$ *and* $l \in \mathbb{N}$. *Assume that* $\log \sigma_l(PT) \prec \log \sigma_l(P) + \log \sigma_l(T)$, *and* $I \subset [-\infty, \infty)$ *is an interval set containing* $\log \sigma_1(P) + \log \sigma_1(T), \log \sigma_l(PT)$. *Let* $h : I \to \mathbb{R}$ *be a convex function. Then*

$$\sum_{i=1}^{l} h(\log \sigma_i(PT)) \le \sum_{i=1}^{l} h(\log \sigma_i(P) + \log \sigma_i(T)). \qquad (5.4.8)$$

Corollary 5.4.8 *Let* $\mathbf{U}, \mathbf{V}, \mathbf{W}$ *be inner product spaces. Assume that* $T \in \mathrm{L}(\mathbf{V}, \mathbf{U}), P \in \mathrm{L}(\mathbf{U}, \mathbf{W})$ *and* $l \in \mathbb{N}$. *Then for any* $t > 0$

$$\sum_{i=1}^{l} \sigma_i(PT)^t \le \sum_{i=1}^{l} \sigma_i(P)^t \sigma_i(T)^t. \qquad (5.4.9)$$

Equality holds if and only if one has equality sign in (5.4.3) *for* $k = 1, \ldots, l$.

Proof. Observe that the function $h : [-\infty, \infty)^l_{\searrow} \to \mathbb{R}$ given by $h((x_1, \ldots, x_l)) = \sum_{i=1}^{l} e^{tx_i}$ is strictly strongly Schur order preserving for any $t > 0$. $\qquad \square$

The following theorem improves the results of Theorem 4.11.12.

Theorem 5.4.9 *Let* $\mathbf{V}$ *be an* n-*dimensional IPS over* $\mathbb{C}$ *and assume that* $T \in \mathrm{L}(\mathbf{V})$. *Let* $\lambda_1(T), \ldots, \lambda_n(T) \in \mathbb{C}$ *be the eigenvalues of* T *counted with their multiplicities and arranged in order* $|\lambda_1(T)| \ge \cdots \ge |\lambda_n(T)|$. *Let* $\boldsymbol{\lambda}_a(T) := (|\lambda_1(T)|, \ldots, |\lambda_n(T)|)$ *and* $\boldsymbol{\lambda}_{a,k}(T) := (|\lambda_1(T)|, \ldots, |\lambda_k(T)|)$ *for* $k = 1, \ldots, n$. *Then*

$$\prod_{i=1}^{l} |\lambda_i(T)| \le \prod_{i=1}^{l} \sigma_i(T) \text{ for } l = 1, \ldots, n-1, \quad \text{and}$$

$$\prod_{i=1}^{n} |\lambda_i(T)| = \prod_{i=1}^{n} \sigma_i(T). \qquad (5.4.10)$$

For $l = 1, \ldots, k \le n$ *equalities hold in the above inequalities if and only if the conditions* 1 *and* 2 *of Theorem 4.11.12 hold.*

In particular, $\log \boldsymbol{\lambda}_{a,k}(T) \preceq \log \boldsymbol{\sigma}_k(T)$ *for* $k = 1, \ldots, n-1$ *and* $\log \boldsymbol{\lambda}_a(T) \prec \log \boldsymbol{\sigma}(T)$.

Proof. By Theorem 4.11.12 $|\lambda_1(\wedge^l T)| \le \sigma_1(\wedge^l T)$. Use Problem 8 below and Proposition 5.4.2 to deduce the inequalities in (5.4.10). The equality $\prod_{i=1}^{n} |\lambda_i(T)| = \prod_{i=1}^{l} \sigma_i(T)$ is equivalent to the identity $|\det T|^2 = \det TT^*$.

Suppose that for $l = 1, \ldots, k \le n$ equalities hold in (5.4.10). Then $|\lambda_i(T)| = \sigma_i(T)$ for $i = 1, \ldots, k$. Hence equality holds in

(4.11.14). Theorem 4.11.12 implies that conditions 1 and 2 hold. Vice versa, assume that the conditions 1 and 2 of Theorem 4.11.12 hold. Then from the proof of Theorem 4.11.12 it follows that $|\lambda_i(T)| = \sigma_i(T)$ for $i = 1, \ldots, k$. Hence, for $l = 1, \ldots, k$ equalities hold in (5.4.10). $\qquad\qquad\qquad\qquad\qquad\qquad\qquad\qquad\qquad\qquad\qquad$ □

Corollary 5.4.10 *Let* **V** *be an n-dimensional IPS. Assume that $T \in L(V)$.*

1. *Assume that $k \in [n-1]$ and $D \subset [-\infty, \infty)^k_\searrow$ is a strong Schur set containing $\log \boldsymbol{\sigma}_k(T)$. Let $h : D \to \mathbb{R}$ be a strong Schur order preserving function. Then $h(\log \boldsymbol{\lambda}_k(T)) \leq h(\log \boldsymbol{\sigma}_k(T))$. Suppose furthermore that h is strict strong Schur order preserving. Then equality holds in the above inequality if and only if equality holds in (5.4.10) for $l = 1, \ldots, k$.*

2. *Let $I \subset [-\infty, \infty)$ be an interval containing $\log \sigma_1(T)$, $\log \sigma_k(T), \log |\lambda_k(T)|$. Assume that $f : I \to \mathbb{R}$ is a convex nondecreasing function.   Then $\sum_{i=1}^k f(\log |\lambda_i(T)|) \leq \sum_{i=1}^k f(\log |\sigma_i(T)|)$. If f is a strictly convex increasing function on I then equality holds if and only if equality holds in (5.4.10) for $l = 1, \ldots, k$. In particular, for any $t > 0$*

$$\sum_{i=1}^k |\lambda_i(T)|^t \leq \sum_{i=1}^k \sigma_i(T)^t. \qquad (5.4.11)$$

Equality holds if and only if equality holds in (5.4.10) for $l = 1, \ldots, k$.

3. *Assume that $D \subset [-\infty, \infty)^n_\searrow$ is a Schur set containing $\log \boldsymbol{\sigma}_n(T)$. Let $h : D \to \mathbb{R}$ be a Schur order preserving function. Then $h(\log \boldsymbol{\lambda}_a(T)) \leq h(\log \boldsymbol{\sigma}_n(T))$. Suppose furthermore that h is strict Schur order preserving. Then equality holds in the above inequality if and only if equality holds T is a normal operator.*

Problems

1. Let $\mathbf{V}_i$ be an n_i-dimensional vector space with the inner product $\langle \cdot, \cdot \rangle_i$ for $i = 1, \ldots, k$.

   (a) Let $\mathbf{e}_{1,i}, \ldots, \mathbf{e}_{n_i,i}$ be an orthonormal basis of $\mathbf{V}_i$ with respect $\langle \cdot, \cdot \rangle_i$ for $i = 1, \ldots, k$. Let $\langle \cdot, \cdot \rangle$ be the inner product in $\mathbf{Y} := \otimes_{i=1}^k \mathbf{V}_i$ such that $\otimes_{i=1}^k \mathbf{e}_{j_i,i}$ where $j_i = 1, \ldots, n_i, i = 1, \ldots, k$ is an orthonormal basis of $\mathbf{Y}$. Show that (5.4.1) holds.

   (b) Prove that there exists a unique inner product on $\mathbf{Y}$ satisfying (5.4.1).

2. Prove Proposition 5.4.1.

3. Let $\mathbf{V}$ be an n-dimensional IPS with an orthonormal basis $\mathbf{e}_1, \ldots, \mathbf{e}_n$. Let $\mathbf{Y} := \otimes_{i=1}^k \mathbf{V}_i, \mathbf{V}_1 = \cdots = \mathbf{V}_k = \mathbf{V}$ be inner product spaces with the canonical inner product $\langle \cdot, \cdot \rangle_{\mathbf{Y}}$. Show

   (a) Let $k \in [n]$. Then the subspace $\bigwedge^k \mathbf{V}$ of $\mathbf{Y}$ has an orthonormal basis

   $$\frac{1}{\sqrt{k!}} \mathbf{e}_{i_1} \wedge \cdots \wedge \mathbf{e}_{i_k}, 1 \le i_1 < i_2 < \cdots < i_k \le n.$$

   (b) Let $k \in \mathbb{N}$. Then the subspace $\mathrm{Sym}^k \mathbf{V}$ of $\mathbf{Y}$ has an orthonomal basis $\alpha(i_1, \ldots, i_k) \mathrm{sym}^k (\mathbf{e}_{i_1}, \ldots, \mathbf{e}_{i_k}), 1 \le i_1 \le \cdots \le i_k \le n$. The coefficient $\alpha(i_1, \ldots, i_k)$ is given as follows. Assume that $i_1 = \cdots = i_{l_1} < i_{l_1+1} = \cdots = i_{l_1+l_2} < \cdots < i_{l_1+\cdots+l_{r-1}+1} = \cdots = i_{l_1+\cdots+l_r}$, where $l_1 + \cdots + l_r = k$. Then $\alpha(i_1, \ldots, i_k) = \frac{1}{\sqrt{k! l_1! \ldots l_r!}}$.

4. (a) Prove Proposition 5.4.2.

   (b) Prove Corollary 5.4.3.

5. Let $\mathbf{U}, \mathbf{V}$ be inner product spaces of dimensions n and m respectively. Let $T \in \mathrm{L}(\mathbf{V}, \mathbf{U})$ and assume that we chose

orthonormal bases $[\mathbf{c}_1, \ldots, \mathbf{c}_n], [\mathbf{d}_1, \ldots, \mathbf{d}_m]$ of $\mathbf{V}, \mathbf{U}$ respectively satisfying (4.10.5). Suppose furthermore that

$$\sigma_1(T) = \cdots = \sigma_{l_1}(T) > \sigma_{l_1+1}(T) = \cdots = \sigma_{l_2}(T) > \cdots >$$

$$\tag{5.4.12}$$

$$\sigma_{l_{p-1}+1}(T) = \cdots = \sigma_{l_p}(T) > 0, \ 1 \le l_1 < \cdots < l_p = \text{rank } T.$$

Let

$$\mathbf{V}_i := \text{span} \, (\mathbf{c}_{l_{i-1}+1}, \ldots, \mathbf{c}_{l_i}), \quad i = 1, \ldots, p, \ l_0 := 0. \quad (5.4.13)$$

(a) Let $k = l_i$ for some $i \in [1, p]$. Show that $\sigma_1(\wedge^k T) > \sigma_2(\wedge^k T)$. Furthermore the vector $\mathbf{c}_1 \wedge \cdots \wedge \mathbf{c}_k$ is a unique right singular vector, (up to a multiplication by scalar), of $\wedge^k T$ corresponding to $\sigma_1(\wedge^k T)$. Equivalently, the one-dimensional subspace spanned by the the right singular vectors of $\wedge^k T$ is given by $\wedge^k \oplus_{j=1}^i \mathbf{V}_i$.

(b) Assume that $l_i - l_{i-1} \ge 2$ and $l_{i-1} < k < l_i$ for some $1 \le i \le p$. Show that

$$\sigma_1(\wedge^k T) = \cdots = \sigma_{\binom{l_i-l_{i-1}}{k-l_{i-1}}}(T) > \sigma_{\binom{l_i-l_{i-1}}{k-l_{i-1}}+1}(T).$$

$$\tag{5.4.14}$$

The subspace spanned by all right singular vectors of $\wedge^k T$ corresponding to $\sigma_1(\wedge^k T)$ is given by the subspace:

$$\left(\bigwedge^{l_{i-1}} \oplus_{j=1}^{i-1} \mathbf{V}_j \right) \bigwedge \left(\bigwedge^{k-l_{i-1}} \mathbf{V}_i \right).$$

6. Let $I := [-\infty, a), a \in \mathbb{R}$ and assume that $f : I \to \mathbb{R}$ is continuous. f is called convex on I if $f(tb + (1-t)c) \le tf(b) + (1-t)f(c)$ for any $b, c \in I$ and $t \in (0, 1)$. We assume that $t(-\infty) = -\infty$ for any $t > 0$. Show that if f is convex on I if and only f is a convex nondecreasing bounded below function on I_o.

7. $D \subset [-\infty, \infty)^n$ such that $D' := D \cap \mathbb{R}^n$ is nonempty. Assume that $f : D \to \mathbb{R}$ is continuous. Show

   (a) D is a Schur set if and only if D' is a Schur set.

   (b) f is Schur order preserving if and only if $f|D'$ is Schur order preserving.

   (c) f is strict Schur order preserving if and only if $f|D'$ is strict Schur order preserving.

   (d) f is strong Schur order preserving if and only if $f|D'$ is strong Schur order preserving.

   (e) f is strict strong Schur order preserving if and only if $f|D'$ is strictly strong Schur order preserving.

8. Let- the assumptions of Theorem 5.4.4 hold. Assume that rank P = rank T = rank PT. Let k = rank P. Show that the arguments of the proof of Theorem 5.4.5 implies that $\prod_{i=1}^{k} \sigma_i(PT) = \prod_{i=1}^{k} \sigma_i(P)\sigma_i(T)$. Hence, $\log \boldsymbol{\sigma}_k(PT) \prec \log \boldsymbol{\sigma}_k(P) + \log \boldsymbol{\sigma}_k(T)$.

9. Prove Theorem 5.4.6 using the results of §4.8.

10. Show that under the assumptions of Theorem 4.11.8 one has the inequality $\sum_{i=1}^{l} \sigma_i(S^*T)^t \leq \sum_{i=1}^{l} \sigma_i(S)^t \sigma_i(T)^t$ for any $l \in \mathbb{N}$ and $t > 0$.

11. (a) Let the assumptions of Theorem 5.4.9 hold. Show that (5.4.10) imply that $\lambda_i(T) = 0$ for $i > $ rank T.

    (b) Let $\mathbf{V}$ be a finite dimensional vector field over the algebraically closed field $\mathbb{F}$. Let $T \in L(\mathbf{V})$. Show that the number of nonzero eigenvalues, counted with their multiplicities, does not exceed rank T. (*Hint*: Use the Jordan canonical form of T.)

5.5   Matrix Exponents

For a matrix $C \in \mathbb{C}^{n \times n}$ with all eigenvalues of C are in the disk $|z - 1| < 1$ we define

$$\log C := \log(I + (C - I)) = \sum_{i=1}^{\infty} \frac{(-1)^{i-1}}{i}(C - I)^i, \quad (5.5.1)$$

$$C^z := e^{z \log C}, \text{ where } z \in C \text{ and } \rho(C - I) < 1. \quad (5.5.2)$$

(See §3.1.)

Proposition 5.5.1 *Let* $a > 0$, $A(t) : (-a, a) \to \mathbb{C}^{n \times n}$, *assume that*

$$\lim_{t \to 0} \frac{A(t)}{t} = B. \quad (5.5.3)$$

Then for any $s \in \mathbb{C}$

$$\lim_{t \to 0}(I + A(t))^{\frac{s}{t}} = e^{sB}. \quad (5.5.4)$$

Proof. The assumption (5.5.3) yields that $B(t) := \frac{1}{t}A(t)$, $t \neq 0, B(0) := B$ is continuous at $t = 0$. Hence, $\|B(t)\|_2 = \sigma_1(B(t)) \leq c$ for $|t| \leq \delta'$. Let $\delta := \min(\delta', \frac{1}{2c})$. Denote $C(t) = I + A(t) = I + tB(t)$. Hence

$$\rho(C(t) - I) \leq \sigma_1(C(t) - I) = \|A(t)\|_2 \leq |t|c < \frac{1}{2} \text{ for } |t| < \delta.$$

Assume that $t \in (-\delta, \delta) \setminus \{0\}$. Use (5.5.1) to obtain

$$\frac{s}{t}\log(I + A(t)) = s\sum_{i=1}^{\infty} \frac{(-t)^{i-1}}{i}B(t)^i = sB(t) + \sum_{i=2}^{\infty} \frac{s(-t)^{i-1}}{i}B(t)^i.$$

Recall that

$$\left\|\sum_{i=2}^{\infty} \frac{s(-t)^{i-1}}{i}B(t)^i\right\|_2 \leq \sum_{i=2}^{\infty} \left\|\frac{s(-t)^{i-1}}{i}B(t)^i\right\|_2$$

$$\leq \sum_{i=2}^{\infty} \frac{|s||t|^{i-1}\|B(t)\|_2^i}{i} \leq \frac{|s|(-|t|c - \log(1 - |t|c)}{|t|}.$$

Hence

$$\lim_{t \to 0}(I + A(t))^{\frac{s}{t}} = \exp(\lim_{t \to 0} \frac{s}{t} \log(I + A(t)) = e^{sB}.$$

Theorem 5.5.2 (*The Lie–Trotter formula*) *Let* $A_1, \ldots, A_k \in \mathbb{C}^{n \times n}$. *Then for any* $s \in \mathbb{C}$

$$\lim_{t \to 0}(e^{tA_1} e^{tA_2} \cdots e^{tA_k})^{\frac{s}{t}} = e^{s \sum_{i=1}^{k} A_i}. \tag{5.5.5}$$

In particular

$$\lim_{N \to \infty, N \in \mathbb{N}}(e^{\frac{1}{N} A_1} e^{\frac{1}{N} A_2} \cdots e^{\frac{1}{N} A_k})^N = e^{\sum_{i=1}^{k} A_i}. \tag{5.5.6}$$

Proof. Clearly $e^{tA_i} = I + tA_i + |t|^2 O(1)$ for $|t| \leq 1$. Hence

$$A(t) := e^{tA_1} e^{tA_2} \cdots e^{tA_k} - I = t \sum_{i=1}^{k} A_i + |t|^2 O(1).$$

Therefore, the condition (5.5.3) holds, where $B = \sum_{i=1}^{k} A_i$. Apply (5.5.4) to deduce (5.5.5). Choose a sequence $t_N = \frac{1}{N}$ for $N = 1, 2, \ldots$, and $s = 1$ in (5.5.5) to deduce (5.5.6). $\qquad \square$

Proposition 5.5.3 *Let* $\mathbf{V}_i$ *be an* n_i-*dimensional vector space for* $i = 1, \ldots, k$ *over* $\mathbb{F} = \mathbb{R}, \mathbb{C}$. *Let* $\mathbf{Y} := \otimes_{i=1}^{k} \mathbf{V}_i$. *Assume that* $A_i \in L(\mathbf{V}_i), i = 1, \ldots$ *Then*

$$\otimes_{i=1}^{k} e^{tA_i} = e^{t(A_1, \ldots, A_k)_\otimes} \in L(\otimes_{i=1}^{k} \mathbf{V}_i), \tag{5.5.7}$$

where

$$(A_1, \ldots, A_k)_\otimes := \sum_{i=1}^{k} I_{n_1} \otimes \cdots \otimes I_{n_{i-1}} \otimes A_{n_i} \otimes I_{n_{i+1}}$$

$$\otimes \cdots \otimes I_{n_k} \in L(\otimes_{i=1}^{k} \mathbf{V}_i),$$

for any $t \in \mathbb{F}$.

See Problem 1.

Definition 5.5.4 *Let* $\mathbf{V}$ *be an* n-*dimensional vector space over* $\mathbb{F}$. *Assume that* $A \in L(\mathbf{V})$. *Denote by* $A_{\wedge k}$ *the restriction of* $\underbrace{(A, \ldots, A)}_{k}_\otimes$ *to* $\wedge^k \mathbf{V}$.

Corollary 5.5.5 *Let the assumptions of Definition 5.5.4 hold for* $\mathbb{F} = \mathbb{R}, \mathbb{C}$. *Then* $\wedge^k e^{tA} = e^{tA}\wedge^k$ *for any* $t \in \mathbb{F}$.

Definition 5.5.6 *A subspace* $\mathbf{U} \subset \mathbf{H}_n$ *is called a commuting subspace if any two matrices* $A, B \in \mathbf{U}$ *commute.*

Note that if $A, B \in \mathbf{H}_n$ then each eigenvalue of $e^A e^B$ are positive. (See Problem 4.)

Recall that a function $f : D \to \mathbb{R}$, where $D \subset \mathbb{R}^N$ is a convex set is called *affine* if f and $-f$ are convex functions on D.

Theorem 5.5.7 *Let*

$$f_k : \mathbf{H}_n \times \mathbf{H}_n \to \mathbb{R}, \quad f_k(A, B) := \sum_{i=1}^{k} \log \lambda_i(e^A e^B), \ k = 1, \ldots, n.$$

$$(5.5.8)$$

(The eigenvalues of $e^A e^B$ *are arranged in a decreasing order.) Then* $f_n(A, B) = \operatorname{tr}(A + B)$ *is an affine function on* $\mathbf{H}_n \times \mathbf{H}_n$. *Assume that* $\mathbf{U}, \mathbf{V} \subset \mathbf{H}_n$ *be two commuting subspaces. Then the functions* $f_k : \mathbf{U} \times \mathbf{V}$ *are convex functions on* $\mathbf{U} \times \mathbf{V}$.

Proof. Clearly

$$f_n(A, B) = \log(\det e^A e^B) = \log\left((\det e^A)(\det e^B)\right)$$
$$= \log \det e^A + \log \det e^B = \operatorname{tr} A + \operatorname{tr} B = \operatorname{tr}(A + B).$$

Hence, $f_n(A, B)$ is an affine function on $\mathbf{H}_n \times \mathbf{H}_n$.

Assume that $k \in [n-1]$. Since $e^A e^B$ has positive eigenvalues for all pairs $A, B \in \mathbf{H}_n$ it follows that each f_k is a continuous function on $\mathbf{U} \times \mathbf{V}$. Hence, it is enough to show that

$$f_k\left(\frac{1}{2}(A_1 + A_2), \frac{1}{2}(B_1 + B_2)\right) \leq \frac{1}{2}(f_k(A_1, B_1) + f_k(A_2, B_2)),$$

$$\text{for } k = 1, \ldots, n-1, \quad (5.5.9)$$

and for any $A_1, A_2 \in \mathbf{U}, B_1, B_2 \in \mathbf{V}$. (See Problem 5.)

We first consider the case $k = 1$. Since $A_1 A_2 = A_2 A_1, B_1 B_2 = B_2 B_1$ it follows that

$$e^{\frac{1}{2}(A_1 + A_2)} e^{\frac{1}{2}(B_1 + B_2)} = e^{\frac{1}{2}A_2} e^{\frac{1}{2}A_1} e^{\frac{1}{2}B_1} e^{\frac{1}{2}B_2}.$$

Observe next that

$$e^{-\frac{1}{2}A_2}(e^{\frac{1}{2}A_2} e^{\frac{1}{2}A_1} e^{\frac{1}{2}B_1} e^{\frac{1}{2}B_2}) e^{\frac{1}{2}A_2} = e^{\frac{1}{2}A_1} e^{\frac{1}{2}B_1} e^{\frac{1}{2}B_2} e^{\frac{1}{2}A_2} \Rightarrow$$

$$f_1\left(\frac{1}{2}(A_1 + A_2), \frac{1}{2}(B_1 + B_2)\right) = \lambda_1(e^{\frac{1}{2}A_1} e^{\frac{1}{2}B_1} e^{\frac{1}{2}B_2} e^{\frac{1}{2}A_2}).$$

Hence

$$\lambda_1\left(e^{\frac{1}{2}A_1} e^{\frac{1}{2}B_1} e^{\frac{1}{2}B_2} e^{\frac{1}{2}A_2}\right)$$

$$\leq \sigma_1(e^{\frac{1}{2}A_1} e^{\frac{1}{2}B_1} e^{\frac{1}{2}B_2} e^{\frac{1}{2}A_2})$$

$$\leq \sigma_1(e^{\frac{1}{2}A_1} e^{\frac{1}{2}B_1}) \sigma_1(e^{\frac{1}{2}B_2} e^{\frac{1}{2}A_2})$$

$$= \lambda_1(e^{\frac{1}{2}A_1} e^{\frac{1}{2}B_1} e^{\frac{1}{2}B_1} e^{\frac{1}{2}A_1})^{\frac{1}{2}} \lambda_1(e^{\frac{1}{2}A_2} e^{\frac{1}{2}B_2} e^{\frac{1}{2}B_2} e^{\frac{1}{2}A_2})^{\frac{1}{2}}$$

$$= \lambda_1(e^{\frac{1}{2}A_1} e^{\frac{1}{2}A_1} e^{\frac{1}{2}B_1} e^{\frac{1}{2}B_1})^{\frac{1}{2}} \lambda_1(e^{\frac{1}{2}A_2} e^{\frac{1}{2}A_2} e^{\frac{1}{2}B_2} e^{\frac{1}{2}B_2})^{\frac{1}{2}}$$

$$= \lambda_1(e^{A_1} e^{B_1})^{\frac{1}{2}} \lambda_1(e^{A_2} e^{B_2})^{\frac{1}{2}}. \tag{5.5.10}$$

This proves the convexity of f_1.

We now show the convexity of f_k. Use Problem 6 to deduce that we may assume that $\mathbf{U} = U\mathbf{D}_n(\mathbb{R})U^*, \mathbf{V} = V\mathbf{D}_n(\mathbb{R})V^*$ for some $U, V \in \mathbf{U}(n)$. Let $\mathbf{U}_k, \mathbf{V}_k \subset \mathbf{H}_{\binom{n}{k}}$ be two commuting subspaces defined in Problem 6(d). The above result imply that $g : \mathbf{U}_k \times \mathbf{V}_k \to \mathbb{R}$ given by $g(C, D) = \log(e^C e^D)$ is convex. Hence

$$g\left(\frac{1}{2}((A_1)_{\wedge^k} + (A_2)_{\wedge^k}), \frac{1}{2}((B_1)_{\wedge^k} + (B_2)_{\wedge^k})\right)$$

$$\leq \frac{1}{2}(g((A_1)_{\wedge^k}, (B_1)_{\wedge^k}) + g((A_1)_{\wedge^k}, (B_1)_{\wedge^k})).$$

The definitions of $\underbrace{(A, \ldots, A)}_{k}{}_{\otimes}$ and $A_{\wedge^k}$ yield the equality $\frac{1}{2}(A_{\wedge^k} + B_{\wedge^k}) = (\frac{1}{2}(A + B))_{\wedge^k}$. Use Problem 3 to deduce that the convexity of g implies the convexity of f_k. $\qquad\square$

Theorem 5.5.8 *Let $A, B \in \mathbf{H}_n, k \in [n]$. Assume that $f_k(tA,$ $tB), t \in \mathbb{R}$ is defined as in (5.5.8). Then the function $\frac{f_k(tA,tB)}{t}$ non-decreases on $(0, \infty)$. In particular*

$$\sum_{i=1}^{k} \lambda_i(A + B) \le \sum_{i=1}^{k} \log \lambda_i(e^A e^B), \quad k = 1, \ldots, n, \qquad (5.5.11)$$

$$\mathrm{tr}\, e^{A+B} \le \mathrm{tr}(e^A e^B). \quad (Golden\text{--}Thompson \ inequality)$$
$$(5.5.12)$$

Equality in (5.5.12) holds if and only if $AB = BA$. Furthermore, for $k \in [n-1]$ equality holds in (5.5.11) if and only if the following condition holds: There exists a common invariant subspace $\mathbf{W} \subset \mathbb{C}^n$ of A and B of dimension k which is spanned by the first k-eigenvectors of $A + B$ and $e^A e^B$ respectively.

Proof. Theorem 5.5.7 yields that $g_k(t) := f_k(tA, tB)$ is convex on $\mathbb{R}$. (Assume $\mathbf{U} = \mathrm{span}\,(A), \mathbf{V} = \mathrm{span}\,(B)$ and $(t, \tau) \in \mathbb{R}^2$ corresponds to $(tA, \tau B) \in \mathbf{U} \times \mathbf{V}$.) Note that $g_k(0) = 0$. Problem 8 implies that $\frac{g_k(t)}{t}$ nondecreasing on $(0, \infty)$. Problem 9 implies that $\lim_{t \searrow 0} \frac{g_k(t)}{t} = \sum_{i=1}^{k} \lambda_i(A+B)$. Hence, (5.5.11) holds. Theorem 5.5.7 yields equality in (5.5.11) for $k = n$. Hence, (5.5.11) is equivalent to

$$\boldsymbol{\lambda}(A + B) \prec \log \boldsymbol{\lambda}(e^A e^B). \qquad (5.5.13)$$

Apply the convex function e^x to this relation to deduce (5.5.12).

We now show that equality holds in (5.5.12) if and only if $AB = BA$. Clearly if $AB = BA$ then $e^{tA} e^{tB} = e^{t(A+B)}$, hence we have equality in (5.5.12).

It is left to show the claim that equality in (5.5.12) implies that A and B commutes. Since e^x is strictly convex it follows that equality in (5.5.12) yields equalities in (5.5.11). That is, $\boldsymbol{\lambda}(A + B) = \log \boldsymbol{\lambda}(e^A e^B)$. In particular, $\lambda_1(A + B) = \log \lambda_1(e^A e^B)$. Hence, $\frac{g_k(t)}{t}$ is a constant function on $(0, 1]$ for $k \in [n]$. Consider first the equality

$2g_1(\frac{1}{2}) = g_1(1)$:

$$\lambda_1(e^{\frac{1}{2}A}e^{\frac{1}{2}B})^2 = \lambda_1(e^A e^B).$$

Recall that

$$\lambda_1(e^{\frac{1}{2}A}e^{\frac{1}{2}B})^2 \leq \sigma_1(e^{\frac{1}{2}A}e^{\frac{1}{2}B})^2 = \lambda_1(e^{\frac{1}{2}A}e^B e^{\frac{1}{2}A}) = \lambda_1(e^A e^B).$$

Hence, we have the equality $\lambda_1(e^{\frac{1}{2}A}e^{\frac{1}{2}B}) = \sigma_1(e^{\frac{1}{2}A}e^{\frac{1}{2}B})$. Similarly we conclude the equality

$$\prod_{i=1}^{k} \lambda_i(e^{\frac{1}{2}A}e^{\frac{1}{2}B}) = \prod_{i=1}^{k} \sigma_i(e^{\frac{1}{2}A}e^{\frac{1}{2}B}) \qquad (5.5.14)$$

for each $k \in [n]$. Therefore $\lambda_k(e^{\frac{1}{2}A}e^{\frac{1}{2}B}) = \sigma_k(e^{\frac{1}{2}A}e^{\frac{1}{2}B})$ for each $k \in [n]$. Theorem 4.11.12 yields that $e^{\frac{1}{2}A}e^{\frac{1}{2}B}$ is a normal matrix. Hence

$$e^{\frac{1}{2}A}e^{\frac{1}{2}B} = (e^{\frac{1}{2}A}e^{\frac{1}{2}B})^* = e^{\frac{1}{2}B^*}e^{\frac{1}{2}A^*} = e^{\frac{1}{2}A}e^{\frac{1}{2}B}.$$

So $e^{\frac{1}{2}A}$ and $e^{\frac{1}{2}B}$ commute. As both matrices are hermitian it follows that there exists an orthonormal basis $\mathbf{x}_1, \ldots, \mathbf{x}_n$ which are eigenvectors of $e^{\frac{1}{2}A}$ and $e^{\frac{1}{2}B}$. Clearly, $\mathbf{x}_1, \ldots, \mathbf{x}_n$ are eigenvectors of A and B. Hence, $AB = BA$.

Assume now that equality holds in (5.5.11) for $k = 1$. Hence, (5.5.14) holds for $k = 1$. Let $F := e^{\frac{1}{2}A}, G := e^{\frac{1}{2}B}$. Let $\mathbf{U}, \mathbf{V} \subset \mathbb{C}^n$ be the invariant subspaces of FG and GF corresponding to $\lambda_1(FG) = \lambda_1(GF)$ respectively. Since $F, G \in \mathbf{GL}(n, \mathbb{C})$ it follows that $G\mathbf{U} = \mathbf{V}$, $F\mathbf{V} = \mathbf{U}$. Assume that $FG\mathbf{x} = \lambda_1(FG)\mathbf{x}$, where $\|\mathbf{x}\|_2 = 1$. So

$$\lambda_1(FG) = \|FG\mathbf{x}\|_2 \leq \sigma_1(FG) = \lambda_1(FG).$$

The arguments of the proof of Theorem 4.11.12 yield that $GF\mathbf{x} = (FG)^*\mathbf{x} = \lambda_1(FG)\mathbf{x}$. Hence, $\mathbf{U} = \mathbf{V}$. So $F\mathbf{U} = G\mathbf{U} = \mathbf{U}$. Let A_1, B_1, F_1, G_1, I_1 be the restrictions of A, B, F, G, I_n to $\mathbf{U}$. Note that $F_1G_1 = G_1F_1 = \lambda_1(FG)I_1$. F_1 and G_1 commute. Since F_1, G_1 are hermitian, it follows that $\mathbf{U}$ has an orthonormal basis $\mathbf{x}_1, \ldots, \mathbf{x}_m$

consisting of eigenvectors of F_1, G_1 respectively. Hence $\mathbf{x}_1, \ldots, \mathbf{x}_m$ are eigenvectors of A_1, B_1 respectively. Assume that

$$A\mathbf{x}_i = \alpha_i \mathbf{x}_i, \quad B\mathbf{x}_i = \beta_i \mathbf{x}_i, \quad i = 1, \ldots, m.$$

As $F_1 G_1 = \lambda_1(FG) I_1$ it follows that $\alpha_i + \beta_i = \gamma$ for $i = 1, \ldots, m$. Furthermore, $e^A e^B|_{\mathbf{U}}$ is $e^\gamma I_1$. So $g_1(1) = \gamma = \lambda_1(A + B)$ and $\lambda_1(e^A e^B) = e^\gamma$. Let $\mathbf{W} = \text{span}(\mathbf{x}_1)$. So $\mathbf{W}$ is an invariant one-dimensional subspace of A and B, such that it corresponds to the first eigenvalue of $A + B$ and $e^A e^B$.

Assume now that equality holds in (5.5.11) for $k \in [n-1] \setminus \{1\}$. As we showed above, this equality is equivalent to the equality (5.5.14). Let $A_2 := A_{\wedge k}, B_2 := B_{\wedge k}$. Then (5.5.14) is equivalent to $\lambda_1(e^{\frac{1}{2}A_2} e^{\frac{1}{2}B_2}) = \sigma_1(e^{\frac{1}{2}A_2} e^{\frac{1}{2}B_2})$. Our results for the case $k = 1$ imply that A_2 and B_2 have a common eigenvector in $\mathbb{C}^{\binom{n}{k}}$, which corresponds to $\lambda_1(A_2 + B_2)$ and $\lambda_1(e^{A_2} e^{B_2})$ respectively. This statement is equivalent to the existence of a common invariant subspace $\mathbf{W} \subset \mathbb{C}^n$ of A and B of dimension k which is spanned by the first k-eigenvectors of $A + B$ and $e^A e^B$ respectively.

Suppose finally that there exists a common invariant subspace $\mathbf{W} \subset \mathbb{C}^n$ of A and B of dimension k which is spanned by the first k-eigenvectors of $A + B$ and $e^A e^B$ respectively. Let A_1, B_1 be the restrictions of A, B to $\mathbf{W}$ respectively. Theorem 5.5.7 implies that $\text{tr}(A_1 + B_1) = \sum_{i=1}^k \log \lambda_i(e^{A_1} e^{B_1})$. As $\mathbf{W}$ is spanned by the first k-eigenvectors of $A + B$ and $e^A e^B$ respectively, it follows that

$$\text{tr}(A_1 + B_1) = \sum_{i=1}^k \lambda_i(A + B), \quad \sum_{i=1}^k \log \lambda_i(e^{A_1} e^{B_1}) = \sum_{i=1}^k \log \lambda_i(e^A e^B).$$

The proof of the theorem is completed. $\square$

Let

$$C(t) := \frac{1}{t} \log e^{\frac{1}{2}tA} e^{tB} e^{\frac{1}{2}tA} \in \mathbf{H}_n, \quad t \in \mathbb{R} \setminus \{0\}. \qquad (5.5.15)$$

$tC(t)$ is the unique hermitian logarithm of a positive definite hermitian matrix $e^{\frac{1}{2}tA} e^{tB} e^{\frac{1}{2}tA}$, which is similar to $e^{tA} e^{tB}$. Proposition

5.5.1 yields

$$\lim_{t \to 0} C(t) = C(0) := A + B. \tag{5.5.16}$$

(See Problem 11.) In what follows we give a complementary formula to (5.5.16).

Theorem 5.5.9 *Let $A, B \in \mathbf{H}_n$ and assume that $C(t)$ is be the hermitian matrix defined as (5.5.15). Then $\sum_{i=1}^{k} \lambda_i(C(t))$ are non-decreasing functions on $[0, \infty)$ for $k = 1, \ldots, n$ satisfying*

$$\boldsymbol{\lambda}(C(t)) \prec \boldsymbol{\lambda}(A) + \boldsymbol{\lambda}(B). \tag{5.5.17}$$

Moreover, there exists $C \in \mathbf{H}_n$ such that

$$\lim_{t \to \infty} C(t) = C, \tag{5.5.18}$$

and C commutes with A. Furthermore, there exist two permutations ϕ, ψ on $\{1, \ldots, n\}$ such that

$$\lambda_i(C) = \lambda_{\phi(i)}(A) + \lambda_{\psi(i)}(B), \quad i = 1, \ldots, n. \tag{5.5.19}$$

Proof. Assume that $t > 0$ and let $\lambda_i(t) = e^{t\lambda_i(C(t))}, i = 1, \ldots, n$ be the eigenvalues of $G(t) := e^{\frac{1}{2}tA}e^{tB}e^{\frac{1}{2}tA}$. Clearly

$$\lambda_1(t) = \|e^{\frac{1}{2}tA}e^{tB}e^{\frac{1}{2}tA}\|_2 \le \|e^{\frac{1}{2}tA}\|_2\|e^{tB}\|_2\|e^{\frac{1}{2}tA}\|_2 = e^{t(\lambda_1(A)+\lambda_1(B))}.$$

By considering $\wedge^k G(t)$ we deduce

$$\prod_{i=1}^{k} \lambda_i(t) \le e^{t\sum_{i=1}^{k} \lambda_i(A)+\lambda_i(B)}, \quad k = 1, \ldots, n, \quad t > 0.$$

Theorem 5.5.7 implies that for $k = n$ equality holds. Hence, (5.5.17) holds. Let $g_k(t)$ be defined as in the proof of Theorem 5.5.8. Clearly $\frac{g_k(t)}{t} = \sum_{i=1}^{k} \lambda_i(C(t))$. Since $\frac{g_k(t)}{t}$ is nondecreasing we deduce that $\sum_{i=1}^{k} \lambda_i(C(t))$ is nondecreasing on $[0, \infty)$. Furthermore (5.5.17) shows that $\frac{g_k(t)}{t}$ is bounded. Hence, $\lim_{t\to\infty} \frac{g_k(t)}{t}$ exists for each $k = 1, \ldots, n$, which is equivalent to

$$\lim_{t\to\infty} \lambda_i(C(t)) = \omega_i, \quad i = 1, \ldots, n. \qquad (5.5.20)$$

Let

$$\omega_1 = \cdots = \omega_{n_1} > \omega_{n_1+1} = \cdots = \omega_{n_2} > \cdots > \omega_{n_{l-1}+1}$$

$$= \cdots = \omega_{n_r}, \quad n_0 = 0 < n_1 < \cdots < n_r = n. \qquad (5.5.21)$$

Let $\omega_0 := \omega_1 + 1, \omega_{n+1} = \omega_n - 1$. Hence, for $t > T$ the open interval $(\frac{\omega_i+\omega_{i+1}}{2}, \frac{\omega_{i-1}+\omega_i}{2})$ contains exactly $n_i - n_{i-1}$ eigenvalues of $C(t)$ for $i = 1, \ldots, r$. In what follows we assume that $t > T$. Let $P_i(t) \in \mathbf{H}_n$ be the orthogonal projection on the eigenspace of $C(t)$ corresponding to the eigenvalues $\lambda_{n_{i-1}+1}(C(t)), \ldots, \lambda_{n_i}(C(t))$ for $i = 1, \ldots, r$. Observe that $P_i(t)$ is the orthogonal projection on the eigenspace of $G(t)$ the eigenspace corresponding to the eigenvalues $\lambda_{n_{i-1}+1}(t), \ldots, \lambda_{n_i}(t)$ for $i = 1, \ldots, r$. The equality (5.5.18) is equivalent to

$$\lim_{t\to\infty} P_i(t) = P_i, \quad i = 1, \ldots, r. \qquad (5.5.22)$$

The claim that $CA = AC$ is equivalent to the claim that $AP_i = P_iA$ for $i = 1, \ldots, r$.

We first show these claims for $i = 1$. Assume that the eigenvalues of A and B are of the form

$$\lambda_1(A) = \ldots = \lambda_{l_1}(A) = \alpha_1 > \lambda_{l_1+1}(A) = \cdots$$

$$= \lambda_{l_2}(A) = \alpha_2 > \ldots > \lambda_{l_{p-1}+1}(A) = \cdots = \lambda_{l_p}(A) = \alpha_p,$$

$$\lambda_1(B) = \ldots = \lambda_{m_1}(B) = \beta_1 > \lambda_{m_1+1}(B) = \cdots = \lambda_{m_2}(B)$$

$$= \beta_2 > \ldots > \lambda_{m_{q-1}+1}(B) = \cdots = \lambda_{m_q}(B) = \beta_q,$$

$$l_0 = 0 < l_1 < \cdots < l_p = n, \quad m_0 = 0 < m_1 < \cdots < m_q = n.$$

$$(5.5.23)$$

Note that if either $p = 1$ or $q = 1$, i.e. either A or B is of the form aI, then the theorem trivially holds. Assume that $p, q > 1$.

Let Q_i, R_j be the orthogonal projections on the eigenspaces of A and B corresponding to the eigenvalues α_i and β_j respectively, for $i = 1, \ldots, p, j = 1, \ldots, q$. So

$$e^{\frac{1}{2}At} = \sum_{i=1}^{p} e^{\frac{1}{2}\alpha_i t} Q_i, \quad e^{Bt} = \sum_{j=1}^{q} e^{\beta_j t} R_j,$$

$$G(t) = \sum_{i_1=i_2=j=1}^{p,p,q} e^{\frac{1}{2}(\alpha_{i_1}+\alpha_{i_2}+\beta_j)t} Q_{i_1} R_j Q_{i_2}.$$

Observe next that

$$\mathcal{K} := \{(i,j) \in \{1,\ldots,p\} \times \{1,\ldots,q\}, \ Q_i R_j \neq 0\},$$

$$I = \left(\sum_{i=1}^{p} Q_i\right)\left(\sum_{j=1}^{q} R_j\right) = \sum_{i,j=1}^{p,q} Q_i R_j = \sum_{(i,j)\in\mathcal{K}} Q_i R_j,$$

$$\text{rank } Q_i R_j = \text{rank } (Q_i R_j)^* = \text{rank } R_j Q_i = \text{rank } (Q_i R_j)(Q_i R_j)^*$$
$$= \text{rank } Q_i R_j^2 Q_i, \ Q_{i_1} R_j Q_{i_2} = (Q_{i_1} R_j)(R_j Q_{i_2}) \neq 0$$
$$\Rightarrow Q_{i_1} R_j Q_{i_1} \neq 0, Q_{i_2} R_j Q_{i_2} \neq 0.$$

(See Problem 14.) Let

$$\gamma_1 := \max_{(i,j)\in\mathcal{K}} \alpha_i + \beta_j, \ \mathcal{K}_1 := \{(i,j) \in \mathcal{K}, \alpha_i + \beta_j = \gamma_1\},$$

$$n_1' = \sum_{(i,j)\in\mathcal{K}_1} \text{rank } Q_i R_j Q_i, \ \gamma_1' = \max_{(i,j)\in\mathcal{K}\setminus\mathcal{K}_1} \alpha_i + \beta_j. \quad (5.5.24)$$

From the above equalities we deduce that $\mathcal{K}_1 \neq \emptyset$. Assume that $(i,j), (i',j') \in \mathcal{K}_1$ are distinct pairs. From the maximality of γ_1 and the definition of $\mathcal{K}_1$ it follows that $i \neq i', j \neq j'$. Hence, $Q_i R_j Q_i (Q_{i'} R_{j'} Q_{i'}) = 0$. Furthermore γ_1' is well defined and $\gamma_1' < \gamma_1$.

Let

$$D_1(t) := \sum_{(i,j) \in \mathcal{K} \setminus \mathcal{K}_1} e^{(\alpha_i + \beta_j - \gamma_1)t} Q_i R_j Q_i$$

$$+ \sum_{(i_1,j),(i_2,j) \in \mathcal{K}, i_1 \neq i_2} e^{\frac{1}{2}(\alpha_{i_1} + \alpha_{i_2} + 2\beta_j - 2\gamma_1)t} Q_{i_1} R_j Q_{i_2}.$$

$$D = \sum_{(i,j) \in \mathcal{K}_1} Q_i R_j Q_i, \quad D(t) = D + D_1(t). \qquad (5.5.25)$$

Then $n_1' = \text{rank } D$. (See Problem 15(b).) We claim that

$$\omega_1 = \gamma_1, \quad n_1 = n_1'. \qquad (5.5.26)$$

From the above equalities and definitions we deduce $G(t) = e^{\gamma_1 t} D(t)$. Hence, $\lambda_i(t) = e^{\gamma_1 t} \lambda_i(D(t))$. As each term $e^{\frac{1}{2}(\alpha_{i_1} + \alpha_{i_2} + 2\beta_j - 2\gamma_1)t}$ appearing in $D_1(t)$ is bounded above by $e^{-\frac{1}{2}(\gamma_1 - \gamma_1')t}$ we deduce $D_1(t) = e^{\frac{1}{2}(\gamma_1' - \gamma_1)t} D_2(t)$ and $\leq \|D_2(t)\|_2 \leq K$. Hence, $\lim_{t \to \infty} D(t) = D$. Since rank $D = n_1'$ we deduce that we have $\frac{1}{2}\lambda_i(D) < \lambda_i(D(t)) \leq 2\lambda_i(D)$ for $i = 1, \ldots, n_1'$. If $n_1' < n$ then from Theorem 4.4.6 we obtain that

$$\lambda_i(D(t)) = \lambda_i(D + D_1(t)) \leq \lambda_i(D) + \lambda_1(D_1(t))$$

$$= \lambda_1(D_1(t)) \leq e^{\frac{1}{2}(\gamma_1' - \gamma_1)t} K,$$

for $i = n_1 + 1, \ldots, n$. Hence

$$\omega_i = \gamma_1, \ i = 1, \ldots, n_1', \quad \omega_i \leq \frac{1}{2}(\gamma + \gamma_1'), \ i = n_1' + 1, \ldots, n,$$

which shows (5.5.26). Furthermore $\lim_{t \to \infty} P_1(t) = P_1$, where P_1 is the projection on $D\mathbb{C}^n$. Since $Q_i Q_{i'} = \delta_{ii'} Q_i$ it follows that $Q_{i'} D = D Q_{i'}$ for $i' = 1, \ldots, p$. Hence, $AD = DA \Rightarrow AP_1 = P_1 A$. Furthermore $P_1 \mathbb{C}^n$ is a direct sum of the orthogonal subspaces $Q_i P_j Q_j \mathbb{C}^n, (i,j) \in \mathcal{K}_1$, which are the eigen-subspaces of A corresponding to $\lambda_i(A)$ for $(i,j) \in \mathcal{K}_1$.

We now define partially the permutations ϕ, ψ. Assume that $\mathcal{K}_1 = \{(i_1, j_1), \ldots, (i_o, j_o)\}$. Then $\omega_1 = \gamma_1 = \alpha_{i_k} + \beta_{j_k}$ for $k = 1,$

$\ldots, o$. Let $e_0 = 0$ and $e_k = e_{k-1} + \text{rank } Q_{i_k} R_{j_k} Q_{i_k}$ for $k = 1, \ldots, o$. Note that $e_o = n'_1$. Define

$$\phi(s) = l_{i_k-1} + s - e_{k-1}, \quad \psi(s) = m_{j_k-1} + s - e_{k-1},$$

$$\text{for } s = e_{k-1} + 1, \ldots, e_k, \quad k = 1, \ldots, o. \tag{5.5.27}$$

Then $\omega_1 = \omega_s = \lambda_{\phi(s)}(A) + \lambda_{\psi(s)}(B)$ for $s = 1, \ldots, n_1$.

Next we consider the matrix

$$G_2 = \wedge^{n_1+1} G(t) = \wedge^{n_1+1} e^{\frac{1}{2}At} e^{Bt} e^{\frac{1}{2}At}$$

$$= e^{\frac{1}{2}A_{\wedge^{n_1+1}}t} e^{B_{\wedge^{n_1+1}}t} e^{\frac{1}{2}A_{\wedge^{n_1+1}}t}.$$

So $\lambda_1(G_2(t)) = \prod_{i=1}^{n_1+1} \lambda_i(t)$ and more generally all the eigenvalues of $G_2(t)$ are of the form

$$\prod_{i=1}^{n_1+1} \lambda_{j_1}(t) \ldots \lambda_{j_{n_1+1}}(t), \ 1 \le j_1 < j_2 < \cdots < j_{n_1+1} \le n.$$

Since we already showed that $\lim_{t \to \infty} \frac{\log \lambda_i(t)}{t} = \omega_i$ for $i = 1, \ldots n$ we deduce that

$$\lim_{t \to \infty} \left(\prod_{i=1}^{n_1+1} \lambda_{j_1}(t) \ldots \lambda_{j_{n_1+1}}(t) \right)^{\frac{1}{t}} = e^{\sum_{i=1}^{n_1+1} \omega_{j_i}}.$$

Hence, all the eigenvalues of $G_2(t)^{\frac{1}{t}}$ converge to the above values for all choices of $1 \le j_1 < j_2 < \cdots < j_{n_1+1} \le n$. The limit of the maximal eigenvalue of $G_2(t)^{\frac{1}{t}}$ is equal to $e^{\omega_1 + \cdots + \omega_{n_1} + \omega_i}$ for $i = n_1+1, \ldots, n_2$, which is of multiplicity $n_2 - n_1$. Let $P_{2,1}(t)$ be the projection on the eigenspace of $G_2(t)$ spanned by the first $n_2 - n_1$ eigenvalues of $G_2(t)$. Our results for $G(t)$ yield that $\lim_{t \to \infty} P_{2,1}(t) = P_{2,1}$, where $P_{2,1}$ is the projection on a direct sum of eigen-subspaces of $A_{\wedge^{n_1+1}}$. Let $\mathbf{W}_2(t) = P_1(t)\mathbb{C}^n + P_2(t)\mathbb{C}^n$ be a subspace of dimension n_2 spanned by the eigenvectors of $G(t)$ corresponding to $\lambda_1(t), \ldots, \lambda_{n_2}(t)$. Then $P_{2,1}(t) \wedge^{n_1+1} \mathbb{C}^n$ is the the subspace of the form $(\wedge^{n_1} P_1(t)\mathbb{C}^n) \wedge (P_2(t)\mathbb{C}^n)$. Since $\lim_{t \to \infty} P_1(t)\mathbb{C}^n = P_1\mathbb{C}^n$ and $\lim_{t \to \infty} P_{2,1}(t) \wedge^{n_1+1} \mathbb{C}^n = P_{2,1} \wedge^{n_1+1} \mathbb{C}^n$ we deduce that $\lim_{t \to \infty} P_2(t)\mathbb{C}^n = \mathbf{W}_2$ for some subspace of dimension $n_2 - n_1$ which is orthogonal to $P_1\mathbb{C}^n$. Let P_2 be the orthogonal projection on $\mathbf{W}_2$. Hence, $\lim_{t\to\infty} P_2(t) = P_2$. (See for details Problem 12.)

We now show that that there exists two permutations ϕ, ψ on $\{1,\ldots,n\}$ satisfying (5.5.27) such that $\omega_i = \alpha_{\phi(i)} + \beta_{\psi(i)}$ for $i = n_1 + 1, \ldots, n_2$. Furthermore $AP_2 = P_2A$. To do that we need to apply carefully our results for $\omega_1, \ldots, \omega_{n_1}$. The logarithm of the first $n_2 - n_1$ limit eigenvalues of $G_2(t)^{\frac{1}{t}}$ has to be of the form $\lambda_a(A_{\wedge n_1+1})) + \lambda_b(B_{\wedge n_1+1})$. The values of indices a and b can be identified as follows. Recall that the indices $\phi(i), i = 1, \ldots, n_1$ in $\omega_i = \lambda_{\phi(i)}(A) + \lambda_{\psi(i)}(B)$ can be determined from the projection P_1, where P_1 is viewed as the sum of the projections on the orthogonal eigen-subspaces $Q_iR_j\mathbb{C}^n, (i,j) \in \mathcal{K}_1$. Recall that $P_{2,1} \wedge^{n_1+1} \mathbb{C}^n$ is of the from $\wedge^{n_1}(P_1\mathbb{C}^n) \wedge (P_2\mathbb{C}^n)$. Since $P_2\mathbb{C}^n$ is orthogonal to $P_1\mathbb{C}^n$ and $\wedge^{n_1}(P_1\mathbb{C}^n) \wedge (P_2\mathbb{C}^n)$ is an invariant subspace of $A_{\wedge n_1+1}$ it follows that $P_2\mathbb{C}^n$ is an invariant subspace of A. It is spanned by eigenvectors of A, which are orthogonal to $P_1\mathbb{C}^n$ spanned by the eigenvectors corresponding $\lambda_{\phi(i)}(A), i = 1, \ldots, n_1$. Hence, the eigenvalues of the eigenvectors spanning $P_2\mathbb{C}^n$ are of the form $\lambda_k(A)$ for $k \in \mathcal{I}_2$, where $\mathcal{I}_2 \subset \{1,\ldots,n\}\backslash\{\phi(1),\ldots,\phi(n_1)\}$ is a set of cardinality $n_2 - n_1$. Therefore, $P_2Q_i = Q_iP_2, i = 1, \ldots, p$, which implies that $P_2A = AP_2$.

Note that $\lambda_a((A_{\wedge n_1+1})) = \sum_{j=1}^{n_1}\lambda_{\phi(j)}(A) + \lambda_k(A)$ for $k \in \mathcal{I}_2$. Since $G_1(t) = e^{\frac{1}{2}At}e^{Bt}e^{\frac{1}{2}At}$ is similar to the matrix $H_1(t) := e^{\frac{1}{2}Bt}$ $e^{At}e^{\frac{1}{2}At}$ we can apply the same arguments of $H_2(t) := \wedge^{n_1+1}H_1(t)$. We conclude that there exists a set $\mathcal{J}_2 \subset \{1,\ldots,n\}\backslash\{\psi(1),$ $\ldots, \psi(n_1)\}$, which is a set of cardinality $n_2 - n_1$ such that $\lambda_b((B_{\wedge n_1+1})) = \sum_{j=1}^{n_1}\lambda_{\psi(j)}(B) + \lambda_{k'}(B)$ for $k' \in \mathcal{J}_2$. Hence, the logarithm of the limit value of the largest eigenvalue of $G_2(t)^{\frac{1}{t}}$ which is equal to $\omega_1 + \cdots + \omega_{n_1} + \omega_{n_1+1}$ is given by $n_2 - n_1$ the sum of the pairs $\lambda_a(A_{\wedge n_1+1})) + \lambda_b(B_{\wedge n_1+1})$. The pairing (a,b) induces the pairing (k,k') in $\mathcal{I}_2 \times \mathcal{J}_2$. Choose any permutation ϕ such that $\phi(1),\ldots,\phi(n_1)$ defined as above and $\{\phi(n_1 + 1),\ldots\phi(n_2)\} = \mathcal{I}_2$. We deduce the existence of a permutation ψ, where $\psi(1), \ldots, \psi(n_1)$ be defined as above, $\{\psi(n_1 + 1),\ldots,\psi(n_2)\} = \mathcal{J}_2$, and $(\phi(i), \psi(i))$ is the pairing (k,k') for $i = n_1 + 1, \ldots, n_2$. This shows that

$\omega_i = \lambda_{\phi(i)}(A) + \lambda_{\psi(i)}(B)$ for $i = n_1 + 1, \ldots, n_2$. By considering the matrices $\wedge^{n_i+1} G(t)$ for $i = 2, \ldots, r$ we deduce the theorem. $\square$

Problems

1. Prove Proposition 5.5.3. (*Hint*: Show that the left-hand side of (5.5.7) is one parameter group in t with the generator $(A_1, \ldots, A_k)_\otimes$.)

2. Let the assumptions of Proposition 5.5.3 hold. Assume that $\boldsymbol{\lambda}(A_i) = (\lambda_1(A_i), \ldots, \lambda_{n_i}(A_i))$ for $i = 1, \ldots, k$. Show that the $n_1, \ldots, n_k$ eigenvalues of $(A_1, \ldots, A_k)_\otimes$ are of the form $\sum_{j=1}^{k} \lambda_{i_j}(A_i)$, where $j_i = 1, \ldots, n_i, i = 1, \ldots k$.

   (*Hint*: Recall that the eigenvalues of $\otimes_{i=1}^{k} e^{tA_i}$ are of the form $\prod_{i=1}^{k} e^{t\lambda_{j_i}(A_i)}$.)

3. Let the assumptions of Definition 5.5.4 hold for $\mathbb{F} = \mathbb{C}$. Assume that $\boldsymbol{\lambda}(A) = (\lambda_1, \ldots, \lambda_n)$. Show that the $\binom{n}{k}$ eigenvalues of $A_{\wedge^k}$ are $\lambda_{i_1} + \cdots + \lambda_{i_k}$ for all $1 \le i_1 < \cdots < i_k \le n$.

4. Let $A, B \in \mathbb{C}^{n \times n}$.

   (a) Show that $e^A e^B$ is similar to $e^{\frac{1}{2}A} e^B e^{\frac{1}{2}A}$.

   (b) Show that if $A, B \in \mathbf{H}_n$ then all the eigenvalues of $e^A e^B$ are real and positive.

5. Let $g \in C[a, b]$. Show that the following are equivalent

   (a) $g(\frac{1}{2}(x_1 + x_2)) \le \frac{1}{2}(g(x_1) + g(x_2))$ for all $x_1, x_2 \in [a, b]$.

   (b) $g(t_1 x_1 + t_2 x_2) \le t_1 g(x_1) + t_2 g(x_2)$ for all $t_1, t_2 \in [0, 1], t_1 + t_2 = 1$ and $x_1, x_2 \in [a, b]$.

   Hint: Fix $x_1, x_2 \in [a, b]$. First show that (a)$\Rightarrow$(b) for any $t_1, t_2 \in [0, 1]$ which have finite binary expansions. Use the continuity to deduce that (a)$\Rightarrow$(b).

6. (a) Let $\mathbf{D}_n(\mathbb{R}) \subset \mathbf{H}_n$ be the subspace of diagonal matrices. Show that $\mathbf{D}_n(\mathbb{R})$ is a maximal commuting subspace.

(b) Let $\mathbf{U} \subset \mathbf{H}_n$ be a commuting subspace. Show that there exists a unitary matrix $U \in \mathbf{U}(n)$ such that $\mathbf{U}$ is a subspace of a maximal commuting subspace $U\mathbf{D}_n(\mathbb{R})U^*$.

(c) Show that a commuting subspace $\mathbf{U} \subset \mathbf{H}_n$ is maximal if and only if $\mathbf{U}$ contains A with n distinct eigenvalues.

(d) Let $\mathbf{U} \subset \mathbf{H}_n$ be a commuting subspace. Show that for each $k \in [n]$ the subspace $\mathbf{U}_k := \mathrm{span}\,(A_{\wedge k} : A \in \mathbf{U})$ is a commuting subspace of $\mathbf{H}_{\binom{n}{k}}$.

7. Let $A, B \in \mathbf{H}_n$ and assume that $f_k(A, B)$ is defined as in (5.5.8). Prove or give a counterexample to the following claim: The function $f_k : \mathbf{H}_n \times \mathbf{H}_n \to \mathbb{R}$ is a convex function for $k = 1, \ldots, n-1$. (**We suspect that the claim is false.**)

8. Let $g : [0, \infty) \to \mathbb{R}$ be a continuous convex function. Show that if $g(0) = 0$ the the function $\frac{g(t)}{t}$ nondecreasing on $(0, \infty)$. (*Hint:* Observe that $g(x) \le \frac{x}{y}g(y) + (1 - \frac{x}{y})g(0)$ for any $0 < x < y$.)

9. (a) Show that one can assume in (5.5.5) that $t \in \mathbb{C}$.

(b) Show that if $A, B \in \mathbf{H}_n$ then $\lim_{t \searrow 0} \frac{1}{t} \log \lambda_i(e^{tA}e^{tB}) = \lambda_i(A + B)$ for $i = 1, \ldots, n$.

10. Let $A, B \in \mathbf{H}_n$. Prove that $\mathrm{tr}\, e^{A+B} = \mathrm{tr}\, e^A e^B$ implies that $AB = BA$ using the conditions for equalities in (5.5.11) stated in the last part of Theorem 5.5.8.

11. Let $C(t)$ be defined by (5.5.15).

(a) Show that $C(-t) = C(t)$ for any $t \ne 0$.

(b) Show the equality (5.5.16).

12. Let $\mathbf{V}$ be an n-dimensional inner product space over $\mathbb{F} = \mathbb{R}, \mathbb{C}$, with the inner product $\langle \cdot, \cdot \rangle$. Let

$$S(\mathbf{U}) := \{\mathbf{u} \in \mathbf{U}, \langle \mathbf{u}, \mathbf{u} \rangle = 1\}, \quad \mathbf{U} \text{ is a subspace of } \mathbf{V} \tag{5.5.28}$$

be the unit sphere in $\mathbf{U}$. $(S(\{\mathbf{0}\}) = \emptyset.)$ For two subspaces of $\mathbf{U}, \mathbf{W} \subseteq \mathbf{V}$ the distance $\mathrm{dist}(\mathbf{U}, \mathbf{V})$ is defined to be the Hausdorff distance between the unit spheres in $\mathbf{U}, \mathbf{W}$:

$$\mathrm{dist}(\mathbf{U}, \mathbf{V})$$

$$:= \max \left(\max_{\mathbf{u} \in S(\mathbf{U})} \min_{\mathbf{v} \in S(\mathbf{V})} \|\mathbf{u} - \mathbf{v}\|, \max_{\mathbf{v} \in S(\mathbf{V})} \min_{\mathbf{u} \in S(\mathbf{U})} \|\mathbf{v} - \mathbf{u}\| \right),$$

$$\mathrm{dist}(\{\mathbf{0}\}, \{\mathbf{0}\}) \tag{5.5.29}$$

$$= 0, \mathrm{dist}(\{\mathbf{0}\}, \mathbf{W}) = \mathrm{dist}(\mathbf{W}, \{\mathbf{0}\}) = 1 \text{ if } \dim \mathbf{W} \geq 1.$$

(a) Let $\dim \mathbf{U}, \dim \mathbf{V} \geq 1$. Show that $\mathrm{dist}(\mathbf{U}, \mathbf{V}) \leq 2$. Equality holds if either $\mathbf{U} \cap (\mathbf{V})^{\perp}$ or $\mathbf{U}^{\perp} \cap \mathbf{V}$ are nontrivial subspaces. In particular, $\mathrm{dist}(\mathbf{U}, \mathbf{V}) = 2$ if $\dim \mathbf{U} \neq \dim \mathbf{V}$.

(b) Show that dist is a metric on $\mathrm{Gr}(\mathbf{V}) := \cup_{m=0}^{n} \mathrm{Gr}(m, \mathbf{V})$.

(c) Show that $\mathrm{Gr}(m, \mathbf{V})$ is a compact connected space with respect to the metric $\mathrm{dist}(\cdot, \cdot)$ for $m = 0, 1 \ldots, n$. (That is, for each sequence of m-dimensional subspaces $\mathbf{U}_i, i \in \mathbb{N}$ one can choose a subsequence $i_j, j \in \mathbb{N}$ such that $\mathbf{U}_{i_j}, j \in \mathbb{N}$ converges in the metric dist to $\mathbf{U} \in \mathrm{Gr}(m, \mathbf{V})$. *Hint*: Choose an orthonormal basis in each $\mathbf{U}_i$.)

(d) Show that $\mathrm{Gr}(\mathbf{V})$ is a compact space with the metric dist.

(e) Let $\mathbf{U}, \mathbf{U}_i \in \mathrm{Gr}(m, \mathbf{V}), i \in \mathbb{N}, 1 \leq m < n$. Let $P, P_i \in \mathbf{S}(\mathbf{V})$ be the orthogonal projection on $\mathbf{U}, \mathbf{U}_i$ respectively. Show that $\lim_{i \to \infty} \mathrm{dist}(\mathbf{U}_i, \mathbf{U}) = 0$ if and only if $\lim_{i \to \infty} P_i = P$.

(f) Let $\mathbf{U}_i \in \mathrm{Gr}(m, \mathbf{V}), \mathbf{W}_i \in \mathrm{Gr}(l, \mathbf{V}), 1 \leq m, l$ and $\mathbf{U}_i \perp \mathbf{W}_i$ for $i \in \mathbb{N}$. Assume that $\lim_{i \to \infty} \mathrm{dist}(\mathbf{U}_i, \mathbf{U}) = 0$ and $\mathrm{dist}((\bigwedge^m \mathbf{U}_i) \bigwedge \mathbf{W}_i, \mathbf{X}) = 0$ for some subspaces $\mathbf{U} \in \mathrm{Gr}(m, \mathbf{V})$, $\mathbf{X} \in \mathrm{Gr}(l, \bigwedge^{m+1} \mathbf{V})$. Show that there exists $\mathbf{W} \in \mathrm{Gr}(l, \mathbf{V})$ orthogonal to $\mathbf{U}$ such that $\lim_{i \to \infty} \mathrm{dist}(\mathbf{W}_i, \mathbf{W}) = 0$.

13. Let $\mathbf{V}$ be an n-dimensional inner product space over $\mathbb{F} = \mathbb{R}, \mathbb{C}$, with the inner product $\langle \cdot, \cdot \rangle$. Let $m \in [n-1] \cap \mathbb{N}$ and assume that $\mathbf{U}, \mathbf{W} \in \mathrm{Gr}(m, \mathbf{V})$. Choose orthonormal bases $\{\mathbf{u}_1, \ldots, \mathbf{u}_m\}, \{\mathbf{w}_1, \ldots, \mathbf{w}_m\}$ in $\mathbf{U}, \mathbf{W}$ respectively. Show

(a) $\det \left(\langle \mathbf{u}_i, \mathbf{w}_j \rangle \right)_{i,j=1}^{m} = \overline{\det \left(\langle \mathbf{w}_j, \mathbf{u}_i \rangle \right)_{i,j=1}^{m}}.$

(b) Let $\mathbf{x}_1, \ldots, \mathbf{x}_m$ another orthonormal basis in $\mathbf{U}$, i.e. $\mathbf{x}_i = \sum_{k=1}^m q_{ki} \mathbf{u}_k, i = 1, \ldots, m$ where $Q = (q_{ki}) \in \mathbb{F}^{m \times m}$ is orthogonal for $\mathbb{F} = \mathbb{R}$ and unitary for $\mathbb{F} = \mathbb{C}$. Then $\det(\langle \mathbf{x}_i, \mathbf{w}_j \rangle)_{i,j=1}^m = \det Q \det(\langle \mathbf{u}_k, \mathbf{w}_j \rangle)_{k,j=1}^m$.

(c) $[\mathbf{U}, \mathbf{W}] := |\det(\langle \mathbf{u}_i, \mathbf{w}_j \rangle)_{i,j=1}^m|$ is independent of the choices of orthonormal bases in $\mathbf{U}, \mathbf{W}$. Furthermore $[\mathbf{U}, \mathbf{W}] = [\mathbf{W}, \mathbf{U}]$.

(d) Fix an orthonormal basis in $\{\mathbf{w}_1, \ldots, \mathbf{w}_m\}$ in $\mathbf{W}$. Then there exists an orthonormal basis $\{\mathbf{u}_1, \ldots, \mathbf{u}_m\}$ in $\mathbf{U}$ such that the matrix $(\langle \mathbf{u}_i, \mathbf{w}_j \rangle)_{i,j=1}^m$ is upper triangular. *Hint*: Let $\mathbf{W}_i = \mathrm{span}\,(\mathbf{w}_{i+1}, \ldots, \mathbf{w}_n)$ for $i = 1, \ldots, m-1$. Consider $\mathrm{span}\,(\mathbf{w}_1)^\perp \cap \mathbf{U}$ which has dimension $m-1$ at least. Let $\mathbf{U}_1$ be an $m-1$ dimensional subspace of $\mathrm{span}\,(\mathbf{w}_1)^\perp \cap \mathbf{U}$. Let $\mathbf{u}_1 \in S(\mathbf{U}) \cap \mathbf{U}_1^\perp$. Use $\mathbf{U}_1, \mathbf{V}_1$ to define an $m-2$ dimensional subspace $\mathbf{U}_2 \subset \mathbf{U}_1$ and $\mathbf{u}_2 \in S(\mathbf{U}_1) \cap \mathbf{U}_2^\perp$ as above. Continue in this manner to find an orthonormal basis $\{\mathbf{u}_1, \ldots, \mathbf{u}_m\}$.

(e) $[\mathbf{U}, \mathbf{W}] \leq 1$. ($[\mathbf{U}, \mathbf{W}]$ *is called the cosine of the angle between* $\mathbf{U}, \mathbf{W}$.)

(f) $[\mathbf{U}, \mathbf{V}] = 0 \iff \mathbf{U}^\perp \cap \mathbf{V} \neq \{0\} \iff \mathbf{U} \cap \mathbf{V}^\perp \neq \{0\}$. *Hint*: Use (d) and induction.

14. Let $\mathbf{V}$ be an n-dimensional inner product space over $\mathbb{F} = \mathbb{R}, \mathbb{C}$, with the inner product $\langle \cdot, \cdot \rangle$. Let $l, m \in [n] \cap \mathbb{N}$ and assume that $\mathbf{U} \in \mathrm{Gr}(l, \mathbf{V}), \mathbf{W} \in \mathrm{Gr}(m, \mathbf{V})$. Let $P, Q \in S(\mathbf{V})$ be the orthogonal projections on $\mathbf{U}, \mathbf{W}$ respectively. Show

(a) $\mathbf{U} + \mathbf{W} = \mathbf{U} \cap \mathbf{W} \oplus \mathbf{U} \cap (\mathbf{U} \cap \mathbf{W})^\perp \oplus \mathbf{W} \cap (\mathbf{U} \cap \mathbf{W})^\perp$.

(b) $\mathrm{rank}\, PQ = \mathrm{rank}\, QP = \mathrm{rank}\, PQP = \mathrm{rank}\, QPQ$.

(c) $\mathrm{rank}\, PQ = \dim \mathbf{W} - \dim \mathbf{W} \cap \mathbf{U}^\perp, \mathrm{rank}\, QP = \dim \mathbf{U} - \dim \mathbf{U} \cap \mathbf{W}^\perp$.

15. Let $\mathbf{V}$ be an n-dimensional inner product space over $\mathbb{F} = \mathbb{R}, \mathbb{C}$, with the inner product $\langle \cdot, \cdot \rangle$. Assume that $\mathbf{V} = \oplus_{i=1}^l \mathbf{U}_i = \oplus_{j=1}^m \mathbf{W}_j$ be two decompositions of $\mathbf{V}$ to nontrivial orthogonal

subspaces:

$$\dim \mathbf{U}_i = l_i - l_{i-1}, \ i = 1, \ldots, p, \ \dim \mathbf{W}_j = m_j - m_{j-1},$$
$$j = 1, \ldots, q, \quad 0 = l_0 < l_1 < \cdots < l_p = n,$$
$$0 = m_0 < m_1 < \cdots < m_q = n.$$

Let $Q_i, R_j \in S(\mathbf{V})$ be the orthogonal projections on $\mathbf{U}_i, \mathbf{W}_j$ respectively for $i = 1, \ldots, p, j = 1, \ldots, q$. Let $n_{ij} :=$ rank $Q_i R_j, i = 1, \ldots, p, j = 1, \ldots, q$.

Denote $\mathcal{K} := \{(i,j) \in \{1, \ldots, p\} \times \{1, \ldots, q\} : Q_i R_j \neq 0\}$. For $i \in \{1, \ldots, p\}, j \in \{1, \ldots, q\}$ let

$$\mathcal{J}_i := \{j' \in \{1, \ldots, q\}, \ (i, j') \in \mathcal{K}\},$$
$$\mathcal{I}_j := \{i' \in \{1, \ldots, p\}, \ (i', j) \in \mathcal{K}\}.$$

Show

(a) $Q_i = \sum_{j \in \mathcal{J}_i} Q_i R_j, \ i = 1, \ldots, p.$

(b) Let $(i_1, j_1), \ldots, (i_s, j_s) \in \mathcal{K}$ and assume that $i_a \neq i_b, j_a \neq j_b$. Then

$$\operatorname{rank} \sum_{a=1}^{s} Q_{i_a} R_{j_a} = \operatorname{rank} \left(\sum_{a=1}^{s} Q_{i_a} R_{j_a} \right) \left(\sum_{a=1}^{s} Q_{i_a} R_{j_a} \right)^*$$

$$= \operatorname{rank} \sum_{a=1}^{s} Q_{i_a} R_{j_a} Q_{i_a}$$

$$= \sum_{a=1}^{s} \operatorname{rank} Q_{i_a} R_{j_a} Q_{i_a} = \sum_{a=1}^{s} \operatorname{rank} Q_{i_a} R_{j_a}.$$

(c) rank $P_i \leq \sum_{j \in \mathcal{J}_i} n_{ij}$, where strict inequality may hold.

(d) $\mathbf{U}_i = \sum_{j \in \mathcal{J}_i} \mathbf{U}_{ij}$, where $\mathbf{U}_{ij} := P_i \mathbf{W}_j, \dim \mathbf{U}_{ij} = n_{ij}$ for $i = 1, \ldots, p, \ j = 1, \ldots, q.$

(e) $Q_j = \sum_{i \in \mathcal{I}_j} P_i Q_j, \ j = 1, \ldots, q.$

(f) rank $Q_j \leq \sum_{i \in \mathcal{I}_j} n_{ij}$, where strict inequality may hold.

(g) $\mathbf{W}_j = \sum_{i \in \mathcal{I}_j} \mathbf{W}_{ji}$, where $\mathbf{W}_{ji} = Q_j \mathbf{U}_i, \dim \mathbf{W}_{ji} = n_{ij}$ for $j = 1, \ldots, q, \ i = 1, \ldots, p.$

5.6   Historical Remarks

§5.1 and 5.2 are standard. The results of §5.3 are taken from Brualdi–Friedland–Pothen [BFP95]. The first part of §5.4 is standard. The second part of §5.4, which starts from Theorem 5.4.6, continues our results from Chapter 4. The results of §5.5 are well known to the experts. Identity 5.5.6 is called the Lie–Trotter formula [Tro59]. Theorem 5.5.7 is taken from Cohen–Friedand–Kato–Kelly [CFKK82]. Inequality (5.5.12) is called Golden–Thompson inequality [Gol57, Tho65]. See also [Pet88, So92]. Theorem 5.5.9 is taken from Friedland–So [FrS94] and Friedland–Porta [FrP04].

Chapter 6

Non-Negative Matrices

6.1 Graphs

6.1.1 Undirected graphs

An *undirected graph* is denoted by $G = (V, E)$. It consists of *vertices* $v \in V$, and edges which are *unordered* set of pairs (u, v), where $u, v \in V$, and $u \neq v$, which are called *edges* of G. u and v are called the end points of (u, v). The set of edges in G is denoted by E. Let $n = \#V$ be the cardinality of V, i.e. V has n vertices. It is useful to identify V with $[n]$. For example, the graph $G = ([4], \{(1, 2), (1, 4), (2, 3), (2, 4), (3, 4)\})$ has four vertices and five edges.

In what follows we assume that $G = (V, E)$ unless stated otherwise. A graph $H = (W, F)$ is called a subgraph of $G = (V, E)$ if W is a subset of V and any edge in F is an edge in E. Given a subset W of V then $E(W) = \{(u, v) \in E, u, v \in W\}$ is the set of edges in G *induced* by W. The graph $G(W) := (W, E(W))$ is called the subgraph *induced* by W. Given a subset F of E, then $V(F)$ is the set of vertices which are end points of F. The graph $G(F) = (V(F), F)$ is called the subgraph *induced* by F.

The *degree* of v, denoted by $\deg v$ is the number of edges that has v as its vertex. Since each edge has two different vertices

$$\sum_{v \in V} \deg v = 2\#E, \tag{6.1.1}$$

where $\#E$ is the number of edges in E. $v \in V$ is called an *isolated* vertex if $\deg v = 0$. Note that $V(E)$ is the set of nonisolated vertices in G, and $G(E) = (V(E), E)$ the subgraph of G obtained by deleting isolated vertices in G.

The complete graph on n vertices is the graph with all possible edges. It is denoted by $K_n = ([n], \mathcal{E}_n)$, where $\mathcal{E}_n = \{(1, 2), \dots, (1, n), (2, 3), \dots, (n-1, n)\}$. For example, K_3 is called a *triangle*. Note that for any graph on n vertices $G = ([n], E)$ is a subgraph of K_n, obtained by erasing some of edges in K_n, but not the vertices! That is, $E \subset \mathcal{E}_n$.

$G = (V, E)$ is called *biparite* if V is a union of two disjoint sets of vertices $V_1 \cup V_2$ so that each edge in E connects some vertex in V_1 to some vertex in E_2. Thus $E \subset V_1 \times V_2 := \{(v, w), v \in V_1, w \in V_2\}$. So any bipartite graph $D = (V_1 \cup V_2, E)$ is a subgraph of the complete bipartite graph $K_{V_1, V_2} := (V_1 \cup V_2, V_1 \times V_2)$. For positive integers l, m the complete bipartite graph on l, m vertices is denoted by $K_{l,m} := ([l] \cup [m], [l] \times [m])$. Note that $K_{l,m}$ has $l + m$ vertices and lm edges.

6.1.2   Directed graphs

A *directed graph*, abbreviated as *digraph*, is denoted by $D = (V, E)$. V is the set of vertices and E is the set of directed edges, abbreviated as diedges, in G. So E is a subset of $V \times V = \{(v, w), v, w \in V$. Thus $(v, w) \in E$ is a directed edge from v to W. For example, the graph $D = ([4], \{(1, 2), (2, 1), (2, 3), (2, 4), (3, 3), (3, 4), (4, 1)\})$ has four vertices and seven diedges.

The diedge $(v, v) \in E$ is called a *loop*, or selfloop.

$$\deg_{in} v := \#\{(w, v) \in E\}, \quad \deg_{out} v := \#\{(v, w) \in E\},$$

the number of diedges to v and out of v in D. $\deg_{in}, \deg_{out}$ are called the *in* or *out* degrees. Clearly we have the analog of (6.1.1)

$$\sum_{v \in V} \deg_{in} v = \sum_{v \in V} \deg_{out} v = \#E. \tag{6.1.2}$$

A subdigraph $H = (W, F)$ of $D = (V, E)$ and the induced subdigraphs $D(W) = (W, E(W)), D(F) = (V(F), F)$ are defined as in §6.1.1. $v \in V$ is called isolated if $\deg_{in}(v) = \deg_{out}(v) = 0$.

6.1.3 Multigraphs and multidigraphs

A *multigraph* $G = (V, E)$ has undirected edges, which may be multiple, and may have multiple loops. A multidigraph $D = (V, E)$ may have multiple diedges.

Each multidigraph $D = (V, E)$ induces an undirected multigraph $G(D) = (V, E')$, where each diedge $(u, v) \in E$ is viewed as undirected edge $(u, v) \in E'$. (Each loop $(u, u) \in E$ will appear twice in E'.) Vice versa, a multigraph $G = (V, E')$ induces a multidigraph $D(G) = (V, E)$, where each undirected edge (u, v) induces diedges (u, v) and (v, u), when $u \neq v$. The loop (u, u) appears p times in $D(G)$ if it appears p times in G.

Most of the following notions are the same for graphs, digraphs, multigraphs or multidigraphs, unless stated otherwise. We state these notions for directed multidigraphs $D = (V, E)$ mostly.

Definition 6.1.1

1. *A walk in $D = (V, E)$ is given by $v_0 v_1 \ldots v_p$, where $(v_{i-1}, v_i) \in E$ for $i = 1, \ldots, p$. One views it as a walk that starts at v_0 and ends at v_p. The* length *of the walk p, is the number of edges in the walk.*

2. *A* path *is a walk where $v_i \neq v_j$ for $i \neq j$.*

3. *A* closed *walk is walk where $v_p = v_0$.*

4. *A* cycle *is a closed walk where $v_i \neq v_j$ for $0 \leq i < j < p$. A loop $(v, v) \in E$ is considered a cycle of length 1. Note that a closed walk vwv, where $v \neq w$, is considered as a cycle of length 2 in a digraph, but not a cycle in undirected multigraph!*

5. *D is called a* diforest *if D does not have cycles. (A multigraph with no cycles is called forest.)*

6. *Let $D = (V, E)$ be a diforest. Then the height of $v \in V$, denoted by* height(v) *is the length of the longest path ending at v.*

7. *Two vertices $v, w \in V, v \neq w$ are called strongly connected if there exist two walks in D, the first starts at v and ends in w, and the second starts in w and ends in v. For multigraphs $G = (V, E)$ the corresponding notion is u, v are connected.*

8. *A multidigraph $D = ([n], E)$ is called strongly connected if either $n = 1$, or $n > 1$ and any two vertices in D are strongly connected.*

9. *A multigraph $G = (V, E)$ is called connected if either $n = 1$, or $n > 1$ and any two vertices in G are connected.*

10. *Assume that a multidigraph $D = (V, E)$ is strongly connected. Then D is called* primitive *if there exists $k \geq 1$ such that for any two vertices $u, v \in V$ there exists a walk of length k which connects u and v. For a primitive multidigraph D, the minimal such k is called the* index of primitivity, *and denoted by* indprim(D). *A strongly connected multidigraph which is not primitive is called* imprimitive.

11. *For $W \subset V$, the multidisubgraph $D(W) = (W, E(W))$ is called a* strongly connected component *of D if $D(W)$ is strongly connected, and for any $W \subsetneq U \subset V$ the induced subgraph $D(U) = (U, E(U))$ is not strongly connected.*

12. *For $W \subset V$, the subgraph $G(W) = (W, E(W))$, of multigraph $G = (V, E)$, is called a* connected component *of G if $G(W)$ is connected, and for any $W \subsetneq U \subset V$ the induced subgraph $G(U) = (U, E(U))$ is not connected.*

13. *A forest $G = (V, E)$ is called a tree if it is connected.*

14. *A diforest $D = (V, E)$ is called a ditree if the induced multigraph $G(D)$ is a tree.*

15. *Let $D = (V, E)$ be a multidigraph. The reduced (simple) digraph $D_r = (V_r, E_r)$ is defined as follows. Let $D(V_i), i = 1, \ldots, k$ be*

all strongly connected components of D. *Let* $V_0 = V \setminus (\cup_{i=1}^{k} V_i)$ *be all vertices in* D *which do not belong to any of strongly connected components of* D. *(It is possible that either* V_0 *is an empty set or* $k = 0$, *i.e.* D *does not have connected components, and the two conditions are mutually exclusive.) Then* $V_r = (\cup_{v \in V_0} \{v\}) \cup_{i=1}^{k} \{V_i\}$, *i.e.* V_r *is the set of all vertices in* V *which do not belong to any connected component and the new* k *vertices named* $\{V_1\}, \ldots, \{V_k\}$. *A vertex* $u' \in V_r$ *is viewed as either a set consisting of one vertex* $v \in V_0$ *or the set* V_i *for some* $i = 1, \ldots, k$. *Then* E_r *does not contain loops. Furthermore* $(s, t) \in E_r$, *if there exists an edge from* $(u, v) \in E$, *where* u *and* v *are in the set of vertices represented by* s *and* t *in* V, *respectively.*

16. *Two multidigraphs* $D_1 = (V_1, E_1), D_2 = (V_2, E_2)$ *are called iso-morphic if there exists a bijection* $\phi : V_1 \to V_2$ *which induces a bijection* $\hat{\phi} : E_1 \to E_2$. *That is if* $(u_1, v_1) \in E_1$ *is a diedge of multiplicity* k *in* E_1 *then* $(\phi(u_1), \phi(v_1)) \in E_2$ *is a diedge of multiplicity* k *and vice versa.*

Proposition 6.1.2 *Let* $G = (V, E)$ *be a multigraph. Then* G *is a disjoint union of its connected components. That is, there is a unique decomposition of* V *to* $\cup_{i=1}^{k} V_i$, *up to relabeling of* $V_1, \ldots, V_k$, *such that the following conditions hold:*

1. $V_1, \ldots, V_k$ *are nonempty and mutually disjoint.*

2. *Each* $G(V_i) = (V_i, E(V_i))$ *is a connected component of* G.

3. $E = \cup_{i=1}^{k} V_i$.

Proof. We introduce the following relation $\sim$ on V. First, we assume that $v \sim v$ for each $v \in V$. Second, for $v, w \in V, v \neq w$ we say that $v \sim w$ if v is connected to w. It is straightforward to show that $\sim$ is an *equivalence* relation. Let $V_1, \ldots, V_k$ be the equivalence classes in V. That is $v, w \in V_i$ if and only if v and w are connected. The rest of the proposition follows straightforward. $\square$

Proposition 6.1.3 *Let $D = (E, V)$ be a multidigraph. Then the reduced digraph D_r is a diforest.*

See Problem 6 for proof.

Proposition 6.1.4 *Let $D = (V, E)$ be a multidigraph. Then D is diforest if and only if it is isomorphic to a digraph $D_1 = ([n], E_1)$ such that if $(i, j) \in E_1$ then $i < j$.*

Proof. Clearly, D_1 cannot have a cycle. So if D is isomorphic to D_1 then D is a diforest. Assume now that $D = (V, E)$ is a diforest. Let V_i be all vertices in V having height i for $i = 0, \ldots, k \geq 0$, where k is the maximal height of all vertices in D. Observe that from the definition of height it follows that if $(u, v) \in D$, where $u \in V_i, w \in V_j$ then $i < j$. Rename the vertices of V such that $V_i = \{n_i + 1, \ldots, n_{i+1}\}$ where $0 = n_0 < n_1 < \cdots < n_{k+1} = n := \#V$. Then one obtains the isomorphic graph $D_1 = ([n], E_1)$, such that if $(i, j) \in E_1$ then $i < j$. $\qquad\qquad$ $\square$

Theorem 6.1.5 *Let $D = (V, E)$ be as strongly connected multidigraph. Assume that $\#V > 1$. Let ℓ be the g.c.d., (the greatest common divisor), of lengths of all cycles in D. Then exactly one of the following conditions hold.*

1. $\ell = 1$. *Then D is primitive. Let s be the length of the shortest cycle in D. Then* $\text{indprim}(D) \leq \#V + s(\#V - 2)$.

2. $\ell > 1$. *Then D is imprimitive. Furthermore, it is possible to divide V to ℓ disjoint nonempty subsets $V_1, \ldots, V_\ell$ such $E \subset \cup_{i=1}^{\ell} V_i \times V_{i+1}$, where $V_{\ell+1} := V_1$.*

   Define $D_i = (V_i, E_i)$ to be the following digraph. $(v, w) \in E_i$ if there is a path or cycle of length ℓ from v to w in D, for $i = 1, \ldots, l$. Then each D_i is strongly connected and primitive.

The proof of this theorem is given in §6.3. If D is a strongly connected imprimitive multidigraph, then $\ell > 1$ given in (2) is called the *index of imprimitivity* of D.

6.1.4 Matrices and graphs

Denote by $\mathbb{R}_+ \supset \mathbb{Z}_+$ the set of non-negative real numbers and non-negative integers respecively. Let $\mathcal{S} \subset \mathbb{C}$. By $S_n(\mathcal{S}) \subset \mathcal{S}^{n \times n}$ denote the set of all symmetric matrices $A = [a_{ij}], a_{ij} = a_{ji}$ with entries in $\mathcal{S}$. Assume that $0 \in \mathcal{S}$. Then by $S_{n,0}(\mathcal{S}) \subset S_n(\mathcal{S})$ the subset of all symmetric matrices with entries in $\mathcal{S}$ and zero diagonal. Denote by $\mathbf{1} = (1, \ldots, 1)^\top \in \mathbb{R}^n$ the vector of length n whose all coordinates are 1. For any $t \in \mathbb{R}$, we let $\operatorname{sign} t = 0$ if $t = 0$ and $\operatorname{sign} t = \frac{t}{|t|}$ if $t \neq 0$. For $A \in \mathbb{R}^{m \times n}$ we denote $A \geq 0, A \gneq 0, A > 0$ if A is a non-negative matrix, a non-negative nonzero matrix, and a positive matrix respectively. For $A, B \in \mathbb{R}^{m \times n}$ we denote $B \geq A, B \gneq A, B > A$ if $B - A \geq 0, B - A \gneq 0, B - A > 0$.

Let $D = (V, E)$ be a multidigraph. Assume that $\#V = n$ and label the vertices of V as $1, \ldots, n$. We have a bijection $\phi_1 : V \to [n]$. This bijection induces an isomorphic graph $D_1 = ([n], E_1)$. With D_1 we associate the following matrix $A(D_1) = [a_{ij}]_{i,j=1}^n \in \mathbb{Z}_+^{n \times n}$. Then a_{ij} is the number of directed edges from the vertex i to the vertex j. (If $a_{ij} = 0$ then there are no diedges from i to j.) When no confusion arises we let $A(D) := A(D_1)$, and we call $A(D)$ the *adjacency matrix* of D. Note that a different bijection $\phi_2 : V \to [n]$ gives rise to a different $A(D_2)$, where $A(D_2) = P^\top A(D_1) P$ for some permutation matrix $P \in \mathcal{P}_n$. See Problem 9.

If D is a simple digraph then $A(D) \in \{0, 1\}^{n \times n}$. If D is a multidigraph, then $a_{ij} \in \mathbb{Z}_+$ is the number of diedges from i to j. Hence $A(G) \in \mathbb{Z}_+^{n \times n}$. If G is a multigraph then $A(G) = A(D(G)) \in S_n(\mathbb{Z}_+)$. If G is a simple graph then $A(G) \in S_{n,0}(\{0, 1\})$.

Proposition 6.1.6 *Let $D = (V, E)$ be a multidigraph on n vertices. Let $A(D)$ be a representation matrix of D. For an integer $k \geq 1$ let $A(D)^k = [a_{ij}^{(k)}] \in \mathbb{Z}_+^{n \times n}$. Then $a_{ij}^{(k)}$ is the number of walks of length k from the vertex i to the vertex j. In particular, $\mathbf{1}^\top A(D)^k, \mathbf{1}^\top A(D)^k \mathbf{1}$ and $\operatorname{tr} A(D)^k$ are the total number of walks and the total number of closed walks of length k in D.*

Proof. For $k = 1$ the proposition is obvious. Assume that $k > 1$. Recall that

$$a_{ij}^{(k)} = \sum_{i_1,\dots,i_{k-1} \in [n]} a_{ii_1} a_{i_1 i_2} \cdots a_{i_{k-1} j}. \tag{6.1.3}$$

The summand $a_{ii_1} a_{i_1 i_2} \cdots a_{i_{k-1} j}$ gives the number of walks of the form $i_0 i_1 i_2 \dots i_{k-1} i_k$, where $i_0 = i, i_k = j$. Indeed if one of the terms in this product is zero, i.e. there is no diedge (i_p, i_{p+1}) then the product is zero. Otherwise each positive integer $a_{i_p i_{p+1}}$ counts the number of diedges (i_p, i_{p+1}). Hence, $a_{ii_1} a_{i_1 i_2} \cdots a_{i_{k-1} j}$ is the number of walks of the form $i_0 i_1 i_2 \dots i_{k-1} i_k$. The total number of walks from $i = i_0$ to $j = i_k$ of length k is the sum given by (6.1.3). To find out the total number of walks in D of length k is $\sum_{i=j=1}^{n} a_{ij}^{(k)} = \mathbf{1}^\top A(D)^k \mathbf{1}$. The total number of closed walks in D of length k is $\sum_{i=1}^{k} a_{ii}^{(k)} = \operatorname{tr} A(D)^k$. $\square$

With a multibipartite graph $G = (V_1 \cup V_2, E)$, where $\#V_1 = m, \#V_2 = n$, we associate a representation matrix $B(G) = [b_{ij}]_{i=j=1}^{m,n}$ as follows. Let $\psi_1 : V_1 \to [m], \phi_1 : V_2 \to [m]$ be bijections. This bijection induces an isomorphic graph $D_1 = ([m] \cup [n], E_1)$. Then b_{ij} is the number of edges connecting $i \in [m]$ to $j \in [n]$ in D_1.

A non-negative matrix $A = [a_{ij}]_{i=j=1}^{n} \in \mathbb{R}_+^{n \times n}$ induces the following digraph $D(A) = ([n], E)$. The diedge (i, j) is in E if and only if $a_{ij} > 0$. Note that of $A(D(A)) = [\operatorname{sign} a_{ij}] \in \{0,1\}^{n \times n}$. We have the following definitions.

Definition 6.1.7

1. $A = [a_{ij}] \in \mathbb{R}^{n \times n}$ is combinatorially symmetric *if* $\operatorname{sign} a_{ij} = \operatorname{sign} a_{ji}$ *for* $i, j = 1, \dots, n$.

2. *Assume that* $A \in \mathbb{R}_+^{n \times n}$. *Then* A *is irreducible if* $D(A)$ *is strongly connected. Otherwise* A *is called reducible.*

3. $A \in \mathbb{R}_+^{n \times n}$ *is primitive if* A^k *is a positive matrix for some integer* $k \geq 1$.

4. *Assume that $A \in \mathbb{R}_+^{n \times n}$ is primitive. Then the smallest positive integer k such that $A^k > 0$ is called the index of primitivity of A, and is denoted by* indprim(A).

5. *$A \in \mathbb{R}_+^{n \times n}$ is imprimitive if A is irreducible but not primitive.*

Proposition 6.1.8 *Let $D = ([n], E)$ be a multidigraph. Then D is strongly connected if and only if $(I + A(D))^{n-1} > 0$. In particular, a non-negative matrix $A \in \mathbb{R}_+^{n \times n}$ is irreducible if and only if $(I + A)^{n-1} > 0$.*

Proof. Apply the Newton binomial theorem for $(1+t)^{n-1}$ to the matrix $(I + A(D))^{n-1}$

$$(I + A(D))^{n-1} = \sum_{p=0}^{n-1} \binom{n-1}{p} A(D)^p.$$

Recall that all the binomial coefficients $\binom{n-1}{p}$ are positive for $p = 0, \ldots, n-1$. Assume first that $(I + A(D))^{n-1} > 0$. That is for any $i, j \in [n]$ the (i, j) entry of $(I + A(D))^{n-1}$ is positive. Hence, the (i, j) entry of $A(D)^p$ is positive for some $p = p(i, j)$. Let $i \neq j$. Since $A(D)^0 = I$, we deduce that $p(i, j) > 0$. Use Proposition 6.1.6 to deduce that there is a walk of length p from the vertex i to the vertex j.

Suppose that D is strongly connected. Then for each $i \neq j$ we must have a path of length $p \in [1, n-1]$ which connects i and j, see Problem 1. Hence, all off-diagonal entries of $(I + A(D))^{n-1}$ are positive. Clearly, $(I + A(D))^{n-1} \geq I$. Hence, $(I + A(D))^{n-1} > 0$.

Let $A \in \mathbb{R}_+^{n \times n}$. Then the (i, j) entry of $(I + A)^{n-1}$ is positive if and only if the (i, j) entry of $(I + A(D(A)))^{n-1}$ is positive. Hence, A is irreducible if and only if $(I + A)^{n-1} > 0$. $\qquad\square$

Problems

1. Assume $v_1 \ldots v_p$ is a walk in multidigraph $D = (V, E)$. Show that it is possible to subdivide this walk to walks

$v_{n_{i-1}+1} \ldots v_{n_i}$, $i = 1, \ldots, q$, where $n_0 = 0 < n_1 < \cdots < n_q = p$, and each walk is either a cycle, or a maximal path.

Erase all cycles in $v_1 \ldots v_p$ and apply the above statement to the new walk. Conclude that a walk can be "decomposed" to a union of cycles and at most one path.

2. Let D be a multidigraph. Assume that there exists a walk from v to w. Show that

    (a) if $v \neq w$ then there exists a path from v to w of length $\#V - 1$ at most;

    (b) if $v = w$ there exists a cycle which which contains v, of length $\#V$ at most.

3. Let $G = (V, E)$ be a multigraph. Show that the following are equivalent.

    (a) G is bipartite;

    (b) all cycles in G have even length.

4. Assume that $G = (V, E)$ is a connected multigraph. Show that the following are equivalent.

    (a) G is imprimitive;

    (b) G is bipartite.

5. Let $D = (V, E)$ be a multidigraph. Assume that the reduced graph D_r of D has two vertices. List all possible D_r up to the isomorphism, and describe the structure of all possible corresponding D.

6. Prove Proposition 6.1.3.

7. Let $A(D) \in \mathbb{Z}_+^{n \times n}$ be the representation matrix of the multidigraph $D = ([n], E)$. Show that $A(D) + A(D)^\top$ is the representation matrix of the multigraph $G(D) = ([n], E')$ induced by D.

8. Let $G = ([n], E')$ be an multigraph, with the representation matrix $A(G) \in S_n(\mathbb{Z}_+)$. Show that $A(G)$ is the representation matrix of the induced multidigraph $D(G)$. In particular, if G is a (simple) graph, then $D(G)$ is a (simple) digraph with no loops.

9. Let $D = (V, E), D_1 = (V_1, E_1)$ be two multidigraphs with the same number of vertices. Show that D and D_1 are isomorphic if and only if $A(D_1) = P^\top A(D)P$ for some permutation matrix.

10. Let $G = (V_1 \cup V_2, E)$ be a bipartite multigraph. Assume that $\#V_1 = m, \#V_2 = n$ and $B(G) \in \mathbb{Z}_+^{m \times n}$ is a representation matrix of G. Show that a full representation matrix of G is of the form
$$A(G) = \begin{bmatrix} 0_{m \times m} & B(G) \\ B(G)^\top & 0_{n \times n} \end{bmatrix}.$$

6.2 Perron–Frobenius Theorem

The aim of this section is to prove the Perron–Frobenius theorem.

Theorem 6.2.1 *Let $A \in \mathbb{R}_+^{n \times n}$ be an irreducible matrix. Assume that $n > 1$. Then*

1. *The spectral radius of A, $\rho(A)$, is a positive eigenvalue of A.*

2. *$\rho(A)$ is an algebraically simple eigenvalue of A.*

3. *To $\rho(A)$ corresponds a positive eigenvector $0 < \mathbf{u} \in \mathbb{R}^n$, i.e. $A\mathbf{u} = \rho(A)\mathbf{u}$. ($\mathbf{u}$ is called the Perron–Frobenius vector of A.)*

4. *All other eigenvalues of λ of A satisfy the inequality $|\lambda| < \rho(A)$ if and only if A is primitive, i.e. $A^k > 0$ for some integer $k \geq 1$.*

5. *Assume that A is imprimitive, i.e. not primitive. Then there exists exactly $h - 1 \geq 1$ distinct eigenvalues $\lambda_1, \ldots, \lambda_{h-1}$ different from $\rho(A)$ and satisfying $|\lambda_i| = \rho(A)$. Furthermore, the following conditions hold.*

    (a) *λ_i is an algebraically simple eigenvalue of A for $i = 1, \ldots, h - 1$.*

(b) *The complex numbers $\frac{\lambda_i}{\rho(A)}, i = 1, \ldots, h - 1$ and 1 are all h roots of unity, i.e. $\lambda_i = \rho(A)e^{\frac{2\pi\sqrt{-1}i}{h}}$ for $i = 1, \ldots, h - 1$. Furthermore, if $A\mathbf{z}_i = \lambda_i \mathbf{z}_i, \mathbf{z}_i \neq \mathbf{0}$ then $|\mathbf{z}_i| = \mathbf{u} > 0$, the Perron–Frobenius eigenvector $\mathbf{u}$ given in 3.*

(c) *Let ζ be any h-root of 1, i.e. $\zeta^h = 1$. Then the matrix ζA is similar to A. Hence, if λ is an eigenvalue of A then $\zeta\lambda$ is an eigenvalue of A having the same algebraic and geometric multiplicity as λ.*

(d) *There exists a permutation matrix $P \in \mathcal{P}_n$ such that $P^\top AP = B$ has a block h-circulant form*

$$B = \begin{bmatrix} 0 & B_{12} & 0 & 0 & \cdots & 0 & 0 \\ 0 & 0 & B_{23} & 0 & \cdots & 0 & 0 \\ \vdots & \vdots & \vdots & \vdots & \cdots & \vdots & \vdots \\ 0 & 0 & 0 & 0 & \vdots & 0 & B_{(h-1)h} \\ B_{h1} & 0 & 0 & 0 & \vdots & 0 & 0 \end{bmatrix},$$

$B_{i(i+1)} \in \mathbb{R}^{n_i \times n_{i+1}}, i = 1, \ldots, h, B_{h(h+1)} = B_{h1},$

$n_{h+1} = n_1, n_1 + \cdots + n_h = n.$

Furthermore, the diagonal blocks of B^h are all irreducible primitive matrices, i.e.

$$C_i := B_{i(i+1)} \cdots B_{(h-1)h} B_{h1} \cdots B_{(i-1)i}$$

$$\in \mathbb{R}_+^{n_i \times n_i}, \ i = 1, \ldots, h, \qquad (6.2.1)$$

are irreducible and primitive.

Note that the 1×1 zero matrix is irreducible. Hence, Theorem 1 excludes the case $n = 1$.

Our proof follows closely the proof of H. Wielandt [Wie50]. For a non-negative matrix $A = [a_{ij}] \in \mathbb{R}_+^{n \times n}$ define

$$r(\mathbf{x}) := \min_{i, x_i > 0} \frac{(A\mathbf{x})_i}{x_i}, \quad \text{where } \mathbf{x} = (x_1, \ldots, x_n)^\top \gneq \mathbf{0}. \qquad (6.2.2)$$

It is straightforward to show, e.g. Problem 1, that

$$r(\mathbf{x}) = \max\{s \geq 0, s\mathbf{x} \leq A\mathbf{x}\}. \tag{6.2.3}$$

Theorem 6.2.2 (*Wielandt's characterization*) *Let $n > 1$ and $A = [a_{ij}] \in \mathbb{R}_+^{n \times n}$ be irreducible. Then*

$$\max_{\mathbf{x} \geq 0} r(\mathbf{x}) = \max_{\mathbf{x} = (x_1, \ldots, x_n)^\top \geq 0} \min_{i, x_i > 0} \frac{(A\mathbf{x})_i}{x_i} = \rho(A) > 0. \tag{6.2.4}$$

The maximum in the above characterization is achieved exactly for all $\mathbf{x} > 0$ of the form $\mathbf{x} = a\mathbf{u}$, where $a > 0$ and $\mathbf{u} = (u_1, \ldots, u_n)^\top > 0$ is the unique positive probability vector satisfying $A\mathbf{u} = \rho(A)\mathbf{u}$. Moreover, $\rho(A)$ is a geometrically simple eigenvalue.

Proof. Let $r(A) := \sup_{\mathbf{x} \geq 0} r(\mathbf{x})$. So $r(A) \geq r(\mathbf{1}) = \min_i \sum_{j=1} a_{ij}$. Since an irreducible A cannot have a zero row, e.g. Problem 2, it follows that $r(A) \geq r(\mathbf{1}) > 0$. Clearly, for any $\mathbf{x} \geq 0$ and $a > 0$ we have $r(a\mathbf{x}) = r(\mathbf{x})$. Hence

$$r(A) = \sup_{\mathbf{x} \geq 0} r(\mathbf{x}) = \sup_{\mathbf{x} \in \Pi_n} r(\mathbf{x}). \tag{6.2.5}$$

Since A is irreducible, $(I + A)^{n-1} > 0$. Hence, for any $\mathbf{x} \in \Pi_n$ $\mathbf{y} = (I + A)^{n-1}\mathbf{x} > 0$. (See Problem 3(a).) As $r(\mathbf{y})$ is a continuous function on $(I + A)^{n-1}\Pi_n$, and Π_n is a compact set, it follows that $r(\mathbf{y})$ achieves its maximum on $(I + A)^{n-1}\Pi_n$

$$r_1(A) := \max_{\mathbf{y} \in (I+A)^{n-1}\Pi_n} = r(\mathbf{v}), \text{ for some } \mathbf{v} \text{ in } (I + A)^{n-1}\Pi_n.$$

$r(A)$ is defined as the supremum of $r(\mathbf{x})$ on the set of all $\mathbf{x} \geq 0$ it follows that $r(A) \geq r_1(A)$. We now show the reversed inequality $r(A) \leq r_1(A)$ which is equivalent to $r(\mathbf{x}) \leq r_1(A)$ for any $\mathbf{x} \geq 0$.

One has the basic inequality

$$r(\mathbf{x}) \leq r((I + A)^{n-1}\mathbf{x}), \mathbf{x} \geq 0, \quad \text{with equality iff}$$

$$A\mathbf{x} = r(\mathbf{x})\mathbf{x}, \tag{6.2.6}$$

see Problem 3(d). For $\mathbf{x} \in \Pi_n$ we have $r(\mathbf{x}) \leq r((I + A)^{n-1}\mathbf{x}) \leq r_1(A)$. In view of (6.2.5) we have $r(A) \leq r_1(A)$. Hence, $r(A) = r_1(A)$.

Suppose that $r(\mathbf{x}) = r(A), \mathbf{x} \gneqq \mathbf{0}$. Then the definition of $r(A)$ (6.2.5) and (6.2.6) yields that $r(\mathbf{x}) = r((I + A)^{n-1}\mathbf{x})$. The equality case in (6.2.6) yields that $A\mathbf{x} = r(A)\mathbf{x}$. Hence, $(1 + r(A))^{n-1}\mathbf{x} = (I + A)^{n-1}\mathbf{x} > \mathbf{0}$, which yields that $\mathbf{x}$ is a positive eigenvector corresponding to the eigenvalue $r(A)$. So $\mathbf{x} = a\mathbf{u}, a > 0$ for the corresponding probability eigenvector $\mathbf{u} = (u_1, \ldots, u_n)^\top$, $A\mathbf{u} = r(A)\mathbf{u}$.

Suppose that $r(\mathbf{z}) = r(A)$ for some vector $\mathbf{z} = (z_1, \ldots, z_n)^\top \gneqq \mathbf{0}$. So $\mathbf{z} > \mathbf{0}$ and $A\mathbf{z} = r(A)\mathbf{z}$. Let $b = \min_i \frac{z_i}{u_i}$. We claim that $\mathbf{z} = b\mathbf{u}$. Otherwise $\mathbf{w} := \mathbf{z} - b\mathbf{u} \gneqq \mathbf{0}$, $\mathbf{w}$ has at least one coordinate equal to zero, and $A\mathbf{w} = r(A)\mathbf{w}$. So $r(\mathbf{w}) = r(A)$. This is impossible since we showed above that $\mathbf{w} > \mathbf{0}$! Hence, $\mathbf{z} = b\mathbf{u}$. Assume now that $\mathbf{y} \in \mathbb{R}^n$ is an eigenvector of A corresponding to $r(A)$. So $A\mathbf{y} = r(A)\mathbf{y}$. There exists a big positive number c such that $\mathbf{z} = \mathbf{y} + c\mathbf{u} > \mathbf{0}$. Clearly $A\mathbf{z} = r(A)\mathbf{z}$. Hence, $r(\mathbf{z}) = r(A)$ and we showed above that $\mathbf{z} = b\mathbf{u}$. So $\mathbf{y} = (b - c)\mathbf{u}$. Hence, $r(A)$ is a geometrically simple eigenvalue of A.

We now show that $r(A) = \rho(A)$. Let $\lambda \neq r(A)$ be another eigenvalue of A, which may be complex valued. Then

$$(\lambda \mathbf{z})_i = \lambda \mathbf{z}_i = (A\mathbf{z})_i = \sum_{j=1}^n a_{ij} z_j, \ i = 1, \ldots, n,$$

where $\mathbf{0} \neq \mathbf{z} = (z_1, \ldots, z_n)^\top \in \mathbb{C}^n$ is the corresponding eigenvector of A. Take the absolute values in the above equality, and use the triangle inequality, and the fact that A is non-negative matrix to obtain

$$|\lambda| \, |z_i| \leq \sum_{j=1}^n a_{ij} |z_j|, \ i = 1, \ldots, n.$$

Let $|\mathbf{z}| := (|z_1|, \ldots, |z_n|)^\top \gneqq \mathbf{0}$. Then the above inequality is equivalent to $|\lambda| \, |\mathbf{z}| \leq A|\mathbf{z}|$. Use (6.2.3) to deduce that $|\lambda| \leq r(|\mathbf{z}|)$. Since $r(|\mathbf{z}|) \leq r(A)$ we deduce that $|\lambda| \leq r(A)$. Hence, $\rho(A) = r(A)$, which yields (6.2.4). $\square$

Lemma 6.2.3 *Let $A \in \mathbb{R}_+^{n \times n}$ be an irreducible matrix. Then $\rho(A)$ is an algebraically simple eigenvalue.*

Proof. Clearly, we may assume that $n > 1$. Theorem 6.2.2 implies that $\rho(A)$ is geometrically simple, i.e. $\mathrm{nul}\,(\rho(A)I - A) = 1$. Hence, rank $(\rho(A)I - A) = n - 1$. Hence, adj $(\rho(A)I - A) = t\mathbf{u}\mathbf{v}^\top$, where $A\mathbf{u} = \rho(A)\mathbf{u}, A^\top\mathbf{v} = \rho(A)\mathbf{v}, \mathbf{u}, \mathbf{v} > \mathbf{0}$ and $0 \neq t \in \mathbb{R}$. Note that $\mathbf{u}\mathbf{v}^\top$ is a positive matrix, hence $\mathrm{tr}\,\mathbf{u}\mathbf{v}^\top = \mathbf{v}^\top\mathbf{u} > 0$. Since

$$(\det\,(\lambda I - A))'(\lambda = \rho(A)) = \mathrm{tr}\,\mathrm{adj}\,(\rho(A)I - A) = t(\mathbf{v}^\top\mathbf{u}) \neq 0,$$

we deduce that $\rho(A)$ is a simple root of the characteristic polynomial of A. $\qquad\square$

As usual, denote by $S^1 := \{z \in \mathbb{C}, |z| = 1\}$ the unit circle in the complex plane.

Lemma 6.2.4 *Let $A \in \mathbb{R}_+^{n\times n}$ be irreducible, $C \in \mathbb{C}^{n\times n}$. Assume that $|C| \leq A$. Then $\rho(C) \leq \rho(A)$. Equality holds, i.e. there exists $\lambda \in \mathrm{spec}\,C$, such that $\lambda = \zeta\rho(A)$ for some $\zeta \in S^1$, if and only if there exists a complex diagonal matrix $D \in \mathbb{C}^{n\times n}$, whose diagonal entries are equal to 1, such that $C = \zeta DAD^{-1}$. The matrix D is unique up to a multiplication by $t \in S^1$.*

Proof. We can assume that $n > 1$. Assume that $A = [a_{ij}]$, $C = [c_{ij}]$. Let $\mathbf{z} = (z_1, \ldots, z_n)^\top \neq \mathbf{0}$ be an eigenvector of C corresponding to an eigenvalue λ, i.e. $\lambda\mathbf{z} = C\mathbf{z}$. The arguments of the proof of Theorem 6.2.2 yield that $|\lambda|\,|\mathbf{z}| \leq |C|\,|\mathbf{z}|$. Hence $|\lambda|\,|\mathbf{z}| \leq |A|\,|\mathbf{z}|$, which implies that $|\lambda| \leq r(|\mathbf{z}|) \leq r(A) = \rho(A)$.

Suppose that $\rho(C) = \rho(A)$. So there exists $\lambda \in \mathrm{spec}\,C$, such that $|\lambda| = \rho(A)$. So $\lambda = \zeta\rho(A)$ for some $\zeta \in S^1$. Furthermore, for the corresponding eigenvector $\mathbf{z}$ we have the equalities

$$|\lambda|\,|\mathbf{z}| = |C\mathbf{z}| = |C|\,|\mathbf{z}| = A|\mathbf{z}| = r(A)|\mathbf{z}|.$$

Theorem 6.2.2 yields that $|\mathbf{z}|$ is a positive vector. Let $z_i = d_i|z_i|, |d_i| = 1$ for $i = 1, \ldots, n$. The equality $|C\mathbf{z}| = |C|\,|\mathbf{z}| = A|\mathbf{z}|$ combined with the triangle inequality and $|C| \leq A$, yields first that $|C| = A$. Furthermore for each fixed i the nonzero complex numbers $c_{i1}z_1, \ldots, c_{in}z_n$ have the same argument, i.e. $c_{ij} = \zeta_i a_{ij}\bar{d}_j$ for $j = 1, \ldots, n$ and some complex number ζ_j, where $|\zeta_i| = 1$. Recall that $\lambda z_i = (C\mathbf{z})_i$. Hence $\zeta_i = \zeta d_i$ for $i = 1, \ldots, n$. Thus $C = \zeta DAD^{-1}$, where $D = \text{diag}(d_1, \ldots, d_n)$. It is straightforward to see that D is unique up tD for any $t \in S^1$.

Suppose now that for $D = \text{diag}(d_1, \ldots, d_n)$, where $|d_1| = \cdots = |d_n| = 1$ and $|\zeta| = 1$ we have that $C = \zeta DAD^{-1}$. Then $\lambda_i(C) = \zeta \lambda_i(A)$ for $i = 1, \ldots, n$. So $\rho(C) = \rho(A)$. Furthermore $c_{ij} = \zeta d_i c_{ij} \bar{d}_j, i, j = 1, \ldots, n$. So $|C| = A$. $\square$

Lemma 6.2.5 *Let* $\zeta_1, \ldots, \zeta_h \in S^1$ *be* h *distinct complex numbers which form a multiplicative semi-group, i.e. for any integers* $i, j \in [1, h]$ $\zeta_i \zeta_j \in \{\zeta_1, \ldots, \zeta_h\}$. *Then the set* $\{\zeta_1, \ldots, \zeta_h\}$ *is the set, (the group), of all* h *roots of* $1 : e^{\frac{2\pi i \sqrt{-1}}{h}}, i = 1, \ldots, h$.

Proof. Let $\zeta \in \mathcal{T} := \{\zeta_1, \ldots \zeta_h\}$. Consider the sequence $\zeta^i, i = 1, \ldots$ Since $\zeta^{i+1} = \zeta \zeta^i$ for $i = 1, \ldots$, and $\mathcal{T}$ is a semigroup, it follows that each ζ^i is in $\mathcal{T}$. Since $\mathcal{T}$ is a finite set, we must have two positive integers such that $\zeta^k = \zeta^l$ for $k < l$. Assume that k and l are the smallest possible positive integers. So $\zeta^p = 1$, where $p = l - k \geq 1$, and $\mathcal{T}_p := \{\zeta, \zeta^2, \ldots, \zeta^{p-1}, \zeta^p = 1\}$ are all p roots of 1. ζ is called a p-primitive root of 1. That is, $\zeta = e^{\frac{2\pi p_1 \sqrt{-1}}{p}}$ where p_1 is an positive integer less than p. Furthermore p_1 and p are *coprime*, which is denoted by $(p_1, p) = 1$. Note that $\zeta^i \in \mathcal{T}$ for any integer i.

Next we choose $\zeta \in \mathcal{T}$, such that ζ is a primitive p-root of 1 of the maximal possible order. We claim that $p = h$, which is equivalent to the equality $\mathcal{T} = \mathcal{T}_p$. Assume to the contrary that $\mathcal{T}_p \subsetneq \mathcal{T}$. Let $\eta \in \mathcal{T} \backslash \mathcal{T}_p$. The previous arguments show that $\eta = $ is a q-primitive root of 1. So $\mathcal{T}_q \subset \mathcal{T}$, and $\mathcal{T}_q \subsetneq \mathcal{T}_p$. So q cannot divide p. Also the maximality of p yields that $q \leq p$. Let $(p, q) = r$ be the g.c.d., the greatest common divisor of p and q. So $1 \leq r < q$. Recall that Euclid algorithm, which is applied to the division of p by q with a residue, yields that there exists two integers i, j such that $ip + jq = r$. Let $l := \frac{pq}{r} > p$ be the least common multiplier of p and q. Observe that $\zeta' = e^{\frac{2\pi\sqrt{-1}}{p}} \in \mathcal{T}_p, \eta' = e^{\frac{2\pi\sqrt{-1}}{q}} \in \mathcal{T}_q$. So

$$\xi := (\eta')^i (\zeta')^j = e^{\frac{2\pi(ip+jq)\sqrt{-1}}{pq}} = e^{\frac{2\pi\sqrt{-1}}{l}} \in \mathcal{T}.$$

As ξ is an l-primitive root of 1, we obtain a contradiction to the maximality of p. So $p = h$ and $\mathcal{T}$ is the set of all h-roots of unity. $\qquad\square$

Lemma 6.2.6 *Let $A \in \mathbb{R}_+^{n \times n}$ be irreducible, and assume that for a positive integer $h \geq 2$, A has $h-1$ distinct eigenvalues $\lambda_1, \ldots, \lambda_{h-1}$, which are distinct from $\rho(A)$, such that $|\lambda_1| = \ldots = |\lambda_{h-1}| = \rho(A)$. Then the conditions (5a–5c) of Theorem 6.2.1 hold. Moreover, $P^\top A P = B$, where B is of the form given in (5d) and P is a permutation matrix.*

Proof. Assume that $\zeta_i := \frac{\lambda_i}{\rho(A)} \in S^1$ for $i = 1, \ldots, h-1$ and $\zeta_h = 1$. Apply Lemma 6.2.4 to $C = A$ and $\lambda = \zeta_i \rho(A)$ to deduce that $A = \zeta_i D_i A D_i^{-1}$ where D_i is a diagonal matrix such that $|D| = I$ for $i = 1, \ldots, h$. Hence, if λ is an eigenvalue of A then $\zeta_i \lambda$ is an eigenvalue of A, with an algebraic and geometrical multiplicity as λ. In particular, since $\rho(A)$ is an algebraically simple eigenvalue of A, $\lambda_i = \zeta_i \rho(A)$ is an algebraically simple of A for $i = 1, \ldots, h-1$. This establishes (5a).

Let $\mathcal{T} = \{\zeta_1, \ldots, \zeta_h\}$. Note that

$$A = \zeta_i D_i A D_i^{-1} = \zeta_i D_i (\zeta_j D_j A D_j^{-1}) D_i^{-1} = (\zeta_i \zeta_j)(D_i D_j) A (D_i D_j)^{-1}. \tag{6.2.7}$$

So $\zeta_i \zeta_j \rho(A)$ is an eigenvalue of A. Hence $\zeta_i \zeta_j \in \mathcal{T}$, i.e. $\mathcal{T}$ is a semigroup. Lemma 6.2.5 yields that $\{\zeta_1, \ldots, \zeta_n\}$ are all h roots of 1. Note that if $A\mathbf{z}_i = \lambda_i \mathbf{z}_i$, $\mathbf{z}_i \neq \mathbf{0}$, then $\mathbf{z}_i = t D_i \mathbf{u}$ for some $0 \neq t \in \mathbb{C}$, where $\mathbf{u} > \mathbf{0}$ is the Perron–Frobenius vector given in Theorem 6.2.1. This establishes (5b) of Theorem 6.2.1.

Let $\zeta = e^{\frac{2\pi\sqrt{-1}}{h}} \in \mathcal{T}$. Then $A = \zeta D A D^{-1}$, where D is a diagonal matrix $D = (d_1, \ldots, d_n), |D| = I$. Since D can be replaced by $\bar{d}_1 D$, we can assume that $d_1 = 1$. (6.2.7) yields that $A = \zeta^h D^h A D^{-h} = I A I^{-1}$. Lemma 6.2.4 yields that $D^h = \mathrm{diag}(d_1^h, \ldots, d_n^h) = tI$. Since $d_1 = 1$ it follows that $D^h = I$. So all the diagonal entries of D are h-roots of unity. Let $P \in \mathcal{P}_n$ be a permutation matrix such that the diagonal matrix $E = P^\top DP$ is of the following block diagonal form

$$E = I_{n_1} \oplus \mu_1 I_{n_2} \oplus \cdots \oplus \mu_{l-1} I_{n_l}, \ \mu_i = e^{\frac{2\pi k_i \sqrt{-1}}{h}},$$

$$i = 1, \ldots, l-1, \ 1 \leq k_1 < k_2 < \cdots < k_{l-1} \leq h - 1.$$

Note that $l \leq h$ and equality holds if and only if $k_i = i$. Let $\mu_0 = 1$.

Let $B = P^\top AP$. Partition B to a block matrix $[B_{ij}]_{i=j=1}^l$ where $B_{ij} \in \mathbb{R}_+^{n_i \times n_j}$ for $i, j = 1, \ldots, l$. Then the equality $A = \zeta DAD^{-1}$ yields $B = \zeta EBE^{-1}$. The structure of B and E implies the equalities

$$B_{ij} = \zeta \frac{\mu_{i-1}}{\mu_{j-1}} B_{ij}, \quad i, j = 1, \ldots, l.$$

Since all the entries of B_{ij} are non-negative we obtain that $B_{ij} = 0$ if $\zeta \frac{\mu_{i-1}}{\mu_{j-1}} \neq 1$. Hence $B_{ii} = 0$ for $i = 1, \ldots, l$. Since B is irreducible it follows that not all $B_{i1}, \ldots, B_{il}$ are zero matrices for each $i = 1, \ldots, l$. First start with $i = 1$. Since $\mu_0 = 1$ and $j_1 \geq 1$ it follows that $\mu_j \neq \zeta$ for $j > 1$. So $B_{1j} = 0$ for $j = 3, \ldots, l$. Hence, $B_{12} \neq 0$, which implies that $\mu_1 = \zeta$, i.e. $k_1 = 1$. Now let $i = 2$ and consider $j = 1, \ldots, l$. As $k_i \in [k_1 + 1, h - 1]$ for $i > 1$, it follows that $B_{2j} = 0$ for $j \neq 3$. Hence, $B_{23} \neq 0$ which yields that $k_2 = 2$. Applying these arguments for $i = 3, \ldots, l-1$ we deduce that $B_{ij} = 0$ for $j \neq i+1$, $B_{i(i+1)} \neq 0, k_i = i$ for $i = 1, \ldots, l-1$. It is left to consider $i = l$. Note that

$$\frac{\zeta \mu_{l-1}}{\mu_{j-1}} = \frac{\zeta^l}{\zeta^{j-1}} = \zeta^{l-(j-1)}, \text{ which is different from 1 for } j \in [2, l].$$

Hence, $B_{lj} = 0$ for $j > 1$. Since B is irreducible, $B_{11} \neq 0$. So $\zeta^l = 1$. Since $l \leq h$ we deduce that $l = h$. Hence, B has the block form given in (5d). $\qquad \square$

Proposition 6.2.7 *Let $A \in \mathbb{R}_+^{n \times n}$ be irreducible. Suppose that $0 \lneq \mathbf{w} \in \mathbb{R}_+^n$ is an eigenvector of A, i.e. $A\mathbf{w} = \lambda \mathbf{w}$. Then $\lambda = \rho(A)$ and $\mathbf{w} > 0$.*

Proof. Let $\mathbf{v} > \mathbf{0}$ be the Perron–Frobenius vector of $A^\top$, i.e. $A^\top \mathbf{v} = \rho(A)\mathbf{v}$. Then

$$\mathbf{v}^\top A\mathbf{w} = \mathbf{v}^\top(\lambda\mathbf{w}) = \rho(A)\mathbf{v}^\top\mathbf{w} \Rightarrow (\rho(A) - \lambda)\mathbf{v}^\top\mathbf{w} = 0.$$

If $\rho(A) \neq \lambda$ we deduce that $\mathbf{v}^\top\mathbf{w} = 0$, which is impossible, since $\mathbf{v} > \mathbf{0}$ and $\mathbf{w} \gneq \mathbf{0}$. Hence $\lambda = \rho(A)$. Then $\mathbf{w}$ is the Perron–Frobenius eigenvector and $\mathbf{w} > \mathbf{0}$. $\square$

Lemma 6.2.8 *Let* $A \in \mathbb{R}_+^{n\times}$ *be irreducible. Then* A *is primitive if and only if one of the following conditions hold:*

1. $n = 1$ *and* $A > 0$.

2. $n > 1$ *and each eigenvalue* λ *of* A *different from* $\rho(A)$ *satisfies the inequality* $|\lambda| < \rho(A)$. *That is, condition* (4) *of Theorem* 6.2.2 *holds.*

Proof. If $n = 1$ then A is primitive if and only if $A > 0$. Assume now that $n > 1$. So $\rho(A) > 0$. By considering $B = \frac{1}{\rho(A)}A$, it is enough to consider the case $\rho(A) = 1$. Assume first that if $\lambda \neq 1$ is an eigenvalue of A then $|\lambda| < 1$. Theorem 6.2.2 implies $A\mathbf{u} = \mathbf{u}, A^\top\mathbf{w} = \mathbf{w}$ for some $\mathbf{u}, \mathbf{w} > \mathbf{0}$. So $\mathbf{w}^\top\mathbf{u} > 0$. Let $\mathbf{v} := (\mathbf{w}^\top\mathbf{u})^{-1}\mathbf{w}$. Then $A^\top\mathbf{v} = \mathbf{v}$ and $\mathbf{v}^\top\mathbf{u} = 1$. The results of §3.3 yield that $\lim_{k\to\infty} A^k = \mathbf{u}\mathbf{v}^\top > 0$. So there exists integer $k_0 \geq 1$, such that $A^k > 0$ for $k \geq k_0$, i.e. A is primitive.

Assume now A has exactly $h > 1$ distinct eigenvalues λ satisfying $|\lambda| = 1$. Lemma 6.2.6 implies that there exists a permutation matrix P such that $B = P^\top AP$ is of the form (5d) of Theorem 6.2.1. Note that B^h is a block diagonal matrix. Hence $B^{hj} = (B^h)^j$ is a block diagonal matrix for $j = 1, \ldots$. Hence, B^{hj} is never a positive matrix, so A^{hj} is never a positive matrix. In view of Problem 4, A is not primitive. $\square$

Lemma 6.2.9 *Let* $B \in \mathbb{R}_+^{n\times n}$ *be an irreducible, imprimitive matrix, having* $h > 1$ *distinct eigenvalues* λ *satisfying* $|\lambda| = \rho(B)$. *Suppose furthermore that* B *has the form* (5d) *of Theorem* 6.2.1.

Then B^h is a block diagonal matrix, where each diagonal block is an irreducible primitive matrix whose spectral radius is $\rho(B)^h$. In particular, the last claim of (5d) of Theorem 6.2.1 holds.

Proof. Let $D(B) = ([n], E)$ be the digraph associated with B. Let $p_0 = 0, p_1 = p_0 + n_1, \ldots, p_h = p_{h-1} + n_h = n$. Denote $V_i = \{p_{i-1} + 1, \ldots, p_i\}$ for $i = 1, \ldots, h$, and let $V_{h+1} := V_1$. So $[n] = \cup_{i=1}^h V_i$. The form of B implies that $E \subset \cup_{i=1}^h V_i \times V_{i+1}$. Thus, any walk that connects vertices $j, k \in V_i$ must be divisible by h. Observe next that $B^h = \mathrm{diag}(C_1, \ldots, C_h)$, where $C_1 = [c_{jk}^{(1)}]_{j=k=1}^{n_1}, \ldots, C_h = [c_{jk}^{(h)}]_{j=k=1}^{n_h}$ are defined in (6.2.1). Let $D(C_i) = (V_i, E_i)$ be the digraph associated with C_i for $i = 1, \ldots, h$. Then there exists a path of length h from j to k in V_i if and only if $c_{jk}^{(i)} > 0$. Since B is irreducible, $D(B)$ is strongly connected. Hence, each $D(C_i)$ is strongly connected. Thus, each C_i is irreducible.

Recall that $B\mathbf{u} = \rho(B)\mathbf{u}$ for the Perron–Frobenius vector $\mathbf{u}^\top = (\mathbf{u}_1^\top, \ldots, \mathbf{u}_h^\top) > \mathbf{0}^\top, \mathbf{u}_i \in \mathbb{R}_+^{n_i}, i = 1, \ldots, h$. Thus, $B^h\mathbf{u} = \rho(B)^h\mathbf{u}$, which implies that $C_i\mathbf{u}_i = \rho(B)^h\mathbf{u}_i, i = 1, \ldots, h$. Since $\mathbf{u}_i > \mathbf{0}$ Proposition 6.2.7 yields that $\rho(C_i) = \rho(B)^h, i = 1, \ldots, h$. Recall that the eigenvalues of B^h are the h power of the eigenvalues of B, i.e. $\lambda_i(B^h) = \lambda_i(B)^h$ for $i = 1, \ldots, n$. Furthermore, B has h simple eigenvalues $\rho(B)e^{\frac{2\pi i\sqrt{-1}}{h}}, i = 1, \ldots, h$ with $|\lambda| = \rho(B)$, and all other eigenvalues satisfy $|\lambda| < \rho(B)$. Hence, B^h has one eigenvalues $\rho(B)^h$ of an algebraic multiplicity h and all other eigenvalues μ satisfy $|\mu| < \rho(B)^h$.

Since $B^h = \mathrm{diag}(C_1, \ldots, C_h)$, we deduce that the eigenvalues of B^h are the eigenvalues of $C_1, \ldots, C_h$. As C_i is irreducible and $\rho(C_i) = \rho(B)^h$, we deduce that all other eigenvalues μ of C_i satisfy $|\mu| < \rho(C_i)$. Lemma 6.2.8 yields that C_i is primitive. $\square$

Problems

1. Prove equality (6.2.3).

2. Show that if $n > 1$ and $A \in \mathbb{R}_+^{n \times n}$ is irreducible then A does not have a zero row or column.

3. Assume that $A \in \mathbb{R}_+^{n \times n}$ is irreducible. Show

   (a) For each $\mathbf{x} \in \Pi_n$ the vector $(I + A)^{n-1}\mathbf{x}$ is positive.

   (b) The set $(I + A)^{n-1}\Pi_n := \{\mathbf{y} = (I + A)^{n-1}\mathbf{x}, \mathbf{x} \in \Pi_n\}$ is a compact set of positive vectors.

   (c) Show that $r(\mathbf{y})$ is a continuous function on $(I + A)^{n-1}\Pi_n$.

   (d) Show (6.2.6). *Hint*: use that $(A + I)^{n-1}(A\mathbf{x} - r(\mathbf{x})\mathbf{x})$ is a positive vector, unless $A\mathbf{x} = r(\mathbf{x})\mathbf{x}$.

4. Assume that $A \in \mathbb{R}_+^{n \times n}$ is irreducible. Show the following are equivalent

   (a) A is primitive.

   (b) There exists a positive integer k_0 such that for any integer $k \geq k_0$ $A^k > 0$.

5. Let $D = ([h], E)$ be the cycle $1 \to 2 \to \cdots \to h - 1 \to h \to 1$.

   (a) Show that representation matrix $A(D)$ is a permutation matrix, which has the form of B given in (5d) of Theorem 6.2.1, where each nonzero block is 1×1 matrix [1]. $A(D)$ is called a *circulant* matrix.

   (b) Find all the eigenvalues and the corresponding eigenvectors of $A(D)$.

6. Let the assumptions of Lemma 6.2.9 hold. Assume the notation of the proof of Lemma 6.2.9.

   (a) Show that the length of any closed walk in $D(B)$ is divisible by h.

   (b) Show that a length of any walk from a vertex in V_i to a vertex V_j, such that $1 \leq i < j \leq h$, minus $j - i$ is divisible by h.

   (c) What can you say on a length of any walk from a vertex in V_i to a vertex V_j, such that $1 \leq j < i \leq h$?

   (d) Show that each C_i is irreducible.

7. Let $D = (V, E)$ be a digraph and assume that V is a disjoint union of h nonempty sets $V_1, \ldots, V_h$. Denote $V_{h+1} := V_1$. Assume that $E \subset \cup_{i=1}^h V_i \times V_{i+1}$. Let $D_i = (V_i, E_i)$ be the following digraph. The diedge $(v, w) \in E_i$, if there is a path of length h in D from v to w in D.

   (a) Show that D is strongly connected if and only if D_i is strongly connected for $i = 1, \ldots, h$.

   (b) Assume that D is strongly connected. Let $1 \leq i < j \leq h$. Then D_i is primitive if and only if D_j is primitive.

8. Let $B \in \mathbb{R}_+^{n \times n}$ be a block matrix of the form given in (5d) of Theorem 6.2.1.

   (a) Show that B^h is a block diagonal matrix $\operatorname{diag}(C_1, \ldots, C_h)$, where C_i is given (6.2.1).

   (b) Show that B is irreducible if and only if C_i is irreducible for $i = 1, \ldots, h$.

   (c) Assume that B is irreducible.

      i. Let $1 \leq i < j \leq h$. Then C_i is primitive if and only if C_j is primitive.

      ii. B has h distinct eigenvalues on the circle $|z| = \rho(B)$ if and only if some C_i is primitive.

9. Assume the assumptions of Lemma 6.2.6. Let $A\mathbf{u} = \rho(A)\mathbf{u}, \mathbf{u} = (u_1, \ldots, u_n)^\top > \mathbf{0}$. Assume that η is an h-root of unity, and suppose that $A\mathbf{z} = \eta\mathbf{z}, \mathbf{z} = (z_1, \ldots, z_n)$, such that $|\mathbf{z}| = \mathbf{u}$. Assume that $z_i = u_i$ for a given $i \in [n]$. (This is always possible by considering $\frac{\bar{z}_i}{|z_i|}\mathbf{z}$.) Show $z_j = \eta^{k(j)}u_j$, for a suitable integer $k(j)$, for $j = 1, \ldots, n$. Furthermore, given an integer k then there exists $j \in [n]$ such that $z_j = \eta^k u_j$.

   Hint: Use the proof of Lemma 6.2.6.

10. Let $B \in \mathbb{R}_+^{n \times n}$ be an irreducible block matrix of the form given in (5d) of Theorem 6.2.1. Let $C_1, \ldots, C_h$ be defined in (6.2.1). Suppose that B has more than h distinct eigenvalues on the circle $|z| = \rho(B)$. Then the following are equivalent,

abbreviated as TFAE:

(a) B has qh eigenvalues the circle $|z| = \rho(B)$, for some $q > 1$.

(b) Each C_i has $q > 1$ distinct eigenvalues on $|z| = \rho(C_i) = \rho(B)^h$.

(c) Some C_i has $q > 1$ distinct eigenvalues on $|z| = \rho(C_i) = \rho(B)^h$.

(d) Let $D(B) = ([n], E)$ and $V_i = \{p_{i-1} + 1, \ldots, p_i\}$ for $i = 1, \ldots, h$ be defined as in the proof of Lemma 6.2.9. Then each V_i is a disjoint union of q nonempty sets $W_i, W_{i+h}, \ldots, W_{i+(q-1)h}$ for $i = 1, \ldots, h$, such that $E \subset \cup_{j=1}^{qh} W_j \times W_{j+1}$, where $W_{qh+1} := W_1$. Let $H_j = (W_j, F_j), F_j \subset W_j \times W_j$ be the following digraph. The diedge (v, w) is in F_j, if and only if there is a path of length qh in $D(B)$ from v to w in W_j. Then each digraph H_j is strongly connected and primitive.

Hint: Use the structure of the eigenvalues λ of B on the circle $|\lambda| = \rho(B)$, and the corresponding eigenvector $\mathbf{z}$ to λ given in (5b) of Theorem 6.2.1.

11. For $A \in R_+^{n \times n}$ and $0 < \mathbf{x} = (x_1, \ldots, x_n)^\top \in R_+^n$ let $R(\mathbf{x}) = \max_{i \in [n]} \frac{(A\mathbf{x})_i}{x_i}$. Assume that A is irreducible. Show that $\inf_{\mathbf{x}>0} R(\mathbf{x}) = \rho(A)$. That is,

$$\min_{\mathbf{x}=(x_1,\ldots,x_n)>0} \max_{i \in [n]} \frac{(A\mathbf{x})_i}{x_i} = \rho(A). \qquad (6.2.8)$$

Furthermore, $R(\mathbf{x}) = \rho(A)$ if and only if $A\mathbf{x} = \rho(A)\mathbf{x}$.

Hint: Mimic the proof of Theorem 6.2.2.

12. Let $n > 1$ and $D = ([n], E)$ be a strongly connected digraph. Show

(a) If D has exactly one cycle, it must be a Hamiltonian cycle, i.e. the length of of this cycle is n. Then D is not primitive.

(b) Suppose that D has exactly two directed cycles. Then the shortest cycle has length $n - 1$ if and only if it is possible to rename the vertices so that the shortest cycle is of the form $1 \to 2 \to \ldots \to n - 1 \to 1$ and the second cycle is a Hamiltonian cycle $1 \to 2 \to \ldots \to n - 1 \to n \to 1$. In this case D is primitive. Moreover, $A(D)^k > 0$ if and only if $k \geq n^2 - 2n + 2$.

(c) Assume that D is primitive. Show that the shortest cycle of D has at most length $n - 1$.

6.3 Index of Primitivity

Theorem 6.3.1 *Let $A \in \mathbb{R}_+^{n \times n}$ be a primitive matrix. Let $s \geq 1$ be the length of the shortest cycle in the digraph $D(A) = ([n], E)$. Then $A^{s(n-2)+n} > 0$. In particular $A^{(n-1)^2+1} > 0$.*

Proof. For $n = 1$ we have that $s = 1$ and the theorem is trivial. Assume that $n > 1$. Note that since A is primitive $s \leq n - 1$. (See Problem 6.2.12(c).)

Suppose first that $s = 1$. So $D(A)$ contains a loop. Relabel the vertices of $D(A)$ to assume that $(1,1) \in E$. That is, we can assume that $A = [a_{ij}]$ and $a_{11} > 0$. Recall that from 1 to $j > 1$ there exists a path of length $1 \leq l(j) \leq n - 1$. By looping at 1 first $n - 1 - l(j)$ times we deduce the existence of a walk of length $n - 1$ from 1 to $j > 1$. Clearly, there exists a walk of length $n - 1$ from 1 to 1: $1 \to 1 \to \cdots \to 1$. Similarly, for each $j > 1$ there exists a walk of length $n - 1$ from j to 1. Hence, the first row and the column of A^{n-1} is positive. Thus, $A^{2(n-1)} = A^{n-1}A^{n-1}$ is a positive matrix.

Assume now that $s \geq 2$. Relabel the vertices of $D(A)$ such that one has the cycle on vertices $c := \{1, 2, \ldots, s\}$: $1 \to \cdots \to s \to 1$. Then the first s diagonal entries of A^s are positive. Since A was primitive, Lemma 6.2.8 implies that A^s is primitive. Our previous arguments show that $(A^s)^{n-1}$ has the first s rows and columns positive. Let

$$A^{n-s} = \begin{bmatrix} F_{11} & F_{12} \\ F_{21} & F_{22} \end{bmatrix}, F_{11} \in \mathbb{R}_+^{s \times s}, F_{12},$$

$$\times F_{21}^{\top} \in \mathbb{R}_+^{s \times (n-s)}, F_{22} \in \mathbb{R}_+^{(n-s) \times (n-s)}.$$

Clearly, $F_{11} \geq ([a_{ij}]_{i=j=1}^{s})^{n-s}$. Since $D(A)$ contains a cycle of length s on $[s]$ it follows that each row and column of F_{11} is not zero. Clearly,

$$A^{s(n-2)+n} = A^{(n-s)} A^{s(n-1)}. \tag{6.3.1}$$

Hence, the first s rows of $A^{s(n-2)+n}$ are positive. We claim that each row of F_{21} is nonzero. Indeed, take the shortest walk from $j \in U := \{s+1, \ldots, n\}$ to the set of vertices $V := [s]$. This shortest walk is a path which can contain at most $n - s$ vertices in U, before it ends in $i \in V$. Hence, the length of this path is $m(j) \leq n - s$. After that continue take a walk on the cycle c of length $n - s - m(j)$, to deduce that there is a walk of length $n - s$ from j to V. Hence, the $j - s$ row of F_{21} is nonzero. Use (6.3.1) and the fact that the first s rows of $(A^s)^{n-1}$ positive to deduce that $A^{s(n-2)+n} > 0$. $\qquad \square$

Proof of Theorem 6.1.5. Problem 6.1.1 yields that the length L of any closed walk in D is a sum of lengthes of a number of cycles in D. Hence, ℓ divides L. Assume that D is primitive, i.e. $A^k > 0$ for any $k \geq k_0$. Hence, for each $k \geq k_0$ there exists a closed walk in D of length k. Therefore $\ell = 1$.

Suppose now that $D = (V, E)$ is imprimitive, i.e. $A(D)$ is imprimitive. (5d) of Theorem 6.2.1 yields that there V decomposes to a nonempty disjoint sets $V_1, \ldots, V_h$, where $h > 1$. Moreover $E \subset \cup_{i=1}^{h} V_i \times V_{i+1}$, where $V_{h+1} = V_1$. So any closed walk must be divisible by $h > 1$. In particular, the length of each cycle is divisible by h. Thus $\ell \geq h > 1$. Hence, D is primitive if and only if $\ell = 1$. Suppose that D is primite. Theorem 6.3.1 yields that $A(D)^{s(n-2)+n} > 0$, where s is the length of the shortest cycle. This proves part 1 of Theorem 6.1.5.

Assume now that D is imprimitive. So $A(D)$ has $h > 1$ distinct eigenvalues of modulus $\rho(A(D))$. Relabel the vertices of D so that $A(D)$ is of the form B given in (5d) of Theorem 6.2.1. As we pointed out, each cycle in D is divisible by h. It is left to show that the $\ell = h$. Let $D_i = (V_i, E_i)$ be defined as in the proof of Lemma 6.2.9. It is straightforward to see that each cycle in D_i corresponds of length L to a cycle in D of length hL. Since C_i is primitive, it follows from the first part of the proof, that the g.c.d of lengths of all cycles in C_i is 1. Hence, the g.c.d. of lengths of the corresponding cycles in D is h.   □

Problems

1. Show that the matrix $A := \begin{bmatrix} 0 & 1 & 0 & \cdots & 0 \\ 0 & 0 & 1 & \cdots & 0 \\ \vdots & \vdots & \vdots & \cdots & \vdots \\ 0 & 0 & 0 & \cdots & 1 \\ 1 & 1 & 0 & \cdots & 0 \end{bmatrix}$ is a primitive

matrix such that $A^{(n-1)^2}$ is not a positive matrix. That is, the last part of Theorem 6.3.1 is sharp [Wie50].

2. Let $A \in \mathbb{R}_+^{n \times n}$ be irreducible. Assume that A has $d \in [n]$ positive diagonal elements. Show that $A^{2n-d-1} > 0$ [HoV58].

6.4   Reducible Matrices

Theorem 6.4.1 *Let $A \in \mathbb{R}_+^{n \times n}$. Then $\rho(A)$, the spectral radius of A, is an eigenvalue of A. There exists a probability vector $\mathbf{x} \in \Pi_n$ such that $A\mathbf{x} = \rho(A)\mathbf{x}$.*

Proof. Let $J_n \in \{1\}^{n \times n}$ be a matrix whose entries are 1. For $\varepsilon > 0$ let $A(\varepsilon) = A + \varepsilon J_n$. Then $A(\varepsilon) > 0$. Hence,

$$\rho(A(\varepsilon)) \in \text{spec}\,(A(\varepsilon)) \quad \text{and} \quad A(\varepsilon)\mathbf{x}(\varepsilon), \quad \mathbf{0} < \mathbf{x}(\varepsilon) \in \Pi_n \quad \text{for } \varepsilon > 0.$$
$$(6.4.1)$$

Since the coefficients of the characteristic polynomial of $A(\varepsilon)$ are polynomial in ε, it follows that the eigenvalues of $A(\varepsilon)$ are continuous function of ε. Hence

$$\lim_{\varepsilon \to 0} \text{spec}\,(A(\varepsilon)) = \text{spec}\,A, \quad \lim_{\varepsilon \to 0} \rho(A(\varepsilon)) = \rho(A).$$

Combine that with (6.4.1) to deduce that $\rho(A) \in \text{spec}\,A$. Choose $\varepsilon_k := \frac{1}{k}, k = 1, \ldots$ Since Π_n is a compact set, there exists a subsequence $1 < k_1 < k_2 < \cdots$ such that $\lim_{j\to\infty} \mathbf{x}(\varepsilon_{k_j}) = \mathbf{x} \in \Pi_n$. The second equality of (6.4.1) yields that $A\mathbf{x} = \rho(A)\mathbf{x}$.   □

It is easy to have examples where $\rho(A) = 0$ for some $A \in \mathbb{R}_+^{n\times n}$, and $\rho(A)$ is not a geometrically simple eigenvalue. (That is, the Jordan canonical form of A contains a Jordan block of order greater than one with the eigenvalue $\rho(A)$.)

Proposition 6.4.2 *Let* $A \in \mathbb{R}_+^{n\times n}$.

1. *Assume that* $C \in \mathbb{R}_+^{n\times n}$ *and* $A \geq C$. *Then* $\rho(A) \geq \rho(C)$. *If either* A *or* C *are irreducible then* $\rho(A) = \rho(C)$ *if and only if* $A = C$.

2. *Assume that* $B \in \mathbb{R}_+^{m\times m}, 1 \leq m < n$ *is a principle submatrix of* A, *obtained by deleting* $n - m$ *rows and columns of* A *from a subset* $J \subset [n]$ *of cardinality* $n - m$. *Then* $\rho(B) \leq \rho(A)$. *If* A *is irreducible then* $\rho(B) < \rho(A)$.

Proof. 1. Suppose first that A is irreducible. Then Lemma 6.2.4 yields that $\rho(A) \geq \rho(C)$. Equality holds if and only if $A = C$. Suppose next that C is irreducible. Then A is irreducible. Hence $\rho(A) \geq \rho(C)$, and equality holds if and only if $C = A$.

Assume now that A is reducible. Let $A(\varepsilon), C(\varepsilon)$ be defined as in the proof of Theorem 6.4.1. For $\varepsilon > 0$ the above arguments show that $\rho(A(\varepsilon)) \geq \rho(C(\varepsilon))$. Letting $\epsilon \searrow 0$ we deduce that $\rho(A) \geq \rho(C)$.

2. By considering a matrix $A_1 = PAP^\top$ for a corresponding $P \in \mathcal{P}_n$ we may assume that $A_1 = \begin{bmatrix} A_{11} & A_{12} \\ A_{21} & A_{22} \end{bmatrix}$, where $B = A_{11}$. Clearly, $\rho(A_1) = \rho(A)$, and A_1 irreducible if and only if A irreducible. Let $C = \begin{bmatrix} B & 0_{m\times(n-m)} \\ 0_{(n-m)\times m} & 0_{(n-m)\times(n-m)} \end{bmatrix}$. Then $A_1 \geq C$. Part 1 yields that $\rho(A) = \rho(A_1) \geq \rho(C)$. Suppose that A_1 is irreducible. Since C is reducible, $A_1 \neq C$. Hence, $\rho(C) < \rho(A_1) = \rho(A)$.   □

Lemma 6.4.3 *Let* $A \in \mathbb{R}_+^{n\times n}$. *Assume that* $t > \rho(A)$. *Then* $(tI - A)^{-1} \geq 0$. *Furthermore,* $(tI - A)^{-1} > 0$ *if and only if* A *is irreducible.*

Proof. Since $t > \rho(A)$ it follows that $\det (tI - A) \neq 0$. (Actually, $\det (tI - A) > 0$. See Problem 1.) So $(tI - A)^{-1}$ exists and $(tI - A)^{-1} = \frac{1}{t}(I - \frac{1}{t}A)^{-1}$. Since $\rho(\frac{1}{t}A) < 1$ we have the the Neumann series [Neu77] (3.4.14):

$$(tI - A)^{-1} = \sum_{k=0}^{\infty} \frac{1}{t^{k+1}} A^k, \quad \text{for } |t| > \rho(A). \tag{6.4.2}$$

Since $A^k \geq 0$ we deduce that $(tI - A)^{-1} \geq 0$. Let $A^k = [a_{ij}^{(k)}]$. The the (i, j) entry of $(tI - A)^{-1}$ is positive, if and only if $a_{ij}^{(k)} > 0$ for some $k = k(i, j)$. This shows that $(tI - A)^{-1} > 0$ if and only if A is irreducible. $\qquad\square$

Theorem 6.4.4 *Let $A \in \mathbb{R}_+^{n \times n}$ be a non-negative matrix. Then there exists a permutation matrix $P \in \mathcal{P}_n$ such that $B = PAP^\top$ has the Frobenius normal form:*

$$
B =
\begin{bmatrix}
B_{11} & B_{12} & \cdots & B_{1t} & B_{1(t+1)} & B_{1(t+2)} & \cdots & B_{1(t+f)} \\
0 & B_{22} & \cdots & B_{2t} & B_{2(t+1)} & B_{1(t+2)} & \cdots & B_{2(t+f)} \\
\vdots & \vdots & \vdots & \vdots & \vdots & \vdots & \vdots & \vdots \\
0 & 0 & \cdots & B_{tt} & B_{t(t+1)} & B_{t(t+2)} & \cdots & B_{t(t+f)} \\
0 & 0 & \cdots & 0 & B_{(t+1)(t+1)} & 0 & \cdots & 0 \\
\vdots & \vdots & \vdots & \vdots & \cdots & \vdots & \vdots & \vdots \\
0 & 0 & 0 & 0 & \vdots & 0 & 0 & B_{(t+f)(t+f)}
\end{bmatrix},
$$

$$B_{ij} \in \mathbb{R}^{n_i \times n_j}, \ i, j = 1, \ldots, t + f, \ n_1 + \ldots + n_{t+f}$$

$$= n, \ t \geq 0, \ f \geq 1. \tag{6.4.3}$$

Each B_{ii} is irreducible, and the submatrix $B' := [B_{ij}]_{i=j=t+1}^{t+f}$ is block diagonal. If $t = 0$ then B is a block diagonal. If $t \geq 1$ then for each $i = 1, \ldots, t$ not all the matrices $B_{i(i+1)}, \ldots, B_{i(i+f)}$ are zero matrices.

Proof. Let $D_r = (W, F)$ be the reduced graph of $D(A) = ([n], E)$. Then D_r is a diforest. Let $\ell \geq 1$ be the length of the longest path in the digraph D_r. For a given vertex $w \in \mathbb{W}$ let $\ell(w)$ be the length of the longest path in D_r from w. So $\ell(w) \in [0, \ell]$. For $j \in \{0, \ldots, \ell\}$ denote by W_j the set of of all vertices in W such that $\ell(w) = j$. Since D_r is diforest, it follows that $W_\ell, \ldots, W_0$ is a decomposition of W to nonempty set. Note if there is a diedge in D_r from W_i to W_j then $i > j$. Also we have always at least one diedge from W_i to W_{i-1} for $i = \ell, \ldots, 1$, if $\ell > 0$.

Assume that $\#W_j = m_{1+\ell-j}$ for $j = 0, \ldots, \ell$. Let $M_0 = 0$ and $M_j = \sum_{i=1}^{j} m_i$ for $j = 1, \ldots, \ell$. Then we name the vertices of W_j as $\{M_{\ell-j} + 1, \ldots, M_{\ell-j} + m_{1+\ell-j}\}$ for $j = 0, \ldots, \ell$. Let $f := \#W_0 = m_{\ell+1}$ and $t := \#(\cup_{j=1}^{\ell} W_j) = \sum_{j=1}^{\ell} m_i$. Note that $f \geq 1$ and $t = 0$ if and only if $\ell = 0$. Hence, the representation matrix $A(D_r)$ is strictly upper triangular. Furthermore the last f rows of $A(D_r)$ are zero rows.

Recall that each vertex in W corresponds to a maximal strongly connected component of $D(A)$. That is, to each $i \in W = [t+f]$ one has a nonempty subset $V_i \subset [n]$, which correspond to the maximal connected component of $D(A)$. Let $n_i := \#V_i$ for $i = 1, \ldots, t+f$. Let $N_0 = 0$ and $N_i = \sum_{j=1}^{i} n_i, i = 1, \ldots, t+s$. Rename vertices of $D(A)$ to satisfy $V_i = \{N_{i-1} + 1, \ldots, N_{i-1} + n_i\}$. Then $PAP^\top$ is of the form (6.4.3). Furthermore, the digraph induced by B_{ii} is a strongly connected component of $D(A)$. Hence, B_{ii} is irreducible. Note $B_{ij} = 0$, for $j > i$ if and only if there is no biedge from the vertex i to the vertex j in D_r. Recall that for $i \leq t$, the vertex i represents a vertex in W_k, for some $k \geq 1$. Hence, for $i \leq t$ there exists $j > i$ such that $B_{ij} \neq 0$. $\qquad \square$

Theorem 6.4.5 *Let $A \in \mathbb{R}_+^{n \times n}$. Then there exists a positive eigenvector $\mathbf{x} > 0$ such that $A\mathbf{x} = r\mathbf{x}$ if and only if the following conditions hold: Let B be the Frobenius normal form of A given in Theorem 6.4.4. Then*

1. $\rho(B_{(t+1)(t+1)}) = \cdots = \rho(B_{(t+f)(t+f)}) = r$;

2. $\rho(B_{ii}) < r$ for $i = 1, \ldots, t$;

3. $r = \rho(A)$.

Proof. Clearly, A has a positive eigenvector corresponding to r, if and only if B has a positive eigenvector corresponding to r. Thus we may assume that $A = B$. Suppose first that $Bx = rx$ for $x > 0$. Let $x^\top = (u_1^\top, \ldots, u_{t+f}^\top)$, where $0 < u_i \in \mathbb{R}^{n_i}$ for $i = 1, \ldots, t+f$. Since $B' = [B_{ij}]_{i=j=t+1}^{t+f}$ is a block diagonal matrix we deduce that $B_{ii} u_i = r u_i$ for $i = t+1, \ldots, t+f$. Proposition 6.2.7 yields the equality $r = \rho(B_{(t+1)(t+1)}) = \ldots = \rho(B_{(t+f)(t+f)})$. Hence, 1 holds. Furthermore,

$$B_{ii} u_i + \sum_{j=i+1}^{t+f} B_{ij} u_j = r u_i, \quad i = 1, \ldots, t. \tag{6.4.4}$$

Since for each $i \in [t]$ there exists an integer $j(i) \in \{i+1, \ldots, t+f\}$ such that $B_{ij} \gneqq 0$ we deduce that $B_{ii} u_i \lneqq r u_i$ for each $i \in [1, t]$. Use Problem 6.2.11 to deduce that $\rho(B_{ii}) < r$ for $i \in [t]$. Hence, 2 holds. As B is block upper triangular then $\rho(B) = \max_{i \in [t+f]} \rho(B_{ii}) = r$. Hence, 3. holds.

Assume now that 1–3 hold. Consider the equality (6.4.4) for $i = t$. Rewrite it as

$$v_t = \sum_{j=t+1}^{t+f} B_{tj} u_j = (rI - B_{tt}) u_i.$$

Since some $B_{tj} \gneqq 0$ it follows that $v_j \gneqq 0$. As $r > \rho(B_{tt})$ and B_{tt} is irreducible, Lemma 6.4.3 implies that $(rI - B_{tt})^{-1} > 0$. Hence, $u_t := (rI - B_{tt})^{-1} v_t > 0$. Thus we showed that there exists $u_t > 0$ so that equality (6.4.4) holds for $i = t$. Suppose we already showed that there exists $u_t, \ldots, u_k > 0$ such that (6.4.4) holds for $i = t, t-1, \ldots, k$. Consider the equality (6.4.4) for $i = k - 1$. Viewing this equality as a system of equations in u_{k-1}, the above arguments show the existence of unique solution $u_{k-1} > 0$. Let $x^\top = (u_1^\top, \ldots, u_{t+f}^\top)$. Then $Bx = rx$. $\qquad\square$

Corollary 6.4.6 *Let $A \in \mathbb{R}_+^{n \times n}$. Then $Ax = \rho(A)x$, $A^\top y = \rho(A)y$, where $x, y > 0$ if and only if:*

1. *The Frobenius normal form of A, given by (6.4.3), is block diagonal, i.e. $t = 0$.*

2. $\rho(B_{11}) = \ldots = \rho(B_{ff})$.

Theorem 6.4.7 *Let $A \in \mathbb{R}_+^{n \times n}$. Assume that $Ax = x$ for some $\mathbf{x} > 0$. $B = [B_{ij}]_{i=j=1}^{t+f}$ be the Frobenius normal form of A given by (6.4.3). Denote $B^k = [B_{ij}^{(k)}]_{i=j=1}^{t+f}$ for $k = 1, 2, \ldots$ Then the block matrix form of B^k is of the form (6.4.3). Furthermore, the following conditions hold.*

1. $\lim_{k \to \infty} B_{ij}^{(k)} = 0$ *for $i, j = 1, \ldots, t$.*

2. $A^k, k = 1, 2, \ldots,$ *converge to a limit if and only if the matrices B_{ii} are primitive for $i = t+1, \ldots, t+f$.*

3. *Assume that B_{ii} are primitive and $B_{ii}\mathbf{u}_i = \mathbf{u}_i$, $B_{ii}^\top \mathbf{v}_i = \mathbf{v}_i$, $\mathbf{u}_i, \mathbf{v}_i > \mathbf{0}$, $\mathbf{v}_i^\top \mathbf{u}_i = 1$ for $i = t+1, \ldots, t+f$. Then $\lim_{k \to \infty} B^k = E = [E_{ij}]_{i=j=1}^{f+t} \geq 0$, where E has the block matrix form (6.4.3). Furthermore*

   (a) *E is a non-negative projection, i.e. $E^2 = E$.*

   (b) *$E_{ii} = \mathbf{u}_i \mathbf{v}_i^\top$ for $i = t+1, \ldots, t+f$, and $E_{ij} = 0$ for $i, j = 1, \ldots, t$.*

   (c) *Let*

$$E = \begin{bmatrix} 0 & \tilde{E}_{12} \\ 0 & \tilde{E}_{22} \end{bmatrix}, \quad \tilde{E}_{12} = [E_{ij}]_{i=1, j=t+1}^{t, t+f}, \quad (6.4.5)$$

$$\tilde{E}_{22} = \mathrm{diag}(E_{(t+1)(t+1)}, \ldots, E_{(t+f)(t+f)}).$$

   Then $\tilde{E}_{12}\tilde{E}_{22} = \tilde{E}_{12}$. In particular, each row r in a matrix E_{ij}, where $i \in [t]$ and $j \in \{t+1, \ldots, t+f\}$, is of the form $c_{r,ij} \mathbf{v}_j^\top$ for some $c_{r,ij} \geq 0$ for $j = t+1, \ldots, t+f$. Moreover, there exists $j = j(r) > t$, such that $c_{r,ij} > 0$.

Proof. Since B is a block upper triangular, it follows that B^k is block upper triangular. Let $B = \begin{bmatrix} \tilde{B}_{11} & \tilde{B}_{12} \\ 0 & \tilde{B}_{22} \end{bmatrix}$. Assume that $\tilde{B}_{22}$ is block diagonal $\text{diag}(B_{(t+1)(t+1)}, \ldots, B_{(t+f)(t+f)})$. Hence, $\rho(B_{(t+i)(t+i)}) = 1$ for $i = 1, \ldots, f$. Furthermore $\tilde{B}_{11}$ is upper triangular and $\rho(\tilde{B}_{11}) < 1$. Clearly,

$$B^k = \begin{bmatrix} \tilde{B}_{11}^k & \tilde{B}_{12}^{(k)} \\ 0 & \tilde{B}_{22}^k \end{bmatrix},$$

$$\tilde{B}_{11}^k = [B_{ij}^{(k)}]_{i=j=1}^t, \quad \tilde{B}_{22}^k = \text{diag}(B_{(t+1)(t+1)}^k, \ldots, B_{(t+f)(t+f)}^k).$$

$$(6.4.6)$$

Hence, B^k is of the form (6.4.3). As $\rho(\tilde{B}_{11}) < 1$ Theorem 3.3.2 yields that $\lim_{k \to \infty} \tilde{B}_{11}^k = 0$. This implies 1.

Clearly, $A^k, k = 1, 2, \ldots$ converges if and only if the sequence $B^k, k = 1, 2, \ldots$ converges. Assume first that the second sequence converges. In particular, the sequence as $B_{ii}^{(k)} = B_{ii}^k$ for $k = t + 1, \ldots, t + f$, we deduce that the sequences $B_{ii}^k, k = 1, 2, \ldots$, converge for $i = [t + 1, t + f]$. Theorem 3.3.2 yields that the only eigenvalue of B_{ii} on the unit circle is 1. Since B_{ii} is irreducible and $\rho(B_{ii}) = 1$ for $i \in [t + 1, t + f]$ we deduce that each B_{ii} is primitive.

Assume now that each B_{ii} is primitive. Hence, the algebraic multiplicity of the eigenvalue of 1 of B is f. We claim that the geometric multiplicity of 1 is also f. Let $\mathbf{u}_{t+1}, \ldots, \mathbf{u}_{t+f}$ be defined as in 3. For any $a_{t+1}, \ldots, a_{t+f} > 0$ we have that $B_{ii}(a_i \mathbf{u}_i) = a_i \mathbf{u}_i$ for $i = t + 1, \ldots, t + f$. From the proof of Theorem 6.4.5 it follows that B has a positive eigenvector $\mathbf{x}$, $B\mathbf{x} = \mathbf{x}$, such that $\mathbf{x}^\top = (\mathbf{x}_1^\top, \ldots, \mathbf{x}_t^\top, a_{t+1}\mathbf{u}_{t+1}^\top, \ldots, a_{t+f}\mathbf{u}_{t+f}), \mathbf{x}_i \in \mathbb{R}_+^{n_i}, i = 1, \ldots, t$. Hence the subspace of eigenvectors of B corresponding to the eigenvalue 1 has dimension f at least f. Since the algebraic multiplicity of 1 is f it follows that the geometric multiplicity of 1 is f. As all other eigenvalues λ of B satisfy $|\lambda| < 1$ Theorem 3.3.2 yields that $\lim_{k \to \infty} B^k = E$. This implies 2. Furthermore, $E \geq 0$ is a projection. This shows 3(a).

Assume that $i \in \{t + 1, \ldots, t + f\}$. As B_{ii} is primitive and $\rho(B_{ii}) = 1$ Theorem 3.3.2 yields $\lim_{k \to \infty} B_{ii}^k = E_{ii}$. $E_{ii} \geq 0$ is a projection with only one eigenvalue equal to 1, which is simple. So rank $E_{ii} = 1$. Clearly $E_{ii}\mathbf{u}_i = \mathbf{u}_i, E_{ii}^\top \mathbf{v}_i = \mathbf{v}_i$. Hence, $E_{ii} = t_i \mathbf{u}_i \mathbf{v}_i^\top$.

As $E_{ii}^2 = E_{ii}$ we deduce $t_i^2 = t_i$, i.e. $t_i \in \{0, 1\}$. Since rank $E_{ii} = 1$ it follows that $t_i = 1$. 1 yields that $E_{ij} = 0$ for $i, j \in [t]$. This shows 3(b).

Let $i \in [t]$. Recall that $E_{ij} = 0$ for $j \in [t]$. Since $E\mathbf{x} = \mathbf{x}$ it follows that $\sum_{j=t+1}^{t+f} E_{ij}\mathbf{u}_i = \mathbf{u}_j$. Hence, for each $i \in [t]$ and a given row r in matrices $E_{i(t+1)}, \ldots, E_{i(t+f)}$, there exists $j = j(r) > t$, such that E_{ij} has a positive element in row r. Assume that E is of the form (6.4.5). As $E^2 = E$ we deduce that $\tilde{E}_{12}\tilde{E}_{22} = \tilde{E}_{12}$. As each row of $\tilde{E}_{12}$ non-negative and nonzero, it follows that each row of $\tilde{E}_{12}$ is a non-negative eigenvector of $\tilde{E}_{22}$ corresponding to the eigenvalue 1. This shows 3(c). □

The following result characterize non-negative projections:

Theorem 6.4.8 *Let $F \in \mathbb{R}_+^{n \times n}$ be a projection, i.e. $F^2 = F$. Then F is permutationally similar to a non-negative projection G, i.e. $G = PFP^\top$ for some $P \in \mathcal{P}_n$, which has one of the following forms:*

1. $G_0 = 0_{n \times n}$.

2. (a) $G_1 = \mathrm{diag}(\mathbf{u}_1\mathbf{v}_1^\top, \ldots, \mathbf{u}_t\mathbf{v}_t^\top)$, *where* $0 < \mathbf{u}_i, \mathbf{v}_i \in \mathbb{R}_+^{n_i}, \mathbf{v}_i^\top\mathbf{u}_i = 1, i = 1, \ldots, t$.

   (b) $G_2 = \begin{bmatrix} 0 & R \\ 0 & G_1 \end{bmatrix}$, *where $G_1 \in \mathbb{R}^{m \times m}$, $m \in [n-1]$, is of the form given in 2(a), and each k-row of R is of the form $(r_{k1}\mathbf{v}_1^\top, \ldots, r_{kt}\mathbf{v}_k^\top)$, where $(r_{k1}, \ldots, r_{kt}) \gneq \mathbf{0}^\top$.*

   (c) $G_3 = \begin{bmatrix} T & H \\ 0 & 0 \end{bmatrix}$, *where $T \in \mathbb{R}_+^{m \times m}$, $m \in [n-1]$, has the form either 2(a) or 2(b), and each column of H is either a zero vector or a non-negative eigenvector of T corresponding to the eigenvalue 1.*

Proof. If $F = 0$ then we have the case 1. Assume that $F \neq 0$. Since $FF = F$ it follows that each column of F is either zero or a non-negative eigenvector of F corresponding to the eigenvalue 1.

Assume first that F does not have a zero row. Let $\mathbf{x}$ be the sum of all columns of F. Then $\mathbf{x} > \mathbf{0}$ and $F\mathbf{x} = \mathbf{x}$. Hence, F satisfies the assumption of Theorem 6.4.7. Let $G = PFP^\top$ be the Frobenius normal form of F. Clearly, $G^k = G$. Hence $G = E$, where E is the projection given in Theorem 6.4.7. If $t = 0$ then we have the case 2(a). If $t \geq 1$ then G has the form 2(b).

Assume now that F has exactly $n - m$ zero rows for some $m \in [n-1]$. Let $F_1 = P_1 F P_1^\top$ such that the last $n - m$ rows of F_1 are zero for some $P \in \mathcal{P}_n$. Let $\mathbf{x}$ be the sum of all columns of P_1. Then $\mathbf{x}^\top = (\mathbf{x}_1^\top, \mathbf{0}_{n-m}^\top)$, $\mathbf{0} < \mathbf{x}_1 \in \mathbb{R}_+^m$ and $F_1 \mathbf{x} = \mathbf{x}$. Let $F_1 = \begin{bmatrix} F_{11} & F_{12} \\ 0 & 0 \end{bmatrix}$. Then

$$F_{11}\mathbf{x}_1 = \mathbf{x}_1, \quad F_{11}^2 = F_{11}, \quad F_{11}F_{12} = F_{12}.$$

Hence, F_{11} satisfies the assumption of Theorem 6.4.7. Let $G_3 = P_2 F_1 P_2^\top$, where $P_2 \in \mathcal{P}_n$ is a permutation that acts as identity on the last $n - m$ coordinates of a vector $\mathbf{y}$, such that T is the Frobenius normal form of F_{11}. Hence, T has the form either 2(a) or 2(b). Furthermore $TH = H$. Let $\mathbf{y}$ be a column of H. Then $T\mathbf{y} = \mathbf{y}$. Hence, either $\mathbf{y} = \mathbf{0}$ or $\mathbf{y}$ is a non-negative eigenvector of T corresponding to the eigenvalue 1. $\qquad \square$

Theorem 6.4.9 *Let* $A = \mathbb{R}_+^{n \times n}$. *Then*

$$\rho(A) = \limsup_{m \to \infty} (\operatorname{tr} A^m)^{\frac{1}{m}}. \tag{6.4.7}$$

Proof. Clearly, for any $B \in \mathbb{C}^{n \times n}$, $|\operatorname{tr} B| = |\sum_{i=1}^n \lambda_i(B)|$. Hence $|\operatorname{tr} B| \leq n\rho(B)$. Therefore, $\operatorname{tr} A^m = |\operatorname{tr} A^m| \leq n\rho(A^m) = n\rho(A)^m$. Thus, $(\operatorname{tr} A^m)^{\frac{1}{m}} \leq n^{\frac{1}{m}}\rho(A)$. Therefore, $\limsup_{m \to \infty} (\operatorname{tr} A^m)^{\frac{1}{m}} \leq \rho(A)$. It is left to show the opposite inequality.

Assume first that A is an irreducible and primitive. Let $A\mathbf{u} = \rho(A)\mathbf{u}$, $A^\top \mathbf{v} = \rho(A)\mathbf{v}$, $\mathbf{0} < \mathbf{u}, \mathbf{v}, \mathbf{v}^\top \mathbf{u} = 1$. Theorem 6.4.7 yields that $\lim_{m \to \infty} \frac{1}{\rho(A)^m} A^m = \mathbf{u}\mathbf{v}^\top$. Hence, for $m > M$

we have:

$$\operatorname{tr} A^m \geq \rho(A)^m \frac{1}{2} \operatorname{tr} \mathbf{u}\mathbf{v}^\top = \frac{\rho(A)^m}{2} \Rightarrow \lim_{m \to \infty} (\operatorname{tr} A^m)^{\frac{1}{m}} \geq \rho(A).$$

Assume that A is an irreducible and imprimitive. If A 1×1 zero matrix, then (6.4.7) trivially holds. Assume that $n > 1$. Without loss of generality we can assume that A is of the form given in Theorem 6.2.1 part 5(d). Then $A^h = \operatorname{diag}(B_1, \ldots, B_h)$, where each B_j is primitive and $\rho(B_j) = \rho(A)^h$, see Lemma 6.2.9. So $\operatorname{tr} A^{hk} = \sum_{j=1}^{h} \operatorname{tr} B_j^k$. Since each B_j is primitive and irreducible, we deduce from the previous arguments that $\lim_{k \to \infty} (\operatorname{tr} A^{hk})^{\frac{1}{hk}} = \rho(A)$. Hence, (6.4.7) holds in this case too.

Assume now that A is not irreducible. Without loss of generality we can assume that A is in the Frobenius form (6.4.3). Then there exists $i \in [t + f]$ such that $\rho(A) = \rho(B_{ii})$. Clearly

$$(\operatorname{tr} A^m)^{\frac{1}{m}} = \left(\sum_{j=1}^{t+f} \operatorname{tr} B_{jj}^m \right)^{\frac{1}{m}} \geq (\operatorname{tr} B_{ii}^m)^{\frac{1}{m}} \Rightarrow$$

$$\limsup_{m \to \infty} (\operatorname{tr} A^m)^{\frac{1}{m}} \geq \limsup_{m \to \infty} (\operatorname{tr} B_{ii}^m)^{\frac{1}{m}} = \rho(B_{ii}) = \rho(A).$$

$\square$

Problems

1. Let $A \in \mathbb{R}^{n \times n}$. Show that $\det (tI - A) > 0$ for $t > \rho(A)$.

6.5 Stochastic Matrices and Markov Chains

Definition 6.5.1 *A matrix S is called a stochastic matrix if $S \in \mathbb{R}_+^{n \times n}$, for some integer $n \geq 1$, and $S\mathbf{1} = \mathbf{1}$. Denote by $\mathcal{S}_n \subset \mathbb{R}_+^{n \times n}$ the set of $n \times n$ stochastic matrices.*

Note that $A \in \mathbb{R}^{n \times n}$ is a stochastic matrix, if and only if each row of A is a probability vector. Furthermore, $\mathcal{S}_n$ is compact semigroup with respect to the product of matrices, see Problem 1.

Definition 6.5.2 *Let $A \in \mathcal{S}_n$.*

1. *Let B be the Frobenius normal form of A given by (6.4.3). Denoted by $V_1, \ldots, V_{t+f}$ the partition of $[n]$ corresponding the partition of B. $i \in V_j$ is called a final state if $j \in \{t+1, \ldots, t+f\}$ and a transient state if $j \in [t]$. $V_{t+1}, \ldots, V_{t+f}$ and $V_1, \ldots, V_t$ are called the final and the transient classes of A respectively.*

2. *$\boldsymbol{\pi} \in \Pi_n$ is called a stationary distribution of A if $\boldsymbol{\pi}^\top A = \boldsymbol{\pi}^\top$.*

Theorem 6.5.3 *Let A be a stochastic matrix. Then*

1. *$\rho(A) = 1$.*

2. *The algebraic multiplicity of the eigenvalue 1 is equal to f: the number of final classes of A. The geometric multiplicity of 1 is f.*

3. *The set of stationary distributions of A is a convex set spanned by the stationary distributions corresponding to the final classes of A. In particular, the stationary distribution is unique, if and only if $f = 1$.*

4. *Assume that λ is an eigenvalue of S such that $|\lambda| = 1$ and $\lambda \neq 1$. Then*

   (a) *λ is an m-th root of unity, where m divides the cardinality of some final class of A.*

   (b) *All other mth roots if unity are eigenvalues of A.*

   (c) *Let $m(\lambda)$ be the algebraic multiplicity of λ. Then $m(\lambda) \leq f$. Furthermore, the geometric multiplicity of λ is $m(\lambda)$.*

Proof. In what follows we apply Theorem 6.4.5. As $A\mathbf{1} = \mathbf{1}$ it follows that $\rho(A) = 1$.

Assume that B is the Frobenius normal of A given by (6.4.3). Since B is stochastic and similar to A by a permutation matrix, it is enough to prove the other statements of the theorem for $A = B$. Recall that $\rho(B_{ii}) < 1$ for $i \in [t]$ and $\rho(B_{ii}) = 1$ for $i = t+1, \ldots, t+f$.

As each B_{ii} is irreducible it follows that 1 is a simple eigenvalue of B_{ii} for $i = t+1, \ldots, t+f$. Hence, the algebraic multiplicity of $\lambda = 1$ is f.

Recall that the geometric multiplicity λ of an eigenvalue λ of B is equal to the geometric multiplicity of λ in $B^\top$. (It is equal to the number of Jordan blocks corresponding to λ in the Jordan canonical form of B.) Recall that $B_{ii} \in \mathbb{R}^{n_i \times n_i}$ is irreducible for $i \in [t+f]$ and stochastic for $i > t$. Hence, B_{ii} has a unique stationary distribution $\boldsymbol{\pi}_i \in \Pi_{n_i}$ for $i > t$. Let

$$\mathbf{x}_i := (\mathbf{0}_{n_1}^\top, \ldots, \mathbf{0}_{n_{i-1}}^\top, \boldsymbol{\pi}_i^\top, \mathbf{0}_{n_{i+1}}, \ldots, \mathbf{0}_{n_{t+f}}^\top)^\top, \quad i = t+1, \ldots, t+f.$$

$$(6.5.1)$$

Then $\mathbf{x}_i$ is a stationary distribution of B. We call $\mathbf{x}_i$ the stationary distribution of B, or A, induced by the final class V_i. Clearly, $\mathbf{x}_{t+1}, \ldots, \mathbf{x}_{t+f}$ are f linearly independent eigenvectors of $B^\top$ corresponding to $\lambda = 1$. Hence, the geometric multiplicity of $\lambda = 1$.

Assume that $\boldsymbol{\pi}$ is a stationary distribution of B. So $B^\top \boldsymbol{\pi} = \boldsymbol{\pi}$. Hence, $\boldsymbol{\pi} = \sum_{i=t+1}^{f} c_i \mathbf{x}_i$. Since $\boldsymbol{\pi} \in \Pi_n$ and $\mathbf{x}_{t+1}, \ldots, \mathbf{x}_{t+f}$ are of the form (6.5.1) it follows that $c_i \geq$ for $i = t+1, \ldots, t+f$ and $c_{t+1} + \cdots + c_{t+f} = 1$. That is, $\boldsymbol{\pi}$ is in the convex set spanned by stationary distributions of B corresponding to final classes of B. Vice versa, if $\boldsymbol{\pi}$ is in the convex set spanned by stationary distributions of B corresponding to final classes of B then $\boldsymbol{\pi}$ is a stationary distribution of B.

Assume that λ is an eigenvalue of B, such that $|\lambda| = 1$ and $\lambda \neq 1$. Hence, λ is an eigenvalue of some $B_{ii} \in \mathbb{R}^{n_i \times n_i}$ for $i \in \{t+1, \ldots, t+f\}$. Since B_{ii} is an irreducible stochastic matrix Perron–Frobenius theorem yield that that all eigenvalues of B_{ii} are mth roots of unity, where m divides $n_i = \#V_i$. Furthermore, λ is a simple eigenvalue of B_{ii}. Hence, $m(\lambda) \leq f$. The proof that $m(\lambda)$ is the geometric multiplicity of λ is similar to the proof that the geometric multiplicity of $\lambda = 1$ is f. $\qquad \square$

The following lemma is straightforward, see Problem 2.

Lemma 6.5.4 *Let $A \in \mathbb{R}_+^{n \times n}$. Then $A = DSD^{-1}$ for some $S \in \mathcal{S}_n$ and a diagonal matrix $D \in \mathbb{R}_+^{n \times n}$ with positive diagonal entries if and only if $A\mathbf{x} = \mathbf{x}$ for some positive $\mathbf{x} \in \mathbb{R}^n$.*

Definition 6.5.5 *Let $S \in \mathcal{S}_n$ be irreducible. We will assume the normalization that the eigenvector of S and $S^\top$ corresponding to the eigenvalue 1 are of the form $\mathbf{1} = \mathbf{1}_n \in \mathbb{R}_+^n$ and $\boldsymbol{\pi} \in \Pi_n$, the stationary distribution of A respectively, unless stated otherwise.*

Theorem 6.4.7 yields:

Theorem 6.5.6 *Let $A \in \mathcal{S}_n$. Denote by $B = [B_{ij}]_{i=j=1}^{t+f}$ the Frobenius normal form of A given by (6.4.3). Then $B, B_{(t+1)(t+1)}, \ldots, B_{(t+f)(t+f)}$ are stochastic. The sequence $A^k, k \in \mathbb{N}$ converges if and only if $B_{(t+1)(t+1)}, \ldots, B_{(t+f)(t+f)}$ are primitive. Assume that $B_{(t+1)(t+1)}, \ldots, B_{(t+f)(t+f)}$ are primitive. Then $\lim_{k \to \infty} B^k = E$, where E is stochastic matrix satisfying the conditions 3 of Theorem 6.4.7. Furthermore, in condition $3(b)$ we can assume that $\mathbf{u}_i = \mathbf{1}_{n_i}, \mathbf{v}_i \in \Pi_{n_i}$ for $i = t + 1, \ldots, t + f$.*

Definition 6.5.7 *Let $C = [c_{ij}]_{i=j=1}^n \in \mathbb{R}_+^{n \times n}$. Denote by $r_i(C) := \sum_{j=1}^n c_{ij}$ the ith row sum of C for $i \in [n]$. Let $R(C) = \mathrm{diag}(r_1(C), \ldots, r_n(C))$. Assume that $r_i(C) > 0$ for $i \in [n]$. Then $A := R(C)^{-1}C$ is a stochastic matrix, called the stochastic matrix induced by C. If C is a symmetric matrix then A is called a symmetrically induced stochastic matrix.*

Lemma 6.5.8 *Let $C \in \mathbb{R}_+^{n \times n}$ be a matrix with no zero row and $A = [a_{ij}]_{i=j=1}^n \in \mathcal{S}_n$. Then*

1. *A is induced by C if and only $C = DA$, where D is a diagonal matrix with positive diagonal entries.*

2. *Assume that C is symmetric. Then*

$$\boldsymbol{\pi}(C) := \frac{1}{\sum_{i=1}^n r_i(C)}(r_1(C), \ldots, r_n(C))^\top$$

is a stationary distribution of stochastic A induced by C.

3. *Assume that A is symmetrically induced. Then the Frobenius normal form of A is* $A_1 = \mathrm{diag}(B_{11}, \ldots, B_{ff})$ *where each* $B_{ii} \in \mathcal{S}_{n_i}$ *is a symmetrically induced irreducible stochastic matrix. Let* $\boldsymbol{\pi}_i = (\pi_{1,i}, \ldots, \pi_{n_i,i})^\top \in \Pi_{n_i}$ *be the unique stationary distribution of* B_{ii}. *Let* $D(\boldsymbol{\pi}_i) = \mathrm{diag}(\pi_{1,i}, \ldots, \pi_{n_i,i})$. *Then a nonnegative symmetric matrix* C_1 *induces* A_1 *if and only if* $C_1 = \mathrm{diag}(a_1 D(\boldsymbol{\pi}_1) B_{11}, \ldots, a_f D(\boldsymbol{\pi}_f) B_{ff})$ *for some* $a_1, \ldots, a_f > 0$.

4. *A is symmetrically induced if and only if the following conditions hold: There exists a positive stationary distribution* $\boldsymbol{\pi} = (\pi_1, \ldots, \pi_n)^\top$ *of A such that* $\pi_i a_{ij} = \pi_j a_{ji}$ *for all* $i, j \in [n]$.

Proof. 1. is straightforward. 2. Assume that C is symmetric with no zero row. So the sum of the ith column is $r_i(C)$. Hence,

$$\gamma \mathbf{1}^\top C = \boldsymbol{\pi}(C)^\top = \boldsymbol{\pi}(C) A, \quad \gamma := \frac{1}{\sum_{i=1}^n r_i(C)}.$$

3. Assume that A is induced by a symmetric C. Then $D(A) = D(C)$. Clearly, $D(A)$ is combinatorially symmetric. $D(C)$ induces a simple graph $G(C)$, where we ignore the loops. So $G(C)$ is a union of f connected simple graphs. This shows that the Frobenius normal form of A is $B = \mathrm{diag}(B_{11}, \ldots, B_{ff})$, where each B_{ii} is irreducible and combinatorially symmetric. Without loss of generality we may assume that $A = B$. Then $C = \mathrm{diag}(C_1, \ldots, C_f)$ and $C_i = [c_{pq,i}]_{p=q=1}^{n_i}$ induces $B_{ii} \in \mathcal{S}_{n_i}$ for $i \in [f]$. Let $\boldsymbol{\pi}(C_i) = (\pi_{1,i}, \ldots, \pi_{n_i,i})^\top$ be the unique stationary distribution of B_{ii} as given in 2. Then $D(\boldsymbol{\pi}_i) B_{ii} = \gamma C_i$. Suppose that $F_i = [f_{pq,i}]_{p=q=1}^{n_i} \in \mathbb{R}_+^{n_i \times n_i}$ is a symmetric matrix that induces B_{ii}. 1 yields that there exists diagonal matrix $D_i = \mathrm{diag}(d_{1,i}, \ldots, d_{n_i,i})$ with positive diagonal entries so that $F_i = D_i \gamma D(\boldsymbol{\pi}_i) B_{ii} = D_i C_i$. It is left to show that $d_{1,i} = \ldots = d_{n_i,1}$. Suppose that $c_{pq,i} > 0$ for $p \neq q$. Since F_i is symmetric it follows that $d_{p,i} = d_{q,i}$. As the induced simple graph $G(C_i)$ is connected it follows that $d_{1,i} = \ldots = d_{n_i,i} = b_i$. Suppose that a symmetric F induces A. Then $F = \mathrm{diag}(F_1, \ldots, F_s)$. The previous argument shows that $F_i = b_i C_i$ for some $b_i > 0$. Hence, $F = \mathrm{diag}(a_1 \gamma C_1, \ldots, a_f \gamma C_g)$ as claimed.

4. Assume that A is induced by a symmetric C. Then $A_1 = PAP^\top = \mathrm{diag}(B_{11}, \ldots, B_{ff})$ for some permutation matrix P, and $D(\pi(B_{ii}))B_{ii}$ are symmetric for $i \in [f]$. 2 shows that $\pi(C)$ is a positive stationary distribution of A. Part 3. of Theorem 6.5.3 yields that $\pi(C) = (\pi_1, \ldots, \pi_n)^\top = (c_1\pi(B_{11}^\top), \ldots, c_f\pi(B_{ff})^\top)^\top$ for some $c_1, \ldots, c_f > 0$, which satisfy $\sum_{i=1}^f c_i = 1$. Since each $D(\pi(B_{ii}))B_{ii}$ is symmetric it follows that $\pi_p a_{pq} = \pi_q a_{qp}$ if $p, q \in V_i$, where $V_i \subset [n]$ induces the finite class corresponding to B_{ii}. If p and q belong to different finite classes then $a_{pq} = a_{qp} = 0$. Hence the conditions of 4 hold.

Assume now that the conditions (4) hold. So $C = D(\pi)A$ is a symmetric matrix. Clearly, C induces A. $\qquad\square$

We now recall the classical connection between the stochastic matrices and Markov chains. To each probability vector $\pi = (\pi_1, \ldots, \pi_n)^\top \in \Pi_n$ we associated a random variable X, which takes values in the set $[n]$, such that $P(X = i) = \pi_i$ for $i = 1, \ldots, n$. Then $\pi(X) := \pi$ is called the *distribution* of X.

Assume that we are given a sequence of random variables $X_0, X_1, \ldots$ each taking values in the set $[n]$. Let $s_{ij}^{(k)}$ be the conditional probability of $X_k = j$ given that $X_{k-1} = i$:

$$s_{ij}^{(k)} := P(X_k = j | X_{k-1} = i), \quad i, j = 1, \ldots, n, \quad k = 1, \ldots \quad (6.5.2)$$

Clearly, $S_k = [s_{ij}^{(k)}]_{i=j=1}^n, k = 1, 2, \ldots$ is a stochastic matrix for $k = 1, \ldots$

Definition 6.5.9 *Let $X_0, X_1, \ldots$, be a sequence of random variables taking values in $[n]$. Then*

1. *$X_0, X_1, \ldots$ is called a homogeneous Markov chain if*

$$P(X_k = j_k | X_{k-1} = j_{k-1}, \ldots, X_0 = j_0) = P(X_1 = j_k | X_0 = j_{k-1})$$

   for $k = 1, 2, \ldots$

2. $X_0, X_1, \ldots$ *is called a nonhomogeneous Markov chain if*

$$\mathrm{P}(X_k = j_k | X_{k-1} = j_{k-1}, \ldots, X_0 = j_0)$$
$$= \mathrm{P}(X_k = j_k | X_{k-1} = j_{k-1})$$

for $k = 1, 2, \ldots$ *(Note that a homogeneous Markov chain is a special case of nonhomogeneous Markov chain.)*

3. *A nonhomogeneous Markov chain is called reversible if there exists* $\boldsymbol{\pi} = (\pi_1, \ldots, \pi_n) \in \Pi_n$ *such that*

$$\pi_i \mathrm{P}(X_k = j | X_{k-1} = i) = \pi_j \mathrm{P}(X_k = i | X_{k-1} = j)$$

for $k = 1, 2, \ldots$.

4. *Assume that* $X_0, X_1, \ldots$ *is a nonhomogeneous Markov chain.* $\boldsymbol{\pi} \in \Pi_n$ *is called a stationary distribution if the assumption that* $\boldsymbol{\pi}(X_0) = \boldsymbol{\pi}$ *yields that* $\boldsymbol{\pi}(X_k) = \boldsymbol{\pi}$ *for* $k \in \mathbb{N}$. *Markov chain is said to have a limiting distribution if the limit* $\boldsymbol{\pi}_\infty := \lim_{k \to \infty} \boldsymbol{\pi}(X_k)$ *exists for some* $\boldsymbol{\pi}(X_0)$. *If* $\boldsymbol{\pi}_\infty$ *exists for each* $\boldsymbol{\pi}(X_0) \in \Pi_n$ *and does not depend on* $\boldsymbol{\pi}(X_0)$ *then* $\boldsymbol{\pi}_\infty$ *is called the unique limiting distribution of the Markov chain.*

The following lemma is straightforward, see Problem 4.

Lemma 6.5.10 *Let* $X_0, X_1, \ldots$ *be a sequence of random variables taking values in* $[n]$. *Let* $\boldsymbol{\pi}_k := \boldsymbol{\pi}(X_k)$ *be the distribution of* X_k *for* $k = 0, 1, \ldots$ *Assume that* $X_0, X_1, \ldots$, *is a nonhomogeneous Markov chain. Then*

1. $\boldsymbol{\pi}_k^\top = \boldsymbol{\pi}_0^\top S_1 \ldots S_k$ *for* $k = 1, \ldots$, *where* S_k *are defined by* (6.5.2).

2. *If* $X_0, X_1, \ldots$, *is a homogeneous Markov process, i.e.* $S_k = S, k = 1, 2, \ldots$, *then* $\boldsymbol{\pi}_k^\top = \boldsymbol{\pi}_0^\top S^k$ *for* $k = 1, \ldots$

3. *Assume that* $X_0, X_1, \ldots$ *is a reversible Markov chain with respect to* $\boldsymbol{\pi}$. *Then* π *is a stationary distribution of* S_k *for* $k = 1, \ldots$ *Furthermore* $\pi_i s_{ij}^{(k)} = \pi_j s_{ji}^{(k)}$ *for* $i, j \in [n]$ *and* $k \in \mathbb{N}$.

Theorem 6.5.11 *Let $X_0, X_1, \ldots$, be a homogeneous Markov chain on $[n]$, given by a stochastic matrix $S = [s_{ij}] \in \mathcal{S}_n$. Let $D(S)$ and $D_r(S)$ be the digraph and the reduced digraph corresponding to S. Label the vertices of the reduced graph $D_r(S)$ by $\{1, \ldots, t + f\}$. Let $V_1, \ldots, V_{t+f}$ be the decomposition of $[n]$ to the strongly connected components of the digraph $D(S)$. Assume that $B = PSP^\top, P \in \mathcal{P}_n$ is given by the form (6.4.3). The vertices, (states), in $\cup_{i=1}^t V_i$ are called the transient vertices, (states). (Note that if $t = 0$ then no transient vertices exist.) The sets $V_1, \ldots, V_t$ are called the transient classes. The vertices, (states), in $\cup_{i=t+1}^{t+f} V_i$ are called the final vertices, (states). $V_{t+1}, \ldots, V_{t+f}$ are called the final classes. Furthermore the following conditions hold.*

1. *The Markov chain is reversible if and only if B_{ii} is symmetrically induced for some $i \in \{t + 1, \ldots, t + f\}$.*

2. *The Markov chain is reversible with respect to a positive distribution if and only if A is symmetrically induced.*

3. *For each $i \in \cup_{j=1}^t V_j$ $\lim_{k \to \infty} \mathrm{P}(X_k = i) = 0$.*

4. *Any stationary distribution $\boldsymbol{\pi}$ of S is a limiting distribution for $\boldsymbol{\pi}(X_0) = \boldsymbol{\pi}$.*

5. *The set of limiting distributions of the Markov chain is the set of stationary distributions.*

6. *For each $\boldsymbol{\pi}(X_0)$ the limiting distribution exists if and only if each B_{ii} is primitive for $i = t + 1, \ldots, t + f$.*

7. *The Markov chain has a unique limiting distribution if and only if $f = 1$ and $B_{(t+1)(t+1)}$ is primitive.*

Proof. Without loss of generality we may assume that $S = B$. Assume that the Markov chain is reversible with respect to $\boldsymbol{\pi} = (\pi_1, \ldots, \pi_n) \in \Pi_n$. Lemma 6.5.10 yields that $\boldsymbol{\pi}$ is a stationary distribution of B. Theorem 6.5.3 claims that $\boldsymbol{\pi}$ is supported on the final classes $V_{t+1}, \ldots, V_{t+f}$. Suppose that the support of $\boldsymbol{\pi}$ is on $V_{t+i_1}, \ldots, V_{t+i_l}$ for $1 \le i_1 < \cdots < i_l \le f$, i.e. $\pi_i = 0$

for $i \in [n] \setminus (\cup_{j=1}^l V_{t+i_j})$. The proof of Theorem 6.5.3 yields that $\pi_i > 0$ if $i \in V_{t+i_j}$ for $j \in [l]$. Let $B' = [b'_{ij}]_{i=j=1}^n = \operatorname{diag}(B_{(t+i_1)(t+i_1)}, \ldots, B_{(t+i_l)(t+i_l)})$ and let $\boldsymbol{\pi}' = (\pi'_1, \ldots, pi'_{n'})^\top$ be the restriction of $\boldsymbol{\pi}$ to $\cup_{j=1}^l V_{t+i_j}$. Then $D(\boldsymbol{\pi}')B'$ is symmetric. Hence, B' is symmetrically induced. In particular, each $B_{(t+i_j)(t+i_j)}$ is symmetrically induced.

Vice versa, assume that $B_{(t+i)(t+i)}$ is symmetrically induced for some $i \in [f]$. Let $\boldsymbol{\pi}$ be the stationary distribution of B supported on V_{t+i}. It is straightforward to show that B is reversible with respect to $\boldsymbol{\pi}$. This shows 1.

Suppose that the Markov chain is reversible with respect to a positive distribution. Hence, $D(\boldsymbol{\pi})A$ is symmetric and A is symmetrically induced. Vice versa, if A is induced by a symmetric C then $\boldsymbol{\pi}(C)$ is a positive stationary distribution of A. This shows 2.

Let $\boldsymbol{\pi}(X_k)^\top = \boldsymbol{\pi}_k^\top = (\boldsymbol{\pi}_{1,k}^\top, \ldots, \boldsymbol{\pi}_{t+f,k}^\top)$ for $k = 0, 1, \ldots$ From the proof of Theorem 6.4.7 we deduce that

$$\boldsymbol{\pi}_{i,k}^\top = \sum_{j=1}^t \boldsymbol{\pi}_{j,0}^\top B_{ji}^{(k)}, \text{ for } i = 1, \ldots, t.$$

In view of part 1 of Theorem 6.4.7 we deduce that $\lim_{k \to \infty} \boldsymbol{\pi}_{i,k} = \mathbf{0}$ for $i = 1, \ldots, t$. This proves part 3.

4. Follows from Lemma 6.5.10.

Clearly, any limit distribution is a stationary distribution. Combine that with 4 to deduce 5.

Suppose that $\boldsymbol{\pi}_0^\top = (\boldsymbol{\pi}_{1,0}^\top, \ldots, \boldsymbol{\pi}_{t+f,0}^\top)$ is supported only on V_{t+i} for some $i \in [f]$. That is $\boldsymbol{\pi}_{j,0} = 0$ if $j \notin V_{t+i}$. Then each $\boldsymbol{\pi}_k$ is supported on V_{t+i}. Furthermore, $\boldsymbol{\pi}_{t+i,k}^\top = \boldsymbol{\pi}_{t+i,0}^\top B_{(t+i)(t+i)}^k$. Assume that $B_{(t+i)(t+i)}$ is imprimitive. Choose $\boldsymbol{\pi}_{t+i,0}$ to have one nonzero coordinate to be 1 and all other coordinates to be zero. Assuming that $B_{(t+i)(t+i)}$ has the form given in 5(d) of Theorem 6.2.1, we see that there is no limit distribution. Hence, to have the limit distribution for each $\boldsymbol{\pi}_0$ we must assume that $B_{(t+i)(t+i)}$ is primitive for $i = 1, \ldots, f$.

Assume that $B_{(t+i)(t+i)}$ is primitive for $i = 1, \ldots, f$. Then Theorem 6.5.6 implies that $\boldsymbol{\pi}_\infty^\top = \boldsymbol{\pi}_0^\top E$, where $E = \lim_{k\to\infty} B^k$ is the stochastic projection. This establishes 6.

As we pointed out before if we assume that $\boldsymbol{\pi}_0$ is supported only on V_{t+i} then the limit probability is also supported on V_{t+i}. Hence, to have a unique stationary distribution we must have that $f = 1$.

Assume that $f = 1$. Then $\lim_{k\to\infty} B_{(t+1)(t+1)}^k = \mathbf{1}_{n_{t+1}} \boldsymbol{\pi}_{t+1}^\top$. Observe that the limit probability is $\boldsymbol{\pi}_0^\top E = (\boldsymbol{\pi}_0^\top E)E$. Since $\boldsymbol{\pi}_0^\top E$ is supported only on V_{t+1} it follows that $\boldsymbol{\pi}_0^\top E^2 = (\underbrace{\mathbf{0}^\top, \ldots, \mathbf{0}^\top}_{t}, \boldsymbol{\pi}_{t+1}^\top)$,

which is independent of $\boldsymbol{\pi}_0$. This shows 7. $\square$

The proof of the above theorem yields the well-known result.

Corollary 6.5.12 *Let $X_0, X_1, \ldots$, be a homogeneous Markov chain on $[n]$, given by a primitive stochastic matrix $S = [s_{ij}] \in \mathcal{S}_n$. Assume that $\boldsymbol{\pi} \in \Pi_n$ is the unique stationary distribution of S. Then this Markov chain has a unique limiting distribution equal to $\boldsymbol{\pi}$.*

A stronger result is proven in [Fri06].

Theorem 6.5.13 *Let $X_0, X_1, \ldots$, be a nonhomogeneous Markov chain on $[n]$, given by the sequence of stochastic matrices $S_1, S_2, \ldots$, defined in (6.5.2). Assume that $\lim_{k\to\infty} S_k = S$, where S is a primitive stochastic matrix with a stationary distribution $\boldsymbol{\pi}$. Then the given nonhomogeneous Markov chain has a unique limiting distribution equal to $\boldsymbol{\pi}$.*

We close this section with *Google's Page Ranking*. Let n be the current number of Web pages. (Currently around a few billions.) Then $S = [s_{ij}] \in \mathcal{S}_n$ is defined as follows. Let $\mathcal{A}(i) \subset [n]$ be the set of all pages accessible from the Web page i. Assume first that i is a *dangling* Web page, i.e. $\mathcal{A}(i) = \emptyset$. Then $s_{ij} = \frac{1}{n}$ for $j = 1, \ldots, n$. Assume now that $n_i = \#\mathcal{A}(i) \geq 1$. Then $s_{ij} = \frac{1}{n_i}$ if $j \in \mathcal{A}(i)$ and otherwise $s_{ij} = 0$. Let $\mathbf{0} < \boldsymbol{\omega} \in \Pi_n, t \in (0, 1)$. Then the Google positive stochastic matrix is given by

$$G = tS + (1 - t)\mathbf{1}\boldsymbol{\omega}^\top. \tag{6.5.3}$$

It is rumored that $t \sim 0.85$. Then the stationary distribution corresponding to G is given by $G^\top \boldsymbol{\pi} = \boldsymbol{\pi} \in \Pi_n$. The coordinates of $\boldsymbol{\pi} = (\pi_1, \ldots, \pi_n)^\top$ constitute Google's popularity score of each Web page. That is, if $\pi_i > \pi_j$ then Web page i is more popular than Web page j.

A reasonable choice of $\boldsymbol{\omega}$ would be the stationary distribution of yesterday Google stochastic matrix. To find the stationary distribution $\boldsymbol{\pi}$ one can iterate several times the equality

$$\boldsymbol{\pi}_k^\top = \boldsymbol{\pi}_{k-1}^\top G, \quad k = 1, \ldots, N. \tag{6.5.4}$$

Then $\boldsymbol{\pi}_N$ would be a good approximation of $\boldsymbol{\pi}$. One can choose $\boldsymbol{\pi}_0 = \boldsymbol{\omega}$.

Problems

1. Show

   (a) If $S_1, S_2 \in \mathcal{S}_n$ then $S_1 S_2 \in \mathcal{S}_n$.

   (b) $P\mathcal{S}_n = \mathcal{S}_n P = \mathcal{S}_n$ for any $P \in \mathcal{P}_n$.

   (c) $\mathcal{S}_n$ is a compact set in $\mathbb{R}_+^{n \times n}$.

2. Prove Lemma 6.5.4.

3. Let $A, B \in \mathbb{C}^{n \times n}$, and assume that A and B are similar. That is, $A = TBT^{-1}$ for some invertible T. Then the sequence $A^k, k = 1, 2, \ldots$, converges if and only if $B^k, k = 1, 2, \ldots$, converges.

4. Prove Lemma 6.5.10.

6.6 Friedland–Karlin Results

Definition 6.6.1 *Let $B = [b_{ij}]_{i=j=1}^n \in \mathbb{R}^{n \times n}$. B is called a Z-matrix if $b_{ij} \leq 0$ for each $i \neq j$. B is called an M-matrix if*

$B = rI - A$ where $A \in \mathbb{R}_+^{n \times n}$ and $r \geq \rho(A)$. B is called a singular M-matrix if $B = \rho(A)I - A$.

The following result is straightforward, see Problem 1.

Lemma 6.6.2 *Let* $B = [b_{ij}] \in \mathbb{R}^{n \times n}$ *be a* Z-matrix. Let $C = [c_{ij}] \in \mathbb{R}_+^{n \times n}$ *be defined as follows.* $c_{ij} = -b_{ij}$ *for each* $i \neq j$ *and* $c_{ii} = r_0 - b_{ii}$ *for* $i = 1, \ldots, n$, *where* $r_0 = \max_{i \in [n]} b_{ii}$. *Then* $B = r_0 I - C$. *Furthermore,* $B = rI - A$ *for some* $A \in \mathbb{R}_+^{n \times n}$ *if and only if* $r = r_0 + t$, $A = tI + C$ *for some* $t \geq 0$.

Theorem 6.6.3 *Let* $B \in \mathbb{R}^{n \times n}$ *be a* Z-matrix. Then TFAE.

1. B *is an* M-matrix.

2. *All principal minors of* B *are non-negative.*

3. *The sum of all* $k \times k$ *principal minors of* B *are non-negative for* $k = 1, \ldots, n$.

4. *For each* $t > 0$ *there exists* $\mathbf{0} < \mathbf{x} \in \mathbb{R}_+^n$, *which may depend on* t, *such that* $B\mathbf{x} \geq -t\mathbf{x}$.

Proof. $1 \Rightarrow 2$. We first show that $\det B \geq 0$. Let $\lambda_1(A), \ldots, \lambda_n(A)$ be the eigenvalues of A. Assume that $\lambda_i(A)$ is real. Then $r - \lambda_i(A) \geq \rho(A) - \lambda_i(A) \geq 0$. Assume that $\lambda_i(A)$ is complex. Since A is a real valued matrix, $\overline{\lambda_i(A)}$ is also an eigenvalue of A. Hence, $(r - \lambda_i(A))(r - \overline{\lambda_i(A)}) = |r - \lambda_i(A)|^2 > 0$. Since $\det B = \prod_{i=1}^n (r - \lambda_i(A))$, we deduce that $\det B \geq 0$. Let B' be a principal submatrix of B. Then $B' = rI' - A'$, where A' is a corresponding principal submatrix of A and I' is the identity matrix of the corresponding order. Part 2 of Proposition 6.4.2 implies that $\rho(A) \geq \rho(A')$. So B' is an M-matrix. Hence, $\det B' \geq 0$.

$2 \Rightarrow 3$. Trivial.

$3 \Rightarrow 1$. Let $\det(tI + B) = t^n + \sum_{i=k}^n \beta_k t^{n-k}$. Then β_k is the sum of all principal minors of B of order k. Hence, $\beta_k \geq 0$ for $k = 1, \ldots, n$. Therefore, $0 < \det(tI + B) = \det((t+r)I - A)$ for $t > 0$. Recall that det $(\rho(A)I - A) = 0$. Thus $t + r > \rho(A)$ for any $t > 0$. So $r \geq \rho(A)$, i.e. B is an M-matrix.

$1 \Rightarrow 4$. Let $t > 0$. Use the Neumann expansion (6.4.2) to deduce that $(tI + B)^{-1} \geq \frac{1}{t+r}I$. So for any $\mathbf{y} > \mathbf{0}$ $\mathbf{x} := (tI + B)^{-1}\mathbf{y} > \mathbf{0}$. So $\mathbf{y} = (tI + B)\mathbf{x} \geq \mathbf{0}$.

$4 \Rightarrow 1$. By considering $PBP^\top = rI - PAP^\top$ we may assume that $A = [A_{ij}]_{i=j=1}^{t+f}$ is in the Frobenius normal form (6.4.3). Let $B\mathbf{x} \geq -t\mathbf{x}$. Partition $\mathbf{x}^\top = (\mathbf{x}_1^\top, \ldots, \mathbf{x}_{t+f}^\top)$. Hence, $(t+r)\mathbf{x}_i \geq A_{ii}\mathbf{x}_i$. Problem 6.2.11 yields that $t + r \geq \rho(A_{ii})$ for $i = 1, \ldots, t + f$. Hence, $t + r \geq \rho(A)$. Since $t > 0$ was arbitrary we deduce that $r \geq \rho(A)$. $\square$

Corollary 6.6.4 *Let B be a Z-matrix. Assume that there exist $\mathbf{x} > \mathbf{0}$ such that $B\mathbf{x} \geq \mathbf{0}$. Then B is an M-matrix.*

Theorem 4.4.2 yields:

Lemma 6.6.5 *Let $B = [f_{ij}]_{i=j=1}^{n}$ be real symmetric matrix with the eigenvalues $\lambda_1(B) \geq \cdots \geq \lambda_n(B)$. Assume that $\lambda_n(B)$ is a simple eigenvalue, i.e. $\lambda_{n-1}(B) > \lambda_n(B)$. Suppose that $B\mathbf{x} = \lambda_n(B)\mathbf{x}$, where $\mathbf{x} \in \mathbb{R}^n, \mathbf{x}^\top\mathbf{x} = 1$. Let $\mathbf{U} \subset \mathbb{R}^n$ be a subspace which does not contain $\mathbf{x}$. Then*

$$\min_{\mathbf{y}\in\mathbf{U},\mathbf{y}^\top\mathbf{y}=1} \mathbf{y}^\top B\mathbf{y} > \lambda_n(B). \tag{6.6.1}$$

Corollary 6.6.6 *Let $B \in \mathbb{R}^{n\times n}$ be an M-singular symmetric matrix of the form $B = \rho(C)I - C$, where C is a non-negative irreducible symmetric matrix. Let $\mathbf{U} = \{\mathbf{y} \in \mathbb{R}^n, \mathbf{1}^\top\mathbf{y} = \mathbf{0}\}$. Then $\lambda_n(B) = 0$ is a simple eigenvalue, and (6.6.1) hold.*

Theorem 6.6.7 *Let $\mathrm{D} \subset \mathbb{R}^m$ be a bounded domain. Assume that $f \in \mathrm{C}^2(\mathrm{D})$. Suppose that $f|\partial\mathrm{D} = \infty$, i.e. for each sequence $\mathbf{x}_i \in \mathrm{D}, i = 1, \ldots,$ such that $\lim_{i\to\infty}\mathbf{x}_i = \mathbf{x} \in \partial\mathrm{D}$, $\lim_{i\to\infty}f(\mathbf{x}_i) = \infty$. Assume furthermore, that for each critical point $\boldsymbol{\xi} \in \mathrm{D}$, the eigenvalues of the Hessian $H(\boldsymbol{\xi}) = [\frac{\partial^2 f}{\partial x_i \partial x_j}(\boldsymbol{\xi})]_{i=j=1}^n$ are positive. Then f has a unique critical point $\boldsymbol{\xi} \in \mathrm{D}$, which is a global minimum, i.e. $f(\mathbf{x}) > f(\boldsymbol{\xi})$ for any $\mathbf{x} \in \mathrm{D}\backslash\{\boldsymbol{\xi}\}$.*

Proof. Consider the negative gradient flow

$$\frac{d\mathbf{x}(t)}{dt} = -\nabla f(\mathbf{x}(t)), \quad \mathbf{x}(t_0) = \mathbf{x}_0 \in \mathrm{D}. \tag{6.6.2}$$

Clearly, the fixed points of this flow are the critical points of f. Observe next that if $\mathbf{x}_0$ is not a critical point then $f(\mathbf{x}(t))$ decreases, as $\frac{df(\mathbf{x}(t))}{dt} = -\|\nabla f(\mathbf{x}(t))\|^2$. Since $f|\partial \mathrm{D} = \infty$, we deduce that all accumulations points of the flow $\mathbf{x}(t), t \in [t_0, \infty)$ are in D, and are critical points of f. Consider the flow (6.6.2) in the neighborhood of a critical point $\boldsymbol{\xi} \in \mathrm{D}$. Let $\mathbf{x} = \mathbf{y} + \boldsymbol{\xi}, \mathbf{x}_0 = \mathbf{y}_0 + \boldsymbol{\xi}$. Then for $\mathbf{x}_0$ close to $\boldsymbol{\xi}$ the flow (6.6.2) is of the form

$$\frac{d\mathbf{y}(t)}{dt} = -(H(\boldsymbol{\xi})\mathbf{y} + \mathrm{Er}(\mathbf{y})), \quad \mathbf{y}(t_0) = \mathbf{y}_0.$$

For a given $\delta > $ the exists $\varepsilon = \varepsilon(\delta) > 0$ such that for $\|\mathbf{y}\| < \varepsilon$, $\|\mathrm{Er}(\mathbf{y})\| < \delta\|\mathbf{y}\|$. Let $\alpha > 0$ be the smallest positive eigenvalue of $H(\boldsymbol{\xi})$. So $\mathbf{z}^\top H(\boldsymbol{\xi})\mathbf{z} \geq \alpha\|\mathbf{z}\|^2$ for any $\mathbf{z} \in \mathbb{R}^m$. Choose $\varepsilon > 0$ so that $\|\mathrm{Er}(\mathbf{y})\| < \frac{\alpha}{2}\|\mathbf{y}\|$ for $\|\mathbf{y}\| < \varepsilon$. Thus

$$\frac{d\|\mathbf{y}(t)\|^2}{dt} = -2(\mathbf{y}(t)^\top H(\boldsymbol{\xi})\mathbf{y}(t) + \mathbf{y}(t)^\top \mathrm{Er}(\mathbf{y}))|$$

$$\leq -\alpha|\mathbf{y}\|^2 \quad \text{if } |\mathbf{y}(t)| < \varepsilon.$$

This shows that if $\|\mathbf{y}(t_0)\| < \varepsilon$ then for $t \geq t_0$ $\|\mathbf{y}(t)\|$ decreases. Moreover

$$\frac{d \log \|\mathbf{y}(t)\|^2}{dt} \leq -\alpha \quad \text{for } t \geq t_0 \Rightarrow \|\mathbf{y}(t)\|^2$$

$$\leq \|\mathbf{Y}_0\|^2 e^{-\alpha(t-t_0)} \quad \text{for } t \geq t_0.$$

This shows that $\lim_{t\to\infty} \mathbf{y}(t) = \mathbf{0}$. Let $\beta \geq \alpha$ be the maximal eigenvalue of $H(\boldsymbol{\xi})$. Similar estimates show that if $\|\mathbf{y}_0\| < \varepsilon$ then $\|\mathbf{y}(t)\|^2 \geq \|\mathbf{y}_0\|^2 e^{-(2\beta+\alpha)(t-t_0)}$.

These results, combined with the continuous dependence of the flow (6.6.2) on the initial conditions $\mathbf{x}_0$, imply the following facts. Any flow (6.6.2) which starts at a noncritical point $\mathbf{x}_0$ must terminate at $t = \infty$ at some critical point $\boldsymbol{\xi}$, which may depend on $\mathbf{x}_0$. For a critical point $\boldsymbol{\xi}$, denote by the set $\mathcal{A}(\boldsymbol{\xi})$ all points $\mathbf{x}_0$ for which the flow

(6.6.2) terminates at finite or infinite time at $\boldsymbol{\xi}$. (The termination at finite time can happen only if $\mathbf{x}_0 = \boldsymbol{\xi}$.) Then $\mathcal{A}(\boldsymbol{\xi})$ is an open connected set of D.

We claim that $\mathcal{A}(\boldsymbol{\xi}) = $ D. If not, there exists a point $\mathbf{x}_0 \in \partial\mathcal{A}(\boldsymbol{\xi}) \cap$ D. Since $\mathcal{A}(\boldsymbol{\xi})$ is open, $\mathbf{x}_0 \notin \mathcal{A}(\boldsymbol{\xi})$. As we showed above $\mathbf{x}_0 \in \mathcal{A}(\boldsymbol{\xi}')$ for some another critical point $\boldsymbol{\xi}' \neq \boldsymbol{\xi}$. Clearly $\mathcal{A}(\boldsymbol{\xi}) \cap \mathcal{A}(\boldsymbol{\xi}') = \emptyset$. As $\mathcal{A}(\boldsymbol{\xi}')$ is open there exists an open neighborhood of $\mathbf{x}_0$ in D which belongs to $\mathcal{A}(\boldsymbol{\xi}')$. Hence $\mathbf{x}_0$ cannot be a boundary point of $\mathcal{A}(\boldsymbol{\xi})$, which contradicts our assumption. Hence $\mathcal{A}(\boldsymbol{\xi}) = $ D, and $\boldsymbol{\xi}$ is a unique critical point of f in D. Hence, $\boldsymbol{\xi}$ is the unique minimal point of f. $\qquad\square$

Theorem 6.6.8 *Let* $A = [a_{ij}]_{i=j=1}^n \in \mathbb{R}_+^{n \times n}$ *be an irreducible matrix. Suppose furthermore that* $a_{ii} > 0$ *for* $i = 1, \ldots, n$. *Let* $\mathbf{w} = (w_1, \ldots, w_n)^\top > \mathbf{0}$. *Define the following function*

$$f = f_{A,\mathbf{w}} = \sum_{i=1}^n w_i \log \frac{(A\mathbf{x})_i}{x_i}, \quad \mathbf{x} = (x_1, \ldots, x_n)^\top > \mathbf{0}. \quad (6.6.3)$$

Let D *be the interior of* Π_n. *(D can be viewed as an open connected bounded set in* $\mathbb{R}^{n-1}$, *see the proof.) Then f satisfies the assumptions of Theorem 6.6.7. Let* $\mathbf{0} < \boldsymbol{\xi} \in \Pi_n$ *be the unique critical point of f in* D. *Then* $f(\mathbf{x}) \geq f(\boldsymbol{\xi})$ *for any* $\mathbf{x} > \mathbf{0}$. *Equality holds if and only if* $\mathbf{x} = t\boldsymbol{\xi}$ *for some* $t > 0$.

Proof. Observe that any probability vector $\mathbf{p} = (p_1, \ldots, p_n)^\top$ can be written as $\mathbf{p} = \frac{1}{n}\mathbf{1} + \mathbf{y}$ where $\mathbf{y} \in \mathbb{R}^n, \mathbf{1}^\top\mathbf{y} = 0$ and $\mathbf{y} \geq -\frac{1}{n}\mathbf{1}$. Since any $\mathbf{y} \in \mathbb{R}^n, \mathbf{1}^\top\mathbf{y} = 0$ is of the form $\mathbf{y} = u_1(-1, 1, 0 \ldots, 0)^\top + \cdots + u_{n-1}(-1, 0, \ldots, 0, 1)^\top$ we deduce that we can view Π_n as a compact connected set in $\mathbb{R}^{n-1}$, and its interior D, i.e. $\mathbf{0} < \mathbf{p} \in \Pi_n$, is an open connected bounded set in $\mathbb{R}^{n-1}$.

We now claim that $f|\partial\Pi_n = \infty$. Let $\mathbf{0} < \mathbf{p}_k = (p_{1,k}, \ldots, p_{n,k})^\top \in \Pi_n, k = 1, \ldots$, converge to $\mathbf{p} = (p_1, \ldots, p_n)^\top \in \partial\Pi_n$. Let $\emptyset \neq Z(\mathbf{p}) \subset [n]$ be the set of vanishing coordinates of $\mathbf{p}$. Observe first that $\frac{(A\mathbf{x})_i}{x_i} \geq a_{ii} > 0$ for $i = 1, \ldots, n$. Since A is irreducible, it follows that there exists $l \in Z(\mathbf{p}), j \in [n] \backslash Z(\mathbf{p})$ such that $a_{lj} > 0$. Hence

$$\lim_{k \to \infty} \frac{(A\mathbf{p}_k)_l}{p_{l,k}} \geq \lim_{k \to \infty} \frac{a_{lj} p_{j,k}}{p_{l,k}} = \infty.$$

Thus

$$\lim_{k \to \infty} f(\mathbf{p}_k) \geq \lim_{k \to \infty} \log \frac{(A\mathbf{p}_k)_l}{p_{l,k}} + \sum_{i \neq l} \log a_{ii} = \infty.$$

Observe next that $f(\mathbf{x})$ is a homogeneous function of degree 0 on $\mathbf{x} > \mathbf{0}$, i.e. $f(t\mathbf{x}) = f(\mathbf{x})$ for all $t > 0$. Hence, $\frac{df(t\mathbf{x})}{dt} = 0$. Thus

$$\mathbf{x}^\top \nabla f(\mathbf{x}) = 0 \qquad (6.6.4)$$

for all $\mathbf{x} > \mathbf{0}$. Let $\boldsymbol{\xi} \in \mathrm{D}$ be a critical point of $f|\mathrm{D}$. Then $\mathbf{y}^\top \nabla f(\boldsymbol{\xi}) = 0$ for each $\mathbf{y} \in \mathbb{R}^n, \mathbf{1}^\top \mathbf{y} = 0$. Combine this fact with (6.6.4) for $\mathbf{x} = \boldsymbol{\xi}$ to deduce that $\boldsymbol{\xi}$ is a a critical point of f in $\mathbb{R}_+^n$. So $\nabla f(\boldsymbol{\xi}) = \mathbf{0}$. Differentiate (6.6.4) with respect to $x_i, i = 1, \ldots, n$ and evaluate these expressions at $\mathbf{x} = \boldsymbol{\xi}$. Since $\boldsymbol{\xi}$ is a critical point we deduce that $H(\boldsymbol{\xi})\boldsymbol{\xi} = \mathbf{0}$. We claim that $H(\boldsymbol{\xi})$ is a symmetric singular M-matrix. Indeed

$$\frac{\partial f}{\partial x_j}(\mathbf{x}) = -w_j \frac{1}{x_j} + \sum_{i=1}^n w_i \frac{a_{ij}}{(A\mathbf{x})_i}. \qquad (6.6.5)$$

Hence, for $l \neq j$

$$\frac{\partial^2 f}{\partial x_l \partial x_j}(\mathbf{x}) = -\sum_{i=1}^n w_i \frac{a_{ij} a_{il}}{(A\mathbf{x})_i^2} \leq 0.$$

So $H(\mathbf{x})$ is a Z-matrix for any $\mathbf{x} > \mathbf{0}$. Since $H(\boldsymbol{\xi})\boldsymbol{\xi} = \mathbf{0}$ Corollary 6.6.4 yields that $H(\boldsymbol{\xi})$ is a symmetric singular M-matrix. So $H(\boldsymbol{\xi}) = \rho(C)I - C, C = [c_{ij}]_{i=j=1}^n$. We claim that C is an irreducible matrix. Indeed assume that $a_{jl} > 0$. Then

$$c_{jl} = c_{lj} = -\frac{\partial^2 f}{\partial x_l \partial x_j} \geq w_j \frac{a_{jj} a_{jl}}{(A\boldsymbol{\xi})_j^2} > 0.$$

Since A is irreducible C is irreducible. Hence, $0 = \lambda_n(H(\boldsymbol{\xi}))$ is a simple eigenvalue of $H(\boldsymbol{\xi})$. The restriction of the quadratic form corresponding to the Hessian of $f|\Pi_n$ at ξ, corresponds to $\mathbf{y}^\top H(\boldsymbol{\xi})\mathbf{y}$

where $\mathbf{1}^\top \mathbf{y} = 0$. Corollary 6.6.5 implies that there exists $\alpha > 0$ such that $\mathbf{y}^\top H(\boldsymbol{\xi})\mathbf{y} \geq \alpha \|\mathbf{y}\|^2$ for all $\mathbf{1}^\top \mathbf{y} = 0$. Hence the Hessian of $f|\mathrm{D}$ at the critical point $\mathbf{0} < \boldsymbol{\xi} \in \Pi_n$ has positive eigenvalues. Theorem 6.6.7 yields that there exists a unique critical point $\boldsymbol{\xi} \in \mathrm{D}$ of $f|\mathrm{D}$ such that $f(\mathbf{p}) > f(\boldsymbol{\xi})$ for any $\mathbf{p} \in \mathrm{D}\backslash\{\boldsymbol{\xi}\}$. Since $f(\mathbf{x})$ is a homogeneous function of degree 0 we deduce that $f(\mathbf{x}) \geq f(\boldsymbol{\xi})$ for any $\mathbf{x} > \mathbf{0}$. Equality holds if and only if $\mathbf{x} = t\boldsymbol{\xi}$ for some $t > 0$. $\qquad\square$

Theorem 6.6.9 *Let $A \in \mathbb{R}_+^{n\times n}$ and assume that*

$$A\mathbf{u} = \rho(A)\mathbf{u}, \quad A^\top \mathbf{v} = \rho(A)\mathbf{v}, \quad 0 < \rho(A),$$

$$\mathbf{0} < \mathbf{u} = (u_1, \ldots, u_n)^\top, \mathbf{v} = (v_1, \ldots, v_n)^\top, \mathbf{v}^\top \mathbf{u} = 1.$$

Then

$$\sum_{i=1}^n u_i v_i \log \frac{(A\mathbf{x})_i}{x_i} \geq \log \rho(A) \text{ for any } \mathbf{x} = (x_1, \ldots, x_n)^\top > \mathbf{0},$$

$$(6.6.6)$$

$$\rho(DA) \geq \rho(A) \prod_{i=1}^n d_i^{u_i v_i} \text{ for any diagonal } D = \mathrm{diag}(d_1, \ldots, d_n) \geq 0.$$

$$(6.6.7)$$

Equality holds for $\mathbf{x} = t\mathbf{u}$ and $D = sI$, where $t > 0, s \geq 0$, respectively. Assume that A is irreducible and all the diagonal entries of A are positive. Then equality holds in (6.6.6) and (6.6.7) if and only if $\mathbf{x} = t\mathbf{u}, D = sI$ for some $t > 0, s \geq 0$ respectively.

Proof. Assume that $A = [a_{ij}]_{i=j=1}^n \in \mathbb{R}_+^{n\times n}$ be irreducible and $a_{ii} > 0$ for $i = 1, \ldots, n$. Let $\mathbf{w} = (u_1 v_1, \ldots, u_n v_n)^\top$. Define $f(\mathbf{x})$ as in (6.6.3). We claim that $\mathbf{u}$ is a critical point of f. Indeed, (6.6.5) yields

$$\frac{\partial f}{\partial x_j}(\mathbf{u}) = -u_j v_j \frac{1}{u_j} + \sum_{i=1}^n u_i v_i \frac{a_{ij}}{(A\mathbf{u})_i}$$

$$= -v_j + \frac{1}{\rho(A)}(A^\top \mathbf{v})_j = 0, \ j = 1, \ldots, n.$$

Similarly, $t\mathbf{u}$ is a critical point of f for any $t > 0$. In particular, $\xi = t\mathbf{u} \in \Pi_n$ is a critical point of f in D. Theorem 6.6.8 implies that $f(\mathbf{x}) \geq f(\mathbf{u}) = \log \rho(A)$ and equality holds if and only if $\mathbf{x} = t\mathbf{u}$ for some $t > 0$.

Let D be a diagonal matrix with positive diagonal entries. Then DA is irreducible, and $DA\mathbf{x} = \rho(DA)\mathbf{x}$ for some $\mathbf{x} = (x_1, \ldots, x_n)^\top > \mathbf{0}$. Note that since $f(\mathbf{u}) \leq f(\mathbf{x})$ we deduce

$$\log \rho(A) \leq \sum_{i=1}^n u_i v_i \log \frac{(A\mathbf{x})_i}{x_i} = \log \rho(DA) - \sum_{i=1}^n u_i v_i \log d_i.$$

The above inequality yields (6.6.7). Suppose that equality holds in (6.6.7). Then $\mathbf{x} = t\mathbf{u}$, which yields that $D = sI$ for some $s > 0$. Suppose that $D \geq 0$ and D has at least one zero diagonal element. Then the right-hand side of (6.6.7) is zero. Clearly $\rho(DA) \geq 0$. Since $d_i a_{ii}$ is a principle submatrix of DA, Lemma 6.4.2 yields that $\rho(DA) \geq \max_{i \in [n]} d_i a_{ii}$. Hence, $\rho(DA) = 0$ if and only if $D = 0$. These arguments prove the theorem when A is irreducible with positive diagonal entries.

Let us now consider the general case. For $\varepsilon > 0$ let $A(\varepsilon) := A + \varepsilon \mathbf{u}\mathbf{v}^\top$. Then $A(\varepsilon) > 0$ and $A(\varepsilon)\mathbf{u} = (\rho(A) + \epsilon)\mathbf{u}$, $A(\varepsilon)^\top \mathbf{v} = (\rho(A)+\epsilon)\mathbf{v}$. Hence, inequalities (6.6.6) and (6.6.7) hold for $A(\varepsilon)$ and fixed $x > \mathbf{0}, D \geq 0$. Let $\varepsilon \searrow 0$ to deduce (6.6.6) and (6.6.7). For $\mathbf{x} = t\mathbf{u}, D = sI$, where $t > 0, s \geq 0$ one has equality.   □

Corollary 6.6.10 *Let the assumptions of Theorem 6.6.9 hold. Then*

$$\sum_{i=1}^n u_i v_i \frac{(A\mathbf{x})_i}{x_i} \geq \rho(A) \text{ for any } \mathbf{x} = (x_1, \ldots, x_n)^\top > \mathbf{0}. \quad (6.6.8)$$

If A is irreducible and has positive diagonal entries then equality holds if and only if $\mathbf{x} = t\mathbf{u}$ for some $t > 0$.

Proof. Use the arithmetic-geometric inequality $\sum_{i=1}^n p_i c_i \geq \prod_{i=1}^n c_i^{p_i}$ for any $\mathbf{p} = (p_1, \ldots, p_n) \in \Pi_n$ and any $\mathbf{c} = (c_1, \ldots, c_n)^\top \geq \mathbf{0}$.   □

Definition 6.6.11 *Let* $\mathbf{x} = (x_1, \ldots, x_n)^\top, \mathbf{y} = (y_1, \ldots, y_n)^\top \in \mathbb{C}^n$. *Denote by* $e^\mathbf{x} := (e^{x_1}, \ldots, e^{x_n})^\top$, *by* $\mathbf{x}^{-1} := (\frac{1}{x_1}, \ldots, \frac{1}{x_n})^\top$ *for* $\mathbf{x} > \mathbf{0}$, *and by* $\mathbf{x} \circ \mathbf{y}$ *the vector* $(x_1 y_1, \ldots, x_n y_n)^\top$. *For a square diagonal matrix* $D = \mathrm{diag}(d_1, \ldots, d_n) \in \mathbb{C}^{n \times n}$ *denote by* $\mathbf{x}(D)$ *the vector* $(d_1, \ldots, d_n)^\top$.

Theorem 6.6.12 *Let* $\mathbf{0} < \mathbf{u} = (u_1, \ldots, u_n)^\top, \mathbf{v} = (v_1, \ldots, v_n)^\top$. *Let* $\mathbf{0} < \mathbf{w} = \mathbf{u} \circ \mathbf{v}$. *Assume that* $A = [a_{ij}]_{i=j=1}^n \in \mathbb{R}_+^{n \times n}$ *is irreducible. Then there exists two diagonal matrices* $D_1, D_2 \in \mathbb{R}_+^{n \times n}$, *with positive diagonal entries, such that* $D_1 A D_2 \mathbf{u} = \mathbf{u}, (D_1 A D_2)^\top \mathbf{v} = \mathbf{v}$ *if one of the following conditions hold. Under any of this conditions* D_1, D_2 *are unique up to the transformation* $t^{-1} D_1, t D_2$ *for some* $t > 0$.

1. *All diagonal entries of* A *are positive.*

2. *Let* $\mathcal{N} \subset [n]$ *be a nonempty set of all* $j \in [n]$ *such that* $a_{jj} = 0$. *Assume that all off-diagonal entries of* A *are positive and the following inequalities hold.*

$$\sum_{i \in [n] \setminus \{j\}} w_i > w_j \text{ for all } j \in \mathcal{N}. \tag{6.6.9}$$

(For $n = 2$ *and* $\mathcal{N} = [2]$ *the above inequalities are not satisfied by any* $\mathbf{w}$.)

Proof. We first observe that it is enough to consider the case where $\mathbf{u} = \mathbf{1}$, i.e. $B := D_1 A D_2$ is a stochastic matrix. See Problem 2. In this case $\mathbf{w} = \mathbf{v}$.

1. Assume that all diagonal entries of A are positive. Let $f = f_{A,\mathbf{w}}$ be defined as in (6.6.3). The proof of Theorem 6.6.8 yields that f has a unique critical $\mathbf{0} < \boldsymbol{\xi}$ in Π_n. (6.6.5) implies that

$$\sum_{i=1}^n \frac{w_i a_{ij}}{(A\boldsymbol{\xi})_i} = \frac{w_j}{\xi_j}.$$

This is equivalent to the equality $(D(A\boldsymbol{\xi})^{-1} A D(\boldsymbol{\xi}))^\top \mathbf{w} = \mathbf{w}$. A straightforward calculation show that $D(A\boldsymbol{\xi})^{-1} A D(\boldsymbol{\xi}) \mathbf{1} = \mathbf{1}$. Hence $D_1 = D(A\boldsymbol{\xi})^{-1}, D_2 = D(\boldsymbol{\xi})$.

Suppose that D_1, D_2 are diagonal matrices with positive diagonal entries so that $D_1 A D_2 \mathbf{1} = \mathbf{1}$ and $(D_1 A D_2)^\top \mathbf{w} = \mathbf{w}$. Let $\mathbf{u} = D_2 \mathbf{1}$. The equality $D_1 A D_2 \mathbf{1} = \mathbf{1}$ implies that $D_1 = D(A\mathbf{u})^{-1}$. The equality $(D_1 A D_2)^\top \mathbf{w} = \mathbf{w}$ is equivalent to $(D(A(\mathbf{u}))^{-1} A D(\mathbf{u}))^\top \mathbf{w} = \mathbf{w}$. Hence, $\mathbf{u}$ is a critical point of f. Therefore, $\mathbf{u} = t\boldsymbol{\xi}$. So $D_2 = t D(\boldsymbol{\xi})$ and $D_1 = t^{-1} D(A\boldsymbol{\xi})^{-1}$.

2. As in the proof of Theorem 6.6.8, we show that $f = f_{\mathbf{w},A}$ blows up to ∞ as $\mathbf{p}$ approaches $\partial \Pi_n$. Let $\mathbf{0} < \mathbf{p}_k = (p_{1,k}, \ldots, p_{n,k})^\top \in \Pi_n, k = 1, \ldots,$ converge to $\mathbf{p} = (p_1, \ldots, p_n)^\top \in \partial \Pi_n$. Let $\emptyset \neq Z(\mathbf{p}) \subset [n]$ be the set of vanishing coordinates of $\mathbf{p}$. Since all off-diagonal entries of A are positive, it follows that $\lim_{k \to \infty} \frac{(A\mathbf{p}_k)_i}{p_{i,k}} = \infty$ for each $i \in Z(\mathbf{p})$. To show that $\lim_{k \to \infty} f(\mathbf{p}_k) = \infty$ it is enough to consider the case where $\lim_{k \to \infty} \frac{(A\mathbf{p}_k)_m}{p_{m,k}} = 0$ for some $m \notin Z(\mathbf{p})$. In view of the proof of Theorem 6.6.8 we deduce that $m \in \mathcal{N}$. Furthermore, $\#Z(\mathbf{p}) = n - 1$. Hence $\lim_{k \to \infty} p_{m,k} = 1$. Assume for simplicity of notation that $m = 1$, i.e. $Z(\mathbf{p}) = \{2, 3, \ldots, n\}$. Let $s_k = \max_{i \geq 2} p_{i,k}$. So $\lim_{k \to \infty} s_k = 0$. Let $a > 0$ be the value of the minimal off-diagonal entry of A. Then $\frac{(A\mathbf{p}_k)_i}{p_{i,k}} \geq \frac{a p_{1,k}}{s_k}$ for $i \geq 2$. Also $\frac{(A\mathbf{p}_k)_1}{p_{1,k}} \geq \frac{a s_k}{p_{1,k}}$. Thus

$$f(\mathbf{p}_k) \geq w_1 \log \frac{a s_k}{p_{1,k}} + \sum_{i \geq 2} w_i \log \frac{a p_{1,k}}{s_k}$$

$$= \left(\sum_{i=1}^n w_i \right) \log a + \left(-w_1 + \sum_{i \geq 2} w_i \right) \log \frac{p_{1,k}}{s_k},$$

(6.6.9) for $j = 1$ implies that $\lim_{k \to \infty} f(\mathbf{p}_k) = \infty$.

Let $\mathbf{0} < \boldsymbol{\xi} \in \Pi_n$ be a critical point of f. We claim that $H(\boldsymbol{\xi}) = \rho(C) I - C$, and $0 \leq C$ is irreducible. Indeed, for $j \neq l$ $c_{jl} = \sum_{i=1}^n w_i \frac{a_{ij} a_{il}}{(A\boldsymbol{\xi})_i^2}$. Since $n \geq 3$ choose $i \neq j, l$ to deduce that $c_{jl} > 0$. So C has positive off-diagonal entries, hence irreducible. Hence, $\mathbf{0} < \boldsymbol{\xi} \in \Pi_n$ is a unique critical point in Π_n. The arguments for 1 yield the existence of D_1, D_2, which are unique up to scaling. $\qquad \square$

Problems

1. Prove Lemma 6.6.2.

2. Let $B \in \mathbb{R}_+^{n \times n}$ and assume that $Bu = u, B^\top v$, where $0 < u = (u_1, \ldots, u_n)^\top, v = (v_1, \ldots, v_n)^\top$. Then $C := D(u)^{-1} B D(u)$ satisfies the following $C1 = 1, C^\top w = w$, where $w = u \circ v$.

3. $A \in \mathbb{R}_+^{n \times n}$ is called fully indecomposable if there exists $P \in \mathcal{P}_n$ such that PA is irreducible and have positive diagonal elements. Show that that if A is fully indecomposable then there exist diagonal matrices D_1, D_2, with positive diagonal entries, such that $D_1 A D_2$ is doubly stochastic. D_1, D_2 are unique up to a scalar factor $t^{-1} D_1, t D_2$.

4. Let $A \in \mathbb{R}_+^{n \times n}$ and $n > 1$. Show that the following are equivalent:

   (a) A does not have a $k \times (n - k)$ zero submatrix for each $k \in [n - 1]$.

   (b) PA is irreducible for each permutation $P \in \mathcal{P}_n$.

   (c) A is fully indecomposable.

6.7   Log-Convexity

The following results are well known, e.g. [Roc70].

Fact 6.7.1

1. *Let* $D \subset \mathbb{R}^n$ *be a convex set, and* $f : D \to \mathbb{R}$ *a convex function. Then* $f : D^o \to \mathbb{R}$ *is a continuous function. Furthermore, at each* $x \in D^o$, f *has a supporting hyperplane. That is there exists* $p = p(x) \in \mathbb{R}^n$ *such that*

$$f(y) \geq f(x) + p^\top (y - x) \text{ for any } y \in D. \qquad (6.7.1)$$

Assume furthermore that $\dim D = n$. *Then the following conditions are equivalent.*

   (a) f *is differentiable at* $x \in D^o$. *That is, the gradient of* f, ∇f, *at* x *exists and the following equality holds.*

$$\|f(y) - (f(x) + \nabla f(x)^\top (y - x))\| = o(\|y - x\|). \qquad (6.7.2)$$

(b) *f has a unique supporting hyperplane at* **x**.

The set of points $\mathrm{Diff}(f) \subset \mathrm{D}^o$, *where f is differentiable, is a dense set in* D *of the full Lebesgue measure, i.e.* $\mathrm{D}\backslash\mathrm{Diff}(f)$ *has zero Lebesgue measure. Furthermore,* ∇f *is continuous on* $\mathrm{Diff}(f)$.

(a) *If f, g are convex on* D *then* $\max(f, g)$, *where* $\max(f, g)(x)$ $:= \max(f(x), g(x))$, *is convex. Furthermore,* $af + bg$ *is convex for any* $a, b \geq 0$.

(b) *Let* $f_i : \mathrm{D} \to \mathbb{R}$ *for* $i = 1, 2, \ldots$ *Denote by* $f := \limsup_i f_i$ *the function given by* $f(\mathbf{x}) := \limsup_i f(\mathbf{x})$ *for each* $\mathbf{x} \in \mathrm{D}$. *Assume that* $f : \mathrm{D} \to \mathbb{R}$, *i.e. the sequence* $f_i(x), i = 1, \ldots,$ *is bounded for each x. If each* f_i *is convex on* D *then f is convex on* D.

Theorem 6.7.2 *Let* $\mathrm{D} \in \mathbb{R}^m$ *be a convex set. Assume that* $a_{ij} :$ $\mathrm{D} \to \mathbb{R}$ *are log-convex functions for* $i, j = 1, \ldots, n$. *Let* $A(\mathbf{x}) :=$ $[a_{ij}(\mathbf{x})]_{i=j=1}^n$ *be the induced non-negative matrix function on* D. *Then* $\rho(A(\mathbf{x}))$ *is a log-convex function on* D. *Assume furthermore that each* $a_{ij}(\mathbf{x}) \in \mathrm{C}^k(\mathrm{D}^o)$, *for some* $k \geq 1$. *(All partial derivatives of* $a_{ij}(\mathbf{x})$ *of order less or equal to k are continuous in* $\int \mathrm{D}$.) *Suppose furthermore that* $A(\mathbf{x}_0)$ *is irreducible. Then* $0 < \rho(A(\mathbf{x})) \in \mathrm{C}^k(\mathrm{D}^o)$.

Proof. In view of Fact 6.7.1.1(a) each entry of the matrix $A(\mathbf{x})^m$ is log-convex. Hence $\mathrm{tr}\, A(\mathbf{x})^m$ is log-convex, which implies that $(\mathrm{tr}\, A(\mathbf{x})^m)^{\frac{1}{m}}$ is log-convex. Theorem 6.4.9 combined with Fact 6.7.1.1(b) yields that $\rho(A(\mathbf{x}))$ is log-convex.

Assume that $A(\mathbf{x}_0)$ is irreducible for some $\mathbf{x}_0 \in \mathrm{D}^o$. Since each $a_{ij}(\mathbf{x})$ is continuous in D^o, Problem 1 yields that the digraph $D(A(\mathbf{x}))$ is a constant digraph on D^o. Since $D(A(\mathbf{x}_0))$ is strongly connected, it follows that $D(A(\mathbf{x}))$ is strongly connected for each $\mathbf{x} \in \mathrm{D}^o$. Hence, $A(\mathbf{x})$ is irreducible for $\mathbf{x} \in \mathrm{D}^o$ and $\rho(A(\mathbf{x})) > 0$ is a simple root of its characteristic polynomial for $\mathbf{x} \in \mathrm{D}^o$. The implicit function theorem implies that $\rho(A(\mathbf{x})) \in \mathrm{C}^k(\mathrm{D}^o)$. $\square$

Theorem 6.7.3 *Let $A \in \mathbb{R}_+^{n \times n}$. Define $A(\mathbf{x}) = D(e^{\mathbf{x}})A$ for any* $\mathbf{x} \in \mathbb{R}^n$. *Then $\rho(A(\mathbf{x}))$ is a log-convex function. Suppose furthermore that A is irreducible. Then $\log \rho(A(\mathbf{x}))$ is a smooth convex function on $\mathbb{R}^n$, i.e. $\log \rho(A(\mathbf{x})) \in C^\infty(\mathbb{R}^n)$. Let*

$$A(\mathbf{x})\mathbf{u}(\mathbf{x}) = \rho(A(\mathbf{x}))\mathbf{u}(\mathbf{x}), \quad A(\mathbf{x})^\top \mathbf{v}(\mathbf{x}) = \rho(A(\mathbf{x}))\mathbf{v}(x),$$

$$\mathbf{0} < \mathbf{u}(\mathbf{x}), \mathbf{v}(\mathbf{x}), \quad \text{with the normalization}$$

$$\mathbf{w}(\mathbf{x}) =: \mathbf{u}(\mathbf{x}) \circ \mathbf{v}(\mathbf{x}) \in \Pi_n.$$

Then

$$\nabla \log \rho(A(\mathbf{x})) = \frac{1}{\rho(A(\mathbf{x}))} \nabla \rho(A(\mathbf{x})) = \mathbf{w}(\mathbf{x}). \tag{6.7.3}$$

That is, the inequality (6.6.7) corresponds to the standard inequality

$$\log \rho(A(\mathbf{y})) \geq \log \rho(A(\mathbf{0})) + \nabla \log \rho(A(\mathbf{0}))^\top \mathbf{y}, \tag{6.7.4}$$

for smooth convex functions.

Proof. Clearly, the function $f_i(\mathbf{x}) = e^{x_i}$ is a smooth log-convex function for $\mathbf{x} = (x_1, \ldots, x_n)^\top \in \mathbb{R}^n$. Since $A \geq 0$ it follows that each entry of $A(\mathbf{x})$ is a log-convex function. Theorem 6.7.2 yields that $\rho(A(\mathbf{x}))$ is log-convex.

Assume in addition that $A = A(\mathbf{0})$ is irreducible. Theorem 6.7.2 yields that $\log \rho(A(\mathbf{x}))$ is a smooth convex function on $\mathbb{R}^n$. Hence, $\log \rho(A(\mathbf{x}))$ has a unique supporting hyperplane at each $\mathbf{x}$. For $\mathbf{x} = 0$ this supporting hyperplane is given by the right-hand side of (6.7.4). Consider the inequality (6.6.7). By letting $D = D(e^{\mathbf{y}})$ and taking the logarithm of this inequality we obtain that $\log \rho(A) + \mathbf{w}(\mathbf{0})^\top \mathbf{y}$ is also a supporting hyperplane for $\log \rho(A(\mathbf{x}))$ at $\mathbf{x} = \mathbf{0}$. Hence, $\nabla \log \rho(A(\mathbf{0})) = \mathbf{w}(\mathbf{0})$. Similar arguments for any $\mathbf{x}$ proves the equality (6.7.3). $\qquad \square$

Problems

1. Let $D \subset \mathbb{R}^m$ be a convex set.

   (a) Show that if f is a continuous log-convex on D, then either f identically zero function or positive at each $\mathbf{x} \in D$.

(b) Assume that f is positive on D, i.e. $f(\mathbf{x}) > 0$ for each $\mathbf{x} \in$ D. Then f is log-convex on D if and only if $\log f$ is convex on D.

(c) Assume that f is log-convex on D. Then f is continuous on D^o.

2. Let $f : D \to \mathbb{R}$ be a log-convex function show that f is a convex function.

3. Let $D \subset \mathbb{R}^n$ be a convex set.

(a) If f, g are log-convex on D then $\max(f, g)$ is log-convex. Furthermore, $f^a g^b$ and $af + bg$ are log-convex for any $a, b \geq 0$.

(b) Let $f_i : D \to \mathbb{R}$ $i = 1, 2, \ldots$ be log-convex. Assume that $f := \limsup_i f_i : D \to \mathbb{R}$. Then f is log-convex on D.

6.8   Min–Max Characterizations of $\rho(A)$

Theorem 6.8.1 *Let* $\Psi : \mathbb{R} \to \mathbb{R}$ *be a differentiable convex non-decreasing function. Let* $A \in \mathbb{R}_+^{n \times n}$, *and assume that* $\rho(A) > 0$. *Then*

$$
\sup_{\mathbf{p}=(p_1,\ldots,p_n)^\top \in \Pi_n} \inf_{\mathbf{x}=(x_1,\ldots,x_n)>0}
$$

$$
\times \sum_{i=1}^n p_i \Psi \left(\log \frac{(A\mathbf{x})_i}{x_i} \right) = \Psi(\log \rho(A)). \tag{6.8.1}
$$

Suppose that $\Psi'(\log \rho(A)) > 0$ *and* A *has a positive eigenvector* $\mathbf{u}$ *which corresponds to* $\rho(A)$. *If*

$$
\inf_{\mathbf{x}=(x_1,\ldots,x_n)>0} \sum_{i=1}^n p_i \Psi \left(\log \frac{(A\mathbf{x})_i}{x_i} \right) = \Psi(\log \rho(A)) \tag{6.8.2}
$$

then the vector $\mathbf{v} = \mathbf{p} \circ \mathbf{u}^{-1}$ *is a non-negative eigenvector of* $A^\top$ *corresponding to* $\rho(A)$. *In particular, if* A *is irreducible, then* $\mathbf{p}$ *satisfying* (6.8.2) *is unique.*

Proof. Let $\mu(A)$ be the left-hand side of (6.8.1). We first show that $\mu(A) \leq \Psi(\log \rho(A))$. Suppose first that there exists $\mathbf{u} > \mathbf{0}$

such that $A\mathbf{u} = \rho(A)\mathbf{u}$. Then

$$\inf_{\mathbf{x}=(x_1,\ldots,x_n)>0} \sum_{i=1}^n p_i \Psi\left(\log \frac{(A\mathbf{x})_i}{x_i}\right)$$

$$\leq \sum_{i=1}^n p_i \Psi\left(\log \frac{(A\mathbf{u})_i}{u_i}\right) = \Psi(\log \rho(A)),$$

for any $\mathbf{p} \in \Pi_n$. Hence, $\mu(A) \leq \Psi(\log \rho(A))$.

Let $J_n \in \mathbb{R}^{n\times n}$ be the matrix whose all entries are equal to 1. For $\varepsilon > 0$ let $A(\varepsilon) := A + \varepsilon J_n$. As $A(\varepsilon)$ is positive, it has a positive Perron–Frobenius eigenvector. Hence, $\mu(A(\varepsilon)) \leq \Psi(\log \rho(A(\varepsilon)))$. Since Ψ is nondecreasing and $A(\varepsilon) > A$, it follows that $\mu(A) \leq \mu(A(\varepsilon)) \leq \Psi(\log \rho(A(\varepsilon)))$. Let $\varepsilon \searrow 0$, and use the continuity of $\Psi(t)$ to deduce $\mu(A) \leq \Psi(\log \rho(A))$.

Assume now that $A \in \mathbb{R}_+^{n\times n}$ is irreducible. Let $\mathbf{u}, \mathbf{v} > \mathbf{0}$ be the right and the left Perron–Frobenius eigenvectors of A, such that $\mathbf{p}^\star = (p_1^\star, \ldots, p_n^\star)^\top := \mathbf{u} \circ \mathbf{v} \in \Pi_n$. Suppose first that $\Psi(t) = t$. Theorem 6.6.9 yields the equality

$$\min_{\mathbf{x}=(x_1,\ldots,x_n)^\top>0} \sum_{i=1}^n p_i^\star \log \frac{(A\mathbf{x})_i}{x_i} = \log \rho(A).$$

Hence, we deduce that $\mu(A) \geq \log \rho(A)$. Combine that with the previous inequality $\mu(A) \leq \log \rho(A)$ to deduce (6.8.1) in this case.

Suppose next that Ψ is a convex differentiable nondecreasing function on $\mathbb{R}$. Let $s := \Psi'(\log \rho(A))$. So $s \geq 0$, and

$$\Psi(t) \geq \Psi(\log \rho(A)) + (t - \log \rho(A))s, \text{ for any } t \in \mathbb{R}.$$

Thus

$$\sum_{i=1}^n p_i \Psi\left(\log \frac{(A\mathbf{x})_i}{x_i}\right) \geq \Psi(\log \rho(A)) - \log \rho(A))s$$

$$+ s \sum_{i=1}^n p_i \log \frac{(A\mathbf{x})_i}{x_i}.$$

Use the equality (6.8.1) for $\Phi(t) = t$ to deduce that $\mu(A) \geq \Psi(\log \rho(A))$. Combine that with the inequality $\mu(A) \leq \Psi(\log \rho(A))$ to deduce (6.8.1) for any irreducible A.

Suppose next that $A \in \mathbb{R}_+^{n \times n}$ is reducible and $\rho(A) > 0$. By applying a permutational similarity to A, if necessary, we may assume that $A = [a_{ij}]$ and $B = [a_{ij}]_{i=j=1}^m \in \mathbb{R}^{m \times m}, 1 \leq m < n$ is an irreducible submatrix of A with $\rho(B) = \rho(A)$. Clearly, for any $\mathbf{x} > \mathbf{0}$, $(A(x_1, \ldots, x_n)^\top)_i \geq (B(x_1, \ldots, x_m)^\top)_i$ for $i = 1, \ldots, m$. Since Ψ is nondecreasing we obtain the following set of inequalities

$$\mu(A) \geq \sup_{\mathbf{q}^\top \in \Pi_m} \inf_{\mathbf{0} < \mathbf{x} \in \mathbb{R}^n} \sum_{i=1}^m q_i \Psi \left(\log \frac{(A\mathbf{x})_i}{x_i} \right)$$

$$\geq \sup_{\mathbf{q}^\top \in \Pi_m} \inf_{\mathbf{0} < \mathbf{y} \in \mathbb{R}^m} \sum_{i=1}^m q_i \Psi \left(\log \frac{(B\mathbf{y})_i}{y_i} \right) = \Psi(\log \rho(B)).$$

Use the equality $\rho(A) = \rho(B)$ and the inequality $\mu(A) \leq \Psi(\log \rho(A))$ to deduce the theorem.

Assume now that

$$\rho(A) > 0, \ \Psi'(\log \rho(A)) > 0, \ A\mathbf{u} = \rho(A)\mathbf{u}, \ \mathbf{u} > \mathbf{0},$$

and equality (6.8.2) holds. So the infimum is achieved at $\mathbf{x} = \mathbf{u}$. Since $\mathbf{x} = \mathbf{u}$ is a critical point we deduce that $A^\top \mathbf{p} \circ \mathbf{u}^{-1} = \rho(A)\mathbf{p} \circ \mathbf{u}^{-1}$. If A is irreducible then $\mathbf{p}$ is unique. $\qquad \square$

Corollary 6.8.2 *Let $A \in \mathbb{R}_+^{n \times n}$. Then*

$$\sup_{\mathbf{p}=(p_1,\ldots,p_n)^\top \in \Pi_n} \inf_{\mathbf{x}=(x_1,\ldots,x_n)>0} \sum_{i=1}^n p_i \frac{(A\mathbf{x})_i}{x_i} = \rho(A). \qquad (6.8.3)$$

Suppose that $\rho(A) > 0$ and A has a positive eigenvector $\mathbf{u}$ which corresponds to $\rho(A)$. If

$$\inf_{\mathbf{x}=(x_1,\ldots,x_n)>0} \sum_{i=1}^n p_i \frac{(A\mathbf{x})_i}{x_i} = \rho(A), \qquad (6.8.4)$$

then the vector $\mathbf{v} = \mathbf{p} \circ \mathbf{u}^{-1}$ *is a non-negative eigenvector of* $A^\top$ *corresponding to* $\rho(A)$. *In particular, if* A *is irreducible, then* $\mathbf{p}$ *satisfying (6.8.3) is unique.*

Proof. If $\rho(A) > 0$ the corollary follows from Theorem 6.8.1 by letting $\Psi(t) = e^t$. For $\rho(A) = 0$ apply the corollary to $A_1 = A + I$ to deduce the corollary in this case. $\qquad\square$

Theorem 6.8.3 *Let* $\mathcal{D}_{n,+}$ *denote the convex set of all* $n \times n$ *non-negative diagonal matrices. Assume that* $A \in \mathbb{R}_+^{n \times n}$. *Then*

$$\rho(A + tD_1 + (1-t)D_2) \le t\rho(A + D_1) + (1-t)\rho(A + D_2) \quad (6.8.5)$$

for $t \in (0,1)$ *and* $D_1, D_2 \in \mathcal{D}_{n,+}$. *If* A *is irreducible then equality holds if and only if* $D_1 - D_2 = aI$.

Proof. Let $\phi(\mathbf{p}) = \inf_{\mathbf{x} > 0} \sum_{i=1}^n p_i \frac{(A\mathbf{x})_i}{x_i}$ for $\mathbf{p} \in \Pi_n$. Since $\frac{((A+D)\mathbf{x})_i}{x_i} = d_i + \frac{(A\mathbf{x})_i}{x_i}$ for $D = \mathrm{diag}(d_1, \ldots, d_n)$ we deduce that

$$\psi(D, \mathbf{p}) := \inf_{\mathbf{x} > 0} \sum_{i=1}^n p_i \frac{((A+D)\mathbf{x})_i}{x_i} = \sum_{i=1}^n p_i d_i + \phi(\mathbf{p}).$$

Thus $\psi(D, \mathbf{p})$ is an affine function, hence convex on $\mathcal{D}_{n,+}$. Therefore, $\rho(A + D) = \sup_{\mathbf{p} \in \Pi_n} \psi(D, \mathbf{p})$ is a convex function on $\mathcal{D}_{n,+}$. Hence, (6.8.5) holds for any $t \in (0,1)$ and $D_1, D_2 \in \mathcal{D}_{n,+}$.

Suppose that A is irreducible and equality holds in (6.8.5). Since $\rho(A + bI + D) = b + \rho(A + D)$ for any $b > 0$, we may assume without loss of generality that all the diagonal elements of A are positive. Let $A_0 = A + tD_1 + (1-t)D_2$. Since A_0 has a positive diagonal and is irreducible we deduce that $A_0\mathbf{u} = r\mathbf{u}$, $A_0^\top \mathbf{v} = r\mathbf{v}$ where $r > 0, \mathbf{u}, \mathbf{v} > \mathbf{0}, \mathbf{w} := \mathbf{v} \circ \mathbf{u} \in \Pi_n$. Corollary 6.8.2 yields that

$$\rho(A_0) = \psi(tD_1 + (1-t)D_2, \mathbf{w}) = t\psi(D_1, \mathbf{w}) + (1-t)\psi(D_2, \mathbf{w}).$$

Hence, equality in (6.8.5) implies that

$$\rho(A + D_1) = \psi(D_1, \mathbf{w}) = \rho(A_0) + (1 - t) \sum_{i=1}^{n}(d_{1,i} - d_{2,i}),$$

$$\rho(A + D_2) = \psi(D_2, \mathbf{w}) = \rho(A_0) + t \sum_{i=1}^{n}(d_{2,i} - d_{1,i}),$$

$$D_1 = \text{diag}(d_{1,1}, \ldots, d_{n,1}), \quad D_2 = \text{diag}(d_{1,2}, \ldots, d_{n,2}).$$

Furthermore, the infima on $\mathbf{x} > \mathbf{0}$ in the $\psi(D_1, \mathbf{w})$ and $\psi(D_2, \mathbf{w})$ are achieved for $\mathbf{x} = \mathbf{u}$. Corollary 6.8.2 that $\mathbf{u}$ is the Perron–Frobenius eigenvector of $A + D_1$ and $A + D_2$. Hence, $D_1 - D_2 = aI$. Clearly, if $D_1 - D_2 = aI$ then equality holds in (6.8.5). $\square$

Theorem 6.8.4 *Let* $A \in \mathbb{R}_+^{n \times n}$ *be an inverse of an M-matrix. Then*

$$\rho((tD_1 + (1 - t)D_2)A) \le t\rho(D_1 A) + (1 - t)\rho(D_2 A), \qquad (6.8.6)$$

for $t \in (0, 1)$, $D_1, D_2 \in \mathcal{D}_{n,+}$. *If* $A > 0$ *and* D_1, D_2 *have positive diagonal entries then equality holds if and only if* $D_1 = aD_2$.

Proof. Let $A = B^{-1}$, where $B = rI - C, C \in \mathbb{R}_+^{n \times n}$ and $\rho(C) < r$. Use Neumann expansion to deduce that $A = \sum_{i=0}^{\infty} r^{-(i+1)} B^i$. Hence, $A > 0$ if and only if C is irreducible. Assume first that A is positive. Denote by $\mathcal{D}_{n,+}^o$ the set diagonal matrices with positive diagonal, i.e. the interior of $\mathcal{D}_{n,+}$. Clearly, $DA > 0$ for $D \in \mathcal{D}_{n,+}^o$. Thus, $\rho(DA) > 0$ is a simple eigenvalue of $\det(\lambda I - DA)$. Hence, $\rho(DA)$ is an analytic function on $\mathcal{D}_{n,+}^o$. Denote by $\nabla\rho(DA) \in \mathbb{R}^n$ the gradient of $\rho(DA)$ as a function on $\mathcal{D}_{n,+}$. Since $\rho(DA) \in C^2(\mathcal{D}_{n,+}^o)$ it follows that convexity of $\rho(DA)$ on $\mathcal{D}_{n,+}^o$ is equivalent to the following inequality.

$$\rho(D(\mathbf{d})A) \ge \rho(D(\mathbf{d}_0)A) + \nabla\rho(D(\mathbf{d}_0)A)^{\top}(\mathbf{d} - \mathbf{d}_0), \quad \mathbf{d}, \mathbf{d}_0 > \mathbf{0}.$$
$$(6.8.7)$$

See Problem 1. We now show (6.8.7). Let

$$D_0 A \mathbf{u} = \rho(D_0)\mathbf{u}, \mathbf{v}^\top D_0 A = \rho(D_0 A)\mathbf{v}^\top,$$

$$\mathbf{u}, \mathbf{v} > 0, \mathbf{v} \circ \mathbf{u} \in \Pi_n, D_0 = D(\mathbf{d}_0).$$

Theorem 6.7.3 implies that $\nabla \rho(D_0 A) = \rho(D_0 A)\mathbf{v} \circ \mathbf{u} \circ \mathbf{d}_0^{-1}$. Hence, (6.8.7) is equivalent to

$$\rho(D(\mathbf{d})A) \geq \rho(D_0 A)(\mathbf{v} \circ \mathbf{u})^\top(\mathbf{d} \circ \mathbf{d}_0^{-1}).$$

Let $D(\mathbf{d})A\mathbf{w} = \rho(D(\mathbf{d})A)\mathbf{w}, \mathbf{w} > 0$. Then the above inequality follows from the inequality

$$\rho(D_0 A)^{-1} \geq (\mathbf{v} \circ \mathbf{u})^\top(\mathbf{w} \circ (D_0 A\mathbf{w})^{-1}). \tag{6.8.8}$$

This inequality follows from the inf sup characterization of $\rho(D_0 A)^{-1}$. See Problem 2. The equality case in (6.8.7) follows from the similar arguments for the equality case in (6.8.6). Since $\rho(DA)$ is a continuous function on $\mathcal{D}_{n,+}$, the convexity of $\rho(DA)$ on $\mathcal{D}_{n,+}^o$ yields the convexity of $\rho(DA)$ on $\mathcal{D}_{n,+}$.

Consider now the case where $A^{-1} = rI - B, B \in \mathbb{R}_+^{n \times n}, r > \rho(B)$, and B is reducible. Then there exists $b > 0$ such that for $\rho(B + b\mathbf{1}^\top \mathbf{1}) < r$. For $\varepsilon \in (0, b)$ let $A(\varepsilon) := (rI - (B + \varepsilon \mathbf{1}\mathbf{1}^\top))^{-1}$. Then the inequality (6.8.7) holds if A is replaced by $A(\varepsilon)$. Let $\varepsilon \searrow 0$ to deduce (6.8.7). □

Problems

1. (a) Let $f \in C^2(a, b)$. Show that $f''(x) \geq 0$ for $x \in (a, b)$ if and only if $f(x) \geq f(x_0) + f'(x_0)(x - x_0)$ for each $x, x_0 \in (a, b)$.

   (b) Let $D \subset \mathbb{R}^n$ be an open convex set. Assume that $f \in C^2(D)$. Let $H(f)(\mathbf{x}) = [\frac{\partial^2 f}{\partial x_i \partial x_j}](\mathbf{x}) \in \mathbf{S}(n, \mathbb{R})$ for $\mathbf{x}_0 \in D$. Show that $H(f)(\mathbf{x}) \geq 0$ for all $\mathbf{x} \in D$ if and only if $f(\mathbf{x}) \geq f(\mathbf{x}_0) + \nabla f(\mathbf{x}_0)^\top(\mathbf{x} - \mathbf{x}_0)$ for all $\mathbf{x}, \mathbf{x}_0 \in D$. (*Hint*: Restrict f to an interval $(\mathbf{u}, \mathbf{v}) \subset D$ and use part (a) of the problem.)

2. (a) Let $F \in \mathbb{R}_+^{n \times n}$ be an inverse of an M-matrix. Show

$$\frac{1}{\rho(F)} = \inf_{\mathbf{p}=(p_1,\ldots,p_n)^\top \in \Pi_n} \sup_{\mathbf{x}=(x_1,\ldots,x_n)>0} \sum_{i=1}^n p_i \frac{x_i}{(F\mathbf{x})_i}.$$

   Hint: Use Corollary 6.8.2.

   (b) Let $0 < F \in \mathbb{R}_+^{n \times n}$ be an inverse of an M-matrix. Assume that

$$F\mathbf{u} = \rho(F)\mathbf{u}, F^\top \mathbf{v} = \rho(F)\mathbf{v}, \mathbf{u}, \mathbf{v} > \mathbf{0}, \mathbf{v} \circ \mathbf{u} \in \Pi_n.$$

   Show

$$\frac{1}{\rho(F)} = \sup_{\mathbf{x}=(x_1,\ldots,x_n)>0} \sum_{i=1}^n v_i u_i \frac{x_i}{(F\mathbf{x})_i}.$$

   Furthermore, $\frac{1}{\rho(F)} = \sum_{i=1}^n v_i u_i \frac{x_i}{(F\mathbf{x})_i}$ for $\mathbf{x} > 0$ if and only if $F\mathbf{x} = \rho(F)\mathbf{x}$.

   (c) Show (6.8.8). *Hint*: Use Corollary 6.6.4 to show that $A^{-1} D_0^{-1}$ is an M-matrix.

3. Let $P = [\delta_{i(j+1)}] \in \mathcal{P}_n$ be a cyclic permutation matrix. Show that $\rho(D(\mathbf{d})P) = (\prod_{i=1}^n d_i)^{\frac{1}{n}}$ for any $\mathbf{d} \in \mathbb{R}_+^n$.

4. Show that (6.8.7) does not hold for all $A \in \mathbb{R}_+^{n \times n}$.

5. Let $A \in \mathbb{R}_+^{n \times n}$ be an inverse of an M-matrix. Show that the convexity of $\rho(D(e^{\mathbf{x}})A)$ on $\mathbb{R}^n$ is implied by the convexity of $\rho(DA)$ on $\mathcal{D}_{n,+}$. *Hint*: Use the generalized arithmetic-geometric inequality.

6.9   An Application to Cellular Communication

6.9.1   Introduction

Power control is used in cellular and ad hoc networks to provide a high signal-to-noise ratio (SNR) for a reliable connection. A higher SNR also allows a wireless system that uses link adaptation to transmit at a higher data rate, thus leading to a greater spectral efficiency. Transmission rate adaptation by power control is an active research area in communication networks that can be used for both interference management and utility maximization [Sri03].

The motivation of the problems studied in this section comes from maximizing sum rate, (data throughput), in wireless communications. Due to the broadcast nature of radio transmission, data rates in a wireless network are affected by interference. This is particularly true in Code Division Multiple Access (CDMA) systems, where users transmit at the same time over the same frequency bands and their spreading codes are not perfectly orthogonal. Transmit power control is often used to control signal interference to maximize the total transmission rates of all users.

6.9.2 Statement of problems

Consider a wireless network, e.g. cellular network, with L logical transmitter/receiver pairs. Transmit powers are denoted as $p_1, \ldots, p_L$. Let $\mathbf{p} = (p_1, \ldots, p_L)^\top \geq \mathbf{0}$ be the power transmission vector. In many situation we will assume that $\mathbf{p} \leq \bar{\mathbf{p}} := (\bar{p}_1, \ldots, \bar{p}_l)^\top$, where $\bar{p}_l$ is the maximal transmit power of the user l. In the cellular uplink case, all logical receivers may reside in the same physical receiver, i.e. the base station. Let $G = [g_{ij}]_{i,j=1}^L > 0_{L \times L}$ representing the channel gain, where g_{ij} is the channel gain from the jth transmitter to the ith receiver, and n_l is the noise power for the lth receiver be given. The signal-to-interference ratio (SIR) for the lth receiver is denoted by $\gamma_l = \gamma_l(\mathbf{p})$. The map $\mathbf{p} \mapsto \boldsymbol{\gamma}(\mathbf{p})$ is given by

$$\gamma_l(\mathbf{p}) := \frac{g_{ll} p_l}{\sum_{j \neq l} g_{lj} p_j + n_l}, \ l = 1, \ldots, L, \ \boldsymbol{\gamma}(\mathbf{p}) = (\gamma_1(\mathbf{p}), \ldots, \gamma_L(\mathbf{p}))^\top.$$

$$(6.9.1)$$

That is, the power p_l is amplified by the factor g_{ll}, and diminished by other users and the noise, inversely proportional to $\sum_{j \neq l} g_{lj} p_j + n_l$.

Define

$$F = [f_{ij}]_{i,j=1}^L, \text{ where } f_{ij} = \begin{cases} 0, & \text{if } i = j \\ \dfrac{g_{ij}}{g_{ii}}, & \text{if } i \neq j \end{cases} \tag{6.9.2}$$

and

$$\mathbf{g} = (g_{11}, \ldots, g_{LL})^\top, \ \mathbf{n} = (n_1, \ldots, n_L)^\top, \tag{6.9.3}$$

$$\mathbf{s} = (\mathbf{s_1}, \ldots, \mathbf{s_L})^\top := \left(\frac{\mathbf{n_1}}{\mathbf{g_{11}}}, \frac{\mathbf{n_2}}{\mathbf{g_{22}}}, \ldots, \frac{\mathbf{n_L}}{\mathbf{g_{LL}}} \right)^\top.$$

Then

$$\boldsymbol{\gamma}(\mathbf{p}) = \mathbf{p} \circ (F\mathbf{p} + \mathbf{s})^{-1}. \tag{6.9.4}$$

Let

$$\Phi_{\mathbf{w}}(\boldsymbol{\gamma}) := \sum_{i=1}^L w_i \log(1 + \gamma_i), \quad \text{where}$$

$$\mathbf{w} = (w_1, \ldots, w_n)^\top \in \Pi_n, \boldsymbol{\gamma} \in \mathbb{R}_+^L. \tag{6.9.5}$$

The function $\Phi_{\mathbf{w}}(\boldsymbol{\gamma}(\mathbf{p}))$ is the sum rate of the interference-limited channel.

We can study the following optimal problems in the power vector $\mathbf{p}$. The first problem is concerned with finding the optimal power that maximizes the minimal SIR for all users:

$$\max_{\mathbf{p} \in [\mathbf{0}; \bar{\mathbf{p}}]} \min_{i \in [L]} \gamma_l(\mathbf{p}). \tag{6.9.6}$$

Then second, more interesting problem, is the sum rate maximization problem in interference-limited channels

$$\max_{\mathbf{p} \in [\mathbf{0}, \bar{\mathbf{p}}]} \Phi_{\mathbf{w}}(\boldsymbol{\gamma}(\mathbf{p})). \tag{6.9.7}$$

The exact solution to this problem is known to be NP-complete [Luo08]. Note that for a fixed $p_1, \ldots, p_{l-1}, p_{l+1}, \ldots, p_L$ each $\gamma_j(\mathbf{p}), j \neq l$ is a decreasing function of p_l, while $\gamma_l(\mathbf{p})$ is an increasing function of l. Thus, if $w_l = 0$ we can assume that in the maximal problem (6.9.7) we can choose $p_l = 0$. Hence, it is enough to study the maximal problem (6.9.7) in the case $\mathbf{w} > \mathbf{0}$.

6.9.3 Relaxations of optimal problems

In this subsection, we study several relaxed versions of (6.9.6) and (6.9.7). We will assume first that we do not have the restriction $\mathbf{p} \leq \bar{\mathbf{p}}$. Let $\gamma(\mathbf{p}, \mathbf{n})$ be given by (6.9.1). Note that since $\mathbf{n} > 0$ we obtain

$$\gamma(t\mathbf{p}, \mathbf{n}) = \gamma(\mathbf{p}, \frac{1}{t}\mathbf{n}) \Rightarrow \gamma(t\mathbf{p}, \mathbf{n}) > \gamma(\mathbf{p}, \mathbf{n}) \text{ for } t > 1.$$

Thus, to increase the values of the optimal problems in (6.9.6) and (6.9.7), we let $t \to \infty$, which is equivalent to the assumption in this subsection that $\mathbf{n} = \mathbf{0}$.

Theorem 6.9.1 *Let* $F \in \mathbb{R}_+^{L \times L}, L \geq 2$ *be a matrix with positive off-diagonal entries. Let* $F\mathbf{u} = \rho(F)\mathbf{u}$ *for a unique* $0 < \mathbf{u} \in \Pi_L$. *Then*

$$\max_{0 < \mathbf{p} \in \Pi_L} \min_{l \in [L]} \frac{p_l}{\sum_{j=1}^L f_{lj} p_j} = \frac{1}{\rho(F)}, \qquad (6.9.8)$$

which is achieved only for $\mathbf{p} = \mathbf{u}$. *In particular, The value of the optimal problem given in* (6.9.6) *is less than* $\frac{1}{\rho(F)}$.

Proof. Clearly, the left-hand side of (6.9.8) is equal to $(\min_{0 < \mathbf{p}} \max_{l \in [L]} \frac{(F\mathbf{p})_l}{p_l})^{-1}$. Since F is irreducible, our theorem follows from Problem 11.

Clearly $\gamma(\mathbf{p}, \mathbf{n}) < \gamma(\mathbf{p}, \mathbf{0})$. Hence, for $\mathbf{p} > 0$ $\min_{l \in [L]} \gamma(\mathbf{p}, \mathbf{n}) < \min_{l \in [L]} \gamma(\mathbf{p}, \mathbf{0})$. Since $\mathbf{p} \in [0, \bar{\mathbf{p}}] \subset \mathbb{R}_+^L$ we deduce that the value of the optimal problem given in (6.9.6) is less than $\frac{1}{\rho(F)}$. $\square$

We now consider the relaxation problem of (6.9.7). We approximate $\log(1 + x)$ by $\log x$ for $x \geq 0$. Clearly, $\log(1 + x) \geq \log x$. Let

$$\Psi_\mathbf{w}(\boldsymbol{\gamma}) = \sum_{j=1}^L w_i \log \gamma_j, \quad \boldsymbol{\gamma} = (\gamma_1, \dots, \gamma_L)^\top. \qquad (6.9.9)$$

Theorem 6.9.2 *Let* $F = [f_{ij}] \in \mathbb{R}_+^{L \times L}$ *have positive off-diagonal elements and zero diagonal entries. Assume that* $L \geq 3$, $\mathbf{w} = (w_1, \ldots, w_L)^\top > \mathbf{0}$, *and suppose that* $\mathbf{w}$ *satisfies the inequalities (6.6.9) for each* $j \in [L]$, *where* $n = L$. *Let* $D_1 = \mathrm{diag}$ $(d_{1,1}, \ldots, d_{L,1}), D_2 = \mathrm{diag}(d_{1,2}, \ldots, d_{L,2})$ *be two diagonal matrices, with positive diagonal entries, such that* $B = D_1 F D_2, B\mathbf{1} = \mathbf{1}$, $B^\top \mathbf{w} = \mathbf{w}$. *(As given by Theorem 6.6.12.) Then*

$$\max_{\mathbf{p} > 0} \Psi_{\mathbf{w}}(\mathbf{p}) = \sum_{j=1}^{L} w_j \log d_{j,1} d_{j,2}. \qquad (6.9.10)$$

Equality holds if and only if $\mathbf{p} = t D_2^{-1} \mathbf{1}$ *for some* $t > 0$.

Proof. Let $\mathbf{p} = D_2 \mathbf{x}$. Then

$$\Psi_{\mathbf{w}}(D_2 \mathbf{x}) = \sum_{j=1}^{L} w_j \log d_{j,1} d_{j,2} - \sum_{j=1}^{L} w_j \frac{(B\mathbf{x})_j}{x_j}.$$

Use Theorem 6.6.9 to deduce that the above expression is not more than the right-hand side of (6.9.10). For $\mathbf{x} = \mathbf{1}$ equality holds. From the proof of the second part of Theorem 6.6.12 it follows that this minimum is achieved only for $\mathbf{x} = t\mathbf{1}$, which is equivalent to $\mathbf{p} = t D_2^{-1} \mathbf{1}$. $\qquad \square$

6.9.4   Preliminary results

Claim 6.9.3 *Let* $\mathbf{p} \geq \mathbf{0}$ *be a non-negative vector. Assume that* $\boldsymbol{\gamma}(\mathbf{p})$ *is defined by (6.9.1). Then* $\rho(\mathrm{diag}(\boldsymbol{\gamma}(\mathbf{p}))F) < 1$, *where* F *is defined by (6.9.2). Hence, for* $\boldsymbol{\gamma} = \boldsymbol{\gamma}(\mathbf{p})$,

$$\mathbf{p} = P(\boldsymbol{\gamma}) := (I - \mathrm{diag}(\boldsymbol{\gamma})F)^{-1} \mathrm{diag}(\boldsymbol{\gamma})\mathbf{v}. \qquad (6.9.11)$$

Vice versa, if $\boldsymbol{\gamma}$ *is in the set*

$$\boldsymbol{\Gamma} := \{\boldsymbol{\gamma} \geq \mathbf{0}, \ \rho(\mathrm{diag}(\boldsymbol{\gamma})F) < 1\}, \qquad (6.9.12)$$

then the vector $\mathbf{p}$ *defined by (6.9.11) is non-negative. Furthermore,* $\boldsymbol{\gamma}(P(\mathbf{p})) = \boldsymbol{\gamma}$. *That is,* $\boldsymbol{\gamma} : \mathbb{R}_+^L \to \boldsymbol{\Gamma}$, *and* $P : \boldsymbol{\Gamma} \to \mathbb{R}_+^L$ *are inverse mappings.*

Proof. Observe that (6.9.1) is equivalent to the equality

$$\mathbf{p} = \operatorname{diag}(\boldsymbol{\gamma})F\mathbf{p} + \operatorname{diag}(\boldsymbol{\gamma})\mathbf{v}. \tag{6.9.13}$$

Assume first that $\mathbf{p}$ is a positive vector, i.e. $\mathbf{p} > \mathbf{0}$. Hence, $\boldsymbol{\gamma}(\mathbf{p}) > \mathbf{0}$. Since all off-diagonal entries of F are positive it follows that the matrix $\operatorname{diag}(\boldsymbol{\gamma})F$ is irreducible. As $\mathbf{v} > \mathbf{0}$, we deduce that $\max_{i\in[1,n]} \frac{(\operatorname{diag}(\boldsymbol{\gamma})F\mathbf{p})_i}{p_i} < 1$. The min–max characterization of Wielandt of $\rho(\operatorname{diag}(\boldsymbol{\gamma})F)$, (6.2.4) implies $\rho(\operatorname{diag}(\boldsymbol{\gamma})F) < 1$. Hence, $\boldsymbol{\gamma}(\mathbf{p}) \in \boldsymbol{\Gamma}$. Assume now that $\mathbf{p} \geq \mathbf{0}$. Note that $p_i > 0 \iff \gamma_i(\mathbf{p}) > 0$. So $\mathbf{p} = \mathbf{0} \iff \boldsymbol{\gamma}(\mathbf{p}) = \mathbf{0}$. Clearly, $\rho(\boldsymbol{\gamma}(\mathbf{0})F) = \rho(0_{L\times L}) = 0 < 1$. Assume now that $\mathbf{p} \gneq \mathbf{0}$. Let $\mathcal{A} = \{i : p_i > 0\}$. Denote $\boldsymbol{\gamma}(\mathbf{p})(\mathcal{A})$ the vector composed of positive entries of $\boldsymbol{\gamma}(\mathbf{p})$. Let $F(\mathcal{A})$ be the principal submatrix of $\mathbb{F}$ with rows and columns in $\mathcal{A}$. It is straightforward to see that $\rho(\operatorname{diag}(\boldsymbol{\gamma}(\mathbf{p}))F) = \rho(\operatorname{diag}(\boldsymbol{\gamma}(p)(\mathcal{A})F(\mathcal{A}))$. The arguments above imply that

$$\rho(\operatorname{diag}(\boldsymbol{\gamma}(\mathbf{p}))F) = \rho(\operatorname{diag}(\boldsymbol{\gamma}(\mathbf{p})(\mathcal{A})\mathbb{F}(\mathcal{A})) < 1.$$

Assume now that $\boldsymbol{\gamma} \in \boldsymbol{\Gamma}$. Then

$$(I - \operatorname{diag}(\boldsymbol{\gamma})F)^{-1} = \sum_{k=0}^{\infty}(\operatorname{diag}(\boldsymbol{\gamma})F)^k \geq 0_{L\times L}. \tag{6.9.14}$$

Hence, $P(\boldsymbol{\gamma}) \geq \mathbf{0}$. The definition of $P(\boldsymbol{\gamma})$ implies that $\boldsymbol{\gamma}(P(\boldsymbol{\gamma})) = \boldsymbol{\gamma}$. $\qquad\square$

Claim 6.9.4 *The set $\boldsymbol{\Gamma} \subset \mathbb{R}_+^L$ is monotonic with respect to the order "$\geq$". That is if $\boldsymbol{\gamma} \in \boldsymbol{\Gamma}$ and $\boldsymbol{\gamma} \geq \boldsymbol{\beta} \geq \mathbf{0}$ then $\boldsymbol{\beta} \in \boldsymbol{\Gamma}$. Furthermore, the function $P(\boldsymbol{\gamma})$ is monotone on $\boldsymbol{\Gamma}$.*

$$P(\boldsymbol{\gamma}) \geq P(\boldsymbol{\beta}) \text{ if } \boldsymbol{\gamma} \in \boldsymbol{\Gamma} \text{ and } \boldsymbol{\gamma} \geq \boldsymbol{\beta} \geq \mathbf{0}. \tag{6.9.15}$$

Equality holds if and only if $\boldsymbol{\gamma} = \boldsymbol{\beta}$.

Proof. Clearly, if $\boldsymbol{\gamma} \geq \boldsymbol{\beta} \geq \mathbf{0}$ then $\operatorname{diag}(\boldsymbol{\gamma})F \geq \operatorname{diag}(\boldsymbol{\beta})F$ which implies $\rho(\operatorname{diag}(\boldsymbol{\gamma})F) \geq \rho(\operatorname{diag}(\boldsymbol{\beta})F)$. Hence, $\boldsymbol{\Gamma}$ is monotonic. Use the Neumann expansion (6.9.14) to deduce the monotonicity of P. The equality case is straightforward. $\qquad\square$

Note that $\gamma(\mathbf{p})$ is not monotonic in $\mathbf{p}$. Indeed, if one increases only the ith coordinate of $\mathbf{p}$, then one increases the ith coordinate of $\gamma(\mathbf{p})$ and decreases all other coordinates of $\gamma(\mathbf{p})$.

As usual, let $\mathbf{e}_i = (\delta_{i1}, \ldots, \delta_{iL})^\top$, $i = 1, \ldots, L$ be the standard basis in $\mathbb{R}^L$. In what follows, we need the following result.

Theorem 6.9.5 *Let $l \in [L]$ and $a > 0$. Denote $[0, a]_l \times \mathbb{R}_+^{L-1}$ the set of all $\mathbf{p} = (p_1, \ldots, p_L)^\top \in \mathbb{R}_+^L$ satisfying $p_l \le a$. Then the image of the set $[0, a]_l \times \mathbb{R}_+^{L-1}$ by the map γ (6.9.1), is given by*

$$\rho\left(\operatorname{diag}(\boldsymbol{\gamma})\left(F + \frac{1}{a}\mathbf{ve}_l^\top\right)\right) \le 1, \ \mathbf{0} \le \boldsymbol{\gamma}. \tag{6.9.16}$$

Furthermore, $\mathbf{p} = (p_1, \ldots, p_L) \in \mathbb{R}_+^L$ satisfies the condition $p_l = a$ if and only if $\boldsymbol{\gamma} = \boldsymbol{\gamma}(\mathbf{p})$ satisfies

$$\rho\left(\operatorname{diag}(\boldsymbol{\gamma})\left(F + \frac{1}{a}\mathbf{ve}_l^\top\right)\right) = 1. \tag{6.9.17}$$

Proof. Suppose that $\boldsymbol{\gamma}$ satisfies (6.9.16). We claim that $\boldsymbol{\gamma} \in \boldsymbol{\Gamma}$. Suppose first that $\boldsymbol{\gamma} > \mathbf{0}$. Then $\operatorname{diag}(\boldsymbol{\gamma})(F + t_1\mathbf{ve}_l^\top) \lneqq \operatorname{diag}(\boldsymbol{\gamma})(F + t_2\mathbf{ve}_l^\top)$ for any $t_1 < t_2$. Lemma 6.2.4 yields

$$\rho(\operatorname{diag}(\boldsymbol{\gamma})F) < \rho(\operatorname{diag}(\boldsymbol{\gamma})(F + t_1\mathbf{ve}_l^\top)) < \rho(\operatorname{diag}(\boldsymbol{\gamma})(F + t_2\mathbf{ve}_l^\top))$$

$$< \rho\left(\operatorname{diag}(\boldsymbol{\gamma})\left(F + \frac{1}{a}\mathbf{ve}_l^\top\right)\right) \le 1 \quad \text{for } 0 < t_1 < t_2 < \frac{1}{a}. \tag{6.9.18}$$

Thus $\boldsymbol{\gamma} \in \boldsymbol{\Gamma}$. Combine the above argument with the arguments of the proof of Claim 6.9.3 to deduce that $\boldsymbol{\gamma} \in \boldsymbol{\Gamma}$ for $\boldsymbol{\gamma} \ge \mathbf{0}$.

We now show that $P(\boldsymbol{\gamma})_l \le a$. The continuity of P implies that it is enough to consider the case $\boldsymbol{\gamma} > \mathbf{0}$. Combine the Perron–Frobenius theorem with (6.9.18) to deduce

$$0 < \det\left(I - \operatorname{diag}(\boldsymbol{\gamma})(F + t\mathbf{ve}_l^\top)\right) \text{ for } t \in [0, a^{-1}). \tag{6.9.19}$$

We now expand the right-hand side of the above inequality. Let $B = \mathbf{xy}^\top \in \mathbb{R}^{L \times L}$ be a rank one matrix. Then B has $L - 1$ zero eigenvalues and one eigenvalue equal to $\mathbf{y}^\top\mathbf{x}$. Hence, $I - \mathbf{xy}^\top$ has $L-1$ eigenvalues equal to 1 and one eigenvalue is $(1 - \mathbf{y}^\top\mathbf{x})$. Therefore,

$\det (I - \mathbf{x}\mathbf{y}^\top) = 1 - \mathbf{y}^\top\mathbf{x}$. Since $\gamma \in \Gamma$ we get that $(I - \mathrm{diag}(\gamma)F)$ is invertible. Thus, for any $t \in \mathbb{R}$

$$
\begin{aligned}
&\det (I - \mathrm{diag}(\gamma)(F + t\mathbf{v}\mathbf{e}_l^\top)) \\
&= \det (I - \mathrm{diag}(\gamma)F)\det (I - t((I - \mathrm{diag}(\gamma)F)^{-1}\,\mathrm{diag}(\gamma)\mathbf{v})\mathbf{e}_l^\top) \\
&\times \det (I - \mathrm{diag}(\gamma)F)(1 - t\mathbf{e}_l^\top(I - \mathrm{diag}(\gamma)F)^{-1}\,\mathrm{diag}(\gamma)\mathbf{v}).
\end{aligned}
$$

$$(6.9.20)$$

Combine (6.9.19) with the above identity to deduce

$$
1 > t\mathbf{e}_l^\top(I - \mathrm{diag}(\gamma)F)^{-1}\,\mathrm{diag}(\gamma)\mathbf{v} = tP(\gamma)_l \text{ for } t \in [0, a^{-1}).
$$

$$(6.9.21)$$

Letting $t \nearrow a^{-1}$, we deduce that $P(\gamma)_l \le a$. Hence, the set of γ defined by (6.9.16) is a subset of $\gamma([0, a]_l \times \mathbb{R}_+^{L-1})$.

Let $\mathbf{p} \in [0, a]_l \times \mathbb{R}_+^{L-1}$ and denote $\gamma = \gamma(\mathbf{p})$. We show that γ satisfies (6.9.16). Claim 6.9.3 implies that $\rho(\mathrm{diag}(\gamma)F) < 1$. Since $\mathbf{p} = P(\gamma)$ and $p_l \le a$ we deduce (6.9.21). Use (6.9.20) to deduce (6.9.19). As $\rho(\mathrm{diag}(\gamma)F) < 1$, the inequality (6.9.19) implies that $\rho(\mathrm{diag}(\gamma)F + t\mathbf{v}^\top\mathbf{e}_l) < 1$ for $t \in (0, a^{-1})$. Hence, (6.9.16) holds.

It is left to show the condition (6.9.17) holds if and only if $P(\gamma)_l = a$. Assume that $\mathbf{p} = (p_1, \ldots, p_L)^\top \in \mathbb{R}_+^L$, $p_l = a$ and let $\gamma = \gamma(\mathbf{p})$. We claim that equality holds in (6.9.16). Assume to the contrary that $\rho(\mathrm{diag}(\gamma)(F + \frac{1}{a}\mathbf{v}\mathbf{e}_l^\top)) < 1$. Then, there exists $\beta > \gamma$ such that $\rho(\mathrm{diag}(\beta)(F + \frac{1}{a}\mathbf{v}\mathbf{e}_l^\top)) < 1$. Since P is monotonic $P(\beta)_l > p_l = a$. On the other hand, since β satisfies (6.9.16), we deduce that $P(\beta)_l \le a$. This contradiction yields (6.9.17). Similarly, if $\gamma \ge \mathbf{0}$ and (6.9.17) then $P(\gamma)_l = a$. $\qquad\square$

Corollary 6.9.6 *Let $\bar{\mathbf{p}} = (\bar{p}_1, \ldots, \bar{p}_L)^\top > \mathbf{0}$ be a given positive vector. Then $\gamma([\mathbf{0}, \bar{\mathbf{p}}])$, the image of the set $[\mathbf{0}, \bar{\mathbf{p}}]$ by the map γ (6.9.1), is given by*

$$
\rho\left(\mathrm{diag}(\gamma)\left(F + \frac{1}{\bar{p}_l}\mathbf{v}\mathbf{e}_l^\top\right)\right) \le 1, \text{ for } l = 1, \ldots, L, \text{ and } \gamma \in \mathbb{R}_+^L.
$$

$$(6.9.22)$$

In particular, any $\boldsymbol{\gamma} \in \mathbb{R}_+^L$ *satisfying the conditions* (6.9.22) *satisfies the inequalities*

$$\boldsymbol{\gamma} \leq \bar{\boldsymbol{\gamma}} = (\bar{\gamma}_1, \ldots, \bar{\gamma}_L)^\top, \ where \ \bar{\gamma}_l = \frac{\bar{p}_l}{v_l}, \ i = 1, \ldots, L. \qquad (6.9.23)$$

Proof. Theorem 6.9.5 yields that $\boldsymbol{\gamma}([0, \bar{\mathbf{p}}])$ is given by (6.9.22). (6.9.4) yields

$$\gamma_l(\mathbf{p}) = \frac{p_l}{((F\mathbf{p})_l + v_l)} \leq \frac{p_l}{v_l} \leq \frac{\bar{p}_l}{v_l} \ \text{for } \mathbf{p} \in [0, \bar{\mathbf{p}}].$$

Note that equality holds for $\mathbf{p} = \bar{p}_l \mathbf{e}_l$. $\qquad \square$

6.9.5   Reformulation of optimal problems

Theorem 6.9.7 *The maximum problem* (6.9.7) *is equivalent to the maximum problem.*

> *maximize* $\sum_l w_l \log(1 + \gamma_l)$
> *subject to* $\rho(\mathrm{diag}(\boldsymbol{\gamma})(F + (1/\bar{p}_l)\mathbf{v}\mathbf{e}_l^\top)) \leq 1 \ \ \forall l \in [\mathbf{L}],$ $\qquad (6.9.24)$
> *variables:* $\gamma_l, \ \ \forall l.$

$\boldsymbol{\gamma}^\star$ *is a maximal solution of the above problem if and only if* $P(\boldsymbol{\gamma}^\star)$ *is a maximal solution* $p^\star$ *of the problem* (6.9.7). *In particular, any maximal solution* $\boldsymbol{\gamma}^\star$ *satisfies the equality* (6.9.22) *for some integer* $l \in [1, L]$.

We now give the following simple necessary conditions for a maximal solution $\mathbf{p}^\star$ of (6.9.7). We first need the following result, which is obtained by straightforward differentiation.

Lemma 6.9.8 *Denote by*

$$\nabla \Phi_{\mathbf{w}}(\boldsymbol{\gamma}) = \left(\frac{w_1}{1 + \gamma_1}, \ldots, \frac{w_L}{1 + \gamma_L} \right)^\top = \mathbf{w} \circ (1 + \boldsymbol{\gamma})^{-1}$$

the gradient of $\Phi_{\mathbf{w}}$. *Let* $\boldsymbol{\gamma}(\mathbf{p})$ *be defined as in (6.9.1). Then* $H(\mathbf{p}) = [\frac{\partial \gamma_i}{\partial p_j}]_{i=j=1}^L$, *the Hessian matrix of* $\boldsymbol{\gamma}(\mathbf{p})$, *is given by*

$$H(\mathbf{p}) = \mathrm{diag}((F\mathbf{p} + \mathbf{v})^{-1})(-\mathrm{diag}(\boldsymbol{\gamma}(\mathbf{p}))F + I).$$

In particular,

$$\nabla_{\mathbf{p}}\Phi_{\mathbf{w}}(\boldsymbol{\gamma}(\mathbf{p})) = H(\mathbf{p})^\top \nabla\Phi_{\mathbf{w}}(\boldsymbol{\gamma}(\mathbf{p})).$$

Corollary 6.9.9 *Let* $\mathbf{p}^\star = (p_1^\star, \ldots, p_L^\star)^\top$ *be a maximal solution to the problem (6.9.7). Divide the set* $[L] = \{1, \ldots, L\}$ *to the following three disjoint sets* $S_{\max}, S_{\mathrm{in}}, S_0$:

$$S_{\max} = \{i \in [L], \ p_i^\star = \bar{p}_i\}, \quad S_{\mathrm{in}} = \{i \in [L], \ p_i^\star \in (0, \bar{p}_i)\},$$

$$S_0 = \{i \in [L], \ p_i^\star = 0\}.$$

Then the following conditions hold.

$$(H(\mathbf{p}^\star)^\top \nabla\Phi_{\mathbf{w}}(\boldsymbol{\gamma}(\mathbf{p}^\star)))_i \geq 0 \quad \text{for } i \in S_{\max},$$

$$(H(\mathbf{p}^\star)^\top \nabla\Phi_{\mathbf{w}}(\boldsymbol{\gamma}(\mathbf{p}^\star)))_i = 0 \quad \text{for } i \in S_{\mathrm{in}}, \qquad (6.9.25)$$

$$(H(\mathbf{p}^\star)^\top \nabla\Phi_{\mathbf{w}}(\boldsymbol{\gamma}(\mathbf{p}^\star)))_i \leq 0 \quad \text{for } i \in S_0.$$

Proof. Assume that $p_i^\star = \bar{p}_i$. Then $\frac{\partial}{\partial p_i}\Phi_{\mathbf{w}}(\boldsymbol{\gamma}(\mathbf{p}))(\mathbf{p}^\star) \geq 0$. Assume that $0 < p_i^\star < \bar{p}_i$. Then $\frac{\partial}{\partial p_i}\Phi_{\mathbf{w}}(\boldsymbol{\gamma}(\mathbf{p}))(\mathbf{p}^\star) = 0$. Assume that $p_i^\star = 0$. Then $\frac{\partial}{\partial p_i}\Phi_{\mathbf{w}}(\boldsymbol{\gamma}(\mathbf{p}))(\mathbf{p}^\star) \leq 0$. $\qquad\square$

We now show that the maximum problem (6.9.24) can be restated as the maximum problem of convex function on a closed unbounded domain. For $\boldsymbol{\gamma} = (\gamma_1, \ldots, \gamma_L)^\top > 0$ let $\tilde{\boldsymbol{\gamma}} = \log\boldsymbol{\gamma}$, i.e. $\boldsymbol{\gamma} = e^{\tilde{\boldsymbol{\gamma}}}$. Recall that for a non-negative irreducible matrix $B \in \mathbb{R}_+^{L \times L}$ $\log\rho(e^{\mathbf{x}}B)$ is a convex function, Theorem 6.7.3. Furthermore, $\log(1 + e^t)$ is a strict convex function in $t \in \mathbb{R}$. Hence, the maximum problem (6.9.24) is equivalent to the problem

maximize $\sum_l w_l \log(1 + e^{\tilde{\gamma}_l})$

subject to $\log \rho(\text{diag}(e^{\tilde{\gamma}})(F + (1/\bar{p}_l)\mathbf{ve}_l^\top)) \leq \mathbf{0} \quad \forall l \in [\mathbf{L}],$ (6.9.26)

variables: $\tilde{\boldsymbol{\gamma}} = (\tilde{\gamma}_1, \ldots, \tilde{\gamma}_n)^\top \in \mathbb{R}^L.$

The unboundedness of the convex set in (6.9.26) is due to the identity $0 = e^{-\infty}$.

Theorem 6.9.10 *Let* $\mathbf{w} > \mathbf{0}$ *be a probability vector. Consider the maximum problem* (6.9.7). *Then any point* $\mathbf{0} \leq \mathbf{p}^\star \leq \bar{\mathbf{p}}$ *satisfying the conditions* (6.9.25) *is a local maximum.*

Proof. Since $\mathbf{w} > \mathbf{0}$, $\Phi_{\mathbf{w}}(e^{\tilde{\gamma}})$ is a strict convex function in $\tilde{\boldsymbol{\gamma}} \in \mathbb{R}^L$. Hence, the maximum of (6.9.26) is achieved exactly on the extreme points of the closed unbounded set specified in (6.9.26). (It may happen that some coordinate of the extreme point are $-\infty$.) Translating this observation to the maximal problem (6.9.7), we deduce the theorem. $\qquad\square$

We now give simple lower and upper bounds on the value of (6.9.7).

Lemma 6.9.11 *Consider the maximal problem* (6.9.7). *Let* $B_l = (F + (1/\bar{p}_l)\mathbf{ve}_l^\top))$ *for* $l = 1, \ldots, L$. *Denote* $R = \max_{l \in [L]} \rho(B_l)$. *Let* $\bar{\boldsymbol{\gamma}}$ *be defined by* (6.9.23). *Then*

$$\Phi_{\mathbf{w}}((1/R)\mathbf{1}) \leq \max_{\mathbf{p} \in [\mathbf{0}, \bar{\mathbf{p}}]} \Phi_{\mathbf{w}}(\boldsymbol{\gamma}(\mathbf{p}) \leq \Phi_{\mathbf{w}}(\bar{\boldsymbol{\gamma}}).$$

Proof. By Corollary 6.9.6, $\boldsymbol{\gamma}(\mathbf{p}) \leq \bar{\boldsymbol{\gamma}}$ for $\mathbf{p} \in [\mathbf{0}, \bar{\mathbf{p}}]$. Hence, the upper bounds holds. Clearly, for $\boldsymbol{\gamma} = (1/R)\mathbf{1}$, we have that $\rho(\text{diag}(\boldsymbol{\gamma})B_l) \leq 1$ for $l \in [L]$. Then, from Theorem 6.9.7, $\Phi_{\mathbf{w}}((1/R)\mathbf{1})$ yields the lower bound. Equality is achieved in the lower bound when $\mathbf{p}^\star = \mathbf{tx}(\mathbf{B_i})$, where $i = \arg\max_{l \in [L]} \rho(B_l)$, for some $t > 0$. $\qquad\square$

We now show that the substitution $\mathbf{0} < \mathbf{p} = e^{\mathbf{q}}$, i.e. $p_l = e^{q_l}$, $l = 1, \ldots, L$, can be used to find an efficient algorithm to solve the optimal problem (6.9.6). As in §6.9.3 we can consider the inverse of the max–min problem of (6.9.6). It is equivalent to the problem

$$\min_{\mathbf{q} \le \bar{\mathbf{q}}} g(\mathbf{q}), \quad g(\mathbf{q}) = \max_{l \in [L]} s_l e^{-q_l} + \sum_{j=1}^{L} f_{lj} e^{q_j - q_l},$$

$$\bar{\mathbf{q}} = (\log \bar{p}_1, \ldots, \log \bar{p}_L)^\top. \tag{6.9.27}$$

Note that $s_l e^{-q_l} + \sum_{j=1}^{L} f_{lj} e^{q_j - q_l}$ is a convex function. Fact 6.7.1.1a implies that $g(\mathbf{q})$ is a convex function. We have quite a good software and mathematical theory to find fast the minimum of a convex function in a convex set as $\mathbf{q} \le \bar{\mathbf{q}}$, i.e. [NoW99].

6.9.6 Algorithms for sum rate maximization

In this section, we outline three algorithms for finding and estimating the maximal sum rates. As above we assume that $\mathbf{w} > \mathbf{0}$. Theorem 6.9.10 gives rise to the following algorithm, which is the gradient algorithm in the variable $\mathbf{p}$ in the compact polyhedron $[\mathbf{0}, \bar{\mathbf{p}}]$.

Algorithm 6.9.12   1. *Choose* $\mathbf{p}_0 \in [\mathbf{0}, \bar{\mathbf{p}}]$:

(a) *Either at random;*

(b) *or* $\mathbf{p}_0 = \bar{\mathbf{p}}$.

2. *Given* $\mathbf{p}_k = (p_{1,k}, \ldots, p_{L,k})^\top \in [\mathbf{0}, \bar{\mathbf{p}}]$ *for* $k \ge 0$, *compute* $\mathbf{a} = (a_1, \ldots, a_L)^\top = \nabla_\mathbf{p} \Phi_\mathbf{w}(\gamma(\mathbf{p}_k))$. *If* $\mathbf{a}$ *satisfies the conditions* (6.9.25) *for* $\mathbf{p}^\star = \mathbf{p}_k$, *then* $\mathbf{p}_k$ *is the output. Otherwise let* $\mathbf{b} = (b_1, \ldots, b_L)^\top$ *be defined as follows.*

(a) $b_i = 0$ *if* $p_{i,k} = 0$ *and* $a_i < 0$;

(b) $b_i = 0$ *if* $p_{i,k} = \bar{p}_i$ *and* $a_i > 0$;

(c) $b_i = a_i$ *if* $0 < p_i < \bar{p}_i$.

Then $\mathbf{p}_{k+1} = \mathbf{p}_k + t_k \mathbf{b}$, *where* $t_k > 0$ *satisfies the conditions* $\mathbf{p}_{k+1} \in [\mathbf{0}, \bar{\mathbf{p}}]$ *and* $\Phi_\mathbf{w}(\gamma(\mathbf{p}_k + t_k \mathbf{b}_k))$ *increases on the interval* $[0, t_k]$.

The problem with the gradient method, and its variations as a conjugate gradient method is that it is hard to choose the optimal value of t_k in each step, e.g. [Avr03]. We now use the reformulation of the maximal problem given by (6.9.26). Since $\mathbf{w} > \mathbf{0}$, the function $\Phi_{\mathbf{w}}(e^{\tilde{\gamma}})$ is strictly convex. Thus, the maximum is achieved only on the boundary of the convex set

$$D(\{F\}) = \{\tilde{\gamma} \in \mathbb{R}^L, \quad \log \rho(\mathrm{diag}(e^{\tilde{\gamma}})(F + (1/\bar{p}_l)\mathbf{ve}_{\mathbf{l}}^\top)) \leq 0, \ \forall \, \mathbf{l}\}.$$

$$(6.9.28)$$

If one wants to use numerical methods and software for finding the maximum value of convex functions on bounded closed convex sets, e.g. [NoW99], then one needs to consider the maximization problem (6.9.26) with additional constraints:

$$D(\{F\}, K) = \{\tilde{\gamma} \in D(\{F\}), \quad \tilde{\gamma} \geq -K\mathbf{1}\}, \tag{6.9.29}$$

for a suitable $K \gg 1$. Note that the above closed set is compact and convex. The following lemma gives the description of the set $D(\{F\}, K)$.

Lemma 6.9.13 *Let* $\bar{\mathbf{p}} > \mathbf{0}$ *be given and let* R *be defined as in Lemma 6.9.11. Assume that* $K > \log R$. *Let* $\underline{\mathbf{p}} = P(e^{-K}\mathbf{1}) = (e^K I - F)^{-1}\mathbf{v}$. *Then* $D(\{F\}, K) \subseteq \log \boldsymbol{\gamma}([\underline{\mathbf{p}}, \bar{\mathbf{p}}])$.

Proof. From the definition of K, we have that $e^K > R$. Hence, $\rho(e^{-K}B_l) < 1$ for $l = 1, \ldots, L$. Thus $-K\mathbf{1} \in D(\{F\})$. Let $\boldsymbol{\gamma} = e^{-K}\mathbf{1}$. Assume that $\tilde{\gamma} \in D(\{F\}, K)$. Then $\tilde{\gamma} \geq -K\mathbf{1}$. Hence, $\gamma = e^{\tilde{\gamma}} \geq \boldsymbol{\gamma}$. Since $\rho(\mathrm{diag}(\gamma)F) < 1$, Claim 6.9.4 yields that $\mathbf{p} = P(\gamma) \geq P(\boldsymbol{\gamma}) = \underline{\mathbf{p}}$, where P is defined by (6.9.11). The inequality $P(\gamma) \leq \bar{\mathbf{p}}$ follows from Corollary 6.9.6. $\qquad \square$

Thus, we can apply the numerical methods to find the maximum of the strictly convex function $\Phi_{\mathbf{w}}(e^{\tilde{\gamma}})$ on the closed bounded set $D(\{F\}, K)$, e.g. [NoW99]. In particular, we can use the gradient method. It takes the given boundary point $\tilde{\gamma}_k$ to another boundary point of $\tilde{\gamma}_{k+1} \in D(\{F\}, K)$, in the direction induced by the gradient of $\Phi_{\mathbf{w}}(e^{\tilde{\gamma}})$. However, the complicated boundary of $D(\{F\}, K)$ will make any algorithm expensive.

Furthermore, even though the constraint set in (6.9.24) can be transformed into a strict convex set, it is in general difficult to determine precisely the spectral radius of a given matrix [Var63]. To make the problem simpler and to enable fast algorithms, we approximate the convex set $D(\{F\}, K)$ by a bigger polyhedral convex sets as follows. Choose a finite number of points $\zeta_1, \ldots, \zeta_M$ on the boundary of $D(\{F\})$, which preferably lie in $D(\{F\}, K)$. Let $H_1(\boldsymbol{\xi}), \ldots, H_N(\boldsymbol{\xi}), \boldsymbol{\xi} \in \mathbb{R}^L$ be the N supporting hyperplanes of $D(\{F\})$. (Note that we can have more than one supporting hyperplane at ζ_i, and at most L supporting hyperplanes.) So each $\boldsymbol{\xi} \in D(\{F\}, K)$ satisfies the inequality $H_j(\boldsymbol{\xi}) \leq 0$ for $j = 1, \ldots, N$. Let $\bar{\boldsymbol{\gamma}}$ be defined by (6.9.23). Define

$$D(\zeta_1, \ldots, \zeta_M, K) = \{\boldsymbol{\xi} \in \mathbb{R}^L, \ -K\mathbf{1} \leq \boldsymbol{\xi} \leq \log \bar{\boldsymbol{\gamma}},$$

$$H_j(\boldsymbol{\xi}) \leq 0 \quad \text{for } j = 1, \ldots, N\}. \tag{6.9.30}$$

Hence, $D(\zeta_1, \ldots, \zeta_M, K)$ is a polytope which contains $D(\{F\}, K)$. Thus

$$\max_{\tilde{\boldsymbol{\gamma}} \in D(\zeta_1, \ldots, \zeta_M, K)} \Phi_{\mathbf{w}}(e^{\tilde{\boldsymbol{\gamma}}}) \geq \tag{6.9.31}$$

$$\max_{\tilde{\boldsymbol{\gamma}} \in D(\{F\}, K)} \Phi_{\mathbf{w}}(e^{\tilde{\boldsymbol{\gamma}}}). \tag{6.9.32}$$

Since $\Phi_{\mathbf{w}}(e^{\tilde{\boldsymbol{\gamma}}})$ is strictly convex, the maximum in (6.9.31) is achieved only at the extreme points of $D(\zeta_1, \ldots, \zeta_M, K)$. The maximal solution can be found using a variant of a simplex algorithm [Dan63]. More precisely, one starts at some extreme point of $\boldsymbol{\xi} \in D(\zeta_1, \ldots, \zeta_M, K)$. Replace the strictly convex function $\Phi_{\mathbf{w}}(e^{\tilde{\boldsymbol{\gamma}}})$ by its first-order Taylor expansion $\Psi_{\boldsymbol{\xi}}$ at $\boldsymbol{\xi}$. Then we find another extreme point $\boldsymbol{\eta}$ of $D(\zeta_1, \ldots, \zeta_M, K)$, such that $\Psi_{\boldsymbol{\xi}}(\boldsymbol{\eta}) > \Psi_{\boldsymbol{\xi}}(\boldsymbol{\xi}) = \Phi_{\mathbf{w}}(e^{\boldsymbol{\xi}})$. Then we replace $\Phi_{\mathbf{w}}(e^{\tilde{\boldsymbol{\gamma}}})$ by its first-order Taylor expansion $\Psi_{\boldsymbol{\eta}}$ at $\boldsymbol{\eta}$ and continue the algorithm. Our second proposed algorithm for finding an optimal $\tilde{\boldsymbol{\gamma}}^*$ that maximizes (6.9.31) is given as follows.

Algorithm 6.9.14 1. *Choose an arbitrarily extreme point $\boldsymbol{\xi}_0 \in D(\zeta_1, \ldots, \zeta_M, K)$.*

2. *Let* $\Psi_{\boldsymbol{\xi}_k}(\boldsymbol{\xi}) = \Phi_{\mathbf{w}}(e^{\boldsymbol{\xi}_k}) + (\mathbf{w} \circ (1 + e^{\boldsymbol{\xi}_k})^{-1} \circ e^{\boldsymbol{\xi}_k})^\top (\boldsymbol{\xi} - \boldsymbol{\xi}_k)$. *Solve the linear program* $\max_{\boldsymbol{\xi}} \Psi_{\boldsymbol{\xi}_k}(\boldsymbol{\xi})$ *subject to* $\boldsymbol{\xi} \in D(\boldsymbol{\zeta}_1, \ldots, \boldsymbol{\zeta}_M, K)$ *using the simplex algorithm in* [Dan63] *by finding an extreme point* $\boldsymbol{\xi}_{k+1}$ *of* $D(\boldsymbol{\zeta}_1, \ldots, \boldsymbol{\zeta}_M, K)$, *such that* $\Psi_{\boldsymbol{\xi}_k}(\boldsymbol{\xi}_{k+1}) > \Psi_{\boldsymbol{\xi}_k}(\boldsymbol{\xi}_k) = \Phi_{\mathbf{w}}(e^{\boldsymbol{\xi}_k})$.

3. *Compute* $\mathbf{p}_k = P(e^{\boldsymbol{\xi}_{k+1}})$. *If* $\mathbf{p}_k \in [0, \bar{\mathbf{p}}]$, *compute* $\mathbf{a} = (a_1, \ldots, a_L)^\top = \nabla_{\mathbf{p}} \Phi_{\mathbf{w}}(\boldsymbol{\gamma}(\mathbf{p}_k))$. *If* $\mathbf{a}$ *satisfies the conditions* (6.9.25) *for* $\mathbf{p}^\star = \mathbf{p}_k$, *then* $\mathbf{p}_k$ *is the output. Otherwise, go to Step 2 using* $\Psi_{\boldsymbol{\xi}_{k+1}}(\boldsymbol{\xi})$.

As in §6.9.3, it would be useful to consider the following related maximal problem:

$$\max_{\tilde{\boldsymbol{\gamma}} \in D(\boldsymbol{\zeta}_1, \ldots, \boldsymbol{\zeta}_M, K)} \mathbf{w}^\top \tilde{\boldsymbol{\gamma}}. \qquad (6.9.33)$$

This problem given by (6.9.33) is a standard linear program, which can be solved in polynomial time by the classical ellipsoid algorithm [Kha79]. Our third proposed algorithm for finding an optimal $\tilde{\boldsymbol{\gamma}}^\star$ that maximizes (6.9.33) is given as follows. Then $\mathbf{p}^\star = P(e^{\tilde{\boldsymbol{\gamma}}^\star})$.

Algorithm 6.9.15   1. *Solve the linear program* $\max_{\tilde{\boldsymbol{\gamma}}} \mathbf{w}^\top \tilde{\boldsymbol{\gamma}}$ *subject to* $\tilde{\boldsymbol{\gamma}} \in D(\boldsymbol{\zeta}_1, \ldots, \boldsymbol{\zeta}_M, K)$ *using the ellipsoid algorithm in* [Kha79].

2. *Compute* $\mathbf{p} = P(e^{\tilde{\boldsymbol{\gamma}}})$. *If* $\mathbf{p} \in [0, \bar{\mathbf{p}}]$, *then* $\mathbf{p}$ *is the output. Otherwise, project* $\mathbf{p}$ *onto* $[0, \bar{\mathbf{p}}]$.

We note that $\tilde{\boldsymbol{\gamma}} \in D(\boldsymbol{\zeta}_1, \ldots, \boldsymbol{\zeta}_M, K)$ in Algorithm 6.9.15 can be replaced by the set of supporting hyperplane $D(\tilde{F}, K) = \{\tilde{\boldsymbol{\gamma}} \in \rho(\text{diag}(e^{\tilde{\boldsymbol{\gamma}}})\tilde{F}) \leq 1, \ \tilde{\boldsymbol{\gamma}} \geq -K\mathbf{1}\}$ or, if $L \geq 3$ and $\mathbf{w}$ satisfies the conditions (6.6.9), $D(F, K) = \{\tilde{\boldsymbol{\gamma}} \in \rho(\text{diag}(e^{\tilde{\boldsymbol{\gamma}}})F) \leq 1, \ \tilde{\boldsymbol{\gamma}} \geq -K\mathbf{1}\}$ based on the relaxed maximal problems in §4. Then Theorem 6.9.2 quantify the closed-form solution $\tilde{\boldsymbol{\gamma}}$ computed by Algorithm 6.9.15.

We conclude this section by showing how to compute the supporting hyperplanes $H_j, j = 1, \ldots, N$, which define $D(\boldsymbol{\zeta}_1, \ldots, \boldsymbol{\zeta}_M, K)$.

To do that, we give a characterization of supporting hyperplanes of $D(\{F\})$ at a boundary point $\zeta \in \partial D(\{F\})$.

Theorem 6.9.16 *Let* $\bar{\mathbf{p}} = (\bar{p}_1, \ldots, \bar{p}_L)^\top > 0$ *be given. Consider the convex set* (6.9.28). *Let* ζ *be a boundary point of* $\partial D(\{F\})$. *Then* $\zeta = \log \gamma(\mathbf{p})$, *where* $0 \le \mathbf{p} = (p_1, \ldots, p_L)^\top \le \bar{\mathbf{p}}$. *The set* $\mathcal{B} := \{l \in [L], \; p_l = \bar{p}_l\}$ *is nonempty. For each* $B_l = (F + (1/\bar{p}_l)\mathbf{v}\mathbf{e}_l^\top))$ *let* $\mathrm{H}_l(\zeta)$ *be the supporting hyperplane of* $\mathrm{diag}(e^{\mathbf{x}})B_l$ *at* ζ, *defined as in Theorem 6.7.3. Then* $\mathrm{H}_l \le 0$, *for* $l \in \mathcal{B}$, *are the supporting hyperplanes of* $D(\{F\})$ *at* ζ.

Proof. Let $\mathbf{p} = P(e^\zeta)$. Theorem 6.9.5 implies the set $\mathcal{B}$ is nonempty. Furthermore, $\rho(e^\zeta B_l) = 1$ if and only if $p_l = \bar{p}_l$. Hence, ζ lies exactly at the intersection of the hypersurfaces $\log \rho(e^\zeta B_l) = 0, l \in \mathcal{B}$. Theorem 6.7.3 implies that the supporting hyperplanes of $D(\{F\})$ at ζ are $\mathrm{H}_l(\xi) \le 0$ for $l \in \mathcal{B}$. $\qquad\square$

We now show how to choose the boundary points $\zeta_1, \ldots, \zeta_M \in \partial D(\{F\})$ and to compute the supporting hyperplanes of $D(\{F\})$ at each ζ_i. Let $\underline{\mathbf{p}} = P(e^{-K}\mathbf{1}) = (\underline{p}_1, \ldots, \underline{p}_L)^\top$ be defined as in Lemma 6.9.13. Choose $M_i \ge 2$ equidistant points in each interval $[\underline{p}_i, \bar{p}_i]$.

$$p_{j_i, i} = \frac{j_i \underline{p}_i + (M_i - j_i)\bar{p}_i}{M_i} \quad \text{for } j_i = 1, \ldots, M_i, \text{ and } i = 1, \ldots, L.$$

$$(6.9.34)$$

Let

$$\mathcal{P} = \{\mathbf{p}_{j_1, \ldots, j_L} = (p_{j_1, 1}, \ldots, p_{j_L, L})^\top$$

$$\times \min(\bar{p}_1 - p_{j_1, 1}, \ldots, \bar{p}_L - p_{j_L, L}) = 0\}.$$

That is, $\mathbf{p}_{j_1, \ldots, j_L} \in \mathcal{P}$ if and only if $\mathbf{p}_{j_1, \ldots, j_L} \not< \bar{\mathbf{p}}$. Then

$$\{\zeta_1, \ldots, \zeta_M\} = \log \gamma(\mathcal{P}).$$

The supporting hyperplanes of $D(\{F\})$ at each ζ_i are given by Theorem 6.9.16.

6.10   Historical Remarks

§6.1 is standard. §6.2 summarizes Frobenius's results on irreducible matrices [Fro08, Fro09, Fro12], which generalizes Perron's results for positive matrices [Per07]. Our exposition follows Collatz–Wielandt minimax characterization [Col42, Wie50]. §6.3 is well known. The sharp upper bound for primitivity index is due to Wielandt [Wie50]. The case $s = 1$ is due to Holladay–Varga [HoV58]. The general case is due to Sedlacek [Sed59] and Dulmage–Mendelsohn [DM64]. See also [BR91]. Most of the results of §6.4 are well known. The Frobenius normal form of a non-negative matrix is due to Frobenius [Fro12]. Most of the results of §6.5 are well known. The first part of Theorem 6.6.12 is well known: [Sin64, BPS66, FrK75]. The exposition of §6.6 follows closely Friedland–Karlin [FrK75]. The second part of Theorem 6.6.12 is taken from Tan–Friedland–Low [TFL11]. Theorem 6.7.2 is due to Kingman [Kin61]. Other results in §6.7 follow some results in [Fri81a, CFKK82]. §6.8 follows some results in [Fri81a]. Theorem 6.8.1 for $\Psi(x) = e^x$ gives the Donsker–Varadjan characterization for matrices [DV75]. Theorem 6.8.3 is due to J.E. Cohen [Coh79]. §6.9 follows Tan–Friedland–Low [TFL11].

Chapter 7

Various Topics

7.1 Norms over Vector Spaces

In this chapter, we assume that $\mathbb{F} = \mathbb{R}, \mathbb{C}$ unless stated otherwise.

Definition 7.1.1 *Let* $\mathbf{V}$ *be a vector space over* $\mathbb{F}$. *A continuous function* $\|\cdot\| : \mathbf{V} \to [0, \infty)$ *is called a norm if the following conditions are satisfied:*

1. *Positivity:* $\|\mathbf{v}\| = 0$ *if and only if* $\mathbf{v} = \mathbf{0}$.

2. *Homogeneity:* $\|a\mathbf{v}\| = |a|\,\|\mathbf{v}\|$ *for each* $a \in \mathbb{F}$ *and* $\mathbf{v} \in \mathbf{V}$.

3. *Subadditivity:* $\|\mathbf{u} + \mathbf{v}\| \leq \|\mathbf{u}\| + \|\mathbf{v}\|$ *for all* $\mathbf{u}, \mathbf{v} \in \mathbf{V}$.

A continuous function $\|\cdot\| : \mathbf{V} \to [0, \infty)$ *which satisfies the conditions 2 and 3 is called a seminorm. The sets*

$$\mathrm{B}_{\|\cdot\|} := \{\mathbf{v} \in \mathbf{V},\ \|\mathbf{v}\| \leq 1\},\ \mathrm{B}^o_{\|\cdot\|} := \{\mathbf{v} \in \mathbf{V},\ \|\mathbf{v}\| < 1\},$$

$$\mathrm{S}_{\|\cdot\|} := \{\mathbf{v} \in \mathbf{V},\ \|\mathbf{v}\| = 1\},$$

are called the (closed) unit ball, the open unit ball and the unit sphere of the norm respectively. For $\mathbf{a} \in \mathbf{V}$ *and* $r > 0$ *we let*

$$\mathrm{B}_{\|\cdot\|}(\mathbf{a}, r) = \{\mathbf{x} \in \mathbf{V} : \|\mathbf{x} - \mathbf{a}\| \leq r\},$$

$$\mathrm{B}^o_{\|\cdot\|}(\mathbf{a}, r) = \{\mathbf{x} \in \mathbf{V} : \|\mathbf{x} - \mathbf{a}\| < r\}$$

be the closed and the open ball of radius r centered at **a** *respectively. If the norm* $\| \cdot \|$ *is fixed, we use the notation*

$$B(\mathbf{a}, r) = B_{\|\cdot\|}(\mathbf{a}, r), \quad B^o(\mathbf{a}, r) = B^o_{\|\cdot\|}(\mathbf{a}, r).$$

See Problem 2 for the properties of unit balls. The standard norms on $\mathbb{F}^n$ are the l_p norms:

$$\|(x_1, \ldots, x_n)^\top\|_p = \left(\sum_{i=1}^n |x_i|^p \right)^{\frac{1}{p}}, \quad p \in [1, \infty), \qquad (7.1.1)$$

$$\|(x_1, \ldots, x_n)^\top\|_\infty = \max_{1 \le i \le n} |x_i|.$$

See Problem 8.

Definition 7.1.2 *Let* **V** *be a finite dimensional vector space over* $\mathbb{F}$. *Denote by* **V**$'$ *the set of all linear functionals* $f : \mathbf{V} \to \mathbb{F}$. *Assume that* $\| \cdot \|$ *is a norm on* **V**. *The conjugate norm* $\| \cdot \| : \mathbf{V}' \to \mathbb{F}$ *is defined as*

$$\|\mathbf{f}\|^* = \max_{\mathbf{x} \in B_{\|\cdot\|}} |\mathbf{f}(\mathbf{x})|, \; for \; \mathbf{f} \in \mathbf{V}'.$$

For a norm $\| \cdot \|$ *on* $\mathbb{F}^n$ *the conjugate norm* $\| \cdot \|$ *on* $\mathbb{F}^n$ *is given by*

$$\|\mathbf{x}\|^* = \max_{\mathbf{y} \in B_{\|\cdot\|}} |\mathbf{y}^\top \mathbf{x}| \; for \; \mathbf{x} \in \mathbb{F}^n. \qquad (7.1.2)$$

A norm $\| \cdot \|$ *on* **V** *is called strictly convex if for any two distinct points* $\mathbf{x}, \mathbf{y} \in S_{\|\cdot\|}$ *and* $t \in (0, 1)$ *the inequality* $\|t\mathbf{x} + (1 - t)\mathbf{y}\| < 1$ *holds. A norm* $\| \cdot \|$ *on* $\mathbb{F}^n$ *is called* C^k, *for* $k \in \mathbb{N}$, *if the sphere* $S_{\|\cdot\|}$ *is a* C^k *manifold.* $\| \cdot \|$ *is called smooth if it is* C^k *for each* $k \in \mathbb{N}$.

For $\mathbf{x} = (x_1, \ldots, x_n)^\top \in \mathbb{C}^n$ *let* abs $\mathbf{x} = (|x_1|, \ldots, |x_n|)^\top$. *A norm* $\| \cdot \|$ *on* $\mathbb{F}^n$ *is called absolute if* $\|\mathbf{x}\| = \|$abs $\mathbf{x}\|$ *for each* $\mathbf{x} \in \mathbb{F}^n$. *A norm* $\||| \cdot \|||$ *on* $\mathbb{F}^n$ *is called a transform absolute if there exists an absolute norm* $\| \cdot \|$ *on* $\mathbb{F}^n$ *and* $P \in \mathbf{GL}(n, \mathbb{F})$ *such that* $\||\mathbf{x}\|| = \|P\mathbf{x}\|$ *for each* $\mathbf{x} \in \mathbb{F}^n$.

A norm $\| \cdot \|$ *on* $\mathbb{F}^n$ *is called symmetric if the function* $\|(x_1, \ldots, x_n)^\top\|$ *is a symmetric function in* $x_1, \ldots, x_n$. *(That is, for*

each permutation $\pi : [n] \rightarrow [n]$ *and each* $\mathbf{x} = (x_1, \ldots, x_n)^\top \in \mathbb{F}^n$
equality $\|(x_{\pi(1)}, \ldots, x_{\pi(n)})^\top\| = \|(x_1, \ldots, x_n)^\top\|$ *holds.)*

Theorem 7.1.3 *Let* $\| \cdot \|$ *be a norm on* $\mathbb{F}^n$. *Then the following are equivalent.*

1. $\| \cdot \|$ *is an absolute norm.*

2. $\| \cdot \|^*$ *is an absolute norm.*

3. *There exists a compact set* $L \subset \mathbb{F}^n$ *not contained in the hyperplane* $\mathrm{H}_i = \{(y_1, \ldots, y_n)^\top \in \mathbb{F}^n, \ y_i = 0\}$ *for* $i = 1, \ldots, n$, *such that* $\|\mathbf{x}\| = \max_{\mathbf{y} \in L} (\mathrm{abs} \ \mathbf{y})^\top \mathrm{abs} \ \mathbf{x}$ *for each* $\mathbf{x} \in \mathbb{F}^n$.

4. $\|\mathbf{x}\| \leq \|\mathbf{z}\|$ *if* $\mathrm{abs} \ \mathbf{x} \leq \mathrm{abs} \ \mathbf{z}$.

Proof. $1 \Rightarrow 2$. Assume that $\mathbf{x}, \mathbf{y} \in \mathbb{F}^n$. Then there exists $\mathbf{z} \in \mathbb{F}^n$, $\mathrm{abs} \ \mathbf{z} = \mathrm{abs} \ \mathbf{y}$ such that $|\mathbf{z}^\top \mathbf{x}| = (\mathrm{abs} \ \mathbf{y})^\top \mathrm{abs} \ \mathbf{x}$. Since $\| \cdot \|$ is absolute $\|\mathbf{z}\| = \|\mathbf{y}\|$. Clearly $|\mathbf{y}^\top \mathbf{x}| \leq (\mathrm{abs} \ \mathbf{y})^\top \mathrm{abs} \ \mathbf{x}$. The characterization (7.1.2) yields that

$$\|\mathbf{x}\|^* = \max_{\mathbf{y} \in \mathrm{B}_{\|\cdot\|}} (\mathrm{abs} \ \mathbf{y})^\top \mathrm{abs} \ \mathbf{x}. \tag{7.1.3}$$

Clearly $\|\mathbf{x}\| = \|\mathrm{abs} \ \mathbf{x}\|$.

$2 \Rightarrow 3$. The equality $(\| \cdot \|^*)^* = \| \cdot \|$, see Problem 3, and the equality (7.1.3) implies 3 with $L = \mathrm{B}_{\|\cdot\|^*}$. Clearly $\mathrm{B}_{\|\cdot\|^*}$ contains a vector whose all coordinates are different from zero.

$3 \Rightarrow 4$. Assume that $\mathrm{abs} \ \mathbf{x} \leq \mathrm{abs} \ \mathbf{z}$. Then $(\mathrm{abs} \ \mathbf{y})^\top \mathrm{abs} \ \mathbf{x} \leq (\mathrm{abs} \ \mathbf{y})^\top \mathrm{abs} \ \mathbf{z}$ for any $\mathbf{y} \in \mathbb{F}^n$. In view of the characterization of the absolute norm given in 3 we deduce 4.

$4 \Rightarrow 1$. Assume that $\mathrm{abs} \ \mathbf{x} = \mathrm{abs} \ \mathbf{y}$. Since $\mathrm{abs} \ \mathbf{x} \leq \mathrm{abs} \ \mathbf{y}$ we deduce that $\|\mathbf{x}\| \leq \|\mathbf{y}\|$. Similarly $\|\mathbf{x}\| \geq \|\mathbf{y}\|$. Hence, 1 holds. $\square$

Definition 7.1.4 *A set* $L \subset \mathbb{F}^n$ *is called symmetric if for each* $\mathbf{y} = (y_1, \ldots, y_n)^\top$ *in* L *the vector* $(y_{\pi(1)}, \ldots, y_{\pi(n)})^\top$ *is in* L, *for each permutation* $\pi : [n] \rightarrow [n]$.

Corollary 7.1.5 *Let* $\| \cdot \|$ *be a norm on* $\mathbb{F}^n$. *Then the following are equivalent.*

1. $\| \cdot \|$ *is an absolute symmetric norm.*

2. $\| \cdot \|^*$ *is an absolute symmetric norm.*

3. *There exists a compact symmetric set* $L \subset \mathbb{F}^n$, *not contained in the hyperplane* $H_i = \{(y_1, \ldots, y_n)^\top \in \mathbb{F}^n, \ y_i = 0\}$ *for* $i = 1, \ldots, n$, *such that* $\|\mathbf{x}\| = \max_{\mathbf{y} \in L}(\text{abs } \mathbf{y})^\top \text{abs } \mathbf{x}$ *for each* $\mathbf{x} \in \mathbb{F}^n$.

See Problem 6.

Proposition 7.1.6 *Assume that* $\| \cdot \|$ *is a symmetric absolute norm on* $\mathbb{R}^2$. *Then*

$$\|\mathbf{x}\|_2^2 \le \|\mathbf{x}\|^* \|\mathbf{x}\| \le \|\mathbf{x}\|_\infty \|\mathbf{x}\|_1 \text{ for any } \mathbf{x} \in \mathbb{R}^2. \tag{7.1.4}$$

Proof. The lower bound follows from the Problem 9. We claim that for any two points on $\mathbf{x}, \mathbf{y}$ satisfying the condition $\|\mathbf{x}\| = \|\mathbf{y}\| = 1$ the following inequality holds

$$(\|\mathbf{x}\|_1 - \|\mathbf{y}\|_1)(\|\mathbf{x}\|_\infty - \|\mathbf{y}\|_\infty) \le 0. \tag{7.1.5}$$

Since $\| \cdot \|$ is symmetric and absolute it is enough to prove the above inequality in the case that $\mathbf{x} = (x_1, x_2)^\top, x_1 \ge x_2 \ge 0, \mathbf{y} = (y_1, y_2)^\top, y_1 \ge y_2 \ge 0$. View $B_{\|\cdot\|}$ as a convex balanced set in $\mathbb{R}^2$, which is symmetric with respect to the line $z_1 = z_2$. The symmetricity of $\| \cdot \|$ implies that all the points $(z_1, z_2)^\top \in B_{\|\cdot\|}$ satisfy the inequality $z_1 + z_2 \le 2c$, where $\|(c, c)^\top\| = 1, c > 0$. Let C, D be the intersection of $B_{\|\cdot\|}, C_{\|\cdot\|}$ with the octant $K = \{(z_1, z_2)^\top \in \mathbb{R}^2, \ z_1 \ge z_2 \ge 0\}$ respectively. Observe that the line $z_1 + z_2 = 2c$ may intersect D at an interval. However, the line $z_1 + z_2 = 2t$ will intersect D at a unique point $(z_1(t), z_2(t))^\top$ for $t \in [b, c)$, where $\|(2b, 0)\| = 1, b > 0$. Furthermore $z_1(t), -z_2(t)$ are decreasing in (b, c). Hence, if $x_1 + x_2 > y_1 + y_2$ it follows that $y_1 > x_1$. Similarly, $x_1 + x_2 < y_1 + y_2$ it follows that $y_1 < x_1$. This proves (7.1.5).

To show the right-hand side of (7.1.4) we may assume that $\|\mathbf{x}\| = 1$. So $\|\mathbf{x}\|^* = |\mathbf{y}^\top \mathbf{x}|$ for some $\mathbf{y} \in S_{\|\cdot\|}$. Hence, $\|\mathbf{x}\| \, \|\mathbf{x}\|^* = |\mathbf{y}^\top \mathbf{x}|$. Clearly

$$|\mathbf{y}^\top \mathbf{x}| \leq \min(\|\mathbf{x}\|_1 \|\mathbf{y}\|_\infty, \|\mathbf{x}\|_\infty \|\mathbf{y}\|_1).$$

Suppose that $\|\mathbf{y}\|_1 \leq \|\mathbf{x}\|_1$. Then the right-hand side of (7.1.4) follows. Assume that $|\mathbf{y}|_1 > \|\mathbf{x}\|_1$. Then (7.1.5) yields that $\|\mathbf{y}\|_\infty \leq \|\mathbf{x}\|_\infty$ and the right-hand side of (7.1.4) follows. $\qquad\square$

A norm $\|\cdot\| : \mathbb{F}^{m \times n} \to \mathbb{R}_+$ is called a *matrix* norm. A standard example of matrix norm is the *Frobenius* norm of $A = [a_{ij}] \in \mathbb{F}^{m \times n}$:

$$\|A\|_F := \sqrt{\sum_{i=j=1}^{m,n} |a_{ij}|^2}. \tag{7.1.6}$$

Recall that $\|A\|_F = (\sum_{i=1}^m \sigma_i(A)^2)^{\frac{1}{2}}$, where $\sigma_i(A), i = 1, \ldots,$ are the singular values of A. More generally, for each $q \in [1, \infty]$

$$\|A\|_{q,S} := \left(\sum_{i=1}^m \sigma_i(A)^q \right)^{\frac{1}{q}} \tag{7.1.7}$$

is a norm on $\mathbb{F}^{m \times n}$, which is called the q-Schatten norm of A. Furthermore, for any integer $p \in [1, m]$ and $w_1 \geq \cdots \geq w_p > 0$, the function $f(A)$ given in Corollary 4.11.5 is a norm on $\mathbb{F}^{m \times n}$. See Problem 4.11.4. We denote $\|\cdot\|_{\infty,S} = \sigma_1(\cdot)$ as the $\|\cdot\|_2$ operator norm:

$$\|A\|_{2,2} := \sigma_1(A) \text{ for } A \in \mathbb{C}^{m \times n}. \tag{7.1.8}$$

See §7.4.

Definition 7.1.7 *A norm $\|\cdot\|$ on $\mathbb{C}^{m \times n}$ is called a unitary invariant if $\|UAV\| = \|A\|$ for any $A \in \mathbb{C}^{m \times n}$ and unitary $U \in \mathbf{U}(m), V \in \mathbf{V}(n)$.*

Clearly, any p-Schatten norm on $\mathbb{C}^{m \times n}$ is unitary invariant.

Theorem 7.1.8 *For positive integers m, n let $l = \min(m, n)$. For $A \in \mathbb{C}^{m \times n}$ let $\boldsymbol{\sigma}(A) := (\sigma_1(A), \ldots, \sigma_l(A))^\top$. Then $\| \cdot \|$ is a unitary invariant norm on $\mathbb{C}^{m \times n}$ if and only if there exists an absolute symmetric norm $||| \cdot |||$ on $\mathbb{C}^l$ such that $\|A\| = |||\boldsymbol{\sigma}(A)|||$ for any $A \in \mathbb{C}^{m \times n}$.*

Proof. Let $\mathbf{D}(m, n) \subset \mathbb{C}^{m \times n}$ be the subspace of diagonal matrices. Clearly, $\mathbf{D}(m, n)$ is isomorphic to $\mathbb{C}^l$. Each $D \in \mathbf{D}(m, n)$ is of the form $\mathrm{diag}(\mathbf{x}), \mathbf{x} = (x_1, \ldots, x_l)^\top$, where $x_1, \ldots, x_l$ are the diagonal entries of D. Assume that $\| \cdot \|$ is a norm on $\mathbb{C}^{m \times n}$. Then the restriction of $\| \cdot \|$ to $\mathbf{D}(m, n)$ induces a norm $||| \cdot |||$ on $\mathbb{C}^l$ given by $|||\mathbf{x}||| := \| \mathrm{diag}(\mathbf{x})\|$. Assume now that $\| \cdot \|$ is a unitary invariant norm. For a given $\mathbf{x} \in \mathbb{C}^l$, there exists a diagonal unitary matrix such that $U \mathrm{diag}(\mathbf{x}) = \mathrm{diag}(\mathrm{abs}\,\mathbf{x})$. Hence

$$|||\mathbf{x}||| = \| \mathrm{diag}(\mathbf{x})\| = \|U \mathrm{diag}(\mathbf{x})\| = \| \mathrm{diag}(\mathrm{abs}\,\mathbf{x})\| = |||\mathrm{abs}\,\mathbf{x}|||.$$

Let $\pi : [l] \to [l]$ be a permutation. Denote $\mathbf{x}_\pi := (x_{\pi(1)}, \ldots, x_{\pi(l)})^\top$. Clearly there exists two permutation matrices $P \in \mathbf{U}(m), Q \in \mathbf{U}(n)$ such that $\mathrm{diag}(\mathbf{x}_\pi) = U \mathrm{diag}(\mathbf{x})V$. Hence, $|||\mathbf{x}_\pi||| = |||\mathbf{x}|||$, and $||| \cdot |||$ is absolute symmetric. Clearly, there exists unitary U, V such that $A = U \mathrm{diag}(\boldsymbol{\sigma}(A))V$. Hence, $\|A\| = |||\boldsymbol{\sigma}(A)|||$.

Assume now that $||| \cdot |||$ is an absolute symmetric norm on $\mathbb{C}^l$. Set $\|A\| = |||\boldsymbol{\sigma}(A)|||$ for any A. Clearly $\| \cdot \| : \mathbb{C}^{m \times n} \to \mathbb{R}_+$ is a continuous function, which satisfies the properties 1–2 of Definition 7.1.1. it is left to show that $\| \cdot \|$ satisfies the triangle inequality. Since $||| \cdot |||$ is an absolute symmetric and $\sigma_1(A) \geq \cdots \geq \sigma_l(A) \geq 0$, Corollary 7.1.5 yields that

$$\|A\| = \max_{\mathbf{y} = (y_1, \ldots, y_l)^\top \in L, |y_1| \geq \cdots \geq |y_l|} (\mathrm{abs}\,\mathbf{y})^\top \boldsymbol{\sigma}(A) \qquad (7.1.9)$$

for a corresponding compact symmetric set $L \subset \mathbb{C}^n$, not contained in the hyperplane $H_i = \{(y_1, \ldots, y_n)^\top \in \mathbb{F}^n, \ y_i = 0\}$ for $i = 1, \ldots, n$. Use Problem 4.11.6(b) to deduce from (7.1.9) that $\|A + B\| \leq \|A\| + \|B\|$. $\qquad \square$

Definition 7.1.9 *A norm on* $\| \cdot \|$ *on* $\mathbb{F}^{n \times n}$ *is called a spectral dominant norm if* $\|A\| \geq \rho(A)$ *for every* $A \in \mathbb{F}^{n \times n}$.

Since $\sigma_1(A) \geq \rho(A)$, see (4.11.14) for $k = 1$, we deduce that any q-Schatten norm is spectral dominant.

Problems

1. Let $\mathbf{V}$ be a finite dimensional vector space over $\mathbb{F}$. Show that a seminorm $\| \cdot \| : \mathbf{V} \to \mathbb{R}_+$ is a convex function.

2. Let $\mathbf{V}$ be a finite dimensional vector space over $\mathbb{F}$. $X \subseteq \mathbf{V}$ is called balanced if $tX = X$ for every $t \in \mathbb{F}$ such that $|t| = 1$. Identify $\mathbf{V}$ with $\mathbb{F}^{\dim \mathbf{V}}$. Then the topology on $\mathbf{V}$ is the topology induced by open sets in $\mathbb{F}^n$. Assume that $\| \cdot \|$ is a norm on $\mathbf{V}$. Show

   (a) $B_{\|\cdot\|}$ is convex and compact.

   (b) $B_{\|\cdot\|}$ is balanced.

   (c) $\mathbf{0}$ is an interior point of $B_{\|\cdot\|}$.

3. Let $\mathbf{V}$ be a finite dimensional vector space over $\mathbb{F}$. Let $X \subset \mathbf{V}$ be a compact convex set balanced set such $\mathbf{0}$ is its interior point. For each $\mathbf{x} \in \mathbf{V} \backslash \{\mathbf{0}\}$ let $f(\mathbf{x}) = \min\{r > 0 : \frac{1}{r}\mathbf{x} \in X\}$. Set $f(\mathbf{0}) = 0$. Show that f is a norm on $\mathbf{V}$ whose unit ball is X.

4. Let $\mathbf{V}$ be a finite dimensional vector space over $\mathbb{F}$ with a norm $\| \cdot \|$. Show

   (a) $\|\mathbf{f}\|^* = \max_{\mathbf{y} \in S_{\|\cdot\|}} |\mathbf{f}(\mathbf{y})|$ for any $\mathbf{f} \in \mathbf{V}'$.

   (b) Show that for any $\mathbf{x} \in \mathbf{V}$ and $\mathbf{f} \in \mathbf{V}'$ the inequality $|\mathbf{f}(\mathbf{x})| \leq \|\mathbf{f}\|^* \|\mathbf{x}\|$.

   (c) Identify $(\mathbf{V}')'$ with $\mathbf{V}$, i.e. any linear functional on $\lambda : \mathbf{V}' \to \mathbb{F}$ is of the form $\lambda(\mathbf{f}) = f(\mathbf{x})$ for some $\mathbf{x} \in \mathbf{V}$. Then $(\|\mathbf{x}\|^*)^* = \|\mathbf{x}\|$.

   (d) Let $L \subset \mathbf{V}'$ be a compact set which contains a basis of $\mathbf{V}'$. Define $\|\mathbf{x}\|_L = \max_{\mathbf{f} \in L} |\mathbf{f}(\mathbf{x})|$. Then $\|\mathbf{x}\|_L$ is a norm on $\mathbf{V}$.

(e) Show that $\|\cdot\| = \|\cdot\|_L$ for a corresponding compact set $L \subset \mathbf{V}'$. Give a simple choice of L.

5. Let $\mathbf{V}$ be a finite dimensional vector space over $\mathbb{F}$, dim $\mathbf{V} \geq 1$, with a norm $\|\cdot\|$. Show

   (a) ext $(\mathrm{B}_{\|\cdot\|}) \subset \mathrm{S}_{\|\cdot\|}$.

   (b) ext $(\mathrm{B}_{\|\cdot\|}) = \mathrm{S}_{\|\cdot\|}$ if and only if for any $\mathbf{x} \neq \mathbf{y} \in \mathrm{S}_{\|\cdot\|}$ $(\mathbf{x}, \mathbf{y}) \subset \mathrm{B}^o_{\|\cdot\|}$.

   (c) For each $\mathbf{x} \in \mathrm{S}_{\|\cdot\|}$ there exists $\mathbf{f} \in \mathrm{S}_{\|\cdot\|^*}$ such that $1 = \mathbf{f}(\mathbf{x}) \geq |\mathbf{f}(\mathbf{y})|$ for any $\mathbf{y} \in \mathrm{B}_{\|\cdot\|}$.

   (d) Each $\mathbf{f} \in \mathrm{S}_{\|\cdot\|^*}$ is a proper supporting hyperplane of $\mathrm{B}_{\|\cdot\|}$ at some point $\mathbf{x} \in \mathrm{S}_{\|\cdot\|}$.

6. Prove Corollary 7.1.5.

7. Let $\mathbf{V}$ be a finite dimensional vector space over $\mathbb{F}$. Assume that $\|\cdot\|_1, \|\cdot\|_2$ are two norms on $\mathbf{V}$. Show

   (a) $\|\mathbf{x}\|_1 \leq \|\mathbf{x}\|_2$ for all $\mathbf{x} \in \mathbf{V}$ if and only if $\mathrm{B}_{\|\cdot\|_1} \supseteq \mathrm{B}_{\|\cdot\|_2}$.

   (b) Show that there exists $C \geq c > 0$ such that $c\|\mathbf{x}\|_1 \leq \|\mathbf{x}\|_2 \leq C\|\mathbf{x}\|_1$ for all $\mathbf{x} \in \mathbf{V}$.

8. For any $p \in [1, \infty]$ define the conjugate $p^* = q \in [1, \infty]$ to satisfy the equality $\frac{1}{p} + \frac{1}{q} = 1$. Show

   (a) Hölder's inequality: $|\mathbf{y}^*\mathbf{x}| \leq (\text{abs } \mathbf{y})^\top \text{abs } \mathbf{x} \leq \|\mathbf{x}\|_p \|\mathbf{y}\|_{p^*}$ for any $\mathbf{x}, \mathbf{y} \in \mathbb{C}^n \backslash \{\mathbf{0}\}$ and $p \in [1, \infty]$. (For $p = 2$ this inequality is called the *Cauchy–Schwarz* inequality.) Furthermore, equalities hold in all inequalities if and only if $\mathbf{y} = a\mathbf{x}$ for some $a \in \mathbb{C} \backslash \{0\}$. (Prove Hölder's inequality for $\mathbf{x}, \mathbf{y} \in \mathbb{R}^n_+$.)

   (b) $\|\mathbf{x}\|_p$ is a norm on $\mathbb{C}^n$ for $p \in [1, \infty]$.

   (c) $\|\mathbf{x}\|_p$ is strictly convex if and only if $p \in (1, \infty)$.

   (d) For $p \in (1, \infty)$ ext $(\mathrm{B}_{\|\cdot\|_p}) = \mathrm{S}_{\|\cdot\|_p}$.

   (e) Characterize ext $(\mathrm{B}_{\|\cdot\|_p})$ for $p = 1, \infty$ for $\mathbb{F} = \mathbb{R}, \mathbb{C}$.

(f) For each $\mathbf{x} \in \mathbb{C}^n$ the function $\|\mathbf{x}\|_p$ is a nonincreasing function for $p \in [1, \infty]$.

9. Show that for any norm $\| \cdot \|$ on $\mathbb{F}^n$ the inequality

$$\|\mathbf{x}\|_2^2 \leq \min(\|\bar{\mathbf{x}}\|^* \|\mathbf{x}\|, \|\mathbf{x}\|^* \|\bar{\mathbf{x}}\|) \text{ for any } \mathbf{x} \in \mathbb{F}^n.$$

In particular, if $\| \cdot \|$ is absolute then $\|\mathbf{x}\|_2^2 \leq \|\mathbf{x}\|^* \|\mathbf{x}\|$. (*Hint:* Use the equality $\|\mathbf{x}\|_2^2 = \bar{\mathbf{x}}^\top \mathbf{x}$.)

10. Let $L \subset \mathbb{F}^n$ satisfy the assumptions of condition 3 of Theorem 7.1.3. Let $\nu(\mathbf{x}) = \max_{\mathbf{y} \in L}(\text{abs } \mathbf{y})^\top \text{abs } \mathbf{x}$ for each $\mathbf{x} \in \mathbb{F}^n$. Show that $\nu(\mathbf{x})$ is an absolute norm on $\mathbb{F}^n$.

11. Let $\| \cdot \|$ be an absolute norm on $\mathbb{R}^n$. Show that it extends in a unique way to an absolute norm on $\mathbb{C}^n$.

12. Let $\mathbf{V}$ be a finite dimensional vector space over $\mathbb{F} = \mathbb{R}, \mathbb{C}$.

(a) Assume that $\| \cdot \|$ is a seminorm on $\mathbf{V}$. Let $\mathbf{W} := \{\mathbf{x} \in \mathbb{V}, \|\mathbf{x}\| = 0\}$. Show that $\mathbf{W}$ is a subspace of $\mathbf{V}$, and for each $\mathbf{x} \in \mathbb{V}$ the function $\| \cdot \|$ is a constant function on $\mathbf{x} + \mathbf{W}$.

(b) Let $\mathbf{W}$ be defined as above. Let $\hat{\mathbf{U}}$ be the quotient space $\mathbf{V}/\mathbf{W}$. So $\hat{\mathbf{v}} \in \hat{\mathbf{V}}$ is viewed as any $\mathbf{y} \in \mathbf{v} + \mathbf{V}$ for a corresponding $\mathbf{v} \in \mathbf{V}$ Define the function $\|\| \cdot \|\| : \hat{\mathbf{V}} \to \mathbb{R}_+$ by $\|\|\hat{\mathbf{v}}\|\| = \|\mathbf{y}\|$. Show that $\|\| \cdot \|\|$ is a norm on $\hat{\mathbf{V}}$.

(c) Let $\mathbf{U}_1, \mathbf{U}_2$ are finite dimensional vector spaces over $\mathbb{F}$. Assume that $\|\| \cdot \|\| : \mathbf{U}_1 \to \mathbb{R}_+$ is a norm. Let $\mathbf{V} = \mathbf{U}_1 \oplus \mathbf{U}_2$. Define $\|\mathbf{u}_1 \oplus \mathbf{u}_2\| = \|\|\mathbf{u}_1\|\|$ for each $\mathbf{u}_i \in \mathbf{U}_i, i = 1, 2$. Show that $\| \cdot \|$ is a seminorm on $\mathbf{V}$. Furthermore, the subspace $\mathbf{0} \oplus \mathbf{U}_2$ is the set where $\| \cdot \|$ vanishes.

13. Show that for any $A \in \mathbb{C}^{m \times n}$ $\|A\|_{2,2} = \sigma(A) = \max_{\|\mathbf{x}\|_2 = 1} \|A\mathbf{x}\|_2$. (*Hint:* Observe that $\|A\mathbf{x}\|_2^2 = \mathbf{x}^*(A^*A)\mathbf{x}$.)

14. For $\mathbb{F} = \mathbb{R}, \mathbb{C}$, identify $(\mathbb{F}^{m \times n})^*$ with $\mathbb{F}^{m \times n}$ by letting $\phi_A : \mathbb{C}^{m \times n} \to \mathbb{F}$ be $\text{tr}(A^\top X)$ for any $A \in \mathbb{F}^{m \times n}$. Show that for any $p \in [1, \infty]$ the conjugate of the p-Schatten norm $\| \cdot \|_{p,S}$ is the q-Schatten norm on $\mathbb{F}^{m \times n}$, where $\frac{1}{p} + \frac{1}{q} = 1$.

7.2   Numerical Ranges and Radii

Let $S^{2n-1} := \{\mathbf{x} \in \mathbb{C}^n, \mathbf{x}^*\mathbf{x} = 1\}$ be the unit sphere of the ℓ_2 norm on $\mathbb{C}^n$.

Definition 7.2.1 *A map ϕ from S^{2n-1} to $2^{\mathbb{C}^n}$, the set of all subsets of $\mathbb{C}^n$, is called a ν-map, if the following conditions hold.*

1. *For each $\mathbf{x} \in S^{2n-1}$ the set $\phi(\mathbf{x})$ is a nonempty compact set.*

2. *The set $\cup_{\mathbf{x} \in S^{2n-1}} \phi(\mathbf{x})$ is compact.*

3. *Let $\mathbf{x}_k \in S^{2n-1}, \mathbf{y}_k \in \phi(\mathbf{x}_k)$ for $k \in \mathbb{N}$. Assume that $\lim_{k\to\infty} \mathbf{x}_k = \mathbf{x}$ and $\lim_{k\to\infty} \mathbf{y}_k = \mathbf{y}$. (Note that $\mathbf{x} \in S^{2n-1}$.) Then $\mathbf{y} \in \phi(\mathbf{x})$.*

4. *$\mathbf{y}^\top \mathbf{x} = 1$ for each $\mathbf{x} \in S^{2n-1}$ and $\mathbf{y} \in \phi(\mathbf{x})$.*

Assume that ϕ from S^{n-1} to $2^{\mathbb{C}^n}$ is ν-map. Then for $A \in \mathbb{C}^{n\times n}$

$$\omega_\phi(A) := \cup_{\mathbf{x} \in S^{2n-1}} \cup_{\mathbf{y} \in \phi(\mathbf{x})} \{\mathbf{y}^\top A\mathbf{x}\}, \qquad (7.2.1)$$

$$r_\phi(A) = \max_{\mathbf{x} \in \phi(\mathbf{x}), \mathbf{y} \in \phi(\mathbf{x})} |\mathbf{y}^\top A\mathbf{x}| \qquad (7.2.2)$$

are called the ϕ-numerical range and the ϕ-numerical radius respectively.

It is straightforward to show that r_ϕ is a seminorm on $\mathbb{C}^{n\times n}$, see Problem 1.

Lemma 7.2.2 *Let $\phi : S^{n-1} \to 2^{\mathbb{C}^n}$ be a ν-map. Then spec (A), is contained in the ϕ-numerical range of A. In particular $r_\phi(A) \geq \rho(A)$.*

Proof. Let $A \in \mathbb{C}^{n\times n}$ and and assume that λ is an eigenvalue of A. Then there exists an eigenvector $\mathbf{x} \in S^{2n-1}$ such that $A\mathbf{x} = \lambda\mathbf{x}$. Choose $\mathbf{y} \in \phi(\mathbf{x})$. Then $\mathbf{y}^\top A\mathbf{x} = \lambda\mathbf{y}^\top\mathbf{x} = \lambda$. Hence, $\lambda \in \omega_\phi(A)$. Thus $r_\phi(A) \geq |\lambda|$. $\qquad\square$

Lemma 7.2.3 *Let $\| \cdot \| : \mathbb{C}^{n\times n} \to \mathbb{R}_+$ be a seminorm, which is spectral dominant. Then $\| \cdot \|$ is a norm on $\mathbb{C}^{n\times n}$.*

Proof. Assume to the contrary that $\| \cdot \|$ is not a norm. Hence, there exists $0 \neq A \in \mathbb{C}^{n \times n}$ such that $\|A\| = 0$. Since $0 = \|A\| \geq \rho(A)$ we deduce that A is a nonzero nilpotent matrix. Hence, $T^{-1}AT = \oplus_{i=1}^{k} J_i$, where each J_i a nilpotent Jordan block and $T \in \mathbf{GL}(n, \mathbb{C})$. Since $A \neq 0$ we may assume that $J_1 \in \mathbb{C}^{l \times l}$, has an upper diagonal equal to 1, all other entries equal to 0 and $l \geq 2$. Let $B = \oplus_{i=1}^{k} B_i$ where each B_i has the same dimensions as J_i. Assume that B_i are zero matrices for $i > 1$, if $k > 1$. Let $B_1 = [b_{ij,1}] \in \mathbb{C}^{l \times l}$, where $b_{21,1} = 1$ and all other entries of B_1 equal to 0. It is straightforward to show that the matrix $\oplus_{i=1}^{k}(J_i + tB_i)$ has two nonzero eigenvalues $\pm\sqrt{t}$ for $t > 0$. Let $C := T(\oplus_{i=1}^{k} B_i)T^{-1}$. Then $\rho(A + tB) = \sqrt{t}$ for $t > 0$. Hence for $t > 0$ we obtain the inequalities

$$\rho(A + tB) = \sqrt{t} \leq \|A + tB\| \leq \|A\| + \|tB\| = t\|B\| \Rightarrow \|B\| \geq \frac{1}{\sqrt{t}}.$$

The above inequality cannot hold for an arbitrary small positive t. This contradiction implies the lemma. $\square$

Use the above Lemmas and Problem 1 to deduce.

Theorem 7.2.4 *Let* $\phi : \mathrm{S}^{2n-1} \to 2^{\mathbb{C}^n}$ *be a ν-map. Then $r_\phi(\cdot)$ is a spectral dominant norm on $\mathbb{C}^{n \times n}$.*

We now consider a few examples of ν-maps.

Example 7.2.5 *The function* $\phi_2 : \mathrm{S}^{2n-1} \to 2^{\mathbb{C}^n}$ *given by* $\phi_2(\mathbf{x}) := \{\bar{\mathbf{x}}\}$ *is a ν-map. The corresponding numerical range and numerical radius of $A \in \mathbb{C}^{n \times n}$ are given by*

$$\omega_2(A) = \{z = \mathbf{x}^*A\mathbf{x}, \ \text{for all } \mathbf{x} \in \mathbb{C}^n \text{ satisfying } \mathbf{x}^*\mathbf{x} = 1\} \subset \mathbb{C},$$
$$r_2(A) := \max_{\mathbf{x} \in \mathbb{C}^n, \mathbf{x}^*\mathbf{x}=1} |\mathbf{x}^*A\mathbf{x}|.$$

It is called the classical numerical range and numerical radius of A, or simply the numerical range and numerical radius of A.

More generally:

Example 7.2.6 *For $p \in (1, \infty)$ the function* $\phi_p : \mathrm{S}^{2n-1} \to 2^{\mathbb{C}^n}$ *given by* $\phi_p((x_1, \ldots, x_n)^\top) := \{\|\mathbf{x}\|_p^{-p}(|x_1|^{p-2}\bar{x}_1, \ldots, |x_n|^{p-2}\bar{x}_n)^\top\}$

*is a ν-map. The corresponding numerical range and numerical radius
of $A \in \mathbb{C}^{n \times n}$ are denoted by $\omega_p(A)$ and $r_p(A)$ respectively.*

The most general example related to a norm on $\mathbb{C}^n$ is as follows.

Example 7.2.7 *Let $\| \cdot \|$ be a norm on $\mathbb{C}^n$. For each $\mathbf{x} \in S^{2n-1}$
let $\phi_{\|\cdot\|}(\mathbf{x})$ be the set of all $\mathbf{y} \in \mathbb{C}^n$ with the dual norm $\|\mathbf{y}\|^* = \frac{1}{\|\mathbf{x}\|}$ satisfying $\mathbf{y}^\top \mathbf{x} = 1$. Then $\phi_{\|\cdot\|}$ is a ν-map. (See Problem 6.)
The corresponding numerical range $\omega_{\|\cdot\|}(A)$ and the numerical radius
$r_{\|\cdot\|}(A)$ is called the Bauer numerical range and the Bauer numerical
radius respectively of $A \in \mathbb{C}^{n \times n}$.*

Definition 7.2.8 *A norm $\| \cdot \|$ on $\mathbb{C}^{n \times n}$ is called stable if there
exists $K > 0$ such that $\|A^m\| \le K\|A\|^m$ for all $A \in \mathbb{C}^{n \times n}$.*

Clearly, $\| \cdot \|$ is stable if and only if the unit ball $B_{\|\cdot\|} \subset \mathbb{C}^{n \times n}$ is power
bounded, see Definition 3.4.5.

Theorem 7.2.9 *Let $\phi : S^{2n-1} \to 2^{\mathbb{C}^n}$ be a ν-map. Set $c := \max_{U \in \mathbf{U}(n)} r_\phi(U)$. Then*

$$\|(zI - A)^{-1}\|_2 \le \frac{c}{|z| - 1} \text{ for all } |z| > 1, \ r_\phi(A) \le 1. \qquad (7.2.3)$$

In particular, a ϕ-numerical radius is a stable norm.

Proof. Fix $\mathbf{x} \in S^{2n-1}$ We first note that $\|\mathbf{y}\|_2 \le c$ for each
$\mathbf{y} \in \phi(\mathbf{x})$. Let $\mathbf{z} = \frac{1}{\|\mathbf{y}\|_2} \bar{\mathbf{y}} \in S^{2n-1}$. Then there exists $U \in \mathbf{U}(n)$ such
that $U\mathbf{x} = \mathbf{z}$. Hence, $\|\mathbf{y}\|_2 = \mathbf{y}^\top U\mathbf{x} \le r_\phi(U) \le c$. Assume next that
$r_\phi(A) \le 1$. Hence, $\rho(A) \le r_\phi(A) \le 1$. So $(zI - A)^{-1}$ is defined for
$|z| > 1$. Let $\mathbf{v} := (zI - A)^{-1}\mathbf{x}, \mathbf{v}_1 := \frac{1}{\|\mathbf{v}\|_2} \mathbf{v}$. Then for $\mathbf{y} \in \phi(\mathbf{v}_1)$ we
have

$$\frac{|\mathbf{y}^\top \mathbf{x}|}{\|\mathbf{v}\|_2} = |\mathbf{y}^\top (zI - A)\mathbf{v}_1| = |z - \mathbf{y}^\top A\mathbf{v}_1| \ge |z| - 1.$$

On the other hand

$$|\mathbf{y}^\top \mathbf{x}| \le \|\mathbf{y}\|_2 \|\mathbf{x}\|_2 = \|\mathbf{y}\|_2 \le c.$$

Combine the above inequalities to deduce $\|(zI - A)\mathbf{x}\|_2 \leq \frac{c}{|z|-1}$ for all $\|\mathbf{x}\|_2 = 1$. Use Problem 7.1.13 to deduce (7.2.3). Theorem 3.4.9 yields that the unit ball corresponding to the norm $r_\phi(\cdot)$ is a power bounded set, i.e. the norm $r_\phi(\cdot)$ is stable. □

Theorem 7.2.10 *Let* $\|\cdot\|$ *be a norm on* $\mathbb{C}^{n\times n}$. *Then* $\|\cdot\|$ *is stable if and only if it is spectral dominant.*

Proof. Assume first that $\|\cdot\|$ is stable. So $B_{\|\cdot\|}$ is a power bounded set. Theorem 3.3.2 yields that each $A \in B_{\|\cdot\|}$ satisfies $\rho(A) \leq 1$. So if $A \neq 0$ we get that $\rho(\frac{1}{\|A\|}A) \leq 1$, i.e. $\rho(A) \leq \|A\|$ for any $A \neq 0$. Clearly $\rho(0) = \|0\| = 0$. Hence, a stable norm is spectral dominant.

Assume now that $\|\cdot\|$ is a spectral dominant norm on $\mathbb{C}^{n\times n}$. Recall that $B_{\|\cdot\|}$ is a convex compact balanced set, and $\mathbf{0}$ is an interior point. Define a new set

$$\mathcal{A} := \big\{B \in \mathbb{C}^{n\times n}, \ B = (1-a)A + zI, \ a \in [0,1], \ z \in \mathbb{C}, |z| \leq a,$$

$$A \in B_{\|\cdot\|}\big\}.$$

It is straightforward to show that $\mathcal{A}$ is a convex compact balanced set. Note that by choosing $a = 1$ we deduce that $I \in \mathcal{A}$. Furthermore, by choosing $a = 0$ we deduce that $B_{\|\cdot\|} \subseteq \mathcal{A}$. So $\mathbf{0}$ is an interior point of $\mathcal{A}$. Problem 7.1.3 yields that there exists a norm $\|\|\cdot\|\|$ on $\mathbb{C}^{n\times n}$ such that $B_{\|\|\cdot\|\|} = \mathcal{A}$. Since $S_{\|\cdot\|} \subset B_{\|\|\cdot\|\|}$ it follows $\|\|A\|\| \leq \|A\|$ for each $A \in \mathbb{C}^{n\times n}$. We claim that $\|\|\cdot\|\|$ is spectral dominant. Assume that $\|\|B\|\| = 1$. So $B = (1-a)A + zI$ for some $a \in [0,1]$, $z \in \mathbb{C}, |z| \leq a$ and $A \in B_{\|\cdot\|}$. Since $\|\cdot\|$ is spectral dominant it follows that $\rho(A) \leq \|A\| \leq 1$. Note that spec $(B) = (1-a)$spec $(A) + z$. Hence, $\rho(B) \leq (1-a)\rho(A) + |z| \leq (1-a) + a = 1$. So $\|\|\cdot\|\|$ is spectral dominant. Since $\|\|I\|\| \leq 1$ and $\|\|I\|\| \geq \rho(I) = 1$ we deduce that $\|\|I\|\| = 1$. Hence, for any $z \in \mathbb{C}, |z| < 1$ we have $\|\|zI\|\| < 1$.

For $\mathbf{x} \in S^{2n-1}$ let

$$C(\mathbf{x}) = \{\mathbf{u} \in \mathbb{C}^n, \ \mathbf{u} = B\mathbf{x}, \ \|\|B\|\| < 1\}.$$

Clearly $C(\mathbf{x})$ is a convex set in $\mathbb{C}^n$. Since for each $\|\|B\|\| < 1$ we have that $\rho(B) < 1$ it follows that $\mathbf{x} \notin C(\mathbf{x})$. The hyperplane separation

Theorem 4.6.6 implies the existence of $\mathbf{y} \in \mathbb{C}^n$ such that

$$\Re(\mathbf{y}^\top \mathbf{x}) > \Re(\mathbf{y}^\top B \mathbf{x}) \text{ for all } |||B||| < 1. \tag{7.2.4}$$

Substitute in the above inequality $B = zI, |z| < 1$ we deduce that $\Re(\mathbf{y}^\top \mathbf{x}) > \Re(z\mathbf{y}^\top \mathbf{x})$. By choosing an appropriate argument of z we deduce $\Re(\mathbf{y}^\top \mathbf{x}) > |z||\mathbf{y}^\top \mathbf{x}|$. Hence, $\Re(\mathbf{y}^\top \mathbf{x}) \geq |\mathbf{y}^\top \mathbf{x}|$. In view of the strict inequality in (7.2.4) we deduce that $\mathbf{y}^\top \mathbf{x}$ is real and positive. Thus we can renormalize $\mathbf{y}$ so that $\mathbf{y}^\top \mathbf{x}$. Let $\phi(\mathbf{x})$ be the set of all $\mathbf{w} \in \mathbb{C}^n$ such that

$$\mathbf{w}^\top \mathbf{x} = 1, \quad \max_{|||B||| \leq 1} |\mathbf{w}^\top B \mathbf{x}| = 1.$$

Clearly, $\mathbf{y} \in \phi(\mathbf{x})$. It is straightforward to show that $\phi : S^{2n-1} \to 2^{\mathbb{C}^n}$ is a ν-map.

As $\mathbf{w}^\top B \mathbf{x} = \operatorname{tr} B(\mathbf{x}\mathbf{w}^\top)$ we deduce that $|||\mathbf{x}\mathbf{w}^\top|||^* = 1$, where $||| \cdot |||^*$ is the dual norm of $||| \cdot |||$ on $\mathbb{C}^{n \times n}$:

$$|||C|||^* = \max_{B \in B_{|||\cdot|||}} |\operatorname{tr} BC| = \max_{B \in S_{|||\cdot|||}} |\operatorname{tr} BC|. \tag{7.2.5}$$

Let $\mathcal{R}(1, n, n) \subset \mathbb{C}^{n \times n}$ be the variety of all matrices of rank one at most. Clearly, $\mathcal{R}(1, n, n)$ is a closed set consisting of all matrices of rank one and $0_{n \times n}$. Hence, $\mathcal{R}(1, n, n) \cap S_{|||\cdot|||^*}$ is a compact set consisting of all $\mathbf{x}\mathbf{w}^\top$, where $\mathbf{x} \in S^{2n-1}$ and $\mathbf{w} \in \phi(\mathbf{x})$. Since $(||| \cdot |||^*)^* = ||| \cdot |||$ it follows that

$$r_\phi(B) = \max_{\mathbf{x} \in S^{2n-1}, \mathbf{w} \in \phi(\mathbf{x})} |\mathbf{w}^\top B \mathbf{x}|$$

$$= \max_{\mathbf{x} \in S^{2n-1}, \mathbf{w} \in \phi(\mathbf{x})} |\operatorname{tr} B(\mathbf{x}\mathbf{w}^\top)|$$

$$\leq \max_{C \in S_{|||\cdot|||^*}} |\operatorname{tr} BC| = |||B||| \leq \|B\|.$$

Hence, $B_{\|\cdot\|} \subseteq B_{|||\cdot|||} \subseteq B_{r_\phi(\cdot)}$. Theorem 7.2.9 yields that $r_\phi(\cdot)$ is a stable norm. Hence, $\| \cdot \|$ and $||| \cdot |||$ are stable norms. $\qquad\square$

Use Theorem 7.2.10 and Problem 3 to deduce:

Corollary 7.2.11 *Let* $\mathcal{A} \subset \mathbb{C}^{n \times n}$ *be a compact, convex, balanced set, whose interior contains* $\mathbf{0}$. *Then* $\mathcal{A}$ *is stable if and only* $\rho(A) \leq 1$ *for each* $A \in \mathcal{A}$.

Definition 7.2.12 *Let* $\mathbb{F}$ *be field. A subspace* $\mathbf{0} \neq \mathbf{U} \subset \mathbb{F}^{n \times n}$ *is called stable if there exists an integer* $k \in [n]$ *such that the dimension of the subspace* $\mathbf{Ux} \subset \mathbb{F}^n$ *is* k *for any* $\mathbf{0} \neq \mathbf{x} \in \mathbb{F}^n$. $\mathbf{U}$ *is called maximally stable if* $k = n$.

The following result is a generalization of Theorem 7.2.10.

Theorem 7.2.13 *Let* $\mathcal{A} \subset \mathbb{C}^{n \times n}$ *be a compact convex balanced set, (see Problem 7.1.2 for the definition of a convex balanced set), which contains the identity matrix. Assume that* $\mathcal{L} := \text{span } \mathcal{A}$ *is a stable subspace. Then* $\mathcal{A}$ *is a stable set if and only if* $\rho(A) \leq 1$ *for each* $A \in \mathcal{A}$.

Proof. Clearly, if $\mathcal{A}$ is stable, then each $A \in \mathcal{A}$ is power bounded, hence $\rho(A) \leq 1$. Assume now that $\mathcal{A}$ is a compact convex balanced set containing identity such that $\mathcal{L}$ is a stable subspace. Let $\mathbf{x} \in S^{2n-1}$ and consider the subspace $\mathcal{L}\mathbf{x}$ of dimension k. Since $\mathcal{A}$ is a compact convex balanced set it follows that $\mathcal{A}\mathbf{x}$ is a compact convex balanced set in $\mathcal{L}$ since span $\mathcal{A}\mathbf{x} = \mathcal{L}\mathbf{x}$ it follows that ri $\mathcal{A}\mathbf{x}$. Hence, $\mathcal{A}\mathbf{x}$ is a unit ball of the norm $\| \cdot \|_{\mathbf{x}}$ on the subspace $\mathcal{L}\mathbf{x}$. Since $I \in \mathcal{A}$ it follows that $\xi \in \mathcal{A}\mathbf{x}$. We claim that $\|\mathbf{x}\|_{\mathbf{x}} = 1$. Assume to the contrary $\|\mathbf{x}\|_{\mathbf{x}} < 1$. Then $(1 + \varepsilon)\mathbf{x} \in \mathcal{A}\mathbf{x}$ for some $\varepsilon > 0$. So there exists $A \in \mathcal{A}$ such that $A\mathbf{x} = (1+\varepsilon)\mathbf{x}$. Hence, $\rho(A) \geq (1+\varepsilon)$ contrary to our assumptions.

Identify $(\mathcal{L}\mathbf{x})'$ with $\mathcal{L}\mathbf{x}$. So a linear functional $f : \mathcal{L}\mathbf{x} \to \mathbb{C}$ is given by $f(\mathbf{y}) = \mathbf{z}^\top \mathbf{y}$ for some $\mathbf{z} \in \mathcal{L}\mathbf{x}$. Let $\| \cdot \|_{\mathbf{x}}^*$ be the conjugate norm on $\mathcal{L}\mathbf{x}$. Denote by $\mathcal{B}(\mathbf{x}) \subset \mathcal{L}\mathbf{x}$ the unit ball of the norm $\| \cdot \|_{\mathbf{x}}^*$. Since $\|\mathbf{x}\|_{\mathbf{x}}^* = 1$ it follows that there exists $\mathbf{z}(\mathbf{x}) \in \mathcal{L}\mathbf{x}$ such that $\mathbf{z}(\mathbf{x})^\top \mathbf{x} = 1$ and $\|\mathbf{z}(\mathbf{x})\|_{\mathbf{x}}^* = 1$. We claim that $\cup_{\mathbf{x} \in S^{2n-1}} \mathcal{B}(\mathbf{x})$ is a compact set in $\mathbb{C}^n$.

Indeed, since $\mathcal{L}\mathbf{x}$ has a fixed dimension k for each $\mathbf{x} \in \mathbb{C}^n$ we can view $\mathcal{L}\mathbf{x}$ of the form $U(\mathbf{x})\mathbf{W}$ for some fixed subspace $\mathbf{W} \subset \mathbb{C}^n$ of dimension k and a unitary matrix $U(\mathbf{x})$. ($U(\mathbf{x})$ maps an orthonormal basis of $\mathbf{W}$ to an orthonormal basis of $\mathcal{L}\mathbf{x}$.) Hence, the set $\mathcal{C}(\mathbf{x}) := U^*(\mathbf{x})\mathcal{A}\mathbf{x}$ is a compact convex balanced set in $\mathbf{W}$, with $\mathbf{0}$ an interior point. Since $\mathcal{A}$ is compact, it follows that $\mathcal{C}(\mathbf{x})$ varies continuously on $\mathbf{x} \in S^{2n-1}$. Therefore, the set $\mathcal{D}(\mathbf{x}) := U^*\mathcal{B}(\mathbf{x}) \subset \mathbf{W}$ varies continuously with $\mathbf{x} \in S^{2n-1}$. Hence, $\cup_{\mathbf{x} \in S^{2n-1}}\mathcal{D}(\mathbf{x})$ is a compact set in $\mathbf{W}$, which yields that $\cup_{\mathbf{x} \in S^{2n-1}}\mathcal{B}(\mathbf{x})$ is a compact set in $\mathbb{C}^n$. In particular, there exists a constant K such that $\|\mathbf{z}(\mathbf{x})\|_2 \le K$. We now claim that $\mathcal{A}$ satisfies the condition 3.4.13 of Theorem 3.4.9. Indeed, for $\mathbf{x} \in S^{2n1-1}, A \in \mathcal{A}$ we have that $A\mathbf{x} \in \mathcal{A}(\mathbf{x})$, hence $\|A\mathbf{x}\|_\mathbf{x} \le 1$. Hence for $|\lambda| > 1$ we have

$$\|(\lambda I - A)\mathbf{x}\|_2 \ge \frac{\|\mathbf{z}(\mathbf{x})^\top(\lambda I - A)\mathbf{x}\|_2}{K} = \frac{|\lambda - \mathbf{z}(\mathbf{x})^\top A\mathbf{x}|}{K}$$

$$\ge \frac{|\lambda| - |\mathbf{z}(\mathbf{x})^\top A\mathbf{x}|}{K} \ge \frac{|\lambda| - |\mathbf{z}(\mathbf{x})^\top A\mathbf{x}|}{K}$$

$$\ge \frac{|\lambda| - \|A\mathbf{x}\|_\mathbf{x}\|\mathbf{z}(\mathbf{x})\|_\mathbf{x}^*}{K} \ge \frac{|\lambda| - 1}{K} = \frac{|\lambda| - 1}{K}\|\mathbf{x}\|_2.$$

Thus for each $|\lambda| > 1$ and $\mathbf{0} \ne \mathbf{x} \in \mathbb{C}^n$ we have the inequality $\|(\lambda I - A)\mathbf{x}\|_2 \ge \frac{|\lambda|-1}{K}\|\mathbf{x}\|_2$. Choose $\mathbf{x} = (\lambda I - A)^{-1}\mathbf{y}$ to deduce the inequality

$$\frac{\|(\lambda I - A)\mathbf{y}\|_2}{\|\mathbf{y}\|_2} \le \frac{K}{|\lambda| - 1}.$$

So $\sigma_1(\lambda I - A)^{-1}) \le \frac{K}{|\lambda|-1}$. Hence, $\mathcal{A}$ satisfies the condition 3.4.13 of Theorem 3.4.9 with the norm $\sigma_1(\cdot)$. Theorem 3.4.9 yields that $\mathcal{A}$ is stable. $\qquad\square$

Problem 7 shows that in Theorem 7.2.13 the assumption that $\mathcal{L}$ is a stable subspace cannot be dropped. In [Fri84], we show the following result.

Theorem 7.2.14 *Let $n \ge 2, d \in [2n-1, n^2]$ be integers. Then a generic subspace $\mathcal{L}$ of $\mathbb{C}^{n \times n}$ of dimension d is maximally stable.*

Problems

1. Let $\phi : S^{2n-1} \to 2^{\mathbb{C}^n}$ be a ν-map. Show that $r_\phi : \mathbb{C}^{n \times n} \to \mathbb{R}_+$ is a seminorm.

2. Let $\phi : S^{2n-1} \to 2^{\mathbb{C}^n}$ be a ν-map. Show that $r_\phi(I_n) = 1$.

3. Show that for any $p \in (1, \infty)$ the map ϕ_p given in Example 7.2.6 is a ν-map.

4. Show

   (a) For any unitary $U \in \mathbb{C}^{n \times n}$ and $A \in \mathbb{C}^{n \times n}$ $\omega_2(U^*AU) = \omega_2(A)$ and $r_2(U^*AU) = r_2(A)$.

   (b) For a normal $A \in \mathbb{C}^{n \times n}$ the numerical range $\omega_2(A)$ is a convex hull of the eigenvalues of A. In particular $r_2(A) = \rho(A)$ for a normal A.

   (c) $\omega_2(A)$ is a convex set for any $A \in \mathbb{C}^{n \times n}$. (Observe that it is enough to prove this claim only for $n = 2$.)

5. Let $A = J_4(0) \in \mathbb{C}^{4 \times 4}$ be a nilpotent Jordan block of order 4. Show that $r_2(A) < 1, r_2(A^2) = \frac{1}{2}, r_2(A^3) = \frac{1}{2}$. Hence, the inequality $r_2(A^3) \leq r_2(A)r_2(A^2)$ does not hold in general.

6. Show that the map $\phi_{\|\cdot\|} : S^{2n-1} \to 2^{\mathbb{C}^n}$ given in Example 7.2.7 is a ν-map.

7. Let $\| \cdot \|$ be the norm on $\mathbb{C}^{n \times n}$ given by $\|[a_{ij}]_{i=j=1}^n\| := \max_{i,j \in [1,n]} |a_{ij}|$. Denote by $\mathcal{U}_n \subset \mathbb{C}^{n \times n}$ the subspace of upper triangular matrices. For $n \geq 2$ show.

   (a) $\mathcal{U}_n$ is not a stable subspace of $\mathbb{C}^{n \times n}$.

   (b) $\mathcal{U}_n \cap B_{\|\cdot\|}$ is a compact, convex, balanced set.

   (c) $\rho(A) \leq 1$ for each $A \in \mathcal{U}_n \cap B_{\|\cdot\|}$.

   (d) $\mathcal{U}_n \cap B_{\|\cdot\|}$ is not a stable set.

8. Let C be a compact convex set in a finite dimensional space $\mathbf{V}$. Show that C is a polytope if an only if ext C is a finite set.

7.3   Superstable Norms

Definition 7.3.1 *A norm* $\| \cdot \|$ *on* $\mathbb{C}^{n \times n}$ *is called superstable if* $\|A^k\| \le \|A\|^k$ *for* $k = 2, \ldots,$ *and each* $A \in \mathbb{C}^{n \times n}$.

Clearly, any operator norm on $\mathbb{C}^{n \times n}$ is a superstable norm, see §7.4.

Theorem 7.3.2 *The standard numerical radius* $r_2(A) = \max_{\mathbf{x} \in \mathbb{C}^n, \|\mathbf{x}\|_2 = 1} |\mathbf{x}^* A \mathbf{x}|$ *is a superstable norm.*

To prove the theorem we need the following lemma.

Lemma 7.3.3 *Assume that* $A \in \mathbb{C}^{n \times n}, \rho(A) < 1$ *and* $\mathbf{x} \in S^{2n-1}$. *Let* $z_j = e^{\frac{2\pi j \sqrt{-1}}{m}}$ *and assume that* $I_n - z_j A \in \mathbf{GL}(n, \mathbb{C})$ *for* $j = 1, \ldots, m$. *Then*

$$1 - \mathbf{x}^* A^m \mathbf{x} = \frac{1}{m} \sum_{j=1}^{m} \|\mathbf{x}_j\|_2^2 (1 - z_j \mathbf{y}_j^* A \mathbf{y}_j), \qquad (7.3.1)$$

where $\mathbf{x}_j = \left(\prod_{k \in [m] \setminus \{j\}} (1 - z_k A) \right) \mathbf{x}, \ \mathbf{y}_j = \dfrac{1}{\|\mathbf{x}_j\| \mathbf{x}_j}, \ j = 1, \ldots, m.$

Proof. Observe the following two polynomial identities in z variable

$$1 - z^m = \prod_{k=1}^{m} (1 - z_k z), \quad 1 = \frac{1}{m} \sum_{j=1}^{m} \prod_{k \in [m] \setminus \{j\}} (1 - z_k z).$$

Replace the variable z by A obtain the identities

$$I_n - A^m = \prod_{k=1}^{m} (I_n - z_k A), \quad I_n = \frac{1}{m} \sum_{j=1}^{m} \prod_{k \in [m] \setminus \{j\}} (I_n - z_k A).$$

$$(7.3.2)$$

Multiply the second identity from the right by $\mathbf{x}$ to get the identity $\mathbf{x} = \frac{1}{m} \sum_{j=1}^{m} \mathbf{x}_j$. Multiply the first identity by $\mathbf{x}$ from the right respectively to obtain $\mathbf{x} - A^m \mathbf{x} = (\prod_{k=1}^{m} (I_n - z_k A)) \mathbf{x}$. Since $I_n - z_k A$

for $k = 1, \ldots, m$ commute, we deduce that for each k, $\mathbf{x} - A^m\mathbf{x} = (I_n - z_k A)\mathbf{x}_k$. Hence

$$1 - \mathbf{x}^* A^m \mathbf{x} = \mathbf{x}^*(\mathbf{x} - A^m\mathbf{x}) = \frac{1}{m}\sum_{j=1}^{m}\mathbf{x}_j^*(\mathbf{x} - A^m\mathbf{x})$$

$$= \frac{1}{m}\sum_{j=1}^{m}\mathbf{x}_j^*(I_n - z_j A)\mathbf{x}_j = \frac{1}{m}\sum_{j=1}^{m}\|\mathbf{x}_j\|_2^2(1 - z_j\mathbf{y}_j^* A\mathbf{y}_j).$$

$\square$

Proof of Theorem 7.3.2. From the the proof of Lemma 7.3.3 it follows that (7.3.1) holds for any $A \in \mathbb{C}^{n\times n}$ and some $\mathbf{y}_1, \ldots, \mathbf{y}_m \in S^{2n-1}$, since for $\mathbf{x}_j = 0$ in (7.3.1) we can choose any $\mathbf{y}_j \in S^{2n-1}$. Suppose that $r_2(A) = 1$. Let $\zeta \in \mathbb{C}, |\zeta| = 1$. Apply the equality (7.3.1) to ζA and $\mathbf{x} \in S^{2n-1}$ to deduce

$$1 - \zeta^m\mathbf{x}^* A^m\mathbf{x} = \frac{1}{m}\sum_{j=1}^{m}\|\mathbf{u}_j\|_2^2(1 - z_j\zeta\mathbf{w}_j^* A\mathbf{w}_j)$$

for corresponding $\mathbf{w}_1, \ldots, \mathbf{w}_m \in S^{2n-1}$. Choose ζ such that $\zeta^m\mathbf{x}^* A^m\mathbf{x} = |\mathbf{x}^* A^m\mathbf{x}|$. Since $r_2(z_k\zeta A) = r_2(A) = 1$, it follows that $\Re(1 - z_j\zeta\mathbf{w}_j^* A\mathbf{w}_j) \geq 0$. Hence, the above displayed equality yields $1 - |\mathbf{x}^* A^m\mathbf{x}| \geq 0$, i.e. $1 \leq |\mathbf{x}^* A^m\mathbf{x}|$. Since $\mathbf{x} \in S^{2n-1}$ is arbitrary, it follows that $r_2(A^m) \leq 1$ if $r_2(A) = 1$. Hence, $r_2(\cdot)$ is a superstable norm. $\square$

Definition 7.3.4 *For an integer $n \geq 2$ and $p \in [1, \infty]$ let $K_{p,n} \geq 1$ be the smallest constant satisfying $r_p(A^m) \leq K_{p,n}r_p(A)^m$ for all $A \in \mathbb{C}^{n\times n}$.*

Theorem 7.3.2 is equivalent to the equality $K_{2,n} = 1$. Problem 1(b) shows that $K_{1,n} = K_{\infty,n} = 1$. It is an open problem if $\sup_{n\in\mathbb{N}} \max_{p\in\infty[1,\infty]} K_{p,n} < \infty$.

Theorem 7.3.5 *Let $\|\cdot\|$ be a norm on $\mathbb{C}^{n\times n}$ which is invariant under the similarity by unitary matrices, i.e. $\|UAU^{-1}\| = \|A\|$ for each $A \in \mathbb{C}^{n\times n}$ and $U \in \mathbf{U}(n)$. Then $\|\cdot\|$ is spectral dominant if and only if $\|A\| \geq r_2(A)$ for any $A \in \mathbb{C}^{n\times n}$.*

To prove the theorem we bring the following two lemmas which are of independent interest.

Lemma 7.3.6 *Let $\| \cdot \|$ be a norm on $\mathbb{C}^{n \times n}$. Assume that $\| \cdot \|$ is invariant under the similarity by $U \in \mathbf{GL}(n, \mathbb{C})$, i.e. $\|A\| = \|UAU^{-1}\|$ for each $A \in \mathbb{C}^{n \times n}$. Then U is similar to diagonal matrix Λ:*

$$\Lambda = \mathrm{diag}(\lambda_1, \ldots, \lambda_n), \quad |\lambda_1| = \cdots = |\lambda_n| > 0. \qquad (7.3.3)$$

Proof. Let $\lambda, \mu \in \mathbb{C}$ be two distinct eigenvalues of U. So there are two corresponding nonzero vectors $\mathbf{x}, \mathbf{y} \in \mathbb{C}^n$ such that $U\mathbf{x} = \lambda\mathbf{x}, U^\top \mathbf{y} = \mu\mathbf{y}$. For $A = \mathbf{x}\mathbf{y}^\top$ we deduce that $UAU^{-1} = \frac{\lambda}{\nu}A$. Since $\|A\| = \|UAU^{-1}\| > 0$ it follows that $|\lambda| = |\mu|$. Hence, all the eigenvalues of U have the same modulus.

It is left to show that U is diagonable. Assume to the contrary that U is not diagonable. Then there exists an invertible matrix T and upper triangular matrix $V = [v_{ij}]_{i=j=1}^n$ such that $v_{11} = v_{22} = \lambda \neq 0, v_{12} = 1$, and $V = TVT^{-1}$. Choose $A = TBT^{-1}$, where $B = [b_{ij}]_{i=j=1}^n$, where $b_{22} = \frac{1}{\lambda^2}$. Since $\|U^k AU^{-k}\| = \|A\|$ for $k = \in \mathbb{N}$ it follows that the sequence of matrices $U^k AU^{-k}, k \in \mathbb{N}$ is bounded. A straightforward calculation shows that the $(1,2)$ entry of $T^{-1}(U^k AU^{-k})T$ is k^2. Hence, the sequence $U^k AU^{-k}, k \in \mathbb{N}$ is not bounded, contrary to our previous claim. The above contradiction establishes lemma. $\square$

Lemma 7.3.7 *Let $\Lambda = \mathrm{diag}(\lambda_1, \ldots, \lambda_n) \in \mathbb{C}^{n \times n}$ and assume that $|\lambda_1| = \cdots = |\lambda_n| > 0$ and $\lambda_i \neq \lambda_j$ for $i \neq j$. Suppose that $\| \cdot \|$ is a norm on $\mathbb{C}^{n \times n}$ which is invariant under the similarity by Λ. Then*

$$\| \mathrm{diag}(A)\| \leq \|A\|. \qquad (7.3.4)$$

Proof. Λ-similarity invariance implies

$$\left\| \frac{1}{m+1} \sum_{k=0}^m \Lambda^k A\Lambda^{-k} \right\| \leq \frac{1}{m+1} \sum_{k=0}^m \|\Lambda^k A\Lambda^{-k}\| = \|A\|.$$

For $A = [a_{ij}] \in \mathbb{C}^{n \times n}$ let

$$A_m = [a_{ij,m}] = \frac{1}{m+1} \sum_{k=0}^{m} \Lambda^k A \Lambda^{-k},$$

where $a_{ii,m} = a_{ii}$, $a_{ij,m} = a_{ij} \dfrac{1 - \frac{\lambda_i^{m+1}}{\lambda_j^{m+1}}}{(m+1)\left(1 - \frac{\lambda_i}{\lambda_j}\right)}$ for $i \neq j$.

Hence, $\lim_{m \to \infty} A_m = \operatorname{diag}(A)$. Since $\|A_m\| \leq \|A\|$ we deduce the inequality (7.3.4). $\qquad \square$

Proof of Theorem 7.3.5. Assume first that $\|A\| \geq r_2(A)$ for each $A \in \mathbb{C}^{n \times n}$. Clearly, $\|\cdot\|$ is spectral dominant. Assume now that $\|\cdot\|$ is invariant under similarity by any unitary matrix U, and $\|\cdot\|$ is spectral dominant. Since $\|\cdot\|$ is invariant under the similarity by a diagonal matrix $\Lambda = \operatorname{diag}(\lambda_1, \ldots, \lambda_n)$, where $|\lambda_1| = \cdots = |\lambda_n| = 1$ and $\lambda_i \neq \lambda_j$ for $i \neq j$, Lemma 7.3.7 yields (7.3.4). Let $A = [a_{ij}]$. Since $\operatorname{diag}(A) = \operatorname{diag}(a_{11}, \ldots, a_{nn})$ and $\|\cdot\|$ is spectral dominant we obtain that

$$\|A\| \geq \|\operatorname{diag}(A)\| \geq \rho(\operatorname{diag}(A)) = \max_{i \in \langle n \rangle} |a_{ii}|.$$

Let $V \in \mathbf{U}(n)$. Then the first column of V is $\mathbf{x} \in \mathbf{S}^{2n-1}$. Furthermore, the $(1,1)$ entry of $V^* A V$ is $\mathbf{x}^* A \mathbf{x}$. Since $\|\cdot\|$ is invariant under unitary similarity we obtain $\|A\| = \|V^* A V\| \geq |\mathbf{x}^* A \mathbf{x}|$. As for any $\mathbf{x} \in \mathbf{S}^{2n-1}$ there exists a unitary V with the first column $\mathbf{x}$ we deduce that $\|A\| \geq r_2(A)$. $\qquad \square$

Problems

1. (a) Describe the ν-maps $\phi_{\|\cdot\|_1}, \phi_{\|\cdot\|_\infty}$.

   (b) Show that for $p = 1, \infty$ $r_p(A)$ is equal to the operator norm of $\|A\|_p$, for $A \in \mathbb{C}^{n \times n}$ viewed as a linear operator $A : \mathbb{C}^n \to \mathbb{C}^n$. (See §7.4). Hence, $K_{1,n} = K_{\infty,n} = 1$, where $K_{p,n}$ is given Definition 7.3.4.

(c) Show that for each $p \in (1, \infty)$ and integer $n \geq 2$ there exists $A \in \mathbb{C}^{n \times n}$ such that $r_p(A) < \|A\|_p$, where $\|A\|_p$ the operator norm of A.

7.4   Operator Norms

Let $\mathbf{V}_a, \mathbf{V}_b$ be two finite dimensional vector spaces over $\mathbb{F} = \mathbb{R}, \mathbb{C}$. Assume that $\|\cdot\|_a, \|\cdot\|_b$ are norms on $\mathbf{V}_a, \mathbf{V}_b$ respectively. Let $T : \mathbf{V}_a \to \mathbf{V}_b$ be a linear transformation. Then

$$\|T\|_{a,b} := \max_{0 \neq \mathbf{x} \in \mathbf{V}_a} \frac{\|T\mathbf{x}\|_b}{\|\mathbf{x}\|_a}, \tag{7.4.1}$$

is called the *operator norm* of T. Clearly

$$\|T\|_{a,b} = \max_{\|\mathbf{x}\|_a \leq 1} \|T\mathbf{x}\|_b = \max_{\|\mathbf{x}\|_a = 1} \|T\mathbf{x}\|_b. \tag{7.4.2}$$

See Problem 1. Let $\mathbf{V}_c$ be a third finite dimensional vector space over $\mathbb{F}$ with a norm $\|\cdot\|_c$. Assume that $Q : \mathbf{V}_b \to \mathbf{V}_c$ is a linear transformation. The we have the well-known inequality

$$\|QT\|_{a,c} \leq \|Q\|_{b,c} \|T\|_{a,b}. \tag{7.4.3}$$

See Problem 2.

Assume now that $\mathbf{V}_a = \mathbf{V}_b = \mathbf{V}$ and $\|\cdot\|_a = \|\cdot\|_b = \|\cdot\|_c = \|\cdot\|$. We then denote $\|T\| := \|T\|_{a,b}$ and $\|Q\| := \|Q\|_{b,c}$. Let $\mathrm{Id} : \mathbf{V} \to \mathbf{V}$ be the identity operator. Hence

$$\|\mathrm{Id}\| = 1, \quad \|QT\| \leq \|Q\| \, \|T\|, \quad \|T^m\| \leq \|T\|^m \text{ for } m = 2, \ldots. \tag{7.4.4}$$

Assume that $\mathbf{V}_a = \mathbb{F}^n, \mathbf{V}_b = \mathbb{F}^m$. Then $T : \mathbb{F}^n \to \mathbb{F}^m$ is represented by a matrix $A \in \mathbb{F}^{m \times n}$. Thus $\|A\|_{a,b}$ is the operator norm of A. For $m = n$ and $\|\cdot\|_a = \|\cdot\|_b = \|\cdot\|$ we denote by $\|A\|$ the *operator* norm. Assume that $s, t \in [1, \infty]$. Then for $A \in \mathbb{F}^{m \times n}$ we denote by $\|A\|_{s,t}$ the operator norm of A, where $\mathbb{F}^n, \mathbb{F}^m$ are equipped with the norms ℓ_s, ℓ_t respectively. Note that $\|A\|_{2,2} = \sigma_1(A)$, see Problem 7.1.13. For $m = n$ and $s = t = p$ we denote by $\|A\|_p$ the ℓ_p operator norm of A.

Lemma 7.4.1 *Let $A = [a_{ij}] \in \mathbb{F}^{m \times n}$ and $\| \cdot \|_a, \| \cdot \|_b$ be norms on $\mathbb{C}^n, \mathbb{C}^m$ respectively. If $\| \cdot \|_b$ is an absolute norm then*

$$\|A\|_{a,b} \leq \|(\|(a_{11}, \ldots, a_{1n})^\top\|_a^*, \ldots, \|(a_{m1}, \ldots, a_{mn})^\top\|_a^*)^\top\|_b.$$
(7.4.5)

If $\| \cdot \|_a$ is an absolute norm then

$$\|A\|_{a,b} \leq \|(\|(a_{11}, \ldots, a_{m1})^\top\|_b, \ldots, \|(a_{1n}, \ldots, a_{mn})^\top\|_b)^\top\|_a^*.$$
(7.4.6)

In both inequalities, equality holds for matrices of rank one.

Proof. Let $\mathbf{x} \in \mathbb{F}^n$. Then

$$(A\mathbf{x})_i = \sum_{j=1}^n a_{ij} x_j$$

$$\Rightarrow |(A\mathbf{x})_i| \leq \|(a_{i1}, \ldots, a_{in})^\top\|_a^* \|\mathbf{x}\|_a, \quad i = 1, \ldots, m \Rightarrow$$

$$|A\mathbf{x}| \leq \|\mathbf{x}\|_a (\|(a_{11}, \ldots, a_{1n})^\top\|_a^*, \ldots, \|(a_{m1}, \ldots, a_{mn})^\top\|_a^*)^\top.$$

Assume that $\| \cdot \|_b$ is an absolute norm. Then

$$\|A\mathbf{x}\|_b \leq \|\mathbf{x}\|_a \|(\|(a_{11}, \ldots, a_{1n})^\top\|_a^*, \ldots, \|(a_{m1}, \ldots, a_{mn})^\top\|_a^*)^\top\|_b,$$

which yields (7.4.5). Suppose that A is rank one matrix. So $A = \mathbf{u}\mathbf{v}^\top$, where $0 \neq \mathbf{u} \in \mathbb{F}^m, 0 \neq \mathbf{v} \in \mathbb{F}^n$. There exists $0 \neq \mathbf{x} \in \mathbb{F}^n$ such that $\mathbf{v}^\top \mathbf{x} = \|\mathbf{v}\|_a^* \|\mathbf{x}\|_a$. For this $\mathbf{x}$ we have that $|A\mathbf{x}| = \|\mathbf{v}\|_a^* \|\mathbf{x}\|_a |\mathbf{u}|$. Hence, $\frac{\|A\mathbf{x}\|_b}{\|\mathbf{x}\|_a} = \|\mathbf{v}\|_a^* \|\mathbf{u}\|_b$. Thus $\|A\|_{a,b} \geq \|\mathbf{v}\|_a^* \|\mathbf{u}\|_b$. On the other hand the right hand of (7.4.5) is $\|\mathbf{v}\|_a^* \|\mathbf{u}\|_b$. This shows that (7.4.5) is sharp for rank one matrices.

Assume now that $\| \cdot \|_a$ is an absolute norm. Theorem 7.1.3 claims that $\| \cdot \|_a^*$ is an absolute norm. Apply (7.4.5) to $\|A^\top\|_{b*,a*}$ and use Problem 1(e) to deduce the inequality (7.4.6). Assume that A is rank one matrix. Then $A^\top$ is a rank one matrix and equality holds in (7.4.6). $\qquad \square$

Theorem 7.4.2 *Let $m, n \geq 2$ be integers. Assume that $s, t \in [1, \infty]$ and suppose $\mathbb{F}^n, \mathbb{F}^m$ are endowed with Hölder with norms $\| \cdot \|_s, \| \cdot \|_t$ respectively. Let s^* be defined by the equality $\frac{1}{s} + \frac{1}{s^*} = 1$. Then for $A = [a_{ij}]_{i=j=1}^{m,n} \mathbb{F}^{m \times n}$ the following hold.*

$$\|A\|_{s,t} \leq \min \left(\left(\sum_{i=1}^{m} \left(\sum_{j=1}^{n} |a_{ij}|^{s^*} \right)^{\frac{t}{s^*}} \right)^{\frac{1}{t}} , \right.$$

$$\left. \left(\sum_{j=1}^{n} \left(\sum_{i=1}^{m} |a_{ij}|^{t} \right)^{\frac{s^*}{t}} \right)^{\frac{1}{s^*}} \right), \tag{7.4.7}$$

$$\|A\|_{\infty,1} \leq \sum_{i=j=1}^{m,n} |a_{ij}|, \tag{7.4.8}$$

$$\|A\|_{1,1} = \max_{1 \leq j \leq n} \sum_{i=1}^{m} |a_{ij}|, \tag{7.4.9}$$

$$\|A\|_{\infty,\infty} = \max_{1 \leq i \leq m} \sum_{j=1}^{n} |a_{ij}|, \tag{7.4.10}$$

$$\|A\|_{1,\infty} = \max_{1 \leq i \leq m, 1 \leq j \leq n} |a_{ij}|. \tag{7.4.11}$$

Proof. Since $\| \cdot \|_s, \| \cdot \|_t$ are absolute norm the inequalities (7.4.5) and (7.4.6) hold. As $\| \cdot \|_s^* = \| \cdot \|_{s^*}$ we deduce (7.4.7). For $s = \infty$ we have $s^* = 1$, and for $t = 1$ (7.4.7) yields (7.4.8).

Assume that $s = t = 1$. So $s^* = \infty$. The second part of the inequality (7.4.7) yields the inequality $\|A\|_{1,1} \leq \max_{1 \leq j \leq n} \sum_{i=1}^{m} |a_{ij}|$. Let $\mathbf{e}_j = (\delta_{j1}, \ldots, \delta_{jn})^\top$. Clearly, $\|\mathbf{e}_j\|_1 = 1$ and $\|A\mathbf{e}_j\|_1 = \sum_{i=1}^{m} |a_{ij}|$. Hence, $\|A\|_{1,1} \geq \sum_{i=1}^{m} |a_{ij}|$. So $\|A\|_{1,1} \geq \max_{1 \leq j \leq n} \sum_{i=1}^{m} |a_{ij}|$, which yields (7.4.9). Since $\|A\|_{\infty,\infty} = \|A^\top\|_{1,1}$, see Problem 1(e), we deduce (7.4.10) from (7.4.9).

Let $s = 1, t = \infty$. Then (7.4.7) yields the inequality $\|A\|_{1,\infty} \leq \max_{1 \leq i \leq m, 1 \leq j \leq n} |a_{ij}| = |a_{i_1 j_1}|$. Clearly, $\|A\mathbf{e}_{j_1}\|_\infty = |a_{i_1 j_1}|$. Hence, $\|A\|_{1,\infty} \geq |a_{i_1 j_1}|$, which proves (7.4.11). $\qquad \square$

Theorem 7.4.3 *Let* **V** *be a finite dimensional vector space over* $\mathbb{C}$ *with an norm* $\| \cdot \|$. *Let* $\| \cdot \|$ *be the induced operator norm* Hom $(\mathbf{V}, \mathbf{V})$. *Then for* $A \in$ Hom $(\mathbf{V}, \mathbf{V})$ *the inequality* $\rho(A) \leq \|A\|$ *holds.*

Proof. Assume that $\lambda \in$ spec A. Then there exists $\mathbf{0} \neq \mathbf{x} \in \mathbf{V}$ such that $A\mathbf{x} = \lambda \mathbf{x}$. So $\|A\| \geq \frac{\|A\mathbf{x}\|}{\|\mathbf{x}\|} = |\lambda|$. $\qquad\square$

Problems

1. Show

   (a) The equality (7.4.2).

   (b) For each $t > 0$ there exists a vector $\mathbf{y} \in \mathbf{V}_a, \|\mathbf{y}\|_b = t$ such that $\|T\|_{a,b} = \frac{\|T\mathbf{y}\|_b}{\|\mathbf{y}\|_a}$.

   (c) $\|T\|_{a,b} = \max_{\mathbf{f} \in S_{\|\cdot\|^*_b}, \mathbf{x} \in S_{\|\cdot\|}} |\mathbf{f}(T\mathbf{x})|$.

   (d) Denote by $\|T^*\|_{b^*,a^*}$ the operator norm of $T^* : \mathbf{V}^*_b \to \mathbf{V}^*_a$ with respect to the norms $\| \cdot \|^*_b, \| \cdot \|^*_a$ respectively. Show that $\|T^*\|_{b^*,a^*} = \|T\|_{a,b}$.

   (e) $\|A^\top\|_{b^*,a^*} = \|A\|_{a,b}$ for $A \in \mathbb{F}^{m \times n}$.

2. Show the inequality (7.4.3).

3. Let $T : \mathbf{V} \to \mathbf{V}$ be a linear transformation on a finite dimensional vector space $\mathbf{V}$ over $\mathbb{C}$ with a norm $\| \cdot \|$. Show that $\rho(T) \leq \|T\|$.

4. Let $A \in \mathbb{C}^{n \times n}$. Then $A = Q^{-1}(\Lambda + N)Q$, where Λ is a diagonal matrix, N strictly upper triangular and $\Lambda + N$ is the Jordan canonical form of A. Show

   (a) A is similar to $\Lambda + tN$ for any $0 \neq t \in \mathbb{C}$, i.e. $A = Q_t^{-1}(\Lambda + tN)Q_t$. Show that $Q_t = QD_t$ for an appropriate diagonal matrix D_t.

   (b) Let $\varepsilon > 0$ be given. Show that one can choose a norm $\| \cdot \|_t$ on $\mathbb{C}^n$ of the form $\|\mathbf{x}\|_t := \|Q_t\mathbf{x}\|_2$ for $|t|$-small enough such that $\|A\|_t \leq \rho(A) + \varepsilon$. *Hint:* Note that $\|\Lambda + tN\|_2 \leq \|\Lambda\|_2 + |t|\|N\|_2 = \rho(A) + t\|N\|_2$.

(c) If $N = 0$ then $\|A\| = \rho(A)$ where $\|\mathbf{x}\| = |Q\mathbf{x}|_2$.

(d) Suppose that each eigenvalue λ of modulus $\rho(A)$ is geometrically simple. Then there exists $|t|$ small enough such that $\|A\|_t = \rho(A)$.

5. Let $A \in \mathbb{C}^{n \times n}$. Then there exists a norm on $\|\cdot\|$ on $\mathbb{C}^n$ such that $\rho(A) = \|A\|$ if and only if A each eigenvalue λ of modulus $\rho(A)$ is geometrically simple. *Hint:* Note that if $\rho(A) = 1$ and there is an eigenvalue $\lambda, |\lambda| = 1$ which is not geometrically simple, then $A^m, m \in \mathbb{N}$ is not a bounded sequence.

6. Assume that $\|\cdot\|_a, \|\cdot\|_b$ are two absolute norm on $\mathbb{C}^n$ and $\mathbb{C}^m$. Assume that $Q_1 \mathbb{C}^{m \times m}, Q_2 \in \mathbb{C}^{n \times n}$ are two diagonal matrices such that the absolute value of each diagonal entry is 1. Show that for any $A \in \mathbb{C}^{m \times n}$ $\|Q_1 A Q_2\|_{a,b} = \|A\|_{a,b}$.

7. Show

(a) If $A \in \mathbb{R}_+^{m \times n}$ or $-A \in \mathbb{R}_+^{m \times n}$ then $\|A\|_{\infty,1} = \sum_{i=j=1}^{m,n} |a_{ij}|$.

(b) Let $\mathbf{x} = (x_1, \ldots, x_n)^\top, \mathbf{y} = (y_1, \ldots, y_n)^\top \in \mathbb{R}^n$. We say that $\mathbf{y}$ has a weak sign pattern as $\mathbf{x}$ if $y_i = 0$ when $x_i = 0$ and $y_i x_i \geq 0$ for $x_i \neq 0$. Let $A \in \mathbb{R}^{m \times n}$. Assume that there exists $\mathbf{x} \in \mathbb{R}^n$ such that each row $\mathbf{r}_i$ of A either $\mathbf{r}_i$ or $-\mathbf{r}_i$ has a weak sign pattern as $\mathbf{x}$. Then $\|A\|_{\infty,1} = \sum_{i=j=1}^{m,n} |a_{ij}|$.

(c) Let $A = \begin{bmatrix} a_{11} & a_{12} \\ a_{21} & a_{22} \end{bmatrix}$. Assume that $a_{11}, a_{12}, a_{21}, -a_{22} > 0$. Show that $\|A\|_{\infty,1} < a_{11} + a_{12} + a_{21} - a_{22}$.

(d) Generalize the results of Problem 7(c) to $A \in \mathbb{C}^{m \times n}$.

7.5   Tensor Products of Convex Sets

Definition 7.5.1 *Assume that* $\mathbf{V}_i$ *is a finite dimensional vector space over* $\mathbb{F} = \mathbb{R}, \mathbb{C}$ *for* $i = 1, \ldots, m$. *Let* $\mathbf{V} = \otimes_{i=1}^m \mathbf{V}_i$. *Assume* *that* $X_i \subset \mathbf{V}_i$ *for* $i = 1, \ldots, m$. *Denote*

$$\odot_{i=1}^m X_i := \{\otimes_{i=1}^m \mathbf{x}_i, \text{ for all } \mathbf{x}_i \in X_i, \ i = 1, \ldots, m\}.$$

We call $\odot_{i=1}^m X_i$ *a set tensor product of* $X_1, \ldots, X_m$, *or simply a tensor product of* $X_1, \ldots, X_m$.

Lemma 7.5.2 *Let* C_i *be a compact convex set in a finite dimensional vector space* $\mathbf{V}_i$ *for* $i = 1, \ldots, m$. *Then* $\text{conv} \odot_{i=1}^m C_i$ *is a compact convex set in* $\otimes_{i=1}^m \mathbf{V}_i$, *whose extreme points are contained in* $\odot_{i=1}^m \text{ext}(C_i)$. *In particular, if* C_1 *and* C_2 *are polytopes then* $\text{conv} \, C_1 \odot C_2$ *is a polytope.*

Proof. Since C_i is compact for $i = 1, \ldots, m$, it is straightforward to show that $C := \odot_{i=1}^m C_i$ is a compact set in $\mathbf{V} = \otimes_{i=1}^m \mathbf{V}_i$. Hence, $\text{conv} \, C$ is compact. Let $\dim \mathbf{V}_i = d_i$. Since C_i is compact, Theorem 4.6.2 implies that each $\mathbf{x}_i \in C_i$ is of the form $\mathbf{x}_i = \sum_{j_i=1}^{d_i+1} a_{ij_i} \mathbf{y}_{ij_i}$ where $a_{ij_i} \geq 0, \mathbf{y}_{ij_i} \in \text{ext}(C_i)$ for $j_i = 1, \ldots, d_i + 1$, and $\sum_{j_i=1}^{d_i+1} a_{ij_i} = 1$. Hence

$$\otimes_{i=1}^m \mathbf{x}_i = \sum_{j_1 = \cdots = j_m = 1}^{d_1+1, \ldots, d_m+1} \left(\prod_{i=1}^m a_{ij_i} \right) (\otimes_{i=1}^m \mathbf{y}_{ij_i}).$$

Note that $\prod_{i=1}^m a_{ij_i} \geq 0$ and $\sum_{j_1 = \cdots = j_m = 1}^{d_1+1, \ldots, d_m+1} \prod_{i=1}^m a_{ij_i} = 1$. Hence, $C \subset \text{conv} \odot_{i=1}^m \text{ext}(C_i)$, which implies $\text{conv} \, C \subset \text{conv} \odot_{i=1}^m \text{ext}(C_i) \subset \text{conv} \, C$.

Assume that C_1, C_2 are polytopes. Then $\text{conv} \, C_1 \odot C_2$ is nonempty, compact and convex whose set of extreme points is finite. Hence, $\text{conv} \, C_1 \odot C_2$ is a polytope. (See Problem 7.2.8.) $\qquad \square$

See Problem 1 for an example where $\text{ext}(C_1 \odot C_2)$ is strictly contained in $\text{ext}(C_1) \odot \text{ext}(C_2)$. The next two examples give two important cases where $\text{ext}(C_1 \odot C_2) = \text{ext}(C_1) \odot \text{ext}(C_2)$. In these two examples we view $\mathbb{C}^{p \times q} \otimes \mathbb{C}^{m \times n}$ as $\mathbb{C}^{pm \times qn}$, where the tensor product of two matrices $A \otimes B$ is the Kronecker tensor product.

Proposition 7.5.3 *Let $m, n \geq 2$ be integers. Then $\Omega_m \odot \Omega_n \subset$ Ω_{mn}, and $\text{ext} \, (\text{conv} \, \Omega_m \odot \Omega_n) = \mathcal{P}_m \odot \mathcal{P}_n$.*

Proof. Let $A = [a_{ij}] \in \Omega_n, B = [b_{pq}] \in \Omega_n$. The the entries of $A \otimes B = [c_{(i,p)(j,q)}] \in \mathbb{R}_+^{mn \times mn}$, where $c_{(i,p)(j,q)} = a_{ij} b_{pq}$. Clearly

$$\sum_{j=1}^{m} c_{(i,p)(j,q)} = \sum_{j=1}^{m} a_{ij} b_{pq} = b_{pq}, \quad \sum_{i=1}^{m} c_{(i,p)(j,q)} = \sum_{i=1}^{m} a_{ij} b_{pq} = b_{pq},$$

$$\sum_{q=1}^{n} c_{(i,p)(j,q)} = \sum_{q=1}^{n} a_{ij} b_{pq} = a_{ij}, \quad \sum_{p=1}^{n} c_{(i,p)(j,q)} = \sum_{p=1}^{n} a_{ij} b_{pq} = a_{ij}.$$

Hence $A \otimes B \in \Omega_{mn}$, where we identify the set $[m] \times [n]$ with $[mn]$. Since Ω_{mn} is convex it follows that $\text{conv} \, \Omega_m \odot \Omega_n \subset \Omega_{mn}$. Recall that $\text{ext} \, (\Omega_{mn}) = \mathcal{P}_{mn}$. Clearly $\mathcal{P}_m \odot \mathcal{P}_n \subset \mathcal{P}_{mn}$. Problem 4.7.10 yields that $\text{ext} \, (\text{conv} \, \Omega_m \odot \Omega_n) = \mathcal{P}_m \odot \mathcal{P}_n$. $\qquad \square$

Proposition 7.5.4 *Let $m, n \geq 2$ be integers. Then $\mathbf{H}_{m,+,1} \odot$ $\mathbf{H}_{n,+,1} \subset \mathbf{H}_{mn,+,1}$, and $\text{ext} \, (\text{conv} \, \mathbf{H}_{m,+,1} \odot \mathbf{H}_{n,+,1}) = \text{ext} \, (\mathbf{H}_{m,+,1}) \odot$ $\text{ext} \, (\mathbf{H}_{n,+,1})$.*

Proof. Let $A \in \mathbf{H}_m, B \in \mathbf{H}_n$ be non-negative definite hermitian matrices. Then $A \otimes B$ is non-negative definite. Since $\text{tr} \, A \otimes B = (\text{tr} \, A)(\text{tr} \, B)$ it follows that $\mathbf{H}_{m,+,1} \odot \mathbf{H}_{n,+,1} \subset \mathbf{H}_{mn,+,1}$. Hence, $\text{ext} \, (\text{conv} \, \mathbf{H}_{m,+,1} \odot \mathbf{H}_{n,+,1}) \subset \text{ext} \, (\mathbf{H}_{m,+,1}) \odot \text{ext} \, (\mathbf{H}_{n,+,1})$.

Recall that $\text{ext} \, (\mathbf{H}_{m,+,1}), \text{ext} \, (\mathbf{H}_{n,+,1}), \text{ext} \, (\mathbf{H}_{mn,+,1})$ are hermitian rank one matrix of trace 1 of corresponding orders. Since $A \otimes B$ is a rank one matrix if A and B is a rank one matrix it follows that $\text{ext} \, (\mathbf{H}_{m,+,1}) \odot \text{ext} \, (\mathbf{H}_{n,+,1}) \subset \text{ext} \, (\mathbf{H}_{mn,+,1})$. Hence

$$\text{ext} \, (\text{conv} \, \mathbf{H}_{m,+,1} \odot \mathbf{H}_{n,+,1}) = \text{ext} \, (\mathbf{H}_{m,+,1}) \odot \text{ext} \, (\mathbf{H}_{n,+,1}). \qquad \square$$

Problem 7.5.5 *Let C_i be a compact convex set in a finite dimensional space $\mathbf{V}_i$ for $i = 1, 2$. Suppose that $C_i = \cap_{\alpha \in \mathcal{F}_i} H(f_\alpha, \mathbf{x}_\alpha)$, were $\mathcal{F}_i$ is the set of all supporting hyperplanes of C_i which characterize C_i, for $i = 1, 2$. ($\mathcal{F}_i$ may not be finite or countable.) The* *problem is to characterize the set of all supporting hyperplanes of* conv $C_1 \odot C_2$.

Equivalently, suppose we know how to decide if $\mathbf{x}_i$ *belongs or does not belong to* C_i *for* $i = 1, 2$. *How do we determine if* $\mathbf{x}$ *belongs or does not belong* conv $C_1 \odot C_2$?

It seems that the complexity of characterization conv $C_1 \odot C_2$ can be much more complex then the complexity of C_1 and C_2. We will explain this remark in the two examples discussed in Propositions 7.5.3 and 7.5.4.

Consider first $\mathbf{H}_{m,+1}, \odot \mathbf{H}_{n,+,1}$. So any element in $\mathbf{H}_{m,+1}, \odot \mathbf{H}_{n,+,1}$ is of the form $A \otimes B$, where A and B are non-negative definite hermitian matrices of trace one. The matrix $A \otimes B$ is called a *pure state* in quantum mechanics. A matrix $C \in$ conv $\mathbf{H}_{m,+1}, \odot \mathbf{H}_{n,+,1}$ is called a separable state. So conv $\mathbf{H}_{m,+1}, \odot \mathbf{H}_{n,+,1}$ is the convex set of separable states. The set $\mathbf{H}_{mn,+1,1} \setminus$ conv $\mathbf{H}_{m,+1}, \odot \mathbf{H}_{n,+,1}$ the set of entangled states. See for example [BeZ06]. The following result is due to L (Gurvits [Gur03]).

Theorem 7.5.6 *For general positive integers* m, n *and* $A \in \mathbf{H}_{mn,+,1}$ *the problem of decision if* A *is separable, i.e.* $A \in$ conv $\mathbf{H}_{m,+1}, \odot \mathbf{H}_{n,+,1}$, *is NP-Hard.*

On the other hand, given a hermitian matrix $A \in \mathbf{H}_n$, it is well known that one can determine in polynomial time if A belongs or does not belong to $\mathbf{H}_{n,+,1}$. See Problem 3. We will discuss the similar situation for conv $\Omega_m \odot \Omega_n$ in §7.6.

Definition 7.5.7 *Let* $\mathbf{V}_i$ *be a finite dimensional vector space over* $\mathbb{F} = \mathbb{R}, \mathbb{C}$ *with a norm* $\| \cdot \|_i$ *for* $i = 1, \ldots, k$. *Let* $\mathbf{V} := \otimes_{i=1}^k \mathbf{V}_i$ *with the norm* $\| \cdot \|$. *Then* $\| \cdot \|$ *is called a cross norm if*

$$\| \otimes_{i=1}^k \mathbf{x}_i \| = \prod_{i=1}^k \|\mathbf{x}_i\|_i \qquad (7.5.1)$$

for all rank one tensors.

Identify $\otimes_{i=1}^k \mathbf{V}'_i$ *with* $\mathbf{V}'$, *where* $(\otimes_{i=1}^k \mathbf{f}_i)(\otimes_{i=1}^k \mathbf{x}_i) = \prod_{i=1}^k \mathbf{f}_i(\mathbf{x}_i)$. *Then* $\| \cdot \|$ *is called a normal cross norm if the norm* $\| \cdot \|^*$ *on* $\mathbf{V}'$ *is a cross norm with respect to the norms* $\| \cdot \|_i^*$ *on* $\mathbf{V}'_i$ *for* $i = 1, \ldots, k$.

See [Sch50] for properties of the cross norms. We discuss the following known results needed in the sequel.

Theorem 7.5.8 *Let* $\mathbf{V}_i$ *be a finite dimensional vector space over* $\mathbb{F} = \mathbb{R}, \mathbb{C}$ *with a norm* $\| \cdot \|_i$ *for* $i = 1, \ldots, k$. *Let* $\mathbf{V} := \otimes_{i=1}^{k} \mathbf{V}_i$. *Then there exists a norm* $\| \cdot \|$ *on* $\mathbf{V}$ *satisfying* (7.5.7). *Furthermore, there exist unique norms* $\| \cdot \|_{\max}, \| \cdot \|_{\min}$ *satisfying the following properties. First,* $\| \cdot \|_{\max}$ *and* $\| \cdot \|_{\min}$ *are normal cross norms. Moreover* $\|\mathbf{z}\|_{\min} \le \|\mathbf{z}\|_{\max}$ *for all* $\mathbf{z} \in \mathbf{V}$. *Any cross norm* $\| \cdot \|$ *on* $\mathbf{V}$ *satisfies the inequality* $\|\mathbf{z}\| \le \|\mathbf{z}\|_{\max}$ *for all* $\mathbf{z} \in \mathbf{V}$. *Third, assume that* $\| \cdot \|_a$ *on* $\mathbf{V}$ *satisfies the equality*

$$\| \otimes_{i=1}^{k} \mathbf{f}_i \|_a^* = \prod_{i=1}^{k} \|\mathbf{f}_i\|_i^* \text{ for all } \mathbf{f}_i \in \mathbf{V}_i', i = 1, \ldots, k. \qquad (7.5.2)$$

That is $\| \cdot \|_a^*$ *is a cross norm with respect to the norms* $\| \cdot \|_i^*$ *on* $\mathbf{V}_i'$ *for* $i = 1, \ldots, k$. *Then* $\|\mathbf{z}\|_{\min} \le \|\mathbf{z}\|_a$ *for all* $\mathbf{z} \in \mathbf{V}$. *More precisely,*

$$\begin{aligned} \mathrm{B}_{\|\cdot\|_{\max}} &= \mathrm{conv}\, \mathrm{B}_{\|\cdot\|_1} \odot \cdots \odot \mathrm{B}_{\|\cdot\|_k}, \\ \mathrm{B}_{\|\cdot\|_{\min}^*} &= \mathrm{conv}\, \mathrm{B}_{\|\cdot\|_1^*} \odot \cdots \odot \mathrm{B}_{\|\cdot\|_k^*}. \end{aligned} \qquad (7.5.3)$$

Proof. For simplicity of the exposition we let $k = 2$. Define the set $\mathcal{B} := \mathrm{conv}\, \mathrm{B}_{\|\cdot\|_1} \odot \mathrm{B}_{\|\cdot\|_2}$. Clearly, $\mathcal{B}$ is a compact convex balanced that $\mathbf{0}$ is in its interior. Hence, there exists a norm $\| \cdot \|_{\max}$ such that $\mathcal{B} = \mathrm{B}_{\|\cdot\|_{\max}}$. We claim that $\|\mathbf{x} \otimes \mathbf{y}\|_{\max} = \|\mathbf{x}\|_1 \|\mathbf{y}\|_2$. Clearly, to show that it is enough to assume that $\|\mathbf{x}\|_1 = \|\mathbf{y}\|_2 = 1$. Since $\mathbf{x} \otimes \mathbf{y} \in \mathrm{B}_{\|\cdot\|_{\max}}$ we deduce that $\|\mathbf{x} \otimes \mathbf{y}\|_{\max} \le 1$. Problem 7.1.5(c) yields that there exists $\mathbf{f} \in \mathrm{S}_{\|\cdot\|_1^*}, \mathbf{g} \in \mathrm{S}_{\|\cdot\|_2^*}$ such that

$$1 = \mathbf{f}(\mathbf{x}) \ge |\mathbf{f}(\mathbf{x}_1)|, \forall \mathbf{x}_1 \in \mathrm{B}_{\|\cdot\|_1}, \quad 1 = \mathbf{g}(\mathbf{y}) \ge |\mathbf{f}(\mathbf{y}_1)|, \forall \mathbf{y}_1 \in \mathrm{B}_{\|\cdot\|_2}.$$

Hence

$$1 = (\mathbf{f} \otimes \mathbf{g})(\mathbf{x} \otimes \mathbf{y}) \ge |(\mathbf{f} \otimes \mathbf{g})(\mathbf{z})| \text{ for all } \mathbf{z} \in \mathrm{B}_{\|\cdot\|_{\max}}. \qquad (7.5.4)$$

Hence $\mathbf{x} \otimes \mathbf{y} \in \partial \mathrm{B}_{\|\cdot\|_{\max}} = \mathrm{S}_{\|\cdot\|_{\max}}$, i.e. $\|\mathbf{x} \otimes \mathbf{y}\|_{\max} = 1$. Therefore, the norm $\| \cdot \|_{\max}$ satisfies (7.5.1). Let $\| \cdot \|$ be another norm on $\mathbf{V}$ satisfying (7.5.1). Hence $\mathrm{B}_{\|\cdot\|_1} \odot \mathrm{B}_{\|\cdot\|_2} \subset \mathrm{B}_{\|\cdot\|}$, which yields $\mathrm{B}_{\|\cdot\|_{\max}} = \mathrm{conv}\, \mathrm{B}_{\|\cdot\|_1} \odot \mathrm{B}_{\|\cdot\|_2} \subset \mathrm{B}_{\|\cdot\|}$. Therefore, $\|\mathbf{z}\| \le \|\mathbf{z}\|_{\max}$.

We next observe that $\| \cdot \|_c := \| \cdot \|_{\max}^*$ on $\mathbf{V}'$ satisfies the equality

$$\| \mathbf{f} \otimes \mathbf{g} \|_c = \| \mathbf{f} \|_1^* \| \mathbf{g} \|_2^* \text{ for all } \mathbf{f} \in \mathbf{V}_1', \mathbf{g} \in \mathbf{V}_2'. \tag{7.5.5}$$

Recall that $\| \mathbf{f} \otimes \mathbf{g} \|_c = \max_{\mathbf{z} \in B_{\| \cdot \|}} |(\mathbf{f} \otimes \mathbf{g})(\mathbf{z})|$. Use the (7.5.4) to deduce (7.5.5). Hence, $\| \mathbf{z} \|_{\max}$ is a normal cross norm.

Let $\| \cdot \|_b$ be the norm given by the unit ball $\mathrm{conv}\, B_{\| \cdot \|_1^*} \odot B_{\| \cdot \|_2^*}$ on $\mathbf{V}'$. Hence, the above results show that $\| \cdot \|_b$ is a normal cross norm. Recall that for any norm $\| \cdot \|_a$ on $\mathbf{V}'$ satisfying (7.5.5) we showed the inequality $\| \mathbf{h} \|_a^* \le \| \mathbf{h} \|_b$ for any $\mathbf{h} \in \mathbf{V}'$. Define $\| \cdot \|_{\min} := \| \cdot \|_b^*$. Hence, $\| \mathbf{z} \|_{\min} \le \| \mathbf{z} \|_a$. The previous arguments show that $\| \cdot \|_{\min}$ satisfies the equality (7.5.1). Hence, $\| \mathbf{z} \|_{\min} \le \| \mathbf{z} \|_{\max}$ for all $\mathbf{z} \in \mathbf{V}$. Also $\| \mathbf{z} \|_{\min}$ is a normal cross norm. $\qquad \square$

Definition 7.5.9 *Let $\mathbf{V}_i$ be a finite dimensional vector space over $\mathbb{F} = \mathbb{R}, \mathbb{C}$ with a Euclidean norm $\| \cdot \|_i$ for $i = 1, \ldots, k$. Let $\mathbf{V} := \otimes_{i=1}^k \mathbf{V}_i$. Then*

1. *The norm $\| \cdot \|_{\min}$ is called the spectral norm and denoted as $\| \cdot \|_{\mathrm{spec}}$.*

2. *The norm $\| \cdot \|_{\max}$ is called the nuclear norm and denoted as $\| \cdot \|_{\mathrm{nuc}}$.*

Theorem 7.5.10 *Let $\mathbf{V}_i$ be a finite dimensional vector space over $\mathbb{F} = \mathbb{R}, \mathbb{C}$ with a Euclidean norm $\| \cdot \|_i$ for $i = 1, \ldots, k$. Let $\mathbf{V} := \otimes_{i=1}^k \mathbf{V}_i$. Then the spectral and nuclear norms are conjugate. Furthermore*

$$\| \mathbf{z} \|_{\mathrm{spec}} = \max_{\| \mathbf{x}_i \|_i = 1, i \in [k]} |\langle \mathbf{z}, \otimes_{i \in [k]} \mathbf{x}_i \rangle|, \tag{7.5.6}$$

$$\| \mathbf{z} \|_{\mathrm{nuc}} = \min \left\{ \sum_{j \in [N]} \prod_{i \in [k]} \| \mathbf{x}_{j,i} \|_i, \ \sum_{j \in [N]} \otimes_{i \in [k]} \mathbf{x}_{j,i} = \mathbf{z} \right\}. \tag{7.5.7}$$

Moreover, one can choose $N = 1 + \prod_{i \in [k]} \dim V_i$ if $\mathbb{F} = \mathbb{R}$ and $N = 1 + 2 \prod_{i \in [k]} \dim V_i$ if $\mathbb{F} = \mathbb{C}$.

Proof. As $\| \cdot \|_i^* = \| \cdot \|_i$ for $i \in [k]$ (7.5.3) yields that the spectral and nuclear norms are conjugate. Furthermore, (7.5.3) yields (7.5.6).

We now show (7.5.7). Assume the decomposition $\sum_{j\in[N]} \otimes_{i\in[k]}$ $\mathbf{x}_{j,i} = \mathbf{z}$. Since the nuclear norm is a cross norm we deduce

$$\|\mathbf{z}\|_{\mathrm{nuc}} \leq \sum_{j\in[N]} \| \otimes_{j\in[N]} \mathbf{x}_{j,i}\|_{\mathrm{nuc}} = \sum_{j\in[N]} \prod_{i\in[k]} \|\mathbf{x}_{j,i}\|_i.$$

It is left to show that there exists a decomposition of $\mathbf{z}$ for which equality holds. One can assume without loss of generality that $\|\mathbf{z}\|_{\mathrm{nuc}} = 1$. Hence, $\mathbf{z}$ is a convex combination of N extreme points of the unit ball of $B_{\|\cdot\|_{\mathrm{nuc}}}$. (7.5.3) yields that each extreme point of $B_{\|\cdot\|_{\mathrm{nuc}}}$ is $\otimes_{i\in[k]}\mathbf{y}_i$, where $\|\mathbf{y}_i\|_i = 1$ for $i \in [k]$. So

$$\mathbf{z} = \sum_{j\in[N]} a_j \otimes_{i\in[k]} \mathbf{y}_{j,i}$$

$$= \sum_{j\in[N]} \otimes_{i\in[k]} \left(a_j^{\frac{1}{k}} \mathbf{y}_{j,i} \right), a_j > 0, j \in [k], \sum_{j\in[k]} a_j = 1.$$

For this decomposition we have that $1 = \|\mathbf{z}\|_{\mathrm{nuc}} = \sum_{j\in[N]} \prod_{i\in[k]}$ $\|a_j^{\frac{1}{k}}\mathbf{y}_{j,i}\|_i.$

It is left to show the upper bound on N. Let $N' := \dim \mathbf{V} = \prod_{i\in[k]} \dim \mathbf{V}_i$. Assume first that $\mathbb{F} = \mathbb{R}$. Then Carathéodory's theorem yield that $N \leq 1 + N'$. Assume that $\mathbb{F} = \mathbb{C}$. Then the real dimension of $\mathbf{V}$ is $2N'$. Hence, $N \leq 1 + 2N'$. $\square$

Proposition 7.5.11 *Let $k > 1$ be an integer. Assume that $\mathbf{V}_1, \ldots, \mathbf{V}_k$ are inner product spaces over $\mathbb{F} = \mathbb{R}, \mathbb{C}$. Let $\mathbf{V} = \otimes_{i=1}^k \mathbf{V}_i$ with the inner product induced by the inner products on $\mathbf{V}_1, \ldots, \mathbf{V}_k$. Assume that $\| \cdot \|_1, \ldots, \| \cdot \|_k, \| \cdot \|$ are the induced norms by the corresponding inner products on $\mathbf{V}_1, \ldots, \mathbf{V}_k, \mathbf{V}$. Then $\| \cdot \|$ is a normal cross norm. If $\dim \mathbf{V}_i > 1$ for $i = 1, \ldots, k$ then $\| \cdot \|$ is different from $\| \cdot \|_{max}$ and $\| \cdot \|_{min}$.*

See Problem 7.

Theorem 7.5.12 *Let $\mathbf{U}, \mathbf{V}$ be finite dimensional vector spaces over $\mathbb{F} = \mathbb{R}, \mathbb{C}$ with norms $\| \cdot \|, ||| \cdot |||$ respectively. Identify $\mathbf{W} = \mathbf{V} \otimes \mathbf{U}'$ with $\mathrm{Hom}\,(\mathbf{U}, \mathbf{V})$, via isomorphism $\theta : \mathbf{W} \to \mathrm{Hom}\,(\mathbf{U}, \mathbf{V})$,* *where $\theta(\mathbf{v} \otimes \mathbf{f})(\mathbf{u}) = \mathbf{f}(\mathbf{u})\mathbf{v}$ for any $\mathbf{f} \in \mathbf{U}'$. Then the minimal cross norm on $\|\cdot\|_{\min}$ on $\mathbf{W}$ is the operator norm on* Hom $(\mathbf{U}, \mathbf{V})$, *where the norms on $\mathbf{U}'$ and $\mathbf{V}$ are $\|\cdot\|^*$ and $\|\|\cdot\|\|$ respectively. Identify $\mathbf{W}'$ with $\mathbf{V}' \otimes \mathbf{U} \sim$* Hom $(\mathbf{U}', \mathbf{V}')$. *Then the maximal cross norm $\|\cdot\|_{\max}$ on $\mathbf{W}$ is the conjugate to the operator norm on* Hom $(\mathbf{U}', \mathbf{V}')$, *which is identified with $\mathbf{W}'$.*

Proof. Let $T \in$ Hom $(\mathbf{U}, \mathbf{V})$. Then

$$\|T\| = \max_{\mathbf{u} \in S_{\|\cdot\|}} \|\|T(\mathbf{u})\|\| = \max_{\mathbf{g} \in S_{\|\|\cdot\|\|^*}, \mathbf{u} \in S_{\|\cdot\|}} |\mathbf{g}(T(\mathbf{u}))|.$$

Let $\theta^{-1} :$ Hom $(\mathbf{U}, \mathbf{V}) \to \mathbf{V} \otimes \mathbf{U}'$ be the isomorphism given in the theorem. Then $\mathbf{g}(T(\mathbf{u})) = (\mathbf{g} \otimes \mathbf{u})(\theta^{-1}(T))$. Let $\|\cdot\|_b$ be the norm given by the unit ball conv $B_{\|\|\cdot\|\|^*} \odot B_{\|\cdot\|}$ on $\mathbf{V}' \otimes \mathbf{U} \sim \mathbf{W}'$, as in the proof of Theorem 7.5.8. Then

$$\|\theta^{-1}(T)\|_b^* = \max_{\mathbf{g} \in S_{\|\|\cdot\|\|^*}, \mathbf{u} \in S_{\|\cdot\|}} = |(\mathbf{g} \otimes \mathbf{u})(\theta^{-1}(T))|.$$

Use the proof of Theorem 7.5.8 to deduce that $\|T\| = \|\phi(T)\|_{\min}$.

Similar arguments show that the conjugate norm to the operator norm of Hom $(\mathbf{U}', \mathbf{V}')$, identified with $\mathbf{V}' \otimes \mathbf{U} \sim \mathbf{W}'$ gives the norm $\|\cdot\|_{\max}$ on $\mathbf{W}$. $\qquad\square$

Use the above theorem and Problem 7.1.14 to deduce.

Corollary 7.5.13 *Let $\mathbf{U}, \mathbf{V}$ be finite dimensional inner product spaces over $\mathbb{F} = \mathbb{R}, \mathbb{C}$, with the corresponding induced norms. Identify $\mathbf{W} = \mathbf{V} \otimes \mathbf{U}'$ with* Hom $(\mathbf{U}, \mathbf{V})$ *as in Theorem 7.5.12. Then $\|T\|_{\min} = \sigma_1(T)$ and $\|T\|_{\max} = \sum_{i=1}^{\dim \mathbf{V}} \sigma_i(T)$.*

More generally, given finite dimensional vectors spaces $\mathbf{U}_i, \mathbf{V}_i$ over $\mathbb{F} = \mathbb{R}, \mathbb{C}$ for $i = 1, \ldots, k$ we identify the tensor spaces $\otimes_{i=1}^k$ Hom $(\mathbf{U}_i, \mathbf{U}_i)$ with Hom $(\otimes_{i=1}^k \mathbf{U}_i, \otimes_{i=1}^k \mathbf{V}_i)$ using isomorphism

$$\iota : \otimes_{i=1}^k \text{Hom } (\mathbf{U}_i, \mathbf{V}_i) \to \text{Hom } (\otimes_{i=1}^k \mathbf{U}_i, \otimes_{i=1}^k \mathbf{V}_i) \text{ satisfying}$$
$$\iota(\otimes_{i=1}^k T_i)(\otimes_{i=1}^k \mathbf{u}_i) = \otimes_{i=1}^k (T_i \mathbf{u}_i) \qquad (7.5.8)$$
$$\text{where } T_i \in \text{Hom } (\mathbf{U}_i, \mathbf{V}_i), \mathbf{u}_i \in \mathbf{U}_i, i = 1, \ldots, k.$$

Theorem 7.5.14 *Let* $\mathbf{U}_i, \mathbf{V}_i$ *are finite dimensional vector spaces over* $\mathbb{F} = \mathbb{R}, \mathbb{C}$ *with the norms* $\|\cdot\|_i, \|\|\cdot\|\|_i$ *respectively for* $i = 1, \ldots, k$. *Let* $N_i(\cdot)$ *be the operator on* $\mathrm{Hom}\,(\mathbf{U}_i, \mathbf{V}_i)$ *for* $i = 1, \ldots, k$. *Let* $\|\cdot\|_{\max}$ *be the maximal cross norms on* $\mathbf{U} := \otimes_{i=1}^{k} \mathbf{U}_i$ *and* $\|\|\cdot\|\|$ *be any cross norm on* $\mathbf{V} := \otimes_{i=1}^{k} \mathbf{V}_i$. *Then the operator norm* $N(\cdot)$ *on* $\mathrm{Hom}\,(\mathbf{U}, \mathbf{V})$, *identified with* $\otimes_{i=1}^{k} \mathrm{Hom}\,(\mathbf{U}_i, \mathbf{V}_i)$, *is a cross norm with respect to the norms* $N_i(\cdot), i = 1, \ldots, k$.

Proof. Since $B_{\|\cdot\|_{\max}} = \odot_{i=1}^{k} B_{\|\cdot\|_i}$ we deduce that for any $T \in \mathrm{Hom}(\mathbf{U}, \mathbf{V})$ one has

$$N(T) = \max_{\mathbf{u}_i \in B_{\|\cdot\|_i}, i \in \langle k \rangle} \|\|T(\otimes_{i=1}^{k} \mathbf{u}_i)\|\|.$$

Let $T = \otimes_{i=1}^{k} T_i$. Since $\|\|\cdot\|\|$ is a cross norm on $\mathbf{V}$ we deduce

$$N(T) = \max_{\mathbf{u}_i \in B_{\|\cdot\|_i}, i \in \langle k \rangle} \|\| \otimes_{i=1}^{k} T_i(\mathbf{u}_i)\|\|$$

$$= \max_{\mathbf{u}_i \in B_{\|\cdot\|_i}, i \in \langle k \rangle} \prod_{i=1}^{k} \|\|T_i(\mathbf{u}_i)\|\|_i = \prod_{i=1}^{k} N_i(T_i). \qquad \square$$

Problems

1. Let $\mathbf{V}_1, \mathbf{V}_2$ be one-dimensional subspaces with bases $\mathbf{v}_1, \mathbf{v}_2$ respectively. Let $C_1 = [-\mathbf{e}_1, 2\mathbf{e}_1], C_2 = [-\mathbf{e}_2, 3\mathbf{e}_2]$. Show that $C_1 \odot C_2 = [-3(\mathbf{e}_1 \otimes \mathbf{e}_2), 6\mathbf{e}_1 \otimes \mathbf{e}_2]$. Hence, $\mathrm{ext}\,(C_1 \odot C_2)$ is contained strictly in $\mathrm{ext}\,(C_1) \odot \mathrm{ext}\,(C_2)$.

2. Index the entries of $C \in \Omega_{mn}$ as $c_{(i,p),(j,q)}$. Show

    (a) Assume that $C \in \mathrm{conv}\,\Omega_m \odot \Omega_n$. Then the entries of C satisfy

$$\sum_{i=1}^{m} c_{(i,p)(j,q)} = \sum_{j=1}^{m} c_{(i,p)(j,q)} \text{ for each } i, j \in \langle m \rangle, \ p, q \in \langle n \rangle,$$

$$\sum_{p=1}^{n} c_{(i,p)(j,q)} = \sum_{q=1}^{n} c_{(i,p)(j,q)} \text{ for each } i, j \in \langle m \rangle, \ p, q \in \langle n \rangle.$$

(b) For $m = n = 2$, the standard conditions for 4×4 doubly stochastic matrices, and the conditions in part 2(a) characterize the set conv $\Omega_2 \odot \Omega_2$.

(c) For $m = n \geq 4$, the standard conditions for $n^2 \times n^2$ doubly stochastic matrices, and the conditions in part 2(a) give a set which contains strictly conv $\Omega_n \odot \Omega_n$. *Hint*: Consult with [Fri08].

3. Assume that $A = [a_{ij}]_{i=j=1}^n \in \mathbf{H}_n$ and $\det A = 0$. Find $\mathbf{0} \neq \mathbf{x} \in \mathbb{C}^n$ such that $A\mathbf{x} = \mathbf{0}$. Let $\mathbf{x}_n = \frac{1}{\|\mathbf{x}\|}\mathbf{x}$ and complete $\mathbf{x}_n$ to an orthonormal basis $\mathbf{x}_1, \ldots, \mathbf{x}_n$. Let $A_{n-1} = [\mathbf{x}_i^* A\mathbf{x}_j]_{i=j=1}^{n-1}$. Then $A_{n-1} \in \mathbf{H}_{n-1}$. Furthermore, $A\mathbf{H}_{n,+}$ if and only if $A_{n-1}\mathbf{H}_{n-1,+}$.

4. Let $\tau : \mathbb{C}^{n \times n}$ be the transpose map: $\tau(A) = A^\top$. Show

(a) $\tau(A)$ is similar to A for any $A \in \mathbb{C}^{n \times n}$.

(b) τ leaves invariant the following subsets of $\mathbb{C}^{n \times n}$:
$$\mathbb{R}^{n \times n}, \mathrm{S}_n(\mathbb{R}), \mathrm{S}_n(\mathbb{C}), \mathrm{O}(n, \mathbb{R}), \mathrm{O}(n, \mathbb{C}),$$
$$\mathrm{U}(n, \mathbb{R}), \mathrm{N}(n, \mathbb{R}), \mathrm{N}(n, \mathbb{C}), \mathbf{H}_n, \mathbf{H}_{n,+}, \mathbf{H}_{n,+,1}.$$

5. On $\mathbb{C}^{mn \times mn}$, viewed as $\mathbb{C}^{m \times m} \otimes \mathbb{C}^{n \times n}$, we define the partial transpose τ_{par} as follows. Let $C = [c_{(i,p),(j,q)}]_{i=j=p=q=1}^{m,m,n,n} \in \mathbb{C}^{m \times m} \otimes \mathbb{C}^{n \times n}$. Then $\tau_{\mathrm{par}}(C) = [\tilde{c}_{(i,p),(j,q)}]_{i=j=p=q=1}^{m,m,n,n}$, where $\tilde{c}_{(i,p),(j,q)} = c_{(i,q),(j,p)}$ for $i, j \in \langle m \rangle, p, q \in \langle n \rangle$. Equivalently, τ_{par} is uniquely determined by the condition $\tau_{\mathrm{par}}(A \otimes B) = A \otimes B^\top$ for any $A \in \mathbb{C}^{m \times n}, B \in \mathbb{C}^{n \times n}$. Show

(a) τ_{par} leaves the following subsets of $\mathbb{C}^{mn \times mn}$ invariant: $\mathrm{S}_{mn}(\mathbb{R}), \mathbf{H}_{mn}$, and all the set of the form $X \odot Y$, where $X \subset \mathbb{C}^{m \times m}$ and $Y \subset \mathbb{C}^{n \times n}$ are given in Problem 4(b). In particular, the convex set of separable states conv $\mathbf{H}_{m,+,1} \odot \mathbf{H}_{n,+,1}$ is invariant under the partial transpose.

(b) Show that for $m = 2$ and $n = 2, 3$ $C \in \mathbf{H}_{mn}$ is a separable state, i.e. $C \in \mathrm{conv}\,\mathbf{H}_{m,+,1} \odot \mathbf{H}_{n,+,1}$, if and only if $C, \tau_{\mathrm{par}}(C) \in \mathbf{H}_{mn,+,1}$. (This is the Horodecki–Peres condition [Hor96, Per96].)

6. Let the assumptions of Theorem 7.5.8 hold. Show

(a) Each $\mathbf{z} \in \mathbf{V}$ can be decomposed, usually in many ways, as a sum of rank one tensors

$$\mathbf{z} = \sum_{j=1}^{N} \otimes_{i=1}^{k} \mathbf{x}_{j,i}, \ \mathbf{x}_{j,i} \in \mathbf{V}_i, \ i = 1, \ldots, k, \ j = 1, \ldots, N,$$

where $N = \prod_{i=1}^{k} \dim \mathbf{V}_i$. Then $\|\mathbf{z}\|_{\max}$ is the minimum of $\sum_{j=1}^{N} \prod_{i=1}^{k} \|\mathbf{x}_{j,i}\|_i$ over all the above decompositions of $\mathbf{z}$.

(b)

$$\|\mathbf{z}\|_{\min} = \max_{\mathbf{f}_i \in B_{\|\cdot\|_i^*}, i=1,\ldots,k} \left| (\otimes_{i=1}^{k} \mathbf{f}_i)(\mathbf{z}) \right|.$$

7. Prove Proposition 7.5.11. *Hint:* To prove the first part of the problem choose orthonormal bases in $\mathbf{V}_1, \ldots, \mathbf{V}_k$. To prove the second part observe that $\|\cdot\|$ is smooth, while $\|\cdot\|_{\min}, \|\cdot\|_{\max}$ are not smooth if $\dim \mathbf{V}_i > 1$ for $i = 1, \ldots, k > 1$.

7.6   The Complexity of $\operatorname{conv} \Omega_n \odot \Omega_m$

In this section, we show that there is a linear programming problem on $\Omega_{n,m} := \operatorname{conv} \Omega_n \odot \Omega_m$, whose solution gives an answer to the *subgraph isomorphism problem*, that will be stated precisely below. The subgraph isomorphism problem belongs to the class of NP-complete problems [GaJ79]. This shows, in our opinion, that the number of half spaces characterizing $\Omega_{n,m}$ is probably not polynomial in $\max(m, n)$, which is analogous to Theorem 7.5.6.

By graph $G = (V, E)$ in this section we mean an undirected simple graph as in §6.1 on the set of vertices V and the set of edges E. We assume that $\#V = n, \#E = m$. We will identify $[n]$ and $[m]$ with V and E, and not ambiguity will arise. Recall that a vertex $v \in V$ is called *isolated* if $\deg v = 0$. Recall:

Definition 7.6.1 *Let* $G = (V, E), G' = (V', E')$ *be two undirected simple graphs. Then* G *and* G' *are called isomorphic if the following condition hold. There is a bijection* $\phi : V \to V'$ *such that*

$(u, v) \in E$ *if and only if* $(\phi(u), \phi(v)) \in E'$. G' *is called isomorphic to a subgraph of* G *if there exists a subgraph* G_1 *of* G *such that* G' *is isomorphic to* G_1.

We recall that the subgraphs isomorphism problem, abbreviated as SGIP, which asks if G' is isomorphic to a subgraph of G, is an NP-complete problem [GaJ79].

We now relate the SGIP to certain linear programming problems on $\Omega_{m,n}$. Let $A(G) = [a_{ij}]_{i=j=1}^{n} \in \{0,1\}^{n \times n} \in$ be the adjacency matrix of G. So $A(G)$ is a symmetric matrix with zero diagonal. Recall that a different labeling of the elements of V gives rise to the adjacency matrix $A' = PA(G)P^{\top}$ for some permutation matrix $P \in \mathcal{P}_n$. Thus the graph G gives rise to the conjugacy class of matrices $\mathcal{A}(G) = \{PA(G)P^{\top}, \ P \in \mathcal{P}_n\}$. The following result is straightforward, see Problem 1.

Lemma 7.6.2 *Let* $G = (V, E), G' = (V', E')$ *are two undirected graphs. Assume that* $\#V = \#V'$. *Then* G *and* G' *are isomorphic if and only if* $\mathcal{A}(G) = \mathcal{A}(G')$.

We next introduce the notion of the vertex-edge incidence matrix $B(G) \in \{0,1\}^{\#V \times \#E}$. Then $B(G) = [b_{ij}]_{i=j=1}^{n,m} \in \{0,1\}^{n \times m}$, such that $b_{ij} = 1$ if and only the edge j contain the vertex i. A different labeling of V and E gives rise to the vertex-edge incidence matrix $B' = PB(A)Q$ for some $P \in \mathcal{P}_n, Q \in \mathcal{P}_m$. Thus the graph G gives rise to the equivalence class of matrices $\mathcal{B}(G) = \{PB(G)Q, \ P \in \mathcal{P}_n, Q \in \mathcal{P}_m\}$.

Lemma 7.6.3 *Let* $G = (V, E), G' = (V', E')$ *are two undirected graphs. Assume that* $\#V = \#V', \#E = \#E'$. *Then* G *and* G' *are isomorphic if and only if* $\mathcal{B}(G) = \mathcal{B}(G')$.

We now restate the SGIP in terms of bilinear programming on $\Omega_m \times \Omega_n$. It is enough to consider the following case.

Lemma 7.6.4 *Let* $G' = (V', E'), G = (V, E)$ *and assume* $n' := \#V' \leq n := \#V, m' := \#E' \leq m := \#E$. *Let* $B(G') \in \{0,1\}^{n' \times m'}, B(G) \in \{0,1\}^{n \times m}$ *be the vertex-edges incidence matrices of* G' *and* G. *Denote by* $C(G') \in \{0,1\}^{n \times m}$ *the matrix obtained* *from $B(G')$ by adding additional $n - n'$ and $m - m'$ zero rows and columns respectively. Then*

$$\max_{P \in \mathcal{P}_n, Q \in \mathcal{P}_m} \mathrm{tr}(C(G')QB(A)^\top P) \le 2m'. \tag{7.6.1}$$

Equality holds if and only if G' is isomorphic to a subgraph of G.

Proof. Let $B_1 = [b_{ij,1}]_{i=j=1}^{n,m} := P^\top B(A)Q^\top \in \mathcal{B}(G)$. Note that B_1 has exactly the same number of ones as $B(G)$, namely $2m$, since each edge is connected is connected to two vertices. Similarly $C(G') = [c_{ij}]_{i=j=1}^{n,m}$ has exactly the same number of ones as $B(G')$, namely $2m'$. Hence, the $\langle C(G'), B_1 \rangle = \mathrm{tr}(C(G_1)B_1^\top) \le 2m'$. Assume that $\mathrm{tr}(C(G_1)B_1^\top) = 2m'$. So we can delete $2(m - m')$ ones in B_1 to obtain $C(G')$. Note that deleting $2(m-m')$ from B_1, means to delete $m - m'$ edges from the graph G. Indeed, assume $c_{ij} = b_{ij,1} = 1$. So the vertex i is connected to the edge j. Hence, there exists another vertex $i' \ne i$ such that $c_{i'j} = 1$. As $\mathrm{tr}(C(G')B_1^\top) = 2m'$ we deduce that $b_{i'j,1} = 1$. Hence, if we rename the vertices and the edges of G corresponding to B_1 we deduce that G', represented by the matrix $B(G')$, is a subgraph of G. $\qquad\square$

We now show how to translate the maximum in (7.6.1) to linear programming problem on $\Omega_{n,m}$. As in §2.8 for $F \in \mathbb{R}^{m \times n}$ let $\hat{F} \in \mathbb{R}^{mn}$ be a vector composed of the columns of F, i.e. first we have the coordinates of the first column, then the coordinates of the second column, and the last n coordinates are the coordinates of the last column. Hence $\widehat{XFY} = (Y^\top \otimes X)\hat{F}$, where $Y^\top \otimes X$ is the Kronecker tensor product.

Lemma 7.6.5 *Let $C, B \in \mathbb{R}^{n \times m}$. Then*

$$\max_{P \in \mathcal{P}_n, Q \in \mathcal{P}_m} \mathrm{tr}(CQB^\top P) = \max_{Z \in \Omega_{m,n}} (\hat{C})^\top Z\hat{B}. \tag{7.6.2}$$

Proof. Since $\Omega^\top = \Omega$ and $\mathrm{ext}\,(\Omega_n) = \mathcal{P}_n$ we deduce

$$\max_{P \in \mathcal{P}_n, Q \in \mathcal{P}_m} \mathrm{tr}(CQB^\top P) = \max_{X \in \Omega_n, Y \in \Omega_m} \mathrm{tr}(CYB^\top X).$$

Observe next that

$$\text{tr}(CYB^\top X) = \text{tr}(C^\top(X^\top BY^\top)) = (\hat{C})^\top(Y \otimes X^\top)\hat{B}.$$

As $\Omega_{m,n} = \text{conv}\,\Omega_m \odot \Omega_n$ we deduce (7.6.2). $\qquad\square$

In summary we showed that if we can solve exactly the linear programming problem (7.6.2), using Lemma 7.6.4 we can determine if G' is isomorphic to a subgraph of G. Since the SGIP is NP-complete, we believe that this implies that for general m, n the number of half spaces characterizing $\Omega_{n,m}$ cannot be polynomial.

Problems

1. Prove Lemma 7.6.2.

2. Prove Lemma 7.6.3.

7.7 Variation of Tensor Powers and Spectra

Definition 7.7.1 *Let* $\mathbf{V}_1, \mathbf{V}_2$ *be finite dimensional vector spaces over* $\mathbb{F} = \mathbb{R}, \mathbb{C}$ *with norms* $\|\cdot\|_1, \|\cdot\|_2$ *respectively. Let* $\mu : \mathbf{V}_1 \to \mathbf{V}_2$ *be a nonlinear map. The map* μ *has a Fréchet derivative at* $\mathbf{x} \in \mathbf{V}_1$, *or simply differentiable at* $\mathbf{x}$, *if there exists a linear transformation* $T_{\mathbf{x}} \in \text{Hom}\,(\mathbf{V}_1, \mathbf{V}_2)$ *such that*

$$\mu(\mathbf{x} + \mathbf{u}) = \mu(\mathbf{x}) + T_{\mathbf{x}}\mathbf{u} + o(\mathbf{u})\|\mathbf{u}\|_1,$$

where $\|o(\mathbf{u})\|_2 \to 0$ *uniformly as* $\|\mathbf{u}\|_1 \to 0$. *Denote* $\mathrm{D}\mu(\mathbf{x}) := T_{\mathbf{x}}$. μ *is differentiable, if it has the Fréchet derivative at each* $\mathbf{x} \in \mathbf{V}_1$, *and* $\mathrm{D}\mu(\mathbf{x})$ *is continuous on* $\mathbf{V}_1$. *(Note that by choosing fixed bases in* $\mathbf{V}_1, \mathbf{V}_2$ *each* $\mathrm{D}\mu(\mathbf{x})$ *is represented by a matrix* $A(\mathbf{x}) = [a_{ij}(\mathbf{x})] \in \mathbb{F}^{m \times n}$, *where* $n = \dim \mathbf{V}_1, m = \dim \mathbf{V}_2$. *Then* $a_{ij}(\mathbf{x})$ *is continuous on* $\mathbf{V}_1$ *for each* $i \in \langle m \rangle, j \in [n]$.*)*

Since all norms on a finite dimensional vector space are equivalent, it is straightforward to show that the notion of Fréchet derivative depend only on the standard topologies in $\mathbf{V}_1, \mathbf{V}_2$. See Problem 1. For properties of the Fréchet derivative consult with [Die69].

Proposition 7.7.2 *Let $\mathbf{V}_1, \mathbf{V}_2$ be finite dimensional vector spaces over $\mathbb{F} = \mathbb{R}, \mathbb{C}$ with norms $\|\cdot\|_1, \|\cdot\|_2$ respectively. Assume that $\mu : \mathbf{V}_1 \to \mathbf{V}_2$ is differentiable. Then for any $\mathbf{x}, \mathbf{y} \in \mathbf{V}_1$ the following equality and inequalities holds.*

$$\mu(\mathbf{y}) - \mu(\mathbf{x}) = \int_0^1 \mathrm{D}\mu((1-t)\mathbf{x} + t\mathbf{y})(\mathbf{y} - \mathbf{x})dt, \tag{7.7.1}$$

$$\|\mu(\mathbf{y}) - \mu(\mathbf{x})\|_2 \le \|\mathbf{y} - \mathbf{x}\|_1 \int_0^1 \|\mathrm{D}\mu((1-t)\mathbf{x} + t\mathbf{y})\|_{1,2}dt \tag{7.7.2}$$

$$\le \|\mathbf{y} - \mathbf{x}\|_1 \max_{t \in [0,1]} \|\mathrm{D}\mu((1-t)\mathbf{x} + t\mathbf{y})\|_{1,2}.$$

(7.7.1) and (7.7.2) are called here the *mean value theorem* and the *mean value inequalities* respectively.

Proof. Let $\mathbf{x}, \mathbf{u} \in \mathbf{V}_1$ be fixed. Clearly, the function $\mu(\mathbf{x} + t\mathbf{u})$ is a differentiable function from $\mathbb{R}$ to $\mathbf{V}_2$, where

$$\frac{d\mu(\mathbf{x} + t\mathbf{u})}{dt} = \mathrm{D}\mu(\mathbf{x} + t\mathbf{u})\mathbf{u}. \tag{7.7.3}$$

Letting $\mathbf{u} = \mathbf{y} - \mathbf{x}$ and integrating the above inequality for $t \in [0, 1]$ we get (7.7.1). Replacing the integration in (7.7.1) by the limiting summation and using the triangle inequality we obtain

$$\|\mu(\mathbf{y}) - \mu(\mathbf{x})\|_2 \le \int_0^1 \|\mathrm{D}\mu((1-t)\mathbf{x} + t\mathbf{y})(\mathbf{y} - \mathbf{x})\|_1 dt$$

$$\le \int_0^1 \|\mathrm{D}\mu((1-t)\mathbf{x} + t\mathbf{y})\|_{1,2}\|(\mathbf{y} - \mathbf{x})\|_1 dt$$

$$\le \|\mathbf{y} - \mathbf{x}\|_1 \max_{t \in [0,1]} \|\mathrm{D}\mu((1-t)\mathbf{x} + t\mathbf{y})\|_{1,2}. \qquad \square$$

Theorem 7.7.3 *Let $\mathbf{V}$ be a finite dimensional vector space. Let $k \in \mathbb{N}$. Denote $\mathbf{V}^{\otimes k} := \underbrace{\mathbf{V} \otimes \cdots \otimes \mathbf{V}}_{k}$. Consider the map $\delta_k : \mathbf{V} \to$*

$\mathbf{V}^{\otimes k}$, *where* $\delta_k(\mathbf{x}) = \underbrace{\mathbf{x} \otimes \cdots \otimes \mathbf{x}}_{k}$. *Then*

$$D\delta_k(\mathbf{x})(\mathbf{u}) = \mathbf{u} \otimes \underbrace{\mathbf{x} \otimes \cdots \otimes \mathbf{x}}_{k-1}$$

$$+ \mathbf{x} \otimes \mathbf{u} \otimes \underbrace{\mathbf{x} \otimes \cdots \otimes \mathbf{x}}_{k-2} + \cdots + \underbrace{\mathbf{x} \otimes \cdots \otimes \mathbf{x}}_{k-1} \otimes \mathbf{u}. \quad (7.7.4)$$

Let $\| \cdot \|$ *be a norm on* $\mathbf{V}$ *and assume that* $\| \cdot \|_k$ *is a cross norm on* $\mathbf{V}^{\otimes k}$:

$$\|\mathbf{x}_1 \otimes \mathbf{x}_2 \otimes \cdots \otimes \mathbf{x}_k\|_k = \prod_{i=1}^{k} \|\mathbf{x}_i\| \ \textit{for } \mathbf{x}_1, \ldots, \mathbf{x}_k \in \mathbf{V}. \quad (7.7.5)$$

Denote by $N_k(T) := \|T\|_{\|\cdot\|, \|\cdot\|_k}$ *the operator norm of* $T \in$ Hom $(\mathbf{V}, \mathbf{V}^{\otimes k})$. *Then*

$$N_k(D\delta_k(\mathbf{x})) = k\|\mathbf{x}\|^{k-1}. \quad (7.7.6)$$

Proof. Fix $\mathbf{x}, \mathbf{u} \in \mathbf{V}$. For $t \in \mathbb{R}$ expand the vector $\delta_k(\mathbf{x} + t\mathbf{u})$ in powers of t. Then

$$\delta_k(\mathbf{x} + t\mathbf{u}) = \delta_k(\mathbf{x}) + t(\mathbf{u} \otimes \underbrace{\mathbf{x} \otimes \cdots \otimes \mathbf{x}}_{k-1} + \mathbf{x} \otimes \mathbf{u} \otimes \underbrace{\mathbf{x} \otimes \cdots \otimes \mathbf{x}}_{k-2}$$

$$+ \cdots + \underbrace{\mathbf{x} \otimes \cdots \otimes \mathbf{x}}_{k-1} \otimes \mathbf{u}) + \text{ higher order terms in } t.$$

Hence, (7.7.4) holds. Apply the triangle inequality to (7.7.4) and use the assumption that $\| \cdot \|_k$ is a cross norms to deduce the inequality $\|D\delta_k(\mathbf{x})(\mathbf{u})\|_k \le k\|\mathbf{x}\|^{k-1}\|\mathbf{u}\|$. Hence, $N_k(D\delta_k(\mathbf{x})) \le k\|\mathbf{x}\|^{k-1}$. Clearly, equality holds if $\mathbf{x} = \mathbf{0}$. Suppose that $\mathbf{x} \ne \mathbf{0}$. Then $\|D\delta_k(\mathbf{x})(\mathbf{x})\|_k = k\|\mathbf{x}\|^{k}$. Hence $N_k(D\delta_k(\mathbf{x})) \ge k\|\mathbf{x}\|^{k-1}$, which establishes (7.7.6). $\qquad \square$

Theorem 7.7.4 *Let* $\mathbf{U}$ *be a finite dimensional vector space over* $\mathbb{F} = \mathbb{R}, \mathbb{C}$. *For an integer* $k > 1$ *consider the map* $\delta_k :$ Hom $(\mathbf{U}, \mathbf{U}) \to$ Hom $(\mathbf{U}, \mathbf{U})^{\otimes k} \sim$ Hom $(\mathbf{U}^{\otimes k}, \mathbf{U}^{\otimes k})$ *given by* $\delta_k(T) = \underbrace{T \otimes \cdots \otimes T}_{k}$. *Let* $\mathbf{W}_k \subset \mathbf{U}^{\otimes k}$ *be a subspace which is invariant* *for each $\delta_k(T), T \in \text{Hom}(\mathbf{U}, \mathbf{U})$. Denote by $\hat{\delta}_k : \text{Hom}(\mathbf{U}, \mathbf{U}) \to$ Hom $(\mathbf{W}_k, \mathbf{W}_k)$ the restriction map $\delta_k(T)|\mathbf{W}_k$. Assume that $\| \cdot \|$ is a norm on $\mathbf{U}$. Let $\| \cdot \|_k$ be the maximal cross norm $\mathbf{U}^{\otimes k}$. Let $\| \cdot \|, \| \cdot \|_k, \||\cdot\||_k$ be the induced operator norms on $\text{Hom}(\mathbf{U}, \mathbf{U})$, Hom $(\mathbf{U}, \mathbf{U})^{\otimes k}$, Hom $(\mathbf{W}_k, \mathbf{W}_k)$ respectively. Let $N_k(\cdot), \hat{N}_k(\cdot)$ be the operator norm on $\text{Hom}(\text{Hom}(\mathbf{U}, \mathbf{U}), \text{Hom}(\mathbf{U}, \mathbf{U})^{\otimes k})$, Hom (Hom $(\mathbf{U}, \mathbf{U}), \text{Hom}(\mathbf{W}_k, \mathbf{W}_k))$ respectively. Then*

$$N_k(\mathrm{D}\delta_k(T)) = k\|T\|^{k-1}, \quad \hat{N}_k(\mathrm{D}\hat{\delta}_k)(T)) \le k\|T\|^{k-1} \qquad (7.7.7)$$

for any $T \in \text{Hom}(\mathbf{V}, \mathbf{V})$.

Proof. Theorem 7.5.14 yields that the operator norm $\| \cdot \|_k$ on $\text{Hom}(\mathbf{U}^{\otimes k}, \mathbf{U}^{\otimes k})$, identified with $\text{Hom}(\mathbf{U}, \mathbf{U})^{\otimes k}$, is a cross norm with respect to the operator norm on $\text{Hom}(\mathbf{U}, \mathbf{U})$. Theorem 7.7.3 yields the equality in (7.7.7). Observe next that $\mathrm{D}\hat{\delta}_k(T)$ is $\mathrm{D}\delta_k(T)|\mathbf{W}_k$. Hence, $\hat{N}_k(\mathrm{D}\hat{\delta}_k(T)) \le N(\mathrm{D}\delta_k(T))$, which implies the inequality in (7.7.7). $\qquad \square$

A simple example of $\mathbf{W}_k$ is the subspace $\bigwedge^k \mathbf{U}$. See Problem 3.

Theorem 7.7.5 *Let $\mathbf{U}$ be an n-dimensional vector space over $\mathbb{F} = \mathbb{R}, \mathbb{C}$, with a norm $\| \cdot \|$. Denote by $\| \cdot \|$ the induced operator norm on $\text{Hom}(\mathbf{V}, \mathbf{V})$. Then for $A, B \in \text{Hom}(\mathbf{U}, \mathbf{U})$*

$$|\det A - \det B| \le \|A - B\|\frac{\|A\|^n - \|B\|^n}{\|A\| - \|B\|}$$

$$\le n\|A - B\|[\max(\|A\|, \|B\|)]^{n-1}. \qquad (7.7.8)$$

Here $\frac{a^n - a^n}{a - a} := na^{n-1}$ for any $a \in \mathbb{C}$. The first inequality is sharp for $A = aI_n, B = bI_n$ for $a, b \ge 0$. The constant n in the second inequality is sharp.

Proof. In $\mathbf{U}^{\otimes n}$ consider the one-dimensional invariant subspace $\mathbf{W}_n := \bigwedge^n \mathbf{U}$ for each $\delta_n(T), T \in \text{Hom}(\mathbf{U}, \mathbf{U})$. See Problem 3. Let $\mathbf{e}_1, \ldots, \mathbf{e}_n$ be a basis in $\mathbf{U}$. Then $\mathbf{e}_1 \wedge \mathbf{e}_2 \wedge \cdots \wedge \mathbf{e}_n$ is a basis vector in $\bigwedge^n \mathbf{U}$. Furthermore

$$\delta_n(T)(\mathbf{e}_1 \wedge \mathbf{e}_2 \wedge \cdots \wedge \mathbf{e}_n) = (\det T)\mathbf{e}_1 \wedge \mathbf{e}_2 \wedge \cdots \wedge \mathbf{e}_n.$$

See Proposition 5.2.7. Note that $\hat{\delta}_n(T) := \delta_n(T)|\bigwedge^n \mathbf{U}$ is the above operator. Observe next that any $Q \in \mathrm{Hom}\,(\bigwedge^n \mathbf{U}, \bigwedge^n \mathbf{U})$, is of the from

$$Q(\mathbf{e}_1 \wedge \mathbf{e}_2 \wedge \cdots \wedge \mathbf{e}_n) = t\mathbf{e}_1 \wedge \mathbf{e}_2 \wedge \cdots \wedge \mathbf{e}_n.$$

Hence, the operator norm of Q is $|t|$. We now apply Theorem 7.7.4 to this case. The inequality in (7.7.7) yields

$$\hat{\mathrm{N}}_n(\mathrm{D}\hat{\delta}_n(T)) \leq n\|T\|^{n-1}.$$

Next we apply Proposition 7.7.2, where $\mathbf{V}_1 := \mathrm{Hom}\,(\mathbf{U}, \mathbf{U})$ and $\mathbf{V}_2 = \mathrm{Hom}\,(\bigwedge^n \mathbf{U}, \bigwedge^n \mathbf{U})$ equipped with the operator norms, and $\mu(T) = \hat{\delta}_n(T)$. So $\|\mu(A) - \mu(B)\|_2 = |\det A - \det B|$. The inequality (7.7.2) combined with the inequality in (7.7.7) yield

$$|\det A - \det B|$$

$$\leq n\|A - B\| \int_0^1 \|(1-t)A + tB\|^{n-1}dt$$

$$\leq n\|A - B\| \int_0^1 ((1-t)\|A\| + t\|B\|)^{n-1}dt$$

$$= \|A - B\|\frac{\|A\|^n - \|B\|^n}{\|A\| - \|B\|} = \|A - B\|\sum_{i=0}^{n-1}\|A\|^{n-1-i}\|B\|^i$$

$$= n\|A - B\|[\max(\|A\|, \|B\|)]^{n-1}.$$

This shows (7.7.8). Recall that $\|xI\| = |x|$ for any $x \in \mathbb{F}$. Hence, for $A = aI, B = bI$ and $a, b \geq 0$ equality holds in the first inequality of (7.7.8). To show that the constant n cannot be improved let $A = (1+x)I, B = I$, where $x > 0$. Then (7.7.8) is equivalent to the inequality $(1+x)^n - 1 \leq nx(1+x)^{n-1}$. Since $\lim_{x \searrow 0} \frac{(1+x)^n - 1}{x(1+x)^{n-1}} = n$ the constant n cannot be improved. $\square$

Definition 7.7.6 *Let* Σ_n *be the group of permutations* $\sigma : [n] \to$ $[n]$. *Let* $S = \{\lambda_1, \ldots, \lambda_n\}, T = \{\mu_1, \ldots, \mu_n\}$ *be two multisets in* $\mathbb{C}$ *containing* n *elements each. Let*

$$\text{dist}(S, T) = \max_{j \in [n]} \min_{i \in [n]} |\lambda_j - \mu_i|,$$

$$\text{hdist}(S, T) = \max(\text{dist}(S, T), \text{dist}(T, S)),$$

$$\text{pdist}(S, T) = \min_{\sigma \in \Sigma_n} \max_{i \in [n]} |\lambda_i - \mu_{\sigma(i)}|.$$

Note: $\text{dist}(S, T)$ is the distance from S to T, viewed as sets; $\text{hdist}(S, T)$ is the Hausdorff distance between S and T, viewed as sets; $\text{pdist}(S, T)$ is called *permutational* distance between two multisets of cardinality n. Clearly

$$\text{hdist}(S, T) = \text{hdist}(T, S), \quad \text{pdist}(S, T) = \text{pdist}(T, S),$$
$$\text{dist}(S, T) \leq \text{hdist}(S, T) \leq \text{pdist}(S, T). \tag{7.7.9}$$

See Problem 4.

Theorem 7.7.7 *Let* $\mathbf{U}$ *be an* n-*dimensional vector space of* $\mathbb{C}$ *with the norm* $\| \cdot \|$. *Let* $\| \cdot \|$ *be the induced operator norm on* $\text{Hom}(\mathbf{U}, \mathbf{U})$. *For* $A, B \in \text{Hom}(\mathbf{U}, \mathbf{U})$ *let* $S(A), S(B)$ *be the eigenvalue multisets of* A, B *of cardinality* n *respectively. Then*

$$\text{pdist}(S(A), S(B)) \leq 4e^{\frac{1}{2e}} n \|A - B\|^{\frac{1}{n}} [max(\|A\|, \|B\|)]^{\frac{n-1}{n}}. \tag{7.7.10}$$

To prove the theorem we need the following lemma.

Lemma 7.7.8 *Let the assumptions of Theorem 7.7.7 holds. Define*

$$h(A, B) := \max_{t \in [0,1]} \text{dist}(S((1 - t)A + tB), S(B)). \tag{7.7.11}$$

Then

$$\text{pdist}(S(A), S(B)) \leq (2n - 1)h(A, B). \tag{7.7.12}$$

Proof. Let $S(A) = \{\lambda_1(A), \ldots, \lambda_n(A)\}, S(B) = \{\lambda_1(B), \ldots, \lambda_n(B)\}$. Let $D(z, r) := \{w \in \mathbb{C}, |w - z| \leq r\}$. Denote $K_B = \cup_{i=1}^n D(\lambda_i(B), h(A, B))$. Then K_B is a closed compact set, which decomposes as union of a $k \in [n]$ connected components. Let

$A(t) = (1-t)A + tB$. Since $\text{dist}(S(A(t)), S(B)) \le h(A, B)$ we deduce that K_B contains $S(A(t))$ for each $t \in [0, 1]$. As $S(A(t))$ various continuously for $t \in [0, 1]$, each connected component of K_B contains a fixed number of the eigenvalues of $S(A(t))$ counting with their multiplicities. Since $A(1) = B$, each connected component of K_B contains a fixed number of the eigenvalues of A and B counting with their multiplicities. Rename the eigenvalues of B such that indices of the eigenvalues of A and B are the same in each component of K_B.

Let $C = \cup_{i=1}^{p} D(z_i, h(A, B))$ be such a connected component, where $z_1, \ldots, z_p$ are p distinct eigenvalues of B. C contains exactly $q \ge p$ eigenvalues of A and B respectively. We claim that if $\lambda \in S(A) \cap C$ then $\max_{j \in [p]} |\lambda - z_j| \le (2p - 1)h(A, B)$. Consider a simple graph $G = (V, E)$, where $V = [p]$ and $(i, j) \in E$ if and only if $|z_i - z_j| \le 2h(A, B)$. Since C is connected it follows that G is connected hence the maximal distance between two distinct points in G is $p-1$. So $|z_i - z_j| \le 2(p-1)h(A, B)$. Since $|\lambda - z_i| \le h(A, B)$ for some $i \in [p]$, it follows that $|\lambda - z_j| \le (2p - 1)h(A, B) \le (2n - 1)h(A, B)$. Therefore, for this particular renaming of the eigenvalues of B we have the inequality $|\lambda_i(A) - \lambda_i(B)| \le (2n - 1)h(A, B), i = 1, \ldots, n$. $\square$

Problem 5 shows that the inequality (7.7.12) is sharp.

Proof of Theorem 7.7.7. First observe that

$$\text{dist}(S(A), S(B))^n \le \max_{i \in [n]} | \prod_{j=1}^{n} (\lambda_i(A) - \lambda_j(B))|$$

$$= \max_{i \in [n]} |\det\,(\lambda_i(A)I - B) - \det\,(\lambda_i(A)I - A)|$$

$$\le \max_{z \in \mathbb{C}, |z| \le \rho(A)} |\det\,(zI - B) - \det\,(zI - A)|.$$

We now apply (7.7.8) to deduce that for $|z| \le \rho(A) \le \|A\|$ we have

$$|\det\,(zI - B) - \det\,(zI - A)|$$

$$\le n\|A - B\|[\max(\|zI - A\|, \|zI - B\|)]^{n-1}$$

$$\le n\|A - B\|[\max(|z| + \|A\|, (|z| + \|B\|)]^{n-1}$$

$$\le n\|A - B\|[\max(2\|A\|, \|A\| + \|B\|)]^{n-1}.$$

Thus

$$\text{dist}(S(A), S(B)) \le n^{\frac{1}{n}} \|A - B\|^{\frac{1}{n}} [\max(2\|A\|, \|A\| + \|B\|)]^{\frac{n-1}{n}}.$$

$$(7.7.13)$$

We apply the above inequality to $A(t)$ for each $t \in [0, 1]$. Clearly, for $t \in [0, 1]$

$$\|A(t)\| \le (1 - t)\|A\| + t\|B\| \le \max(\|A\|, \|B\|)$$
$$\Rightarrow \max(2\|A(t)\|, \|A(t)\| + \|B\|) \le 2\max(\|A\|, \|B\|).$$

Also $\|A(t) - B\| = (1 - t)\|A - B\| \le \|A - B\|$. Hence, we deduce

$$h(A, B) \le n^{\frac{1}{n}} \|A - B\|^{\frac{1}{n}} [2\max(|A|, \|B\|)]^{\frac{n-1}{n}}. \qquad (7.7.14)$$

Use (7.7.12) to obtain

$$\text{pdist}(S(A), S(B)) \le (2n - 1)n^{\frac{1}{n}} \|A - B\|^{\frac{1}{n}} [2\max(|A|, \|B\|)]^{\frac{n-1}{n}}.$$

$$(7.7.15)$$

Use the inequality

$$(2n - 1)2 \left(\frac{n}{2}\right)^{\frac{1}{n}} \le 4n \left(\frac{n}{2}\right)^{\frac{1}{n}} \le 4ne^{\frac{1}{2e}} \text{ for } n \in \mathbb{N}, \qquad (7.7.16)$$

to deduce (7.7.10). (See Problem 6.) $\square$

The inequality (7.7.10) can be improved by a factor of 2 using the following theorem [EJRS83]. (See Problem 7.)

Theorem 7.7.9 *Let* **U** *be an n-dimensional vector space of* $\mathbb{C}$. *For* $A, B \in \text{Hom}(\mathbf{U}, \mathbf{U})$ *let* $S(A), S(B)$ *be the eigenvalue multisets of* A, B *of cardinality n respectively. Then*

$$\text{pdist}(S(A), S(B)) \le \left(2 \left\lfloor \frac{n+1}{2} \right\rfloor - 1\right) \max(h(A, B), h(B, A)).$$

$$(7.7.17)$$

The above constant is sharp.

Problems

1. Let $\mu : \mathbf{V}_1 \to \mathbf{V}_2$ be a nonlinear map. Show

   (a) Assume that μ has a Fréchet derivative at $\mathbf{x}$ with respect to given two norms $\| \cdot \|_1, \| \cdot \|_2$. Then μ has a Fréchet derivative at $\mathbf{x}$ with respect to any two norms $\| \cdot \|_a, \| \cdot \|_b$.

   (b) Suppose that μ has a Fréchet derivative at $\mathbf{x}$. Then μ is continuous at $\mathbf{x}$ with respect to the standard topologies on $\mathbf{V}_1, \mathbf{V}_2$.

   (c) Assume that μ has a Fréchet derivative at each point of a compact set $O \subset \mathbf{V}_1$. Then $\mu : O \to \mathbf{V}_2$ is uniformly continuous.

2. Let $\mathbf{U}_1, \ldots, \mathbf{U}_k, \mathbf{V}_1, \ldots, \mathbf{V}_k$ be finite dimensional inner product vector space overs $\mathbb{F} = \mathbb{R}, \mathbb{C}$. Assume that $\mathbf{U} := \otimes_{i=1}^k \mathbf{U}_i, \mathbf{V} := \otimes_{i=1}^k \mathbf{V}_i$ have the induced inner product. Identify $\mathrm{Hom}\,(\mathbf{U}, \mathbf{V})$ with

   $\prod_{i=1}^k \mathrm{Hom}\,(\mathbf{U}_i, \mathbf{V}_i)$. Show

   (a) The operator norm on $\mathrm{Hom}\,(\mathbf{U}, \mathbf{V})$, with respect to Hilbert norms, is a normal cross norm with respect to the operator norms on $\mathrm{Hom}\,(\mathbf{U}_i, \mathbf{V}_i)$, the Hilbert norms, for $i = 1, \ldots, k$.

   Hint: Express the operator norm on $\mathrm{Hom}\,(\mathbf{U}_i, \mathbf{V}_i)$ and its conjugate norm in terms of singular values of $T_i \in \mathrm{Hom}\,(\mathbf{U}_i, \mathbf{V}_i)$ for $i = 1, \ldots, k$.

   (b) Assume that $\mathbf{U}_1 = \mathbf{V}_1 = \cdots = \mathbf{U}_k = \mathbf{V}_k$. Let $\delta_k : \mathrm{Hom}\,(\mathbf{U}_1, \mathbf{U}_1) \to \mathrm{Hom}\,(\mathbf{U}, \mathbf{U})$. Then $\mathrm{N}_k(\delta_k(T)) = k\|T\|^{k-1}$, where $\| \cdot \|$ is the operator norm on $\mathrm{Hom}\,(\mathbf{U}_1, \mathbf{U}_1)$.

3. Let $\mathbf{U}$ be a vector space over $\mathbb{F} = \mathbb{R}, \mathbb{C}$ of dimension $n > 1$. Let $k \in [n] \backslash \{1\}$. Show

   (a) $\mathbf{W}_k := \bigwedge^k \mathbf{U}$ is an invariant subspace for each $\delta_k(T)$ given in Theorem 7.7.4.

(b) Assume that $\mathbf{U}$ is an inner product space. Let $T \in$ Hom $(\mathbf{U}, \mathbf{U})$, and denote by $\|T\| = \sigma_1(T) \geq \cdots \geq \sigma_k(T) \geq \cdots$ the singular values of T. Then

$$\hat{N}_k(D\hat{\delta}_k(T)) = \sum_{i=1}^{k} \sigma_1(T) \ldots \sigma_{i-1}(T) \sigma_{i+1}(T) \ldots \sigma_k(T).$$

In particular, $\hat{N}_k(D\hat{\delta}_k(T)) \leq k\sigma_1(T)^{k-1} = k\|T\|^{k-1} = N_k(D\delta_k(T))$. Equality holds if and only if $\sigma_1(T) = \cdots = \sigma_k(T)$. *Hint:* Consult with [BhF81].

4. Prove (7.7.9).

5. Let $A = \operatorname{diag}(0, 2, 4, \ldots, 2n - 2), B = (2n - 1)I_n \in \mathbb{R}^{n \times n}$. Show that in this case equality holds in (7.7.12).

6. Using the fact that $\min_{t \in [0,1]} -t \log t = \frac{1}{e}$ deduce the last part of (7.7.16).

7. Show

(a) Let $n = 2k + 1$ and

$$A = \operatorname{diag}(\underbrace{0, \ldots, 0}_{k+1}, 2, 4, \ldots, 2k),$$

$$B = \operatorname{diag}(1, 3, \ldots, 2k - 1, \underbrace{2k + 1, \ldots, 2k + 1}_{k+1}).$$

Then equality holds in (7.7.17).

(b) Let $n = 2k$ and

$$A = \operatorname{diag}(\underbrace{0, \ldots, 0}_{k}, 2, 4, \ldots, 2k),$$

$$B = \operatorname{diag}(1, 3, \ldots, 2k - 1, \underbrace{2k + 1, \ldots, 2k + 1}_{k}).$$

Then equality holds in (7.7.17).

(c) $\max(h(A, B), h(B, A))$ is bounded above by the right-hand side of (7.7.14).

(d) Deduce from (7.7.17) and the previous part of the problem the improved version of (7.7.10).

$$\text{pdist}(S(A), S(B)) \leq 2e^{\frac{1}{2e}} n \|A - B\|^{\frac{1}{n}} [max(\|A\|, \|B\|)]^{\frac{n-1}{n}}.$$

$$(7.7.18)$$

7.8   Variation of Permanents

Definition 7.8.1 *For* $A = [a_{ij}] \in \mathbb{D}^{n \times n}$ *the permanent of* A, *denoted as* perm A

$$\text{perm } A = \sum_{\sigma \in \Sigma_n} \prod_{i=1}^{n} a_{i\sigma(i)}.$$

The determinant and the permanent share some common properties as multilinear functions on $\mathbb{D}^{n \times n}$, as Laplace expansions. However, from the computational point of view the determinants are easy to compute while permanents are hard to compute over all fields, except the fields of characteristic 2. (Over the field of characteristic 2 perm $A = \det A$.) For $A \in \mathbb{Z}_+^{n \times n}$ the permanent of A has a fundamental importance in combinatorics, and usually is hard to evaluate [Val79]. The main aim of this section is to generalize the inequality (7.7.8) to the permanents of matrices. The analog of (7.7.8) holds for the norms $\ell_p, p \in [1, \infty]$ [BhE90]. However, it was also shown in [BhE90] that the analog of fails for some operator norm on $\mathbb{C}^{n \times n}$.

Theorem 7.8.2 *Let* $\| \cdot \|$ *be a norm on* $\mathbb{C}^n$, *and denote by* $\| \cdot \|$ *the induced operator norm on* $\mathbb{C}^n$. *Let* $A, B \in \mathbb{C}^{n \times n}$. *Then*

$$|\text{perm } A - \text{perm } B| \leq \frac{\|A - B\|(\|A\|^n - \|B\|^n)}{2(\|A\| - \|B\|)}$$

$$+ \frac{\|A^* - B^*\|(\|A^*\|^n - \|B^*\|^n)}{2(\|A^*\| - \|B^*\|)}. \qquad (7.8.1)$$

To prove this theorem we need two lemmas. The first lemma gives the following formula for the standard numerical radius of a square matrix with complex entries.

Lemma 7.8.3 *Let $A \in \mathbb{C}^{n \times n}$. Then*

$$r_2(A) = \max_{|z|=1} \rho \left(\frac{zA + \bar{z}A}{2} \right). \tag{7.8.2}$$

In particular, $r_2(A) \leq \frac{1}{2}(\|A\| + \|A^*\|)$ for any operator norm on $\mathbb{C}^n$.

Proof. Let $z \in S^1, B = zA$. Assume that $\mathbf{x}$ is an eigenvector of $\frac{1}{2}(B + B^*)$ of length one corresponding to the eigenvalue λ. Then

$$|\lambda| = |\Re(\mathbf{x}^*(zA)\mathbf{x})| \leq |\mathbf{x}^*(zA)\mathbf{x}| = |\mathbf{x}^*A\mathbf{x}| \leq r_2(A).$$

Hence, the right-hand side of (7.8.2) is not bigger than its left-hand side. On the other hand, there exists $\mathbf{x} \in \mathbb{C}^n, \mathbf{x}^*\mathbf{x} = 1$ and $z \in \mathbb{C}, |z| = 1$ such that $r_2(A) = |\mathbf{x}^*A\mathbf{x}| = \mathbf{x}^*(zA)\mathbf{x}$. For this value of z we have that

$$r_2(A) \leq \lambda_1 \left(\frac{zA + \bar{z}A}{2} \right) \leq \rho \left(\frac{zA + \bar{z}A}{2} \right).$$

Clearly,

$$\rho \left(\frac{zA + \bar{z}A}{2} \right) \leq \left\| \frac{zA + \bar{z}A}{2} \right\| \leq \frac{\|A\| + \|A\|}{2}.$$

Hence, $r_2(A) \leq \frac{1}{2}(\|A\| + \|A^*\|)$. $\qquad\square$

For $A \in \mathbb{C}^{n \times n}$, view the matrix $\otimes^n A$ as a linear operator on $\otimes^n \mathbb{C}^n$, which is identified with $\mathbb{C}^{n^n}$. $\omega_2(\otimes^n A), r_2(\otimes^n A)$ are the numerical range and the numerical radius of $\otimes^n A$ corresponding to the inner product $\langle \cdot, \cdot \rangle$ on $\otimes^n \mathbb{C}^n$ induced by the standard inner product $\mathbf{y}^*\mathbf{x}$ on $\mathbb{C}^n$.

Lemma 7.8.4 *Let $A \in \mathbb{C}^{n \times n}$. Then* perm $A \in \omega_2(\otimes^n A)$.

Proof. Assume that $\mathbf{e}_i = (\delta_{i1}, \ldots, \delta_{in})^\top, i \in [n]$ is a standard basis in $\mathbb{C}^n$. Then $\otimes_{j=1}^n \mathbf{e}_{i_j}$, where $i_j \in [n], j = 1, \ldots, n$, is the standard basis in $\otimes^n \mathbb{C}^n$. A straightforward calculation shows that

$$\langle \otimes^n A\mathbf{x}, \mathbf{x} \rangle = \text{perm } A, \ \mathbf{x} = \frac{1}{\sqrt{n!}} \sum_{\sigma \in \Sigma_n} \otimes_{i=1}^n \mathbf{e}_{\sigma(i)}, \langle \mathbf{x}, \mathbf{x} \rangle = 1. \tag{7.8.3}$$

(See Problem 1.) Hence, perm $A \in \omega_2(\otimes^n A)$. $\qquad\square$

Proof of Theorem 7.8.2. Since $\mathbf{x}$ given in (7.8.3) does not depend on A we deduce perm $A -$ perm $B \in \omega_2(\otimes^n A - \otimes^n B)$. Let $|||\cdot|||$ the maximal cross norm on $\otimes^n \mathbb{C}^n$ induced by the norm $\|\cdot\|$ on $\mathbb{C}^n$. Denote by $|||\cdot|||$ the operator norm on $|||\cdot|||$ on $\otimes^n \mathbb{C}^{n \times n}$, induced by the norm $|||\cdot|||$ on $\otimes^n \mathbb{C}^n$. Use the definition of $r_2(\otimes^n A - \otimes^n B)$ and Lemma 7.8.3 to deduce

$$|\text{perm } A - \text{perm } B| \leq r_2(\otimes^n A - \otimes^n B)$$

$$\leq \frac{1}{2}(||| \otimes^n A - \otimes^n B||| + ||| \otimes^n A^* - \otimes^n B^*|||).$$

$$(7.8.4)$$

Theorem 7.5.14 implies that the operator norm $|||\cdot|||$ on $\otimes^n \mathbb{C}^{n \times n}$, induced by the norm $|||\cdot|||$ on $\otimes^n \mathbb{C}^n$, is a cross norm with respect to the operator norm $\|\cdot\|$ on $\mathbb{C}^{n \times n}$. Observe next

$$\otimes^n A - \otimes^n B = \sum_{i=0}^{n-1} (\otimes^i B) \otimes (A - B) \otimes^{n-1-i} A.$$

(Here, $\otimes^0 A$ means that this term does not appear at all.) Use the triangular inequality and the fact that the operator norm $|||\cdot|||$ is a cross norm we deduce

$$||| \otimes^n A - \otimes^n B||| \leq \sum_{i=0}^{n-1} |||(\otimes^i B) \otimes (A - B) \otimes^{n-1-i} A|||$$

$$\leq \sum_{i=0}^{n-1-i} \|B\|^i \|A - B\| \|A\|^{n-1-i}$$

$$= \frac{\|A - B\|(\|A\|^n - \|B\|^n)}{\|A\| - \|B\|}.$$

Apply the above inequality to (7.8.4) to deduce the theorem. $\qquad \square$

Problems

1. Prove (7.8.3).

2. Let the assumptions of Theorem 7.8.2 hold. Show that

$$|\text{perm } A - \text{perm } B| \leq \|A - B\| \frac{\|A\|^n - \|B\|^n}{\|A\| - \|B\|} \qquad (7.8.5)$$

in the following cases.

(a) $A = A^*, B = B^*$.

(b) The norm $\|\cdot\|$ on $\mathbb{C}^n$ is $\|\cdot\|_2$.

3. Show that (7.8.5) holds for the norms $\|\cdot\|_1, \|\cdot\|_\infty$ using the following steps.

(a) For $A = [a_{ij}] \in \mathbb{C}^n$ denote $|A| := [|a_{ij}|] \in \mathbb{R}_+^{n \times n}$. Show

$$|\text{perm } A| \le \text{perm } |A| \le \prod_{i=1}^n \sum_{j=1}^n |a_{ji}|.$$

(b) Let $A = [\mathbf{a}_1, \ldots, \mathbf{a}_n], B = [\mathbf{b}_1, \ldots, \mathbf{b}_n] \in \mathbb{C}^{n \times n}$, where $\mathbf{a}_i, \mathbf{b}_i$ are the ith columns of A, B respectively, for $i = 1, \ldots, n$. Let

$$C_0 = [\mathbf{a}_1 - \mathbf{b}_1, \mathbf{a}_2, \ldots, \mathbf{a}_n],$$
$$C_{n-1} = [\mathbf{b}_1, \ldots, \mathbf{b}_{n-1}, \mathbf{a}_n - \mathbf{b}_n],$$
$$C_i = [\mathbf{b}_1, \ldots, \mathbf{b}_i, \mathbf{a}_{i+1} - \mathbf{b}_{i+1}, \mathbf{a}_{i+2}, \ldots, \mathbf{a}_n],$$
$$\text{for } i = 1, \ldots, n-2.$$

Then

$$\text{perm } A - \text{perm } B = \sum_{i=1}^n \text{perm } C_{i-1}$$

$$\Rightarrow |\text{perm } A - \text{perm } B| \le \sum_{i=1}^n \text{perm } |C_{i-1}|.$$

(c) Recall that $\|A\|_1 = \| |A| \| = \max_{i \in [n]} \|\mathbf{a}_i\|_1$. Then

$$\|C_{i-1}\| \le \|\mathbf{a}_i - \mathbf{b}_i\|_1 \|B\|_1^{i-1} \|A\|_1^{n-i}$$
$$\le \|A - B\|_1 \|B\|_1^{i-1} \|A\|_1^{n-i},$$

for $i = 1, \ldots, n$. Hence, (7.8.5) holds for $\|\cdot\|_1$ norm.

(d) Use the equalities perm $A^\top = $ perm $A, \|A\|_\infty = |A^\top|_1$ deduce that (7.8.5) holds for $\|\cdot\|_\infty$ norm.

7.9 Vivanti–Pringsheim Theorem and Applications

We start with the following basic result on the power series in one complex variable, which is usually called the Cauchy–Hadamard formula on power series [Rem98, §4.1].

Theorem 7.9.1 *Let*

$$f(z) = \sum_{i=0}^{\infty} a_i z^i, \quad a_i \in \mathbb{C}, i = 0, 1, \dots, \text{ and } z \in \mathbb{C}, \tag{7.9.1}$$

be power series. Define

$$R = R(f) := \frac{1}{\limsup_i |a_i|^{\frac{1}{i}}} \in [0, \infty]. \tag{7.9.2}$$

(R is called the radius of convergence of the series.) Then

1. *For $R = 0$ the series converge only for $z = 0$.*

2. *For $R = \infty$ the series converge absolutely and uniformly for each $z \in \mathbb{C}$, and $f(z)$ is an entire function, i.e. analytic on $\mathbb{C}$.*

3. *For $R \in (0, \infty)$ the series converge absolutely and uniformly to an analytic function for each $z, |z| < R$, and diverge for each $|z| > R$. Furthermore, there exist $\zeta, |\zeta| = R$, such that $f(z)$ cannot be extended to an analytic function in any neighborhood of ζ. (ζ is called a singular point of f.)*

Consider the Taylor series for the function complex valued $\frac{1}{1-z}$

$$\frac{1}{1-z} = \sum_{i=0}^{\infty} z^i.$$

Then $R = 1$, the function $\frac{1}{1-z}$ is analytic in $\mathbb{C} \backslash \{1\}$, and has a singular point at $z = 1$. Vivanti–Pringsheim theorem is an extension of this example [Viv93, Pri94].

Theorem 7.9.2 *Let the power series $f(z) = \sum_{i=0}^{\infty} a_i z^i$ have positive finite radius of convergence R, and suppose that the sequence $a_i, i = 0, 1, \ldots$, is eventually non-negative. (That is, all but finitely many of its coefficients are real and non-negative.) Then $\zeta := R$ is a singular point of f.*

See [Rem98, §8.1] for a proof. In what follows we need a stronger version of this theorem for rational functions, e.g. [Fri78b, Theorem 2]. Assume that $f(z)$ is a rational function with 0 as a point of analyticity. So f has power series (7.9.1). Assume that f is not polynomial, i.e. $R(f) \in (0, \infty)$. Then f has the following form.

$$f(z) = P(z) + \sum_{i=1}^{N} \sum_{j=1}^{p_i} \frac{b_{j,i}}{(1 - \lambda_i z)^j}, \tag{7.9.3}$$

$$P \in \mathbb{C}[z], \lambda_i, b_{p_i,i} \in \mathbb{C} \setminus \{0\}, \lambda_i \neq \lambda_{i'} \text{ for } i \neq i'.$$

Note that

$$R(f) = \frac{1}{\max_i |\lambda_i|}. \tag{7.9.4}$$

Definition 7.9.3 *Let $f(z)$ be a rational function of the form (7.9.3). Let $p := \max_{|\lambda_i| = R(f)^{-1}} p_i$. Denote*

$$f_{\text{prin}} = \sum_{i: |\lambda_i| = R(f)^{-1} \text{ and } p_i = p} \frac{b_{p,i}}{(1 - \lambda_i z)^p}.$$

f_{prin} *is called the principle part of f, i.e. $f - f_{\text{prin}}$ does not have poles of order p on $|z| = R(f)$.*

Theorem 7.9.4 *Let $f(z)$ be a rational function of the form (7.9.3). Assume that the sequence of coefficients in the power expansion (7.9.1) is eventually non-negative. Then*

1. *The set $\{\lambda_1, \ldots, \lambda_N\}$ is symmetric with respect to $\mathbb{R}$. That is, for each $i \in [N]$ there exists $i' \in [N]$ such that $\bar{\lambda}_i = \lambda_{i'}$. Furthermore $p_i = p_{i'}$, and $\bar{b}_{j,i} = b_{j,i'}$ for $j = 1, \ldots, p_i$ and $i = 1, \ldots, N$.*

2. *After renaming the indices in $[N]$ we have: $\lambda_1 = \frac{1}{R(f)}$, $|\lambda_i| = \lambda_1$ for $i = 2, \ldots, M$, and $|\lambda_i| > \lambda_1$ for $i > M$. (Here, $M \in [N]$.)*

3. *Let $p := p_1$. There exists $L \in [M]$ such that $p_i = p$ for $i \in [L]\backslash\{1\}$, and $p_i < p$ for $i > L$.*

4. *$b_{p,1} > 0$ and there exists $m \in [L]$ such that $|b_{p,i}| = b_{p,1}$ for $i = 1, \ldots, m$ and $|b_{p,i}| < b_{p,1}$ for $i \in [L]\backslash[m]$.*

5. *Let $\zeta = e^{\frac{2\pi\sqrt{-1}}{m}}$. After renaming the indices $2, \ldots, m$, $\lambda_i = \zeta^{i-1}\lambda_1$ for $i = 2, \ldots, m$. Furthermore, there exists $l \in [m]$ such that $b_{p,i} = \zeta^{l(i-1)}b_{p,1}$ for $i = 2, \ldots, m$.*

6. *$f_{\mathrm{prin}}(\zeta z) = \zeta^{-l} f_{\mathrm{prin}}(z)$.*

Proof. We outline the major steps in the proof of this theorem. For all details see the proof of [Fri78b, Theorem 2]. By considering $g(z) = f(z) + P_1$ for some polynomial P_1, we may assume that the MacLaurin coefficients of g are real and non-negative. As $g_{\mathrm{prin}} = f_{\mathrm{prin}}$, without loss of generality we may assume that the MacLaurin coefficients of f real and non-negative. Hence, $\overline{f(\bar{z})} = f(z)$ for each z where f is defined. This shows part 1.

Part 2 follows from Theorem 7.9.2. For simplicity of the exposition assume that $R(f) = 1$. Recall that for each singular point $\lambda_i^{-1} = \bar{\lambda}_i$ of f on the circle $|z| = 1$ we have the equality

$$b_{p_i,i} = \lim_{r \nearrow 1}(1 - r)^{p_i} f(\bar{\lambda}_i r). \tag{7.9.5}$$

In particular $b_{p_1,1} = \lim_{r \nearrow 1}(1 - r)^{p_1} f(r) \geq 0$. Since $b_{p_1,1} \neq 0$ we obtain that $b_{p_1,1} > 0$. Let $p := p_1$. Since all the MacLaurin coefficients of f are non-negative we have the inequality $|f(z)| \leq f(|z|)$ for all $|z| < 1$. Hence, $\limsup_{r \nearrow 1}(1 - r)^p|f(\bar{\lambda}_i r)| \leq b_{p,1}$. This inequality and (7.9.5) implies parts 3 and 4.

For $m = 1$ parts 5 and 6 are trivial. Assume that $m > 1$. Let $b_{p,i} = \eta_i b_{p,1}, |\eta_i| = 1$ for $i = 2, \ldots, m$. In view of the part 1 for each $i \in [m]\backslash\{1\}$ there exists $i' \in [m]\backslash\{1\}$ such that $\bar{\lambda}_i = \lambda_{i'}, \bar{\eta}_i = \eta_{i'}$. Consider the function

$$g(z) = 2f(z) - \eta_i f(\lambda_i z) - \bar{\eta}_i f(\bar{\lambda}_i z) = \sum_{j=0}^{\infty} 2(1 - \Re(\eta_i \lambda_i^j))a_j z^j.$$

So the MacLaurin coefficients of g are non-negative. Clearly $R(g) \geq 1$, and if g has a pole at 1, its order is at most $p - 1$. This implies the equality

$$2f_{\text{prin}}(z) - \eta_i f_{\text{prin}}(\lambda_i z) - \bar{\eta}_i f_{\text{prin}}(\bar{\lambda}_i z) = 0.$$

Therefore, the set $\{\lambda_1, \ldots, \lambda_m\}$ form a multiplicative group of order m. Hence, it is a group of of all m-roots of unity. So we can rename the indices $2, \ldots, m$ such that $\lambda_i = \zeta^{(i-1)}$ for $i = 1, \ldots, m$. Similarly, $\eta_1 = 1, \eta_2, \ldots, \eta_m$ form a multiplicative group, which must be a subgroup of m roots of 1. Furthermore, $\eta_i \mapsto \lambda_i$ is a group homomorphism. This shows part 5. Part 5 straightforward implies part 6.   □

Definition 7.9.5 *Let* $S = \{\lambda_1, \ldots, \lambda_n\} \subset \mathbb{C}$ *be a finite multiset. That is, a point* $z \in S$ *appears exactly* $m(z) \geq 1$ *times in* S. *Denote*

1. $r(S) := \max_{z \in S} |z|$.

2. *For any* $t \geq 0$ *denote by* $S(t)$ *the multiset* $S \cap \{z \in \mathbb{C}, |z| = t\}$.

3. *For an integer* $k \in \mathbb{N}$ *denote by* $s_k(S) := \sum_{i=1}^{n} \lambda_i^k$ *the* k*th moment of* S. *Let* $s_0(S) = n$.

4. *For* $\mathbf{z} = (z_1, \ldots, z_N)^\top \in \mathbb{C}^N$ *denote by* $\sigma_k(\mathbf{z}) = \sum_{1 \leq i_1 < \cdots < i_k \leq N} z_{i_1} \ldots z_{i_N}$, *for* $k = 1, \ldots, N$, *the elementary symmetric polynomial in* $z_1, \ldots, z_N$.

5. *Denote* $\mathbf{z}(S) = (\lambda_1, \ldots, \lambda_n)^\top \in \mathbb{C}^n$. *Then* $\sigma_k(S) := \sigma_k(\mathbf{z}(S))$, *for* $k = 1, \ldots, n$, *are called the elementary symmetric polynomials of* S.

S is called a Frobenius multiset if the following conditions hold.

1. $\bar{S} = S$.

2. $r(S) \in S$.

3. $m(z) = 1$ *for each* $z \in S(r(S))$.

4. *Assume that* $\#S(r(S)) = m$. *Then* $\zeta S = S$ *for* $\zeta = e^{\frac{2\pi\sqrt{-1}}{m}}$.

A simple example of Frobenius multiset is the set of eigenvalues, counted with their multiplicities, of a square non-negative irreducible matrix.

Theorem 7.9.6 *Let* $S \subset \mathbb{C}$ *be a multiset. Assume that the moments* $s_k(S), k \in \mathbb{N}$ *are eventually non-negative. Then the following conditions hold.*

1. $r(S) \in S$.

2. *Denote* $\mu := m(r(S))$. *Assume that* $\lambda \in S(r(S))$. *Then* $m(\lambda) \leq \mu$.

3. *Assume that* $r(S) > 0$ *and suppose that* $\lambda_1 = r(S), \lambda_2, \ldots, \lambda_m$ *are all the distinct elements of* S *satisfying the conditions* $|\lambda_i| = r(S), m(\lambda_i) = \mu$ *for* $i = 1, \ldots, m$. *Then* $\frac{\lambda_i}{r(S)}, i = 1, \ldots, m$ *are the* m *distinct roots of* 1.

4. *Let* $\zeta = e^{\frac{2\pi\sqrt{-1}}{m}}$. *Then* $\zeta S(r(S)) = S(r(S))$.

5. *If* $r(S) > 0, \mu = 1$ *and none of the other elements of* S *are positive, then* S *is a Frobenius multiset.*

Proof. For a finite multiset $S \subset \mathbb{C}$ define

$$f_S(z) := \sum_{\lambda \in S} \frac{1}{1 - \lambda z} = \sum_{k=0}^{\infty} s_k(S) z^k. \tag{7.9.6}$$

Apply Theorem 7.9.4 to deduce the parts 1–4.

Assume that $r(S)$ is the only positive element of S and $\mu = 1$. If $m = 1$ then S is a Frobenius set. Suppose the $m > 1$. Consider the function $g(z) = 2f_S(z) - f_S(\zeta z) - f_S(\bar{\zeta} z)$. We claim that g is the zero function. Suppose to the contrary that $g \neq 0$. In view of 4 we deduce $R(f) < R(g) < \infty$. Since the MacLaurin coefficients of g are eventually non-negative, Theorem 7.9.4 yields that g must have a singular point $\xi > 0$ whose residue at ξ is positive. Since $\rho(A)$ is the only positive eigenvalue of A, all other positive residues of g, coming from $2f_S$ are not located on positive numbers. Hence, the residues of g at its poles located on the positive axes are negative integers. The above contradiction shows that $g = 0$, i.e. S is a Frobenius multiset. $\square$

Let $A \in \mathbb{C}^{n \times n}$ and assume that $S(A)$ is the eigenvalue multiset of A. Then

$$f_{S(A)}(z) = \text{tr}(I - zA)^{-1} = \sum_{i=0}^{\infty} (\text{tr } A^i) z^i. \qquad (7.9.7)$$

Corollary 7.9.7 *Let $A \in \mathbb{C}^{n \times n}$. Denote by S be the multiset consisting of all eigenvalues of A, counted with multiplicities. Assume that the traces of $A^k, k \in \mathbb{N}$ are eventually non-negative. Then the following conditions hold.*

1. *$\rho(A)$ is an eigenvalue of A.*

2. *Assume that the algebraic multiplicity of $\rho(A)$ is μ. Let λ be an eigenvalue of A of multiplicity $m(\lambda)$ satisfying $|\lambda| = \rho(A)$. Then $m(\lambda) \leq \mu$.*

3. *Assume that $\rho(A) > 0$ and suppose that $\lambda_1 = \rho(A), \lambda_2, \ldots, \lambda_m$ are all the distinct eigenvalues of A satisfying the conditions $|\lambda_i| = \rho(A), m(\lambda_i) = \mu$ for $i = 1, \ldots, m$. Then $\frac{\lambda_i}{\rho(A)}, i = 1, \ldots, m$ are the m distinct roots of 1.*

4. *Let $\zeta = e^{\frac{2\pi \sqrt{-1}}{m}}$. Then $\zeta S(\rho(A)) = S$.*

5. *If $\rho(A) > 0$ is an algebraically simple eigenvalue of A, and none of the other eigenvalues of A are positive, then S is a Frobenius multiset.*

Definition 7.9.8 *$A \in \mathbb{R}^{n \times n}$ is called eventually non-negative if $A^k \geq 0$ for all integers $k > N$.*

Lemma 7.9.9 *Let $B \in \mathbb{R}^{n \times n}$. Then there exists a positive integer M with the following property. Assume that $L > M$ is a prime. Suppose that B^L is similar to a non-negative matrix. Then the eigenvalue multiset of B is a union of Frobenius multisets.*

Proof. Associate with the eigenvalues of B the following set $T \subset S^1$. For $0 \neq \lambda \in \operatorname{spec} B$ we let $\frac{\lambda}{|\lambda|} \in T$. For $\lambda \neq \kappa \in \operatorname{spec} B$ satisfying the conditions $|\lambda| = |\kappa| > 0$ we assume that $\frac{\lambda}{\kappa}, \frac{\kappa}{\lambda} \in T$. Let $T_1 \subset T$ be the set of all roots of 1 that are in T. Recall that $\eta \in S^1$ is called a primitive k-root of 1, if $\eta^k = 1$, and $\eta^{k'} \neq 1$ for all integers $k' \in [1, k)$. k is called the primitivity index of η. Let $L > k$ be a prime. Then η^L is a k-primitive root of 1. Furthermore, the map $\eta \mapsto \eta^L$ is an isomorphism of the group of all kth roots of 1, which commutes with conjugation $\eta \mapsto \bar{\eta}$. Clearly, if $\eta \in S$, and η is not a root of unity then η^L is not root of unity. Define $M \in \mathbb{N}$ to be the maximum over all primitivity indices of $\eta \in T_1$. If $T_1 = \emptyset$ then $M = 1$. Let $L > M$ be a prime. Assume that B^L is similar to $C \in \mathbb{R}_+^{n \times n}$. Apply Theorem 6.4.4 and the Perron–Frobenius theorem to each irreducible diagonal block of of the matrix in (6.4.3), to deduce that the eigenvalue multiset of $S(C)$ is $\cup_{j=1}^{t+f} F_j$ and each F_j a Frobenius multiset.

Clearly, $\overline{\operatorname{spec} B} = \operatorname{spec} B$ and $\operatorname{spec} B^L = \operatorname{spec} C$. Observe next that the condition that L is a prime satisfying $L > M$ implies that the map $z \mapsto z^L$ induces a $1 - 1$ and onto map $\phi : \operatorname{spec} B \to \operatorname{spec} C$. Moreover, $\phi^{-1}(r) > 0$ if and only of $r > 0$. Hence, ϕ can be extended to a $1 - 1$ and onto map $\tilde{\phi} : S(B) \to S(C)$. Furthermore, $\phi^{-1}(F_j)$ is a Frobenius set, where the number of distinct points $F_j(r(F_j))$ is equal to the number of points in $\phi^{-1}(F_j)(r(\phi^{-1}(F_j))$. Hence, $S(B) = \cup_{j=1}^{t+f} \phi^{-1}(F_j)$ is a decomposition of $S(B)$ to a union of Frobenius multisets. $\square$

Corollary 7.9.10 *Assume that a matrix $B \in \mathbb{R}^{n \times n}$ is similar to an eventually non-negative matrix. Then the eigenvalue multiset $S(B)$ of B is a union of Frobenius multisets.*

Theorem 7.9.11 *Assume that the eigenvalue multiset $S(B)$ of $B \in \mathbb{R}^{n \times n}$ is a union of Frobenius multisets. Then there an eventually non-negative $A \in \mathbb{R}^{n \times n}$, such that $S(A) = S(B)$.*

Proof. It is enough to show that for a given Frobenius multiset F there exists an eventually $A \in \mathbb{R}^{n \times n}$ such that $S(A) = F$. The claim is trivial if $F = \{0\}$. Assume that $r(F) > 0$. Without loss of generality we can assume that $r(F) = 1$. Suppose first that $F \cap S^1 = \{1\}$. To each real point $\lambda \in F$ of multiplicity $m(\lambda)$ we associate $m(\lambda)$ the diagonal matrix $G(\lambda) = \lambda I_{m(\lambda)} \in \mathbb{R}^{m(\lambda) \times m(\lambda)}$. For nonreal points $\lambda \in F$ of multiplicity $m(\lambda)$ we associate the block diagonal matrix $H(\lambda) = I_{m(\lambda)} \otimes \begin{bmatrix} 2\Re(\lambda) & |\lambda|^2 \\ -1 & 0 \end{bmatrix}$. Note that $H(\bar{\lambda}) = H(\lambda)$. Let

$$C := [1] \oplus_{\lambda \in F \cap \mathbb{R} \setminus \{1\}} G(\lambda) \oplus_{\lambda \in F, \Im \lambda > 0} H(\lambda),$$

$$C = C_0 + C_1, \quad C_0 = [1] \oplus 0_{(n-1) \times (n-1)},$$

$$C_1 = [0] \oplus (\oplus_{\lambda \in F \cap \mathbb{R} \setminus \{1\}} G(\lambda) \oplus_{\lambda \in F, \Im \lambda > 0} H(\lambda)).$$

Clearly,

$$S(C) = F, \quad C_0 C_1 = C_1 C_0 = 0, \quad C_0^m = C_0,$$

$$C^m = C_0^m + C_1^m, \quad \rho(C_1) < 1, \quad \lim_{m \to \infty} C_1^m = 0.$$

Let $X \in \mathbf{GL}(n, \mathbb{R})$ be a matrix such $X\mathbf{1} = X^\top \mathbf{1} = \mathbf{e}_1$. Define

$$A_0 := X^{-1} C_0 X = \tfrac{1}{n} \mathbf{1} \mathbf{1}^\top, \quad A_1 = X^{-1} C_1 X,$$

$$A := A_0 + A_1 = X^{-1} X.$$

So $S(A) = S(C) = F$. Also

$$A^m = A_0^m + A_1^m, \quad A_0^m = A_0, \quad \lim_{m \to \infty} A_1^m \Rightarrow \lim_{m \to \infty} A^m = A_0.$$

So A is eventually positive.

Assume now that F is a Frobenius set with $r(F) = 1$ such that $F \cap S^1$ consist of exactly $m > 1$ roots of unity. Let $\zeta = e^{\frac{2\pi\sqrt{-1}}{m}}$. Recall that $\zeta F = F$. Let $F = F_1 \cup F_0$, where $0 \notin F_1$ and F_0 consists of $m(0)$ copies of 0. $(m(0)) = 0 \iff F_0 = \emptyset$. If $F_0 \neq \emptyset$ then the zero matrix of order $m(0)$ has F_0 as its eigenvalue multiset. Thus it is enough to show that there exists an eventually non-negative matrix B whose eigenvalue multiset is F_1. Clearly, F_1 is a Frobenius set satisfying $r(F_1) = 1$ and $F \cap S^1$ consist of exactly $m > 1$ roots of unity. Assume that all the elements of F_1, counted with their multiplicity are the coordinates of the vector $\mathbf{z} = (z_1, \ldots, z_N)^\top \in \mathbb{C}^N$. Let $\sigma_1(\mathbf{z}), \ldots, \sigma_N(\mathbf{z})$ be the elementary symmetric polynomials in $z_1, \ldots, z_N$. Hence, the multiset F_1 consists of the roots of $P(z) := z^N + \sum_{k=1}^{N}(-1)^k \sigma_k(\mathbf{z})z^{N-k}$. Since $\bar{F}_1 = F_1$ it follows that each $\sigma_k(\mathbf{z})$ is real. As $\zeta F_1 = F_1$ we deduce that $N = mN'$ and $\sigma_k = 0$ if m does not divide k. Let F_2 be the root multiset $Q(z) := z^{N'} + \sum_{k=1}^{N'}(-1)^{mk}\sigma_{km}(\mathbf{z})z^{N'-k}$. Clearly, $\bar{F}_2 = F_2$. Since $1 \in F_1$ it follows that $1 \in F_2$. Furthermore, $F_1 = \phi^{-1}(F_2)$, where $\phi(z) : \mathbb{C} \to \mathbb{C}$ is the map $z \mapsto z^m$. That is, if $z \in F_2$ has multiplicity $m(z)$ then $\phi^{-1}(z)$ consists of m points, each of multiplicity $m(z)$ such that these m-points are all the solutions of $w^m = z$. Hence, $F_2 \cap S^1 = \{1\}$. Therefore, F_2 is a Frobenius set.

According to the previous case there exists an eventually non-negative matrix $A \in \mathbb{R}^{N' \times N'}$ such that F_2 is its eigenvalue multiset. Let $P \in \mathcal{P}_m$ be a permutation matrix corresponding to the cyclic permutation on $\langle m \rangle$ $i \mapsto i + 1$, for $i = 1, \ldots, m$, where $m + 1 \equiv 1$. Consider the matrix $B = P \otimes A$. Then B is eventually non-negative, and the eigenvalue multiset of B is F_1. $\qquad\square$

7.10 Inverse Eigenvalue Problem for Non-Negative Matrices

The following problem is called the *inverse eigenvalue problem for non-negative matrices*, abbreviated an IEPFNM:

Problem 7.10.1 *Let* $S \subset \mathbb{C}$ *be a multiset consisting of n points, (counting with their multiplicities). Find necessary and sufficient conditions such that there exists a non-negative $A \in \mathbb{R}^{n \times n}$ whose eigenvalue multiset is* S.

Proposition 7.10.2 *Let* $A \in \mathbb{R}_+^{n \times n}$. *Then the eigenvalue multiset S satisfies the following conditions.*

1. S *is a union of Frobenius multisets.*

2. *All the moments of $s_k(S) \geq 0$.*

3. $s_{kl}(S) \geq \frac{s_k(S)^l}{n^{l-1}}$ *for each $k, l \in \mathbb{N}$.*

Proof. 1 follows from Theorem 6.4.4 and the Perron–Frobenius theorem applied to each irreducible diagonal block of of the matrix in (6.4.3). Since $A^k \geq 0$ it follows that $\operatorname{tr} A^k \geq 0$. Hence, 2 holds. Since $A^k \geq 0$ it is enough to show the inequality in 3 for $k = 1$. Decompose $A = [a_{ij}]$ as $D + A_0$, where $D = \operatorname{diag}(a_{11}, \ldots, a_{nn})$ and $A_0 := A - D \geq 0$. So $A^l - D^l \geq A_0^l \geq 0$. Hence, $\operatorname{tr} A^l \geq \operatorname{tr} D^l = \sum_{i=1}^{l} a_{ii}^l$. Hólder inequality for $p = l$ yield that $\sum_{i=1}^{n} a_{ii} \leq (\sum_{i=1}^{n} a_{ii}^l)^{\frac{1}{l}} n^{\frac{l-1}{l}}$, which yields 3. $\square$

The following result gives simple sufficient conditions for a multiset S to be the eigenvalue multiset of a non-negative matrix.

Proposition 7.10.3 *Let* $S \subset \mathbb{C}$ *be a multiset containing* n *elements, counting with multiplicities. Assume that the elementary symmetric polynomials corresponding to* S *satisfy* $(-1)^{k-1}\sigma_k(S) \geq 0$ *for* $k = 1, \ldots, n$. *Then there exists* $A \in \mathbb{R}_+^{n \times n}$ *such that* S *is the eigenvalue multiset of* A.

Proof. Note that the companion matrix to the polynomial $P(z) = z^n + \sum_{i=1}^{n}(-1)^i \sigma_i(S) z^{n-i}$ is a non-negative matrix. $\square$

Recall the MacLaurin inequalities [HLP52, p. 52].

Proposition 7.10.4 *Let* $\mathbf{w} = (w_1, \ldots, w_{n-1})^\top \in \mathbb{R}_+^{n-1}$. *Then the sequence* $\left(\frac{\sigma_k(\mathbf{w})}{\binom{n-1}{k}}\right)^{\frac{1}{k}}$ *nonincreasing for* $k = 1, \ldots, n - 1$.

Proposition 7.10.5 *Let* S *be a multiset of real numbers, which contains exactly one positive number. Assume that the sum of all elements in* S *is non-negative. Then* S *satisfies the conditions of Proposition 7.10.3. In particular, there exists* $A \in \mathbb{R}_+^{n \times n}$ *such that* S *is the eigenvalue multiset of* A.

Proof. Without loss of generality we may assume that $S = \{1, -w_1, \ldots, w_{n-1}\}$ where $w_i \geq 0$ for $i = 1, \ldots, n-1$ and $1 \geq \sum_{i=1}^{n-1} w_i$. Denote $\mathbf{z} = (1, -w_1, \ldots, -w_{n-1})^\top$ and $\mathbf{w} = (w_1, \ldots, w_{n-1})^\top$. Clearly, $\sigma_1(\mathbf{z}) \geq 0, (-1)^{n-1}\sigma_n(\mathbf{z}) = \sigma_{n-1}(\mathbf{w}) \geq 0$.

Observe next that

$$\sigma_{k+1}(\mathbf{z}) = (-1)^k(\sigma_k(\mathbf{w}) - \sigma_{k+1}(\mathbf{w})) \quad \text{for } k = 1, \dots, n-2.$$

Thus to prove that $(-1)^k\sigma_{k+1}(\mathbf{z}) \geq 0$ it is enough to show that the sequence $\sigma_i(\mathbf{w}), i = 1, \dots, n-1$ is nonincreasing.

Observe that $\sigma_1(\mathbf{w}) \leq 1$. We now use Use Proposition 7.10.5. First observe that

$$\left(\frac{\sigma_k(\mathbf{w})}{\binom{n-1}{k}}\right)^{\frac{1}{k}} \leq \frac{\sigma_1(\mathbf{w})}{n-1} \leq \frac{1}{n-1} \quad \text{for } k = 1, \dots, n-1.$$

Next

$$\sigma_k(\mathbf{w}) - \sigma_{k+1}(\mathbf{w}) \geq \sigma_k(\mathbf{w}) - \left[\frac{\sigma_k(\mathbf{w})}{\binom{n-1}{k}}\right]^{\frac{k+1}{k}} \cdot \binom{n-1}{k+1}$$

$$= \frac{\sigma_k(\mathbf{w})}{\binom{n-1}{k}}\left[\binom{n-1}{k} - \left(\frac{\sigma_k(\mathbf{w})}{\binom{n-1}{k}}\right)^{\frac{1}{k}}\binom{n-1}{k+1}\right]$$

$$\geq \frac{\sigma_k(\mathbf{w})}{\binom{n-1}{k}}\left[\binom{n-1}{k} - \frac{1}{n-1}\binom{n-1}{k+1}\right] \geq 0.$$

Hence, S satisfies the conditions of Proposition 7.10.3. The last part of Proposition 7.10.3 yields that there exists $A \in \mathbb{R}_+^{n \times n}$ such that S is the eigenvalue multiset of A. $\square$

Example 7.10.6 *Let* S $= \{\sqrt{2}, \sqrt{-1}, -\sqrt{-1}\}$. *Then* S *is a Frobenius set. Furthermore, $s_2(\text{S}) = 0$ and all other moments of* S *are positive. Hence, the condition 3 of Proposition 7.10.2 does not hold for $k = 1, l = 2$. In particular, there is an eventually non-negative matrix $A \in \mathbb{R}_+^{3 \times 3}$, which cannot be non-negative, whose eigenvalue multiset is* S.

Theorem 7.10.7 *Let* S $= \{\lambda_1, \lambda_2, \lambda_3\}$ *be a multiset satisfying the following properties.*

1. $r(\text{S}) \in$ S.

2. $\overline{\text{S}} = $ S.

3. $s_1(S) \geq 0$.

4. $(s_1(S))^2 \leq 3s_2(S)$.

Then there exist $A \in \mathbb{R}_+^{3\times3}$ such that S is the eigenvalue multiset of A.

Proof. Suppose first that $S \subset \mathbb{R}$. It is straightforward to show that S is a union of Frobenius multisets. In that case the theorem can be shown straightforward. See Problem 2. It is left to discuss the following renormalized case $S = \{r, e^{\sqrt{-1}\theta}, e^{-\sqrt{-1}\theta}\}$, where $r \geq 1, \theta \in (0, \pi)$. The condition $s_1(S) \geq 0$ yields that

$$2\cos\theta + r \geq 0. \tag{7.10.1}$$

The condition $(s_1(S))^2 \leq 3s_2(S)$ boils down to

$$\left(r - 2\cos\left(\frac{\pi}{3} + \theta\right)\right)\left(r - 2\cos\left(\frac{\pi}{3} - \theta\right)\right) \geq 0.$$

For $r \geq 1, \theta \in (0, \pi)$ we have $r - 2\cos\left(\frac{\pi}{3} + \theta\right) > 0$. Hence, the condition 4 is equivalent to

$$r - 2\cos\left(\frac{\pi}{3} - \theta\right) \geq 0. \tag{7.10.2}$$

Let U be the orthogonal matrix $\frac{1}{6}\begin{bmatrix} \sqrt{2} & \sqrt{3} & -1 \\ \sqrt{2} & 0 & 2 \\ \sqrt{2} & -\sqrt{3} & -1 \end{bmatrix}$ and $J = 1_3 1_3^\top$.

So $U^\top JU = \mathrm{diag}(3, 0, 0)$. S is the eigenvalue set of $B = \begin{bmatrix} r & 0 & 0 \\ 0 & \cos\theta & \sin\theta \\ 0 & -\sin\theta & \cos\theta \end{bmatrix}$. Then $A := UBU^\top$ is the following matrix

$$\frac{r}{3}\begin{bmatrix} 1 & 1 & 1 \\ 1 & 1 & 1 \\ 1 & 1 & 1 \end{bmatrix} - \frac{2}{3}\begin{bmatrix} -\cos\theta & \cos\left(\frac{\pi}{3} + \theta\right) & \cos\left(\frac{\pi}{3} - \theta\right) \\ \cos\left(\frac{\pi}{3} - \theta\right) & -\cos\theta & \cos\left(\frac{\pi}{3} + \theta\right) \\ \cos\left(\frac{\pi}{3} + \theta\right) & \cos\left(\frac{\pi}{3} - \theta\right) & -\cos\theta \end{bmatrix}.$$

The above inequalities show that $A \geq 0$. $\square$

A weaker version of the solution of Problem 7.10.1 was given in [BoH91].

Theorem 7.10.8 *Let* $T \subset \mathbb{C}\backslash\{0\}$ *be Frobenius mulitiset satisfying the following conditions.*

1. $T(r(T)) = \{r(T)\}$.

2. $s_k(T) \geq 0$ *for* $k \in \mathbb{N}$.

3. *If* $s_k(T) > 0$ *then* $s_{kl}(T) > 0$ *for all* $l \in \mathbb{N}$.

Then there exist a square non-negative primitive matrix A, *whose eigenvalues multiset is a union of* T *and* $m_0 \geq 0$ *copies of* 0.

We prove the above theorem under the stronger assumption

$$s_k(T) > 0 \quad \text{for } k \geq 2, \qquad (7.10.3)$$

following the arguments of [Laf12].

Lemma 7.10.9 *Let* $A_n \in \mathbb{C}^{n \times n}$ *be the following lower Hessenberg matrix*

$$A_n = \begin{bmatrix} s_1 & 1 & 0 & & \cdot & \cdot & \cdot & & 0 \\ s_2 & s_1 & 2 & 0 & & \cdot & \cdot & \cdot & 0 \\ s_3 & s_2 & s_1 & 3 & 0 & & \cdot & \cdot & 0 \\ \cdot & s_3 & & \cdot & \cdot & \cdot & \cdot & & \\ \cdot & & \cdot & & & \cdot & \cdot & \cdot & \\ \cdot & & & & & & \cdot & & \\ & & & & & \cdot & & & \\ s_{n-1} & s_{n-2} & & \cdot & \cdot & \cdot & s_2 & s_1 & n-1 \\ s_n & s_{n-1} & & \cdot & \cdot & \cdot & s_3 & s_2 & s_1 \end{bmatrix}. \qquad (7.10.4)$$

Let $S = \{\lambda_1, \ldots, \lambda_n\} \subset \mathbb{C}$ *be the unique multiset such that* $s_k = s_k(S)$ *for* $k = 1, \ldots, n$. *Let* $\sigma_1, \ldots, \sigma_n$ *be the* n-*elementary symmetric polynomials corresponding to* S. *Then the characteristic polynomial of* A_n *is given by*

$$\det(zI_n - A_n) = z^n + \sum_{i=1}^{n}(-1)^i i!\binom{n}{i}\sigma_i. \qquad (7.10.5)$$

Proof. Recall the Newton identities.

$$s_1 = \sigma_1, \quad s_k = (-1)^{k-1}k\sigma_k + \sum_{i=1}^{k-1}(-1)^{i-1}\sigma_i s_{k-i} \quad \text{for } k = 2, \ldots, n.$$

Let $p(z)$ be the polynomial given by the right-hand side of (7.10.5). Denote by $C(p(z)) \in \mathbb{C}^{n\times n}$ the companion matrix corresponding to $p(z)$. Let $Q = [q_{ij}] \in \mathbb{C}^{n\times n}$ be the following lower triangular matrix.

$$q_{ij} = \frac{(-1)^{i-j}\sigma_{i-j}}{(j-1)!}, \quad j = 1, \ldots, i, \ i = 1, \ldots, n, \text{ where } \sigma_0 := 1.$$

Use the Newton identities to verify the equality $A_n Q = C(p(z))Q$. Hence, A_n is similar to $C(p(z))$, and the characteristic polynomial of A_n is given by (7.10.5). $\qquad\square$

Proof of Theorem 7.10.8 under the assumption 7.10.3. Let $T = \{\lambda_1, \ldots, \lambda_n\}$ be a multiset in $\mathbb{C}$. Denote by $\sigma_1, \ldots, \sigma_n$ the elementary symmetric polynomials corresponding to T. Let $p(z) = z^n + \sum_{i=1}^{n}(-1)^i\sigma_i z^{n-i}$ be the normalized polynomial whose zero set is T. For $m \in \mathbb{N}$ denote $S_m := T \cup \underbrace{\{0, \ldots, 0\}}_{m}$. Let $\sigma_{i,m}$ be the ith elementary symmetric polynomial corresponding to S_m for $i = 1, \ldots, n+m$. Then $\sigma_{i,m} = \sigma_i$ for $i = 1, \ldots, n$ and $\sigma_{i,m} = 0$ for $i = n+1, \ldots, n+m$. The $s_k(T) = s_k(S_m)$ for all $k \in \mathbb{N}$. Denote by $A_{n+m} \in \mathbb{C}^{(n+m)\times(n+m)}$ the matrix (7.10.4), where $s_k = s_k(T)$ for $k = 1, \ldots, n+m$. Observe that

$$\det\left(zI_{n+m} - \frac{1}{n+m}A_{n+m}\right)$$

$$= \left(z^n + \sum_{i=1}^{n}\left(\prod_{j=1}^{i}\left(1 - \frac{j-1}{n+m}\right)\right)(-1)^i\sigma_i z^{n-i}\right)z^m.$$

Let

$$p_m(z) := z^n + \sum_{i=1}^{n}\left(\prod_{j=1}^{i}\left(1 - \frac{j-1}{n+m}\right)^{-1}\right)(-1)^i\sigma_i z^{n-i}. \quad (7.10.6)$$

Denote by $T_m = \{\lambda_{1,m}, \ldots, \lambda_{n,m}\}$ the multiset formed the n zeros of p_m. Since $\lim_{m \to \infty} p_m(z) = p(z)$ we deduce that $\lim_{m \to \infty} \operatorname{pdist}(T_m, T) = 0$. That is, we can rename $\lambda_{1,m}, \ldots, \lambda_{n,m}, m \in \mathbb{N}$ such that $\lim_{m \to \infty} \lambda_{i,m} = \lambda_i$ for $i = 1, \ldots, n$. Let $B_{m+n} \in \mathbb{C}^{(m+n) \times (m+n)}$ be the matrix defined by (7.10.4), where $s_k, k = 1, \ldots, m+n$ are the moments corresponding to T_m. Then $\det\left(z I_{n+m} - \frac{1}{n+m} B_{n+m}\right) = z^m p(z)$. Thus, if the first $n+m$ moments corresponding to T_m are non-negative, it follows that that the multiset S_m is realized as an eigenvalue set of a non-negative matrix.

We now show that the above condition holds for $m \geq N$, if T satisfies the assumption 1 of Theorem 7.10.8 and (7.10.3). It is enough to consider the case where $T = \{\lambda_1 = 1, \lambda_2, \ldots, \lambda_n\}$, where $1 > |\lambda_2| \geq \cdots \geq |\lambda_n|$. Let $\varepsilon := \frac{1-|\lambda_2|}{4}$. First we choose M big enough such that after renaming the elements of the multiset of T_m we have that $|\lambda_{i,m} - \lambda_i| \leq \varepsilon$ for $i = 1, \ldots, n$ and $m \geq M$. Note that since $\bar{T}_m = T_m$ it follows that $\lambda_{1,m} \in \mathbb{R}$ and $\lambda_{1,m} \geq 1 - \varepsilon$ for $m \geq M$. Furthermore, $|\lambda_{i,m}| \leq 1 - 3\varepsilon$ for $i = 2, \ldots, n$. Hence

$$\mathbf{s}_k(T_m) \geq (1 - \varepsilon)^k \left(1 - (n-1)\left(\frac{1 - 3\varepsilon}{1 - \varepsilon}\right)^k\right).$$

Thus for $k > k(\varepsilon) := \frac{\log(n-1)}{\log(1-\varepsilon) - \log(1-3\varepsilon)}$ and $m \geq N$ we have that $s_k(T_m) > 0$. Clearly $s_1(T_m) = s_1(T)$ and $\lim_{m \to \infty} s_k(T_m) = s_k(T)$ for $k = 2, \ldots, \lceil k(\varepsilon) \rceil$. Since $s_k(T) > 0$ for $n > 1$ we deduce the positivity of all $s_k(T_m)$ for all $k > 1$ if $m \geq N \geq M$. □

It is straightforward to generalize this result to a general Frobenius multiset. See [Fri12].

Theorem 7.10.10 *Let $T \subset \mathbb{C} \backslash \{0\}$ be a Frobenius multiset satisfying the following conditions.*

1. $T(r(T)) = \{r(T), \zeta r(T), \ldots, \zeta^{m-1} r(T)\}$ *for* $\zeta = e^{\frac{2\pi \sqrt{-1}}{m}}$ *where $m > 1$ is an integer.*

2. $s_k(T) \geq 0$ *for $k \in \mathbb{N}$.*

3. *If $s_k(T) > 0$ then $s_{kl}(T) > 0$ for all $l \in \mathbb{N}$.*

Then there exist a square non-negative irreducible matrix A, whose eigenvalues multiset is a union of T *and* $m_0 \geq 0$ *copies of* 0.

Proof. Observe first that $s_k(T) = 0$ if $m / |k$. Let $\phi : \mathbb{C} \to \mathbb{C}$ be the map $z \mapsto z^m$. Since $\zeta T = T$ it follows that for $z \in T$ with multiplicity $m(z)$ we obtain the multiplicity z^m in $\phi(T)$ is $m m(z)$. Hence, $\phi(T)$ is a union of m copies of a Frobenius set T_1, where $r(T_1) = r(T)^m$ and $T_1(r(T_1)) = \{r(T_1)\}$. Moreover, $s_{km}(T) = m s_k(T_1)$. Hence, T_1 satisfies the assumptions of Theorem 7.10.8. Thus, there exists a primitive matrix $B \in \mathbb{R}_+^{n \times n}$ whose nonzero eigenvalue multiset is T_1. Let $A = [A_{ij}]_{i=j=1}^m$ be the following non-negative matrix of order mn.

$$A = \begin{bmatrix} 0_{n \times n} & I_n & 0_{n \times n} & 0_{n \times n} & \cdots & 0_{n \times n} \\ 0_{n \times n} & 0_{n \times n} & I_n & 0_{n \times n} & \cdots & 0_{n \times n} \\ \vdots & \vdots & \vdots & \vdots & \vdots & \vdots \\ 0_{n \times n} & 0_{n \times n} & 0_{n \times n} & 0_{n \times n} & \cdots & I_n \\ B & 0_{n \times n} & 0_{n \times n} & 0_{n \times n} & \cdots & 0_{n \times n} \end{bmatrix}. \qquad (7.10.7)$$

Then A is irreducible and the nonzero part of eigenvalue multiset if T. (See Problems 4 and 5.)

Problems

1. **Definition 7.10.11** *Let* $S \subset \mathbb{C}$ *be a finite multiset.* S *is called a semi Frobenius multiset if either* S *has* m *elements all equal to* 0, *or the following conditions hold.*

   (a) $r(S) > 0$, $\bar{S} = S$, $r(S) \in S$.

   (b) $m(z) \leq \mu := m(r(S))$ *for each* $z \in S$ *such that* $|z| = r(S)$.

   (c) *Assume that* S *contains exactly* m *distinct points satisfying* $|z| = r(S), m(z) = \mu$. *Then* $\zeta S = S$ *for* $\zeta = e^{\frac{2\pi \sqrt{-1}}{m}}$.

   S *is called an almost a Frobenius multiset if the number points in* S, *counted with their multiplicities, satisfying* $z \in S, |z| = r(S), m(z) < \mu$ *is strictly less than* $m\mu$.

Let f_S by (7.9.6). Show

(a) $S = \{1, 1, z, \bar{z}\}$, with $|z| = 1, z \neq \pm 1$ is a semi Frobenius multiset, and f_S has non-negative moments.

(b) Let $S = \cup_{i=1}^{j} S_i$, where each S_i is almost a Frobenius multiset. Then f_S has eventually non-negative MacLaurin coeffients.

(c) Assume that the MacLaurin coefficients f_S are eventually non-negative. Then $r(S) \in S$. Suppose furthermore that $0 < \alpha < r(S)$ is the second largest positive number contained in S. Then $S \cap \{z \in \mathbb{C}, \alpha < |z| \leq r(S)\}$ is a semi Frobenius set.

(d) Assume that the MacLaurin coefficients f_S are eventually non-negative. Suppose that S contains only one positive number of multiplicity one. Then S is semi Frobenius.

(e) Assume that the MacLaurin coefficients f_S are eventually non-negative. Suppose that S contains only two positive number of multiplicity one each: $r(S) > \alpha > 0$. Decompose S to $S_1 \cup S_2$, where S_1 is a maximal semi Frobenius set containing $\{z \in \mathbb{C}, \alpha < |z| \leq r(S)\}$. If $\alpha \in S_1$ then $S_2 = \emptyset$. Suppose that $\alpha \in S_2$. Then $S_3 := S_2 \cap \{z \in \mathbb{C}, |z| = \alpha\}$ is a set, i.e. $m(z) = 1$ for each $z \in S_3$. Assume for simplicity of the exposition that $\alpha = 1$. Let $m' \in [1, l)$ be the greatest divisor of $m > 1$, entering in the definition of the Frobenius multiset S_1, such that all m' roots of 1 are in S_3. Let $m'' := \frac{m}{m'} > 1$. Then there exists $r \in \mathbb{N}$ coprime with m'' such that one of the following conditions hold.

   i. If m'' is even then $S_3 = S_4$, where S_4 consists of all $m'r$ roots of 1.

   ii. If m'' is odd then either $S_3 = S_4$ or $S_3 = S_4 \cup S_5$, where

$$S_5 = \cup_{q,k,j=1}^{m',r,m''-1} e^{2\pi\sqrt{-1}((q + \frac{2k-1}{2r} + \frac{j}{m''})\frac{1}{m'})}.$$

Hint: Use the function g in the proof of Theorem 7.9.6, or/and consult with [Fri78b, Theorem 4] and its proof.

2. Let $S = \{\lambda_1, \ldots, \lambda_n\} \subset \mathbb{R}$ be a union of Frobenius multiset. Assume furthermore that $\sum_{i=1}^{n} \lambda_i \geq 0$. Show that if $n \leq 4$ then

there exists a non-negative $n \times n$ matrix whose eigenvalue multiset if S. *Hint*: For $n = 3$ use Proposition 7.10.5. For $n = 4$ and the case where S contains exactly two negative numbers consult with the proof of [LoL78, Theorem 3].

3. Show that for $n \geq 4$ the multiset S $:= \{\sqrt{2}, \sqrt{2}, \sqrt{-1}, -\sqrt{-1}, 0, \ldots, 0\}$ satisfies all the conditions of Proposition 7.10.2. However, there is no $A \in \mathbb{R}_+^{n \times n}$ with the eigenvalue set S.

4. Let $B \in \mathbb{R}_+^{n \times n}$ be a primitive matrix. Show that the matrix $A \in \mathbb{R}_+^{mn \times mn}$ defined (7.10.7) is irreducible for any integer $m > 1$.

5. Let $B \in \mathbb{C}^{n \times n}$. Assume that T is the eigenvalue multiset of B Assume that $A \in \mathbb{C}^{mn \times mn}$ is defined by (7.10.7). Let S be the eigenvalue multiset of A. Show that $w \in$ S if and only if $w^m \in$ T. Furthermore the multiplicity of $0 \neq w \in$ S equals to the multiplicity of w^m in T. The mulitplicity of $0 \in$ S is m times the multiplicity of $0 \in$ T.

7.11   Cones

Let $\mathbf{V}$ be a vector space over $\mathbb{C}$. Then $\mathbf{V}$ is a vector space over $\mathbb{R}$, which we denote by $\mathbb{V}_{\mathbb{R}}$, or simply $\mathbf{V}$ when no ambiguity arises. See Problem 1.

Definition 7.11.1 *Let* $\mathbf{V}$ *be a finite dimensional vector space over* $\mathbb{F} = \mathbb{R}, \mathbb{C}$. *A set* K $\subset \mathbf{V}$ *is called a cone if*

1. K $+$ K $\subset$ K, *i.e.* $\mathbf{x} + \mathbf{y} \in$ K *for each* $\mathbf{x}, \mathbf{y} \in$ K.

2. $\mathbb{R}_+$K $\subset$ K, *i.e.* $a\mathbf{x} \in$ K *for each* $a \in [0, \infty)$ *and* $\mathbf{x} \in$ K.

Assume that K $\subset \mathbf{V}$ *is a cone.* (*Note that* K *is convex set.*) *Then*

1. ri K, dim K, *is the relative interior and the dimension of* K, *viewed as a convex set.*

2. K$' := \{\mathbf{f} \in \mathbf{V}', \Re\mathbf{f}(\mathbf{x}) \geq 0 \text{ for all } \mathbf{x} \in \mathbf{V}\}$ *is called the conjugate cone,* (*in* $\mathbf{V}'$).

3. K *is called pointed if* $K \cap -K = \{\mathbf{0}\}$.

4. K *is called generating if* $K - K = \mathbf{V}$, *i.e. any* $\mathbf{z} \in \mathbf{V}$ *can be represented as* $\mathbf{x} - \mathbf{y}$ *for some* $\mathbf{x}, \mathbf{y} \in K$.

5. K *is called proper if* K *is closed, pointed and generating.*

6. *For* $\mathbf{x}, \mathbf{y} \in \mathbf{V}$ *we denote:* $\mathbf{x} \geq^K \mathbf{y}$ *if* $\mathbf{x} - \mathbf{y} \in K$; $\mathbf{x} \gneq^K \mathbf{y}$ *if* $\mathbf{x} \geq^K \mathbf{y}$ *and* $\mathbf{x} \neq \mathbf{y}$; $\mathbf{x} >^K \mathbf{y}$ *if* $\mathbf{x} - \mathbf{y} \in \mathrm{ri}\, K$.

7. *For* $\mathbf{x} \in \mathbf{V}$ *we call:* $\mathbf{x}$ *non-negative relative to* K *if* $\mathbf{x} \geq^K \mathbf{0}$; $\mathbf{x}$ *is semipositive relative to* K *if* $\mathbf{x} \gneq^K \mathbf{0}$; $\mathbf{x}$ *is positive relative to* K *if* $\mathbf{x} \in \mathrm{ri}\, K$. *When there is no ambiguity about the cone* K *we drop the term relative to* K.

8. $K_1 \subset K$ *is called a subcone of* K *if* K_1 *is a cone in* $\mathbf{V}$. $F \subset K$ *is called a face of* K, *if* F *is a subcone of* K *and the following condition holds: Assume that* $\mathbf{x}, \mathbf{y} \in K$ *and* $\mathbf{x} + \mathbf{y} \in F$. *Then* $\mathbf{x}, \mathbf{y} \in F$. $\dim F$, *the dimension of* F, *is called the dimension of* F. $F = \{\mathbf{0}\}, F = K$ *are called trivial faces,* $(\dim \{\mathbf{0}\} = 0)$. $\mathbf{x} \gneq \mathbf{0}$ *is called an extreme ray if* $\mathbb{R}_+ \mathbf{x}$ *is a face in* K, (*of dimension* 1). *For a set* $X \subset K$, *the face* $F(X)$ *generated by* X, *is the intersections of all faces of* K *containing* X.

9. *Let* $T \in \mathrm{Hom}\,(\mathbf{V}, \mathbf{V})$. *Then:* $T \geq^K 0$, *and* T *is called non-negative with respect to* K, *if* $TK \subset K$; $T \gneq^K 0$, *and* T *is called semipositive with respect to* K, *if* $T \geq^K 0$ *and* $T \neq 0$; $T >^K 0$, *and* T *is called positive with respect to* K, *if* $T(K \backslash \{\mathbf{0}\}) \subset \mathrm{ri}\, K$. $T \geq^K 0$ *is called primitive with respect to* K, *if* F *is a face of* K *satisfying* $TF \subset F$, *i.e.* F *is* T *invariant, then* F *is a trivial face of* K. T *is called eventually positive with respect to* K *if* $T^l >^K 0$ *for all integers* $l \geq L (\geq 1)$. *When there is no ambiguity about the cone* K *we drop the term relative to* K. *Denote by* $\mathrm{Hom}\,(\mathbf{V}, \mathbf{V})^K$ *the set of all* $T \geq^K 0$. *For* $T, S \in \mathrm{Hom}\,(\mathbf{V}, \mathbf{V})$ *we denote:* $T \geq^K S \iff T - S \geq^K 0$, $T \gneq^K S \iff T - S \gneq^K 0, T >^K S \iff T - S >^K 0$.

As pointed out in Problem 3, without loss of generality we can discuss only the cones in real vector spaces. Also, in most of the applications the cones of interest lie in the real vector spaces. Since most of the results we state hold for cones over complex vector spaces, we state our results for cones in real or complex vector spaces, and give a proof only for the real case, when possible.

Lemma 7.11.2 *Let* $\mathbf{V}$ *be a finite dimensional vector space over* $\mathbb{F} = \mathbb{R}, \mathbb{C}$. *Let* K *be a cone in* $\mathbf{V}$. *Then* $\mathbf{V} = \mathrm{K} - \mathrm{K}$, *i.e.* K *is generating, if and only if the interior of* K *is nonempty, i.e.* $\dim \mathrm{K} = \dim_{\mathbb{R}} \mathbf{V}$.

Proof. It is enough to assume that $\mathbf{V}$ is an n-dimensional vector space over $\mathbb{R}$. Let $k = \dim \mathrm{K}$. Then span K is a k-dimensional vector space in $\mathbf{V}$. Assume that $\mathrm{K} - \mathrm{K} = \mathbf{V}$. Since $\mathrm{K} - \mathrm{K} \subset$ span K we deduce that $\dim \mathrm{K} = n$. Hence, K must have an interior.

Assume now that K has an interior. Hence, its interior must contain n linearly independent vectors $\mathbf{x}_1, \ldots, \mathbf{x}_n$ which form a basis in $\mathbf{V}$. So $\sum_{i=1}^{n} a_i \mathbf{x}_i \in \mathrm{K}$ for any $a_1, \ldots, a_n \geq 0$. Since any $\mathbf{z} \in \mathbf{V}$ is of the form $\sum_{i=1}^{n} z_i \mathbf{x}_i$ is of the form $\sum_{z_i \geq 0} z_i x_i - \sum_{z_i < 0} (-z_i) \mathbf{x}_i$ we deduce that $\mathrm{K} - \mathrm{K} = \mathbf{V}$. $\qquad\square$

Theorem 7.11.3 *Let* $\mathrm{K} \subset \mathbf{V}$ *be a proper cone over* $\mathbb{F} = \mathbb{R}, \mathbb{C}$, *where* $\dim \mathbf{V} \in [1, \infty)$. *Then the following conditions holds.*

1. *There exists* $\mathbf{f} \in \mathrm{K}'$ *which is strictly positive, i.e.* $\Re \mathbf{f}(\mathbf{x}) > 0$ *if* $\mathbf{x} \gneq^{\mathrm{K}} \mathbf{0}$.

2. *Every* $\mathbf{x} \geq \mathbf{0}$ *is a non-negative linear combination of at most* $\dim_{\mathbb{R}} \mathbf{U}$ *extreme rays of* K.

3. *The conjugate cone* $\mathrm{K}' \subset \mathbf{V}'$ *is proper.*

Proof. It is enough to assume that $\mathbf{V}$ is a vector space over $\mathbb{R}$. Observe that for any $\mathbf{u} \gneq^{\mathrm{K}} \mathbf{0}$ the set $\mathrm{I}(\mathbf{u}) := \{\mathbf{x} \in \mathbf{V}, \ \mathbf{u} \geq^{\mathrm{K}} \mathbf{x} \geq^{\mathrm{K}} -\mathbf{u}\}$ is a compact set. Clearly, $\mathrm{I}(\mathbf{u})$ is closed. It is left to show that $\mathrm{I}(\mathbf{u})$ is bounded. Fix a norm $\| \cdot \|$ on $\mathbf{V}$. Assume to the contrary that there exists a sequence $\mathbf{0} \neq \mathbf{x}_m \in \mathrm{C}$ such that $\lim_{m \to \infty} \|\mathbf{x}_m\| = \infty$. Let $\mathbf{y}_m = \frac{1}{\|\mathbf{x}_m\|} \mathbf{x}_m, m \in \mathbb{N}$. Since $\|\mathbf{y}_m\| = 1, m \in \mathbb{N}$ it follows that there exists a subsequence $m_k, k \in \mathbb{N}$ such that $\lim_{k \to \infty} \mathbf{y}_{m_k} = \mathbf{y}, \|\mathbf{y}\| = 1$. Since $\mathbf{y}_m \in \mathrm{I}(\frac{1}{\|\mathbf{x}_m\|}\mathbf{u})$ it follows that $\mathbf{y} \in \mathrm{I}(\mathbf{0})$. So $\mathbf{y} \in \mathrm{K} \cap -\mathrm{K} = \{\mathbf{0}\}$ which is impossible. Hence, $\mathrm{I}(\mathbf{u})$ is compact.

Choose $\mathbf{u} \in \mathrm{ri}\,\mathrm{K}$. We claim that $\mathbf{u}$ is an isolated extreme point of $\mathrm{I}(\mathbf{u})$. Since $\mathbf{u} \in \mathrm{ri}\,\mathrm{K}$ it follows that there exist $r > 0$ so that $\mathbf{u} + \mathbf{x} \in \mathrm{K}$ for each $\|\mathbf{x}\| \leq r$. Suppose that there exist $\mathbf{v}, \mathbf{w} \in \mathrm{I}(\mathbf{u})$ such that $t\mathbf{v} + (1-t)\mathbf{w} = \mathbf{u}$ for some $t \in (0,1)$. So $\mathbf{v} = \mathbf{u} - \mathbf{v}_1, \mathbf{w} = \mathbf{u} - \mathbf{v}_2$ for some $\mathbf{v}_1, \mathbf{w}_1 \geq^{\mathrm{K}} \mathbf{0}$. The equality $\mathbf{u} = (1-t)\mathbf{v} + t\mathbf{w}$ yields $\mathbf{0} = (1-t)\mathbf{v}_1 + t\mathbf{w}_1 \geq (1-t)\mathbf{v}_1 \geq \mathbf{0}$. Hence, $\mathbf{v}_1 = 0$. (See Problem 2). Similarly, $\mathbf{w}_1 = 0$. Hence, $\mathbf{u}$ is an extreme point.

We now show that for any $\mathbf{x} \in \mathbf{U}$ such that $\mathbf{x} \geq^{\mathrm{K}} \mathbf{0}, \|\mathbf{x}\| \leq r$ the point $\mathbf{u} - \mathbf{x}$ is not an extreme point of $\mathrm{I}(\mathbf{u})$. Indeed, $\mathbf{u} - \frac{3}{2}\mathbf{x}, \mathbf{u} - \frac{1}{2}\mathbf{x} \in \mathrm{I}(\mathbf{u})$ and $\mathbf{u} - \mathbf{x} = \frac{1}{2}(\mathbf{u} - \frac{3}{2}\mathbf{x}) + \frac{1}{2}(\mathbf{u} - \frac{1}{2})\mathbf{x}$. Since $\mathbf{u}$ is an isolated extreme point, Corollary 4.6.10 yields that $\mathbf{u}$ is exposed. Hence, there exists $\mathbf{f} \in \mathbf{U}^*$ such that $\mathbf{f}(\mathbf{u}) > \mathbf{f}(\mathbf{u} - \mathbf{x})$ for any $\mathbf{x} \geq^{\mathrm{K}} \mathbf{0}$ satisfying $\|\mathbf{x}\| < r$. So $\mathbf{f}(\mathbf{x}) > 0$ for any $\mathbf{x} \geq^{\mathrm{K}} \mathbf{0}$ satisfying $\|\mathbf{x}\| < r$. Hence, $\mathbf{f}(\mathbf{y}) = \frac{\|\mathbf{y}\|}{r} f(\frac{r}{\|\mathbf{y}\|}\mathbf{y}) > 0$ for any $\mathbf{y} \geq^{\mathrm{K}} \mathbf{0}$. This proves the part 1 of the theorem.

Let $\mathrm{C} = \{\mathbf{x} \geq^{\mathrm{K}} \mathbf{0}, \ \mathbf{f}(\mathbf{x}) = 1\}$. Since K is closed, it follows that C is a convex closed set. We claim that C is compact, i.e. bounded. Fix a norm $\|\cdot\|$ on $\mathbf{U}$. Assume to the contrary that there exists a sequence $\mathbf{x}_m \in \mathrm{C}$ such that $\lim_{m \to \infty} \|\mathbf{x}_m\| = \infty$. Let $\mathbf{y}_m = \frac{1}{\|\mathbf{x}_m\|}\mathbf{x}_m, m \in \mathbb{N}$. Since $\|\mathbf{y}_m\| = 1, m \in \mathbb{N}$ it follows that there exists a subsequence $m_k, k \in \mathbb{N}$ such that $\lim_{k \to \infty} \mathbf{y}_{m_k} = \mathbf{y}, \|\mathbf{y}\| = 1$. Since K is closed it follows that $\mathbf{y} \geq^{\mathrm{K}} \mathbf{0}$. Note that

$$\mathbf{f}(\mathbf{y}) = \lim_{k \to \infty} f(\mathbf{y}_{m_k}) = \lim_{k \to \infty} \frac{f(\mathbf{x}_{m_k})}{\|\mathbf{x}_k\|} = \lim_{k \to \infty} \frac{1}{\|\mathbf{x}_{m_k}\|} = 0.$$

This contradicts the assumption that $\mathbf{f}$ is strictly positive on K. Thus C is a convex compact set. We next observe that $\dim \mathrm{C} = n - 1$. First observe that $\mathbf{f}(\mathrm{C} - \mathbf{x}) = 0$ for any $\mathbf{x} \in \mathbb{C}$. Hence $\dim \mathrm{C} \leq n - 1$. Observe next that if $\mathbf{f}(\mathbf{z}) = 0$ and $\|\mathbf{z}\| \leq r$ then $\mathbf{u} + \mathbf{z} \in \mathrm{C}$. Hence $\dim \mathrm{C} = n - 1$. Let $\mathbf{w} \geq^{\mathrm{K}} \mathbf{0}$. Define $\mathbf{w}_1 = \frac{1}{f(\mathbf{w})}\mathbf{w} \in \mathrm{C}$. Carathéodory's theorem claims that $\mathbf{w}_1$ is a convex combination of at most n extreme points of C. This proves part 2 of the theorem.

Let $R := \{\max \|\mathbf{x}\|, \ \mathbf{x} \in C\}$. Let $\mathbf{g} \in \mathbf{U}^*, \|\mathbf{g}\|^* \leq \frac{1}{R}$. Then for $\mathbf{x} \in C$ $|\mathbf{g}(\mathbf{x})| \leq 1$. Hence, $(\mathbf{f} + \mathbf{g})(\mathbf{x}) \geq 1 - |\mathbf{g}(\mathbf{x})| \geq 0$. Thus $\mathbf{f} + \mathbf{g} \in K'$. So $\mathbf{f}$ is an interior point of K'. Clearly K' is a closed and a pointed cone. Hence, part 3 of the theorem hold.   □

Theorem 7.11.4 *Let* V *be a vector space over* $\mathbb{F} = \mathbb{R}, \mathbb{C}$. *Assume that* $K \subset V$ *be a proper cone. Assume that* $T \in$ Hom $(\mathbf{V}, \mathbf{V})^K$. *Let* $S(T) \subset \mathbb{C}$ *be the eigenvalue multiset of* T, *(i.e. the root set of the polynomial* det $(zI - T)$.*) Then*

1. $\rho(T) \in S(T)$.

2. *Let* $\lambda \in S(T)(\rho(T))$. *Then* index $(\lambda, T)) \leq k :=$ index $(\rho(T), T)$.

3. *There exists* $\mathbf{x} \geq^K \mathbf{0}$ *such that* $T\mathbf{x} = \rho(T)\mathbf{x}$, *and* $\mathbf{x} \in (\rho(T)I - T)^{k-1}\mathbf{V}$.

4. *If* $T >^K \mathbf{0}$ *then* $\rho(T) > 0, k = 1, S(T)(\rho(T)) = \{\rho(T)\}$ *and* $\rho(A)$ *is a simple root of the characteristic polynomial of* T. *(This statement can hold only if* $\mathbb{F} = \mathbb{R}$.*)*

Assume in addition that $\rho(T) = 1$. *Let* $P \in$ Hom $(\mathbf{V}, \mathbf{V})$ *be the spectral projection, associated with* T, *on the generalized eigenspace corresponding to* 1. *Then*

$$\lim_{m \to \infty} \frac{k!}{m^k} \sum_{i=0}^{m-1} T^i = (T - I)^{k-1}P \geq^K \mathbf{0}. \tag{7.11.1}$$

Assume finally that $\mathbb{F} = \mathbb{R}$, $\lambda \neq 1, |\lambda| = 1$ *is an eigenvalue of* T *of index* k. *Let* $P(\lambda) \in$ Hom $(\mathbf{V}, \mathbf{V})$ *be the spectral projection, associated with* T, *on the generalized eigenspace corresponding to* λ. *View* $P(\lambda) = P_1 + \sqrt{-1}P_2$, *where* $P_1, P_2 \in$ Hom $(\mathbf{V}, \mathbf{V})$. *Then*

$$|\mathbf{f}((T - \lambda I)^{k-1}P(\lambda)\mathbf{y})| \leq \mathbf{f}((T - I)^{k-1}P\mathbf{y}) \text{ for any } \mathbf{y} \in K, \mathbf{f} \in K'. \tag{7.11.2}$$

Proof. Suppose first that $\rho(T) = 0$, i.e. T is nilpotent. Then parts 1–2 are trivial. Choose $\mathbf{y} \geq^K \mathbf{0}$. Then there exists an integer $j \in [0, k-1]$ so that $T^j\mathbf{y} \geq^K \mathbf{0}$ and $T^{j+1}\mathbf{y} = \mathbf{0}$. Then $\mathbf{x} := T^j\mathbf{y}$ is an eigenvector of A which lies in K. Since $A\mathbf{x} = \mathbf{0}$ it follows that A cannot be positive.

From now on we assume that $\rho(T) > 0$, and without loss of generality we assume that $\rho(T) = 1$. In particular, dim $\mathbf{V} \geq 1$. Choose a basis $\mathbf{b}_1, \ldots, \mathbf{b}_n$ in $\mathbf{V}$. Assume first that $\mathbb{F} = \mathbb{R}$. Then T represented in the basis $\mathbf{b}_1, \ldots, \mathbf{b}_n$ by $A = [a_{ij}] \in \mathbb{R}^{n \times n}$. Consider the matrix $B(z) = (I - zA)^{-1} = [b_{ij}]_{i=j=1}^{n} \mathbb{C}(z)^{n \times n}$. Using the Jordan canonical form of A we deduce that all the singular points of all $b_{ij}(z)$ are of the form $\mu := \frac{1}{\lambda}$ where λ is a nonzero eigenvalue of A. Furthermore, if $0 \neq \lambda \in$ spec (A), and λ has index $l = l(\lambda)$. Then for each i, j, $b_{ij}(z)$ may have a pole at μ of order l at most, and there is at least one entry $b_{ij}(z)$, where $i = i(\lambda), j = j(\lambda)$, such that $b_{ij}(z)$ has a pole of order l exactly. In particular, for each $\mathbf{x}, \mathbf{y} \in \mathbb{R}^n$ the rational function $\mathbf{y}^\top B(z)\mathbf{x}$ may have a pole of order l at most μ. Furthermore, there exists $\mathbf{x}, \mathbf{y} \in \mathbb{R}^n$, $\mathbf{x} = \mathbf{x}(\lambda), \mathbf{y} = \mathbf{y}(\lambda)$ such that $\mathbf{y}^\top B(z)\mathbf{x}$ has a pole at μ of order l. (See Problem 7.)

Let $\hat{K} \in \mathbb{R}^n$ denote the induced cone by K $\subset \mathbf{V}$. Then $\hat{K}$ is a proper cone. Denote by

$$\hat{K}' := \{\mathbf{y} \in \mathbb{R}^n, \mathbf{y}^\top \mathbf{x} \geq 0 \text{ for all } \mathbf{x} \in \hat{K}'\}.$$

Theorem 7.11.3 implies that $\hat{K}'$ is a proper cone. Observe next that $A\hat{K} \subset \hat{K}$. Clearly, we have the following MacLaurin expansion

$$B(z) = (I - zA)^{-1} = \sum_{i=0}^{\infty} z^i A^i, \quad \text{for } |z| < \frac{1}{\rho(A)}, \quad (7.11.3)$$

$$\mathbf{y}^\top B(z)\mathbf{x} = \sum_{i=0}^{\infty} (\mathbf{y}^\top A^i \mathbf{x}) z^i, \quad \text{for } |z| < \frac{1}{\rho(A)}. \quad (7.11.4)$$

Note that $\mathbf{y}^\top B(z)\mathbf{x}$ is a rational function. Denote by $r(\mathbf{x}, \mathbf{y}) \in (0, \infty]$ the convergence radius of $\mathbf{y}^\top B(z)\mathbf{x}$. So $r(\mathbf{x}, \mathbf{y}) = \infty$ if and only if $\mathbf{y}^\top B(z)\mathbf{x}$ is polynomial. Assume first that $\mathbf{x} \in \hat{K}, \mathbf{y} \in \hat{K}'$. Then the MacLaurin coefficients of $\mathbf{y}^\top B(z)\mathbf{x}$ are non-negative. Hence, we can apply the Vivanti–Pringsheim Theorem 7.9.2, i.e. $r(\mathbf{x}, \mathbf{y})$ is a singular point of $\mathbf{y}^\top B(z)\mathbf{x}$. Hence, $\frac{1}{r(\mathbf{x}, \mathbf{y})} \in$ spec (A) if $r(\mathbf{x}, y) < \infty$. Suppose now that $\mathbf{x}, \mathbf{y} \in \mathbb{R}^n$. Since $\hat{K}$ and $\hat{K}'$ are generating it follows

$\mathbf{x} = \mathbf{x}_+ - \mathbf{x}_-, \mathbf{y} = \mathbf{y}_+ - \mathbf{y}_-$ for some $\mathbf{x}_+, \mathbf{x}_- \in \hat{K}, \mathbf{y}_+, \mathbf{y}_- \in \hat{K}'$. So

$$\mathbf{y}^\top B(z)\mathbf{x} = \mathbf{y}_+^\top B(z)\mathbf{x}_+ + \mathbf{y}_-^\top B(z)\mathbf{x}_- - \mathbf{y}_-^\top B(z)\mathbf{x}_+ - \mathbf{y}_+^\top B(z)\mathbf{x}_-.$$

Hence

$$r(\mathbf{x}, \mathbf{y}) \geq r(\mathbf{x}_+, \mathbf{x}_-, \mathbf{y}_+, \mathbf{y}_-)$$
$$:= \min(r(\mathbf{x}_+, \mathbf{y}_+), r(\mathbf{x}_-, \mathbf{y}_-), r(\mathbf{x}_+, \mathbf{y}_-), r(\mathbf{x}_-, \mathbf{y}_+)).$$
$$(7.11.5)$$

Let $\lambda \in \text{spec}(A), |\lambda| = \rho(A)$, and assume that $l = \text{index}(\lambda)$. Choose $\mathbf{x}, \mathbf{y}$ such that $\mu = \frac{1}{\lambda}$ is a pole of $\mathbf{y}^\top B(z)\mathbf{x}$ of order l. Hence, we must have equality in (7.11.5). More precisely, there exists $\mathbf{x}_1 \in \{\mathbf{x}_+, \mathbf{x}_-\}, \mathbf{y}_1 \in \{\mathbf{y}_+, \mathbf{y}_-\}$ such that $r(\mathbf{x}, \mathbf{y}) = r(\mathbf{x}_1, \mathbf{y}_1)$ and $\mathbf{y}_1^\top B(z)\mathbf{x}_1$ has a pole at μ of order l. Vivanti–Pringsheim theorem yields that $r(\mathbf{x}_1, \mathbf{y}_1)$ is pole of order $k' \geq l$ of $\mathbf{y}_1^\top B(z)\mathbf{x}_1$. Hence, $\rho(A) \in \text{spec}(A)$ and index $(\rho(A)) \geq k' \geq l = \text{index}(\lambda)$. This proves parts 1 and 2. Choose $\lambda = \rho(A)$ that satisfies the above assumptions. Hence, $B(z)\mathbf{x}_1$ must have at least one coordinate with a pole at $\rho(A)^{-1}$ of order $k = \text{index}(\rho(A))$. Problem 7 yields that $\lim_{t \nearrow 1} B(t\rho(A)^{-1})\mathbf{x}_1 = \mathbf{u} \neq \mathbf{0}$ such that $A\mathbf{u} = \rho(A)\mathbf{u}$, and $\mathbf{u} \in (\rho(A)I - A)^{k-1}\mathbb{R}^n$. Use the fact that for $z = t\rho(A)^{-1}, t \in (0, 1)$ we have the equality (7.11.3). So $(1 - t)^k B(t\rho(A)^{-1})\mathbf{x}_1 \in \hat{K}$ for each $t \in (0, 1)$. Since $\hat{K}$ is closed we deduce that $\mathbf{u} \in \hat{K}$. This proves part 3.

The equality (7.11.1) follows from the Tauberian theorem given in Problem 8 and it application to the series (7.11.3). (Consult with [Har49] for various Tauberian theorems.)

Assume now that $A >^{\hat{K}} 0$. Observe first that the eigenvector $\mathbf{x} \geq^{\hat{K}} \mathbf{0}, A\mathbf{x} = \mathbf{x}$ satisfies $\mathbf{x} >^{\hat{K}} \mathbf{0}$, i.e. $\mathbf{x} \in \text{ri}\,\hat{K}$. Next we claim that the dimension of the eigenspace $\{\mathbf{y}, (A - I)\mathbf{y} = \mathbf{0}\}$ is 1. Assume to the contrary that $A\mathbf{y} = \mathbf{y}$ and $\mathbf{x}, \mathbf{y}$ are linearly independent. Since $x(s_0) = A\mathbf{x}(s_0) >^{\hat{K}} \mathbf{0}$ we obtain a contradiction. Hence, $\mathbf{x}$ is a geometrically simple eigenvalue.

Next we claim that $k = \text{index}(1) = 1$. Assume to the contrary that $k > 1$. Recall that $\mathbf{x} \in (A - I)^{k-1}\mathbb{R}^n$. So $\mathbf{x} = (A - I)\mathbf{y}$. Hence, $\mathbf{x} = (A - I)(\mathbf{y} + t\mathbf{x})$. Choose $t > 0$ big enough so that $\mathbf{z} = \mathbf{y} + t\mathbf{x} = t(\frac{1}{t}\mathbf{y} + \mathbf{x}) >^{\hat{K}} \mathbf{0}$. Since $\mathbf{x} >^{\hat{K}} \mathbf{0}$ it follows that there exists $s > 0$ such that $(A - I)\mathbf{z} - r\mathbf{z} = \mathbf{x} - r\mathbf{z} >^{\hat{K}} \mathbf{0}$. That is $A\mathbf{z} >^{\hat{K}} (1 + r)\mathbf{z}$. Hence, $A^m\mathbf{z} \geq (1+r)^m\mathbf{z} \Rightarrow (\frac{1}{1+r}A)^m\mathbf{z} \geq^{\hat{K}} \mathbf{z}$. Since $\rho(\frac{1}{1+r}A) = \frac{1}{1+r} < 1$ it follows that $\mathbf{0} = \lim_{m\to\infty}(\frac{1}{1+r}A)^m\mathbf{z} \geq^{\hat{K}} \mathbf{z}^{\hat{K}} >^{\hat{K}} \mathbf{0}$, which is impossible. Hence, index $(1) = 1$.

We now show that if $\lambda \in \operatorname{spec}(A)$ and $|\lambda| = 1$ then $\lambda = 1$. Let $J := \{\mathbf{y} \in \hat{K}, \|\mathbf{y}\|_2 = 1\}$. Since $\hat{K}$ is closed it follows that J is compact set. Since $A >^{\hat{K}} 0$ it follows that $AJ \in \operatorname{ri}\hat{K}$. Hence, there exists $s \in (0,1)$ such that $A\mathbf{y} - s\mathbf{z} >^{\hat{K}} \mathbf{0}$ for any $\mathbf{y} \in J, \mathbf{z} \in \mathbb{R}^n, \|\mathbf{y}\|_2 = 1$. In particular, $A\mathbf{y} - s\mathbf{y} >^{\hat{K}} \mathbf{0}$ for each $\mathbf{y} \in J$. That is $(A - sI)J \in \operatorname{ri}\hat{K}$. Hence, $(A - sI) >^{\hat{K}} 0$. Note that $(A - sI)\mathbf{x} = (1 - s)\mathbf{x} >^{\hat{K}} \mathbf{0}$. So $\rho(A - sI) = 1 - s$. Each eigenvalue of $A - sI$ is $\lambda - s$, where $\lambda \in \operatorname{spec}(A)$. Apply part 1 of the theorem to deduce that $|\lambda - s| \leq 1 - s$. Since for any $\zeta \in S\backslash\{1\}$ we must have that $|\zeta - s| > |\zeta| - s = 1 - s$, we obtain that $S(A)(1) = \{1\}$, which concludes the proof of part 4 for $\mathbb{F} = \mathbb{R}$.

Assume next that $\rho(T) = \rho(A) = 1$, and $k = \operatorname{index}(1)$. (7.11.11) of Problem 9 yields the equality in (7.11.1). Since any sum in the left-hand side of (7.11.1) is non-negative with respect to the cone K it follows that $(A - I)^{k-1}P \geq^K 0$. Let $\lambda_1 = 1$ so $s_1 = k$, see notation of Problem 7. Recall that $(A - I)^{k-1}P$ in the basis $\mathbf{b}_1,\ldots,\mathbf{b}_n$ is represented by the component $Z_{1(k-1)} \neq 0$. Hence, $(A - I)^{k-1}P \gneq^K 0$.

Assume finally that $\lambda \in \operatorname{spec}(T), |\lambda| = 1, \lambda \neq 1, \operatorname{index}(\lambda) = k$. Let $P(\lambda)$ be the spectral projection on the eigenvalue λ. Then (7.11.11) yields

$$\lim_{m\to\infty} \frac{k!}{m^k} \sum_{r=0}^{m-1} \bar{\lambda}^r T^r = \bar{\lambda}^{k-1}(T - \lambda I)^{k-1}P(\lambda). \qquad (7.11.6)$$

Let $\mathbf{y} \in K, \mathbf{f} \in K$. Since $\mathbf{f}(T^r\mathbf{y}) \geq 0$ and $|\lambda| = 1$ we obtain $|\mathbf{f}(\bar{\lambda}^r T^r\mathbf{y})| = \mathbf{f}(T^r\mathbf{y})$. The triangle inequality

$$\left| \frac{k!}{m^k} \sum_{r=0}^{m-1} \mathbf{f}(\bar{\lambda}^r T^r\mathbf{y}) \right| \leq \frac{k!}{m^k} \sum_{r=0}^{m-1} \mathbf{f}(T^r\mathbf{y}).$$

Let $m \to \infty$ and use the equalities (7.11.6) and (7.11.1) to deduce (7.11.2).

We now point out why our results hold for a vector space $\mathbf{V}$ over $\mathbb{C}$. Let $T \in \mathrm{Hom}\,(\mathbf{V}, \mathbf{V})$ and assume that $\mathbf{b}_1, \ldots, \mathbf{b}_n$ is a basis $\mathbf{V}$. Then $\mathbf{V}_{\mathbb{R}}$ has a basis $\mathbf{b}_1, \ldots, \mathbf{b}_n, \sqrt{-1}\mathbf{b}_1, \ldots, \sqrt{-1}\mathbf{b}_n$. Clearly T induces an operator $\tilde{T} \in \mathrm{Hom}\,(\mathbf{V}_{\mathbb{R}}, \mathbf{V}_{\mathbb{R}})$. Let $A \in \mathbb{C}^{n \times n}$ represents T in the basis $\mathbf{b}_1, \ldots, \mathbf{b}_n$. Observe that $A = B + \sqrt{-1}C$, where $B, C \in \mathbb{R}^{n \times n}$. Then $\tilde{T}$ is presented by the matrix $\tilde{A} = \begin{bmatrix} B & -C \\ C & B \end{bmatrix}$ in the basis $\mathbf{b}_1, \ldots, \mathbf{b}_n, \sqrt{-1}\mathbf{b}_1, \ldots, \sqrt{-1}\mathbf{b}_n$. See Problem 10.

Problems

1. Let $\mathbf{V}$ be a vector space of dimension n over $\mathbb{C}$, with a basis $\mathbf{z}_1, \ldots, \mathbf{z}_n$. Show

   (a) $\mathbf{V}$ is a vector space over $\mathbb{R}$ of dimension $2n$, with a basis $\mathbf{z}_1, \sqrt{-1}\mathbf{z}_1, \ldots, \mathbf{z}_n, \sqrt{-1}\mathbf{z}_n$. We denote this real vector space by $\mathbf{V}_{\mathbb{R}}$ and its dimension $\dim{}_{\mathbb{R}}\mathbf{V}$.

   (b) Let the assumptions of 1(a) hold. Then $\mathbf{V}'$ can be identified with $(\mathbf{V}_{\mathbb{R}})^*$ as follows. Each $\mathbf{f} \in \mathbf{V}'$ gives rise to $\hat{\mathbf{f}} \in (\mathbf{V}_{\mathbb{R}})'$ by the formula $\hat{\mathbf{f}}(\mathbf{z}) = \Re \mathbf{f}(\mathbf{z})$. In particular, if $\mathbf{f}_1, \ldots, \mathbf{f}_n$ form a basis in $\mathbf{V}'$ then $\hat{\mathbf{f}}_1, \widehat{\sqrt{-1}\mathbf{f}_1}, \ldots, \hat{\mathbf{f}}_n, \widehat{\sqrt{-1}\mathbf{f}_n}$ is a basis in $(\mathbf{V}_{\mathbb{R}})'$.

2. Let K be a cone. Show that K is pointed if and only the two inequalities $\mathbf{x} \geq^K \mathbf{y}, \mathbf{y} \geq^K \mathbf{x}$ imply that $\mathbf{x} = \mathbf{y}$.

3. Let the assumptions of Problem 1 hold. Assume that $K \subset \mathbf{V}$ is a cone. Denote by $K_{\mathbb{R}}$ the induced cone in $\mathbf{V}_{\mathbb{R}}$. Show

   (a) K is closed if and only if $K_{\mathbb{R}}$ is closed.

   (b) K is pointed if and only if $K_{\mathbb{R}}$ is pointed.

   (c) K is generating if and only if $K_{\mathbb{R}}$ is generating.

   (d) K is pointed if and only if $K_{\mathbb{R}}$ is pointed.

4. Let $\mathbf{U}$ be a real vector space. Denote by $\mathbf{U}_{\mathbb{C}}$ as in Proposition 4.1.2. Assume that $K \subset \mathbf{U}$ is a cone. Let $K_{\mathbb{C}} := \{(\mathbf{x}, \mathbf{y}),\ \mathbf{x},$

$\mathbf{y} \in \mathrm{K}\}$. Show

   (a) $\mathrm{K}_{\mathbb{C}}$ is a cone in $\mathbf{U}_{\mathbb{C}}$.

   (b) K is closed if and only if $\mathrm{K}_{\mathbb{C}}$ is closed.

   (c) K is pointed if and only if $\mathrm{K}_{\mathbb{C}}$ is pointed.

   (d) K is generating if and only if $\mathrm{K}_{\mathbb{C}}$ is generating.

   (e) K is proper if and only if $\mathrm{K}_{\mathbb{C}}$ is proper.

5. Let the assumptions of Problem 4 hold. Assume that $A \in$ Hom $(\mathbf{U}, \mathbf{U})^{\mathrm{K}}$. Define $\hat{A} : \mathbf{U}_{\mathbb{C}} \to \mathbf{U}_{\mathbb{C}}$ by $\hat{A}(\mathbf{x}, \mathbf{y}) = (A\mathbf{x}, A\mathbf{y})$. Show

   (a) $\hat{A} \in$ Hom $(\mathbf{U}_{\mathbb{C}}, \mathbf{U}_{\mathbb{C}})^{\mathrm{K}_{\mathbb{C}}}$.

   (b) $\det (zI - A) = \det (zI - \hat{A})$.

   (c) $\hat{A}$ is not positive with respect to $\mathrm{K}_{\mathbb{C}}$.

6. Let $\mathbf{V}$ be a vector space over $\mathbb{F} = \mathbb{R}, \mathbb{C}$. Assume that $\mathrm{K} \subset \mathbf{V}$ and $A \in$ Hom $(\mathbf{V}, \mathbf{V})^{\mathrm{K}}$. Then $A^* \in$ Hom $(\mathbf{V}', \mathbf{V}')^{\mathrm{K}'}$.

7. Let $A \in \mathbb{C}^{n \times n}$. Assume that $S(A) = \{\lambda_1, \dots, \lambda_n\}$ is the eigenvalue multiset of A. Consider the matrix $B(z) = (I - zA)^{-1} = [b_{ij}]_{i=j=1}^{n} \in \mathbb{C}(z)^{n \times n}$. Show

   (a) Let $Z_{i1}, \dots, Z_{i(s_i - 0)}, i = 1, \dots, \ell$ be all the matrix components of A as in §3.1. Then

      i. Z_{i0} is the spectral projection of A on λ_i. (See §3.4.) Furthermore

$$(A - \lambda_i I)^{s_i - 1} Z_{i0} = Z_{i(s_i - 1)} \text{ for } i = 1, \dots, \ell. \qquad (7.11.7)$$

      (Use (3.1.6).) So $Z_{i(s_i - 1)}\mathbb{C}^n$ is a subspace of eigenvectors of A corresponding to all Jordan blocks of A of order s_i and eigenvalue λ_i.

ii.

$$(I - zA)^{-1} = \sum_{i=1}^{\ell} \sum_{j=0}^{s_i-1} \frac{z^j}{(1-\lambda_i z)^{j+1}} Z_{ij}, \qquad (7.11.8)$$

$$\lim_{t \nearrow 1}(1-t)^{s_i}\left(I - \frac{t}{\lambda_i}A\right)^{-1} = \frac{1}{\lambda_i^{s_i}} Z_{i(s_i-1)}. \qquad (7.11.9)$$

(*Hint*: To show the first equality use (3.4.1) by letting $\lambda = \frac{1}{z}$ and divide (3.4.1) by z.)

(b) All the singular points of all $b_{ij}(z)$ are of the form $\mu := \frac{1}{\lambda}$ where λ is a nonzero eigenvalue of A. Furthermore, if $0 \neq \lambda \in$ spec (A), and λ has index $l = l(\lambda)$. Then for each i, j $b_{ij}(z)$ may have a pole at μ of order l at most, and there is at least one entry $b_{ij}(z)$, where $i = i(\lambda), j = j(\lambda)$, such that $b_{ij}(z)$ has a pole of order l exactly. Furthermore Suppose furthermore, that for $\mathbf{x} \in \mathbb{C}^n$, at least one of the entries of $B(z)\mathbf{x}$ has a pole of order l at μ. Then $\lim_{t \nearrow 1}(1-t)^l B(t\mu)\mathbf{x} = \mathbf{y} \neq \mathbf{0}$, $A\mathbf{y} = \lambda\mathbf{y}$ and $\mathbf{y} \in (\lambda I - A)^{l-1}\mathbb{C}^n$.

(c) Let $\mathbf{e}_i = (\delta_{i1}, \ldots, \delta_{in})^\top, i \in \langle n \rangle$. For each $0 \neq \lambda \in$ spec (A) of index $l = l(\lambda)$ there exists $\mathbf{e}_i, \mathbf{e}_j, i = i(\lambda), j = j(\lambda)$ such that $\mathbf{e}_i^\top B(z)\mathbf{e}_j$ has a pole at $\frac{1}{\lambda}$ of order l exactly.

8. Let $k, l \in \mathbb{N}$ and consider the rational function

$$f(z) = \frac{1}{(1-\mu z)^l} = \sum_{i=0}^{\infty}(-1)^i \binom{-l}{i}\mu^i z^i.$$

Show

(a)

$$\frac{k!}{m^k}\sum_{i=0}^{m-1}(-1)^i\binom{-k}{i} = \frac{k!}{m^k}\sum_{i=0}^{m-1}|\binom{-k}{i}| = 1.$$

Hint: Use the Riemann sums for the integral $\int_0^1 x^{k-1}dx$ to show

$$\lim_{m \to \infty} \frac{k}{n^k}\sum_{i=1}^{m} i^{k-1} = 1.$$

(b)

$$\frac{k!}{m^k} \sum_{i=0}^{m-1} (-1)^i \binom{-l}{i} \mu^i = 0.$$

Under the following assumptions

i. $|\mu| = 1, \mu \neq 1$ and $l = k$. *Hint*: Recall the identity

$$s_m(z) = \sum_{i=0}^{m-1} (-1)^i \binom{-1}{i} z^i = \sum_{i=0}^{m-1} z^i = \frac{1 - z^m}{1 - z}.$$

Show that

$$\frac{1}{(k-1)!} \frac{d^{k-1}}{dz^{k-1}} s_{m+k-1}(z) = \sum_{i=0}^{m-1} (-1)^i \binom{-k}{i} z^i.$$

ii. $|\mu| = 1$ and $l < k$. *Hint*: Sum the absolute values of the corresponding terms and use part 8(a).

iii. $|\mu| < 1$. *Hint*: Use the Cauchy–Hadamard formula to show that $\sum_{i=0}^{m-1} |\binom{-l}{i}| |\mu|^i < \infty$.

9. Let the assumptions of Problem 7 hold.

(a) For $m \geq \max(s_1, \ldots, s_\ell)$

$$\sum_{r=0}^{m-1} z^r A^r = \sum_{i=1}^{\ell} \sum_{j=0}^{s_i-1} z^j \left(\sum_{r=0}^{m-1-j} (-1)^r \binom{-(j+1)}{r} \lambda_i^r z^r Z_{ij} \right).$$

$$(7.11.10)$$

(Use the first m terms of MacLaurin expansion of both sides of (7.11.8).)

(b) Assume furthermore that $\rho(A) = 1$ and k is the maximal index of all eigenvalues λ_i satisfying $|\lambda_i| = 1$. Assume that $\lambda = \lambda_1, |\lambda| = 1$, and $k = \text{index}(\lambda_1) = s_1$. Let $P(\lambda) = Z_{10}$ be the spectral projection on λ and $Z_{1(s_1-1)} = (A-\lambda)^{k-1} P(\lambda_1)$.

Then (7.11.10) and Problem 8 implies.

$$\lim_{m \to \infty} \frac{k!}{m^k} \sum_{r=0}^{m-1} \bar{\lambda}^r A^r = \bar{\lambda}^{k-1}(A - \lambda I)^{k-1} P(\lambda). \quad (7.11.11)$$

10. Let $\mathbf{V}$ be a vector space over $\mathbb{C}$ with a basis $\mathbf{b}_1, \ldots, \mathbf{b}_n$. Then $\mathbf{V}_{\mathbb{R}}$ has a basis $\mathbf{b}_1, \ldots, \mathbf{b}_n, \sqrt{-1}\mathbf{b}_1, \ldots, \sqrt{-1}\mathbf{b}_n$. Let $T \in \mathrm{Hom}\,(\mathbf{V}, \mathbf{V})$. Show Clearly T induces an operator $\tilde{T} \in \mathrm{Hom}\,(\mathbf{V}_{\mathbb{R}}, \mathbf{V}_{\mathbb{R}})$. Let $A \in \mathbb{C}^{n \times n}$ represents T in the basis $\mathbf{b}_1, \ldots, \mathbf{b}_n$. Observe that $A = B + \sqrt{-1}C$, where $B, C \in \mathbb{R}^{n \times n}$. Then $\tilde{T}$ is presented by the matrix $\tilde{A} = \begin{bmatrix} B & -C \\ C & B \end{bmatrix}$ in the basis $\mathbf{b}_1, \ldots, \mathbf{b}_n, \sqrt{-1}\mathbf{b}_1, \ldots, \sqrt{-1}\mathbf{b}_n$

7.12   Historical Remarks

§7.1 is standard. §7.2 follows Friedland–Zenger [FrZ84] and [Fri84]. The results of §7.3 are well known. Theorem 7.3.2 was conjectured by P.R. Halmos, and proved by several authors: [Ber65, Kat65, Pea66, BeS67]. §7.4 is standard. §7.5 follows the basic notion of separable states in quantum mechanics as in [Gur03, BeZ06]. The notion of the cross norm is well known [Sch50]. §7.6 follows some ideas in [Fri08]. §7.7 follows closely [Fri82]. See also [Bha84, BhE90, BhF81, Els82, EJR83]. §7.8 follows closely [Fri90]. §7.9 follows [Fri78b]. See also [Els88]. §7.10 gives a short account on the inverse eigenvalue problem for non-negative matrices non-negative matrices, abbreviated as IEPFNM. See [Fri77] for general account on inverse eigenvalue problems for matrices and [Fri78] on preliminary results IEPFNM. Part 3 of Proposition 7.10.2 is due to Loewy–London and Johnson [LoL78, Joh81]. Theorem 7.10.8 is due to Boyle–Handelman [BoH91]. Our proof follows the arguments of Laffey [Laf12]. Theorem 7.10.10 is due to Friedland [Fri12]. The results of §7.11 are well known, e.g. [BeP79].

Bibliography

[Arn71]   V.I. Arnold, On matrices depending on paramemters, *Usp. Math. Nauk.* 26 (1971), 101–114.

[Avr03]   M. Avriel, *Nonlinear Programming: Analysis and Methods*, Dover Publishing, Mineota, N.Y., 2003.

[BR97]   R.B. Bapat and T.E.S. Raghavan, *Nonnegative Matrices and Applications*, Cambridge University Press, Cambridge, UK, 1997.

[BeZ06]   I. Bengtsson and K. Zyckowski, *Geometry of Quantum States: An Introduction to Quantum Entanglement*, Cambridge University Press, Cambridge, UK, 2006.

[Ber65]   C.A. Berger, On the numerical range of powers of an operator, Abstract No. 625152, *Notices Amer. Math. Soc.* 12 (1965), 590.

[BeS67]   C.A. Berger and J.G. Stampfli, Mapping theorems for the numerican range, *Amer. J. Math.* 89 (1967), 1047–1055.

[BeP79]   A. Berman and R.J. Plemmons, *Nonnegative Matrices in Mathematical Sciences*, Academic Press, New York, 1979.

[Bha84]   R. Bhatia, Variation of symmetric tensors and permanents, *Linear Algebra Appl.* 62 (1984), 268–276.

[BhE90]   R. Bhatia and L. Elsner, On the variation of permanents,
Linear Multilin. Algebra 27 (1990), 105–109.

[BhF81]   R. Bhatia and S. Friedland, Variation of Grassman powers and spectra, *Linear Algebra Appl.* 40 (1981), 1–18.

[Big96]   N. Biggs, *Algebraic Graph Theory*, Cambridge University
Press, Cambridge, second edition, 1996.

[Bir46]   G. Birkhoff, Three observations on linear algebra, *Univ.
Nac. Tacumán Rev. Ser. A* 5 (1946) 147–151.

[BoH91]   M. Boyle and D. Handelman, The spectra of nonnegative
matrices via symbolic dynamics, *Ann. Math.* 133 (1991),
249–316.

[BFP95]   R.A. Brualdi, S. Friedland and A. Pothen, The sparse
basis problem and multilinear algebra, *SIAM J. Matrix
Anal. Appl.* 16 (1995), 1–20.

[BPS66]   R.A. Brualdi, S.V. Parter and H. Schneider, The diagonal equivalence of a nonnegative matrix to a stochastic
matrix, *J. Math. Anal. Appl.* 16 (1966), 31–50.

[BR91]   R.A. Brualdi and H.J. Ryser, *Combinatorial Matrix Theory*, Cambridge University Press, Cambridge, 1991.

[Cal64]   E. Calabi, Linear systems of real quadratic forms, *Proc.
Amer. Math. Soc.* 15 (1964), 844–846.

[Coh79]   J.E. Cohen, Random evolutions and the spectral radius
of a non-negative matrix, *Mat. Proc. Camb. Phil. Soc.* 86
(1979), 345–350.

[CFKK82] J.E. Cohen, S. Friedland, T. Kato and F.P. Kelly, Eigenvalue inequalities for products of matrix exponentials,
Linear Algebra Appl. 45 (1982), 55–95.

[Col42]   L. Collatz, Einschließungssatz für die charakteristischen Zahlen von Matrizen, *Math. Z.* 48 (1942), 221–226.

[Cou20]   R. Courant, Über die eigenwerte bei den differentialgleichungen der mathematischen physik, *Math. Z.* 7 (1920), 1–57.

[CuR62]   C.W. Curtis and I. Reiner, *Representation Theory of Finite Groups and Associative Algebras*, Interscience, New York, 1962.

[Dan63]   G.B. Dantzig, *Linear Programming and Extensions*, Princeton University Press, Princeton, NJ, 1963.

[Dav57]   C. Davis, All convex invariant functions of hermitian matrices, *Arch. Math.* 8 (1957), 276–278.

[Die69]   J. Dieudonné, *Foundations of Modern Analysis*, Academic Press, New York–London, 1969.

[DV75]   M.D. Donsker and S.R.S. Varadhan, On a variational formula for the principal eigenvalue for operators with maximum principle, *Proc. Natl. Acad. Sci. U.S.A.* 72 (1975), 780–783.

[DDG51]   M.A. Drazin, J.W. Dungey and K.W. Gruenberg, Some theorems on commutative matrices, *J. London Math. Soc.* 26 (1951), 221–228.

[DM64]   A.L. Dulmage and N.S. Mendelsohn, Gaps in the exponent of primitive matrices, *Illinois J. Math.* 8 (1964), 642–656.

[Els82]   L. Elsner, On the variation of spectra of matrices, *Linear Algebra Appl.* 47 (1982), 127–132.

[Els88]   L. Elsner, A note on the variation of permanents, *Linear Algebra Appl.* 109 (1988), 37–39.

[EJRS83]   L. Elsner, C. Johnson, J. Ross and J. Schönheim, On a generalized matching problem arising in estimating the eigenvalue variation of two matrices, *Eur. J. Combin.* 4 (1983), 133–136.

[Fad66]   D.K. Fadeev, On the equivalence of systems of integral matrices, *Izv. Akad. Nauk SSSR, Ser. Mat.* 30 (1966), 449–454.

[Fan49]   K. Fan, On a theorem of Weyl concerning eigenvalues of linear transformations, I. *Proc. Nat. Acad. Sci. U.S.A.* 35 (1949), 652–655.

[Fis05]   E. Fischer, Über quadratische formen mit reellen koeffizienten, *Monatshefte Math. Phys.* 16 (1905), 234–249.

[Fri73]   S. Friedland, Extremal eigenvalue problems for convex sets of symmetric matrices and operators, *Israel J. Math.* 15 (1973), 311–331.

[Fri77]   S. Friedland, Inverse eigenvalue problems, *Linear Algebra Appl.* 17 (1977), 15–51.

[Fri78]   S. Friedland, Extremal eigenvalue problems, *Bull. Brazilian Math. Soc.* 9 (1978), 13–40.

[Fri78b]   S. Friedland, On inverse problem for nonnegative and eventually nonnegative matrices, *Israel J. Math.* 29 (1978), 43–60.

[Fri80a]   S. Friedland, On pointwise and analytic similarity of matrices, *Israel J. Math.* 35 (1980), 89–108.

[Fri80b]   S. Friedland, Analytic similarities of matrices, *Lectures in Applied Math., Amer. Math. Soc.* 18 (1980), 43–85 (edited by C.I. Byrnes and C.F. Martin).

[Fri81a]   S. Friedland, A generalization of the Kreiss matrix theorem, *SIAM J. Math. Anal.* 12 (1981), 826–832.

[Fri81b]   S. Friedland, Convex spectral functions, *Linear Multilin. Algebra* 9 (1981), 299–316.

[Fri82]   S. Friedland, Variation of tensor powers and spectra, *Linear Multilin. Algebra* 12 (1982), 81–98.

[Fri83]   S. Friedland, Simultaneous similarity of matrices, *Adv. Math.* 50 (1983), 189–265.

[Fri84]   S. Friedland, Stable convex sets of matrices, *Linear Multilin. Algebra* 16 (1984), 285–294.

[Fri90]   S. Friedland, A remark on the variation of permanents, *Linear Multilin. Algebra* 27 (1990), 101–103.

[Fri00]   S. Friedland, Finite and infinite dimensional generalizations of Klyachkos theorem, *Linear Alg. Appl.* 319 (2000), 3–22.

[Fri06]   S. Friedland, Convergence of products of matrices in projective spaces, *Linear Alg. Appl.* 413 (2006), 247–263.

[Fri08]   S. Friedland, On the graph isomorphism problem, arXiv:0801.0398.

[Fri12]   S. Friedland, A note on the nonzero spectra of irreducible matrices, *Linear Multilin. Algebra* 60 (2012), 1235–1238.

[Fri15]   S. Friedland, Equality in Wielandt's eigenvalue inequality, *Spec. Matrices 3* (2015), 53–57.

[FrK75]   S. Friedland and S. Karlin, Some inequalities for the spectral radius of non-negative matrices and applications, *Duke Math. J.* 42 (3) (1975), 459–490.

[FrL76]   S. Friedland and R. Loewy, Subspaces of symmetric matrices containing matrices with a multiple first eigenvalue, *Pacific J. Math.* 62 (1976), 389–399.

[FMMN11] S. Friedland, V. Mehrmann, A. Miedlar and M. Nkengla, Fast low rank approximations of matrices and tensors, *Electron. J. Linear Algebra* 22 (2011), 1031–1048.

[FrP04]   S. Friedland and G. Porta, The limit of the product of the parameterized exponentials of two operators, *J. Funct. Anal.* 210 (2004), 436–464.

[FrS80]   S. Friedland and H. Schneider, The growth of powers of nonnegative matrix, *SIAM J. Algebraic Discrete Methods* 1 (1980), 185–200.

[FrS94]   S. Friedland and W. So, On the product of matrix exponentials, *Linear Algebra Appl.* 196 (1994), 193–205.

[FrT84]   S. Friedland and E. Tadmor, Optimality of the Lax–Wendroff condition, *Linear Alg. Appl.* 56 (1984), 121–129.

[FriT07] S. Friedland and A. Torokhti. Generalized rank-constrained matrix approximations, *SIAM J. Matrix Anal. Appl.* 29 (2007), 656–659.

[FrZ84]   S. Friedland and C. Zenger, All spectral dominant norms are stable, *Linear Algebra Appl.* 58 (1984), 97–107.

[Fro08]   G. Frobenius, Über Matrizen aus positiven Elementen, 1, *S.-B. Preuss. Akad. Wiss.* Berlin 1908, 471–476.

[Fro09]   G. Frobenius, Über Matrizen aus positiven Elementen, 2, *S.-B. Preuss. Akad. Wiss.* Berlin 1909, 514–518.

[Fro12]   G. Frobenius, Über Matrizen aus nicht negativen Elementen, *S.-B. Preuss. Akad. Wiss.* Berlin 1912, 456–477.

[Ful00]   W. Fulton, Eigenvalues of majorized Hermitian matrices and Littlewood–Richardson coefficients, *Linear Alg. Appl.* 319 (2000), 23–36.

[Gan59]   F.R. Gantmacher, *The Theory of Matrices*, Vols. I and II, Chelsea Publ. Co., New York, 1959.

[GaJ79]   M.R. Garey and D.S. Johnson, *Computers and Intractability: A Guide to the Theory of NP-Completeness*, W. H. Freeman, 1979.

[GaB77]   M.A. Gauger and C.I. Byrnes, Characteristic free, improved decidability criteria for the similarity problem, *Lin. Multilin. Alg.* 5 (1977), 153–158.

[Gin78]   H. Gingold, A method for global block diagonalization for matrix-valued functions, *SIAM J. Math. Anal.* 9 (1978), 1076–1082.

[Gol57]   S. Golden, Statistical theory of many-electron systems; discrete bases of representation, *Phys. Rev.* 107 (1957), 1283–1290.

[GolV96]   G.H. Golub and C.F. Van Loan, Matrix Computation, John Hopkins University Press, Baltimore, 3rd editon, 1996.

[GorT01]   S.A. Goreinov and E.E. Tyrtyshnikov, The maximum-volume concept in approximation by low-rank matrices, *Contemporary Mathematics* 280 (2001), 47–51.

[GorTZ95]   S.A. Goreinov, E.E. Tyrtyshnikov and N.L. Zamarashkin, Pseudo-skeleton approximations of matrices, *Reports of the Russian Academy of Sciences* 343(2) (1995), 151–152.

[GorTZ97]   S.A. Goreinov, E.E. Tyrtyshnikov and N.L. Zamarashkin, A theory of pseudo-skeleton approximations of matrices, *Linear Algebra Appl.* 261 (1997), 1–21.

[GrH78]   P. Griffiths and J. Harris, *Principles of Algebraic Geometry*, J. Wiley, New York, 1978.

[GuR65]   R. Gunning and H. Rossi, *Analytic Functions of Several Complex Variables*, Prentice-Hall, New Jersey, 1965.

[Gur80]   R.M. Guralnick, A note on the local–global prinicple for similarity of matrices, *Lin. ALg. Appl.* 30 (1980), 241–245.

[Gur81]   R.M. Guralnick, Similarity of matrices over local rings, *Linear Algebra Appl.* 41 (1981), 161–174.

[Gur03]   L. Gurvits, Classical deterministic complexity of Edmond's problem and quantum entanglement, In: *Proc. 35th STOC* 2003, pp. 10–19.

[Hal35]   P. Hall, On representation of subsets, *J. London Math. Soc.* 10 (1935), 26–30.

[Har49]   G.H. Hardy, *Divergent Series*, Clarendon Press, Oxford, 1949.

[HLP52]   G.H. Hardy, J.E. Littlewood and G. Pólya, *Inequalities*, Cambridge University Press, Cambridge, second edition, 1952.

[Hel40]   O. Helmer, Divisibility properties of integral functions, *Duke Math. J.* 6 (1940), 345–356.

[Hel43]   O. Helmer, The elementary divisor theorems for certain rings without chain conditions, *Bull. Amer. Math. Soc.* 49 (1943), 225–236.

[HW53]   A.J. Hoffman and H.W. Wielandt, The variation of the spectrum of a normal matrix, *Duke Math. J.* 20 (1953), 37–39.

[HoV58]   J.C. Holladay and R.S. Varga, On powers of non-negative matrices, *Proc. Amer. Math. Soc.* 9 (1958), 631–634.

[Hor62]   A. Horn, Eigenvalues of sums of Hermitian matrices, *Pacific J. Math.* 12 (1962), 225–241.

[Hor96]   M. Horodecki, P. Horodecki and R. Horodecki, Separability of mixed states: Necessary and sufficient conditions, *Phys. Lett. A* 223 (1996), 1–8.

[HJ88]   R.A. Horn and C.R. Johnson, *Matrix Analysis*, Cambridge University Press, New York, 1988.

[Joh81]   C. R. Johnson, Row stochastic matrices similar to doubly stochastic matrices, *Lin. Multilin. Alg.* 10 (1981), 113–130.

[Kap49]   I. Kaplansky, Elementary divisors and modules, *Trans. Amer. Math. Soc.* 66 (1949), 464–491.

[Kar84]   N.K. Karmakar, A new polynomial agorithm for linear programming, *Combinatorica* 4 (1984), 373–395.

[Kat65]   T. Kato, Some mapping theorems for the numerical range, *Proc. Japan Acad.* 41 (1965), 652–655.

[Kat80]   T. Kato, *Perturbation Theory for Linear Operators*, Springer-Verlag, New York, 2nd edition, 1980.

[Kat82]   T. Kato, *A Short Introduction to Perturbation Theory for Linear Operators*, Springer-Verlag, New York, 2nd edition, 1982.

[Katz70]   M. Katz, On the extreme points of a certain convex polytope, *J. Combinatorial Theory* 8 (1970), 417–423.

[Kha79]   L.G. Khachian, A polynomial algorithm in linear programming, *Dokl. Akad. Nauk SSSR* 244 (1979), 1093–1096. English translation in *Soviet Math. Dokl.* 20 (1979), 191–194.

[Kin61]   J.F.C. Kingman, A convexity property of positive matrices, *Quart. J. Math. Oxford* 12 (1961), 283–284.

[Kly98]   A.A. Klyachko, Stable bundles, representation theory and Hermitian operators, *Selecta Math. (N.S.)* 4 (1998), 419–445.

[Knu98]   D.E. Knuth, *The Art of Computer Programming, vol. 3: Sorting and Searching*, Addison-Wesley, second edition, 1998.

[KT99]   A. Knutson and T. Tao, The honeycomb model of $GL_n(\mathbb{C})$ tensor products, I. Proof of the saturation conjecture, *J. Amer. Math. Soc.* 12 (1999), 1055–1090.

[Kre62]   H.O. Kreiss, Über die stabilitätsdefinition für differenzgleichungen die partielle differentiel gleichungen approximieren, *Nordisk Tidskr., Informations-Behandling* 2 (1962), 153–181.

[Kro90]   L. Kronecker, Algebraische reduction der schaaren bilinear formen, *S-B Akad. Berlin* (1890), 763–778.

[Laf12]   T.J. Laffey, A constructive version of the Boyle–Handelman theorem on the spectra of nonnegative matrices, *Linear Algebra Appl.* 436 (2012), 1701–1709.

[Lanc50]   C. Lanczos, An iteration method for the solution of the eigenvalue problem of linear differential and integral operators, *J. Res. Nat. Bureau Standards, Sec. B* 45 (1950), 255–282.

[Lan58]   S. Lang, *Introduction to Algebraic Geometry*, Interscience, New York, 1958.

[Lan67]   S. Lang, *Algebra*, Addison-Wesley, MA, 1967.

[Lea48]   W.G. Leavitt, A normal form for matrices whose elements are holomorphic functions, *Duke Math. J.* 15 (1948), 463–472.

[Lid50]   V.B. Lidskii, On the characteristic numbers of the sum and product of symmetric matrices. *Doklady Akad. Nauk SSSR (N.S.)* 75 (1950), 769–772.

[LoL78]   R. Loewy and D. London, A note on an inverse problem for nonnegative matrices, *Linear Multilin. Algebra* 6 (1978), 83–90.

[Lov86]   L. Lovász, *An Algorithmic Theory of Numbers, Graphs and Convexity*, CBMS-NSF Regional Conference Series in Applied Mathematics, Vol. 50, SIAM, Philadelphia, PA, 1986.

[Luo08]   Z.-Q. Luo and Z. Zhang, Dynamic spectrum management: Complexity and duality, *IEEE J. Selected Areas in Signal Processing* 2(1) (2008), 57–73.

[McD33]   C.C. MacDuffee, *The Theory of Matrices*, Springer, Berlin, 1933.

[MaM64]   M. Marcus and H. Minc, *A Survey of Matrix Theory and Matrix Inequalities*, Prindle, Weber & Schmidt, Boston, 1964.

[MOA11]   A.W. Marshall, I. Olkin and B. Arnold, *Inequalities: Theory of Majorization and Its Applications*, Springer, New York, 2011.

[MoF80]   N. Moiseyev and S. Friedland, The association of resonance states with incomplete spectrum of finite complex scaled Hamiltonian matrices, *Phys. Rev. A* 22 (1980), 619–624.

[Moo20]   E.H. Moore, On the reciprocal of the general algebraic matrix, *Bul. Amer. Math. Soc.* 26 (1920), 394–395.

[MoT52]   T.S. Motzkin and O. Taussky, Pairs of matrices with property L, *Trans. Amer. Math. Soc.* 73 (1952), 108–114.

[MoT55]   T.S. Motzkin and O. Taussky, Pairs of matrices with property L. II, *Trans. Amer. Math. Soc.* 80 (1955), 387–401.

[Neu77]   C. Neumann, Untersuchungen über das logarithmische Newtonsche potential, Teubner Leipzig, 1877.

[NoW99]   J. Nocedal and S.J. Wright, *Numerical Optimization*, Springer, New York, 1999.

[Olv74]   F.W.J. Olver, *Asymptotic and Special Functions*, Academic Press, New York, 1974.

[Ost66]   A. Ostrowski, *Solutions of Equations and Systems of Equations*, Academic Press, New York, second edition, 1966.

[Pea66]   C. Pearcy, An elementary proof of the power inequality for the numerical radius, *Michigan Math. J.* 13 (1966), 289–291.

[Pen55]   R. Penrose, A generalized inverse for matrices, *Proc. Cambridge Phil. Soc.* 51 (1955), 406–413.

[Per96]   A. Peres, Separability criterion for density matrices, *Phys. Rev. Lett.* 77 (1996), 1413–1415.

[Per07]   O. Perron, Zur theorie der matrices, *Math. Ann.* 64 (2) (1907), 248–263.

[Pet88]   D. Petz, A variational expression for the relative entropy, *Commun. Math. Phys.* 114 (1988), 345–349.

[Pin09]   M.A. Pinsky, *Introduction to Fourier Analysis and Wavelets*, American Mathematical Society, 2009.

[Poi90]   H. Poincaré, Sur les équations aux derivées partielle de la physique mathématiques, *Amer. J. Math.* 12 (1890), 211–294.

[PS54]   G. Pólya and M. Schiffer, Convexity of functionals by transplantation, *J. Analyse Math.* 3 (1954), 245–346.

[Pri94]   A. Pringsheim, Über Functionen, welche in gewissen
Punkten endliche differentialquotienten jeder endlichen
Ordnung, aber keine Taylor'sche Reihenentwickeling
besitzen, *Math. Annalen* 44 (1894), 41–56.

[Ray73]   Lord Rayleigh, Some general theorems relating to variations, *Proc. London Math. Soc.* 4 (1873), 357–368.

[Rel37]   F. Rellich, Störungstheorie der Spektralzerlegung. I,
Math. Ann. 113 (1937), 600–619.

[Rel69]   F. Rellich, *Perturbation Theory of Eigenvalue Problems*,
Gordon & Breach, New York, 1969.

[Rem98]   R. Remmert, *Theory of Complex Functions*, Springer-
Verlag, New York, 4th edition, 1998.

[Roc70]   R.T. Rockafeller, *Convex Analysis*, Princeton University
Press, Princeton, 1970.

[Rot81]   U. Rothblum, Expansions of sums of matrix powers,
SIAM Rev. 23 (1981), 143–164.

[Rot52]   W.E. Roth, On the equaiton $AX - YB = C$ and $AX - XB = C$ in matrices, *Proc. Amer. Math. Soc.* 3 (1952),
392–396.

[Rud74]   W. Rudin, *Real and Complex Analysis*, McGraw Hill,
New York, 1974.

[Sch50]   R. Schatten, A theory of cross spaces, *Ann. Math. Studies*
26 (1950).

[Schu09]   I. Schur, On the characteristic roots of a linear substitution with an application to the theory of integral equations, *Math. Ann.* 66 (1909), 488–510.

[Sed59]   O. Sedláček, On incedenčnich matiach orientirovaných
grafò, *Časop. Pěst. Mat.* 84 (1959), 303–316.

[Sin64]   R.A. Sinkhorn, A relationship between arbitrary positive matrices and doubly stochastic matrices, *Ann. Math.
Statist.* 35 (1964), 876–879.

[Sen81]   E. Seneta, *Non-Negative Matrices and Markov Chains*,
Springer, New York, second edition, 1981.

[Smi68]   M.F. Smiley, Inequalities related to Lidskii's, *Proc.
Amer. Math. Soc.* 19 (1968), 1029–1034.

[So92]   W. So, Equality cases in matrix exponential inequalities,
SIAM J. Matrix Anal. Appl. 13 (1992), 1154–1158.

[Sri03]   R. Srikant, *The Mathematics of Internet Congestion
Control*, Birkhauser, Boston, 2003.

[Ste93]   G.W. Stewart, On the early history of the singular value
decomposition, *SIAM Rev.* 35 (1993), 551–566.

[Str35]   S. Straszewicz, Über exponierte punkte abgeschlossener
punktmengen, *Fund. Math.* 24 (1935), 139–143.

[Tad81]   E. Tadmor, The equivalence of L_2-stability, the resolvent
condition, and strict H-stability, *Linear Alg. Appl.* 41
(1981), 151–159.

[TFL11]   C.W. Tan, S. Friedland and S. Low, Non-negative matrix
inequalities and their application to nonconvex power
control optimization, *SIAM J. Matrix Anal. Appl.* 32
(2011), 1030–1055.

[TeD07]   F. de Terán and F.M. Dopico, Low rank perturbation of
Kronecker structure without full rank, *SIAM J. Matrix
Anal. Appl.* 29 (2007), 496–529.

[Tho65]   C.J. Thompson, Inequality with applications in statistical mechanics, *J. Math. Phys.* 6 (1965), 1812–1813.

[Tro59]   H.F. Trotter, On the product of semi-groups of operators,
Proc. Amer. Math. Soc. 10 (1959), 545–551.

[Val79]   L.G. Valiant, The complexity of computing the perma-
           nent, *Theor. Comput. Sci.* 8 (1979), 189–201.

[vdW59]   B.L. Van der Waerden, *Algebra*, Vols. I and II, Springer,
           Berlin, 1959.

[Var63]   R.S. Varga, *Matrix Iterative Analysis*, Prentice Hall,
           1963.

[Viv93]   G. Vivanti, Sulle serie di potenze, *Rivista di Matematica*
           3 (1893), 111–114.

[vNe37]   J. von Neumann, Some matrix-inequalities and metriza-
           tion of matrix-space, *Tomsk Univ. Rev.* 1 (1937),
           286–300.

[vNe53]   J. von Neumann, A certain zero-sum two-person game
           equivalent to an optimal assignment problem, *Ann.
           Math. Studies* 28 (1953), 5–12.

[Was63]   W. Wasow, On holomorphically similar matrices, *J.
           Math. Anal. Appl.* 4 (1963), 202–206.

[Was77]   W. Wasow, Arnold's canonical matrices and asymptotic
           simplifcation of ordinary differential equations, *Lin. Alg.
           Appl.* 18 (1977), 163–170.

[Was78]   W. Wasow, *Topics in Theory of Linear Differential Equa-
           tions Having Singularities with respect to a Parameter*,
           IRMA, Univ. L. Paster, Strassbourg, 1978.

[Wei67]   K. Weierstrass, Zur theorie der bilinearen un quadratis-
           chen formen, *Monatsch. Akad. Wiss. Berlin* (1867),
           310–338.

[Wey12]   H. Weyl, Das asymptotische verteilungsgesetz der eigen-
           werte linearer partieller differentialgleichungen, *Math.
           Ann.* 71 (1912), 441–479.

[Wie50]   H. Wielandt, Unzerlegbare nicht-negative Matrizen, *Math. Z.* 52 (1950), 642–648.

[Wie56]   H. Wielandt, An extremum property of sums of eigenvalues, *Proc. Amer. Math. Soc.* 6 (1955), 106–110.

[Wie67]   H. Wielandt, *Topics in Analytic Theory of Matrices*, Lecture Notes, University of Wisconsin, Madison, 1967.

[Wim86]   H.K. Wimmer, Rellich's perturbation theorem on hermitian matrices of holomorphic functions, *J. Math. Anal. Appl.* 114 (1986), 52–54.

Index of Symbols

Symbol	Description
$\mathbf{1}$	column vector of whose all coordinates are 1
$a\|b$	a divides b
$(a_1, \ldots, a_n)$	g.c.d. of $a_1, \ldots, a_n$
$A^\top$	transpose of matrix A
A^*	$\bar{A}^\top$ for complex-valued matrix A
$A[\alpha, \beta]$	submatrix of A formed by the rows α and columns β
$A(\alpha, \beta)$	submatrix of A obtained by deleting the rows α and columns β
adj A	adjoint matrix of A
$A \sim_l B$	A is left equivalent to B
$A \sim_r B$	A is right equivalent to B
$A \sim B$	A is equivalent to B
$A \approx B$	A is similar to B
$\hat{A}$	augmented coefficient matrix
$A \geq 0, A \gneq 0,$ $A > 0$	non-negative, nonzero non-negative and positive matrix
$A(D)$	adjacency matrix of digraph D
$A \oplus B$	block diagonal matrix $\begin{bmatrix} A & 0 \\ 0 & B \end{bmatrix}$
$\oplus_{i=1}^{k} A_i$	block diagonal matrix $\mathrm{diag}(A_1, \ldots, A_k)$

571

$\mathbf{AS(V)}$	all anti self-adjoint operators
$\mathbf{AS}(n, \mathbb{D})$	all $n \times n$ skew-symmetric matrices with entries in $\mathbb{D}$
$\mathbf{AH}_n$	all $n \times n$ skew hermitian matrices
$\mathbb{C}$	the field of complex numbers
$\mathbb{C}^n$	n-dimensional vector space over $\mathbb{C}$
C(D)	the set of continuous function in D
$C^k(D)$	the set of continuous function in D with k continuous derivatives
Cl $\mathcal{T}$	closure of $\mathcal{T}$
conv S	all convex combinations of elements in S
conv$_{j-1}$ S	all convex combinations of at most j elements in S
$\mathbb{D}$	integral domain
$\mathbb{D}_B$	Bezout domain (BD)
$\mathbb{D}_E$	Euclidean domain (ED)
$\mathbb{D}_{ED}$	elementary divisor domain (EDD)
$\mathbb{D}_G$	greatest common divisor (GCD) domain
$\mathbb{D}_P$	principal ideal domain (PID)
$\mathbb{D}_U$	unique factorization domain (UFD)
$\mathbb{D}^{m \times n}$	ring of $m \times n$ matrices with entries in $\mathbb{D}$
$\mathbb{D}^{m_1 \times \cdots \times m_k}$	all k-mode $m_1 \times \cdots \times m_k$ tensors with entries in $\mathbb{D}$
$\mathbb{D}[x_1, \ldots, x_n]$	ring of polynomials in $x_1, \ldots, x_n$ with coefficients in $\mathbb{D}$
$D = (V, E)$	digraph with V vertices and E diedges
D^o	the interior of D
$D(p)$	discriminant of polynomial p
$D(\mathbf{x})$	diagonal matrix with diagonal entries equal to coordinates of $\mathbf{x}$
$\mathbf{D}(n, \mathbb{D})$	all $n \times n$ diagonal matrices with entries in $\mathbb{D}$
$\mathbf{DO}(n, \mathbb{D})$	all $n \times n$ diagonal orthogonal matrices with entries in $\mathbb{D}$
$\mathbf{DU}_n$	all $n \times n$ diagonal unitary matrices
$\mathcal{D}_{n,+}$	the set of $n \times n$ diagonal matrices with non-negative entries

$\deg p$	degree of polynomial p
$\deg v$	degree of a vertex v
$\deg_{in} v, \deg_{out} v$	in and out degree of a vertex v
$\det$	determinant
$\delta_k(A)$	g.c.d. of all $k \times k$ minors of A
$\mathrm{diag}(A)$	the diagonal matrix with the diagonal entries of A
$\mathbf{e}_i$	$(\delta_{1i}, \ldots, \delta_{ni})^\top$
$\mathrm{ext}\, C$	the set of extreme points of a convex set C
$e^{\mathbf{x}}$	$(e^{x_1}, \ldots, e^{x_n})^\top$ for $\mathbf{x} = (x_1, \ldots, x_n)^\top \in \mathbb{C}^n$
$\eta(A)$	degree of maximal local invariant polynomial of $A(z)$
$\mathbb{F}$	field
$\mathrm{Fr}(k, \mathbf{V})$	all orthonormal k-frames in $\mathbf{V}$
$\mathbf{GL}(n, \mathbb{D})$	group of $n \times n$ invertible matrices in $\mathbb{D}^{n \times n}$
$\mathrm{Gr}(m, \mathbf{V})$	Grassmannian of m-dimensional subspaces in $\mathbf{V}$
$G = (V, E)$	undirected graph with V vertices and E edges
g.c.d.	greatest common divisor
$\mathbb{H}$	ring of quaternions
$\mathbf{H}_n$	all $n \times n$ hermitian matrices
$\mathbf{H}_{n,+}$	all $n \times n$ non-negative definite hermitian matrices
$\mathbf{H}_{n,+,1}$	all $n \times n$ non-negative definite hermitian matrices with trace one
$\mathrm{H}(\Omega)$	ring of analytic functions in $\Omega \subset \mathbb{C}^n$
H_ζ	is $\mathrm{H}(\{\zeta\})$ for $\zeta \in \mathbb{C}^n$
$\mathrm{Hom}(\mathbf{N}, \mathbf{M})$	all homomorphisms from $\mathbf{N}$ to $\mathbf{M}$
$\mathrm{Hom}(\mathbf{M})$	all homomorphisms from $\mathbf{M}$ to $\mathbf{M}$
$H(A)$	$\begin{bmatrix} 0 & A \\ A^* & 0 \end{bmatrix}$ for $A \in \mathbb{C}^{m \times n}$
$i_k(A)$	k-invariant factor of A

Ω_n	all $n \times n$ doubly stochastic matrices
$\Omega_{n,s}$	all $n \times n$ symmetric doubly stochastic matrices
p_π	principal homogeneous part of polynomial p
$\mathcal{P}_n$	all $n \times n$ permutation matrices
$\mathcal{P}_{n,s}$	all $n \times n$ matrices of the form $\frac{1}{2}(P + P^\top)$ where $P \in \mathcal{P}_n$
Π_n	the set of probability vectors with n coordinates
$\mathbb{R}$	the field of real numbers
$\mathbb{R}_+$	the set of all non-negative real numbers
$\overline{\mathbb{R}}$	$[-\infty, \infty]$
rank A	the rank of a matrix A
$r(A, B)$	rank of $I_n \otimes A - B^\top \otimes I_m$
$r_\phi(A)$	ϕ-numerical radius of $A \in \mathbb{C}^{n \times n}$
$r_{\|\cdot\|}(A)$	Bauer numerical radius
$r_p(A)$	p-numerical radius
R	a ring
$R(p, q)$	resultant of polynomials p and q
$\mathcal{R}_{m,n,k}(\mathbb{F})$	all $m \times n$ matrices of at most rank k with entries in $\mathbb{F}$
Range T	range of a homomorphism T
Rank τ	rank of a tensor τ
$\rho(A)$	spectral radius of A
sign σ	the sign of permutation σ
S^1	the unit circle $|z| = 1$ in the complex plane
S^{2n-1}	the $2n - 1$ dimensional sphere
$\mathbf{S}(\mathbf{V})$	all self-adjoint operators acting on IPS $\mathbf{V}$
$\mathbf{S}_+(\mathbf{V})$	all self-adjoint non-negative definite operators
$\mathbf{S}_{+,1}(\mathbf{V})$	all self-adjoint non-negative definite operators with trace one

$\mathbf{S}_+(\mathbf{V})^o$	all self-adjoint positive definite operators
$S^{m \times n}$	all $m \times n$ matrices with entries in S
$\mathcal{S}_n$	the set of $n \times n$ stochastic matrices
$\mathbf{S}(n, \mathbb{D})$	all $n \times n$ symmetric matrices with entries in $\mathbb{D}$
$\mathbf{SO}(n, \mathbb{D})$	all $n \times n$ special orthogonal matrices with entries in $\mathbb{D}$
$\mathbf{SU}_n$	all $n \times n$ special unitary matrices
$\operatorname{spec}(T)$	spectrum of T
$\operatorname{spec}_{\mathrm{peri}}(T)$	peripheral spectrum of T
$\sigma_1(T) \geq \sigma_2(T) \geq \cdots$	singular values of T
$\boldsymbol{\sigma}(A)$	$(\sigma_1(A), \ldots, \sigma_{\min(m,n)}(A))^\top$ for $A \in \mathbb{C}^{m \times n}$
$\boldsymbol{\sigma}_{(p)}(T)$	$(\sigma_1(T), \ldots, \sigma_p(T))^\top$
Σ_n	the group of permutations of $[n]$
$T \succ S,$ $(T \succeq S)$	$T - S$ is positive (non-negative) definite
$\operatorname{tr} T$	trace of T
$\operatorname{tr}_k T$	sum of the first k eigenvalues of self-adjoint T
$\mathbf{U}(\mathbf{V})$	all unitary operators
$\mathbf{U}_n$	all $n \times n$ unitary matrices
$\mathbf{UT}(m, \mathbb{D})$	all $m \times m$ upper triangular matrices with entries in $\mathbb{D}$
$\mathbf{V}$	a vector space
$\mathbf{V}'$	the dual vector space $\operatorname{Hom}(\mathbf{V}, \mathbb{F})$
$\mathbf{x}(D)$	$(d_1, \ldots, d_n)^\top$ for $D = \operatorname{diag}(d_1, \ldots, d_n)$
$\mathbf{x}^{-1}$	$(x_1^{-1}, \ldots, x_n^{-1})^\top$ for $\mathbf{0} < \mathbf{x} = (x_1, \ldots, x_n)^\top$
$\mathbf{x} \circ \mathbf{y}$	$(x_1 y_1, \ldots, x_n y_n)^\top$ for $\mathbf{x}, \mathbf{y} \in \mathbb{C}^n$
$\mathbf{x} \prec \mathbf{y} \ (\mathbf{x} \preceq \mathbf{y})$	$\mathbf{x}$ (weakly) majorized by $\mathbf{y}$
$(\mathbf{x}, \mathbf{y}), [\mathbf{x}, \mathbf{y}]$	open and closed intervals spanned by $\mathbf{x}, \mathbf{y}$

$\hat{X}, \mu(X)$	vector in $\mathbb{D}^{mn}$ corresponding to matrix $X \in \mathbb{D}^{m \times n}$
$\bar{z}$	complex conjugate of z
$\mathbb{Z}$	all integers
$\mathbb{Z}_+$	all non-negative integers
$\#J$	cardinality of a set J
$\otimes_{i=1}^{k} \mathbf{V}_i$	tensor product of $\mathbf{V}_1, \ldots, \mathbf{V}_k$
$\otimes^k \mathbf{V} = \mathbf{V}^{\otimes k}$	k-tensor power of $\mathbf{V}$
$\odot_{i=1}^{m} \mathrm{X}_i$	a set tensor product of $\mathrm{X}_1, \ldots, \mathrm{X}_m$
$\mathrm{Sym}^k \mathbf{V}$	k-symmetric power of $\mathbf{V}$
$\bigwedge^k \mathbf{V}$	k-exterior power or wedge product of $\mathbf{V}$
$\otimes_{i=1}^{k} \mathbf{x}_i$	k-decomposble tensor, rank one tensor if each $\mathbf{x_i}$ is nonzero
$\wedge_{i=1}^{k} \mathbf{x}_i$	k-wedge product
$\otimes_{i=1}^{k} A_i$	Kronecker tensor product of matrices $A_1, \ldots, A_k$
$\wedge^k A$	kth compound matrix of A
$(A_1, \ldots, A_k)_{\otimes}$	the generator of the group $\otimes_{i=1}^{k} e^{tA_i}$
$A_{\wedge^k}$	the generator of the group $\wedge^k e^{tA}$

Index

Printed in the United States
By Bookmasters